W0071803

Modular Chemistry

NATO ASI Series

Advanced Science Institutes Series

A Series presenting the results of activities sponsored by the NATO Science Committee, which aims at the dissemination of advanced scientific and technological knowledge, with a view to strengthening links between scientific communities.

The Series is published by an international board of publishers in conjunction with the NATO Scientific Affairs Division

A Life Sciences **B Physics**	Plenum Publishing Corporation London and New York
C Mathematical and Physical Sciences **D Behavioural and Social Sciences** **E Applied Sciences**	Kluwer Academic Publishers Dordrecht, Boston and London
F Computer and Systems Sciences **G Ecological Sciences** **H Cell Biology** **I Global Environmental Change**	Springer-Verlag Berlin, Heidelberg, New York, London, Paris and Tokyo

PARTNERSHIP SUB-SERIES

1. Disarmament Technologies	Kluwer Academic Publishers
2. Environment	Springer-Verlag / Kluwer Academic Publishers
3. High Technology	Kluwer Academic Publishers
4. Science and Technology Policy	Kluwer Academic Publishers
5. Computer Networking	Kluwer Academic Publishers

The Partnership Sub-Series incorporates activities undertaken in collaboration with NATO's Cooperation Partners, the countries of the CIS and Central and Eastern Europe, in Priority Areas of concern to those countries.

NATO-PCO-DATA BASE

The electronic index to the NATO ASI Series provides full bibliographical references (with keywords and/or abstracts) to more than 50000 contributions from international scientists published in all sections of the NATO ASI Series.
Access to the NATO-PCO-DATA BASE is possible in two ways:

– via online FILE 128 (NATO-PCO-DATA BASE) hosted by ESRIN, Via Galileo Galilei, I-00044 Frascati, Italy.

– via CD-ROM "NATO-PCO-DATA BASE" with user-friendly retrieval software in English, French and German (© WTV GmbH and DATAWARE Technologies Inc. 1989).

The CD-ROM can be ordered through any member of the Board of Publishers or through NATO-PCO, Overijse, Belgium.

Series C: Mathematical and Physical Sciences – Vol. 499

Modular Chemistry

edited by

Josef Michl
Department of Chemistry & Biochemistry,
University of Colorado,
Boulder, CO, U.S.A.

Springer-Science+Business Media, B.V.

Proceedings of the NATO Advanced Research Workshop on
Modular Chemistry
Aspen Lodge at Estes Park, Colorado, U.S.A.
September 9–12, 1995

A C.I.P. Catalogue record for this book is available from the Library of Congress

ISBN 978-94-010-6353-1 ISBN 978-94-011-5582-3 (eBook)
DOI 10.1007/978-94-011-5582-3

Printed on acid-free paper

This book contains the proceedings of a NATO Advanced Research Workshop held within the programme of activities of the NATO Special Programme on Supramolecular Chemistry as part of the activities of the NATO Science Committee.

Other books previously published as a result of the activities of the Special Programme are:

WIPFF, G. (Ed.), *Computational Approaches in Supramolecular Chemistry.* (ASIC 426) 1994. ISBN 0-7923-2767-5

FLEISCHAKER, G.R., COLONNA, S. and LUISI, P.L. (Eds.), *Self-Production of Supramolecular Structures.* From Synthetic Structures to Models of Minimal Living Systems. (ASIC 446) 1994. ISBN 0-7923-3163-X

FABBRIZZI, L., POGGI, A. (Eds.), *Transition Metals in Supramolecular Chemistry.* (ASIC 448) 1994. ISBN 0-7923-3196-6

BECHER, J. and SCHAUMBURG, K. (Eds.), *Molecular Engineering for Advanced Materials.* (ASIC 456) 1995. ISBN 0-7923-3347-0

LA MAR, G.N. (Ed.), *Nuclear Magnetic Resonance of Paramagnetic Macromolecules.* (ASIC 457) 1995. ISBN 0-7923-3348-9

SIEGEL, JAY S. (Ed.), *Supramolecular Stereochemistry.* (ASIC 473) 1995. ISBN 0-7923-3702-6

WILCOX, C.S. and HAMILTON A.D. (Eds.), *Molecular Design and Bioorganic Catalysis.* (ASIC 478) 1996. ISBN 0-7923-4024-8

MEUNIER, B. (Ed.), *DNA and RNA Cleavers and Chemotherapy of Cancer and Viral Diseases.* (ASIC 479) 1996. ISBN 0-7923-4025-6

KAHN, O. (Ed.), *Magnetism: A Supramolecular Function.* (ASIC 484) 1996. ISBN 0-7923-4153-8

ECHEGOYEN, L., KAIFER ANGEL E. (Eds.), *Physical Supramolecular Chemistry.* (ASIC 485) 1996. ISBN 0-7923-4181-3

DESVERGNE J.P., CZARNIK A.W. (Eds.), *Chemosensors of Ion and Molecule Recognition.* (ASIC 492) 1997. ISBN 0-7923-4555-X

Table of Contents

POSTER PRESENTATIONS

PREFACE

Modular Chemistry: the First Steps

In recent years, there has been increasing interest among chemists, physicists, materials scientists, biologists, engineers, and others in the assembly of well defined, relatively large functional structures from repetitive units that themselves are molecules of some complexity. Using the dictionary definition of a module (a detachable section, compartment, or unit with a specific purpose or function, and in electronics, a compact assembly functioning as a component of a larger unit) [1], we feel that this newly emerging field of endeavor could be called "modular chemistry" [2].

The NATO Advanced Research Workshop on Modular Chemistry that was held on September 9 to 12, 1995, at Aspen Lodge near Estes Park, Colorado, was meant to bring together prominent contributors to modular chemistry as it is being born, and to examine the associated birth pangs. It was concluded that although real, these are not nearly as bad as giving birth to a hedgehog tail first, and that the ultimate rewards were likely to be far more satisfying in terms of new ideas and enabling methodology. The level of excitement about the possibilities that are opening up for modular chemists, and also the challenge involved, are perhaps best documented by noting that the planned discussion periods at the workshop were as long as the oral presentation periods, and yet, each discussion ran over the allocated time.

Perhaps the most striking, and most useful, aspect of the workshop was its interdisciplinary nature. The participants came from a variety of subdisciplines that seem to be combining to create the new field of modular chemistry. In spite of these widely differing backgrounds, and perhaps precisely because of them, the participants clearly relished listening to each other and debating thereafter.

The two most visible sources from which modular chemistry originates are supramolecular ("self-assembly") chemistry, which exploits non-covalent intermolecular forces among modules to form larger assemblies, and polymer chemistry, which is based on attaching modules to each other through covalent bonds. However, many of the participating practitioners of modular chemistry consider themselves neither supramolecular nor polymer chemists. Some are primarily crystallographers, others are solid-state chemists or physicists, small-molecule organic synthetic chemists, inorganic coordination or main group chemists, photochemists or photophysicists, surface chemists, electrochemists, carbon or semiconductor cluster chemists, biochemists, biomimetic chemists, biomineral chemists, materials scientists, microscopists, or theoreticians, while still others work in "molecular electronics", in non-linear optics, with Langmuir-Blodgett or self-assembled monolayers, or with liquid crystals.

The size of the modules also varies enormously, from less than one nanometer across to nearly macroscopic. The concept of a hierarchy of structures of increasing size was referred to repeatedly, and comparisons were inevitably made with the organization of living matter. There is clearly much that is being learnt from biology, yet we also heard much from participants whose research programs consciously focus on structures that have no known analogues in living matter.

The present volume represents an attempt to capture the tantalizing atmosphere of the workshop as faithfully as possible, and to provide a record of what modular chemistry looked like close to the time of its birth. The authors were encouraged to be just as daring, provocative, and speculative in their written contributions as they had been in their oral or poster presentations at the workshop. They were asked not to provide reviews of previously published material but to concentrate on the newest ongoing research in their laboratories, even if it is still incomplete and the results not quite understood. They were asked to include their views on where modular chemists stand, where they are going, and what they propose to do once they get there. All of the articles were peer-reviewed by at least one independent reviewer.

Perhaps the accounts of the participants' current thoughts on their work in modular chemistry, and of the ensuing discussions, will entice additional scientists to enter the field and make it flourish. This would be the best possible reward to the contributors to this volume, whom I thank for their painstaking labor, and to the editor.

References

1. (1979) in *Webster's New Universal Unabridged Dictionary*, Simon & Schuster, Cleveland, OH, p. 1156.
2. Michl, J. (1996) Synthesis of Giant Modular Structures, in C. Chatgilialoglu and V. Snieckus (eds.), *Chemical Synthesis: Gnosis to Prognosis*, Kluwer Academic Publishers, Dordrecht, p. 429.

JOSEF MICHL
Department of Chemistry and Biochemistry,
University of Colorado
Boulder, CO 80309-0215, U.S.A.

March 8, 1997

ACKNOWLEDGEMENTS

The Advanced Research Workshop was financed by a grant from NATO, and the participants are grateful to the NATO Scientific Affairs Division. Special thanks go to Ms. Susan Robeck for tireless effort invested into the production of the symposium volume.

Modular Chemistry Participants

1. Kuki
2. Seddon
3. Coleman
4. Scuseria
5. Billups
6. Janata
7. Bunz
8. Whitesell
9. Ward
10. Port
11. Guard
12. Lindsey
13. Diederich
14. Seeman
15. Tomalia
16. Smalley
17. Stupp
18. Prinzbach
19. Chidsey
20. Newkome
21. M. Grimme
22. Yaghi
23. R. Müllen
24. Rampi
25. Harada
26. Mingotaud
27. Kubiak
28. Mallouk
29. Moore
30. Hensel
31. Banaszak-Holl
32. Bein
33. Tilley
34. Mortimer
35. Schauer
36. Wegner
37. Rajca
38. Schlüter
39. K. Müllen
40. Kaszynski
41. Ding
42. Stoddart
43. Balzani
44. Miller
45. Sita
46. Kahn
47. Nuzzo
48. Ozin
49. Palacin
50. Zaworotko
51. Meijer
52. Wuest
53. Michl
54. Grimme

TOWARDS DESIGNER SOLIDS

An Approach to Tinkertoy-like Molecular Grids and Scaffolds

ROBIN M. HARRISON, THOMAS F. MAGNERA, JAROSLAV VACEK, AND JOSEF MICHL
Department of Chemistry and Biochemistry,
University of Colorado
Boulder, CO 80309-0215
U.S.A.

ABSTRACT. We describe work with molecular construction kits, one of the subareas of modular chemistry. Ultimately, we wish to learn how these kits can be used to produce thin layers of designer solids, materials in which arbitrary preselected substructures have been imbedded in a rigid or a mobile way into preselected positions into a scaffolding in a controlled, and if desired, totally aperiodic fashion. Our present long-term goal is the production of designer solids whose structure is periodic in two long dimensions and controlled in an arbitrary fashion in the third thin one. We describe efforts to reach the immediate synthetic goal, a single grid layer, and computer simulation of the response of a turbine-shaped molecule mounted on the grid to a stream of gas.

1. Introduction

An important subarea of modular chemistry [1] are molecular construction kits [2][3], the preparation and use of artificial structurally well defined large molecules built from fairly rigid repetitive modules through strong chemical bonds in a way that allows the modules to be still clearly recognized. The modules and the assembly are reminiscent of children's construction kits.

There are two types of target structures for synthesis with molecular construction kits, soluble (free-floating) structures, and surface-anchored structures, and both offer much excitement. We have chosen to work on the latter. This choice has the advantage that solubility of the target macromolecules is not an issue, and the disadvantage that many standard methods of analysis, such as solution NMR, are inapplicable. The more demanding and more expensive analytical tools of surface science have to be used instead.

J. Michl (ed.), Modular Chemistry, 1–16.

2. The Overall Plan

2.1 THE TARGET STRUCTURES

We hope to synthesize thin layers of covalent designer solids by building layers on a scaffolding one at a time. The scaffolding will consist of connectors and rods, which may be thick enough to touch sideways if desired, for instance for conductivity or increased mechanical strength. These properties might also be achieved by filling the finished scaffolding with a polymerizable material and curing, by electrodepositing a metal or a semiconductor into it, etc.

Fig. 1 shows examples of trigonal, tetragonal, and hexagonal scaffoldings. For clarity, it shows only two layers, although in reality we would like to be able to prepare solids with dozens or hundreds of such layers. Since they are to be synthesized on top of the existing layers one at a time, each one can carry different active groups if desired. This design feature would require using different rods and/or connectors in the addition of each layer, of dimensions such as to still make epitaxial matching possible.

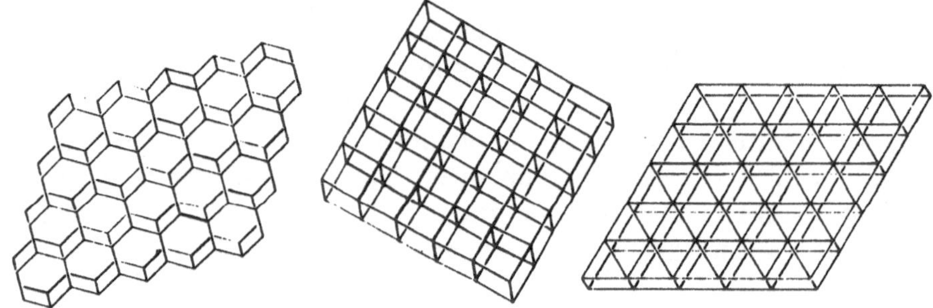

Figure 1. Examples of hexagonal, square, and trigonal covalent molecular scaffolds.

For simplicity, no active groups are shown in Fig. 1. We envisage attaching them to the rods or the connectors directly or by tethers, or incorporating them in an integral way into the structure of the rods or the connectors themselves. The active groups could be absorbers or emitters of light, and donors or acceptors of electrons. They could be charged, dipolar or magnetic, they could have non-linear polarizability, specific chemical affinity for specified partners, they could be installed rigidly or could be mobile, etc.

Our immediate target is a polymer molecule consisting of a single layer of this type, i.e. a regular firmly bonded two-dimensional grid of two very large dimensions and the thickness of a few Å, with controlled trigonal, tetragonal, or hexagonal structure, with preselected rod length and thickness, and preferably permitting subsequent attachment of functionalities at the rods or the connectors, useful for future epitaxial extension by covalent construction of additional layers.

2.2 THE BASIC MODULES

Our modules need to be straight rods and connectors to which the rods can be attached at fixed regular angles of 60°, 90°, or 120°. The construction kit that these building elements resemble the most is the "Tinkertoy" kit [4] common in the U.S.A. It is possible to build with crooked beams, and we have admired many half-timbered houses in Northern Europe that prove it. At least initially, however, we think that it is easier to build with straight beams.

Although we choose rods that prefer to be straight at equilibrium, they can still bend relatively easily and should be thought of as analogous to rubber night sticks rather than beams of wood or steel. This flexibility is an advantage, as it ought to increase the mechanical resilience of the designer solids.

The length of rods that we are most interested in ranges from a few Å to a few dozen Å, with increments of perhaps 1 Å or so. Eventually, we hope to have at our disposal a collection of rods varying in properties such as length, rigidity, polarizability, absorptivity, and electrical conductivity.

We foresee two fundamentally different modes of coupling the modules (Fig. 2). In one, the basic monomeric unit to be polymerized into a grid is a "star" connector carrying six, four, or three half-rod arms. The polymerization reaction consists of linear (180°) coupling of arm ends to each other directly or through a linear coupler, which could also be an additional rod. The reaction thus converts pairs of part-rods on adjacent monomers into full-length rods.

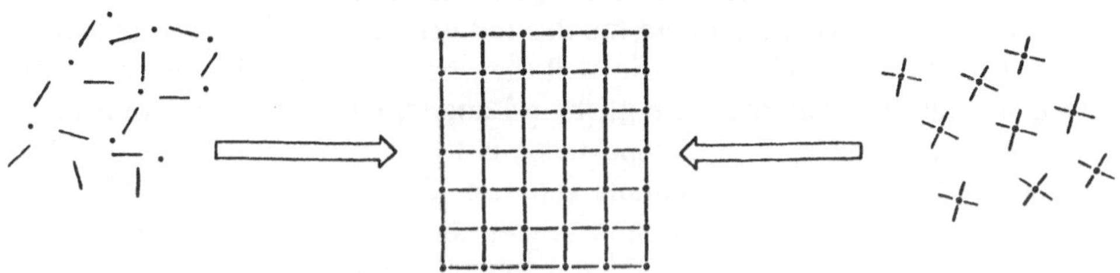

Figure 2. Modes of polymerization into a two-dimensional grid.

The other mode uses the full rods themselves as the basic monomeric units. The polymerization reaction consists of angular (60°, 90°, or 120°) coupling of six, four, or three rods to "point" connectors. It is therefore characterized by the formation of three or more geminal bonds, whereas the coupling reaction of the star connectors only involved the formation of at most two geminal bonds. Yet, as Fig. 2 shows, the final geometry is the same.

2.3 GRID ASSEMBLY

It is clear that ordinary solution polymerization of either type of monomer with a coupling agent has essentially zero chance of producing anything even remotely resembling the desired huge flat grid. Instead, a three-dimensionally cross-linked polymer is most likely to result. The way in which we plan to force our monomers to form a two-dimensional regular polymer is to confine their motion to two dimensions. We plan to effect this constraint by mounting the monomers on the top of oriented pedestals that are free to translate and rotate on a surface, but are not free to leave it or to turn over. All reactions of the arm ends of the monomers, including the coupling processes described above, will then be forced to occur within a thin segment of space parallel to the confining surface and separated from it by the height of the pedestal. In order to guarantee translational and rotational mobility for the pedestals, we initially use a liquid surface, which has the additional advantage of being atomically flat.

We need to force the pedestals to stay confined to the surface in the proper orientation, with the pillar and thus the monomer held on the top, away from the surface, and not hindered by it sterically. To accomplish this kind of orientation, the pedestal must consist of a pillar erected on a platform carrying three or more tentacles that have large affinity for the surface. The pillar permits the attachment of the monomer on top. The tentacles are chosen to be so sticky that they can virtually never leave the surface once thay attach themselves to it, unless chemically transformed by the action of a suitable reagent. They are firmly attached to the platform at the bottom of the pedestal. Ideally, they should be detachable from it chemically upon demand.

Once the polymerization to the desired grid occurs, the tentacles are to be deactivated or cut off. The grid will then have no special affinity for the surface and can be manipulated, e.g., by a Langmuir-Blodgett type transfer.

An issue of the utmost importance is the prevention of defects. The simplest way to minimize their number is to use "thermodynamic control", i.e., coupling reactions that are reversible under the reaction conditions and yet produce links that are strong under the intended conditions of use. Under reversible coupling conditions, it might be possible to anneal most initially formed defects in a process similar to zone refining, producing large two-dimensional "single crystal" domains from the presumably small initial domains. Ultimately, it might be possible to develop methods for the conversion of a "reversible" link into an "irreversible" one without disturbing the grid pattern, forming grids that are sturdy under all conditions of practical interest.

Components of a construction kit for a molecular grid are shown in Fig. 3. They are (A) axially functionalized rods, (B) point and/or star connectors, and (C) tentacled pedestals suitable for mounting the connectors or the rods. In

(D), we show schematically the side and top views of three complete assembled modules ready for the coupling reaction. The pedestal-mounted star connector on the left has its four arms held parallel to the surface, facilitating their end-to-end coupling. The point connector in the center is ready to couple with three rods, and the star one on the right has three arms pointing obliquely away from the surface to permit attachment of subsequent layers into a diamond-like lattice.

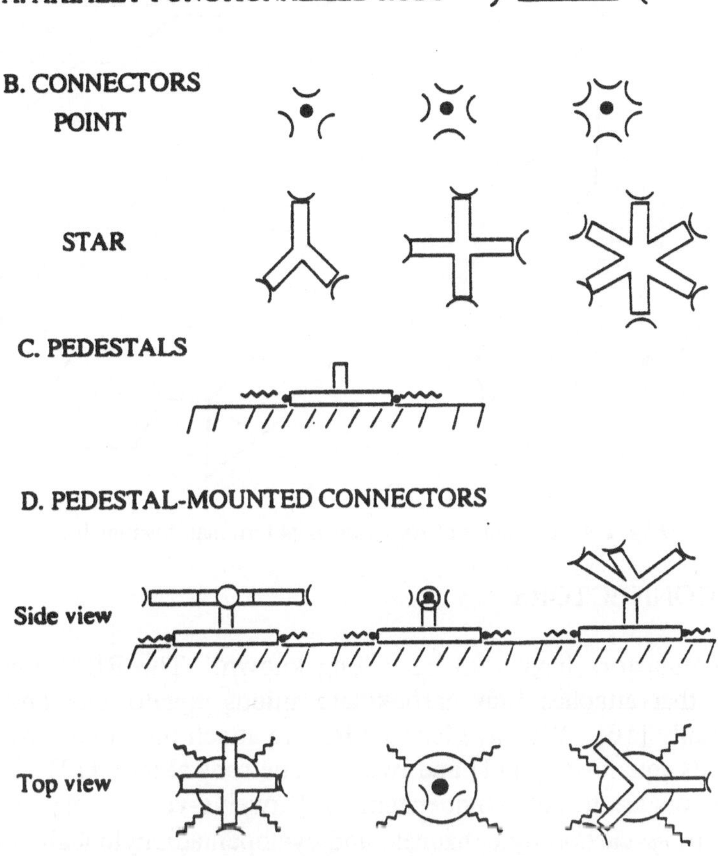

Figure 3. Components of a molecular-size construction kit.

3. Specific Structures

3.1 THE RODS

Some axially functionalized rod-like molecules have been known for a long time, and many more were made recently (Fig. 4). We have worked on [n]staffanes [3] and 10-vertex and 12-vertex carborods [5]. Important contributions were made by Szeimies [6], Eaton [7], Hawthorne [8], and Zimmerman [9].

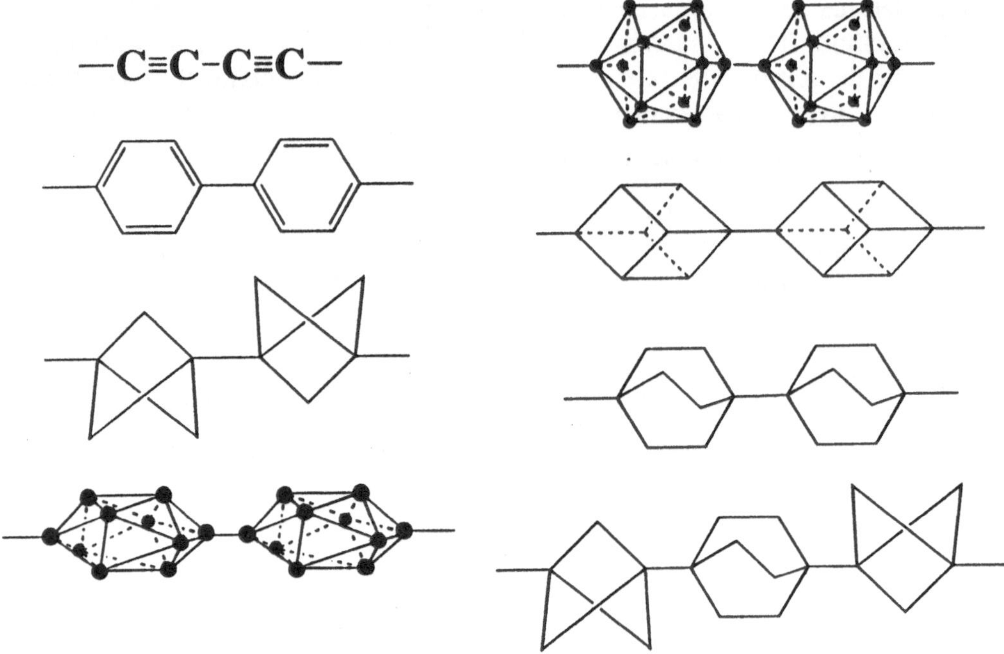

Figure 4. Examples of rods with axial terminal functionalities.

3.2 THE CONNECTORS

Suitable connectors have also been long known. The Rh_2^{4+} ion is a point connector that attaches four carboxylate anions equatorially and two other ligands axially [10]. We have learned how to attach one, two, three, and four staffane rods equatorially [11] and two staffane rods axially [12]. Examples of star connectors are hexaethynylbenzene [13], twelve-vertex [14] and ten-vertex [15] 1,3,5-tris-*p*-carboranylbenzenes, and cyclopentadienylcobalt complexes of tetrasubstituted cyclobutadienes [16].

3.3 PEDESTAL-MOUNTED STAR CONNECTORS

The modules that we have worked with were metal sandwich complexes. One deck is the tentacle-carrying bottom, one is the arms-carrying top, and the metal is the pillar. An example with carbethoxy groups at arms ends is the cyclobutadiene complex **1** (Fig. 5). We have also prepared modules with other terminal functionalities.

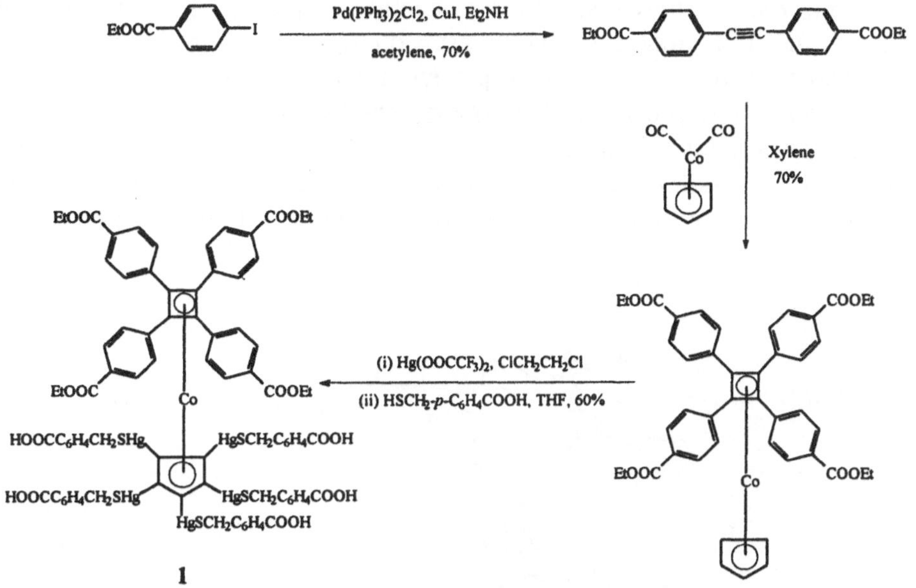

Figure 5. Synthesis of the pedestal-mounted connector **1**.

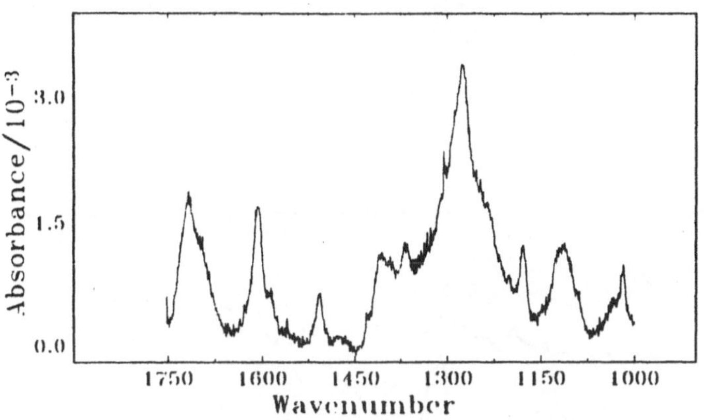

Figure 6. Reflectance IR of **1** on Hg surface.

Aromatic carboxyls have high affinity for a clean Hg surface, but desorb at low pH. Once the carboxyl-terminated tentacles are attached to the surface, the functional groups at the arms ends on the top of the module are available for coupling. Fig. 6 shows the IR spectrum of **1** adsorbed on mercury. The signal-to-noise ratio suggests that it will be possible to follow its chemical reactions by IR spectroscopy, and this is presently under examination.

Fig. 7 shows the IR spectrum of another module (**2**) [17] adsorbed on Hg. Here, the sticky tentacles and the arms on top are both pyridine rings. A comparison with the ordinary isotropic spectra of the Zn salts of a typical porphyrin, *meso*-tetra(*p*-carboxyphenyl)porphyrin (**3**), and of the isolated single deck, *meso*-tetrapyridylporphyrin (**4**), shows that **2** resides on mercury with the porphyrin rings parallel to the surface (the in-plane 1003 cm^{-1} vibration of the porphyrin rings is absent and the out-of-plane 870 cm^{-1} vibration is present).

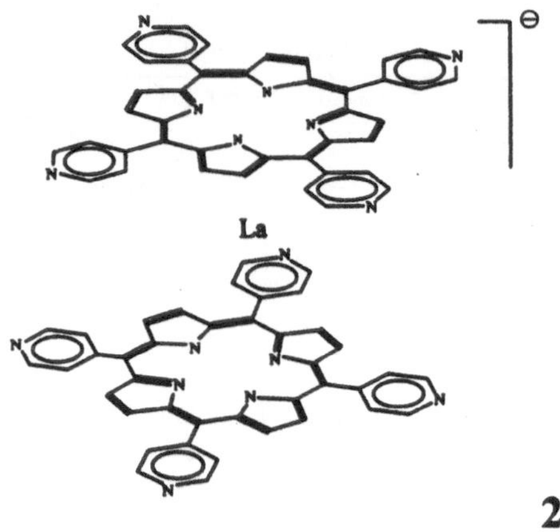

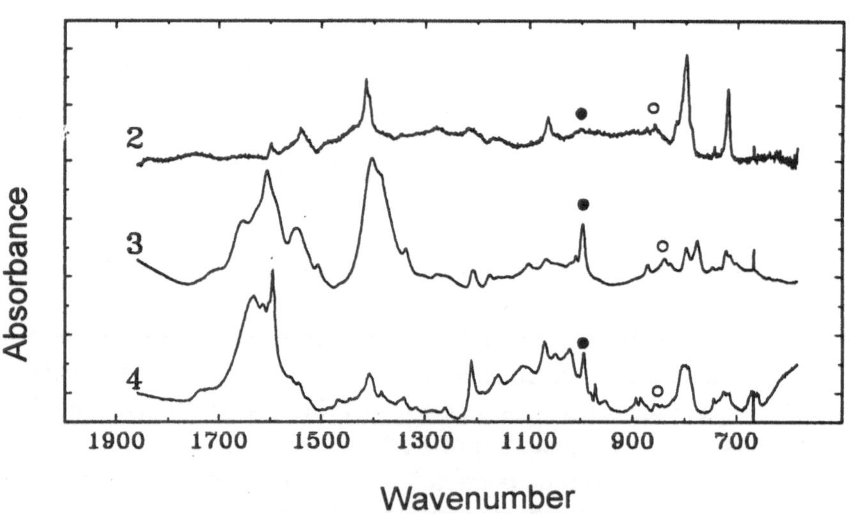

Figure 7. Reflectance IR of **2** on Hg surface and IR of the Zn salts of **3** and **4** in KBr pellet.

4. Two-Dimensional Polymerization

4.1 REACTIONS OF SURFACE-MOUNTED CONNECTORS

It is hoped that at suitable surface concentrations, coupling reactions of the arms ends on top of our modules will permit the formation of a two-dimensional grid under essentially reversible conditions. An example of the planned transformation in shown schematically in Fig. 8 and refers specifically to the coupling of pyridine arms ends through MX_2 metal dihalide or dicyanide point connector units, such as $NiCl_2$. We are examining reversible reaction conditions first, using both IR and Raman, since such reactions will be useful for testing annealing procedures and defect minimization. Although surface spectroscopy permits us to follow the chemical transformations, it provides no information on the presence or absence of long-range order. We plan to use a scanning tunneling microscope and low-energy electron diffraction to search for such order, for local grid formation, and for domain size. Ultimately, we wish to identify coupling reactions that are reversible in the presence of the coupling solution, but produce bonds that are sturdy in its absence.

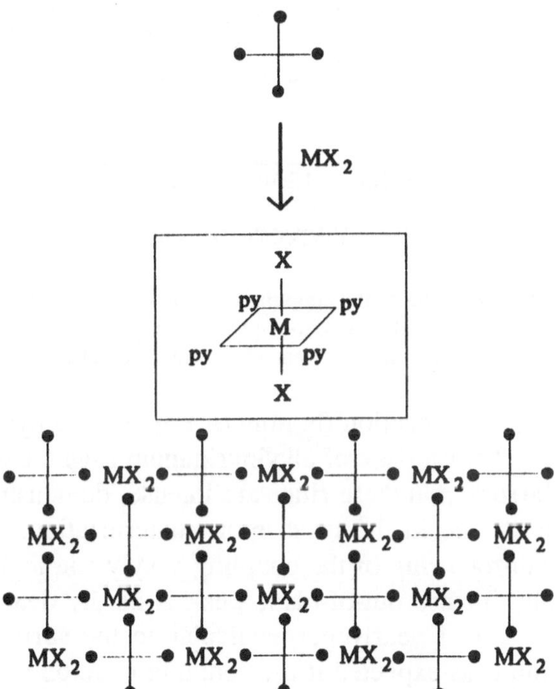

Figure 8. Proposed coupling of module **2** into a two-dimensional square grid by metal-mediated pyridine-to-pyridine links.

We have already started to examine some irreversible coupling conditions, using the module **2**. The initially used coupling agent, *p*-xylylene dibromide, is expected to quaternize the pyridine rings and form a kinked -CH_2-C_6H_4-CH_2- link connecting them pairwise. It is not ideal, in that the connections are not really linear, but it is highly reactive and easy to follow by IR spectroscopy. The top part of Fig. 9 shows the assignment of transitions in the IR spectrum of **2** on Hg surface. It contains both in-plane and out-of-plane vibrations of the pyridine rings, indicating that they are somewhat twisted relative to the porphyrin plane, but are not orthogonal to it. As already pointed out, the in-plane porphyrin vibration is nearly invisible.

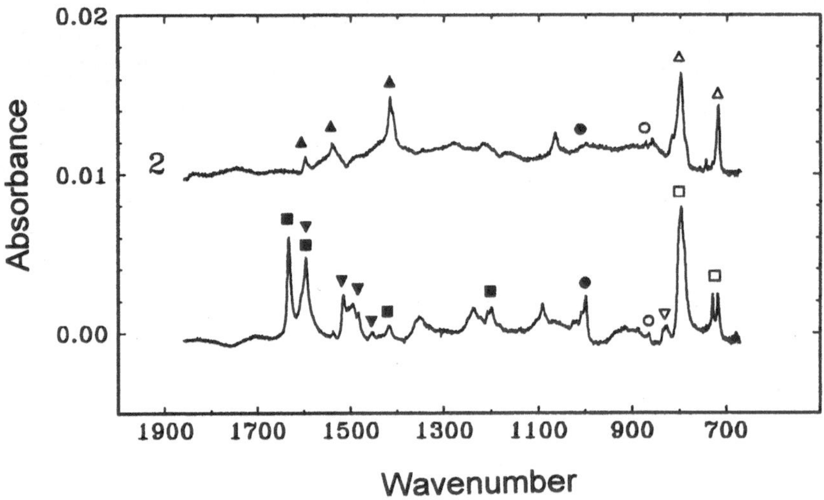

Figure 9. Reflectance IR monitoring of the coupling reaction of **2** with *p*-xylylene dibromide on Hg surface. Important in-plane (black) and out-of-plane (white) vibrations of porphyrin (●,○), pyridine (▲,△), pyridinium (■,□), and *p*-xylylene (▼,▽) rings are labeled.

After the coupling reaction (bottom of Fig. 9), the pyridine vibrations are replaced by those characteristic of alkylpyridinium rings. Once again, peaks of both types of polarization in these rings are intense, demonstrating that there has not been much change in the twist upon quaternization. In-plane peaks attributable to the benzene ring of the coupling *p*-xylylene moiety also appear clearly. Its normally intense out-of-plane peak is hardly detectable, so these benzene rings are oriented nearly perpendicular to the surface. All of this structural information is as expected if the coupling reaction worked perfectly.

However, the in-plane porphyrin vibration now also appears, with variable intensity depending on the coupling conditions. The estimated average angle between the porphyrin ring and the Hg surface is 15 - 40°, presumably due to irregularities in the coupling process. We are still optimizing the reaction conditions. As pointed out above, even if the IR spectra end up perfectly compatible with the expected chemical structure, we will not know whether a regular grid has formed, until we use complementary tools [18].

5. Speculation on Possible Uses of Designer Solids

Suppose for a moment that the scaffolding-based designer solids described here can indeed be synthesized. What might they be good for?

We believe that we are close enough to preparing at least small domains of the ground floor that it is not a total waste of time to speculate about its possible uses. We refrain from speculating about the uses of multilayer designer solids, whose synthesis seems to be farther off in the future.

Two types of uses of a single-layer grid might be worth exploring.

5.1 TRIVIAL USES

First, various functions might be performed better with a designer grid than with existing materials. For example, today's porous membranes permit the separation of gaseous or dissolved molecules by size, but the rate of the process is limited by the relatively slow rate of diffusion through the membrane. Also, the crooked "tunnels" through the membrane are not all of equal size, making the separation imperfect. The process is a little like separating thin earthworms from fat ones by waiting for them to crawl spontaneously through a layer of earth.

The rate of separation of molecules by the grid of the type we are attempting to synthesize, mounted on a fine mesh or a coarsely porous polymer for mechanical strength, would be limited by the much faster rate of diffusion through solution or gas to the grid, and the passage through the grid itself would be essentially instantaneous. This would be like allowing bumblebees but not hummingbirds to escape from a birdcage. A demonstration of this principle on an LB multilayer made of a calixarene derivative ("perforated membrane") has been published [19]; the designer grid would be much thinner and sturdier.

The separation by passage through a grid could be most simply based on the smallest cross-section, but it could also be based on the total length of a molecule, say a linear polymer. This would be like separating long from short worms by shaking them through a coarse sieve.

In addition to size, chemical affinity could be utilized. The rims of the regular openings in the grid could be provided with suitable functionalities that would preferentially recognize certain substrates. One or both faces of the grid could also be functionalized in ways that promote biocompatibility for applications such as enzyme reactors or kidney dialysis. The rods or the connectors could be provided with flexible or rigid hairs that lock upon demand, blocking passage through the openings. For example, terminal thiol groups could be reversibly oxidized to disulfides to build a web across the openings in a grid.

5.2 NOVEL USES

The second kind of uses is even more appealing. These are the applications of designer solids that no existing materials can perform. We have not been able to think of many of these, both because we know little solid state physics and because we have spent little time thinking about the issue. Once designer solids have been synthesized, it will be reasonable to spend time on such speculation.

One of the few possibilities that have occurred to us is to use a regular array of rotatable charges, dipoles, or magnets attached to a grid for various purposes. If the rotatable elements had the shape of turbines or paddlewheels and could be driven by a stream of gas, they might mediate an interconversion of electromagnetic energy and the energy of gas flow. This possibility has fascinated us sufficiently that we have performed an actual computer simulation.

6. Computer Modeling of a Molecular Windmill

We have simulated the motion of a single molecular turbine, represented by a dipolar derivative of the bis(phenanthroline)(3-cyanobicyclo[1.1.1]pentylmalonic dialdehyde) complex of rhenium(I) **5**, with a positive and a negative charge at the extremities of the phenanthroline "blades".

The turbine was held in position either artificially, by fixing the atoms of the cyano group, or in some runs, more realistically, by mounting on to a square grid of twelve terminally carboxylated [4]staffane rods linked through nine dirhodium connectors. The cyano group of the turbine, carried by the outer bridgehead position of the bicyclo[1.1.1]pentane axis, was attached to an axial site of one of the dirhodium connectors. All the other axial sites were saturated with acetonitrile molecules, and the twelve equatorial sites left over at the edges were saturated with [2]staffanecarboxylate groups. Although sturdier arrangements can be envisaged, this one seemed adequate for the simulation.

In the simulation, the grid was exposed to a flow of helium gas from a supersonic nozzle, using a universal force field (UFF) [20] and Newton's laws of motion [21], ignoring radiative losses for the time being. All atoms in the model (with the realistic grid, over a thousand) were allowed to move except for a few at the edges, used to hold the grid in place. In some runs, a Morse curve instead of a harmonic potential was used for the bond holding the cyano group to the dirhodium ion in order to permit dissociation, but on the time scale of the simulation, it made no difference. The integration steps were 2.3 fs long and the motion was usually followed for a period of a little over 100 ps, and sometimes as long as 500 ps.

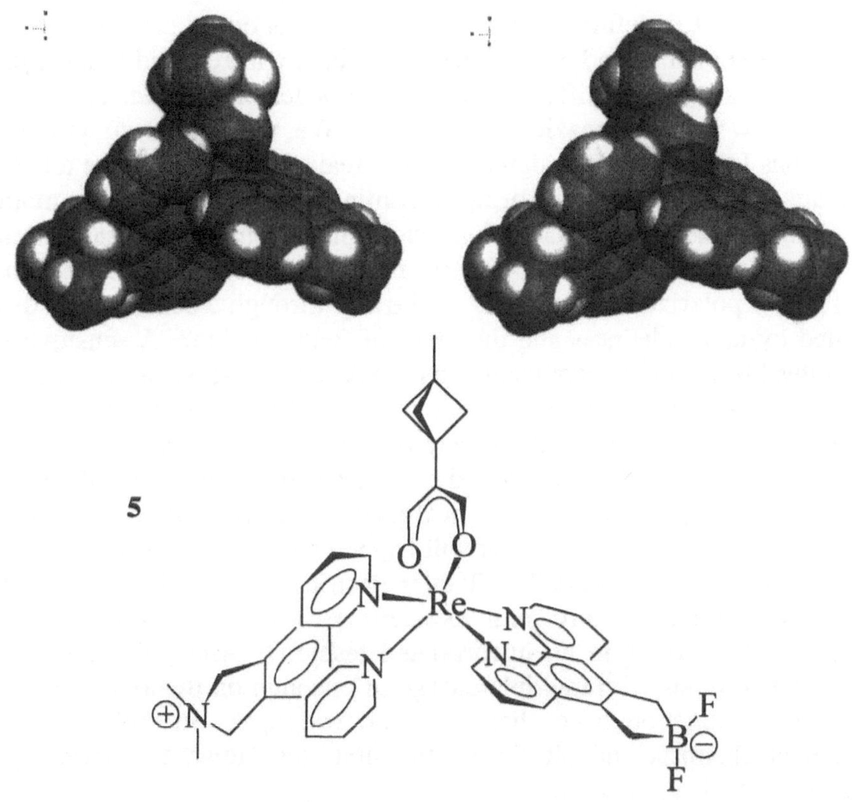

The question was, will the impact of the cold but relatively fast helium atoms on the blades cause the turbine to rotate, or will it just cause it to shake violently, while the energy of the collisions is lost to irregular thermal motion of the grid? Regardless of the initial conditions and of the mounting mode, we found that the turbine begins to rotate in the intuitively expected direction within a few ps and continues to rotate at a quite stable rate that depends on the density and the velocity of the gas, typically one turn every few dozen ps. The rotation is not perfectly steady, but reflects the violent collisions with the gas atoms. These also cause the turbine to rock irregularly. Since the kinetic energy of the helium atoms corresponds to room temperature or only a little above, it is not a surprise that the grid with the turbine warms up to this temperature after a while, independently of the temperature at which it was started. Under these conditions, the chemical bonds that hold the grid and the turbine together will be stable essentially indefinitely.

There seems to be little doubt about the qualitative outcome: this turbine will indeed rotate in the expected sense, with an average kinetic energy that greatly exceeds *kT*, defined by the irregular motions of the grid and the effluent gas. We next plan to find out whether the turbine will respond in a similar way to a mere gas pressure difference on the two sides of the grid, or whether the laminar flow from a nozzle is essential. We would like to pursue these simulations further with a design that is still realistic but permits the turbine axle to be attached at both ends in order to eliminate the violent rocking motion.

Subsequently, we plan to introduce time-dependent external electric field into the simulation. Will the turbine act as a molecular submarine when exposed to circularly polarized microwaves, and travel through a solution in a direction dictated by its handedness and the sense of field rotation? A sensitive way to detect this kind of movement would be to look for optical resolution of a racemic mixture of turbines. Finding a solvent with acceptably low dielectric losses that still dissolves the dipolar turbine may be hard, but perhaps supercritical CO_2 might work. We have recently found out that a simulation of a similar molecular submarine concept has been performed totally independently by Rabitz [22].

The next question is, what will happen when a large number of dipolar turbines are attached to a grid? Will their electrical and possibly also mechanical interactions cause them to rotate in lockstep? How far apart would they have to be before they rotate in an uncorrelated fashion? And how about a set of multipolar turbines with multiple charge of one sign on the axle, say positive, and compensating opposite charges on all wingtips? In collaboration with theoretical chemists and physicists, the first steps towards answering these questions have been taken [23].

Once the answers are understood, it will be fascinating to consider whether such an array of dipolar turbines could be driven by externally applied circularly polarized microwave radiation, and whether it would then pump gas and generate pressure waves (acting as a loudspeaker, or more likely, a softspeaker?). Would it be levitated in a gas or liquid? Better still, could a flow of gas directed at a grid of dipolar turbines be used to generate microwave radiation, permitting it to act as a microphone without a power supply? Could an array of such grids at adjustable separations act as a gas-driven tunable cavity? Would a grid of multipolar turbines produce a detectable magnetic field? Could temporal variation in the gas flow and thus in the magnetic field induce a detectable current in pick-up loops wound along the periphery of the grid?

Presently, we do not know the answers. The questions illustrate our conviction that the availability of designer grids, and ultimately, designer solids, will offer novel possibilities limited by little else than our imagination. Perhaps chemists are now on their way to fulfill Richard Feynman's old dream [24].

Acknowledgement. Our work was supported initially by the U.S. National Science Foundation (CHE-9213399, CHE-9412767) and subsequently by the U.S. Department of Energy (FG03-94ER12141). Technical assistance by Mr. Kevin Vanderveen is gratefully acknowledged.

7. References

1. Michl, J. (1996) Synthesis of Giant Modular Structures, in C. Chatgilialoglu and V. Snieckus (eds.), *Chemical Synthesis: Gnosis to Prognosis*, Kluwer Academic Publishers, Dordrecht, p. 429.
2. Kaszynski, P. and Michl, J. (1988) [n]Staffanes: A Molecular-Size `Tinkertoy' Construction Set for Nanotechnology. Preparation of End-functionalized Telomers and a Polymer of [1.1.1]Propellane, *J. Am. Chem. Soc.*, **110**, 5225; Michl, J., Kaszynski, P., Friedli, A.C., Murthy, G.S., Yang, H.-C., Robinson, R.E., McMurdie, N.D., and Kim, T. (1989) Harnessing Strain: From [1.1.1]Propellanes to Tinkertoys, in A. de Meijere and S. Blechert (eds.), *Strain and Its Implications in Organic Chemistry*, Kluwer Academic Publishers, Dordrecht, p 463; Michl, J. (1995) The 'Molecular Tinkertoy' Approach to Materials, in J.F. Harrod and R.M. Laine (eds.), *Applications of Organometallic Chemistry in the Preparation and Processing of Advanced Materials*, Kluwer Academic Publishers, Dordrecht, p. 243.
3. Kaszynski, P., Friedli, A.C., and Michl, J. (1992) Towards a Molecular-Size 'Tinkertoy' Construction Set. Methyl [*n*]Staffane-3-carboxylates, Dimethyl [*n*]Staffane-3,$3^{(n-1)}$-dicarboxylates, and [*n*]Staffane-3,$3^{(n-1)}$-dithiols from [1.1.1]Propellane, *J. Am. Chem. Soc.*, **114**, 601.
4. Tinkertoy is a trademark of Playskool, Inc., Pawtucket, RI 02862, and designates a children's toy construction set consisting of straight wooden sticks and other simple elements insertable into spool-like connectors.
5. Müller, J., Base, K., Magnera, T.F., and Michl, J. (1992) Rigid-Rod Oligo-*p*-Carboranes for Molecular Tinkertoys, *J. Am. Chem. Soc.*, **114**, 9721.
6. Szeimies, G. (1989) From Bicyclo[1.1.0.] Butanes to [n.1.1] Propellanes, in A. de Meijere and S. Blechert (eds) *Strain and Its Implications in Organic Chemistry*, Kluwer Academic Publishers, Dordrecht, p 361.
7. Gilardi, R., Maggini, M., and Eaton, P.E. (1988) X-ray Structures of Cubylcubane and 2-*tert*-Butylcubylcubane: Short Cage-Cage Bonds, *J. Am. Chem. Soc.*, **110**, 7232; Eaton, P.E. and Tsanaktsidis, J. (1990) The Reactions of 1,4-Dihalocubanes with Organolithiums. The Case for 1,4-Cubadiyl, *J. Am. Chem. Soc.*, **112**, 876.
8. Yang, X., Jiang, W., Knobler, C.B., and Hawthorne, M.F. (1992) Rigid-Rod Molecules: Carborods. Synthesis of Tetrameric *p*-Carboranes and the Crystal Structure of Bis(tri-*n*-butylsilyl)tetra-*p*-carborane, *J. Am. Chem. Soc.*, **114**, 9719.
9. Zimmerman, H.E., King, R.K., and Meinhardt, M.B. (1992) Molecular Rods: Synthesis and Properties, *J. Org. Chem.*, **57**, 5484.

10. Cotton, F.A. and Walton, R.A. (1993) *Multiple Bonds between Metal Atoms*, Clarendon Press, Oxford, Chap 7.
11. Ibrahim, M. A., King, B.T., and Michl, J., unpublished results.
12. Janecki, T., Shi, S., Kaszynski, P., and Michl, J. (1993) [n]Staffanes with Terminal Nitrile and Isonitrile Functionalities and Their Metal Complexes, *Collect. Czech. Chem. Commun.*, **58**, 89.
13. Diercks, R., Armstrong, J.C., Boese, R., Vollhardt, K.P.C. (1986) Hexaethynylbenzene, *Angew. Chem. Int. Ed. Engl.* **25**, 268.
14. Schöberl, U.; Magnera, T. F.; Harrison, R.; Fleischer, F.; Pflug, J. L.; Schwab, P. F. H.; Meng, X.; Lipiak, D.; Noll, B. C.; Allured, V. S.; Rudalevige, T.; Lee, S.; Michl, J. Towards a Hexagonal Grid Polymer: Synthesis, Coupling, and Chemically Reversible Surface-Pinning of the Star Connectors, 1,3,5-$C_6H_3(CB_{10}H_{10}CX)_3$, *J. Am. Chem. Soc.*, in press.
15. Janoušek, Z., Harrison, R.M., Grüner, B., Magnera, T.F., and Michl, J., unpublished results.
16. Harrison, R. M.; Brotin, T.; Noll, B. C.; Michl, J. Towards a Square Grid Polymer: Synthesis and Structure of Pedestal-Mounted Tetragonal Star Connectors, C_4R_4-Co-C_5Y_5, *Organometallics*, in press.
17. Magnera, T. F.; Peslherbe, L. M.; Körblová, E.; Michl, J. The Organometallic 'Molecular Tinkertoy' Approach to Planar Grid Polymers, *J. Organomet. Chem.*, in press.
18. ADDED IN PROOF: In the time since the conference, we have indeed succeeded in transferring pieces of the grid to a graphite surface and obtaining their STM images. They were 0.7 nm thin, as expected, and up to about 150 nm × 150 nm wide, and were clearly composed of the expected 2.5 nm × 2.5 nm hollow squares. The long-range order was poor, presumably due to the irreversible nature of the coupling reaction used: Magnera, T. F., Pecka, J., and Michl, J., unpublished results. Presented at the 4th International Conference on Frontiers of Polymers and Advanced Materials, Cairo, Egypt, Jan. 4 - 9, 1997 (Book of Abstracts, p. 19).
19. Conner, M., Janout, V., and Regen, S.L. (1993) Molecular Sieving by a Perforated Langmuir-Blodgett Film, *J. Am. Chem. Soc.* **115**, 1178. Lee, W., Hendel, R.A., Dedek, P., Janout, V., Regen, S.L. (1995) Unusual Pressure Effects on the Permeation Properties of a Langmuir-Blodgett Composite Membrane, *J. Am. Chem. Soc.* **117**, 10599.
20. Rappé, A.K., Casewit, C.J., Colwell, K.S., Goddard III, W.A., and Skiff, W.M. (1992) UFF, a Full Periodic Table Force Field for Molecular Mechanics and Molecular Dynamics Simulations, *J. Am. Chem. Soc.*, **114**, 10024. Rappé, A.K., Goddard III, W.A. (1991) Charge Equilibration for Molecular Dynamics Simulations, *J. Phys. Chem.* **95**, 3358. We are grateful to Prof. Rappé for several subroutines.
21. We used an extensively modified form of the programs Moil and Moil-View, which were kindly provided by Prof. Elber (Hebrew University).
22. Space, B., Rabitz, H., Lörincz, A., and Moore, P. (1996) Feasibility of Using Photophoresis to Create a Concentration Gradient of Solvated Molecules, *J. Chem. Phys.* **105**, 9515.
23. DeLeeuw, S. W., Solvaeson, D., Ratner, M. A., Michl, J. (1996) Molecular Dipole Chains: Excitations and Dissipation, submitted for publication.
24. Feynman, R.P. (1960) The Wonders that Await a Micro-Microscope, *Saturday Rev.* **43**, April 2, 45; Feynman, R. P. (1961) There's Plenty of Room at the Bottom, in H.P. Gilbert (ed), *Miniaturization*, Reinhold, New York, NY, p. 282.

Tetraethynylethenes: Versatile Carbon-Rich Building Blocks for Two-Dimensional Acetylenic Scaffolding

François Diederich
Department of Chemistry
Swiss Federal Institute of Technology
ETH-Zentrum, CH-8092 Zürich

ABSTRACT. Derivatives of tetraethynylethene (TEE, 3,4-diethynyl-hex-3-ene-1,5-diyne) constitute a versatile "molecular construction kit" for acetylenic molecular scaffolding. TEEs have been introduced into multinanometer-sized functional molecular and polymeric materials with stable, extended carbon cores that exhibit unusual electronic and optical properties. In addition, the planar TEE carbon frame is a basic repeat unit for the construction of two-dimensional crystalline all-carbon networks. Starting from *cis*-bis-deprotected TEEs, annulenes were prepared as macrocyclic precursors to such networks. The challenges encountered in the formation of extended regular carbon networks by oxidative acetylenic coupling are discussed, and techniques from supramolecular chemistry are proposed to overcome the difficulties that prevented their preparation so far. One approach consists in the self-assembly of metal-acetylenic networks under thermodynamic control and error checking, followed by reductive elimination of the metal centers to the all-carbon net. Expanded radialenes represent another class of stable, extended carbon-rich compounds which were prepared for the first time starting from TEE precursors. *Trans*-bis-deprotected TEEs provided access to rod-like oligomers and polymers with the novel polytriacetylene (PTA) backbone. The redox-properties of these remarkably stable materials are discussed. Tetrakis(phenylethynyl)ethene was shown to form highly ordered charge-transfer complexes with π-acceptors in the solid state and in solution. By attaching *p*-donor and *p*-acceptor substituted phenyl rings to TEEs, novel NLO materials were obtained. It was shown for a large class of TEEs that donor/acceptor substitutions and fully two-dimensional conjugation strongly enhance the third-order nonlinear optical properties. The relevance of the results obtained from studies of extended unsaturated carbon-rich materials for carbon allotropy in general is discussed.

1. Introduction

In 1989, graduate student Yves Rubin and I planned the preparation of regular two-dimensional carbon networks such as **2** and **3** containing the carbon frame of tetraethynylethene (**1**, TEE, 3,4-diethynyl-hex-3-ene-1,5-diyne) as the infinite repeat unit [1]. As a synthetic approach to these carbon nets, we proposed, in a rather naive way (from today's perspective), the oxidative polymerization of **1**. A literature survey rapidly showed that **1** was unknown; only the tetraphenyl derivative as well as a few persilylated and peralkylated derivatives of **1** had been described [2]. The first synthesis of **1** was

J. Michl (ed.), Modular Chemistry, 17–31.

accomplished by Yves Rubin at the end of 1990 and communicated shortly after in *Angewandte Chemie* [3]. Since then, synthetic routes to tetraethynylethenes with essentially any desired substitution and protection pattern have been worked out and published in full detail [4,5]. This has provided a unique "molecular construction kit" for the preparation of novel, multinanometer-sized functional molecular and polymeric materials with extended carbon cores [6-8]. In this paper, I discuss the properties of these extended TEE scaffolds, which represent end-capped conjugated carbon sheets and rods. Also, new approaches towards all-carbon networks such as **2** and **3** are presented [9,10]. As a fundamental result, comparisons of the properties of the extended TEE scaffolds to those of the fullerenes provided new insight into the electronic properties of unsaturated carbon matter in general and thus profoundly enhanced the present knowledge on the allotropy of carbon. It will become clear from these discussions that tetraethynylethenes rival fullerenes as novel molecular components for future materials with technological applications.

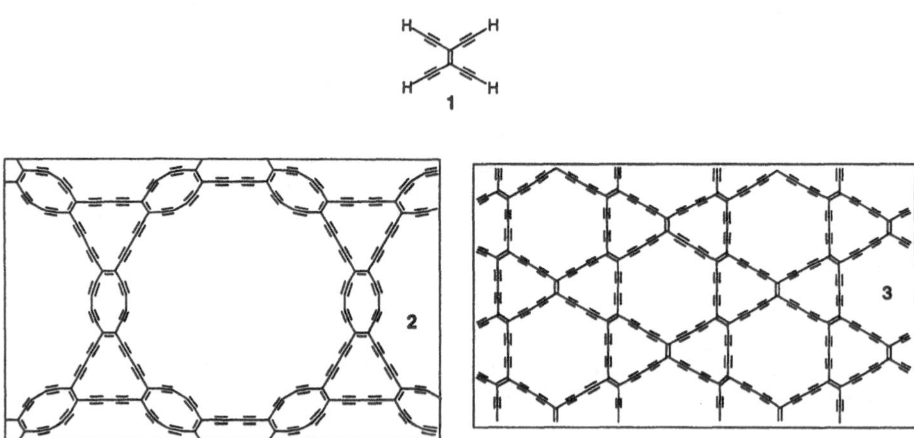

2. Acyclic and Macrocyclic Tetraethynylethene Molecular Scaffolding

Oxidative polymerization (CuCl, *N,N,N',N'*-tetramethylethylenediamine (TMEDA), O_2) of *trans*-bis-deprotected **4** yielded, after end-capping with phenylacetylene, the remarkably stable, high melting, and readily soluble oligomers **5-9** with the hitherto unknown polytriacetylene backbone (Scheme 1A). The perfectly planar dimeric **6** showed a remarkable supramolecular structure in the solid state (Figure 1). In the crystal, molecules of **6** pack in chains composed of three parallel-stacked rows of conjugated polytriacetylene (PTA) backbone insulated by the (i-Pr)$_3$Si groups. There is electronic communication between the molecules in the three rows of each chain, with the phenyl rings of one molecule stacking directly above the central C=C bonds of its neighbors' tetraethynylethene fragments. The centroid-centroid distance between the stacking subunits is 3.73 Å. Thus, each chain can be seen as an insulated 'wire'.

Compounds **5-9** underwent facile one-electron reductions with the number of reversible reduction steps being equal to the number of TEE moieties in each molecular rod. Thus, the first reduction of **5** occurred at $E^o = -1.57$ V (*vs.* the ferrocene/ferricinium (Fc/Fc^+) couple; in THF) whereas the first reduction of **9** was much facilitated and occurred at –1.07 V.

When the oxidative polymerization of **4** was carried out in 1,2-dichlorobenzene at 65 °C in the presence of 3,5-bis(tert-butyl)phenylacetylene as end-capping reagent, an air-stable longer-chain PTA (**10**) was obtained (Scheme 1B) for which the number-averaged molecular weight was determined as $M_n = 9600$ (degree of oligomerization $X_n = 22$) [8]. Deep red-brown **10** was soluble in hot chloroform and in 1,2-dichlorobenzene above 65 °C and showed an optical gap of $E_g = 2.0$ eV, *i.e.* in the range of values measured for polydiacetylenes (PDAs) [11,12]. Polymer **10** underwent a reversible one-electron reduction at ca. – 0.7 V (*vs.* Fc/Fc^+) in THF.

Scheme 1A

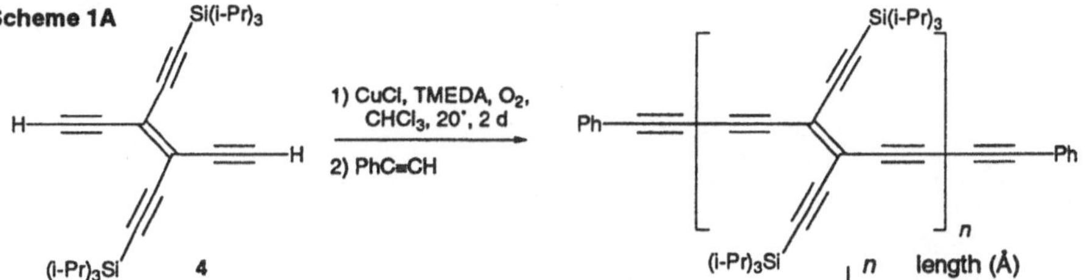

	n	length (Å)
5	1	19.4
6	2	26.8
7	3	34.3
8	4	41.8
9	5	49.2

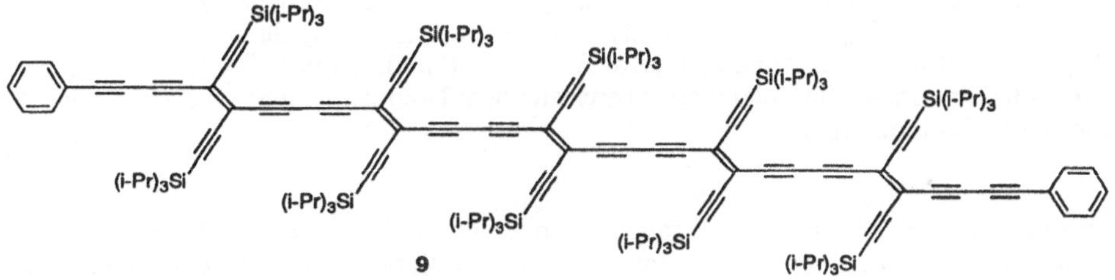

Scheme 1B

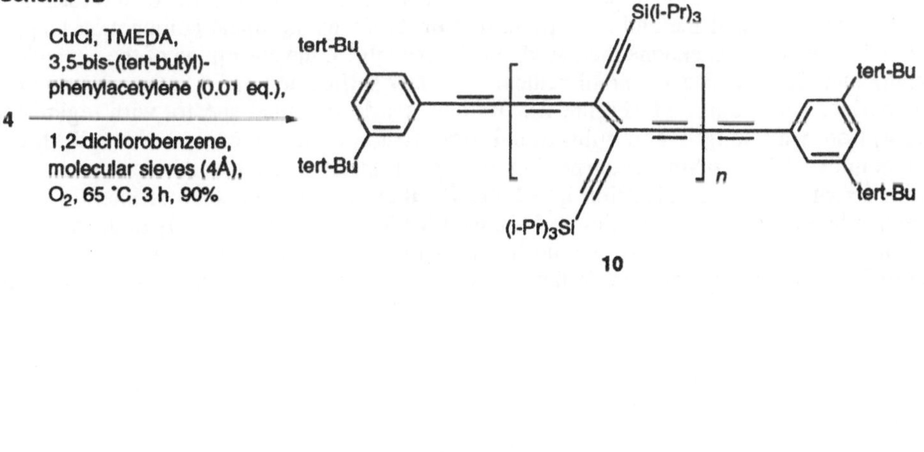

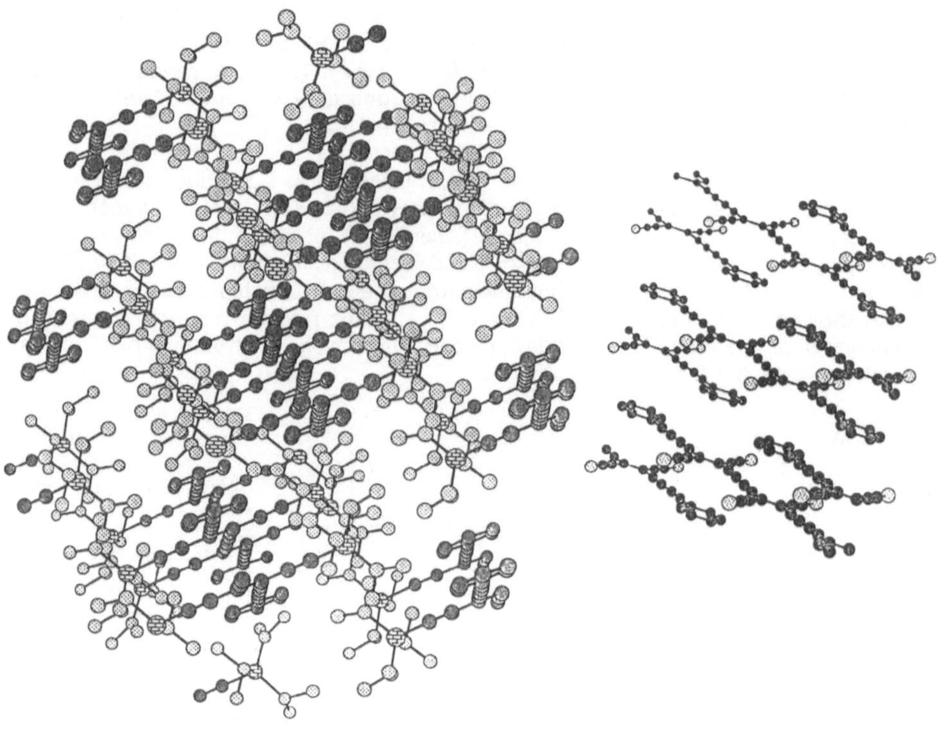

Figure 1. Left: Crystal packing of **6** showing the chains formed by three parallel-stacked rows of conjugated backbone (dark atoms) surrounded by the insulating $(i\text{-Pr})_3Si$ groups (light atoms; different shading for Si and C atoms). Right: Crystal packing of **6** showing the intermolecular interactions between phenyl and tetraethynylethene moieties (C atoms dark, Si atoms light).

By other routes, the cross-conjugated expanded dendralenes **12** and **13** were prepared which, similar to **5-9**, were stable to moisture, air, and light, and remained unchanged for months on the laboratory bench [6,13]. The electronic absorption spectra in the series **11-13** are remarkable. In the series **5-9**, an increasing bathochromic shift of the longest-wavelength band and the end absorption was observed as the linear conjugation length along the molecular backbone increased. In sharp contrast, the absorption of the expanded dendralenes **12** and **13** is not at all bathochromically shifted compared to the spectrum of comparison compound **11** (Figure 2). Rather, nearly identical spectra with regard to absorption wavelength and vibrational fine structure are measured for all three compounds. We explain the surprising absorption spectra in the dendralene series by limited π-electron delocalization in **12** and **13**, that appears to extend efficiently only through the longest linearly conjugated fragment which corresponds exactly to the dodeca-3,9-diene-1,5,7,11-tetrayne backbone in **11**. Apparently, cross-conjugation across the central bifurcating tetraethynylethene moieties in **12** and **13** is not a very efficient mechanism for π-electron delocalization. However, steric reasons for the reduced π-electron conjugation, namely distortion of the chromophores out of planarity due to repulsion between the bulky interior (i-Pr)$_3$Si groups cannot be fully excluded.

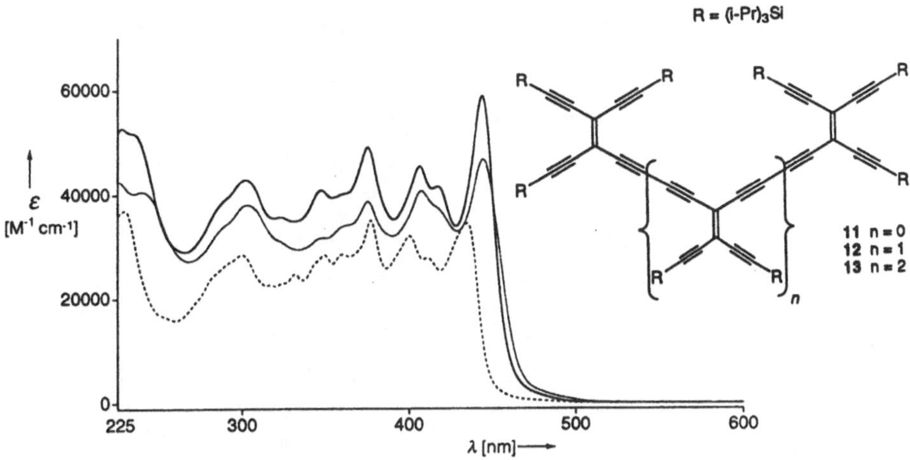

Figure 2. Electronic absorption spectra of **11**(-------), **12** (———), and **13** (·······) in hexane. $T = 298$ K.

A limited extent of π-electron delocalization, presumably due to inefficient cross-conjugation, was also observed in the electronic absorption spectra of the first series of expanded radialenes **14-16** [7,14]. These end-capped, nanometer-sized carbon sheets with diameters, not including the (i-Pr)$_3$Si groups, of *ca.* 17 (**14**), 19 (**15**), and 22 Å (**16**) are amazingly stable compounds with melting points above 220 °C and can be viewed as persilylated C_{40}, C_{50}, and C_{60} isomers, respectively. Despite their high degree of unsaturation, **14-16** are yellow colored and, similar to the series **11-13**, their UV/Vis spectral end absorption remains almost constant. Again, π-electron delocalization extends efficiently only through the longest linearly conjugated fragment (which is equal to the conjugated π-backbone in **11**), mainly due to inefficient macrocyclic cross-conjugation. This explanation is further supported by recent studies [15] of expanded radialenes related to **14-16** but without the bulky (i-Pr)$_3$Si groups which could induce nonplanarity of the macrorings. Compounds **14-16** show 2, 3, and 4 one-electron reduction waves, respectively [7]; the fact that the first reversible reduction occurs in all three compounds at similar potentials [$E^o = -1.08$ (**14**), -1.35 (**15**), and -1.27 (**16**) *vs.* Fc/Fc$^+$ in THF] is also in accord with π-conjugation being limited by inefficient cross-conjugation and possibly to some extent also by nonplanarity.

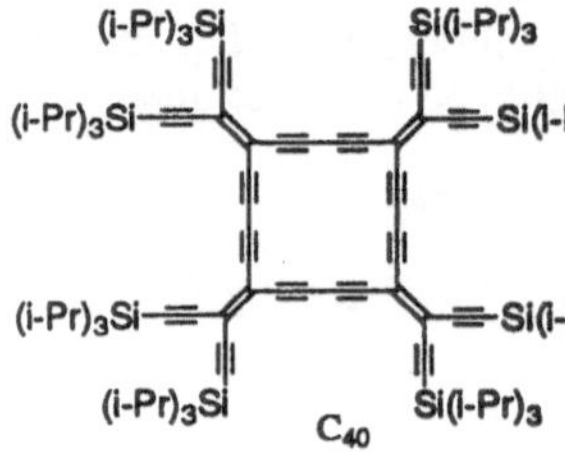

14

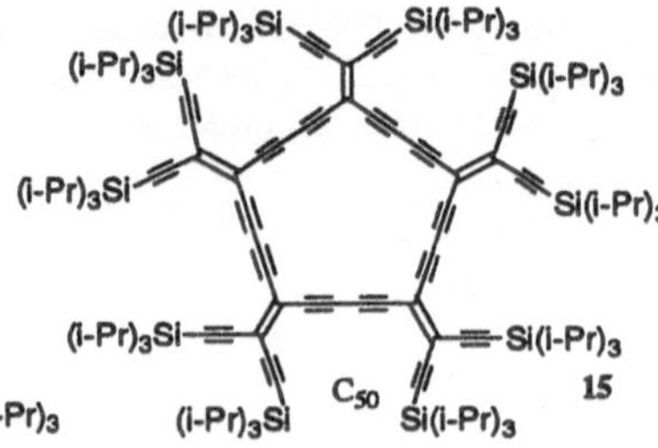

15

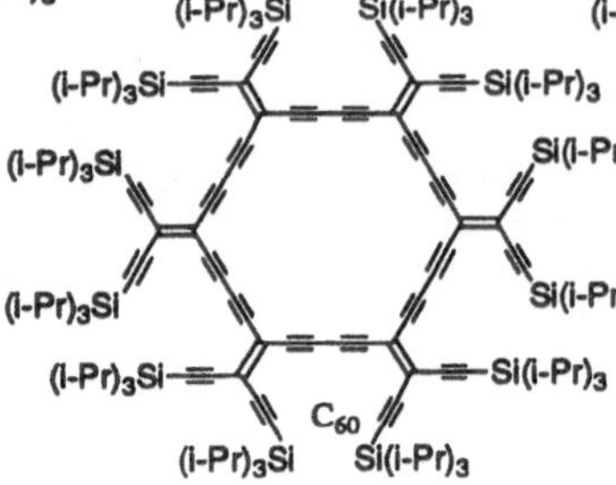

16

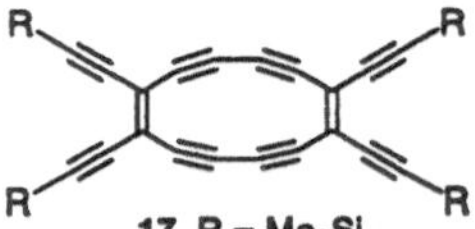

17 R = Me_3Si
18 R = (i-Pr)$_3$Si
19 R = H

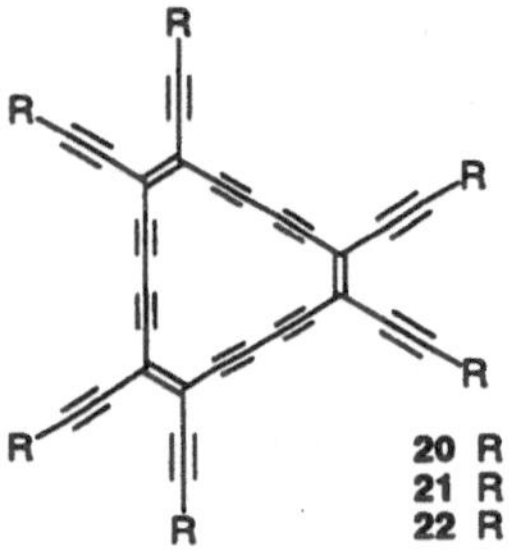

20 R = Me_3Si
21 R = (i-Pr)$_3$Si
22 R = H

Whereas the cross-conjugated expanded radialenes are highly stable compounds, the annulenic precursors **17-22** to the all-carbon networks **2** and **3** with extended linear macrocyclic π-conjugation are much more sensitive [7]. X-ray crystal structures demonstrated perfect planarity for (i-Pr)$_3$Si-protected octadehydro[12]annulene **18** and Me_3Si-protected dodecadehydro[18]annulene **20**. Both ^{1}H-NMR and electronic absorption spectral comparisons provided strong evidence that the macrorings in the deep-purple-colored **17** and **18** are antiaromatic (4n π-electrons) whereas those in yellow **20** and **21** are aromatic [(4n+2) π-electrons]. In agreement with these findings, **18** undergoes two stepwise one-electron reductions ($E^o = -0.99$ and -1.46 V *vs.* Fc/Fc$^+$ in THF) more readily than **21** ($E^o = -1.12$ V and -1.52 V). This redox behavior is best explained by the formation of an aromatic (4n+2) π-electron dianion from antiaromatic **18**, whereas **21** loses its aromaticity upon reduction. The 18–π electron systems expectedly showed higher stability than the 12–π electron systems which, in addition to the electronic destabilization, show considerable strain in the 12-membered ring as expressed by the strong bending in the butadiyne moieties of **18**, with C≡C–C angles as low as 164.5°.

Despite antiaromaticity and strain, **18** was quite stable. At room temperature, dilute pentane solutions remained unchanged over weeks, and crystals were stable to air and light for months. However, if a solution of **18** was concentrated without crystallization, significant decomposition occurred. The remarkable stability of **18** (m.p. 200 °C (dec.)) clearly originates from the way individual molecules arrange in the crystal. An examination of the crystal lattice of **18** (Figure 3) showed that the delicate annulenic cycles are offset in stacking, so that they are completely surrounded by the bulky, inert (i-Pr)$_3$Si groups. This insulating matrix-type effect of (i-Pr)$_3$Si groups has now been observed in several X-ray crystal structures of extended acetylenic compounds and seems to represent a more general mode to stabilize these compounds in the solid state [5-7].

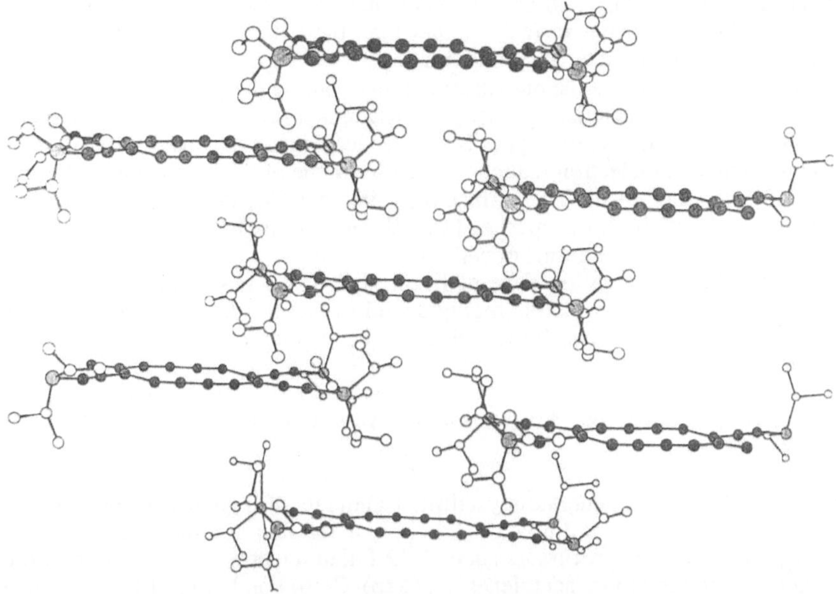

Figure 3. Crystal lattice of **18** showing the conjugated C_{20} chromophores (dark) fully surrounded by the bulky (i-Pr)$_3$Si groups.

The Me$_3$Si-protecting groups of **17** and **20** could be removed by treatment with sodium tetraborate (borax) in MeOH/THF, yielding **19** and **22** as very unstable compounds. Whereas dilute solutions of **22** decomposed only slowly in the dark at –20 °C, rapid decomposition of **19** even prevented its characterization. Attempts to obtain characterizable structures related to network **3** by oxidative polymerization of **22** have failed thus far.

From the work on carbon-rich acetylenic acyclic and macrocyclic scaffolds constructed from TEE building blocks, the following conclusions emerge for unsaturated carbon-based matter in general [6,7]:

i) Cross-conjugation drastically reduces the extent of π-electron delocalization, and, by this compartmentalization of the π-electrons, stabilizes the molecules in terms of chemical reactivity. This principle may serve as a useful and previously unrecognized guideline in the synthetic targeting of large unsaturated all-carbon molecules and networks. Buckminsterfullerene C_{60} is best described electronically as a cross-conjugated molecule with [5]radialene substructures [16]. Also, the electronic structure of graphite may be viewed as cross-conjugated.

ii) The high solubility and high stability of the expanded radialenes with their large all-carbon cores raises great hope that much larger carbon surfaces can be prepared and handled as long as the peripheral valences contain stabilizing and solubilizing groups such as the $(i\text{-}Pr)_3Si$ groups in **14-16**. In the preparation of **14** and **16**, we isolated small quantities of even larger expanded radialenes with central C_{80}, C_{100}, and C_{120}-cores, which were found to be perfectly stable compounds [7]. A further expansion of the molecular dimensions will yield compounds which resemble large graphite or diamond fragments in that the carbon surfaces, rather than the heterofunctions at the peripheral valences, determine their materials properties.

iii) Facile, multiple reversible one-electron reductions paired with difficult oxidizability seems to represent a characteristic property of unsaturated acetylenic carbon-rich molecules and presumably also carbon allotropes. Both acyclic and macrocyclic scaffolds underwent multiple reversible one-electron reductions whereas none of the compounds was oxidized below + 1.0 V (*vs.* Fc/Fc^+ in THF). Since the fullerenes C_{60} and C_{70} as well as many of their adducts exhibit similar properties [16,17], such redox behavior may be a general phenomenon for unsaturated forms of carbon.

iv) Finally, laser-desorption time-of-flight (LD-TOF) mass spectrometry with or without matrix assistance, which is highly useful in the analytics of fullerene materials [18], is also an extremely powerful technique in the characterization of large carbon-rich acetylenic scaffolds.

3. A Metal-Template Approach to the Construction of All-Carbon Networks

Numerous attempts to generate crystalline fragments of regular two-dimensional all-carbon networks such as **2** or **3** by oxidative polymerization of either tetraethynylethene (**1**) or larger macrocyclic precursors such as **22** failed under a variety of conditions. In the oxidative coupling of two acetylenes, an C(sp)–C(sp) bond of *ca.* 130 kcal mol^{-1} that joins two acetylenes is formed irreversibly [19]. Any error made in a conventional oxidative acetylenic polymerization targeting a regular two-dimensional network will therefore be irreversible, thus lowering the crystallinity of the resulting material. Such errors are unavoidable in oxidative polymerizations; therefore we believe that a metal-template approach will be a superior technique to construct regular networks. In this approach, judicious choice of transition metal and ligands should allow the reversible formation of σ-bis(acetylide) substructures [20] leading to the assembly of a metal-acetylenic network such as **23** under thermodynamic control with error checking [21]. In a second step, the all-carbon network (*e.g.* **24**) could be generated by reductive elimination at the metal centers [22]. Metal-acetylenic networks are by themselves of considerable novelty and interest since the metal centered *d*-orbitals may guarantee some electronic delocalization within the plane.

In research toward these aims, we prepared compounds **25a-c**, **26**, and **27** and proved that tetraethynylethene units may function as η^1-ligands in transition metal complexes [23]. The electronic absorption spectra of **25a-c** and **27** provided clear evidence for transmission of electron density across the metal centers via metal-to-ligand charge-transfer (MLCT) leading to electronic delocalization over the entire planar π-systems. Platinum centers however do not seem suitable for the construction of the targeted regular metal-acetylide networks, since the σ-acetylide bonds are too stable, preventing assembly under thermodynamic control. Decomplexation of the TEE moieties in **2c** could only be achieved under very special conditions. Therefore, in continuation of this work, we now explore other transition metals for the assembly of networks such as **23**, in collaboration with Prof. H. Berke from the University of Zürich.

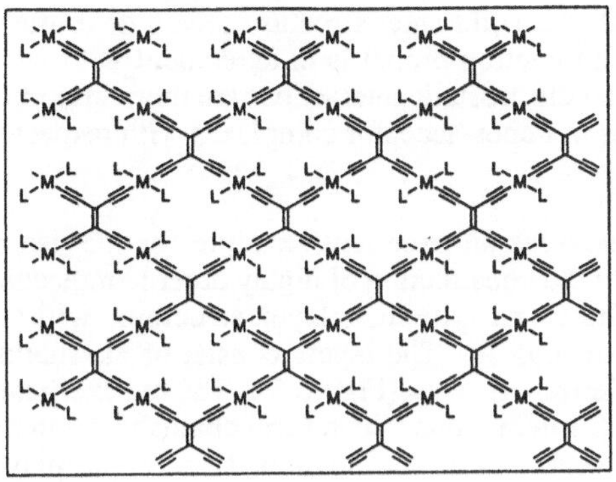

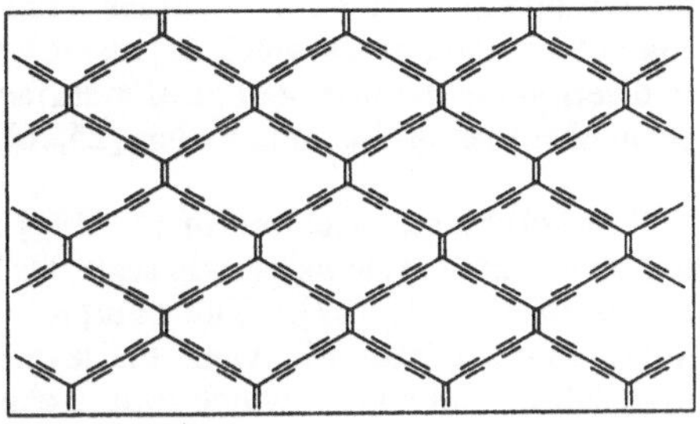

24

23

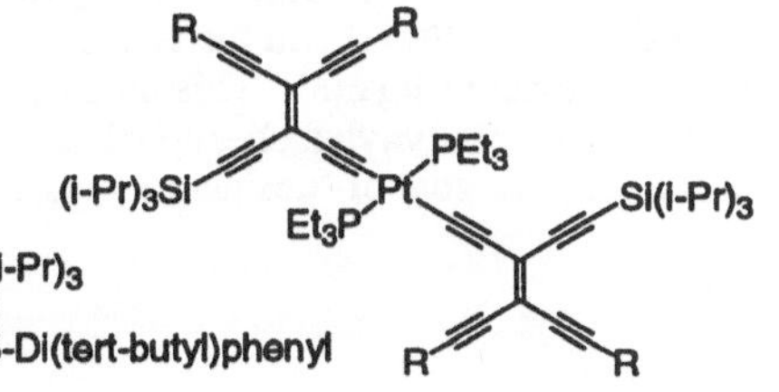

25a: R = Si(i-Pr)$_3$
25b: R = Ph
25c: R = 3,5-Di(tert-butyl)phenyl

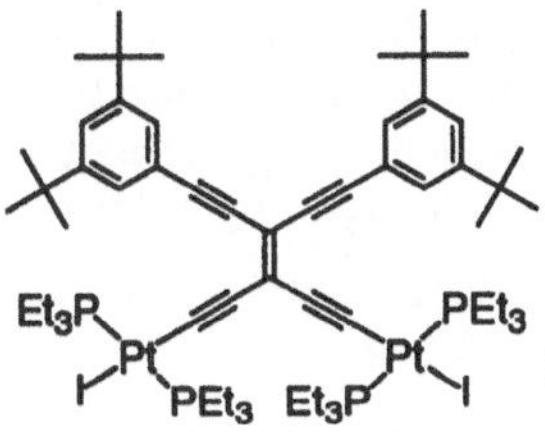

26

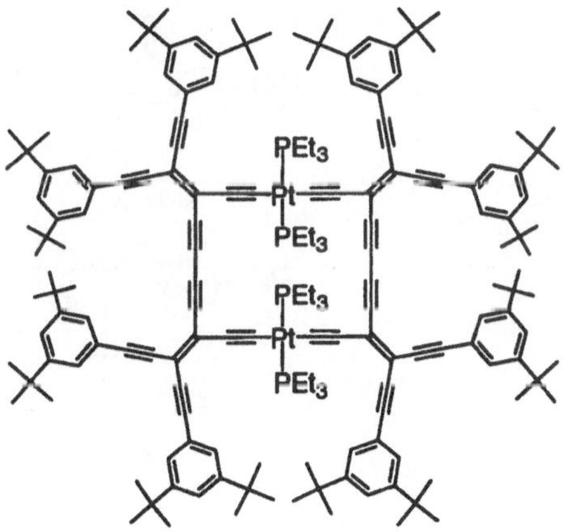

27

4. Crystalline Charge-Transfer Complexes of Tetrakis(phenylethynyl)-ethene

Tetrakis(phenylethynyl)ethene (**28**) and the π-acceptors 2,4,7-trinitrofluoren-9-one (**29**) and (2,4,7-trinitrofluoren-9-ylidine)malononitrile (**30**) form highly ordered donor-acceptor π-complexes having 1:2 stoichiometry in the solid state (Figure 4) whereas in solution, relatively weak 1:1 complexes are formed (formation free energies: $\Delta G° \approx -1.4$ kcal mol^{-1}) [24]. The orientation of the donor with respect to the two acceptors in the solid state complexes showed a good correlation with the atom-centered point charges on each component. Since there is a very poor correlation between the orbital coefficients of the donor HOMO and the acceptor LUMO, the solid state structure seems determined largely by electrostatic interactions. This interaction model is in agreement with work from other laboratories which has asserted that electrostatic interactions are more important in determining the orientations of π-surfaces in donor-acceptor complexes with respect to each other than orbital interactions [25,26].

The solid state structure of [**28**·**30**$_2$] provides an excellent example for the use of acetylene-based systems as rigid scaffolds for the construction of highly ordered molecular arrays in crystals [27]. The complex shows an extended layer structure, with an interplanar separation between the layers of 3.33 Å. The layers consist of alternating molecules of **28** and **30** which form a checkerboard array (Figure 5). The checkerboard pattern is stabilized by three sets of C–H····X interactions which form chains through the layers. The CH_2Cl_2 solvent molecules, incorporated in a stoichiometric amount, play a pivotal role in the stabilization of the layered structure. In addition to the 'in layer' contacts between the chlorine atoms of the CH_2Cl_2 molecules and H-atoms on both **28** and **30**, there are also short contacts between the H-atoms of the CH_2Cl_2 molecules and both the aromatic rings of **28** (centroid···H distance 2.98 Å) and the N-atoms of the nitrile groups of **30** (N····H distance 2.67 Å) in the adjacent layers. This study clearly shows the great potential of rigid frameworks based on tetraethynylethene, coupled with relatively weak π-π interactions, for achieving the challenging goal of 'designer solids'.

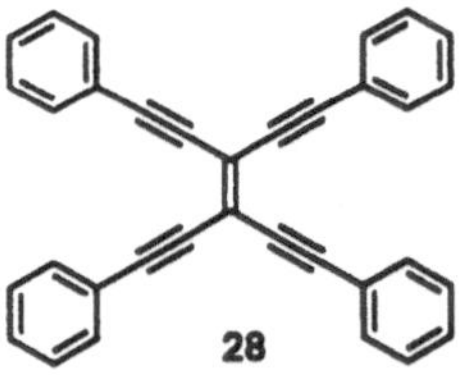

28

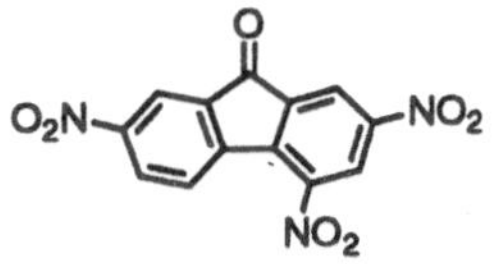

O

O_2N NO_2 NO_2 29

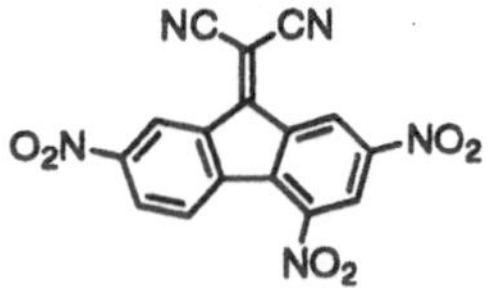

NC CN

O_2N NO_2 NO_2 30

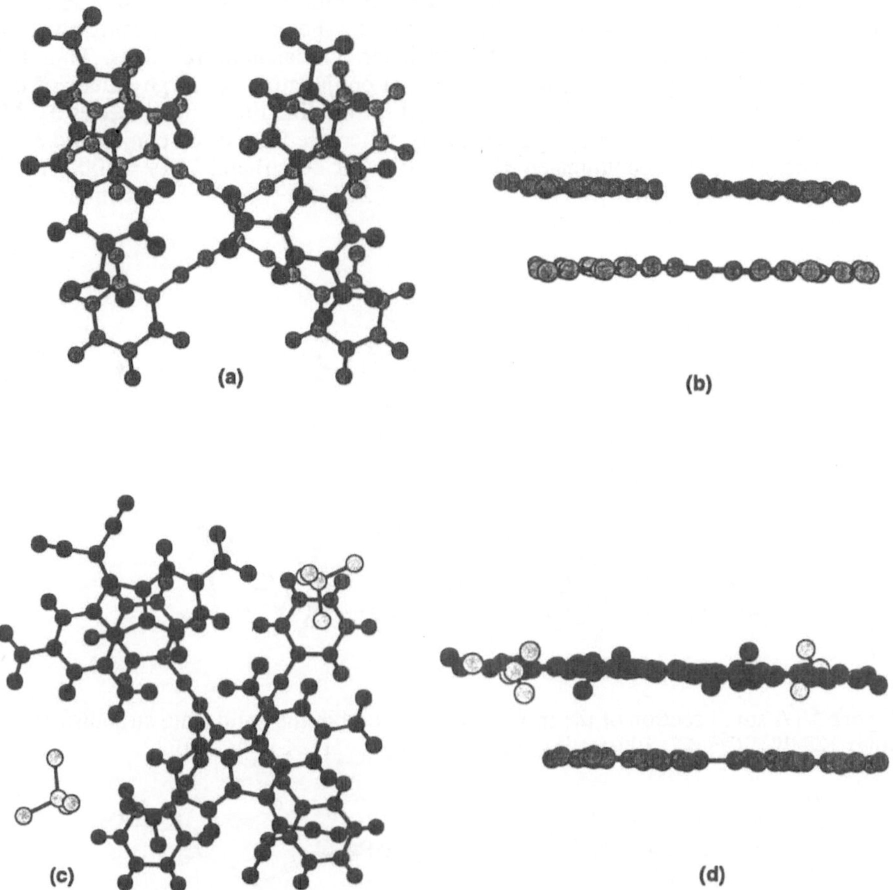

Figure 4. Ball-and stick representations of the solid state structure of complexes [**28**·**29**$_2$] (a, b) and [**28**·**30**$_2$] (c, d). (a) and (c) are views perpendicular to the mean plane of **28** and (b) and (d) are views parallel to the mean plane of **28**. In the structure of [**28**·**30**$_2$], the solvent CH_2Cl_2, which is included in stoichiometric amount in two equally populated orientations, is shown.

5. Donor-Acceptor Substituted Tetraethynylethenes for Nonlinear Optical Applications

In most recent work [28,29] we prepared a comprehensive series of donor-acceptor substituted TEEs such as **31**-**34** (Figure 6) by palladium catalyzed acetylenic coupling. Several X-ray crystal structures demonstrated that the conjugated C-atom scaffolds of these remarkably stable compounds are planar, including the substituted phenyl rings. The absorption spectra showed that intramolecular donor-acceptor interactions via linear conjugation paths are substantially more effective than by cross-conjugation paths. The third-order nonlinear optical properties [30] were subsequently studied by using third-harmonic generation under non-resonant conditions in chloroform solution for a series of 16 donor/acceptor substituted TEEs [29], and it could be shown that significant structure-property relationships exist in this novel nonlinear optical material class. Thus the importance of acentricity and full two-dimensional conjugation for large values of the third-order nonlinearity were clearly demonstrated for the first time. In the series **31-34**, the values for the second-order hyperpolarizability γ increase from cross-conjugated **31** ($\gamma = 59 \cdot 10^{-36}$ esu) to linearly conjugated **32** ($\gamma = 1100 \cdot 10^{-36}$ esu) and **33** ($\gamma = 1000 \cdot 10^{-36}$ esu), to fully two-dimensionally conjugated **34** ($\gamma = 3200 \cdot 10^{-36}$ esu) which contains four linear and two cross-conjugation paths.

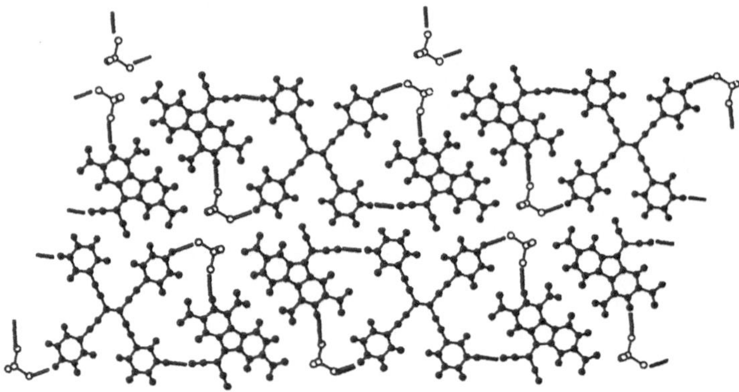

Figure 5. A small section of the infinite layers found in the solid state structure of the complex [**28**·**30**$_2$]·CH_2Cl_2 illustrating the stabilizing C–H···X interactions.

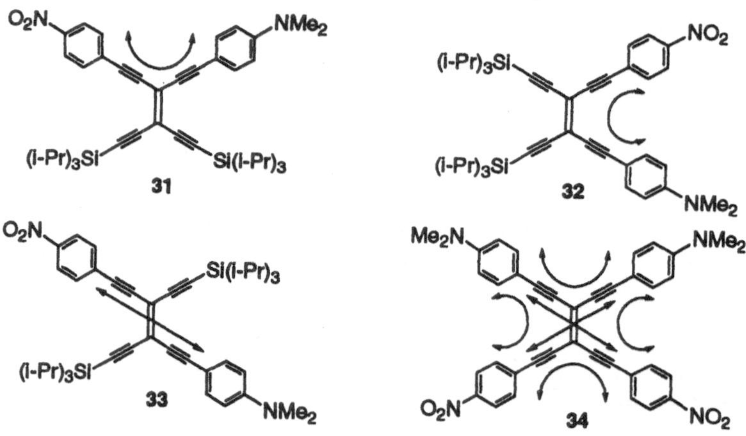

Figure 6. Donor-acceptor substituted tetraethynylethenes with conjugation paths.

Future work will take advantage of the fact that compound **33**, after removal of the silyl protecting groups, can be easily polymerized to give laterally donor-acceptor substituted PTAs suitable for thin film fabrication and study. Also, investigations of the structure-property relationship with respect to the first-order hyperpolarizability β are projected.

6. Conclusions

Although convenient syntheses of tetraethynylethenes (TEEs) have only been available for a few years, their importance for carbon-rich molecular scaffolding targeting a broadest variety of structures with interesting electronic and optical properties has already been fully established. TEEs are without doubt among the most convenient and versatile molecular building blocks for nanoscale and supramolecular construction known today. It is clear that only the small tip of a large iceberg in structural diversity and function has been reached today. We see future applications of TEE-derived molecules, polymers, and networks in molecular wiring and in optical processing. Tunable nonlinear optical properties, light-emitting diodes, and photorefractive polymers are targets of this research. Supramolecular construction such as the design of solid state structures based on non-covalent interactions using the rigid TEE π-skeleton will be increasingly pursued. Hybrid fullerene-tetraethynylethene molecular and polymeric scaffolding is under intense investigation. In addition to providing perspectives for technological applications, this research enhances the understanding of carbon-rich systems and ultimately, the allotropy of carbon. Finally, there remains the challenge of developing fully π-conjugated building blocks for three-dimensional scaffolding of similar efficiency and use to the TEEs. One such system could be tetraethynylallene [10], a compound that is currently under construction in our laboratory.

7. Acknowledgments

This work was supported by the Swiss National Science Foundation, the ETH research council, the U.S. Office of Naval Research and, in its beginning, the U.S. National Science Foundation.

8. References

1. Rubin, Y. (1991) 'Novel Carbon Allotropes: The Assembly of Carbon Atoms into Rings, Spheres, and Nets', Ph. D. Thesis, University of California at Los Angeles.
2. Hopf, H.; Kreutzer, M.; Jones, P. G. (1991) 'Zur Darstellung und Struktur von Tetrakis(phenylethinyl)ethen', *Chem. Ber.* **124**, 1471-1475.
3. Rubin, Y.; Knobler, C. B.; Diederich F. (1991) 'Tetraethynylethene', *Angew. Chem. Int. Ed. Engl.* **30**, 698-700.
4. Diederich, F. (1995) 'Oligoacetylenes' in '*Modern Acetylene Chemistry*', Stang, P. J.; Diederich, F., Eds., VCH, Weinheim, pp. 443-471.
5. Anthony, J.; Boldi, A. M.; Rubin, Y.; Hobi, M.; Gramlich, V.; Knobler, C. B.; Seiler, P.; Diederich, F. (1995) 'Tetraethynylethenes: Fully Cross-Conjugated π-Electron Chromophores and Molecular Scaffolds for All-Carbon Networks and Carbon-Rich Nanomaterials', *Helv. Chim. Acta* **78**, 13-45.

6. Boldi, A. M.; Anthony, J.; Gramlich, V.; Knobler, C. B.; Boudon, C.; Gisselbrecht, J.-P.; Gross, M.; Diederich, F. (1995) 'Acyclic Tetraethynylethene Molecular Scaffolding: Multinanometer-Sized Linearly Conjugated Rods with the Poly(triacetylene) Backbone and Cross-Conjugated Expanded Dendralenes', *Helv. Chim. Acta* **78**, 779-796.
7. Anthony, J.; Boldi, A. M.; Boudon, C.; Gisselbrecht, J.-P.; Gross, M.; Seiler, P.; Knobler, C. B.; Diederich, F. (1995) 'Macrocyclic Tetraethynylethene Molecular Scaffolding: Perethynylated Aromatic Dodecadehydro[18]annulenes, Antiaromatic Octadehydro[12]annulenes, and Expanded Radialenes', *Helv. Chim. Acta* **78**, 797-817.
8. Schreiber, M.; Anthony, J.; Diederich, F.; Spahr, M. E.; Nesper, R.; Hubrich, M.; Bommeli, F.; DeGiorgi, L.; Wachter, P.; Kaatz, P.; Bosshard, Ch.; Günter, P.; Colussi, M.; Suter, U. W.; Boudon, C.; Gisselbrecht, J.-P.; Gross, M. (1994) 'Polytriacetylenes: Conjugated Polymers with a Novel All-Carbon Backbone', *Adv. Mater.* **6**, 786-790.
9. Diederich, F.; Rubin, Y. (1992) 'Synthetic Approaches toward Molecular and Polymeric Carbon Allotropes', *Angew. Chem. Int. Ed. Engl.* **31**, 1101-1123.
10. Diederich, F. (1994) 'Carbon Scaffolding: Building Acetylenic All-carbon and Carbon-rich Compounds', *Nature* **369**, 199-207.
11. Baughman, R. H.; Brédas, J. L.; Chance, R. R.; Elsenbaumer, R. L.; Shacklette, L. W. (1982) 'Structural Basis for Semiconducting and Metallic Polymer/Dopant Systems', *Chem. Rev.* **82**, 209-222.
12. Wegner, G. (1969) 'Polymerisation von Derivaten des 2,4-Hexadiin-1,6-diols im kristallinen Zustand', *Z. Naturf.* **24b**, 824-832.
13. Hopf, H. (1984) 'The Dendralenes - a Neglected Group of Highly Unsaturated Hydrocarbons', *Angew. Chem. Int. Ed. Engl.* **23**, 948-960.
14. Hopf, H.; Maas, G. (1992) 'Preparation and Properties, Reactions, and Applications of Radialenes', *Angew. Chem. Int. Ed. Engl.* **31**, 931-954.
15. Tykwinski, R. R.; Schreiber, M.; Diederich, F. unpublished results.
16. Hirsch, A. (1994) '*The Chemistry of the Fullerenes*', Thieme, Stuttgart.
17. Boudon, C.; Gisselbrecht, J.-P.; Gross, M.; Isaacs, L.; Anderson, H. L.; Faust, R.; Diederich F. (1995) 'Electrochemistry of Mono- through Hexa-Adducts of C_{60}', *Helv. Chim. Acta* **78**, 1334-1344.
18. Isaacs, L.; Seiler, P.; Diederich, F. (1995) 'Solubilized Derivatives of C_{195} and C_{260}: The First Members of a New Class of Carbon Allotropes $C_{n(60+5)}$', *Angew. Chem. Int. Ed. Engl.* **34**, 1466-1469.
19. Diederich, F.; Rubin, Y.; Knobler, C. B.; Whetten, R. L.; Schriver, K. E.; Houk, K. N.; Li, Y. (1989) 'All-Carbon Molecules: Evidence for the Generation of Cyclo[18]carbon from a Stable Organic Precursor', *Science* **245**, 1088-1090.
20. Beck, W.; Niemer, B.; Wieser, M. (1993) 'Methods for the Synthesis of μ-Transition Metal Complexes without Metal-Metal Bonds', *Angew. Chem. Int. Ed. Engl.* **32**, 923-949.
21. Whitesides, G. M.; Mathias, J. P.; Seto, C. T. (1991) 'Molecular Self-Assembly and Nanochemistry: A Chemical Strategy for the Synthesis of Nanostructures', *Science* **254**, 1312-1319.
22. Low, J. J.; Goddard, W. A., III (1986) 'Theoretical Studies of Oxidative Addition and Reductive Elimination. 3. C–H and C–C Reductive Coupling from Palladium and Platinum Bis(phosphine) Complexes', *J. Am. Chem. Soc.* **108**, 6115-6128.
23. Faust, R.; Diederich, F.; Gramlich, V.; Seiler. P. (1995) 'Linear and Cyclic Platinum σ-Acetylide Complexes of Tetraethynylethene', *Chem. Eur. J.* **1**, 111-117.

24. Philp, D.; Gramlich, V; Seiler, P.; Diederich, F. (1995) 'π-Complexes Incorporating Tetrakis(phenylethynyl)ethene', *J. Chem. Soc. Perkin Trans. 2*, 875-886.
25. Hunter, C. A.; Sanders, J. K. M. (1990) 'The Nature of π-π Interactions', *J. Am. Chem. Soc.* **112**, 5525-5534.
26. Cozzi, F.; Cinquini, M.; Annuziata, R.; Siegel, J. S. (1993) 'Dominance of Polar/π over Charge-Transfer Effects in Stacked Phenyl Interactions', *J. Am. Chem. Soc.* **115**, 5330-5331
27. Young, J. K.; Moore, J. S. (1995) 'Acetylenes in Nanostructures' in *'Modern Acetylene Chemistry'*, Stang, P. J.; Diederich, F., Eds., VCH, Weinheim, pp. 415-442.
28. Tykwinski, R. R.; Schreiber, M.; Gramlich, V.; Seiler, P.; Diederich, F. (1995) 'Donor-Acceptor Substituted Tetraethynylethenes', *Adv. Mater.* **8**, 226-230.
29. Bosshard, Ch.; Spreiter, R.; Günter, P.; Tykwinski, R. R.; Schreiber, M.; Diederich, F. (1995) 'Structure-Property Relationships in Nonlinear Optical Tetraethynylethenes', Adv. Mater. **8**, 231-234.
30. Nalwa, H. S. (1993) 'Organic Materials for Third-Order Nonlinear Optics', *Adv. Mater.* **5**, 341-358.

DISCUSSION OF THE MICHL AND DIEDERICH LECTURES

HORST PRINZBACH
Institut für Organische Chemie und Biochemie
Lehrstuhl für Organische Chemie
Albert-Ludwigs-Universität
Albertstrasse 21
79104 Freiburg, Germany

H. PRINZBACH, discussion leader, introduced himself as one of the senior participants of the workshop - who over the years has witnessed, or had active part in, research topics such as "organic conductors" and "solar energy conversion". This relatively recent past has taught lasting lessons to keep in mind when it comes to the evaluation of new research trends, to the definition of new targets, of applications, and to the generation of "new words". As a (loose) tie to the workshop theme he gave the background of his poster "EN ROUTE TO LINEAR C_{20} OLIGOMERS? - ELECTRON DELOCALIZATION IN HYDROCARBON CAGES". In short, this describes the utilization of unsaturated, spherical dodecahedranes (e.g. $C_{20}H_{10}$; $C_{20}X_{16}$) as "modules" for the construction of $C_{40}H_x$ and $C_{60}H_y$ tubular carbon skeletons by thermal or photochemical [2+2] cycloaddition/cycloreversion (metathesis) reactions.

1. Questions directed to Prof. Michl

RAJCA (Nebraska): When the H-bonded carboxylic acid dimer lattice is transformed to a Rh_2-linked lattice, the surface area occupied by the lattice appears to shrink, according to your slide.

MICHL: Yes.

RAJCA: If the lattice indeed shrinks by the factor of four, it is not likely that a continuous fragment of the lattice, extending from one end to the other end, will not form; that is, critical concentration (Pc) for penetration in a square lattice is significantly greater than 1/4 (bonds and sites). What is going to be your product?

J. Michl (ed.), Modular Chemistry, 33–39.

MICHL: As the coupling proceeds, the surface area will be continuously adjusted in an L-B trough to accommodate changes in the surface area per monomer.

RAJCA: Are you implying cooperativity over macroscopic distances?

MICHL: Ultimately, I am hoping for it when the coupling reaction is reversible. I want a two-dimensional crystal.

KUKI (Alanex): I want to return to the subject of Tinkertoys and coherence lengths. The discussion need not focus only on materials design. If I want to make a wall, I would not use Tinkertoys; I would use bricks. But there may be other important non-material applications of grids and nets of one or two hundred Å size. In your list I think you alluded to enzymatic reactivity at designed functional sites, and other functions.

MICHL: I agree that non-materials applications may turn out to be just as important as the materials applications. At this stage, we are still just trying to demonstrate the concept. To build a wall from tinkertoys one just needs to use thick rods and let those on one floor touch those on the next floor. The catalytic reactivity could perhaps be built in by attaching the right chemically active groups at the right places in the scaffolding as it is being built.

WEGNER (Mainz): (i) Why is it important to aim for two-dimensional periodic structures when one wants to synthesize a monomolecular layer composed of rods and connectors? Wouldn't aperiodic structures do just as well?

(ii) What about the problem of commensurability between the layered structure to be formed on a substrate (possibly dressed with pedestals) and the periodic structure of the substrate itself?

(iii) When constructing 3-dimensional solids from 2-dimensional nets which have been synthesized separately, the problem of commensurability arises again or it could probably arise. Why could one not allow for disorder on a local level in order to overcome this problem. Note also that structures exhibiting long range rather perfect order could contain short range disorder and that would not deteriorate or actually could enhance the properties.

This remark started an extensive discussion on perfection and imperfections in materials and on all kinds of definitions characterizing "order" in materials.

MICHL: (i) For many purposes, an irregular structure is useful, but for others, a two-dimensionally periodic structure will have advantages. Even for the simplest application, molecular separation by size, it will be best if all the openings in a grid are exactly alike. In more advanced applications that will involve electronic, optical, magnetic or similar properties, a large coherence length will frequently be desirable.

(ii) This problem does not come up since we use a liquid metal substrate, and this has no permanent periodic structure. The pedestals will not dress the surface in a static fashion, they will glide on it along with the monomers that they will carry.

(iii) The "upper" layers will not be prepared separately and then added onto the "floor" layer. Rather, we anticipate growing them epitaxically one by one, ultimately perhaps in a machine similar to an oligonucleotide synthesizer. The rod lengths and connectivity pattern on the connectors have to be similar in each layer to produce the type of scaffolding that I have in mind, but they do not need to be exactly identical. Since the connector-to-connector separations will presumably be on the order of 5-25 Å, a slight bending of a rod would accommodate small differences in the length of rods or size of connectors in different layers. For some applications local disorder between layers might not matter but for many it will. For instance, if I wish to provide long uninterrupted "tunnels" for moving atoms or other particles, or for building wires, I need perfect epitaxy between layers.

MÜLLEN (Mainz): When a pillared connector, held to the metal surface by the S-metal interaction, is subjected to coupling or insertion reactions at the carboxy functions, the question will arise as to the mobility of the pillar and thus to the commensurate distances in the ground layer and the second layer.

MICHL: We assure nearly free translation and rotation of our pillar (monomer) on the metal surface by using a liquid metal. The issue therefore does not arise when we build the ground layer. It will arise when we attempt to build scaffoldings consisting of two or more grid layers that are different from each other. Then, clearly, rod lengths in the various layers need to be approximately equal. They do not need to be exactly equal since the rods are not perfectly rigid and can accommodate some strain by bending.

SEEMAN (New York): You state that you deesterify your pedestal molecule after it is fixed in the substrate. Can you exclude any inversion of the molecule following hydrolysis, given that both the top and bottom of the molecule now contain carboxyls?

MICHL: The molecule is actually fixed **on** the substrate (liquid metal), not **in** it. Yes, this is the type of problem we are concerned with all the time. We need to monitor carefully every step of the procedure and we use the usual tools of surface science to do this. In this case, the aliphatic carboxylic acid groups of the tentacles are distinct in the IR from the aromatic carboxylic acid groups of the arms attached to the cyclobutadiene ring and we can follow their behavior separately.

KAHN (Bordeaux): I have been very interested in your transparency entitled "speculation on possible uses". It was quite useful to suggest some challenges in this field of modular chemistry. However, a very important issue was missing, namely "bistability". Molecular bistability may be defined as the ability of a molecular system to exhibit two stable states in a given range of perturbations. Bistability leads to the phenomenon of hysteresis, which confers a memory effect on the system. Bistability is directly related to the interaction between the active sites within the molecular assembly. When the interactions are large enough, the system may acquire the property of cooperativity giving rise to the bistability. For me, to design such cooperative molecular assemblies from modules (or bricks) is one of the most important issues in the field of modular chemistry.

MICHL: Yes, I only listed a few of the many applications. I agree that cooperative phenomena will be extremely interesting and have indeed begun a collaborative effort with Prof. Mark Ratner (Northwestern University) on the theory of coupled motion of a planar assembly of rotating dipoles, with obvious connection to the molecular machine I talked about.

Questions directed to Prof. Diederich

SCHLÜTER (FU Berlin): Please make a comment on the problems associated with the synthesis of two-dimensional networks from your building blocks. Do you believe that 2D molecules can actually be prepared by using standard flask-type chemistry?

DIEDERICH: We tried several approaches to get crystalline all-carbon network materials by oxidative polymerization of tetraethynylethene or macrocyclic annulenic precursors. These experiments involved also electropolymerization on an electrode surface as well as attempts on graphite where we tried to monitor conversions by STM. Now we have turned to the following approaches: a) end-capping monomers by long-chain alkyl groups and orienting them for crystallinity-generating polymerization. Langmuir-Blodgett techniques should be suitable to provide the desired orientation of the end-capped monomers. b) metal-template assembly looks particularly promising, since the ligands or the metal could help enhance solubility of the metal-acetylenic nets. I am quite optimistic this approach will be successful.

LINDSEY (N. Carolina State): Can you comment on the molecular design features that give rise to high solubility in your compounds? In those cases where you now have problems of insolubility, how do you anticipate imparting solubility?

DIEDERICH: Solubility is crucial to developing large fully characterizable nanostructures. In our work, the TIPS-group (triisopropylsilyl) is ideal for protecting the highly reactive terminal alkynes. It provides three distinct advantages, 1) solubility, 2) stability, 3) crystallinity suitable for X-ray analysis. I am confident we should soon construct planar carbon nets as large as C_{240} provided we have peripheral $Si(i\text{-}Pr)_3$ end-capping groups. Already we have stable C_{120} sheets, end-capped by TIPS groups, in hand.

PRINZBACH: When you speak of 'stability' of your polyynes - what do you mean? Thermodynamic, kinetic stability?

DIEDERICH: First, I mean kinetic stability. Our systems derived from tetraethynylethene do not undergo Bergman cyclization, since the terminal C atoms of the cis-enediyne moieties are at too large a distance. Kinetic stability is high if all terminal alkynes are end-capped; terminal acetylenic hydrogen atoms significantly decrease stability.

But there might also be a thermodynamic advantage present in our cross-conjugated systems, as compared to linearly conjugated ones of similar size. In short conjugated π systems, linear conjugation provides more stabilization than cross-conjugation; this is well documented. However, it is also well known that linear conjugation becomes less effective (and correspondingly systems less stable and more reactive) when it becomes increased in length. This is illustrated nicely by the polyacetylenes which have linear conjugation lengths of 20-25 C atoms. Cross-conjugation limits electron delocalization in extended π systems and, in our view, in such systems is thermodynamically favorable to extended linear conjugation. Although accurate computing will be difficult on very large molecules, we hope to demonstrate this in computational work. We like to speak of "compartmentalization of π-electron delocalization" provided by cross-conjugation in our systems. This compartmentalization, which others would call bond localization over finite atom arrays, generates shorter linear conjugation paths and we believe that this is at the origin of the exceptional stability seen in our extended radialenes and possibly also in fullerenes, which can be described as cross-conjugated [5]radialene electronic structures.

I like to compare this electron compartmentalization by cross-conjugation to that seen in polybenzenoid hydrocarbons such as kekulene. In kekulene, we find all-phenanthrene-type subunits instead of macrocyclic annulenoid π-electron delocalization. Obviously, the formation of six electronic sextets (in Clar's terms) makes the molecule much more stable. The X-ray data we have so far show distinct bond length alternation in all tetraethynylethene containing chromophores, which provides some additional support for reduced bond delocalization.

STUPP (Illinois): Can you tell us if the conjugated structures are resistant to photochemical damage when exposed to the high laser intensities necessary in third order sum frequency experiments as opposed to four wave mixing experiments?

DIEDERICH: The TIPS groups provide enormous stability to our NLO materials and prevent ready damage from the laser. The laser is at 1.9 microns which give 634 nm as frequency-tripled light. Experiments were interference third-harmonic generation, and $\chi^{(3)}$ values varied between 10^{-10} 10^{-12} esu.

WUEST (Montréal): How expensive are your building blocks?

DIEDERICH: I do not consider the price of starting materials as a priority design criterion in the development of novel materials. This would be too limiting, besides, prices change. Our TIPS acetylene building block costs today 1/20 of the price four years ago.

STODDART (Birmingham): Referring to the donor and acceptor molecules (work of D. Philp) based on the tetraethylethene framework, I wonder what price would have to be paid in energy terms - I am thinking of the enthalpic difference - when two molecules in the CT complex are distorted as far as possible away from the preferred geometry reflecting the point-charge distributions.

DIEDERICH: As an answer to your specific question, I would say that turning two aromatic systems away from their optimal interacting geometry, which is governed by the electrostatic interactions between atoms of opposite partial charge ($\delta+$ - δ -), would easily prevent an intermolecular association from taking place.

The understanding of noncovalent interactions has been increased in a dramatic way over the past 10 years. Among others, this is illustrated by progress in rational drug design in the pharmaceutical industry.

KAHN: I am wondering whether the kind of networks you showed us could exhibit some cooperativity. Cooperativity occurs when the interaction between adjacent active sites of the network is very large, larger than a threshold value. Cooperativity usually leads to phase transitions and hysteresis effects, which confer a memory effect on the system. My feeling is that most of the supramolecular assemblies described so far do not exhibit this type of behavior. Sometimes, they exhibit some additivity, but not cooperativity. Could you comment on this?

DIEDERICH: There is cooperativity in many supramolecular phenomena. An example is the two-dimensional checkerboard structure I showed - single C-H...N and C-H...Cl interactions and their contributions to the free energy are still poorly defined, perhaps not so much in terms of enthalpy as in Gibbs free energy. But they are for sure very weak and only through the cooperative effect of these very weak interactions does the final net constitute.

MODULAR ASSEMBLY OF SURFACE HETEROSTRUCTURES FROM INORGANIC CLUSTERS AND POLYELECTROLYTES

DANIEL L. FELDHEIM, HYUK-NYUN KIM, HUN-GI HONG[†], STEVEN W. KELLER[‡], KATHERINE C. GRABAR, MICHAEL J. NATAN, AND THOMAS E. MALLOUK

Departments of Chemistry, The Pennsylvania State University, University Park, PA 16802 USA, [†]Sejong University, Seoul 133-747, Korea, and [‡]The University of Missouri, Columbia, MO 65211 USA

ABSTRACT. Layered inorganic compounds, such as metal phosphates and ternary oxides, can be exfoliated by ion exchange and acid-base reactions to produce unilamellar suspensions. Multilayer thin films can then be grown on both planar and high surface area supports by alternately adsorbing these anionic sheets with spherical, inorganic oligocations, such as $Al_{13}O_4(OH)_{12}(H_2O)_{24}^{7+}$ or ethylene diamine-modified C_{60}. Proteins such as cytochrome c, myoglobin, and hemoglobin are other examples of monodisperse and roughly spherical oligocations that can be incorporated into lamellar heterostructures by this technique. By terminating these inorganic/organic stacks with cationic polyelectrolyte layers, such as poly(allylamine) hydrochloride, it is also possible to prepare surfaces that bind (2.5±1.5) nm diameter anionic gold colloid clusters. Insulator-cluster-insulator heterostructures of nanometer dimensions, grown between a gold substrate and a polypyrrole overlayer, show coulomb blockade effects in their current-voltage characteristics. The 0.3 - 0.4 V coulomb gap observed in this case is consistent with the dimensions of the gold clusters and the insulator layers.

1. Introduction

The modular, or "tinkertoy" approach to chemical synthesis provides a convenient route to supramolecular complexes and extended solids. In general, this synthetic strategy involves joining atoms and larger molecular building blocks through covalent or non-covalent bonds. The size, shape, symmetry, and chemical affinity of the synthons used are chosen judiciously to give a desired supramolecular structure or framework topology, through reactions that proceed under mild conditions. The literature in this area is already vast, and the chemistry encompasses a tremendous variety of synthons, from metal ions and organic ligands to larger entities such as oligonucleotides, proteins, dendrimers, micelles and vesicles.

One of the goals of modular chemistry is to provide a rational synthetic approach to complex hierarchical structures with interesting properties and functions. These include physical effects -- for example, current rectification or electronic signal amplification, logical operations, and photonic switching -- as well as more intricate biomimetic

J. Michl (ed.), Modular Chemistry, 41–51.

functions such as artificial photosynthesis, conversion of chemical to mechanical energy, and template-directed synthesis. These complex and seemingly far-off targets may require structures that are well ordered over large distances in one or two dimensions, but still aperiodic, i.e., fabricated in a specific molecular sequence or in certain macroscopic shapes. The concurrent requirements of supramolecular or long-range order and aperiodic assembly represent a serious problem if the "tinkertoy" chemist uses modules of atomic or molecular dimensions. In these syntheses it is difficult to achieve long-range order if strong chemical bonds are formed irreversibly between building units. The assembly proceeds through growth of a product phase from many different nuclei, and annealing of the resulting microdomains into larger well-ordered structures cannot occur [1]. Alternatively, one can design materials held together by weaker coordinate covalent or non-covalent bonds. In this case it is possible to make relatively perfect crystals with predictable (and sometimes even more interesting unpredictable) linking topologies [2]. However, this approach always gives periodic frameworks and therefore precludes the possibility of assembling hierarchical structures in serial fashion.

This paper describes a hybrid approach, namely, taking apart well-ordered, periodic crystals to make solutions or suspensions of extended building blocks, and then putting them back together through a serial, irreversible adsorption process. The assembly relies primarily on electrostatic forces between modules, and is closely related to a strategy developed earlier by Decher and coworkers for growing thin films of organic polyelectrolytes [3]. One important advantage of the electrostatic approach is that it does not rely on specific chemical functionality, and therefore allows one to mix and match a very large variety of charged inorganic and organic modules. While the individual anion-cation interactions are weaker than covalent bonds, between macroscopic building blocks they are collectively strong and therefore lead to the formation of robust heterostructures.

Inorganic solids that are prepared in high temperature solid-solid reactions, or in lower temperature hydrothermal reactions, are usually crystallographically well ordered in three dimensions. In many cases, however, the bonding forces are weak in one or two directions in the crystal. Asbestos and graphite are everyday examples of solids in which strong bonds define chains and sheets in the structure, respectively. Weaker ionic or van der Waals bonds between these infinite building units are easily broken, and this effect accounts for the morphology and cleavage of these solids along certain crystal planes. Taking the cleavage operation one step further, it is often possible to decompose the crystals of anisotropic solids to suspensions of isolated chains or sheets through wet chemical reactions [4]. For example, $Li_2Mo_6Se_6$ crystals dissociate in high dielectric solvents to Li^+ ions and anionic $[Mo_6Se_6]^{2-}$ chains of nanometer width and micron length [5]. A more familiar example is the smectite clays, which exfoliate in water to give unilamellar suspensions. We have used mild acid-base and ion-exchange reactions to exfoliate lamellar metal phosphates and alkali transition metal oxides. The resulting suspensions contain sheets of molecular width (8-25 Å), which are macroscopic (nm to mm, depending on the dimensions of the starting crystals) in the two remaining dimensions. These sheets adsorb irreversibly as monolayers or bilayers onto cationic surfaces. Subsequent adsorption reactions in which the anionic sheets are interleaved with organic polycations give rise to thin films that resemble inorganic intercalation compounds, except that the layering can be done in any desired sequence, i.e., in aperiodic fashion [6].

While these films appear, from ellipsometry and from diffraction data, to be well ordered in the stacking direction, there is no evidence (or expectation) that the polydisperse and flexible organic polycations, which charge compensate the anionic sheets, can form ordered structures in the lateral directions. As a first step towards preparing thin films with at least local in-plane order, we are investigating the properties of films and bulk solids made from infinite inorganic sheets and monodisperse, oligomeric "balls" (inorganic polyoxocations, fullerenes, proteins, and metal clusters). This paper reports the synthesis and charaterization of these films, and presents evidence for quantum electron transfer effects in some new insulator-nanocluster-insulator heterostructures.

2. Materials and Methods

2.1. MATERIALS AND SYNTHESIS

Microcrystalline α-zirconium phosphate, (α-ZrP) was prepared by a modification of Alberti's method [7]. The gel prepared by mixing solutions of $ZrOCl_2{\cdot}8H_2O$ and excess phosphoric acid was dissolved in water by dropwise addition of 48% aqueous HF, and the solution was slowly evaporated at 80-100 °C by means of an air stream. Water was periodically added to maintain the volume (approximately 50 mL per gram zirconium phosphate), and after one week the suspension was filtered to yield microcrystalline α-$Zr(HPO_4)_2{\cdot}H_2O$. Semi-crystalline α-ZrP was prepared by the method of Clearfield et al. [8].

Both microcrystalline and semi-crystalline α-ZrP were exfoliated by reaction with aqueous tetrabutylammonium hydroxide (TBA^+OH^-), as described in a previous communication [6]. This solid was suspended in water and titrated to pH 8.5 with 0.5 M TBA^+OH^-; typically 0.20-0.25 equivalents were needed. The resulting turbid suspension (0.7 meq/L α-ZrP) was centrifuged, and the translucent supernatant was used in film growth experiments. $K_4Nb_6O_{17}{\cdot}3H_2O$ and $CsTi_2NbO_7$ were prepared by reaction of the appropriate carbonates and oxides (K_2CO_3, Cs_2CO_3, Nb_2O_5, and TiO_2) in stoichiometric proportions, at 1100°C in alumina crucibles [9]. They were ion-exchanged overnight with 2 M HCl, and then filtered, washed with deionized water, and dried in air to produce solids of approximate composition $H_2K_2Nb_6O_{17}{\cdot}xH_2O$ and $HTi_2NbO_7{\cdot}yH_2O$. In a typical exfoliation procedure, 0.32 g of $H_2K_2Nb_6O_{17}{\cdot}xH_2O$ was suspended in 80 mL water, and 1.0 mL of 0.50 M TBA^+OH^- was added. After stirring overnight, the pH of the suspension was adjusted to 8.0 using 0.10 M HCl, and water was added to make the total volume 100 mL. Unreacted solid particles were allowed to settle, and the supernatant was diluted to double its volume before use in film growth. The final concentration of $K_2Nb_6O_{17}^{2-}$ was approximately 3.5 meq/L.

Solutions of $Al_{13}O_4(OH)_{12}(H_2O)_{24}^{7+}$ (as the chloride), and polyallylamine hydrochloride (PAH) were prepared as described previously [6]. The water-soluble adduct of C_{60} with ethylenediamine (C_{60}-en_x) was synthesized as described by Wudl and coworkers [10]. Infrared spectra (KBr pellet) were similar to those reported in the literature [10], and the FABMS spectrum showed predominantly peaks at 780 ($C_{60}(NH_2C_2H_4NH_2)$) and 720 (C_{60}) amu. 20 mg of the fullerene adduct was dissolved in 10 mL deionized water, and the yellow-brown solution was acidified to pH 6.0 by addition of 1.0 M HCl. Horse skeletal myoglobin (Mb) and beef blood hemoglobin (Hb) were used as received from Sigma Chemical Co. Aqueous protein (Mb and Hb) solutions were prepared at 25 μM concentration in 50 mM phosphate buffer solution (pH 4.1). Solutions of 2.5 ± 1.5 nm diameter gold colloid particles were made by borohydride reduction of $HAuCl_4$ solutions in the presence of sodium citrate [11].

2.2. SUBSTRATE PREPARATION AND FILM GROWTH

Single crystal silicon wafers (100), optical quartz, and gold-coated silicon substrates were prepared, cleaned, and primed with the appropriate cationic monolayer as described elsewhere [6]. For silicon and quartz substrates the primer layer was (4-aminobutyl)dimethylmethoxysilane, whereas for polycrystalline gold substrates it was 2-mercaptoethylamine. The derivatized substrates were immersed in the appropriate inorganic suspension (α-ZrP, $Ti_2NbO_7^-$, or $H_2Nb_6O_{17}^{2-}$) for 15-30 min. at room temperature, and then rinsed thoroughly with deionized water. They were then immersed in the oligocation ($Al_{13}O_4(OH)_{12}(H_2O)_{24}^{7+}$, Hb, or C_{60}-en_x) solution for 30-40 min. In the case of Hb, the samples were rinsed with 50 mM phosphate buffer (pH 4.1), whereas the $Al_{13}O_4(OH)_{12}(H_2O)_{24}^{7+}$ and fullerene samples were rinsed with deionized water. PAH layers were deposited similarly [6]. The anion-cation adsorption steps were repeated as desired to make multilayer films.

Anionic colloidal gold particles were deposited onto PAH-terminated assemblies in 12 h. adsorption steps, followed by rinsing with water. These clusters have a high affinity for amine functionalities such as those contained in PAH [11]. A second layer of PAH was adsorbed onto the colloidal gold monolayer, followed by sequential adsorption steps of the appropriate anionic sheets and PAH. These insulator-cluster-insulator assemblies were then overcoated with a thin conducting polymer layer by first depositing a film of $FeCl_3$ (from a 1 M solution in acetone), allowing the solvent to dry, and then exposing the film to pyrrole vapor. A doped polypyrrole layer, which presumably contains $FeCl_4^-$ anions, forms on top of the $FeCl_3$ layer, and the latter is subsequently removed by dissolving in water. This procedure brings the conductive polypyrrole and insulator-cluster-insulator films into mechanical contact.

2.3. ELLIPSOMETRIC AND SPECTROSCOPIC CHARACTERIZATION OF THIN FILM HETEROSTRUCTURES

Ellipsometric measurements were made using a Gaertner model L2W256D ellipsometer with a 6328 Å (He-Ne) laser source and a rotating analyzer. The incident angle was 70°, and the polarizer was set at 45°. The real part of the refractive index of the film was typically fixed at 1.54, and the imaginary part set to zero. For gold substrates, the refractive index was determined experimentally prior to film growth. For Si substrates, the literature value of the complex refractive index (n = 3.875, k = -0.018) was used in the calculation of film thickness from the observed Ψ and Δ values. Film thicknesses were generally measured at 8-10 different positions for each sample and averaged. The range of these measurements for good films was typically 1-2 Å.

Visible absorption spectra were acquired in transmission mode at normal incidence, using planar quartz substrates, with a Hewlett Packard 8452A diode array spectrophotometer. The spectral resolution was 2 nm. All spectra were referenced to a blank quartz substrate.

3. Results and Discussion

3.1. LAYER-BY-LAYER GROWTH OF INTERCALATION COMPOUNDS FROM INORGANIC POLYANION SHEETS AND ORGANIC POLYCATIONS

High charge density inorganic sheets, such as α-ZrP and $(Ti_2NbO_7)_n^{n-}$, adsorb onto cationic surfaces as monolayers from unilamellar colloidal suspensions. Ellipsometric measurements show that this adsorption reaction occurs within a few seconds on an amine-terminated silicon surface. Figure 1, which shows ellipsometric data for alternating inorganic sheets and poly(allylamine) hydrochloride (PAH), illustrates the regularity of layer growth in this system. Adsorption of a monolayer of PAH, at low ionic strength, results in a thickness change of 5 Å, which corresponds well to the expected van der Waals thickness of this linear polymer in its uncoiled form. Adsorption of an α-ZrP sheet causes a thickness change of 10 Å, which is close to the average of the crystallographic layer spacings of α-ZrP and its TBA^+ intercalation compound (7.6 and 14.6 Å, respectively). Note the regular "sawtooth" pattern (Figure 1), which is characteristic of very reproducible layer growth. X-ray reflectometry provides a check of the ellipsometric thickness measurements for PAH/α-ZrP multilayers. The observation of a Bragg diffraction peak from a 13-layer sample demonstrates that the multilayer stack is well-ordered in the growth direction [12].

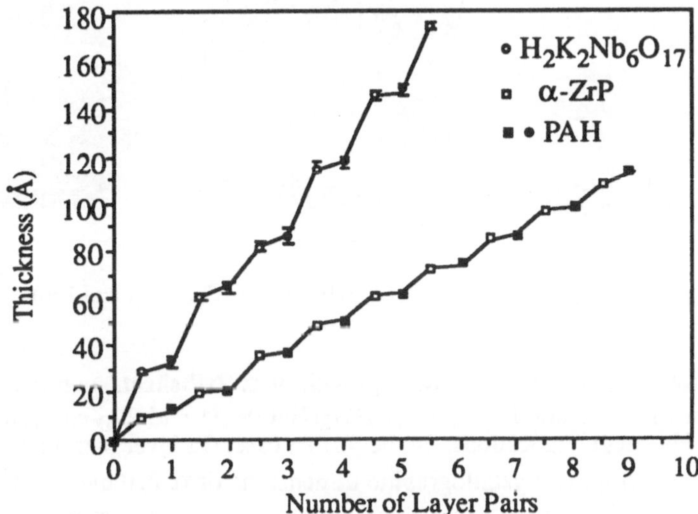

Figure 1. Ellipsometric data for growth of alternating anion sheets ($K_2H_2Nb_6O_{17}$ or α-ZrP) and PAH on Si. "Zero" thickness corresponds to native oxide + amine primer.

The mica-like solid $K_4H_2Nb_6O_{17}{\cdot}3H_2O$ is an interesting special case, because it has alternating, chemically distinct interlayers [13]. Proton exchange to an approximate stoichiometry $K_2H_2Nb_6O_{17}$ yields a solid that can be exfoliated with TBA^+OH^- to give a colloid in which two niobate layers sandwich a layer of K^+. Each anion adsorption step then results in a thickness change (Figure 1) of 22 Å, again in excellent agreement with the expected dimensions of the niobate-K^+-niobate sandwich.

3.2. SERIAL LAYER GROWTH OF INORGANIC SHEETS, CLUSTERS, FULLERENE CATIONS, AND PROTEINS

The serial adsorption technique described above is very general, and may be useful in a number of modular chemistry applications. But the structure of the thin films that are grown in this way, and the process whereby they are formed, are poorly understood at present. As an illustration of this point, we note that multilayer films made from PAH and α-ZrP grow as single, alternating organic and inorganic layers (see Figure 1). On the other hand, alternate adsorption of exfoliated hectorite (a smectite clay) with poly(diallyldimethylammonium) chloride, PDDA, is subtly different. X-ray diffraction shows that while the surface structure is similar to the bulk hectorite/PDDA intercalation compound, not one but *two* layers of the organic/inorganic composite grow per adsorption cycle [14]. Significant structural rearrangement of the polycation layer must attend adsorption of hectorite in the latter case. In order to better understand the factors controlling layer growth and structural integrity, and expand to the scope of the technique, we have conducted adsorption experiments with the cationic and anionic "balls" shown schematically in Figure 2.

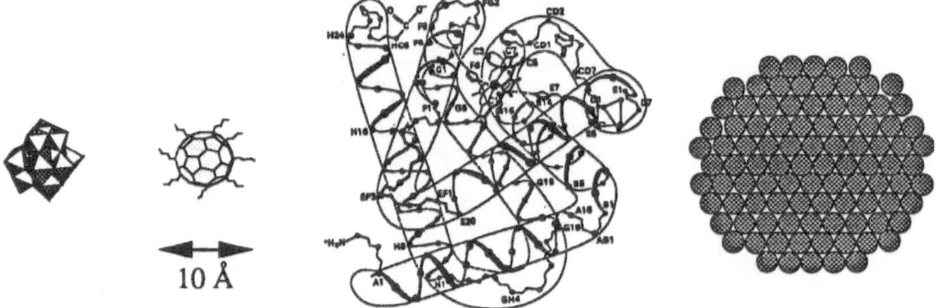

Figure 2. Oligomeric cations ($Al_{13}O_4(OH)_{12}(H_2O)_{24}{}^{7+}$, C_{60}-en_x, and Mb) and anions (Au colloids).

Typical ellipsometric data for layerwise growth of α-ZrP sheets, cationic clusters and proteins in shown in Figure 3. $Al_{13}O_4(OH)_{12}(H_2O)_{24}{}^{7+}$ and C_{60}-en_x can be regarded here as relatively small oligocations. In the former case, the layer pair thickness (16 Å) is nicely consistent with the crystallographic dimensions of α-ZrP and the Keggin cation [15]. The average monolayer thicknesses of C_{60}-en_x and α-ZrP are respectively 11.1(±1.8) and 9.6(±1.9) Å. As in the case of α-ZrP/PAH, the thickness change for adsorption of the inorganic sheets reflects the fact that they carry with them a monolayer of TBA+. The thickness observed upon adsorption of oligo- or polycations reflects the displacement of TBA+. Assuming that α-ZrP accounts for 7.6 Å of the average layer pair thickness, the thickness of C_{60}-en$_x$ in the layers is therefore 13.1 Å. The difference between this and the theoretical size of a fully extended C_{60}-en$_x$ adduct (approximately 15 Å) is easily rationalized by considering the small average number (ca. 6) of ethylenediamine units attached to the fullerene.

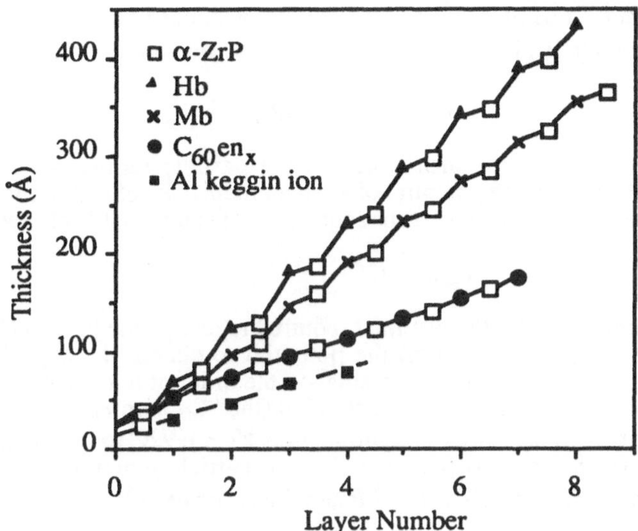

Figure 3. Ellipsometric data for alternate layer growth of α-ZrP and spherical oligocations ($Al_{13}O_4(OH)_{12}(H_2O)_{24}{}^{7+}$, C_{60}-en$_x$, Mb, and Hb).

Since Hb has an isoelectric point of 6.8, it is positively charged at the pH (4.1) of the phosphate buffer solution used in the adsorption experiments. Again the average thickness change per α-ZrP adsorption step is consistently 9.8 Å, whereas that for Hb is 41.9 Å. Accounting in the same way for the change that attends loss of TBA+, we arrive at an average thickness of 44 Å for Hb in the film. This is significantly less than the molecular dimensions (50x55x60Å) obtained from crystallographic data [16], and suggests that somewhat incomplete monolayers form from these dilute (25 μM) solutions, or possibly that denaturation of the protein occurs. Similar results were obtained with cytochrome c and Mb layers.

3.3. THIN FILM METAL-INSULATOR-NANOCLUSTER-INSULATOR-METAL (MINIM) STRUCTURES

The ability to control the transfer of a single electron per time interval in solid state devices has potential applications in precision electrometers, memory devices, and high speed, low power, and highly miniaturized transistors. The simplest structures of this kind that can be prepared by the techniques described above are two-terminal devices with nanometer-size clusters between metallic top and bottom contacts. In this case, however, it is apparent that Keggin ions and cationic proteins are inappropriate, and one must include between insulating layers materials that contain mobile hole or electron carriers; that is, the clusters should be metallic or semiconducting. We report here preliminary data on the synthesis, structure, and electronic properties of MINIM devices containing 2.5 nm diameter gold clusters sandwiched between insulating sheets of α-ZrP/PAH.

3.3.1. *Single Electron Diodes.* Classically, the capacitance of a metal-insulator-spherical metal particle junction is given by (1), where ε_0 is the vacuum permittivity constant, ε is

$$C = 4\pi\varepsilon_0\varepsilon r(1 + r/2l) \qquad (1)$$

the dielectric constant of the insulator, r is the particle radius and l is the junction thickness [17]. The charging potential of such a junction is related to the capacitance by (2), where e is the electron charge. From equations (1) and (2) it is obvious that as the

$$V_{gap} = e/2C \qquad (2)$$

capacitance becomes smaller the potential required to charge the junction becomes larger. In the small-particle limit (<10 nm) the potential to charge the capacitor by a single electron becomes experimentally observable–on the order of tens to hundreds of mV [17-21]. This leads to a high resistance current offset (the "Coulomb gap") in the i-V curve of the junction. However, one further requirement is $kT < e^2/2C$ to avoid thermal tunneling of electrons across the gap. Typical small metal particle junctions constructed to date have contained capacitances of ca. 10^{-17} F and thus operate below 77 K.

3.2.2. *Preparation and Electrical Characterization of MINIM Devices.* Double-tunnel junction single-electron diodes were constructed on gold surfaces by stepwise growth of the desired number (2-6) of α-ZrP/PAH bilayers, followed by adsorption of the negatively charged Au clusters, followed by adsorption of PAH/α-ZrP bilayers. Ellipsometric data are shown in Figure 4 for fabrication of a typical two-terminal device. While the apparent thickness of the gold layer exceeds the size of the particles, it is thought that this is an artifact of the ellipsometric measurement of a gold layer on a gold substrate. Scanning probe microscopy experiments, which are currently in progress, should provide a check of this layer thickness. While measurement of the nanocluster layer thickness is problematic, we note that the insulators above and below show normal growth characteristics. A top contact was made by depositing a thick, doped polypyrrole layer, and then attaching to it a copper wire via silver epoxy or a metal clip. Typically, the area of the device was 1 cm^2.

An i-V curve for a (2x2) MINIM device is shown in Figure 5. A high resistance region is observed for ca. 150 mV on both sides of 0 V. This region is termed the "coulomb gap" and represents the charging potential (e/2C) of each particle of the double junction array by a single electron. Once enough energy is supplied to the junction, electrons tunnel through the junction, resulting in the exponential current rise on either side of the gap. These data were obtained at room temperature and were reproducible as the potential was scanned from -1 V to +1 V. By varying the number of (α-ZrP/PAH) layers flanking the

Au nanoclusters, the magnitude of the gap was varied. Coulomb gap potentials for three such devices with 30, 80, and 90 Å junction thicknesses were 275, 300, and 400 mV. These values are in good agreement with those calculated from equations (1) and (2) (320, 360, and 360 mV, respectively), especially when one considers the polydispersity

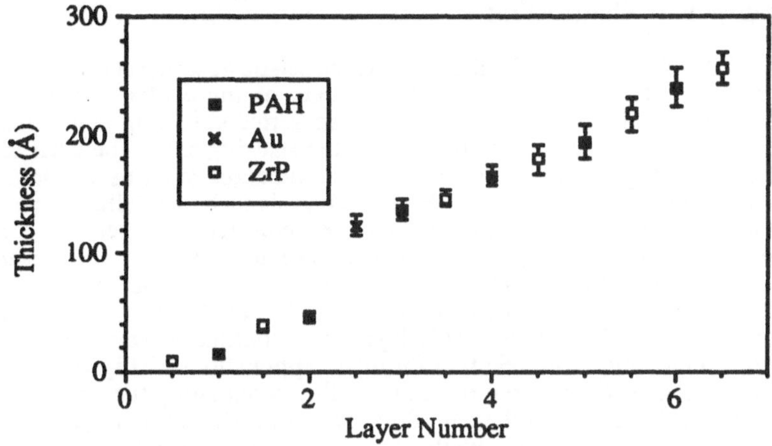

Figure 4. Thickness vs. layer number for a (α-ZrP/PAH)$_2$-Au-(PAH/α-ZrP)$_6$, or (2x6), MINIM device.

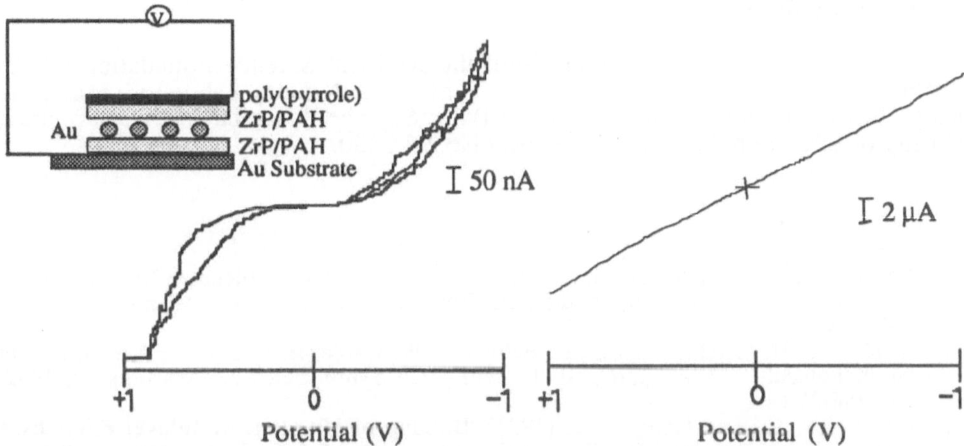

Figure 5. Left: i-V curve for a (2x2) MINIM device. Right: i-V curve for control (2x2) device, in which the Au adsorption step was eliminated. Scan rate = 100 mV/s.

of the Au particles (2.5 ± 1.5 nm), and uncertainties in the values of the layer thickness and dielectric constant of the insulator. Figure 5 also shows the i-V behavior of a (2x2) device prepared without gold nanoclusters. The linear curve is typical of ohmic devices, and is an important result since poly(pyrrole) has been shown previously to act as an ohmic, non-ohmic or rectifying contact depending on the polymerization conditions, film thickness, and dopant concentration [22].

4. Summary and Conclusions

Exfoliation of lamellar solids and subsequent electrostatic adsorption provides a modular route to surface heterostructures in which sheets of molecular dimensions (metal phosphates, silicates, titanates, and niobates, and semiconducting and metallic metal disulfides) and cationic "spacers" alternate along the stacking axis. The latter can include polymers, oligomers, clusters, and biological molecules such as cationic proteins, deposited as in any desired sequence. Effectively, this technique allows one to prepare structures that are surface analogs of intercalation compounds. Because it is a serial deposition technique, however, complex layer sequences with interesting device applications are possible. MINIM structures are one example, in which both the deposition sequence and the dimensions of the individual layers are reflected in the physical properties of the device. This layering technique is very general and may ultimately be an enabling synthetic tool for many applications. Ellipsometry and other spectroscopic probes show that the films are well-ordered along the stacking direction. However, the lateral structure of the thin films that are grown in this way, and the process whereby they are formed, are not completely understood at present. Future experiments will focus on characterization of this lateral structure, the dynamics of layer formation, and further applications of the technique.

5. Acknowledgment

This work was supported by grants from the National Science Foundation (CHE-9396243) and the Office of Naval Research (Lawrence Berkeley Laboratory Molecular Design Institute), and also in part by the Basic Science Research Institute Program, Ministry of Education, Korea, 1995, Project No. 94-3430.

6. References

[1] Fagan, P. J., Ward, M. D., and Calabrese, J. C. (1989) 'Molecular Engineering of Solid-State Materials: Organometallic Building Blocks' J. Am. Chem. Soc., **111**, 1698-1719.

[2] R. Robson, B. F. Abrahams, S. R. Batten, R. W. Gable, B. F. Hoskins, and J. Liu, in 'Supramolecular Architecture', ed. T. Bein, ACS Symp. Ser. **499**, Washington, 1992, pp. 256-273.

[3] (a) Decher, G., and Hong, J. D. (1991) 'Buildup of Ultrathin Multilayer Films by a Self-Assembly Process: II. Consecutive adsorption of Anionic and Cationic Bipolar Amphiphiles and Polyelectrolytes on Charged Surfaces' Ber. Bunsen-Ges. Phys. Chem. **95**, 1430-4; (b) Y. Lvov, G. Decher, and H. Möhwald (1993) 'Assembly, Structural Characterization, and Thermal Behavior of Layer-by-Layer Deposited Ultrathin Films of Poly(vinyl sulfate) and Poly(allylamine)' Langmuir, **9**, 481-486.

[4] A. J. Jacobson (1994) 'Colloidal Dispersions of Compounds with Layer and Chain Structures' Mater. Sci. Forum, **152-153**, 1-12.

[5] Tarascon, J. M., DiSalvo, F. J.,Chen, C. H., Carroll, P. J., Walsh, M., and Rupp, L. (1985) 'First Example of Monodispersed $(Mo_3Se_3)^{1-}$ Clusters' J. Solid State Chem., **58**, 290-7.

[6] Keller, S. W., Kim, H.-N., and Mallouk, T. E. (1994) 'Layer-by-Layer Assembly of Intercalation Compounds and Heterostructures on Surfaces: Towards Molecular "Beaker" Epitaxy'J. Am. Chem. Soc., **116**, 8817-8.

[7] Alberti, G., Costantino, U., and Giulietti, R. (1980) 'Preparation of Large Crystals of α-$Zr(HPO_4)_2.H_2O$' J. Inorg. Nucl. Chem., **42**, 1062-3.

[8] Clearfield, A., Oskarsson, A., and Oskarsson, C. (1972) 'Mechanism of Ion Exchange in Crystalline Zirconium Phosphates.VI. Effect of Crystallinity of the Exchanger on Sodium Ion/Proton Exchange.' Ion Exch. Membranes, **1**, 91.

[9] Kim, Y.-I., Atherton, S. J., Brigham, E. S., and Mallouk, T. E. (1993) 'Sensitized Layered Metal Oxide Particles for Photochemical Hydrogen Evolution from Nonsacrificial Electron Donors' J. Phys. Chem., **97**, 11802-10.

[10] Wudl, F., Hirch, A., Khemani, K. C., Suzuki, T., Allemand, P.-M., Koch, A., Eckert, H., Srdanov, H. G., and Webb, H. (1992) 'Survey of Chemical Reactivity of C_{60}, Electrophile and Dieno-polarophile Par Excellence' in *Fullerenes: Synthesis, Properties, and Chemistry of Large Carbon Clusters*, Hammond, G. S. and Kuck, V. J., Eds., ACS Symp. Ser. **481**, 161-175.

[11] R. G. Freeman, K. C. Grabar, K. J. Allison, R. M. Bright, J. A. Davis, A. P. Guthrie, M. B. Hommer, M. A. Jackson, P. C. Smith, D. G. Walter, and M. J. Natan (1995) 'Self-Assembled Metal Colloid Monolayers: An Approach to SERS Substrates' Science, **267**, 1629.

[12] Kim, H.-N., Keller, S. W., Louder, D., Parkinson, B., Schmitt, J., Decher, G., and Mallouk, T. E., unpublished results.

[13] Gasperin, M., and Le Bihan, M.-T. (1980) 'Rubidium Niobate of a New Structural Type: $Rb_4Nb_6O_{17}{\cdot}3H_2O$' J. Solid State Chem., **33**, 83-9.

[14] E. R. Kleinfeld and G. S. Ferguson, (1994) 'Stepwise Formation of Multilayered Nanostructural Films from Macromolecular Precursors' Science, **265**, 370-3.

[15] Johansson, G.; Lundgren, G.; Sillen, L. G.; Soderquist, R. (1960) 'The Cyrstal Structure of a Basic Aluminum Sulfate and the Corresponding Selenate' Acta Chem. Scand., **14**, 769-71.

[16] Perutz, M., Muirhead, H., Cox, J., Goaman, L., Mathews, L., McGandy, E., and Webb, L. (1968) 'Three-Dimensional Fourier Synthesis of Horse Oxyhemoglobin at 2.8 Å Resolution' Nature, **219**, 29-32.

[17] Barner, J. B. and Ruggiero, S. T. (1987) 'Observation of Incremental Charging of Ag Particles by Single Electrons' Phys. Rev. Lett., **59**, 807-10.

[18] Mullen, K., Ben-Jacob, E., Jaklevic, R. C., and Schuss, Z. (1988) 'I-V Characteristics of Coupled Ultrasmall Tunnel Junctions' Phys. Rev. B, **37**, 98-105.

[19] Alder, J. G. and Straus, J. (1976) 'Observation of Localized States in Barrier Regions of Metal-Insulator-Metal Tunnel Junctions' Phys. Rev. B, **13**, 1377.

[20] Lambe, J. and Jaklevic, R. C. (1969) 'Charge Quantization Studies Using a Tunnel Capacitor' Phys. Rev. Lett. **22**, 1371-4.

[21] Fulton, T. A. and Dolan, G. J. (1987) 'Observation of Single Electron Charging Effects in Small Tunnel Junctions' Phys. Rev. Lett., **59**, 109-12.

[22] Inganas, O. and Lundstrom, I. (1984) 'Electronic Properties of Metal-Polypyrrole Junctions' Synth. Met., **10**, 5-11.

PHOTOPATTERNING TO CREATE NEW STRUCTURES ON SURFACES

M.A. FOX,* M.O. WOLF, AND G.M. STEWART
Department of Chemistry and Biochemistry
University of Texas at Austin
Austin, TX 78712 USA

Abstract. Ultraviolet irradiation of self-assembled monolayers of *cis*- and *trans*-4-cyano-4'-(10-thiodecoxy)-stilbene on a polycrystalline gold surface results in visual surface patterning as a consequence of pronounced changes in the surface hydrophobicity induced by photochemical geometric isomerization. The metal surface influences the efficiency of the photoconversion but does not completely quench excited state reactivity.

A novel series of dendrimer segments bearing functionalized aryl chromophores (such as pyrene and naphthalene) at the periphery has been synthesized using a convergent-growth methodology. Selective excitation of the naphthyl substituted dendrons shows no intramolecular excimer formation, although substantial excimer emission is observed with the pyrenyl substituted dendrons. Fluorescence quantum yields are used to define energy migration through the dendritic backbone.

1. Introduction

If high levels of efficiency are to be attained in photoresponsive molecular electronic devices, a means must exist to produce long term charge separation over distances that are large on a molecular scale. Furthermore, the direction of charge carrier migration must be non-random, ideally with electrons and holes moving in opposite, and predictable, directions along a gradient in a spatially well-defined array [1]. Such directional migration can be observed in macroscopically self-ordered solids and/or liquid crystals, and our recent observations of significant photovoltaic effects [2] and of long term information storage [3] in micron thick solid layers of stacked metalloporphyrins attest to the validity of the general principle of charge carrier separation in ordered media as a key component of materials useful for the next generation of sensors, imaging devices, and recording materials.

J. Michl (ed.), Modular Chemistry, 53–67.

Photoinduced electron transfer [4] has proven to be a convenient way to initiate long distance charge separation, provided that appropriate electron relays are precisely positioned in a structurally rigid environment. Such sequential electron transfers constitute, for example, the primary photoprocesses taking place within the reaction center in natural photosynthesis. When an analogous photoinduced electron transfer is initiated along a synthetic rigid macromolecular framework bearing an energetically graded series of redox relays, efficient charge separation may ensue, providing both a useful technique for defining the structure of the polymer scaffold and an interesting means by which the consequences of long distance charge separation can be exploited [5]. Metal and semiconductor surfaces provide a convenient base for the construction of such arrays, and modified electrodes, in particular, offer the additional possibility of field-dependent interrogation of the electron migration event and of electrochemical monitoring of the structured environment near the electrode surface [6]. Thus, the construction of rigid polymers with well-defined backbone structures and high molar absorptivities on smooth electrode surfaces would allow for interesting model systems on which further fundamental principles relevant to long distance vectorial electron transfer could be studied [7,8].

For such studies to succeed, an excited state or exciton must persist on the chemically modified surface for a period long enough for the desired electron transfer sequence to compete with radiationless decay. That is, the supporting surface must not completely quench the expected excited state chemistry emanating from an attached photophysical probe. Many examples exist in the literature that describe the efficient quenching of many emissive molecules when chemisorbed or physisorbed to the surface of a metal. The conventional explanation for this observation rests on the high conductivity of such metals, which is afforded by a continuum of accessible states that can provide for rapid and reversible electron injection to and from an adsorbed excited state molecule. Whether this efficient quenching also pertains to chemically bound reagents, particularly those tethered at the outer edge of a potentially insulating layer, is currently not well known.

An effective means for examining whether this quenching can be suppressed is accomplished by studying the photoreactivity of organic molecules attached to an appropriate metal surface as self-assembled monolayers. If the probe molecule is chosen such that the corresponding solution state photoreactivity is very well characterized, a comparison of this known solution-phase photochemistry with that observed within the covalently bound monolayer will provide a definitive evaluation about whether metals can be used as supports for such complex arrays. The first problem addressed in this article, therefore, is whether "normal" photoreactivity can be observed in photoactive probes tethered to a self-assembled monolayer on a conductive metal surface, and whether the proximity to the metal surface catastrophically interferes with the ability to employ photophysical probes as a means for characterizing appended photochemically activated redox couples.

An additional problem associated with modified electrodes bearing photosensitizers as key components of macromolecular redox relays is in providing sufficiently high concentrations of the absorptive chromophores near the surface of the electrode without promoting efficient self-quenching. Such self-quenching has been a serious problem in traditional attempts to expand the wavelength responsiveness of modified semiconductor electrodes [9], and recent successes in improving the efficiency of photosensitization of porous TiO_2 electrodes [10] have relied instead on monolayer coverages of a highly porous metal oxide surfaces. It would be of interest, therefore, to explore other architectures for providing high local concentrations of physically isolated, but electronically interactive, photosensitizers within the proposed macromolecular arrays. For an efficient photoresponse, in other words, it is likely to be necessary to arrange the appended chromophores in an architecture that can permit directional funneling of the excitation energy to the arranged scaffolding through which efficient electronic coupling might then ensue.

Recent advances in supramolecular photochemistry [11] have provided a plethora of new spatial arrangements for the arrangement of highly absorptive chromophores. One particularly appealing approach would be to exploit recent synthetic success in the preparation of three-dimensional dendrimers [12], especially those bearing chosen chromophores at the periphery of the molecule. By examining the relevant excited state interactions that take place within a dendrimer segment (called a "dendron"), we might reasonably expect to model local light harvesting arrays that could be attached to an electrode surface. The second problem addressed in this article, therefore, deals with a photophysical characterization of intramolecular interactions between aromatic probe molecules attached to the periphery of a dendron.

Much progress has been made in preparing dimensionally ordered arrays on metal and metal oxide surfaces, but relatively little attention has been paid to characterizing photochemistry in these self-assembled layers. Comparisons with solution phase and solid state photoreactivity provide interesting means for learning about local organization and about the character of electronic interactions between attached molecules and the surface. This article will discuss two molecular architectures that can be used as construction units on chemically modified surfaces and can be probed by photophysical methods.

2. Photochemically Reactive Molecules Tethered as Self-Assembled Monolayers on Gold

A particularly convenient means for providing a highly ordered monolayer on a metal electrode surface is to permit functionalized alkylthiols to self-assemble on metals freshly deposited on atomically flat supports. For example, densely packed insulating monolayers are produced on gold, silver, copper, or mercury by simply allowing a solution of the alkylthiol to contact the clean metal surface [13]. It is also known that such ordered arrays can be patterned by exposure to ultraviolet light in the presence of oxygen by a process that removes the absorbate at the irradiated portions from the organized monolayer [14]. Although mechanistic details are not generally available for these ablative processes [15], even less is known about the course of possible photoreactions of molecules present at the outer edge of such monolayers.

It can be arguably asserted that more is known about the excited state behavior of stilbene [16] than that of any other molecule. In particular, stilbene is known to undergo an efficient cis-trans isomerization in which the steady state equilibrium position is wavelength dependent, eqn 1. When

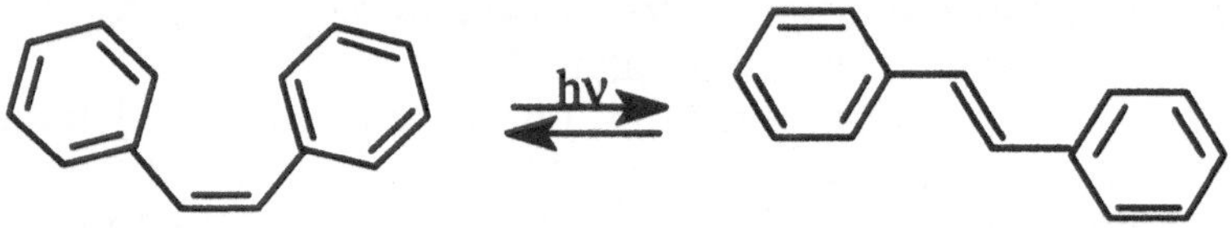

hν (1)

such irradiations are conducted at high local concentrations in solution or in the solid state, a slower [2+2] photochemical cyclodimerization can also be observed.

Seeking to explore the effect of a proximate metal surface of such a photoreaction, we have studied the photochemistry of a functionalized stilbene constrained to a self-assembled monolayer on gold as a probe for excited state chemistry near a conductive metal surface. We have investigated the photoreactions of the two geometric isomers **1** and **2** of a

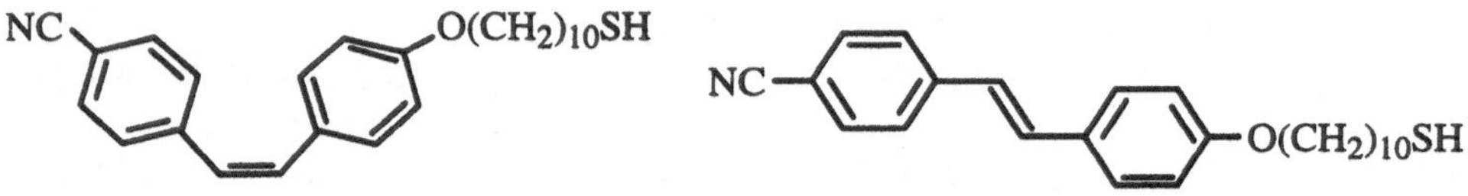

NC O$(CH_2)_{10}$SH

cis

1

NC O$(CH_2)_{10}$SH

trans

2

cyano-functionalized stilbene derivative attached through an ether-linkage to the alcohol end of an ω-hydroxylalkylthiol [16]. As expected by analogy with the unsubstituted parents, the absorption spectra of these compounds show divergent intensities, Figure 1, permitting preferential excitation of one isomer in the presence of the other.

Such compounds form densely packed monolayers, as shown by the electrochemical blocking behavior exhibited by such layers. With the cis isomer, an overpotential of about 0.4 V is observed for the oxidation of ferricyanide present in the contacting electrolyte, with an even greater

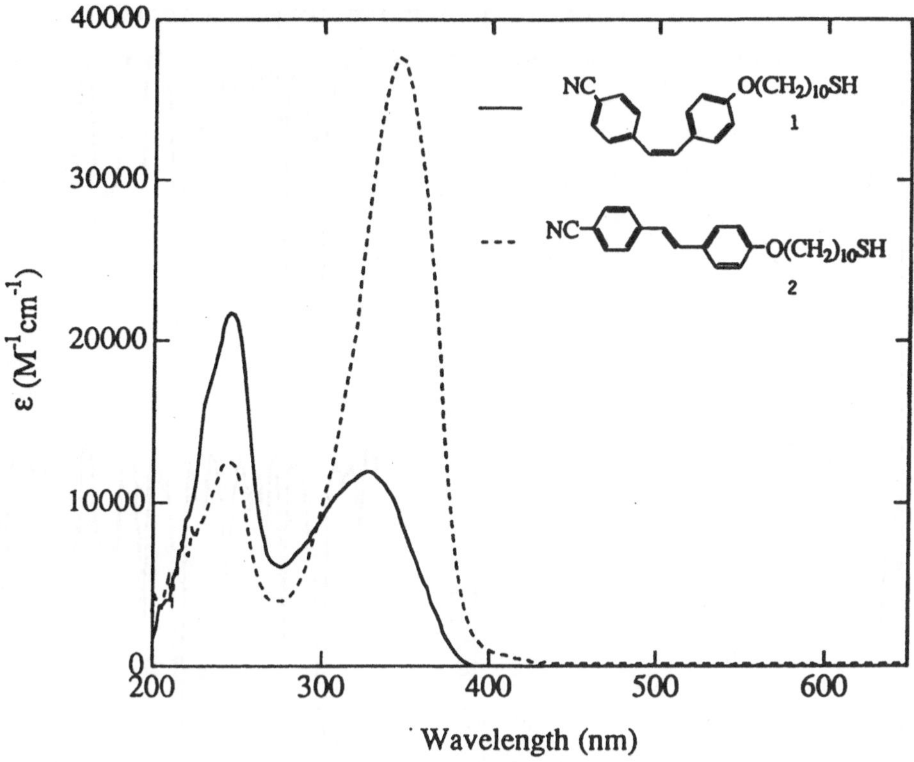

Figure 1. Absorption spectra of **1** and **2** in deaerated methylene chloride.

overpotential being observed for the trans isomer [16]. Surface Fourier-transform infrared spectra of these compounds, Figure 2, similarly demonstrate the presence of these two molecules in discernibly different environments, as compared with the same molecules present in a pressed pellet. Moreover, molecular mechanics (MM-2) calculations indicate that the polar nitrile functional group would protrude differently in the attached cis and trans isomers, Figure 3, rendering the outer surfaces of these two monolayers quite different in surface polarity. In particular, the cis isomer **1** is expected to display a hydrocarbon-like hydrophobic surface, whereas the external surface presented by trans isomer **2** should be appreciably more hydrophilic by virtue of the exposed and highly polarizable nitrile group. Better surface packing is also anticipated with monolayer **2** where, in fact, substantial π,π interactions may enforce the structural order imposed by attachment of the sulfide to the surface gold atoms and by hydrophobic packing of the long alkyl chains. Consistent with this expectation, contact angles with water produced by these two monolayers differ by about 16°, the larger value being observed for the more hydrophobic cis isomer **1**.

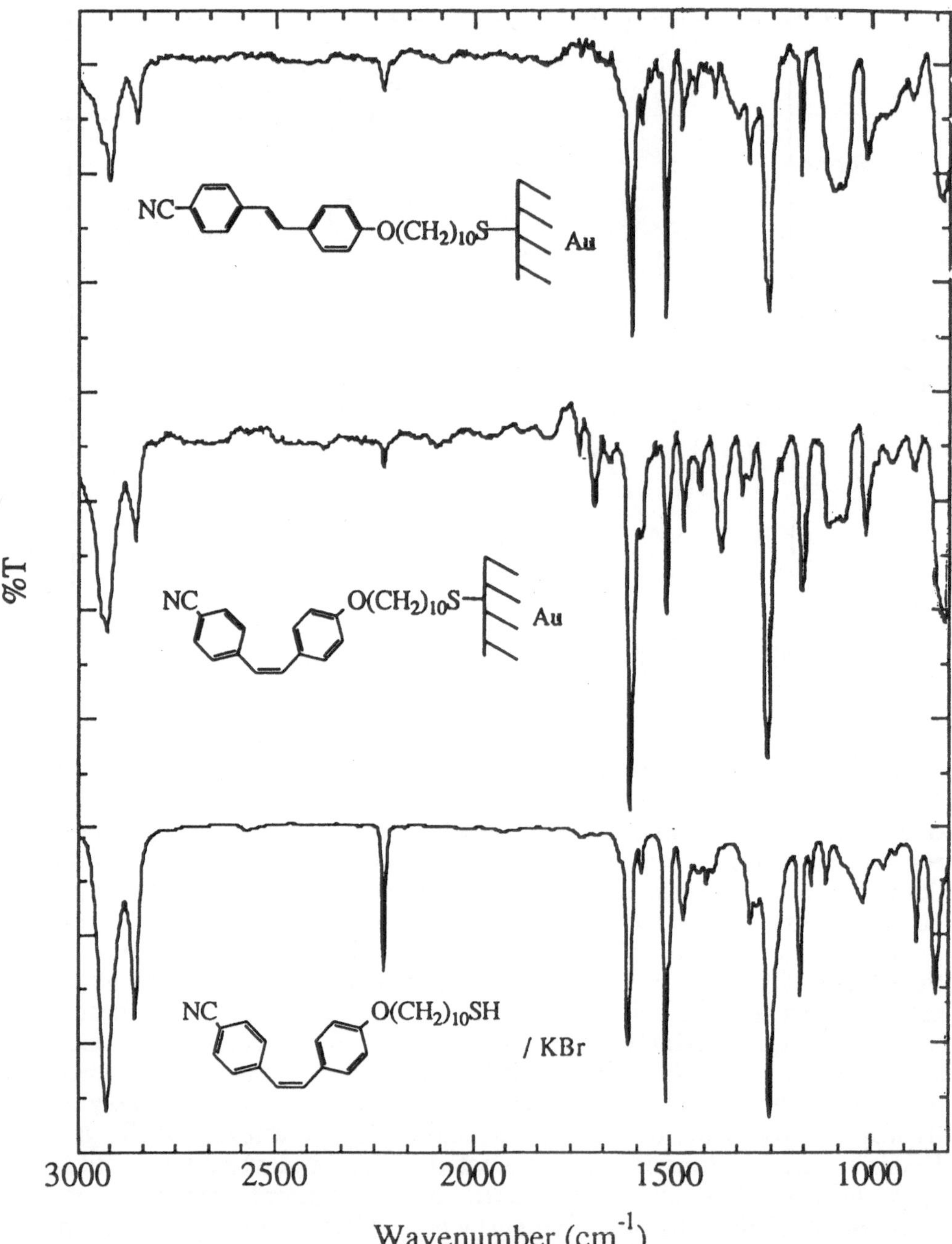

Figure 2. Fourier-transform infrared spectra of **1** and **2** as self-assembled monolayers on freshly deposited gold (unannealed) on a silicon wafer and of **1** as a KBr pellet.

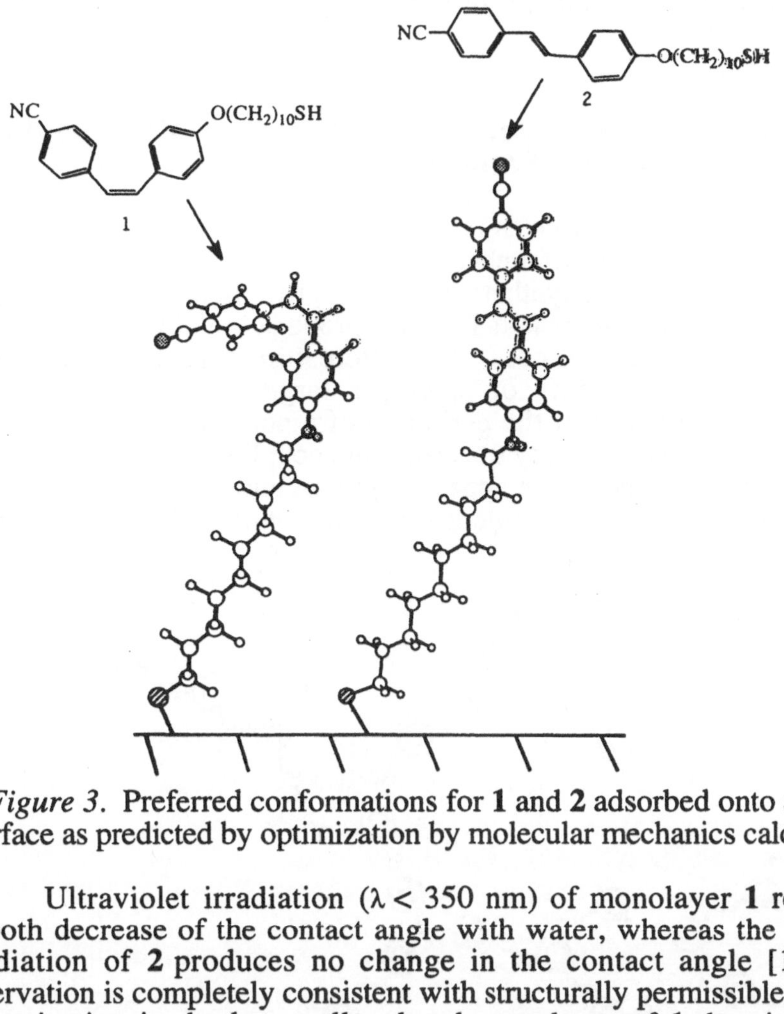

NC-[stilbene]-O(CH$_2$)$_{10}$SH 2

NC ... O(CH$_2$)$_{10}$SH 1

Figure 3. Preferred conformations for **1** and **2** adsorbed onto a metal surface as predicted by optimization by molecular mechanics calculations.

Ultraviolet irradiation ($\lambda < 350$ nm) of monolayer **1** results in a smooth decrease of the contact angle with water, whereas the analogous irradiation of **2** produces no change in the contact angle [16]. This observation is completely consistent with structurally permissible geometric isomerization in the less well-ordered monolayer of **1** than in the more highly ordered monolayer of **2**. This differential photoreactivity can be used as the basis for a new method of non-silver based imaging, as this difference in contact angle in a (masked) photochemically isomerized surface become visible to the naked human eye upon exposure of a chilled plate to moist air.

The observed photochemistry shows that the metal surface per se does not quench completely the normal photoreactivity of appended organic molecules, although the surface does perturb normal excited state partitioning between competing reaction channels. Both the efficiency of the geometric isomerization and the subsequent photochemical cyclodimerization [17] are influenced by the local order, and additional photochemical control is possible in these layers than is attainable in either homogeneous solution, in melts, or in solid films of these same molecules [17].

3. Intramolecular Interactions between Chromophores Appended to Dendrimer Segments

Dendritic architectures provide an interesting three dimensional alternative to the more common linear, branched, and cross-linked architectures more commonly encountered in synthetic polymers [18]. In a typical dendrimer, a hyper-branched structure results from the propagation of multiple reaction sites from a functionalized core. Our interest in dendritic architectures derives from the possibility of placing a large number of absorptive chromophores inside and at the periphery of a dendrimer and employing photophysical methods to study the electronic coupling (either by energy or electron transfer) between these appended groups and a central probe.

A possible target molecule is shown in Figure 4, where a sequence

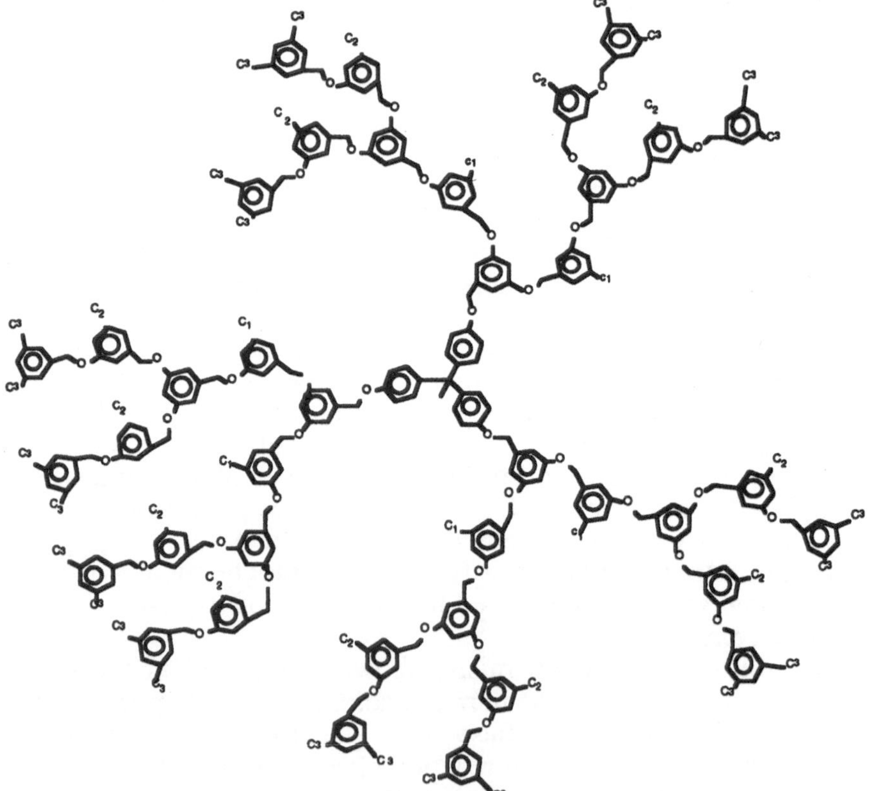

Figure 4. A possible functionalized dendrimer bearing graded chromophores or redox relays C_1, C_2, and C_3.

of appended chromophores C_1, C_2, and C_3 which can act as electron or energy donors or acceptors can be attached at defined positions along the progressively larger support. Steady state emission measurements are employed here to characterize these interactions.

There are several reasons to use functionalized dendrimers or dendrons (dendrimer segments) as probe arrays for studying long range photoinduced electron transfer. These include: a variable surface-to-volume ratio as a function of sequential generations of dendritic growth; easy suppression of possible chain entanglement which is frequently observed in conformationally flexible polymers; typically high solubility in organic solvents; and precise synthetic control of the relevant donor / acceptor ratio and of the placement of specific absorbers or relays. In addition, the dendron segments prepared as intermediates in a converging synthetic pathway may also provide interesting simple models for exploring intramolecular multichromophore interactions.

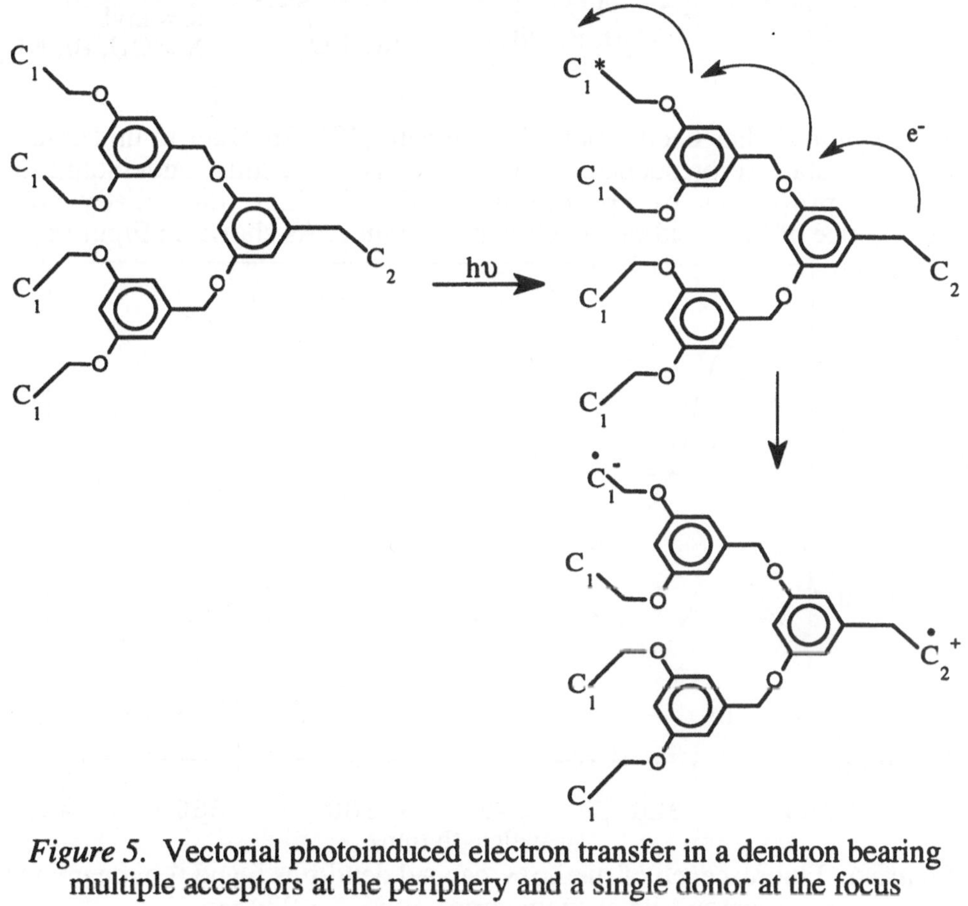

Figure 5. Vectorial photoinduced electron transfer in a dendron bearing multiple acceptors at the periphery and a single donor at the focus

The desired directional electron transfer in such dendrons is attained by several hops through the backbone relay, as shown in Figure 5. As in natural photosynthesis, these sequential hops achieve spatial separation of positive and negative charges.

Such effects have been examined in our target dendrons **3** and **4**,

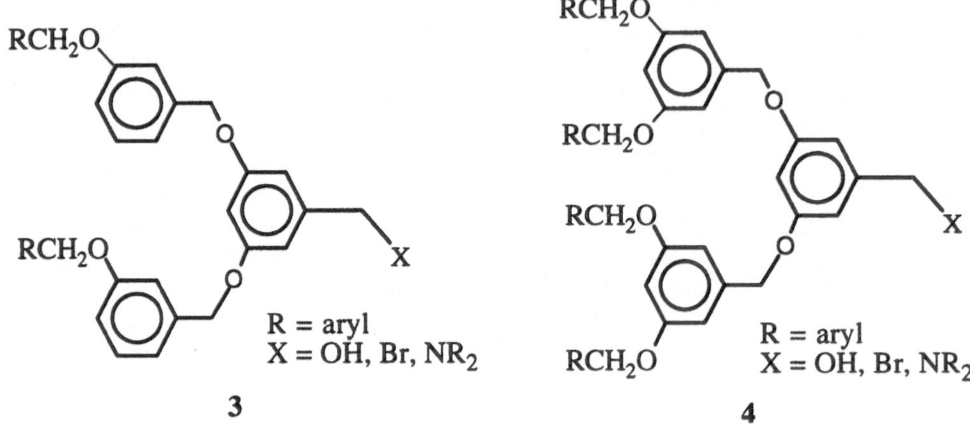

whose synthesis has been reported elsewhere [19]. In order to understand the steady state fluorescence of these dendrons, we must understand the emission properties of appropriate models. First, the steady state fluorescence of the backbone itself was examined. As shown in Figure 6,

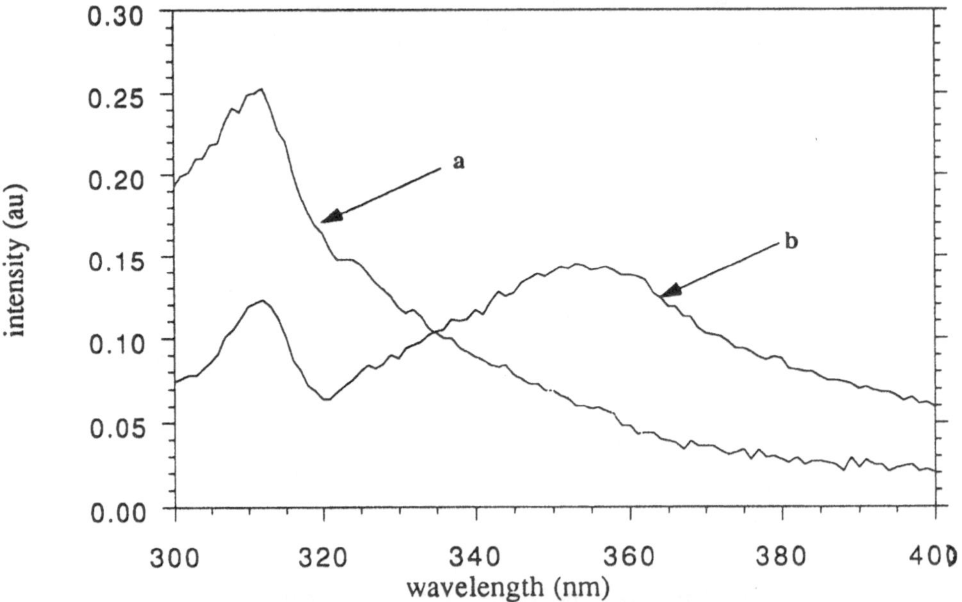

Figure 6. Fluorescence of methoxy-capped dendrons **5** and **6** in deareated acetonitrile at room temperature, λ = 284 nm.

only very weak, structureless emission is observed for the alkyl-substituted skeletons being either an alcohol group (**5**) or a dimethylanilinophenoxy group (**6**) at the focal point.

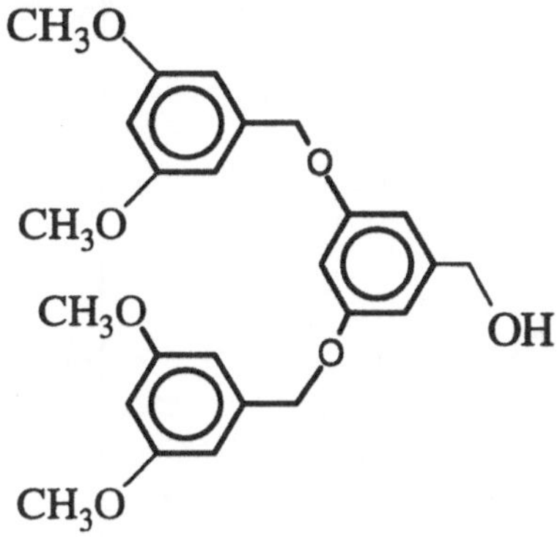

CH_3O CH_3O CH_3O CH_3O OH

CH_3O CH_3O CH_3O CH_3O $N(CH_3)_2$

5 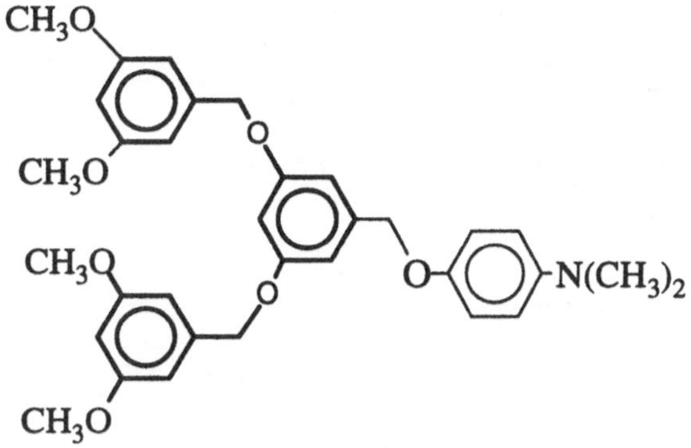 **6**

In dendrons **7** and **8**, which bear different degrees of peripheral absorbing substituents, partial self-quenching is evident from the contrasting fluorescence yields observed upon excitation at $\lambda = 284$ nm, respectively 0.06 and 0.02. The observed emission spectrum, however, closely

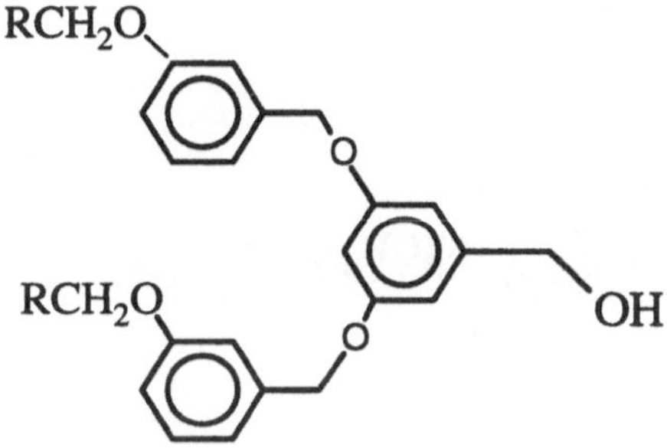

RCH_2O RCH_2O OH

R = 2-Np

7

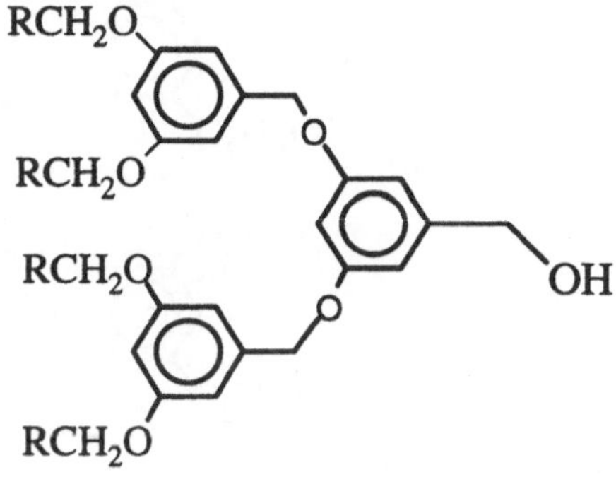

RCH_2O RCH_2O RCH_2O RCH_2O OH

R = 2-Np

8

resembles that expected for isolated naphthalene units, and no evidence of intramolecular excimer emission can be found. In the presence of a distant (about 15 Å) dimethylaminophenoxy group, substantial further quenching can be observed, Figure 7.

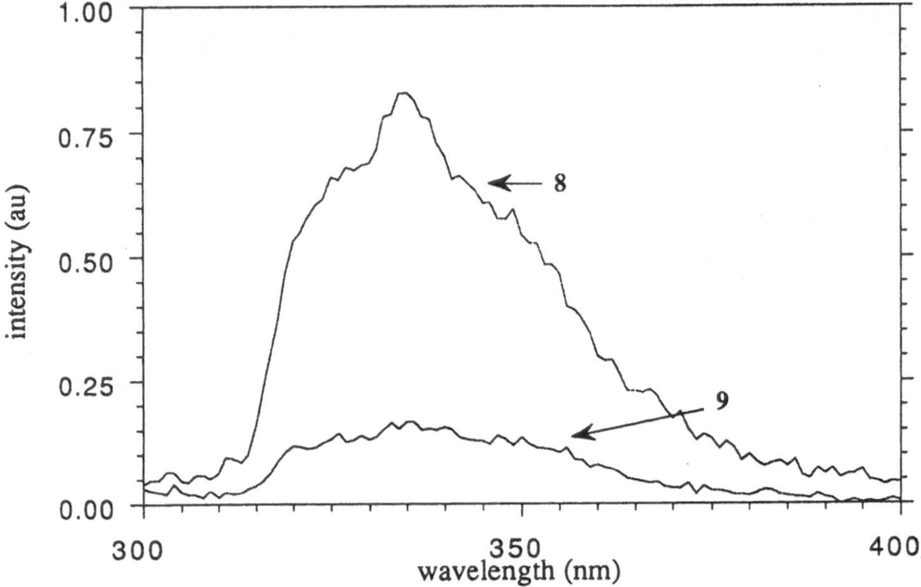

Figure 7. Fluorescence of naphthyl-capped dendrons as a terminal alcohol **8** and a dimethylaminophenoxy quencher **9** in deaerated acetonitrile at room temperature, λ_{ex} = 284 nm.

RCH_2O RCH_2O RCH_2O RCH_2O O O O $N(CH_3)_2$

R = 2-Np

9

In contrast, the observed emission is dominated by excimer emission in the analogous pyrenyl-capped dendrons. Compared with the fluorescence observed in pyrenylmethanol **10** or in a simple model bearing a single pyrenyl group **11**, the dendrons **12** and **13** exhibited significantly red-shifted emission, assigned to an intramolecular excimer, Figure 8. In the presence of an intramolecular electron donor, i.e., the dimethylaminophenoxy group, this emission is extensively quenched.

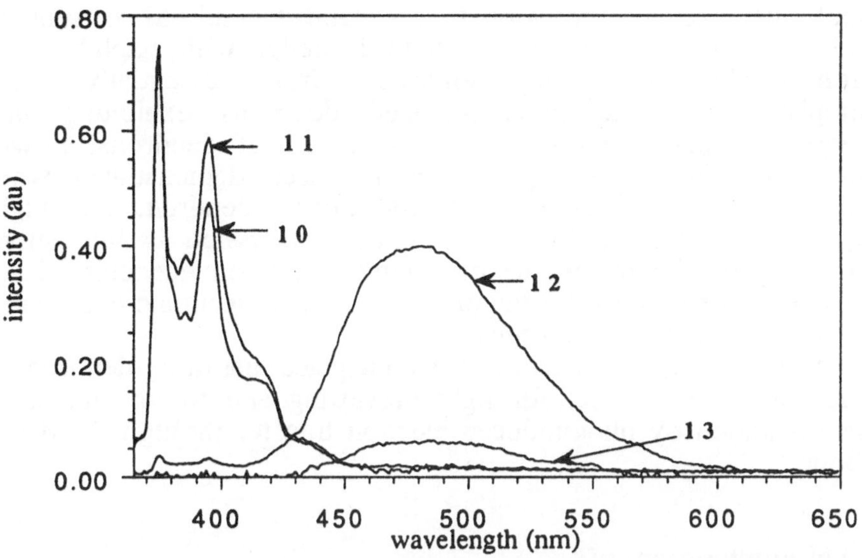

Figure 8. Fluorescence of pyrenyl methanol **10** and a monopyrenyl model **11** of the pyrenyl-capped dendrons **12** and **13** in deaerated acetonitrile at room temperature, λ_{ex} = 344 nm.

CH_2OH CH_2O OH

10 **11**

RCH_2O RCH_2O O OH RCH_2O RCH_2O O O O $N(CH_3)_2$

R = 1-Pyr R = 1-Pyr

12 **13**

From such studies, we conclude that the significance of

intramolecular excimer formation in aryl-capped dendrons is sensitively dependent on the size of the appended arene, with naphthyl-capped dendrons exhibiting mainly emission from essentially isolated chromophores and the pyrenyl-capped dendrons exhibiting mainly intramolecular excimer emission. In both sets of dendrons, however, intramolecular fluorescence quenching has been demonstrated when a suitable electron donor is affixed at the focus of the dendron. Although the absorption maximum of the pyrenyl excimer emission is insensitive to solvent polarity, the magnitude of excimer emission quenching is quite responsive, as expected from the proposed strong electronic coupling as a means to directional electron transfer.

When affixed to surfaces, it is anticipated that these dendrons may provide interesting vehicles for light harvesting and for the initiation of charge separation by photoinduced electron transfer through the dendritic backbone.

4. Acknowledgements

This work was supported by the U.S. Department of Energy, Office of Basic Energy Sciences. MOW thanks the National Science and Engineering Research Council of Canada for a postdoctoral fellowship.

5. References

1. Fox, M.A. (1992) Polymeric and Supramolecular Arrays for Directional Energy and Electron Transport over Macroscopic Distances, *Accts. Chem. Res.* **25**, 569-574.
2. Gregg, B.A., Fox, M.A., and Bard, A.J. (1990) Photovoltaic Effect in Symmetrical Cells of a Liquid Crystal Porphyrin, *J. Phys. Chem.* **94**, 1586-1598.
3. Liu, C.Y., Pan, H.L., Fox, M.A., and Bard, A.J. (1993) High Density Nanosecond Charge Trapping in Thin Films of the Photoconductor Zinc-octakis (b-decoxyethyl) porphyrin, *Science* **261**, 897-899.
4. Fox, M.A. and Chanon, M. (1984) *Photoinduced Electron Transfer*, Elsevier, Amsterdam, and references cited therein.
5. Watkins, D.M. and Fox, M.A. (1995) Synthesis and Photophysical Characterization of Aryl-substituted Polynorbornenediol Acetal and Ketal Multiblock Copolymers, *Macromolecules* **28**, 4939-4950.
6. Fox, M.A., Nobs, F. and Voynick, V. (1980) Covalent Attachment of Arenes to SnO_2 Semiconductor Electrodes, *J. Amer. Chem. Soc.* **102**, 4029-4036.
7. Hong, B. and Fox, M.A. (1994) Arene-Functionalized Polyisocyanides: A Kinetic Study of Polymerization to Prepare Homopolymers and Block Copolymers, *Macromolecules* **27**, 5311-5317.

8. Fox, H.H. and Fox, M.A. (1995) Fluorescence and Singlet Energy Migration in Rationally Designed Acrylate Polymers Bearing Pendant Chromophores, *Macromolecules* **28**, 4570-4576.
9. Gerischer, H. and Willig, F. (1976) Reaction of Excited Dye Molecules at Electrodes, *Top. Curr. Chem.* ***61***, 33-82.
10. Nazeeruddin, M.K., Kay, A., Rodicio, I., Humphry-Baker, R., Müller, E., Liska, P., Vlachopoulos, N., and Grätzel, M. (1993) Conversion of Light to Electricity by *cis*-X_2 Bis(2,2'-bipyridyl-4,4'-dicarboxylate)ruthenium(II) Charge-Transfer Sensitizers (X = Cl^-, Br^-, I^-, CN^-, and SCN^-) on Nanocrystalline TiO_2 Electrodes, *J. Am. Chem. Soc.* **115**, 6382-6390.
11. (a) see Schneider, H.J. and Dürr, H., Eds. (1991) *Frontiers in Supramolecular Organic Chemistry and Photochemistry*, VCH Publishers, Weinheim for an overview of supramolecular assemblies. (b) Möbius, D. (1990) Organized Monolayers and Monolayer Assemblies as Potential Components of Molecular Devices, *Can. J. Phys.* **68**, 992-998. (c) Möbius, D. and Kuhn, H. (1988) Energy Transfer i nMonolayers with Cyanine Dye Scheibe Aggregates, *J. Appl. Phys.* **64**, 5138-5146.
12. Tomalia, D.A. and Durst, H.D. (1993) Genealogically Directed Synthesis: Starburst/Cascade Dendrimers and Hyperbranched Structures, *Top. Curr. Chem.* **165**, 93-138.
13. Bain, C.D., Troughton, E.B., Tao, Y.T., Evall, J., Whitesides, G.M., and Nuzzo, R.G. (1989) Formation of Monolayer Films by the Spontaneous Assembly of Organic Thiols from Solution onto Gold, *J. Am. Chem. Soc.* **111**, 321-335.
14. Rozsnyai, L.F. and Wrighton, M.S. (1994) Selective Electrochemical Deposition of Polyaniline via Photopatterning of a Monolayer-Modified Substrate, *J. Am. Chem. Soc.* **116**, 5993-5994.
15. Lewis, M., Tarlov, M., and Carron, K. (1995) Study of the Photooxidation Process of Self-Assembled Alkanethiol Monolayers, *J. Am. Chem. Soc.* **117**, 9574-9575.
16. Wolf, M.O. and Fox, M.A. (1995) Photochemistry and Surface Properties of Self-Assembled Monolayers of *cis*- and *trans*-4-Cyano-4'-(10-thiodecoxy)stilbene on Polycrystalline Gold, *J. Am. Chem. Soc.* **117**, 1845-1846.
17. Wolf, M.O. and Fox, M.A. (1996) Photoisomerization and Photodimerization in Self-Assembled Monolayers of *cis*- and *trans*-4-cyano-4'-(10-thiodecoxy)stilbene on Gold, *Langmuir* **12**, 955-962.
18. Hawker, C.J. and Frechet, J.M.J. (1992) Unusual Macromolecular Architectures: The Convergent Growth Approach to Dendritic Polyesters and Novel Block Copolymers, *J. Am. Chem. Soc.* **114**, 8405-8413.
19. Stewart, G. and Fox, M.A. (1996) Chromophore-labeled Dendrons as Light Harvesting Antennae, *J. Am. Chem. Soc.* ***118***, 4354-4360.

DISCUSSION OF THE MALLOUK AND FOX LECTURES

CHRISTOPHER E. D. CHIDSEY
Department of Chemistry
Stanford University
Stanford, CA 94305-5080
U.S.A.

A general theme of the Session II talks by Professors Mallouk and Fox was the formation, structure and properties of layered systems. In the context of modular chemistry, an extended layer may be thought of as a two-dimensional network of modules composed of atoms, molecules, ions, polymers, or, as illustrated in Prof. Mallouk's talk, the exfoliated leaflets of a layered solid. Though novel bulk solids can in principle be built up by sequential formation of such layers, this may be a very time-consuming process, as Prof. Mallouk points out below in response to a question. On the other hand, as single layers or the components of a heterostructure of several layers, extended layers have a large and potentially significant role to play in the separation and interconnection of functional modules and phases. Prof. Chidsey began the discussion by commenting on some of the possibly important structural features of extended layers and some of the implications for a modular chemistry in two dimensions.

He argued that various degrees of lateral "order" are possible and valuable in a two-dimensional network of modules. The most obvious, 2-dimensional crystalline order, is possible and is achieved in some self-assembled monolayers and in many inorganic layers on crystalline substrates. Such crystalline layers will usually be composed of multiple domains separated by domain boundaries, and it is important to consider what role may be played by the defects at these domain boundaries. Prof. Chidsey argued that multiple domains are a very common and hard-to-avoid consequence of forming extended, crystalline 2-dimensional arrays of modules starting from many different places independently. Less ordered 2-dimensional networks of modules including glassy and liquid-like arrangements may be adequate or even preferable for some uses. In fact in some cases, the modules may usefully be arranged in a dilute way in two dimensions in what could be called a 2-dimensional gas.

J. Michl (ed.), Modular Chemistry, 69–73.

Prof. Chidsey also argued that the transverse structure of these extended layers could have various characteristics as required by their function. For instance, to serve as a barrier, a layer must presumably be densely packed and of sufficient thickness. To serve as a mechanical joint between phases or other layers, only the strength of bonding may be important. Finally, to mediate chemical events across layers as described by Professors Mallouk and Fox in their talks, the lateral structure must allow for the required energy or electron transfer, and provide the appropriate degree of steric flexibility or constraint to promote the desired chemistry.

A broad ranging exploration of the systems described by Professors Mallouk and Fox followed. Unfortunately, the recording of this session was lost. What follows is a partial reconstruction by the participants of many of the comments and questions.

Comments and questions directed to Prof. Mallouk

WARD (Minnesota): Is there any evidence for ordered, commensurate arrangements of monomolecular cations on the exfoliated sheets? It should be possible to bridge zirconium phosphate sheets with multivalent cations, and the density of the cations can be controlled by varying their charge. This may be important in the optical/electron transfer properties if the cation is a chromophore or electron acceptor.

MALLOUK: There are, to my knowledge, no detailed studies of this kind, but it is a great idea. We should examine these molecular interactions by AFM to see if there are specific sites where, for example, $R\text{-}NH_3^+$ groups attach by hydrogen bonds to triangularly disposed phosphate oxygen atoms on the phosphate sheets. If there are commensurate structures that are stable, it would provide some guidance in the design of polyelectrolytes, and would enable us to optimize lateral distances in the electron and energy transfer cascades.

KUBIAK (Purdue): Tom, it's my understanding from Tinkham's semiclassical model for single electron tunneling and "Coulomb Blockade", that the model consists of two capacitances: C_1, C_2 and two resistances, R_1, R_2. Could you point out in the layered heterostructure that showed Coulomb Blockade behavior: first, how exactly the "device" is constructed and which parts you think act like R_1, R_2, C_1, C_2? How is the electrical connection made?

MALLOUK: Thank you for drawing attention to my contribution to this volume. The experimental details you request are in the experimental section of my chapter.

OZIN (Toronto): Why not perform TEM to study your layered materials? Embedding and sectioning is relatively straightforward.

MALLOUK: It's a good idea. This is one of several techniques we should use to elucidate 3-D structure in these materials. We are currently using monodisperse silica spheres as TEM-friendly substrates.

OZIN: Your synthetic approach to layered materials can be applied to make functionally gradient materials, with interesting mechanical properties, in the way that Nature creates these architectures, and currently materials science fabricates using processing methods.

MALLOUK: Our technique may provide a straightforward route to such gradient materials. The caveat is that we grow layers one at a time, and each one takes ~10 minutes. So making structures on micron or larger length scales in the stacking direction may be very time-consuming unless we use thicker building blocks.

MORTIMER (UCLA): A general comment: with regard to materials science, the degree of ordering in a system need only be enough to fulfill its function, but when talking about making things in a modular fashion, the better ordered, the better.

MALLOUK: That is true.

MORTIMER: There is a type of disorder in films which I don't think has been mentioned yet. There are certain clay minerals -- smectites? -- where the layers are folded and crumpled like a tissue. Do you observe any behavior of that type?

MALLOUK: The layers are very flexible, so buckling is possible. We need to apply direct imaging tools to see if this is the case. My guess is that we could find conditions of temperature and ionic strength where such defects could be annealed away.

MORTIMER: Is it possible to pick up small monolayers on an AFM tip and build up layered structures like that? It might avoid the overlap problem with successive layers.

MALLOUK: That might be a good way to tile the surface efficiently, although I am not convinced that we are not already tiling it well. The fact that the layer thicknesses are very reproducible, and correspond to the dimensions of related bulk intercalation compounds, means that we are making relatively complete layers. The disadvantage of the approach you suggest is that it would be hard to cover macroscopic areas in a reasonable period of time.

Comments and questions directed to Prof. Fox

MALLOUK (to Prof. Fox or others with dendrimer expertise, e.g. Professors Newkome and Tomalia): Our MM2 calculations show that the Frechet convergent synthesis you are using gives largely planar wedges. Are there other connection strategies, which might be similarly accessible in the synthetic sense, that would give more 3-dimensional wedges? This might help to minimize excimer formation with molecules like pyrene at the periphery.

FOX: It may be questionable whether MM2 provides a good picture of three dimensional structure in compounds with many aryl/aryl interactions. It fails, for example, to predict observed crystal packing in simple arenes. The rigidity of the intervening units, or the possibility of including other additives within dedritic holes, can alternatively influence the attainable structure, we hope.

TOLBERT (Georgia Tech): Given that surface of a sphere goes as r^2 and ET rate goes as r^{-3}, isn't the efficiency of light harvesting limited as you go to larger dendrimers?

FOX: Light absorption is incomplete in a monolayer so the density of chromophores overcomes inherent geometric inefficiency. Further, it will depend on the magnitude of the electronic coupling through the dendritic backbone and on the disposition of the chromophores appended (they will likely be dispersed at an angle to the surface of the sphere and will not be strictly limited by the geometric area).

BALZANI (Bologna): What is the rate of the photoinduced electron-transfer process in the dendrimers carrying donor and acceptor moieties?

FOX: For the naphthyl and pyrenyl-substituted second generation dendrimers, it is approximately 10^{10} s^{-1}.

BALZANI: Is there any sizable electronic interaction between the (electron donor) dimethoxybenzene units contained in the backbone of the dendrimer and the appended electron acceptor moieties?

FOX: There is evidence of small charge-transfer shifts in the absorption and emission spectra of both sets of dendrimers.

KUBIAK: Marye Anne, my comment concerns this issue of energy and electron transfer quenching of molecular excited states by metals. I think in your case that it's important to remember that the molecular excited states are attached to the metal surface by an intervening SAM of an alkyl thiol on gold. In the limit of strong adsorption of the thiols on gold, it is almost a certainty that the Fermi level of gold is "pinned" within the HOMO/LUMO gap of the dithiol/chromophore. In this case, energy and electron transfer quenching should not be as efficient as from a metal with a "continuous density of states."

FOX: Yes, that's right, I should have pointed that out.

Ag(I)···NC-R COORDINATION NETWORKS

KEITH A. HIRSCH, PENG ZHANG, AND JEFFREY S. MOORE
Departments of Chemistry and Materials Science & Engineering,
University of Illinois
Urbana, IL 61801
U.S.A.

GEOFFREY B. GARDNER AND STEPHEN LEE
Department of Chemistry,
University of Michigan
Ann Arbor, MI 48109-1055
U.S.A.

ABSTRACT. Crystal structures of di-, tri-, and tetratopic nitriles with silver(I) triflate ($AgCF_3SO_3$) and silver(I) hexafluorophosphate ($AgPF_6$) are described. From this study, two porous coordination networks involving tritopic ligands and $AgCF_3SO_3$ have been identified and experiments conducted to prove their porosity are discussed. Also, an empirical set of rules to predict network connectivity for complexes formed with $AgCF_3SO_3$ is provided. These rules derive from the observation that triflate coordination limits the number of network-forming ligands that may enter the coordination sphere of silver(I).

1. Introduction

The rational design of extended solids *via* directed interactions has gained considerable attention in the literature recently. For example, many interesting hydrogen bonded networks have been achieved through the use of appropriately functionalized, rigid organic building blocks [1-4]. The coordination bond is another directional interaction which can be used to construct infinite frameworks [5-9]. Coordination networks are realized through the binary combination of a multitopic organic ligand [10] and a transition metal salt. In order to reliably predict the topology of the ensuing coordination network, the interactions between these chemical units must be understood. Since the geometry and functionality of the ligand may be controlled through synthesis, the rational formation of such networks lies in control of the coordination geometry of the transition metal. Factors affecting the coordinating propensity of the metal include the nature of the counterion and the crystallization solvent. Towards this end, we have investigated the coordination of multitopic nitriles with silver(I) salts. The counterions employed are triflate ($CF_3SO_3^-$) which is considered to be weakly coordinating [11] and hexafluorophosphate (PF_6^-) which is

J. Michl (ed.), Modular Chemistry, 75–94.

generally non-coordinating [12]. This review summarizes our efforts to predictably construct coordination networks comprised of Ag(I)···NC-R interactions and to apply this knowledge to the formation of porous materials.

2. Coordination Networks of Multitopic Nitriles with Silver(I) Salts

Table 1 outlines the multitopic nitriles used for this study and briefly summarizes the results of crystallization of these ligands with both silver(I) triflate ($AgCF_3SO_3$) and silver(I) hexafluorophosphate ($AgPF_6$). Illustrations of the coordination networks characterized are shown in Figure 1. A discussion of the crystal structures obtained follows.

2.1 COORDINATION NETWORKS OF DITOPIC NITRILES

The network topology observed with ditopic nitriles is varied. Crystallization of 1,4-dicyanobenzene with either $AgCF_3SO_3$ or $AgPF_6$ (complexes **1** and **2**; Figures 1(a) and 1(b), respectively) results in the formation of one-dimensional structures in which 1,4-dicyanobenzene units bridge silver(I) ions [13]. Specifically, with $AgCF_3SO_3$, neighboring infinite chains are linked by adventitious molecules of water which form the rungs of a ladder. Triflate counterions fill the space between the rungs. In the case of $AgPF_6$, adjacent chains are bridged by two molecules of water. The void space created by this assembly is filled by hexafluorophosphate counterions.

The influence of the counterion on network topology is illustrated by comparison of the structures resulting from crystallization of 4,4'-biphenyldicarbonitrile with both $AgCF_3SO_3$ and $AgPF_6$ (complexes **3** and **4**; Figures 1(c) and 1(d), respectively) [7]. The structure with $AgCF_3SO_3$, grown from benzene, is that of infinite chains of alternating 4,4'-biphenyldicarbonitrile molecules and silver(I) ions. The coordination sphere of silver(I) is completed through weak coordination to a benzene molecule *via* Ag-π interactions and a bond to an oxygen of the triflate counterion. Triflate coordination presumably prevents the formation of a network of higher dimensionality. However, with $AgPF_6$, a ninefold interpenetrated diamondoid network is obtained from either ethanol or toluene. Silver(I) is tetrahedral and coordinates to four 4,4'-biphenyldicarbonitrile ligands. The non-coordinating propensity of the hexafluorophosphate counterion allows silver(I) to achieve its maximal coordination number (with respect to the ligand) of four [14]. Disordered hexafluorophosphate counterions fill channels along the *c*-axis of this structure which are maintained despite the high level of interpenetration.

Crystallization of 4,4'-dicyanodiphenylacetylene with $AgPF_6$ in xylenes is accompanied by counterion hydrolysis to the relatively unususal $PO_2F_2^-$ species [15]. The likely source of water is the $AgPF_6$ used, which is known to be very hygroscopic. The resulting network (complex **5**; Figure 1(e)) consists of undulating chains of 4,4'-dicyanodiphenylacetylene molecules coordinated to silver(I) ions. Proximal chains are bridged through counterion oxygen atoms [16].

2.2 COORDINATION NETWORKS OF TRI- AND TETRATOPIC NITRILES

With tritopic nitriles, coordination networks have been obtained most often through crystallization with $AgCF_3SO_3$. The exception is 1,3,5-tricyanobenzene which has been crystallized with both $AgCF_3SO_3$ [8] and $AgPF_6$ [17] (complexes **6** and **7**; Figures 1(f) and 1(g), respectively). With $AgCF_3SO_3$, two-dimensional, planar sheets of the AlB_2-type [18] are obtained which consist of alternating 1,3,5-tricyanobenzene and silver(I) units. Silver(I) is four-coordinate, bonding to three ligand nitrogen atoms and an oxygen of triflate. Void space created within a single sheet is filled by triflate counterions from adjacent sheets. With $AgPF_6$, undulating sheets of alternating 1,3,5-tricyanobenzene and silver(I) moieties are formed from toluene. In this structure, as the hexafluorophosphate counterion does not bond, silver(I) coordinates to a molecule of toluene in an η^2 fashion. Hexafluorophosphate counterions fill space created within the layers.

Crystallization of 1,3,5-tris(4-ethynylbenzonitrile)benzene (TEB) with $AgCF_3SO_3$ in benzene results in the *controlled* formation of two polymorphs. The first, polymorph A (complex **8**; Figure 1(h)) [8], is obtained by dissolving TEB and $AgCF_3SO_3$ in benzene at 100 °C followed by cooling to room temperature at a rate of 1.2 °C/h. The resulting structure is a three-dimensional network of alternating TEB ligands and silver(I) ions and resembles $ThSi_2$ [18]. The triflate counterion coordinates to silver(I) through oxygen. A significant portion of the void space created by a single $ThSi_2$-type net is filled by five more identical nets (a sixfold interpenetrated structure). Despite this interpenetration, *ca.* 15 Å x 22 Å channels along [1 0 0], which are filled with benzene molecules, are maintained. This polymorph has been proven to be porous by solvent exchange (see Section 3.1) [8].

The second polymorph involving TEB and $AgCF_3SO_3$, polymorph B (complex **9**; Figure 1(i)) [19], is obtained by initially dissolving the two components in benzene at 100 °C and cooling at a rate of 1.2 °C/h. Performing this heating/cooling cycle three more times leads to complete conversion of polymorph A to polymorph B as confirmed by X-ray powder diffraction. The structure of polymorph B consists of undulating sheets of alternating TEB and silver(I) units and is reminiscent of AlB_2. The triflate counterion bonds to silver(I) through oxygen and, similar to polymorph A, a majority of the void space in a single sheet is filled by five equivalent nets. Remaining void space is filled by benzene molecules. However, unlike the $ThSi_2$-type structure, polymorph B is not porous.

With $AgCF_3SO_3$, crystallization of 1,3,5-tris(3-ethynylbenzonitrile)benzene (*m*-TEB) in benzene yields a single polymorph (complex **10**; Figure 1(j)) [20]. The structure is that of planar sheets of alternating *m*-TEB and silver(I) moieties whose repeat unit is [12]annulene-like. As before, triflate coordinates to silver(I) through oxygen. Face-to-face π-π stacking of the sheets in an ···ABCD··· manner is observed in this structure. This arrangement creates channels at an oblique angle to [1 0 1] which are filled with benzene. There are sixteen benzene molecules per

TABLE 1. Results of crystallization of multitopic nitriles with silver(I) salts

	Silver(I) Salts		
Ligand	$AgCF_3SO_3$	$AgPF_6$	Reference
ditopic ligands			
NC CN	•complex **1** •1-D chains •M:L stoichiometry 1:1 •grown from benzene	•complex **2** •1-D chains •M:L stoichiometry 1:1 •grown from ethanol	13
NC CN	•complex **3** •1-D chains •M:L stoichiometry 1:1 •grown from benzene	•complex **4** •9-fold interpenetrated diamondoid structure •M:L stoichiometry 1:2 •suitable crystals grown from either ethanol or toluene	7
NC CN	•no suitable crystals yet obtained	•complex **5** •counterion hydrolysis to $PO_2F_2^-$ •1-D chains with bridging counterion •M:L stoichiometry 1:1 •grown from xylenes	16
tritopic ligands			
CN NC CN	•complex **6** •2-D sheets (AlB_2 prototype) •M:L stoichiometry 1:1 •grown from benzene	•complex **7** •undulating 2-D sheets •M:L stoichiometry 1:1 •Ag(I) coordinated to η^2–toluene •grown from toluene	8 ($CF_3SO_3^-$) 17 (PF_6^-)

TABLE 1. (continued)

Ligand	Silver(I) Salts		Reference
	$AgCF_3SO_3$	$AgPF_6$	
tritopic ligands (continued)			
	polymorph A •complex **8** •3-D net ($ThSi_2$ prototype) •6-fold interpenetrated •M:L stoichiometry 1:1 •grown from benzene polymorph B •complex **9** •undulating sheets (AlB_2 prototype) •6-fold interpenetrated •M:L stoichiometry 1:1 •grown from benzene	•no suitable crystals yet obtained	8 (A) 19 (B)
	•complex **10** •planar, 2-D sheets •M:L stoichiometry 1:1 •grown from benzene	•no suitable crystals yet obtained	20

TABLE 1. (continued)

	Silver(I) Salts		
Ligand	$AgCF_3SO_3$	$AgPF_6$	Reference
tritopic ligands (continued)			
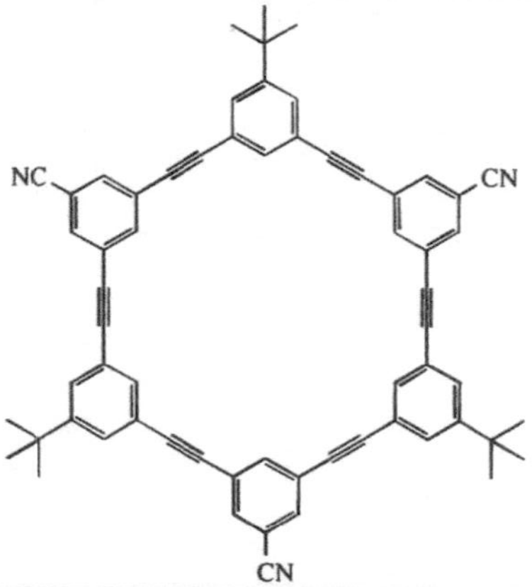 NC, CN, CN	•no suitable crystals yet obtained	•no suitable crystals yet obtained	-
tetratopic ligands			
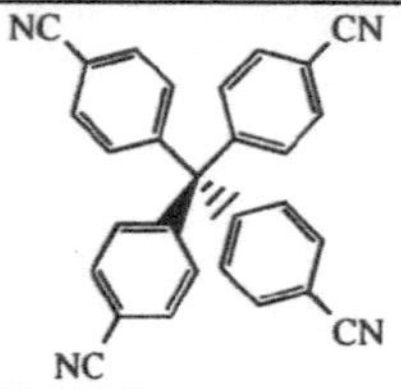 NC, CN, NC, CN	•complex **11** •2-D sheets (AlB_2 prototype) •M:L stoichiometry 1:1(one nitrile uncoordinated) •grown from benzene	•no suitable crystals yet obtained	19

TABLE 1. (continued)

Ligand	Silver(I) Salts		Reference
	$AgCF_3SO_3$	$AgPF_6$	
tetratopic ligands (continued)			
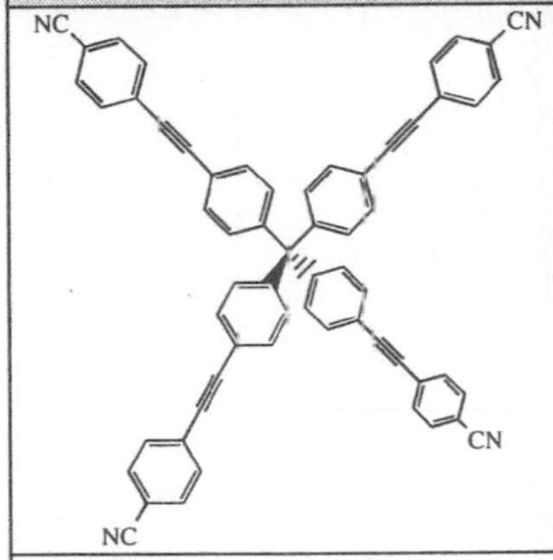	•small needles grown from toluene	•no suitable crystals yet obtained	-
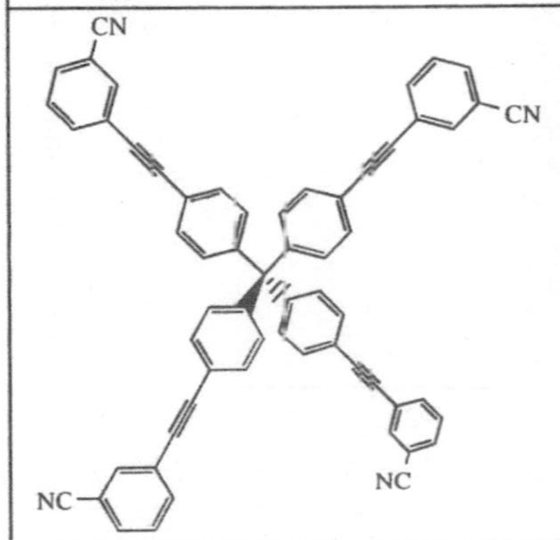	•small and weakly diffracting crystals grown from either toluene or xylenes	•no suitable crystals yet obtained	-

(a) [Ag(1,4-dicyanobenzene)(CF_3SO_3)(H_2O)$_{0.5}$] (**1**)

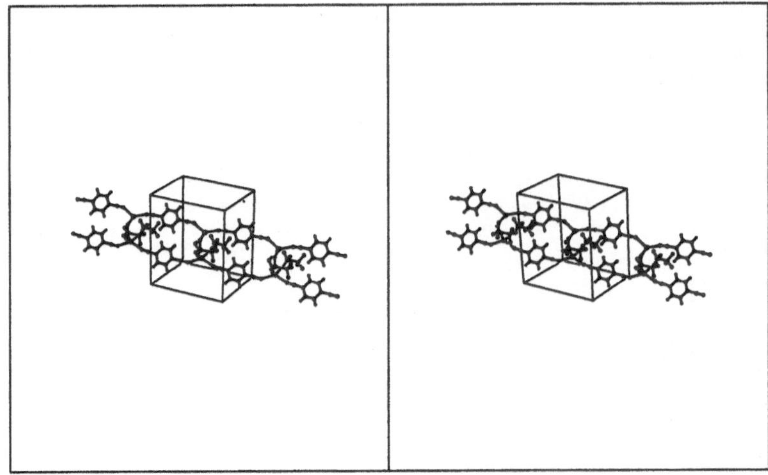

(b) [Ag(1,4-dicyanobenzene)(H_2O)]PF_6 (**2**) (counterions omitted for clarity)

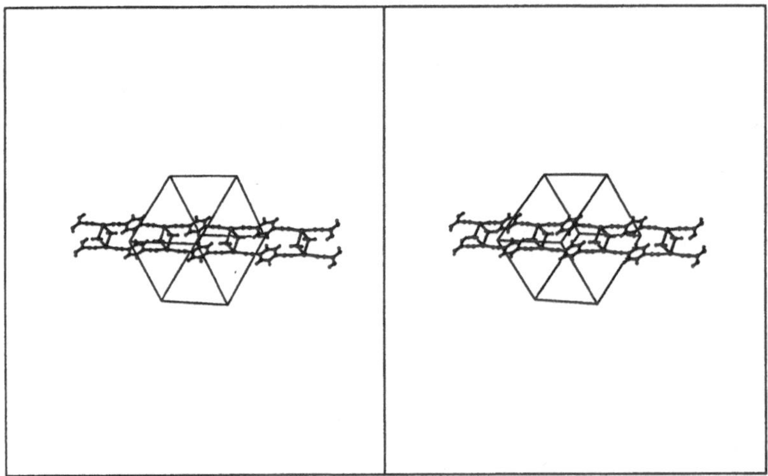

Figure 1. Coordination networks of multitopic nitriles and silver(I) salts.

(c) [Ag(4,4'-biphenyldicarbonitrile)(CF_3SO_3)]·C_6H_6 (**3**) (benzene molecules omitted for clarity)

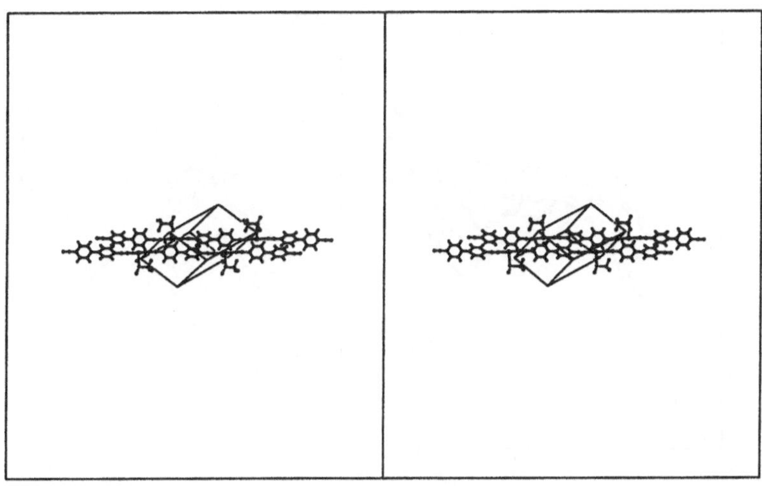

(d) [Ag(4,4'-biphenyldicarbonitrile)$_2$]PF_6 (**4**) (counterions omitted for clarity)

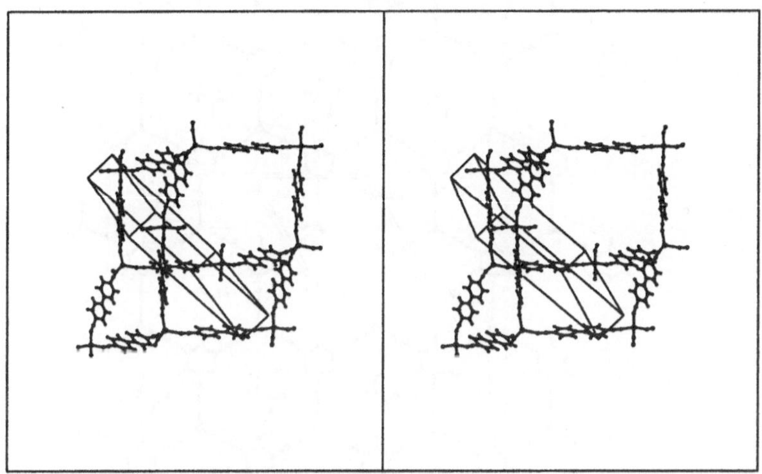

Figure 1. (continued).

(e) [Ag(4,4'-dicyanodiphenylacetylene)(PO_2F_2)] (**5**)

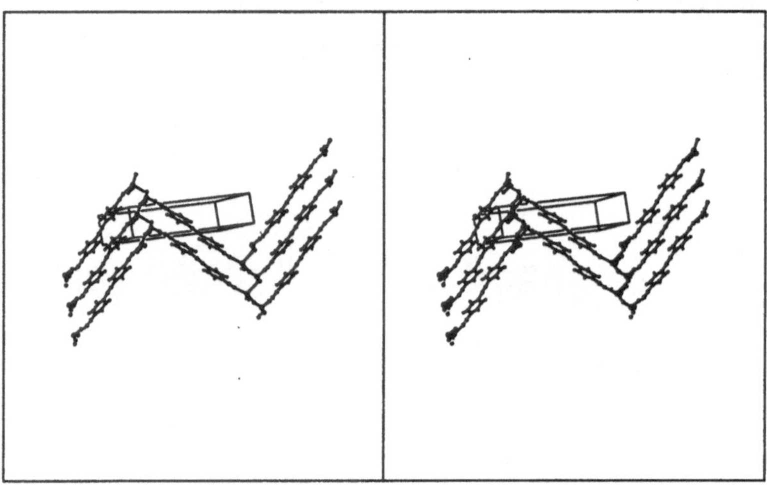

(f) [Ag(1,3,5-tricyanobenzene)(CF_3SO_3)] (**6**)

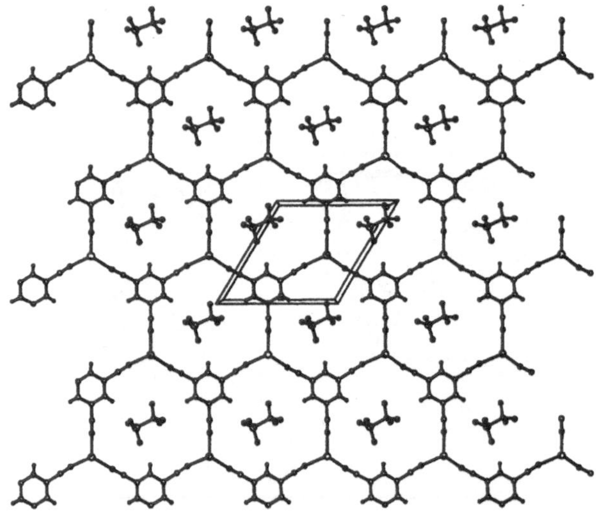

Figure 1. (continued).

(g) [Ag(1,3,5-tricyanobenzene)(η^2-C_7H_8)]PF_6 (**7**) (toluene molecules and counterions omitted for clarity)

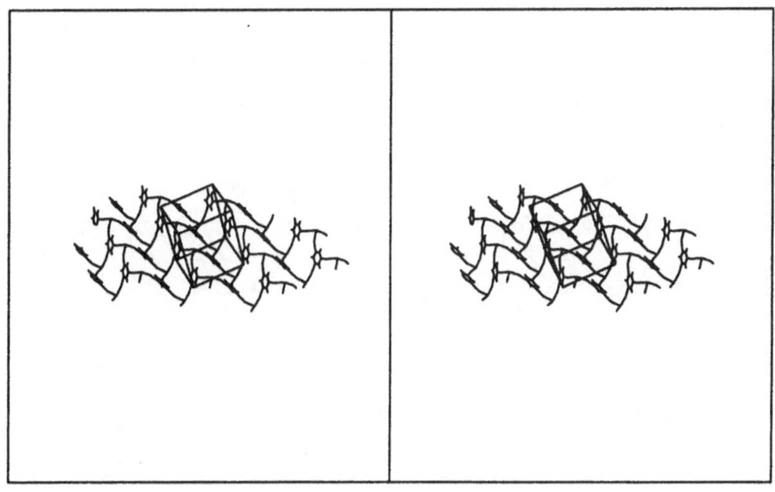

(h) [Ag(1,3,5-tris(4-ethynylbenzonitrile)benzene)(CF_3SO_3)]·$2C_6H_6$ (polymorph A, **8**) (benzene molecules and counterions omitted for clarity)

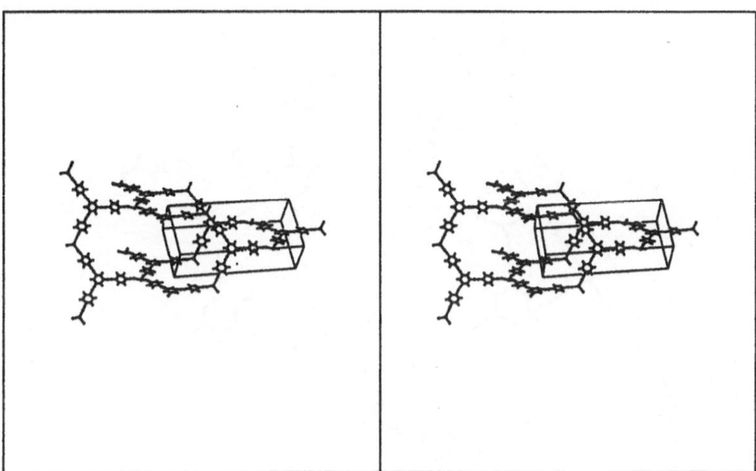

Figure 1. (continued).

(i) [Ag(1,3,5-tris(4-ethynylbenzonitrile)benzene)(CF_3SO_3)]·$2C_6H_6$ (polymorph B, **9**) (benzene molecules and counterions omitted for clarity)

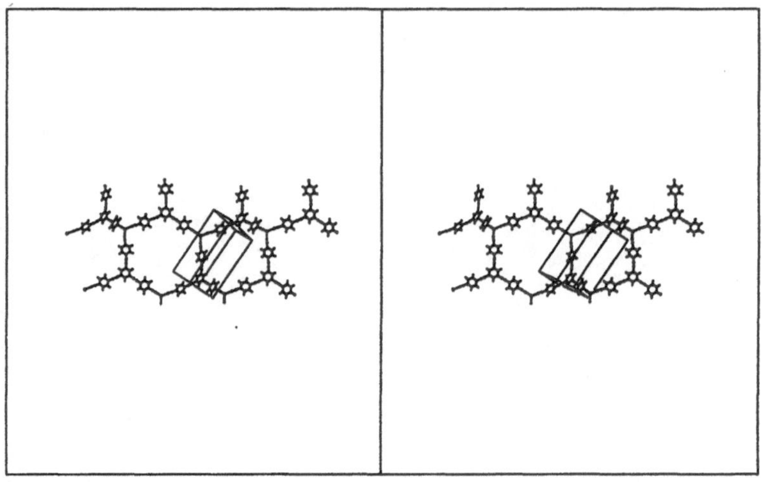

(j) [Ag(1,3,5-tris(3-ethynylbenzonitrile)benzene)(CF_3SO_3)]·$2C_6H_6$ (**10**) (benzene molecules and counterions omitted for clarity)

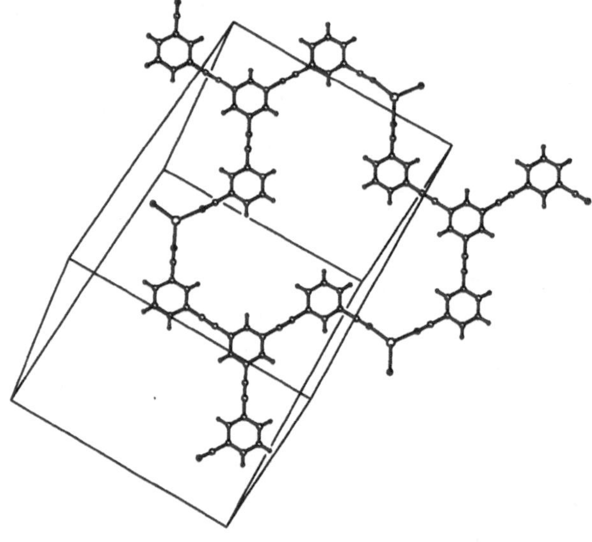

Figure 1. (continued).

(k) [Ag(4,4',4",4'"-tetracyanotetraphenylmethane)(CF_3SO_3)]·$3C_6H_6$ (**11**) (benzene molecules omitted for clarity)

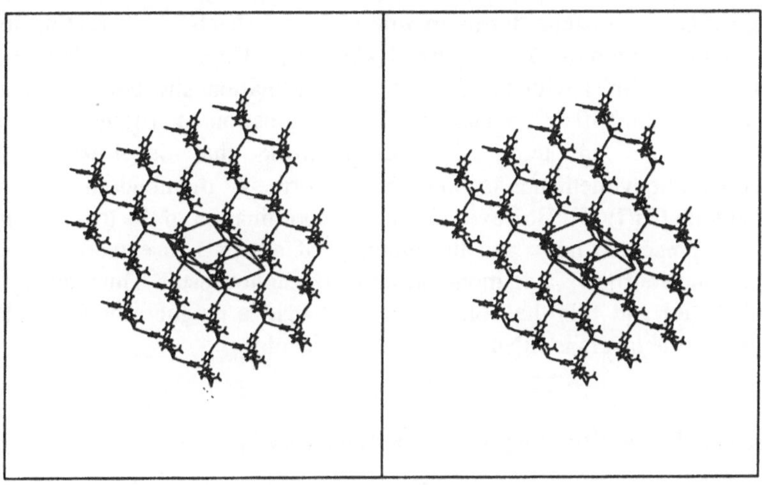

Figure 1. (continued).

unit cell. Four of these are disordered and twelve are well-ordered. This material has proven to be porous and displays unusual thermal properties (see Section 3.3) [20].

The tetratopic ligand, 4,4',4",4'"-tetracyanotetraphenylmethane, when complexed with $AgCF_3SO_3$ in benzene, forms two-dimensional sheets of alternating ligands and silver(I) atoms (complex **11**; Figure 1(k)) [19]. Three of the four nitriles are coordinated to silver(I) with the remaining coordination site being occupied by an oxygen atom of the triflate counterion. The coordination of triflate in this case is of great consequence in terms of the network that is formed. 4,4',4",4'"-tetracyanotetraphenylmethane is known to form a diamondoid network when complexed with Cu(I) [5]. However, due to the coordination of the triflate counterion in the example presented here, and the inability of silver(I) to extend its coordination number beyond four [14], a diamondoid network cannot result. Thus, one nitrile group is uncoordinated and fills the void space created in a neighboring layer. Remaining space between the layers is filled by benzene molecules.

3. Porosity and Stability of Ag(I)···NC-R Coordination Networks

This section describes experiments conducted to prove the porosity of silver(I) triflate complexes of 1,3,5-tris(4-ethynylbenzonitrile)benzene (polymorph A, **8**) and 1,3,5-tris(3-ethynylbenzonitrile)benzene (**10**). Guest exchange experiments with complex **8** and thermal properties of complex **10** are also discussed.

3.1 PROOF OF POROSITY IN [Ag(1,3,5-TRIS(4-ETHYNYLBENZONITRILE)-BENZENE)CF_3SO_3]·$2C_6H_6$ (POLYMORPH A, **8**)

As mentioned in Section 2.2, complex **8** displays *ca.* 15 Å x 22 Å channels along [1 0 0] which are filled with benzene (crystallization solvent). To prove porosity in this material, a non-destructive exchange of benzene with benzene-d_6 was attempted [8]. X-ray powder diffraction of large crystals grown from benzene indicated that the product was quantitatively pure and the same phase as the single crystal structure. Large crystals from the same sample used for powder diffraction were placed in benzene-d_6 of isotopic purity >99.5% at room temperature and subsequently washed with fresh benzene-d_6 several times. These crystals were then placed in benzene-d_6 of isotopic purity >99.96% and washed several times with fresh benzene-d_6 of the same purity. Identical samples were examined under an optical microscope and the morphology of the crystals subjected to similar treatment was monitored over time. Visual inspection showed that the sample changed very little in its overall crystalline morphology while soaking in benzene-d_6. In particular, no microscopic cracks were observed and the dimensions and shape of the crystals remained constant. An X-ray powder pattern obtained after subjecting the crystals to benzene-d_6 of isotopic purity >99.96% showed that the original phase was maintained. ^1H NMR spectroscopy in acetone-d_6 showed, to the limits of detection, no peak at δ 7.34 indicating the absence of the original benzene. Thus, although the crystal appears to be structurally unchanged, the actual composition has changed from [Ag(1,3,5-tris(4-ethynylbenzonitrile)benzene)CF_3SO_3]·$2C_6H_6$ to [Ag(1,3,5-tris(4-ethynylbenzonitrile)benzene)CF_3SO_3]·$2C_6D_6$ *without* dissolution and reformation of the crystal lattice. Therefore, this material is indeed porous to benzene exchange.

3.2 GUEST ABSORPTION IN [Ag(1,3,5-TRIS(4-ETHYNYLBENZONITRILE)-BENZENE)CF_3SO_3]·$2C_6H_6$ (POLYMORPH A, 8)

The exchange studies involving complex **8** have been extended to examine the absorption of other guests from solution [21]. Exchange was confirmed by ^{1}H NMR according to the following general method. Crystals grown from benzene were initially washed with a small amount of toluene and filtered. The crystals were then soaked in a solution of a guest at known concentration at room temperature. After two hours, the crystals were vacuum filtered to dryness. In addition, if the guest was an alcohol, the crystals were further washed with cold toluene (-40 °C) and suction dried. Unit cell parameters were determined by X-ray powder diffraction. The solid was then dissolved in acetone-d_6 and the guest:ligand ratio was determined by integration of the peaks in the ^{1}H NMR spectrum. The results are presented in Table 2.

TABLE 2. Solution phase absorption of guests by host complex **8**

guest	unit cell edges of host/guest complexes (Å)[a] *a*	*b*	*c*	guest:ligand stoichiometry
benzene	11.625(3)	19.110(7)	38.856(15)	-
toluene	11.705(3)	19.351(11)	38.908(16)	1.47:1.00
m-xylene	11.64(6)	19.58(3)	38.50(5)	2.31:1.00
undecane	11.76(2)	19.231(9)	38.85(3)	1.42:1.00
benzyl alcohol	b	22.43(1)	33.98(2)	1.92:1.00
2,6-di-*tert*-butylphenol	b	22.48(4)	32.84(4)	0.80:1.00
sec-phenethyl alcohol	b	22.49(1)	34.18(3)	1.88:1.00

[a] All two-dimensional cells are rectangular and all three-dimensional cells are orthorhombic. [b] Indexes as a two-dimensional unit cell.

In order to verify that the guest uptake was not simply a result of surface adsorption, crystals of the materials obtained after exchange were grown and the guest:ligand ratios were determined [21]. In all cases, the ratios were within experimental error of those obtained for microcrystalline powders. Furthermore, the crystals were monitored by optical microscopy to insure that absorption did not take place simply by dissolution and recrystallization of the host. The morphology remained constant in all cases. As a control, the non-porous complex of 1,3,5-tricyanobenzene and $AgCF_3SO_3$ (**6**) was exposed to the same guests under identical conditions. ^{1}H NMR spectra of these solids obtained in acetone-d_6 indicated that virtually no guests were present (the largest guest:ligand ratio observed was 0.03:1.00).

Guest absorption by the crystalline host was also studied by vapor phase transfer [21]. Initially, a microcrystalline powder of complex **8** was heated to 200 °C at 10 °C·min^{-1} to remove the included benzene (the benzene:ligand ratio after this treatment was determined to be 0.07:1.00 by ^{1}H NMR). This material was then exposed to vapor of various guests in a sealed apparatus at room temperature. The guest:ligand ratio was determined by ^{1}H NMR and/or thermogravimetric analysis (TGA). Unit cell parameters were determined by X-ray powder diffraction.

The results of the vapor phase studies are summarized in Table 3. Note that benzyl alcohol and *sec*-phenethyl alcohol are strongly absorbed. A somewhat lower absorption is observed for non-functionalized aromatics while aliphatics are essentially not absorbed. Also note that exposure of the host to benzene or benzyl alcohol produces crystalline solids with unit cell dimensions that are nearly identical to those values obtained by solution phase exchange (Table 2). However, exposure of the host to vapor of cyclooctane or undecane results in negligible changes of the cell constants from those of the initial solid heated to 200 °C.

3.3 ZEOLITIC AND INCLUSION-LIKE BEHAVIOR OF [Ag(1,3,5-TRIS(3-ETHYNYLBENZONITRILE)BENZENE)CF_3SO_3]·$2C_6H_6$ (**10**)

As discussed in Section 2.2, for complex **10**, sixteen molecules of benzene per unit cell are located in channels, of which four are found to be disordered. Studies of the loss of included benzene have been very informative [20]. TGA shows two discrete mass losses at 110 °C and 145 °C corresponding to the mass percent of four and twelve benzene molecules, respectively (see Figure 2). Differential scanning calorimetry (DSC) and optical microscopy confirm that no phase change is associated with the first mass loss (Figure 2). Unit cell parameters of a single crystal of complex **10** heated to 110 °C for ten minutes remain unchanged within the standard deviation of the original crystal. This behavior is confirmed on bulk powder samples by X-ray powder diffraction. Such an observation, namely the removal of a guest contained within the cavity of a host lattice, without an accompanying phase change, resembles that of zeolites. At 145 °C, a solid-to-solid phase transition occurs concomitant with the loss of the remaining benzene. This behavior, whereby removal of the guest occurs with a phase change, is like that of classical inclusion compounds. The high temperature solid phase undergoes a melting transition upon further heating to 169 °C. Finally, microcrystalline samples heated to 110 °C or 145 °C and subsequently cooled to room temperature re-absorb benzene vapor in an amount corresponding to the mass percent in the original, unheated sample.

TABLE 3. Vapor phase absorption of guests by host complex **8**[a]

guest	exposure time (h)	exposure temp (°C)	unit cell edges of host/guest complexes (Å)[b]			guest:ligand stoichiometry	
			a	*b*	*c*	by TGA	by ^{1}H NMR
none[c]	12	65	d	22.761(8)	36.48(15)	-	0.07:1.00[e]
benzene	48	40	11.363(5)	19.07(3)	39.15(3)	2.55:1.00	3.65:1.00
m-xylene	60	60	d	21.87(5)	34.65(9)	1.29:1.00	-
cyclooctane	36	60	unchanged from original, dry solid			0.06:1.00	-
undecane	60	60	d	22.867(6)	36.533(7)	0.11:1.00	-
benzyl alcohol	22	65	d	22.393(12)	34.138(14)	3.62:1.00	3.40:1.00
sec-phenethyl alcohol	36	60	d	22.319(7)	34.42(2)	3.60:1.00	3.60:1.00

[a] Microcrystalline powder heated to 200 °C prior to exposure to guests. [b] All two-dimensional cells are rectangular and all three-dimensional cells are orthorhombic. [c] Control experiment. Sample heated to 200 °C and then placed in the chamber without a guest for the indicated time. [d] Indexes as a two-dimensional unit cell. [e] Residual benzene.

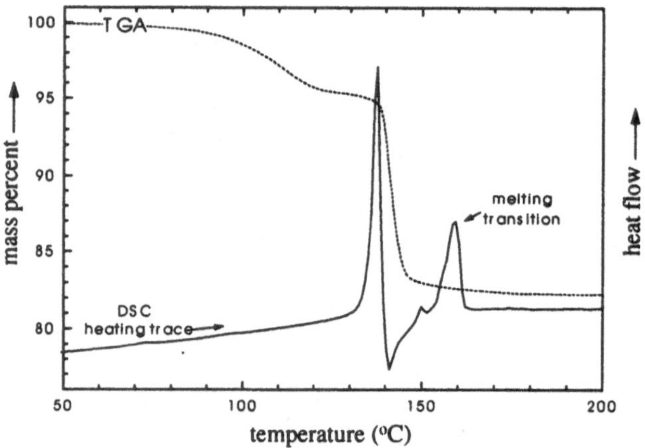

Figure 2. TGA and DSC traces (recorded at a heating rate of 10 °C·min^{-1}) for complex **10** showing the loss of included benzene at two distinct temperatures. The loss at 110 °C is not accompanied by a change in crystalline phase or morphology, a behavior characteristic of zeolites. The loss at 145 °C is accompanied by a phase transition, a behavior characteristic of inclusion compounds.

4. Conclusions

The Ag(I)···NC-R coordination bond is a reliable supramolecular interaction that can be used to prepare a variety of network structures. Although incorporation of the non-coordinating hexafluorophosphate counterion allows diamondoid networks to be accessed, use of the weakly coordinating triflate counterion facilitates the prediction of network connectivity using the rules summarized in Table 4. In the examples presented here with silver(I) triflate, ditopic ligands form two-connected nets while tritopic ligands form three-connected networks. Finally, the coordination of the triflate counterion to silver(I) allows for the rational construction of porous coordination networks involving large tritopic ligands.

TABLE 4. Empirically derived rules of network connectedness for complexes of silver(I) triflate and multitopic nitriles.

ligand topicity	most probable (p,q)-connected network[a,b]	other possible (p,q)-connected networks[a]
ditopic	(2,2)	(2,3)
tritopic	(3,3)	(3,2)
tetratopic	(3,3)	(4,2), (4,3)

[a] The values of p and q correspond to the network connectedness of the ligand and metal ion, respectively (i.e., a p-topic ligand coordinated to a transition metal ion having a network connectivity of q). [b] Assuming M:L stoichiometry of 1:1.

5. Acknowledgments

Contributions by D. Venkataraman, Y.-H. Kiang, A.C. Covey, and A. Asgaonkar are gratefully acknowledged. This work was supported by the National Science Foundation (Grant CHE-94-23121) and the U.S. Department of Energy through the Materials Research Laboratory at the University of Illinois (Grant DEFG02-91-ER45439). A portion of this research was carried out at the Center for Microanalysis of Materials, University of Illinois, which is supported by the U.S. Department of Energy under Grant DEFG02-91-ER45439. J.S.M. acknowledges support from the 3M Co. and the Camille Dreyfus Teacher-Scholar Awards Program. S.L. thanks the J.D. and C.T. MacArthur Foundation (1993-97) and the A.P. Sloan Foundation (1993-95) for fellowships. We thank the School of Chemical Sciences Materials Chemistry Laboratory at the University of Illinois and Dr. Jeffrey W. Kampf at the University of Michigan for single crystal X-ray data collection.

6. References

1. Subramanian, S. and Zaworotko, M.J. (1994) Exploitation of the hydrogen bond: recent developments in the context of crystal engineering, *Coord. Chem. Rev.* **137**, 357-401.
2. Ermer, O. (1988) Fivefold-Diamond Structure of Admantane-1,3,5,7-tetracarboxylic Acid, *J. Am. Chem. Soc.* **110**, 3747-3754.
3. Simard, M., Su, D., and Wuest, J.D. (1991) Use of Hydrogen Bonds to Control Molecular Aggregation. Self-Assembly of Three-Dimensional Networks with Large Chambers, *J. Am. Chem. Soc.* **113**, 4696-4698.
4. Venkataraman, D., Lee, S., Zhang, J., and Moore, J.S. (1994) An organic solid with wide channels based on hydrogen bonding between macrocycles, *Nature* **371**, 591-593.
5. Hoskins, B.F. and Robson, R. (1989) Infinite Polymeric Frameworks Consisting of Three Dimensionally Linked Rod-like Segments, *J. Am. Chem. Soc.* **111**, 5962-5964.
6. Ermer, O. (1991) Sevenfold Diamond Structure and Conductivity of Copper Dicyanoquinonediimines $Cu(DCNQI)_2$, *Adv. Mater.* **3**, 608-611.
7. Hirsch, K.A., Venkataraman, D., Wilson, S.R., Moore, J.S., and Lee, S. (1995) Crystallization of 4,4'-Biphenyldicarbonitrile with Silver(I) Salts: a Change in Topology Concomitant with a Change in Counterion Leading to a Ninefold Diamondoid Network, *J. Chem. Soc., Chem. Commun.*, 2199-2200.
8. Gardner, G.B., Venkataraman, D., Moore, J.S., and Lee, S. (1995) Spontaneous assembly of a hinged coordination network, *Nature* **374**, 792-795.
9. Fujita, M., Kwon, Y.J., Washizu, S., and Ogura, K. (1994) Preparation, Clathration Ability, and Catalysis of a Two-Dimensional Square Network Material Composed of Cadmium(II) and 4,4'-Bipyridine, *J. Am. Chem. Soc.* **116**, 1151-1152.

10. A multitopic ligand is one equipped with multiple coordination sites and, as a consequence of a rigid geometry, is unable to chelate to a metal center. These structural constraints ensure network formation unless counterions and/or solvent molecules coordinate to the metal to the extent that the ligand is unable to coordinate.
11. Lawrance, G.A. (1986) Coordinated Trifluoromethanesulfonate and Fluorosulfate, *Chem. Rev.* **86**, 17-33.
12. Beck, W. and Sünkel, K. (1988) Metal Complexes of Weakly Coordinating Anions. Precursors of Strong Cationic Organometallic Lewis Acids, *Chem. Rev.* **88**, 1405-1421.
13. Venkataraman, D., Gardner, G.B., Covey, A.C., and Moore, J.S. Silver(I) Complexes of 1,4-Dicyanobenzene, Accepted for publication in *Acta Crystallogr. C.*
14. For a rare example of six-coordinate silver(I), see: Carlucci, L., Ciani, G., Proserpio, D.M., and Sironi, A. (1995) Novel Networks of Unusually Coordinated Silver(I) Cations: The Wafer-Like Structure of $[Ag(pyz)_2][Ag_2(pyz)_5](PF_6)_3 \cdot 2G$ and the Simple Cubic Frame of $[Ag(pyz)_3](SbF_6)$, *Angew. Chem., Int. Ed. Engl.* **34**, 1895-1898.
15. Kitagawa, S., Kawata, S., Nozaka, Y., and Munakata, M. (1993) Synthesis and Crystal Structures of Novel Copper(I) Co-ordination Polymers and a Hexacopper(I) Cluster of Quinoline-2-thione, *J. Chem. Soc., Dalton Trans.*, 1399-1403.
16. Hirsch, K.A., Wilson, S.R., and Moore, J.S. Silver(I) Complex of 4,4'-Dicyanodiphenylacetylene with a Difluorophosphate ($PO_2F_2^-$) Counterion, *Acta Crystallogr. C*, In press.
17. Venkataraman, D., Covey, A.C., Lee, S., and Moore, J.S. Manuscript in preparation.
18. Wells, A.F. (1984) *Structural Inorganic Chemistry*, Oxford University Press, New York, 5th ed.
19. Venkataraman, D., Lee, S., Moore, J.S., Zhang, P., Hirsch, K.A., Gardner, G.B., Covey, A.C., and Prentice, C.L. Coordination Networks Based On Multitopic Ligands and Silver(I) Salts: A Study of Network Connectivity and Topology as a Function of Counterion, *Chem. Mater.*, In press.
20. Venkataraman, D., Gardner, G.B., Lee, S., and Moore, J.S. (1995) Zeolite-like Behavior of a Coordination Network, *J. Am. Chem. Soc.* **117**, 11600-11601.
21. Gardner, G.B., Kiang, Y.-H., Lee, S., Asgaonkar, A., and Venkataraman, D. Exchange Properties of the Three-Dimensional Coordination Compound 1,3,5-tris(4-ethynylbenzonitirle)benzene$\cdot AgO_3SCF_3$, *J. Am. Chem. Soc.*, In press.

THE CONTROL OF DNA STRUCTURE
From Topological Modules to Geometrical Modules

NADRIAN C. SEEMAN‡, JING QI‡, XIAOJUN LI‡, XIAOPING YANG‡, NEOCLES B. LEONTIS§, BING LIU‡, YUWEN ZHANG‡, SHOU MING DU‡, and JUNGHUEI CHEN‡

Departments of Chemistry, ‡New York University, New York, NY 10003, USA and *§Bowling Green State University, Bowling Green, OH 43403 USA*

ABSTRACT. The generation of chemical modules that can be used for the construction of nanoscale objects and devices is clearly a major goal of current research in molecular construction. DNA molecules containing fixed branch points are very powerful modules, because their associations can be directed by terminating each double helical arm in a sticky end. It is possible to direct the associations of these modules to form stick figures whose edges consist of double helical DNA, and whose strands form DNA catenanes. Likewise, the double-helical half-turn corresponds to the unit tangle necessary to generate the topology of specific knots. Geometrical control must be added to the topological control in hand if periodic matter is to be built from DNA.

1. Nanoconstruction with Branched DNA Modules

The double helical structure of DNA [1] is fundamental to molecular biotechnology. One of the key features of the DNA molecule is that its helix axis is linear, in the sense that it is unbranched. Nevertheless, DNA molecules in living cells do form ephemeral branched structures, such as Holliday recombination intermediates, in which four double helices surround a branch point. When branched molecules appear in biological systems, they do so through processes that generate sequence symmetry around their branch points; this symmetry renders them unstable, because they can isomerize to produce linear duplex molecules. Synthetic DNA methodology [2] has made it possible to model branched structures through the use of DNA molecules (termed 'immobile junctions') whose branch points are fixed, because they lack the sequence symmetry needed to isomerize [3]. The principles used to minimize the sequence symmetry of branched molecules have been discussed previously [3,4].

The ligation of (topologically) linear DNA molecules containing sticky ends [5] is the fundamental reaction of biotechnology; through this means, it is possible to assemble large DNA molecules of any sequence at all, thereby permitting genetic engineers to clone, manipulate, exploit, and examine genes of interest. It is also possible to add sticky ends to immobile junctions [3]. From a structural standpoint, this combination contributes vertices to the line segments already present in nucleic acid biotechnology. Consequently one can imagine ligating branched junctions to form stick figures; the edges of these figures consist of double helical DNA, and the vertices

J. Michl (ed.), Modular Chemistry, 95–104.

occur at the branch points of the junctions. The concept of connectedness [6] is a critical topological feature for describing such objects: An object is N-connected if each of its vertices is connected through an edge to N other vertices; for example, a regular octahedron is a 4-connected object, and a regular icosahedron is a 5-connected object. The number of double-helical arms that flank an immobile junction limits the connectedness of any object constructed from junctions of that sort. Junctions containing up to six arms have been constructed successfully [7]. Figure 1 illustrates the ligation of four 4-arm junctions into a quadrilateral that could be extended to form a 2-dimensional lattice.

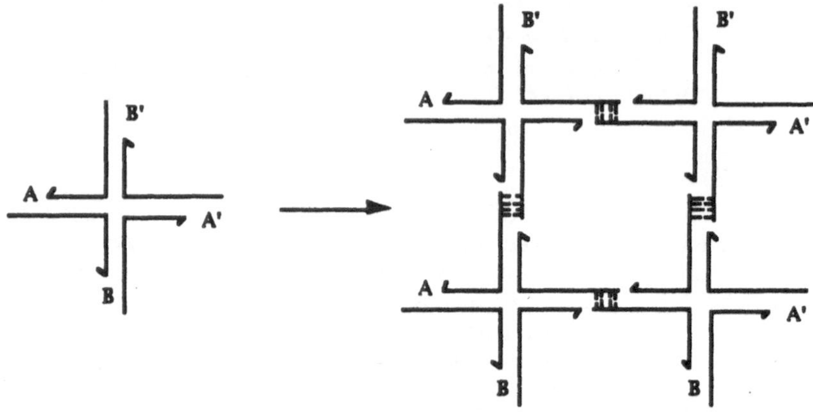

Figure 1. Formation of a Two-Dimensional Lattice from a Junction with Sticky Ends. A is a sticky end and A' is its complement. The same is true of B and B'. Four of the monomers on the left are complexed in parallel orientation to yield the structure on the right. DNA ligase can close the gaps left in the complex. The complex has open valences, so that it can be extended by the addition of more monomers.

When the type of assembly illustrated in Figure 1 is attempted, one discovers that individual junctions are not completely rigid [8,9]. We can expect that the double helical edges of each object are torsionally and flexurally stiff, because the persistence length of DNA is usually taken to be approximately 500 Å [10]: This value leads to a 66.5 Å RMS end-to-end distance for two turns of DNA, nominally 68 Å long. Thus, on a small scale, say using 2-turn edges, the rigidity of linear DNA appears to be adequate for our purposes. However, the 'valence angles' between the double helical edges of each object are flexible [8,9]. If a given junction is oligomerized to yield oligolaterals with two turns between branch points, it readily forms a series of cyclic products: trimers, tetramers, pentamers and so on. Thus, the flexibility of DNA objects corresponds to multiple minima in the potential energy surface; these multiple minima prevent the use of oligomerization to generate a given cyclic polygon, or to generate the lattice suggested by Figure 1.

1.1 THE CONSTRUCTION OF A PLATONIC POLYHEDRON, A CUBE

The problem with multiple minima can be overcome in the construction of discrete objects. One can employ a large set of different sticky ends in order to specify a finite set of edges. We have shown this to be true, by constructing from 3-arm junctions a molecule whose helix axes have the connectivity of a cube [11]. A schematic of the 3-dimensional, 3-connected object is shown in Figure 2.

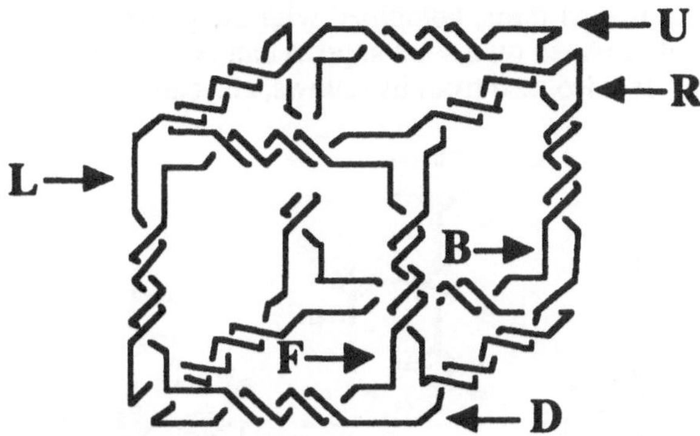

Figure 2. A DNA Molecule Whose Helix Axes Have the Connectivity of a Cube. The molecule shown consists of six cyclic strands that have been catenated together in this particular arrangement. They are labeled by the first letters of their positional designations, Up, Down, Front, Back, Left and Right.

The molecule is depicted as a cube, although we have not characterized the angles between the edges. It contains 12 edges that consist of double helical DNA. Each of the edges contains a unique site for recognition and cleavage by a restriction endonuclease, thereby enabling us to establish the validity of the synthesis. From model building [12], the axis-to-axis distance across a square face appears to be about 100 Å, with a volume (in a cubic configuration) of approximately 1760 nm^3, when the cube is folded so that major grooves form the outsides of the corners; it is markedly smaller when the outside corners are formed from minor grooves. The direction of folding is unknown. The cube was synthesized in solution, and the details of its synthesis have been described previously [11]. The flexibility of the angles between the double helical arms led to a major problem in the first step of the assembly: Two arms ending in sticky ends were ligated simultaneously, to form two edges; it was not possible to ensure that the two arms carrying their complementary sticky ends were both carried by the same unit. Consequently, undesirable purification and reconstitution steps were included in the process.

1.2 THE SOLID-SUPPORT BASED CONSTRUCTION OF AN ARCHIMEDEAN POLYHEDRON, A TRUNCATED OCTAHEDRON

Whereas the synthetic methodology used to construct the cube in solution does not provide enough control to assemble complex objects, we have developed a methodology that is more effective [13]. This methodology is based on the use of a solid support, which permits the convenient removal of reagents and catalysts from the growing product. Each ligation cycle creates an intermediate object that is covalently closed and topologically bonded together. This feature permits exonuclease digestion of incompletely ligated edges, thereby effecting a purification of the growing object during synthesis. A single edge of an object can be built at a time. Control derives from the restriction endonuclease digestion of hairpin loops; the sticky ends exposed will cohere to produce the new edge. Sequences are chosen so that restriction sites are destroyed when the edge forms. In principle, the object added to the growing construct can be a junction, a polygon, a group of polygons, a polyhedron or an array of polyhedra. The solid-support based methodology allows one to overcome the problems generated by the flexibility of double helical arms, combined with the presence of many molecules in solution. A molecule isolated on a solid support can be treated similarly to a molecule on a piece of paper: It is isolated from its fellows, and cannot react with them.

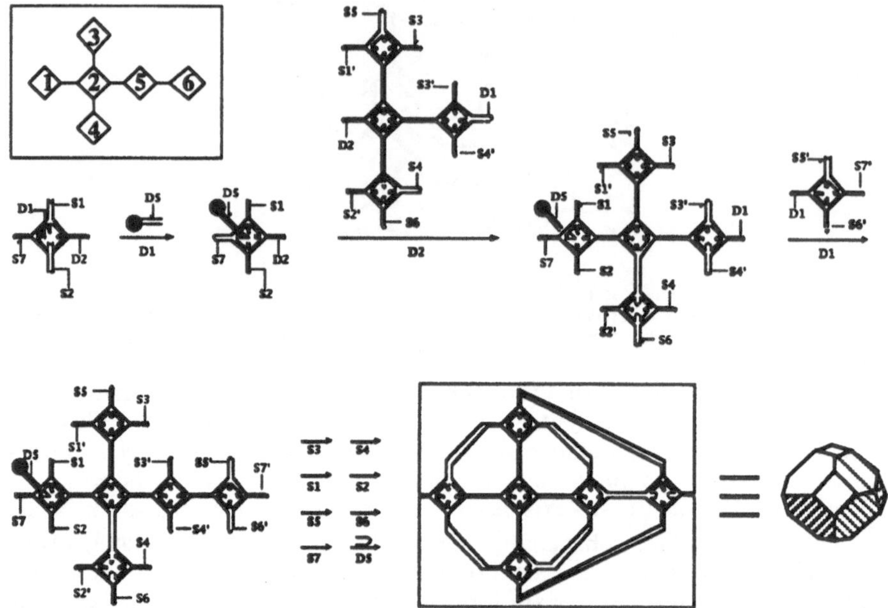

Figure 3. The Synthetic Scheme Used to Synthesize the Truncated Octahedron. The boxed diagram in the upper left indicates the numbering of individual squares. Each square in the rest of the diagram is shown with its restriction sites indicated. Symmetric restriction sites are named 'S', indicated in pairs, with one member primed; restriction sites cut distally are named 'D'. Arms that will eventually combine to form edges are drawn on the outside of each square, and exocyclic arms are drawn on the inside of the square. A reaction is indicated by a line above a restriction site: This means that the restriction enzyme is added, protecting hairpins are removed and then the two sticky ends are ligated together. The product is shown in two forms. On the left, the S1-S6 closures are shown as triple edges, to emphasize their origins; the two strands of the edge formed by the S7 closure are separated to maintain the symmetry of the picture. On the right, a slightly rotated front view of a polyhedral representation of a truncated octahedron is shown without the exocyclic arms; the 432 cubic symmetry of the ideal object is evident from this view.

We have constructed a truncated octahedron from DNA [14] by using this solid support methodology. In this construction, the objects added to the support are squares and square groupings, as indicated in Figure 3. A truncated octahedron is a 3-connected object, but it has been constructed from branched junctions containing four arms apiece. The figure contains six squares and eight hexagons. Each edge is formed from two double helical turns of DNA, so a single strand is associated with each of the fourteen faces.

The structure on the lower left of Figure 3 is a hexasquare complex that is a heptacatenane. The square strands are already intact in this intermediate, and the outer strand is the precursor to all of the hexagons. The hexagons result from successive closures of the sticky ends associated with the restriction enzyme site pairs, S1-

S1'...S7-S7'. Whereas the molecules are isolated from each other on the support, it is possible to expose the sticky ends by using symmetrically-cleaving restriction enzyme pairs that recognize six nucleotide pairs each. The final step in the synthesis involves releasing the structure from the support and annealing it shut with a hairpin. The synthesis is demonstrated in two stages: First, by showing that all six cyclic strands corresponding to the square molecules are in the heptacatenane, and second, by digesting the final product to the tetracatenanes that flank the squares.

2. The Construction of Target Knots from DNA

The cube and the truncated octahedron have been characterized primarily by their topological properties: The cube has been shown to be a hexacatenane and the truncated octahedron has been shown to be a 14-catenane, in which each molecule is linked twice to every one of its neighbors. The low yields we experience and the flexibility of the branched junction components currently restrict characterization of these molecules to the topological level, and prevent structural characterization. A close relationship exists between catenanes and knots [15]: Removal of a node by switching strands, yet maintaining local symmetry and strand polarity, converts a catenane to a knot, and a knot to a catenane (Figure 4). This operation is sometimes called forming a 'zero node.' [15] We have pointed out elsewhere the relationship between this operation and both the solid state approach and the possibilities of cloning DNA objects by embedding them in single-stranded knots [16].

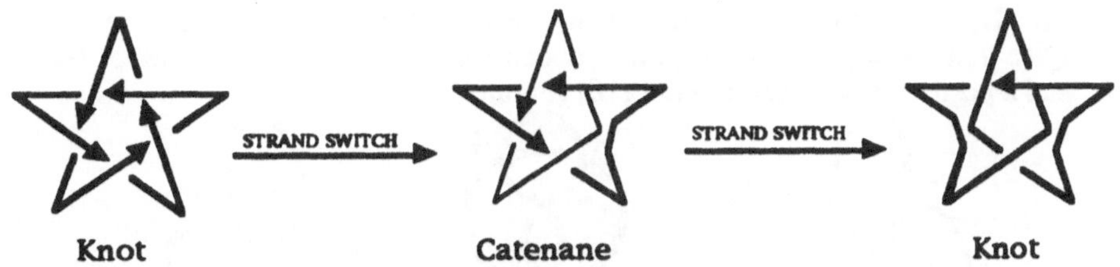

Figure 4. Interconversions of Knots and Catenanes by Switching Strands at a Node. The structure shown on the left is a 5_1 knot. The strand direction is indicated by the arrowheads appearing along the strand. When the two strands entering the lower node on the right exchange outgoing partners, the node disappears, and a 'zero node' is introduced. This converts the knot to a catenane, shown in the middle; the two linked cycles are drawn so as to retain their shapes, but they are drawn with lines of different thicknesses. The lower left node of the catenane undergoes a strand switch, and the structure is converted to a trefoil knot, illustrated on the right. The trefoil knot is one strand, so it is drawn with one thickness.

The fundamental unit of a knotted or catenated structure is a unit tangle. This can be thought of as a structure consisting of two strands that wrap around each other for about a half-turn of their mutual local helix axis; when projected down the line connecting their midpoints, a node is formed, where one strand crosses the other. A trefoil knot with negative nodes is shown in Figure 5; each of its three tangles is surrounded by a dotted square. The arbitrary polarity of the strand is indicated by the arrowheads. The strand of the knot forms the diagonals of each square, which divide the square into four regions, two between parallel strands and two between antiparallel strands. We can make the transition from topology to nucleic acid chemistry by equating each of the tangles with a half-turn of DNA [17]. Base pairs are shown between antiparallel strands, a helix axis (terminating in arrowheads) is drawn normal to them, and the local tangle dyad axis (terminating in lens-shaped objects) is drawn normal to the helix axis.

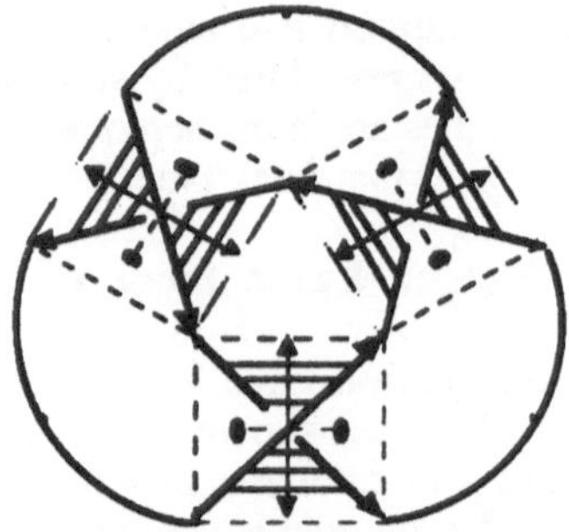

Figure 5. The Relationship Between Nodes and Antiparallel B-DNA Illustrated on a Trefoil Knot. A trefoil knot is drawn with negative nodes. The path is indicated by the arrows and the very thick curved lines connecting them. The nodes are formed by individual arrows drawn at right angles to each other. Each pair of arrows forming a node defines a quadrilateral (a square in this figure), which is drawn in dotted lines. Double-arrowheaded helix axes are shown perpendicular to these lines. The twofold axis that relates the two strands is perpendicular to the helix axis; its ends are indicated by lens-shaped figures. The twofold axis intersects the helix axis and lies halfway between the upper and lower strands. The DNA shown base paired at the center of each tangle corresponds to half a double helical turn, taken here to be six nucleotide pairs.

A negative node corresponds to B-DNA, but a positive node can be built from Z-DNA. By using this methodology, we have constructed from a single strand topologies that correspond to a negative-noded trefoil knot, an amphicheiral figure-8 knot, and a positive-noded trefoil knot [18] (Figure 6).

Figure 6. Three Knots Synthesized from a Single Strand of DNA. From left to right are a trefoil knot with all negative nodes, a figure-8 knot, with two negative and two positive nodes, and a trefoil knot with all positive nodes. The knot on the left is synthesized from pure B-DNA, the one on the right from pure Z-DNA, and the molecule in the middle from a mixture of B-DNA and Z-DNA.

It is possible to add 'topological protecting groups' to any construction, such as a knot or catenane in which the target contains single strands that are not designated to be unstructured material. These protecting groups consist of single strands that pair with the target single strands, but which can be removed from the target by denaturation and purification [16]. Topological protecting groups are more successful with catenanes than with knots. Nevertheless, the topological goals of knot construction are well in hand, and at this point characterization of the products is the factor limiting progress in this area: It is possible to design and synthesize single-stranded DNA knots whose overall number of nodes can be differentiated on gels; however, the signs of the nodes in these knots cannot yet be determined unambiguously, so identities of the products cannot yet be established.

3. The Construction of Periodic Matter: The Search for Rigidity

A key aim of biotechnology and nanotechnology is a rational approach to the construction of new biomaterials, particularly periodic matter [19]. We have seen above that the construction of discrete closed structural entities, such as polyhedra, can be controlled readily, because the symmetry of the molecules can be limited by the molecular design. Thus, the ligation of identical DNA sticky end pairs (to yield an edge) can be separated from each other in time, by protection techniques [13]; the use of sticky ends with unique sequences also provides control over the assembly of finite objects [11]. Likewise, we have seen that purely topological goals, such as the construction of knots or particular catenanes, is reasonably well in hand.

This comfortable situation does not apply to the construction of periodic matter (crystals), where translational symmetry is an inherent characteristic of the system, because the contacts between all unit cells are identical. It is possible to envision deprotection schemes to unmask successively individual polyhedra or polyhedral clusters containing the same sticky ends by means of different restriction enzymes [20]. Likewise, one can imagine the construction of 'pseudocrystals', having the same backbone structure and topology, but differing in sequence at key sites. Such schemes are both cumbersome and expensive. They do not offer a practical means to the assembly of large repetitive constructs, even if one pictures hierarchical assembly of subsections of the target crystal.

There are at least three key elements necessary for the control of three-dimensional structure in molecular construction that involves the high symmetry associated with crystals: (1) The predictable specificity of intermolecular interactions between components; (2) the structural predictability of intermolecular products; and (3) the structural rigidity of the components [21]. We have seen that DNA branched junctions are excellent building blocks from the standpoint of the first two requirements, which are also needed for the construction of individual objects: (1) Ligation directed by Watson-Crick base pairing between sticky ended molecules works well in this system; and (2) the ligated product is double helical B-DNA, whose local structural parameters are well-known. As noted above, the key problem in working with branched DNA as a construction medium is that branched junctions are flexible molecules. We regard a rigid DNA component as one in which *the vectors of double-helix axes (and hence the angles between them) vary within limits of flexibility no greater than those of linear DNA*. Thus, connections between two such components generate predictable and unique structures, *not* a series of products.

In addition to the flexibility of 3-arm and 4-arm branched junctions, 5-arm DNA branched junction has been shown to have no well-defined structure, and a 6-arm DNA branched junction has only a single preferred stacking domain [7]. However, Leontis and his colleagues have shown that a three-arm branched junction containing a bulge of two deoxythymidine nucleotides has a preferred stacking direction [22]. We have explored the possibilities of using bulged three-arm junctions in DNA construction [21]. We have done this by performing ligation-closure experiments similar to those used to ascertain the relative rigidities of 3-arm and 4-arm branched junctions. The bulge provides twelve bonds that allow the arm 3' to the bulge to stack on the arm 3' to it. This stacking interaction indeed provides the dominant structural mode. Figure 7 shows the results of the ligation experiments, which entail cyclization through each strand in junctions with/without the bulge.

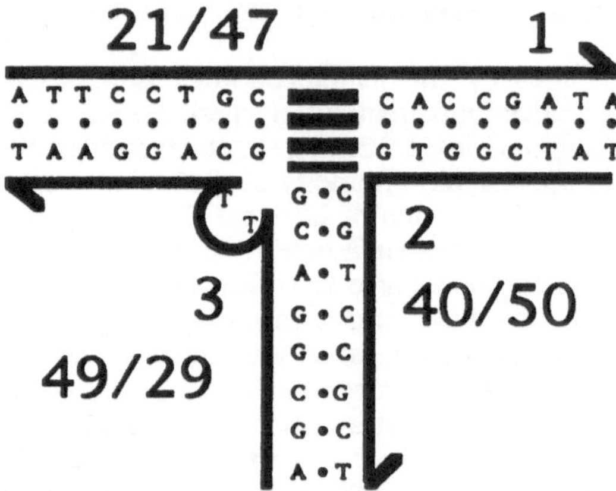

Figure 7. Ligation Results of Bulged Junctions. The core of the junction is shown, with arrowheads on the 3' ends of each strand. Sticky-ends (not shown) have been attached to each arm pairwise. The numbers represent the cyclization percentage seen in the presence/absence of the bulge on strand 3. The striped rectangle represents the stacking interaction permitted by the bulge. The low 21% cyclization percentage on strand 1 indicates that this stacking is clearly the favored interaction in the presence of the bulge.

The results of the ligation experiments with bulged 3-arm branched junctions have suggested that an element of structural preference exists in this molecule that is lacking in previous junctions explored. Details of the ligation experiments (not shown) do not support the notion that the relative orientation of the third arm is well-fixed for purposes of construction [21]. Consequently, we have tried to immobilize the third arm in triangular molecules, such as those shown in Figure 8. In three dimensions, only deltahedra (polyhedra whose faces are all triangles) are rigid polyhedra [23].

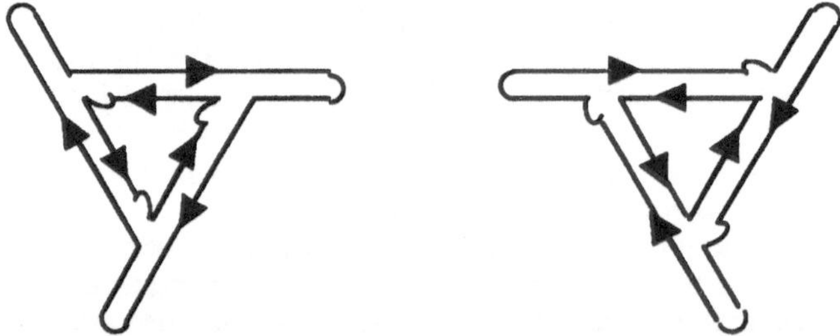

Figure 8. Schematic Drawings of Equilateral Triangles Constructed from Bulged Junctions. The arrowheads indicate the 5' → 3' direction. The edges of the two triangles shown are of equal length. The bulges on the molecule on the left are on the inner strand, so that it would be 6 nucleotides longer than the inner strand of the triangle on the right, where the bulges are on the outer strand.

In addition to the rigidity of the polygonal component, an apparent advantage of the triangular motif is that one of the two alternative stacking structures has been eliminated, because the two arms of each junction that participate in the triangle cannot stack on themselves.

In order to test the rigidity of these triangular components, it is necessary to use both triangles, in alternating fashion, if one is to have a 'reporter strand', whose cyclization will indicate cyclization of the triangular array. Thus, one must alternate triangles with inside and outside bulges. These are indicated by the 'I' and 'O' in Figure 8. The triangles are drawn with the stacking interactions promoted by the presence of the bulges.

Figure 9. Ligated DNA Triangles Containing Bulges on the Inside and Outside Strands. Two DNA triangles are shown. Each contains two turns in each edge, as indicted by the medium thickness strands that form cyclic molecules corresponding to the individual polygons. The triangle with the bulges on the inside is indicated by the label 'I', and the triangle with the bulges on the outside is indicted by the label 'O'. The reporter strand is drawn with the thickest lines.

The angles between the two triangles and the external arms that are ligated to connect them in Figure 9 are 120°. Thus, if the stacking interaction is sufficiently strong as to

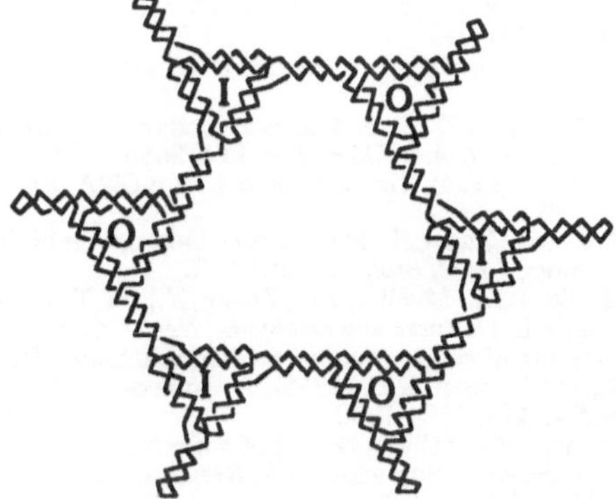

Figure 10. Projected Structure of Ligated Rigid Triangles. The ideal structure would contain six 0° angles, counterclockwise between 'I' and 'O' triangles, and six 120° angles between 'O' and 'I' triangles.

render these triangles 'trigonal valence clusters', the preferred ligation product would in fact be a hexagon, in which three such units form a cyclic structure, as shown in Figure 10.

We have performed this ligation experiment in several different modes [24]. For simplicity, we begin with the linked dimer of triangles, such as the one shown in Figure 9. We have used external arms that when linked together yield connecting helices of 20, 21, or 22 nucleotides. In addition, we have mixed connecting helix lengths. Regardless of which triangles we use, the primary product is the cyclic tetramer of triangles. Thus, triangles built from bulged junctions do not constitute a rigid system.

The search for rigid modules in DNA nanoconstruction continues.

4. Acknowledgments

This research has been supported by ONR grant N00014-89-J-3078 (N.C.S.), NIH grants GM-29554 (N.C.S.) and GM-41454 (N.B.L.) and by Research Corporation Grant C-2314 (N.B.L.).

5. References

1. Watson, J.D. and Crick, F.H. (1953), A Structure for Deoxyribose Nucleic Acid, *Nature (London)* **171**, 737-738.
2. Caruthers, M.H. (1985), Gene synthesis machines, *Science* *230*, 281-285.
3. Seeman, N.C. (1982), Nucleic acid junctions and lattices *J. Theor. Biol.* **99**, 237-247.
4. Seeman, N.C. (1990), *De Novo* design of sequences for nucleic acid engineering, *J. Biomol. Str. & Dyns*. **8**, 573-581.
5. Cohen, S.N., Chang, A.C.Y., Boyer, H.W. and Helling, R.B. (1973), Construction of biologically functional bacterial plasmids *in vitro*, *Proc. Nat. Acad. Sci. (USA)* **70**, 3240-3244.
6. Wells, A. F. (1977), *Three-dimensional Nets and Polyhedra* , John Wiley & Sons, New York.
7. Wang, Y., Muller, J.E., Kemper, B. and Seeman, N.C. (1991),The assembly and characterization of 5-arm and 6-arm DNA branched junctions, *Biochemistry* **30**, 5667-5674.
8. Ma, R.-I., Kallenbach, N.R., Sheardy, R.D., Petrillo, M.L. and Seeman, N.C. (1986), Three arm nucleic acid junctions are flexible, *Nucl. Acids Res*. **14**, 9745-9753.
9. Petrillo, M.L., Newton, C.J., Cunningham, R.P., R.-I. Ma, Kallenbach, N.R., and Seeman, N.C. (1988), Ligation and flexibility of four-arm DNA junctions, *Biopolymers* **27**, 1337-1352.
10. Hagerman, P.J. (1988), Flexibility of DNA, *Ann. Rev. Biophys. & Biophys. Chem* **17**, 265-286.
11. Chen, J. and Seeman, N.C. (1991), Synthesis from DNA of a molecule with the connectivity of a cube, *Nature (London)* **350**, 631-633.
12. Seeman, N.C. (1988), Physical models for exploring DNA topology, *J. Biomol. Str. & Dyns* **5**, 997-1004.
13. Zhang, Y. and Seeman, N.C. (1992), A solid-support methodology for the construction of geometrical objects from DNA, *J. Am. Chem. Soc*. **114**, 2656-2663.
14. Zhang, Y. and Seeman, N.C. (1994), The construction of a DNA truncated octahedron, *J. Am. Chem. Soc.* **116**, 1661-1669.
15. White, J.H., Millett, K.C. and Cozzarelli, N.R. (1987), Description of the topological entanglement of DNA catenanes and knots, *J Mol. Biol.* **197**, 585-603.
16. Seeman, N.C., Chen, J., Du, S.M., Mueller, J.E., Zhang, Y., Fu, T.-J., Wang, Y., Wang, H. and Zhang, S. (1993), Synthetic DNA knots and catenanes, *New J. Chem.* **17**, 739-755.
17. Seeman, N.C. (1992), Design of single-stranded nucleic acid knots, *Mol. Engineering* **2**, 297-307.
18. Du, S.M., Stollar, B.D., and Seeman, N.C. (1995), A synthetic DNA molecule in three knotted topologies, *J.Am.Chem.Soc*. **117**, 1194-1200.
19. Robinson, B.H. and Seeman, N.C. (1987), Design of a biochip, *Prot. Eng*. **1**, 295-300.
20. Seeman, N.C. (1993), Nanoengineering with DNA, *Biomolecular Materials: Materials Res.Soc.Symp. Proc*. **242**, 123-134.
21. Liu, B., Leontis, N.B. and Seeman, N.C. (1995), Bulged 3-arm DNA branched junctions as components for nanoconstruction, *Nanobiol*.**3**, 177-188.
22. Leontis, N.B., Kwok, W. and Newman, J.S. (1991), Stability and structure of three way DNA junctions containing unpaired nucleotides, *Nucl. Acids Res*. **19**, 759-766.
23. Kappraff, J. (1990), *Connections*, McGraw-Hill, New York.
24. Qi, J., Li, X., Yang, X., and Seeman, N.C. (1996), *J. Am. Chem. Soc*. in press.

DISCUSSION OF THE MOORE AND SEEMAN LECTURES

JONATHAN S. LINDSEY
Department of Chemistry
North Carolina State University
Raleigh, NC 27695-8204
U.S.A.

LINDSEY: It is hard to know how to add anything to such beautiful talks. It struck me that both speakers dealt with the issue of how to make compounds of specific shapes, and this reminded me of a game which predates Tinkertoys by many years. This is a Chinese game known as tangrams where tang means Chinese and gram comes from the English word diagram [1]. This is a simple game played with seven modular components in which there are five distinct shapes and two of the shapes appear in duplicate. The object of the game is to use these modular components to create diverse morphologies.

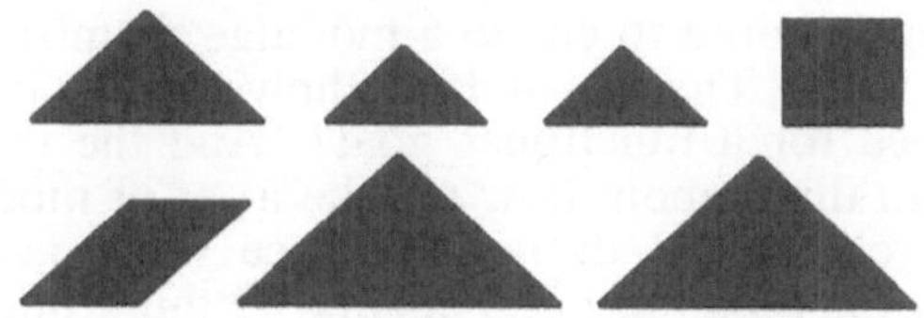

constructing a uniform object

Let me give an example where the goal is to make a large square. Retrosynthetic analysis of a square is very straightforward. We can build that out of the seven components as shown in the diagram.

J. Michl (ed.), Modular Chemistry, 105–122.

One can make a wide variety of target structures (tangrams) based on these seven components. For example, all of the different objects shown below are created in a very straightforward manner. Some of these are of purely architectural interest, others are purely artistic, though none has any functional application.

Some of the issues that come up in thinking about this game are how many components are required to create a modular chemistry, that is, what is the minimal basis set? The second is, how many different modular chemistries do we need for a functional goal? And the third is, can we use computational tools to think about how to take a set of modular components and create diverse architectural designs, or vice versa, going from diverse architectures back to the modular components? I think both of the speakers today may have used some computational tools and I hope they will comment on those.

Jeff Moore showed a slide about building block approaches in biology, and this may be quite a tired slide, but I do want to comment on one aspect of it. Biology uses a variety of different modular chemistries, and the number of components constituting the basis set of each of those is different. For proteins there are 20 amino acids (at least in mammalian biochemistry), there are four (or five) nucleotides in nucleic acids, a handful of sugars make all of the polysaccharides, and for energy storage only one modular component is used, that's acetate, and so forth. The number of modular components for different chemistries can be quite different.

To pose the first question, I'd like to have both Nadrian Seeman and Jeff Moore comment briefly on the different computational tools that they use.

SEEMAN: The only design tools that we use are physical models [2] and a program I wrote some time ago that minimizes the sequence symmetry. The sequence symmetry minimization program, SEQUIN [3], is based on treating each strand of DNA as a series of overlapping N-mers. For example, the first structure we ever designed is a 4-arm junction in which each arm contains eight nucleotide pairs, and is constructed from four 16-mer strands [4]. Each strand is treated as a series of 13 overlapping tetramers. We insist that each of the tetramers be unique, and that the sequence contain no tetramers that complement a tetramer containing a bend. In this way, there is no competition with the designated structure, except at the level of DNA trimers. Of course, larger structures require larger units, such as pentamers or hexamers. This methodology relies on the cooperativity of DNA double helix formation.

MOORE: I'd say the one thing we routinely do is when thinking about new network structures, we build hypothetical periodic models to check for unfavorable non-bonded interactions. Really just the simplest models have been the most valuable for us. For example, where we have identified polytopic ligands that we imagined would form a planar network, but where the ligand's substituents overlap with one another when forced to be in a plane, we obviously don't observe that planar two-dimensional network. A good example from our lab is the complex formed from silver triflate and the nitrogen heterocycle triazine. But as far as any more sophisticated modeling, like what Josef showed us this morning, we haven't done anything quite like that.

KAHN (Bordeaux): Yes - I would like to ask a question of Dr. Moore. I very much like your cellular-containing structures, they are absolutely magnificent. In the meantime I am wondering whether there is anything new in terms of physics or in terms of chemistry arising from the fact that you have a network. Are there any new physical properties, for instance, or is this just a game?

MOORE: I can tell you what our target has been. One of these targets has been the control of free volume in the solid state. We would like to be able to make an organic analog of a zeolite. Not an inclusion compound. That means we would like to be able to reversibly remove those guest molecules that are in the solid without having that lattice reorganized. That means we need networks that are robust like the covalent connectivity typical of zeolites. And so, we are asking for stronger and stronger network bonds while being faced with this problem of making bonds that are reversible so that we can get periodic structures. In this case I think periodic structures are important because periodicity will define a very characteristic shape to these channels that exist. And so that has been the target. Maybe you want to talk about why I think these are interesting solids to be studied? First let me state that there are certainly a lot of questions to address before that application can be realized. For example, "Can we understand how to control supramolecular chemistry to the degree that we know how to control covalent bond chemistry?". This is a completely different issue from the more practical one I just mentioned. One would think that if you have high symmetry ligands that have strongly directional interactions, that is at least a good starting point in an effort to control solid state structure. Network motifs are easily identified and that is why we have been studying them. I don't know if that answers your question or not.

KAHN: Oh that is a very good answer. Do you plan to incorporate different metal ions which would provide more physics?

MOORE: We have only begun recently working with copper(I) ions.

KAHN: By the way, this morning, during one of the discussion sessions, Francois Diederich said that any periodical molecular assembly shows some cooperativity. I'm not sure that it is true, actually, cooperativity is not that easy to achieve. In fact, I'm sure that it is very difficult to get. In phase transition theory, cooperativity appears only when interaction is larger than a critical value. And I mean it. I really think that this is one of the very important issues and once again I wanted to raise this issue

MOORE: Those of us who are working on organic analogs of zeolites aren't necessarily trying to put zeolites out of business, but recognize that this is an interesting state of matter and that organic chemistry does allow the opportunity to add a lot of structure through the building blocks we use, and that may or may not prove to be useful. But we don't know how to control the structure very well to begin with and to a large degree that is the issue on the table.

TILLEY (Berkeley): I think one of the nice things that Jeff is doing is basically developing synthetic guidelines for how to control structure and shape and how things can be linked together in the solid state. Along those lines, I wanted to ask you if you had enough structural data to make any correlations between the role of the anion and how the structure might develop in a templated way.

MOORE: The counterion and solvent from which our networks crystallize, I can say very definitively, do play a role. The one example that I showed, silver (I) complexes with dicyanobiphenyl, is probably the best example that I can point to where the counterion is important. The question of reliability is the second and probably more important issue. I can say that with a constant counterion and a constant solvent from which we crystallize, the coordination geometry is pretty constant for silver nitrile systems. If I have a tri-topic ligand, a tris-nitrile for example, then I usually see with silver triflate crystallized from benzene, in fact in every case of over 10 structures, I find that 3 nitriles and one triflate are always bound to silver. And so the coordination geometry at silver is a trigonal pyramidal-type of coordination geometry, a flattened tetrahedron. The nitrile groups occupy the basal plane with an angle of about 120° between nitrile pairs. At the apical position is the triflate. That coordination geometry holds very well from one structure to the other when we keep the counterion constant as triflate and the solvent which we use in the crystallization, generally benzene.

COLEMAN (Lyon): To be totally provocative, I would like to suggest that you don't try to make organic zeolites with organic Werner complexes. If what you're doing is replacing one of the Werner skeletons with an organic skeleton, then you're trying to do the same things that have been done with Werner complexes for about 150 years. Just sideswap in and out. I would suggest that you look at some more recent results by Iwamoto and some guy at University of Novosibirsk where they have done the full network analysis on how you build up and form all these structures. It's quite nice, they demonstrate the point quite beautifully. So would you like to comment on that?

MOORE: I don't see your point.

COLEMAN: I would like to get your comment on whether or not you're trying to build Werner structures or zeolites.

MOORE: I've already talked about zeolites as a target. Why is this a challenging target? Because nature tends to close-pack in the solid state. I described an approach that will lead to open structures. Can we achieve open structures as a target? That, in and of itself, was really the question that we put on the table. I think everybody has to have a target if you want to learn chemical rules.

MÜLLEN (Mainz): You pointed out how sensitively topologies are reflected by the gel electrophoresis, but to a stranger in the field, how did you then assign a spot to a potential topology?

SEEMAN: What we do is we construct marker molecules that we can characterize. Each of those different bands on the gels that I show corresponds to a specific topology, such as a corner triple catenane or a linear triple catenane. We build those species out of strands that are going to form those topologies, and then run them on gels to establish characteristic mobilities under particular conditions [5]. Gel electrophoresis is actually more sensitive than was evident on the gels shown here. In other experiments, we have demonstrated that we can separate, by gel electrophoresis, singly linked, doubly linked and triply linked double catenanes of strands that are the same length as the cyclic strands (80-mers) that make up the faces of the cube [6].

PORT (Stuttgart): Do you have MALDI-TOF or electrospray mass spectrometry?

SEEMAN: No, in our experience, that doesn't really work well with DNA. At this point, a hundred-mer is about the most you can do with mass spectrometry, and that wouldn't tell you anything new. Gel electrophoresis is more sensitive, because mass spectrometry wouldn't tell you linking topology.

KAHN: Even since its charged you wouldn't see the molecular ion peak by mass spectrometry?

SEEMAN: Let me be very clear. We're doing exactly what old-time organic chemists did. They didn't have NMR, crystallography, or other direct probes of structure. They took their molecule and they broke it down into something that they could recognize. We're doing the same thing here with this chemistry. We have specific breakpoints on every one of these molecules, so that the 12 edges that I pointed out on the cube, for instance, correspond to 12 different places we can cut our hexacatenane, at will. Each of these cuts breaks it down to a tetracatenane. We can get all the predicted products. If we cut on every edge we get a tetracatenane. Our synthetic protocol can be executed in four different ways, and we did it all four ways and got all four predicted products. We are very confident of our results.

EBBESEN (NEC): Can you get crystals of them? That would be great.

SEEMAN: Yes, it would be great. We certainly don't have anything at this point.

EBBESEN: You're trying to do that?

SEEMAN: What we're trying to do is to stop crystallizing, going through the awkward search, blindly stumbling through a series of trials with limited feedback from our failures. Instead, we are actually trying to build the crystals, exactly the same thing that Josef discussed this morning and in some respects the sort of thing that Jeff just discussed, as well.

KAHN: Can't you use STM or some auxiliary methods?

SEEMAN: STM hasn't worked well for us. We've tried a few things and the technique is not very effective. Let me talk about yields here for a second. Our yields are low, which makes it difficult to apply many physical techniques. A typical synthesis of the cube, for example, yields 4 femtomoles. Now for those of you multiplying through by Avogadro's number, that is 2.5 billion molecules. 2.5 billion molecules, when I was born, was enough to give everybody on the planet their own cube. But today every 2 or 2.5 people would have to share a molecule. <Laughter> So there are lots of things we would like to be able to do, but cannot do yet.

MALLOUK (Penn State): I have a question for Jeff. In general, with coordination polymers the problem goes back to Richard Robson's work and others, and that is the problem of collapse when we look at these things or try to get the solvent out. It's also a problem in other areas of molecular sieves science but it's not as severe, let's say, in the zeolites. The reason being that the building blocks in zeolites are rings that go together to make prisms, rather than being single spindly rod. So there's a bundling effect that gives them strength. The individual bonds are not that much stronger, actually, than a good coordination metal link-coordination bond, but there is that reinforcement. It seems like not so difficult a prescription to first make rings and then hang your ligands off those by analogy with zeolites to make secondary building units and then connect those.

MOORE: The best example of that is actually Robson's. He has a cage-like structure which doesn't have extended channels through it but it does have a cage which has a huge amount of solvent trapped in it, and because it doesn't have channels, he has no ability to get the trapped guests out, but he clearly showed that the trapped guests behaved like a liquid.

Now, there are a couple of comments I want to make. First of all, these nets are interpenetrated and a question is, does interpenetration help you in a sense of stability, that is if you have highly interpenetrated structures, does that provide any added stability? That is one point I think that these networks have that you don't see in zeolites. The cage concept is probably a very real and important one that leads to stability. In the data I didn't get to show we actually have one crystal where we can lose the guest species amounting up to 5 mass percent of the complex, reversibly.

KAHN: Without destroying the crystal?

MOORE: Yes. We have obtained rotation photographs after removing this portion of the guests and the samples are still single crystalline. The unit cell indexes to the same cell before guest removal. It remains optically transparent and uniformly birefringent. In many respects, this is our first example that suggests the feasibility of the zeolite concept. And I think that Omar Yaghi has some nice data which also supports this concept.

MALLOUK: Well that's true, both you guys have examples, I wasn't contesting that, I'm just saying as a general design strategy, most of them are going to fall apart. There are people, for example Mark Davis I think is doing this, that are making a real silicate-like secondary building material. But it seems like this is an opportunity for coordination chemistry to design something like that.

YAGHI (Arizona State): I would like to add to that because Tom touches on a very important point that has been a bit hidden in the literature- it is the fact that these frameworks are destroyed when the guest leaves. I think one of the ways to create more stable ones is to aggregate bulkier building units, which we have shown in at least two examples. So I think that this point is still a sticky point in the area where there are still challenges on trying to make a stable framework but at the same time have large voids in the structure. The Robson compound that you refer to, the cage compound, I think that is almost irrelevant in this discussion because the guest is actually a hostage.

MOORE: Right, that's what I said.

YAGHI: It's a hostage situation and that is no different than the rest of the hydrated halide clathrates that exist or have existed for a long time. I just have one more point about the professor from Lyon, Professor Coleman. The people who work in this area are not dissociating themselves from the Hofmann-type clathrates. I think only recently have people very carefully tried to find out what factors are important in building these structures.

MOORE: In many respects, the coordination bond is no different than a hydrogen bond. It's a different type of supramolecular interaction, that's all. I don't think that its an important issue whether coordination chemistry has been around for 100 years. I don't think anyone said that at all.

COLEMAN: Right, I just saw two weeks ago the analysis of some Werner compounds which remarkably resembled what you presented.

SEDDON (Belfast): Two-part question: If you're trying to control structure, why did you use copper and silver? They probably have the most flexible coordination geometries of any element.

MOORE: Right, that's a good point.

SEDDON: Secondly, related to that, do you see any signs of polymorphism?

MOORE: Yes, now back to the first question. Does one want the most rigid structure that doesn't have any means of adopting the geometry to allow those variations in packing, or does one want a softer center that will allow some flexibility? Do you see my point?

SEDDON: Yes, I see your question, but surely you also want to create the ability to control structure?

MOORE: Which approach works best? Is one component, say the polytopic ligand, the leader and the other, say the transition metal, the follower? That's the sort of the approach of using a malleable coordination sphere. The second question about polymorphism: Yes, we've seen polymorphism in one of the crystal structures that I've shown. The one that has the three-dimensional thorium disilicide network has been crystallized in a two-dimensional network as well. It's a two-dimensional graphite structure.

MICHL (Colorado): I'd like to address a similar thing, this problem of trying to make "organic zeolites". I think there are several points to make. First, possibly some of these crystals actually would not collapse if one removed the solvent under the conditions under which aerogels are made, that is above the critical temperature, avoiding problems of capillary forces pulling at the pores and destroying the crystals. This probably should be tried as a matter of course each time one tries to desolvate a crystal of that kind. Maybe the percentage of success would only go from 1 to 2%, but that still would be a good advance. That's just a suggestion. Second, what we are trying to build, I think, are really frameworks or scaffoldings, that's the word that I used in my presentation and it seems to me to be more general than "organic zeolites", for example, because zeolites are very specific structures and we don't necessarily try to imitate those, nor Werner complexes. Werner complexes are simply one way to establish bonding of the kind that's needed to produce a framework or a scaffolding. One could use hydrogen bonding instead, or use covalent bonds, or other kinds of bonds. So I don't see any conflict between trying to make frameworks or scaffolding using one or another type of complex or bonding. And finally, the idea of bundling that has been suggested by Tom Mallouk and also mentioned earlier this morning, I think that is ultimately going to be very important, and I believe that Professor Wegner mentioned I-shaped beams and things like that. I believe people will address these problems after they have overcome the ones that we are faced with right now.

MOORE: The only remarks I want to make are first I agree with everything you said, and second from our target-oriented approach to this project, I can easily make a single statement about organic analogs to zeolites so that everyone immediately recognizes what we are interested in. It's in this light that I made the comment about an analogy to zeolites.

DIEDERICH (ETH Zürich): I have two questions - First of all, you mentioned that the phenylacetylene linkage involves a barrierless torsion. The barrier for rotation about the phenylacetylene $C(sp^2)$-$C(sp)$ single bond has been of interest ever since the first studies published in 1951 when Leslie Orgel worked as a postdoc with Jack Dunitz at that time at Caltech. Theoretical work at various levels on this topic has come to controversial results.

MOORE: Nearly a barrierless torsion.

DIEDERICH: How much is nearly barrierless?

MOORE: It is less than 1 kcal/mol.

DIEDERICH: Do you know that so precisely? Does it come from theory?

MOORE: This wasn't a theory, this was an experimental measurement.

DIEDERICH: How do you measure a kcal barrier?

MOORE and MICHL: I think it was supersonic jet, electronic spectroscopy.

LINDSEY: If I could interject, we have measured the rotational barrier in our diphenylethyne-linked porphyrins and the rotational barrier is 900 calories measured by NMR in solution. Furthermore, it's true that ethynes are very rigid in the linear direction but ethynes bend, and they bend because the carbon distance is 1.2 Å and at that distance the 2s orbitals, which are spherically symmetrical, play a significant role in the bonding. The bending force constant of ethynes is less than that of a corresponding alkane, and so, for example, in our porphyrin systems the average out of plane deformation of adjacent porphyrin rings is 29° [7]. In Jeff Moore's case, that's probably not the case that you get a lot of flexibility because many of those compounds are macrocyclic, and if one ethyne bends then all of them would have to bend in compensation.

MOORE: They might have some sort of coupled motion.

LINDSEY: But if all you have is a single ethyne you can anticipate significant bending at the ethyne.

DIEDERICH: The other question is whether the perylene absorption is really going to zero?

MOORE: Well, there's definitely a finite absorption.

DIEDERICH: Can you quantify how much of your energy goes into your system by an antenna effect and how much goes directly to your perylene?

MOORE: I won't tell you an exact number but it is greater than 100:1 antenna absorption to perylene absorption at 310 nm. If you normalize our emission spectrum per mole of perylene, and you look at our antenna molecules versus ethynylperylene, you see emission from the antenna molecule that is almost a thousand times greater than that from perylene itself.

DIEDERICH: The supramolecular nucleic acid structures that you presented, are those present in homogeneous form or are they in an equilibrium exchange? An example for the latter that comes to my mind are the 4-helix bundle proteins, which were for a long time postulated by their proponents to be homogeneous, whereas later experimentation showed that the 4-helix bundles were in equilibrium with 5- and 6-helix bundles. How is it in your systems?

SEEMAN: Ours are topologically unique structures. I would like to re-emphasize that we can only characterize topology. Insofar as topologies are concerned, they are invariant. You can't change the DNA topology without a topoisomerase. In fact, I didn't talk very much about that in the knot work, but we can create conditions whereby we can get topoisomerases to interconvert particular species. In that case where I showed four different topologies for a single strand, one of those topological species can be made more favorable in particular solution conditions than the topology in which the strand happens to be locked. We can then take a topoisomerase, something that will catalyze strand passage operations, and actually convert one topological form into another [8]. But without the catalysis of the topoisomerase, the transformation cannot happen on its own.

KUKI (Alanex): With regards to the large phenylacetylene constructs, there is the obvious idea of making an extended two- dimensional "super-graphite" based on the tri-functional monomer unit. Perhaps the monomers could be supported on a metal surface by a pedestal as Josef Michl was describing. This "super-graphite" would have a phenyl ring in place of every carbon vertex of regular graphite, and an ethynyl linker in place of every bond. If you make a single layer, it could be a terrific membrane. Is that possible?

MOORE: The difficulty is dealing there with irreversible bonding. I'm not sure we know how to control that without introducing many defects and so, it would be hard to make it truly two-dimensional super-graphite. The whole point of going to a system like metal coordination chemistry is that we have reversible bonds and we can have error checking that Francois mentioned this morning so that we can get more perfect structures. Irreversible bonds are going to lead to highly defective irregular structures. I don't know of a way of controlling that in a way to get a perfect crystal.

WUEST (Montréal): I've noticed before that chemists interested in zeolites and other inorganic porous solids do not like to hear about "organic zeolites". When I talk about organic zeolites, I mean organic compounds that have properties of selected porosity, the potential of compartmentalization of guests, and the possibility of internal catalysis. Whether or not you can actually remove guests from inside them reversibly is an important question but not for me the most important. I think the term "organic zeolites" presents a challenge to the zeolite chemists, who may fear that some of this potential will be realized and that their materials will no longer be unique.

I have a question. The talks of Ned and Jeff indicated that there are two limiting approaches to the construction of networks with interesting architectures. One is a synthesis-intensive approach in which a molecule with a great deal of information is built, sometimes heroically, and then is allowed to associate to create a structure. That is the approach that I have taken, that Jeff Moore has taken in the first part of his talk, and that Ned Seeman has taken. The other approach, which is favored by people who are less good at synthesis or who are lazier or who are smarter, possibly, is to use materials that can be bought or can be obtained simply and then to mix them together in some clever way. That's the approach that Jeff Moore used in the second part of his talk. This suggests the potential of a combinatorial approach. Someone can mix things together and see if interesting networks can be obtained by crystallization. That could be an effective approach. You indicate some ambivalence about which approach you favor. However, if we have to make materials that incorporate more elaborate structures, such as tubes or reinforcing I-beams, then we will clearly have to do more synthesis.

MOORE: Absolutely.

WUEST: How far can we get with compounds we can buy from Aldrich? Which approach do you favor right now?

MOORE: Part of my group pretty much takes the approach you say, that is working hard can get you after 6 months this complicated macrotetracycle; while working for a week on crystallization and then doing an X-ray structure can get you the exact same topology in an infinite network. So the working hard, working smart thing is what my group faces right now. Where do I stand? You're right. If you want something more complex in your solid like this mechanical device that I showed early on, I would say that it's not beyond the possibility of putting it together with simple components, but it's not obvious and I don't think we know how to do it today.

SEEMAN: I'd like to comment also on the issue of using cheap materials. One of the things that we're aiming for, which is one of the reasons for the work that we've done with knots, is that it's possible, in principle, to embed a polyhedron within a knot [9]. The nice thing about a knot is that it's a single strand. It is not possible to clone a branch in DNA, but it is possible to clone the sequence of a knot in which a polyhedron containing branches has been embedded. It's a bit further down the pike, but in principle we may be able to create a cheap source of DNA polyhedra. If we can make these molecules from a single strand, then one could get a bug to make the single strand cheaply instead of us having to go through all the effort that we do now. Of course, you have to be sure that you can control the folding of the strand. Once you got the single strand to fold properly, you could restrict off the garbage DNA that you don't need for your application.

TOMALIA (MMI): I have a question for you, Jeff. Have you done any density measurements? What kind of excluded volume are we talking about here?

MOORE: Densities on what materials?

TOMALIA: On any of the rigid symmetrical systems.

MOORE: Yes, they have normal densities, 1.4 g/cc. First of all, density is really hard to make an accurate measurement of, when you have things that can come in and out, and people that solve protein structures don't normally report density, or if they do, they have to use something besides the do flotation methods that we use.

KAHN: Does the crystal structure give you the density?

MOORE: Well yes, but, ultimately, if your structure is ready you should be able to experimentally obtain that same density. If you account for everything. But the problem is crystallographically we often don't see everything. We see the skeleton, sometimes we see the guest molecules, but sometimes they're too disordered, in which case making density measurements is very difficult to do just because you run the risk of changing the composition while you make the density measurement.

TOMALIA: I have a question for Ned. Is it possible that any of your topologies are found in Nature?

SEEMAN: Of the ones that we're talking about today, it's highly unlikely.

LINDSEY: I would like to pose a question - it seems to me that one of the problems that you're struggling with is that the genetic code is so degenerate, and I wonder whether you wouldn't be better off if you could incorporate other sites, such as minor or variant bases for recognition.

SEEMAN: Well, first there are the Benner bases. Steve Benner (ETH) has enumerated a dozen possible bases with three hydrogen bonding groups [10]. However, I don't think we're struggling a lot against that problem, because of the algorithm I described earlier, that is employed by SEQUIN. That's the way these molecules are designed and the whole question is whether we can get the molecules to hydrogen-bond uniquely. Certainly, we have thought about using other bases. If we had a ready source of them, there is no question that we would try them; we'd much prefer to have a dozen bases to use, rather than four. The nice thing about the cooperativity of DNA is that the individual bases are not completely independent of each other; as a consequence you can actually treat a group of 4 or 5 or 6 as a tetramer, pentamer or hexamer, and they will behave as a single unit. They behave sufficiently like single units that mispairing and lack of specificity hasn't bothered us very much so far. If we had phosphoramidites of the exotic bases, and had a specific reason for using them, we'd be happy to test them out.

LINDSEY: The other question is that in Nature DNA is stabilized by lots of other species, such as polyamines and very basic proteins, such as histones and have you thought of exploiting those kinds of structures for helping to guide the types of motifs you want to put into your molecules?

SEEMAN: Right now, the motifs we want involve strictly DNA. If you toss histones at our structures, they will very likely try to wrap them into a nucleosome, that is, they will take the DNA and wrap it around it and that's the last thing we need. We're operating on what is for Nature a very small scale. The edges of everything that we've been making so far are two turns in DNA. That's about 70 Å, roughly a seventh of a persistence length. We would like their helix axes to be as close to linear as possible. I should say the branched structure that we started off with was not something that we cooked up. This is a stable analog of an ephemeral structure found in the process of genetic recombination. The three-arm junctions are a non-disjoint version of a replication fork; there are other structures found in Nature, for example, the bulged junctions are found in RNA structures. There are other nucleic acid structures out there that are likely to be useful, as well. Thus, rather than dumping in proteins whose effects are unpredictable, and possibly antithetical to our purposes, what we have been doing is seeking more and more of the unusual structures found in Nature.

If they are found in Nature, this tells us that they are somewhat stable, so there may be some merit in trying to use them in our constructions.

LINDSEY: One other thing that makes it possible for you to do this chemistry is the powerful techniques from molecular biology.

SEEMAN: That is correct.

LINDSEY: And at the same time, you have maintained that one can envisage using your DNA molecules as scaffolding for other types of things. Do you think you're going to be able to use the molecular biology techniques when you have non-DNA molecules bound to the DNA by some means?

SEEMAN: The molecular biological approach we are using utilizes restriction enzymes and DNA ligase. We are experimenting now with non-enzymatic ligation procedures, which are somewhat effective, but they're not competitive with enzymatic ligation yet. What we would be doing when we included non-biological molecules would be to exploit structures we already knew were relatively stable so we wouldn't, for example, have to use restriction analysis. In the picture I showed you, the non-biological molecule is hanging off the bottom, so, if needed for some reason, you could put a restriction site on the top, and possibly the enzyme could tolerate its presence. The chemistry is usually done at the sub-picomole scale, because the enzymes are expensive.

STODDART (Birmingham): I want to ask Jeff something abut organic zeolites. Are you taking into consideration sufficiently the molecular recognition that you might have between the zeolitic structure you are creating, or hoping to create, and the guest species that you wish to have included in the structure? In raising the question, I would like to refer to the first communication we published on our cyclobis(paraquat-*p*-phenylene) cyclophane back in 1988 [11]. In it, we mentioned that this tetracationic cyclophane crystallizes to form a kind of 'chimney' stack in which the paraquat units 'coat' opposite sides of the 'chimney' oriented parallel to its direction. The cyclophane tetracations are cemented together by four hexafluorophosphate counterions and two acetonitrile molecules. Whereas your zeolitic structure relies upon coordinative bonding, ours is held together by electrostatic forces including hydrogen bonding. Since we have two pi-acceptors on opposite sides of each tetracationic cyclophane, we find we can locate pi-donors within each cyclophane up and down the stack. We find, for example, that 1,2-dimethoxybenzene is included with 1:1 stoichiometry with respect to each cyclophane tetracation. The inclusion complex is deep orange in color. Under high vacuum, we can remove the 1,2-dimethoxybenzene molecules from the 'chimney' stacks and the crystals become colorless. When these crystals were returned to liquid dimethoxybenzene they became colored once again. And so, in a sense, this salt is behaving like a zeolite that exhibits molecular recognition. The point behind this story is - do you have enough molecular recognition in your system to achieve selectivity?

MOORE: One of the biggest problems for selectivity to be achieved in the materials that I talked about today is that they've got very large channels, and large channels aren't going to be very selective. The channels that you refer to are much smaller and they had a shape that was better defined with respect to a smaller molecule than the ones we have, which are more liquid-like, and obviously not going to show selectivity.

STODDART: So if you use a bigger guest, you might achieve better selectivity.

MOORE: We could use either a bigger guest or a smaller building block. It depends on what we're interested in achieving selectivity for.

STODDART: Do you feel you have enough molecular recognition information directed into the channels or cavities from the periphery? Or putting it another way, don't you think you might have to put in some more information in the form of hydrogen bonding or pi-stacking interactions?

MOORE: I think you're right. If you have a robust framework, there is hope that the same structure will result when you start to make those modifications. If that space is really open and filled with liquid-like guest molecules then you might guess that you can put some amount of structured recognition-defining groups directed into those channels and still maintain that same structure.

STODDART: You would know you're making progress if your solvent, whatever the guest, was not disordered but instead was highly ordered and oriented in a specific manner because of these interactions.

MOORE: Right. And right now we're nowhere near that.

NUZZO (Illinois): Jeff, you mentioned the structure of the nine-fold interpenetrated network. Your student showed me a picture of that. I think that is an absolutely amazing structure. Is that the world's all-time record for continuously interpenetrating networks?

MOORE: There are a lot of examples of interpenetrated networks. I don't know of anything larger than nine. But there are many, many examples of interpenetrated networks. It's pretty amazing that it knows to copy itself nine times, and by doing so space filling is very efficient.

NUZZO: That can't be a kinetic structure, can it?

MOORE: We have obtained the same structure from at least two different crystallization solvents, actually two very different crystallization solvents, ethanol and toluene. The same structure resulted, except in one case, one molecule of ethanol was incorporated in and the other one, maybe the toluene is there, but we haven't found it.

NUZZO: Do enzymes change their activity based on the topology of the DNA? With the kind of molecules you work with, for example, I would have thought that unwinding a helix would be sensitive to the fact that your molecules have short lengths and corners and all the funny stuff at the edges. When you want to snip DNA, you have something that recognizes a particular sequence and you're going to cut it in a certain place. So then, does the topology of the DNA you're working with have any effects?

SEEMAN: No, these are not really great substrates for the enzymes, for reasons that I can imagine, but that we haven't checked out. They're not great substrates, but we do get specificity, which is the real question. Do we do it with the amount that is recommended in the catalog? No! We use 10 times as much, maybe 100. But certainly we get the results that are expected. We don't get any extra restriction sites, for example. In fact, when we were first doing the cube, we wondered about exactly that question: There was a point when we had a single band of material, and when we cleaved it, not only did we get the linear triple catenane that I showed you, we also got a double catenane. My student (Junghuei Chen) and I had a long series of debates about whether or not the enzymes were changing their specificity. Ultimately, we decided to explore other gel conditions, which is part of the black art in this business. We found that, in fact, the single band was two bands, one corresponding to the hexacatenane (the cube) and one corresponding to the pentacatenane having a single failure of ligation. Once we bagged the pentacatenane, there was no more of the double-catenane present as a digestion product. We've had a fair number of failures inside the lab, that we haven't reported, that have convinced us that we've got it right, and that things are working the way they're supposed to work.

WARD (Minnesota): Jeff, you talked about density before- you can couch density in terms of packing fractions. You didn't mention any numbers. Using as an example the structure you mentioned which can lose 5% of its solvent volume without collapse, can you give us an idea of how much space is filled before and after solvent removal?

MOORE: Are you asking how much volume fraction the channel occupies?

WARD: I'm asking what would be the packing fraction of the network without the solvent, and then what is the amount of space filled by the solvent?

MOORE: Do you mean the volume fraction occupied by the guest molecules that we are able to remove?

WARD: No, I am asking what space is filled in the lattice in terms of the packing fraction.

MOORE: In the example you mentioned I don't know the number, but I can tell you the number for some of the other ones. In that hexagonal structure of that macrobicycle, the void volume is about 40 to 50% of the unit cell. In the example you brought up where we could reversibly remove that 5%, I should point out that not all of the guests are removed when 5% of the mass is lost. The 5% mass loss corresponds to about a quarter of the guest species that we've been able to locate crystallographically. There are two very distinct mass losses by TGA. One corresponds to something that is reversible without a phase change, and the other corresponds to a loss with a phase change. (A calculation of the packing coefficient was done following the meeting. The skeleton atoms account for 43% of the unit cell volume fraction in this particular example).

WARD: You're just talking about the volume of void that is left, not the empty space. Because usually organics will have a packing fraction of 0.65 or 0.7.

MOORE: OK, now I see what you are getting at. I don't know what the packing coefficient is for this example.

WARD: You're just talking about the volume of void that is left, not the empty space. Because usually organics will have a packing fraction of 0.65 or 0.7.

PRINZBACH (Freiburg): I have a general problem. It is beautiful work, definitely, but why do you put on a label like "organic zeolites". What do you want to insinuate, analogous functions? Your labeling reminds me a little bit of the story of "enzyme models". When is a model a model? You don't have to use wrong labels.

MOORE: Mainly because, as chemists, we often work in terms of analogies. We think of what we know and we consider new things in light of what we know.

PRINZBACH: In function or structure or why? Is it popular, is it trendy? For what reason? I don't see why do you want to put a label organic zeolites when it has no function. I don't see really what's the analogy with your beautiful beautiful work and zeolites?

MOORE: It's no more than in the sense of what I just said. Everyone can connect very easily to the concept of a zeolite. It's an analogy which I make, and I always use the word "analogy."

PRINZBACH: But you put yourself into a competition that you cannot win.

MOORE: No, I disagree with you. I don't think this story's over yet. How can you say that we're not going to win? We've basically been working on a problem that hasn't been fully investigated.

STODDART: I think chemistry is at a very early stage of development. There is no saying what can be achieved or not achieved by chemists in the fullness of time.

PRINZBACH: I think we must identify what we can do, and what we cannot do.

References

1. Read, R.C. (1965) *Tangrams - 330 Puzzles*, Dover Publications, Inc., New York.
2. Seeman, N.C. (1988) Physical models for exploring DNA topology, *J. Biomol. Str. Dyn.* **5**, 997-1004.
3. Seeman, N.C. (1990) De novo design of sequences for nucleic acid structure engineering, *J. Biomol. Str. Dyn.* **8**, 573-581.
4. Seeman, N.C. and Kallenbach, N.R. (1983) Design of immobile nucleic acid junctions, *Biophys. J.* **44**, 201-209.
5. Chen, J. and Seeman, N.C. (1991) The electrophoretic properties of a DNA cube and its sub-structure catenanes, *Electrophoresis* **12**, 607-611.
6. Fu, T-J., Tse-Dinh, Y.-C., and Seeman, N.C. (1994) Holliday junction crossover topology, *J. Mol. Biol.* **236**, 91-105.
7. Bothner-By, A.A., Dadok, J., Johnson, T.E., and Lindsey, J.S. (1996) Molecular Dynamics of covalently-linked multi-porphyrin arrays, *J. Phys. Chem.* **100**, 17551-17557.
8. Du, S.M., Wang, H., Tse-Dinh, Y.C., and Seeman, N.C. (1995) Topological transformations of synthetic DNA knots, *Biochemistry* **34**, 673-682.
9. Seeman, N.C. (1991) The construction of 3-D stick figures from branched DNA, *DNA Cell Biol.* **10**, 475-486.
10. Piccirilli, J.A., Krauch, T., Moroney, S.E., and Benner, S.A. (1990) Enzymatic incorporation of a new base pair into DNA and RNA extends the genetic alphabet, *Nature (London)* **343**, 33-37.
11. Ashton, P.R., Odell, B., Reddington, M.V., Slawin, A.M.Z., Stoddart, J.F., and Williams, D.J. (1988) Isostructural, Alternatively-Charged Receptor Stacks. The Inclusion Complexes of Hydroquinone and Catechol Dimethyl Ethers with Cyclobis(paraquat-*p*-phenylene), *Angew. Chem. Int. Ed. Engl.* **27**, 1550-1553.

WIRING-UP NANOSTRUCTURES

LAREN M. TOLBERT* AND XIAODONG ZHAO
School of Chemistry and Biochemistry
Georgia Institute of Technology
Atlanta, GA 30332-0400 USA

ABSTRACT. The concept of "molecular wires", in which an unsaturated chain of atoms acts as a conduit for electrons, has been a central theme in the emerging field of molecular electronics. The simplest examples are provided by the polyenes. However, the way in which such chains couple to the end groups has a profound effect on the nature of the conductive mechanism. In general, all such one-dimensional molecules undergo a Peierls distortion, making charge transport between chain ends dependent upon thermal (phonon) processes and limiting the rate of such charge transport. A classification of such systems based upon the nature of the termini is proposed.

1. Introduction.

The use of linear unsaturated molecules has become a motif in the creation of molecular "wires" for application in redox systems and nanoscale devices[1,2,3] following some initial concepts proposed by Aviram and Ratner in 1974 for the creation of molecular "rectifiers"[4]. Although the past twenty years have seen an exponential growth, both in theory and experiment[5], the subject has been one of continuing controversy. The most fundamental limitation concerns the difficulties in addressing a single molecule. However, technical advances, particularly exemplified by modern atomic probe microscopy, are finally making available the tools and means to study individual molecules. Thus the molecular issues involved in connecting redox-active nanostructures must now be addressed.

2. Polyenes as molecular wires.

Since the early proposals of Aviram and Ratner, the intervention of intramolecular electron transfer has formed the basis for model systems directed toward molecular "electronics"[6]. Polyenes bearing donor and/or acceptor groups at the termini, in which a discrete segment of polyacetylene acts as the electron shuttle, have served as the major system for exploring the possibility of molecular wiring and logic memory devices. Among these, two types of polyenes have been the focus of vigorous studies.

The first type involves symmetrical substitution of acceptor moieties, which requires a formal reduction of the polyene chain to introduce the charge carrier and is represented by Lehn's so-called "caroviologens"[1]. Another type of polyene involves

J. Michl (ed.), Modular Chemistry, 123–134.

the use of donor-acceptor (or "push-pull") polyenes, e.g., merocyanines, in which photoexcitation provides the driving force for the separation of the metastable state[7]. Recently, Marder and co-workers have pointed out the relationship between the nonlinear optical (NLO) properties and bond alternation in the polyene chain and have demonstrated that the bond alternation can be perturbed by suitable choice of solvent[8]. Variations of such polyenes are the mixed valence binuclear complexes bridged by polyenes, in which intervalence transfer facilitates the long range electron transfer[9].

Unfortunately, for all such approaches, charge injection is accompanied by the occupation of high-lying antibonding orbitals which couple poorly into the Fermi levels of typical electroactive, e. g., metal or semiconductor, surfaces. Polyenes, the prototypical molecular wires, exhibit band-gap behavior which extrapolates to the (insulating) polyacetylene limit. Only upon doping does polyacetylene develop the charge carriers, i. e., solitons, necessary for conduction. Thus the proper molecular wire should take full advantage of these charge carriers (see Figure 1)[10].

The use of solitonic moieties as charge carriers in molecular wires is among the initial proposals for molecular electronics.[5] Solitons have also been suggested for use in molecular gates or switches, in which passage of a soliton switches the bond alternation pattern and blocks charge propagation in a transverse direction. Thus a full examination of the electronic connection of nanostructures via polyenes should appeal to the electronic structures of such systems as a function of charge and end group.

3. Classification of polyenes.

The classification of one dimensional conjugated systems was first proposed by Dähne and Radeglia[11]. They modified the Lewis-Calvin-Regel rule for the absorption behavior of polyenes and suggested that the wavelength of the first absorption band of homologous polymethines, polyenes and merocyanines, could be described by Equation 1, in which n_π is the number of pairs of π-electrons involved in conjugation, q is an increment responding to additional terminal groups, and the constant k describes the effective absorption wavelength of a π-electron pair. The exponent a reveals the polyene-like or polymethine-like nature of the conjugated systems, ranging from 1.0 (polymethines) to 0.37 (polyenes).

$$\lambda = k(n_\pi + q)^a \qquad (1)$$

Although this classification is generally accepted and widely used, it is empirical and lacks a clear physical basis. This equation also suggests a zero band gap for polymethines at infinite chain length, which is contrary to the known Peierls distortion at large coherence lengths leading to a non-zero band-gap.

Recently, following the efforts of Brédas, Silbey and others[12], based on the early work of Kuhn[13], we realized that one dimensional conjugated systems could be best depicted by the particle in a box model[10], which originates from modified free electron theory. Such a model predicts that the absorption behavior of one-dimensional conjugated systems can be described by Equation 2.

$$E = hc/\lambda = E_{\infty} + k/(n_{\pi} + q) \tag{2}$$

In Equation 2, as before, n_{π} is the number of π-electrons involved in the conjugation, q is a function of the terminal groups; the combination of n_{π} + q corresponds to the effective conjugation length, n_{eff}. The intercept E_{∞} corresponds to the band gap at infinite chain length. When the lowest optical excitation energy is plotted versus the effective conjugation length, a linear relationship is obtained for one-dimensional conjugated systems. As we shall see, one dimensional conjugated systems can be distinctively divided into three categories based upon the extrapolated intercept at infinite chain length (see Table 1). The extrapolated band gaps of even membered neutral polyenes, which include neutral polyenes and merocyanines, are between 1.4 and 1.7 eV, which correlates well with the experimental or theoretically calculated band gaps of 1.4 eV for *trans*-polyacetylene[14]. The band gaps of polyene-like polymethines, such as diphenylpolyenyl cations and diphenylpolyenyl anions, approach 0.6-0.8 eV at infinite

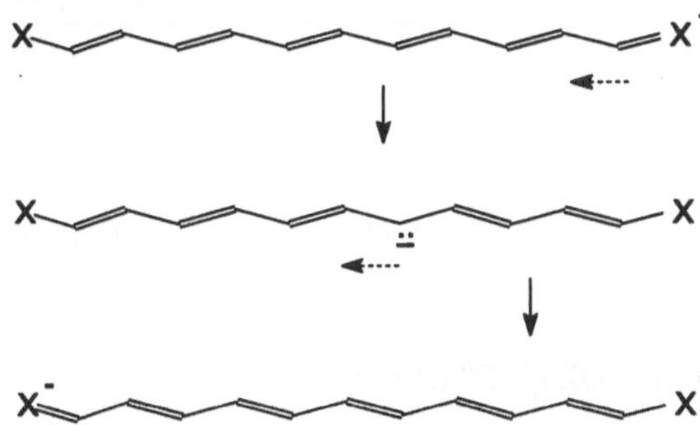

Figure 1. Electron-translocation through soliton migration.

chain length, which corresponds to the 0.7 eV value obtained for polyacetylene doped to high conductivity, in which soliton is proposed to be the absorptive species[15]. This intercept is approximately half of that for polyenes, a fact one would expect from the population of a non-bonding orbital associated with the mid gap Fermi level. In contrast, the optical transition energy of cyanine extrapolates to zero at infinite chain length.

3.1. CASE 1. THE POLYENES.

The polyenes are represented by the α,ω-diphenyl polyenes **DPN** of Boudreaux and coworkers[12] and the α,ω-bis(*t*-butyl) polyenes **(D-t-BuN)** of Schrock and coworkers[15]. The ΔE vs. 1/n plot provides a limiting intercept at 1.8 eV characteristic of undoped polyacetylene (see Figure 2). Such systems are characterized by complete bond alternation, with little change in bond lengths for either double or single bonds as a function of chain length or atomic coordinate.

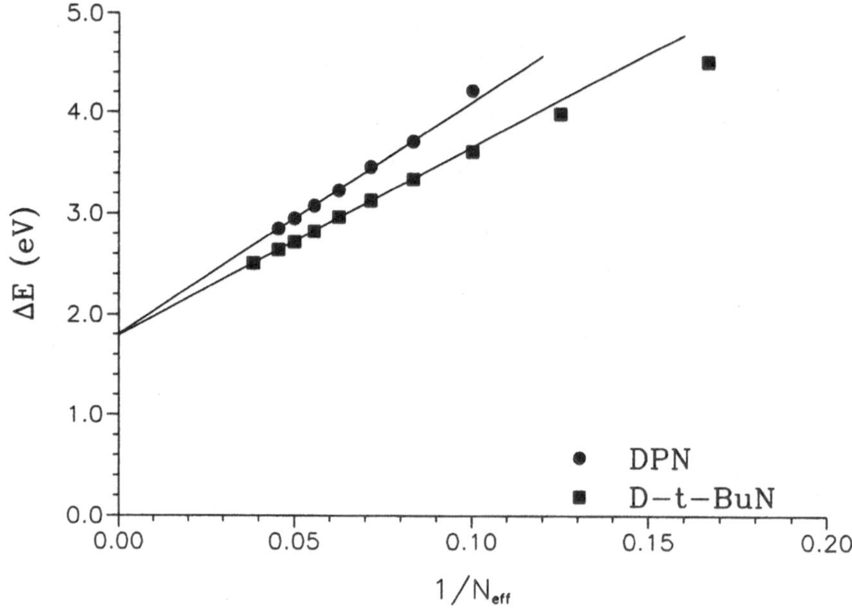

Figure 2. ΔE vs. 1/n plot for linear neutral polyenes.

3.2 CASE 2. THE POLYMETHINES.

When polyacetylene is doped with dissolving metals or oxidants, the material undergoes an insulator to semi-conductive transition. Although the conductivity can approach metallic levels, the conduction mode remains semi-conductive, since thermal activation is required. The charge carrier associated with this process has been identified as a closed-shell charge-density wave called the soliton[16]. Recent investigations in our laboratory on α,ω-diphenylpolyenyl anions $\mathbf{DPN^-}$, i. e., polymethines, have confirmed the charge-density save form of the soliton using ^{13}C nmr spectroscopy[17]. The charge is localized in the center of the extended polymethine chain. Bond alternation increases from zero upon moving away from the center of the soliton "defect", while the charge density decreases. A plot of E vs. 1/n is linear and provides an intercept of 0.6-0.7 eV, analogous to the absorption energy maximum for doped polyacetylene,[16] A similar intercept is observed for the α,ω-diphenylpolyenyl cations $\mathbf{DPN^+}$ (see Figure 3)[18] and α,ω-bisferrocenylpolyenyl cations $\mathbf{DFcN^+}$ (see below).

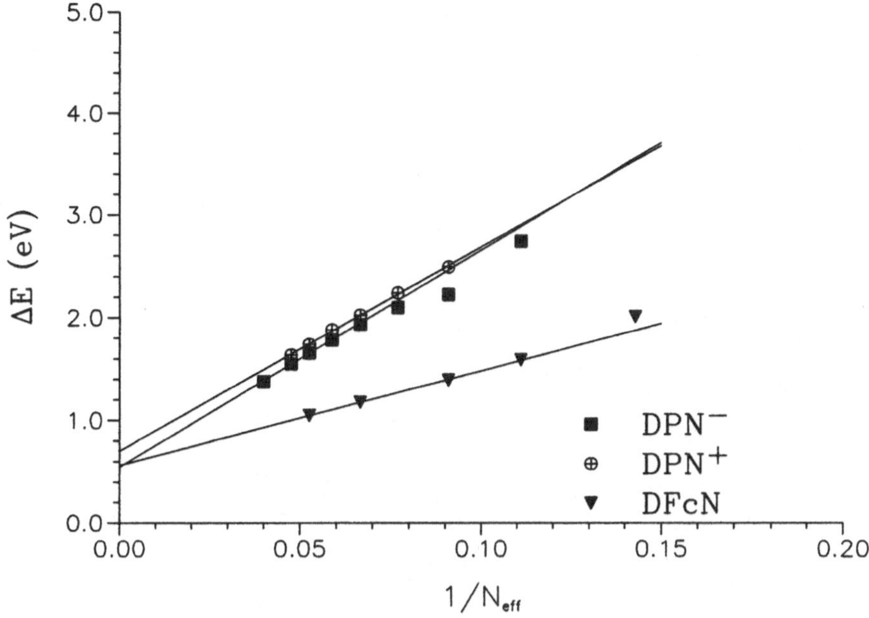

Figure 3. ΔE vs. 1/n plot for linear polymethines.

3.3. CASE 3. THE CYANINES.

Cyanines are characterized by the presence of two elements from group IV, V, or VI at their termini. The electronic structures of cyanines are generally described by two equivalent resonance forms which place charge at either of the heteroatoms, resulting in equal bond lengths and a zero band gap at infinite chain length. When the lowest absorption energies of a series of dipyridocyanines **DPyN** are plotted vs. $(N_{eff})^{-1}$, where the N_{eff} is the number of polyene atoms plus 4 per phenyl, a linear relationship is obtained with zero intercept (Figure 4). The polyoxenols **DCON** also extrapolate to the same intercept. These ions clearly possess a C_{2v} symmetrical ground state and can be best described by classical resonance theory. Recent theoretical treatment of cyanine systems based upon nonalternating bond lengths predicts that through chain conductivity should be exceedingly high relative to soliton transport or interchain hopping mechanisms[19]. However, based on semi-empirical calculations, we[20] and others[19,21,22], have concluded that, for polyenes of sufficient length, a Peierls distortion should obtain, producing a double-minimum potential in the ground state and a barrier for interconversion of the two bond-localized forms.

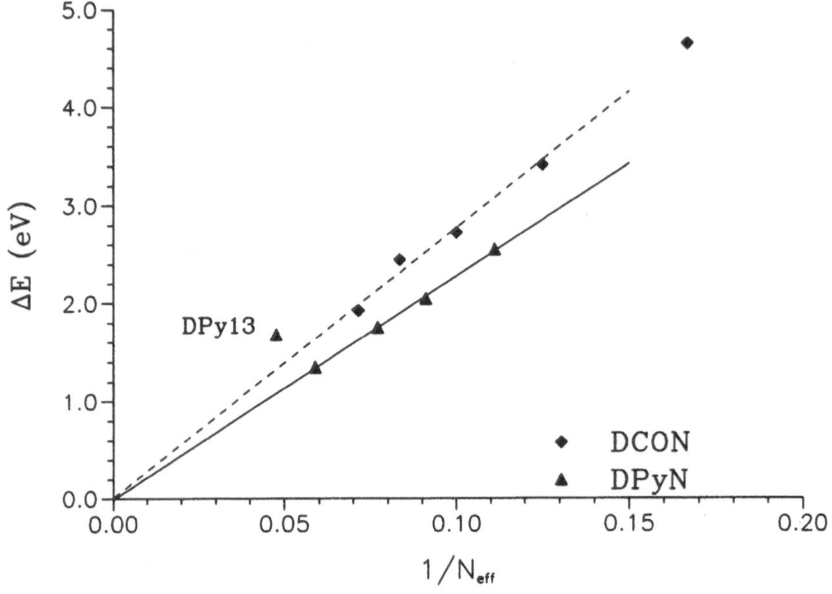

Figure 4. ΔE vs. 1/n plot for linear cyanines.

4. Redox Systems

Although the intervention of solitons in electronic conduction for doped polyacetylene is now widely accepted, direct measurements of solitonic processes in bulk materials have always been limited by the requirement for rate-limiting interchain hopping processes. Thus we turned to polymethines terminated with tightly coupled redox centers.

Mixed-valence molecules are ideal models for the studies of intramolecular electron transfer. In a bridged diferrocenyl system Fc-X-Fc, the oxidation of the ferrocenyl groups may occur either by two distinct one-electron steps or by a single two electron oxidation, depending upon the nature and the length of the connectors[23,24,25]. When the ferrocenyl moieties are connected directly or separated by a single carbon atom, the two ferrocenyl moieties interact effectively. producing separate one-electron waves[26]. However, such through bond interaction shows rapid attenuation upon chain elongation.

4.1. BIS(FERROCENYL)POLYENYL CATIONS

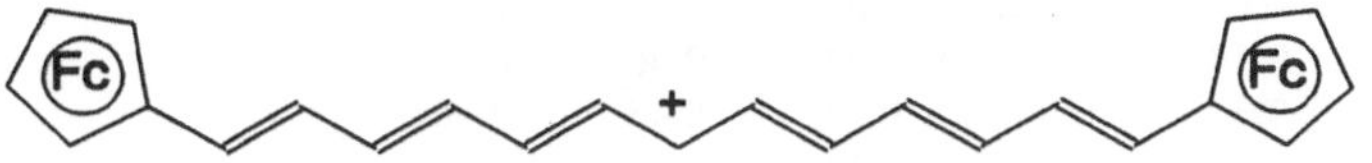

DFc13

Bis(ferrocenyl)polymethine cations $\mathbf{DFcN^+}$ were synthesized using straightforward Wittig methodology[27]. The absorption maxima of this class of polymethines are shown in Table II, along with those of diphenylpolyenyl cations[18], diphenylpolyenyl anions[17], dicyclopentadienylpolyenyl anions[18] and dipyridocyanines[20]. The organometallic polymethines absorb at a much longer wavelength than cyanines or vinylogous carbenium ions and carbanions of analogous length. This lower than usual band gap is presumably the consequence of the strong electronic coupling between the electropositive iron complexes and the HOMO orbitals of the polyene chains.

When the absorption energies of the diferrocenyl polymethines were plotted vs. $(N_{eff})^{-1}$, where the N_{eff} is the number of polyene atoms plus three per ferrocenyl group, a linear relationship was obtained with an intercept of 4400 cm^{-1} or 0.55 eV. This intercept, places this series of ions clearly in the Class 2 polymethine group. The optical spectra of the polymethine cations are fully congruent with those of a system of molecules converging to solitonic conductivity. This suggests, moreover, that the electronic structure of the bisferrocenyl cations involves a true polyenic cation structure with delocalization into the ferrocenyl moieties that vanishes at longer chain lengths.

In Table III are shown the results of electrochemical studies of the bis(ferrocenyl) polymethine cations (**DFcN**). Cyclic voltammetry (CV) of all the polymethine cations exhibited two oxidation waves., The appearance of two oxidation waves even when the two iron atoms were separated by thirteen carbon atoms suggests that in the mixed-valence state the ferrocinium moiety strongly affects the ferrocenyl group at the other end through a distance of 20 Å. Such electronic coupling must be propagated through the soliton-incorporating polymethine chains. This fact strongly supports the notion that the soliton mediated connectors greatly enhance the through bond electronic coupling between the two redox centers, thus making very long range electron transfer feasible.

The variation of electronic coupling (H_{AB}) has been shown to be an exponential function of the distance between the redox centers (l_{AB}), Equation 3.

$$H_{AB} = H_0 \exp(-\gamma l_{AB}) \tag{3}$$

Although this formula is largely empirical, The H_0 and γ values are generally used as a guide for electronic coupling in a bridged system, with γ increasing as electronic coupling diminishes. The γ values lie within the range of 0.4 to 0.6 $Å^{-1}$ for a variety of systems[28,29,30]. Such high γ values would effectively prohibit fast electron transfer when the distance between the donor and the acceptor exceeds 10-15 Å

Table I. Half-wave redox potentials of bis(ferrocenyl)polymethine cations **DFcN⁺**.

	DFc1	**DFc3**	**DFc5**	**DFc9**	**DFc13**
$E_{1/2}(1)^b$	0.39	0.42	0.37	0.36	0.34
$E_{1/2}(2)^c$	0.72	0.60	0.51	0.43	0.38
ΔE	0.33	0.18	0.14	0.07	0.04

and diminish the chances of such systems functioning as molecular electronic devices. However, recent studies on mixed valence ruthenium complexes bridged by dipyridylpolyenes have evoked much interest in the polyene based systems[31,32]. For mixed valence systems of general type, $[(NH_3)_5Ru\text{-}py\text{-}(CH{=}CH)_n\text{-}py\text{-}Ru(NH_3)_5]^{5+}$, the electronic coupling decreases with the increasing chain length with a particularly small exponent, 0.078 $Å^{-1}$.

Since the differences between the two oxidation potentials are presumably the results of the electronic coupling in mixed valence states, it is not surprising that ΔE of **DFc3** to **DFc13** also fits the exponential law analogous to Equation 3, Equation 4:

$$\Delta E = \Delta E_0 \exp - (\gamma' l_{AB}) \tag{4}$$

The least-squares fit provides a γ' value of 0.12 $Å^{-1}$, which is much lower than the 0.5 $Å^{-1}$ value for most of the other systems and is on the same order as the γ value of polyene bridged ruthenium mixed valence system. The exceptionally large long range electronic coupling and the unusual low attenuation displayed by the bis(ferrocenyl) polymethine cations suggest that such solitonic-bridged mixed-valence systems be interesting candidates for molecular wires.

4.2. DIPYRIDOCYANINES.

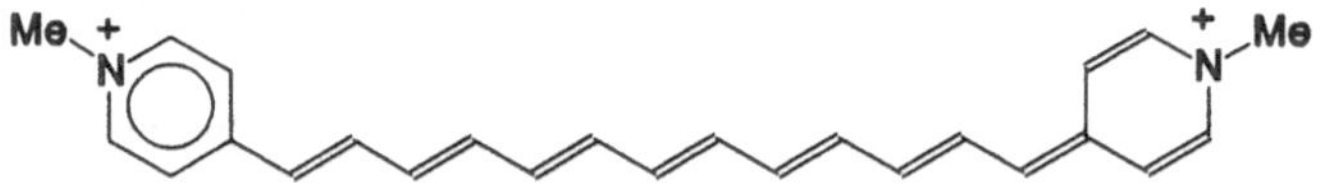

DPy13

In order to test the prediction of symmetry collapse in linear cyanines, we synthesized bis(N-methyl-4-pyridinium)polymethines **DPyN⁺**, which are isoelectronic with the anions **DPN⁻**. The absorption spectra of **DPy1** through **DPy9** exhibited the typical behavior of cyanine dyes, with a sharp, intense absorption and a shoulder (see Figure 5)[33], indicating a minimal change in bond length and bond order upon excitation. The UV-Vis spectra of these classes of cyanines indicate that the electronic structures of the molecules shifted to be more polyene-like upon chain elongation. However, **DPy13** exhibits a remarkable hypsochromic shift relative to **DPy9**, and also exhibits a significant solvatochromic effect, consistent with a symmetry collapse. The structureless and broad absorption band differs radically from the sharp peak absorptions of the cyanine dyes. Similar phenomena in other cyanine systems were first reported by Brooker *et al* in their studies of thiocyanine series[34], who attributed such behaviors to *cis-trans* isomerism. However, the ^{1}H-NMR spectrum of **DPy13** is consistent with the all *trans* configuration.

The ΔE vs 1/n behavior for the dipyrido cyanines are displayed in Figure 4. Significant is the deviation of the energy for **DPy13** from the line, indicating the onset of Peierls distortion and symmetry collapse. Consistent with this behavior, **DPY13** also displays a significant negative solvatochromic effect.

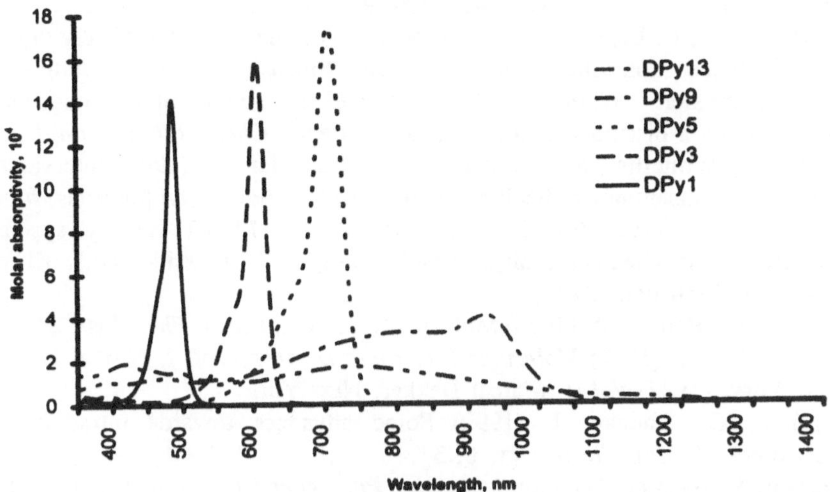

Figure 5. Absorption spectra of dipyridocyanines **DPyN**.

5. Conclusions

As shown in the above classification, the real difference between cyanines and polyene-like polymethines lies in the electron affinity of the end groups; in cyanines these are stronger than in polymethines. Beyond the coherence length of a soliton, the symmetry of a cyanine collapses, and the structure of the charge carrier resembles that of a charge-defect in *trans*-polyacetylene. Thus the possibility of employing cyanines or other polyenes as true molecular "wires" remains remote. What remains is the possibility of activated processes involving solitonic conduction.. However, these processes involve phonon modes in the polymer, and thus are considerably slower at the molecular level than purely electronic conduction.

6. Acknowledgment

Support of this research by the U. S. Department of Energy through grant no. DE-FG05-85ER45194 is gratefully acknowledged.

7. References.

1. Arrhenius, T.; Blanchard-Desce, M.; Dvolaitzky, M.; Lehn, J.-M., Malthete, J. (1986) Molecular devices: Caroviologens as an approach to molecular wires--synthesis and incorporation into vesicle membranes, *Proc. Natl. Acad. Sci. USA*, **83**, 5355-59.
2. Lehn, J.-M. (1990) Perspectives in supramolecular chemistry--from molecular recognition towards molecular information processing and self-organization, *Angew. Chem. Int. Ed. Engl.*, **29**, 1304-19.
3. (a) Bonvoisin, J.; Launay, J.-P.; Rovira, C.; Veciana, J. (1994) Purely organic mixed-valence molecules with nanometric dimensions showing long-range electron transfer. Synthesis, and optical and epr studies of a radical anion derived from a bis(triarylmethyl)diradical, *Angew. Chem. Int. Ed. Engl.*, *33*, 2106. (b) Woitellier, S.; Launay, J.-P.; Spangler, C. (1989) Intervalence transfer in pentaammineruthenium complexes of α,ω-dipyridyl polyenes, *Inorg. Chem.*, **28**, 758-62. (c) Launay, J.-P., (1990) Mixed valency systems: Applications in chemistry, physics and biology, (K. Prassides ed.), Kluwer Academic Publishers, **321**.
4. Aviram, A.; Ratner, M. (1974) Molecular rectifiers, *Chem. Phys. Lett.* **29**, 277.
5. Carter, F. (ed.) (1982) *Molecular Electronic Devices, Vol I,* Marcel Dekker, New York. (1987) *Vol II*, Marcel Dekker, New York.
6. Joachim, C.; Launay, J. (1990) Some advances towards intramolecular electronics, *J. Mol. Electronics*, 6, 37.
7. Balzani, V. (ed.) (1987) *Supramolecular Photochemistry*, Reidel, Dordrecht.
8. Marder, S.; Gorman, C.; Meyers, F.; Perry, J.; Bourhill, G.; Brédas, J.-L.; Pierce, B. (1994) A unified description of linear and nonlinear polarization in organic polymethine dyes, *Science*, **265**, 632. Marder, S.; Beratan, D.; Cheng, L.-T. (1991) Approaches for optimizing the first electronic hyperpolarizability of conjugated organic molecules, *Science*, **252**, 103.
9. Woitellier, S.; Launay, J.; Spangler, C. (1989) Intervalence transfer in pentaammineruthenium complexes of α,ω-dipyridyl polenes, *Inorg. Chem.* **28**, 758; Launay, J. P. (1991) *Mixed valency systems: Applications in chemistry, physics and biology* (Prassides, K. ed.), Kluwer Academic Publishers, **321.**
10. Tolbert, L. M. (1992) Solitons in a box. The organic chemistry of conducting polyenes, *Acc. Chem. Res.*, **25**, 561.
11. Dähne, S.; Radeglia, R. (1971) Revision der Lewis-Calvin-Regel zur charakterisierung vinyloger Polyen und polymethinähnlicher Verbindungen, *Tetrahedron*, **27**, 3673.
12. Brédas, J.-L.; Silbey, R.; Boudreaux, D.; Chance, R. (1983) Chain-length dependence of electronic and electrochemical properties of conjugated systems: Polyacetylene, polyphenylene, polythiophene, and polypyrrole, *J. Am. Chem. Soc.*, **105**, 6555.
13. Kuhn, H. (1949) A quantum-mechanical theory of light absorption of organic dyes and similar compounds, *J. Chem. Phys.*, **17**, 1198.

14. André, J.-M.; Delhalle, J.; Brédas, J.-L. (1991) Quantum Chemistry Aided Design of Organic Polymers, *World Scientific.*
15. Schlund, R.; Schrock, R.; Crowe, W. (1989) Direct polymerization of acetylene to give living polyenes, *J. Am. Chem. Soc.*. **111**, 8004-6.
16. Fisher, A.; Hayes, W.; Wallace, D. (1989) Polarons and solitons, *J. Phys. Condens. Matter*, **1**, 5567-93; Su, W.; Schrieffer J.; Heeger, A. (1979) Solitons in polyacetylene, *Phys. Rev. Lett.*, **42**, 1698; Su, W.; Schrieffer J. R.; Heeger, A. (1980) Soliton excitations in polyacetylene, *Phys. Rev. B.*, **22**, 2099.
17. Tolbert, L.; Ogle, M. (1990) ^{13}C-nmr evidence for a fixed soliton width in anionic polyacetylene, *Mol. Cryst. Liq. Cryst.*, **189**, 279; Tolbert, L.; Ogle, M. (1990) How far can a carbanion delocalize. ^{13}C-nrm studies of soliton model compounds, *J. Am. Chem. Soc.*, **112**, 9519; Tolbert, L.; Ogle, M. (1989)^{13}C-nmr spectroscopy of α,ω-diphenylpolyenyl anions. Confirmation of charge localization in soliton model compounds, *J. Am. Chem. Soc.*, **111**, 5958.
18. Fabian, J.; Zahradnik, R. (1977) PPP-rechnungen zum vinylenshift symmetrischer farbstoffe, *Wiss. Zeit. Tech. Univ. Dresden* **26**, 315.
19. Reimers, J.; Hush, N. (1993) Hole, electron and energy transfer through bridges systems. VIII. Soliton molecular switching in symmetry-broken brooker (polymethinecyanine) cations, *Chem. Phys.*, **176**, 407.
20. Tolbert, L.; Zhao, X. (1993) Extended cyanine dyes, *Syn. Metals*, **55-57**, 4788.
21. Kuhn, C. (1991) Step potential model for non-linear optical properties of polyenes, push-pull polyenes and cyanines and the motion of solitons in long-chain cyanines, *Synth. Metals*, **41-3**, 3681-88.
22. Reimers, J.; Craw, J.; Hush, N.; (1991) Proceedings, U. S. engineering foundation conf. on molecular electronics: Science and technology, St. Thomas, U. S. V. I.; Craw, J.; Reimers, J.; Bacskay, G.; Wong, A.; Hush, N. (1992) Solitons in finite- and infinite-length negative-defect trans-polyacetylene and the corresponding brooker (polymethinecyanine) cations. I. geometry, *Chem. Phys.*, **167**, 77-99; Craw, J.; Reimers, J.; Bacskay, G.; Wong, A.; Hush, N. (1992) Solitons in finite- and infinite-length negative-defect trans-polyacetylene and the corresponding brooker (polymethinecyanine) cations. II. Charge density wave, *Chem. Phys.*, **167**, 101-109.
23. Carugo, O.; De Santis, G.; Fabbrizzi, L.; Licchelli, M.; Monichino A.; Pallavicni, P. (1992) Using platinum(II) as a building block to two-electron redox systems. Crystal structure and redox behavior of cis-[Pt^{II}(3-ferrocenylpyridine)$_2Cl_2$], *Inorg. Chem.*, **31**, 765-9.
24. Silva, M.; Pombeiro, A.; da Silva, J.; Herrmann, R.; Deus, N.; Castiho, T.; Silva, M.; (1991) Redox potential and substituent effects at ferrocene derivatives. Estimates of Hammett σ_p and Taft polar σ substitute constants, *J. Organomet. Chem.*, **421**, 75-90.
25. Miller, T.; Ahmed, K.; Wrighton, M. (1989) Complexes of rhenium carbonyl containing ferrocenyl-derived ligands: Tunable electron density at rhenium by control of the redox state of the ferrocenyl ligand, *Inorg. Chem.*, **28**, 2347-55.

26. Floris, B.; Tagliatesta, P. (1993) One- and two-wave oxidation potentials of some diferrocenyl systems, *J. Chem. Research(S)*, 42-3.

27. Tolbert, L.; Zhao, X.; Ding, Z.; Bottomley, L. (1995) Bis(ferrocenyl)polymethine cations. A prototype molecular wire with redox active end groups, *J. Am. Chem. Soc.*, **117**, 12891.

28. Beratan, D.; Hopfield, J. (1984) Calculation of electron tunneling matrix elements in rigid systems: Mixed-valence dithiaspirocyclobutane molecules, *J. Am. Chem. Soc.*, **106**, 1584-94.

29. Stein, A.; Lewis, N.; Seitz, G. (1982) Long-range intervalence electron tunneling through fully saturated systems, *J. Am. Chem. Soc.*, **104**, 2596-99.

30. Richardson, D.; Taube, H. (1983) Electronic interactions in mixed-valence molecules as mediated by organic bridging groups, *J. Am. Chem. Soc.*, **105**, 40-51.

31. Joachim, C.; Launay, J.; Woitellier, S. (1990) Distance dependence of the effective coupling parameters through conjugated ligands of the polyene type, *Chem. Phys.*, **147**, 131-141.

32. Launay, J.; Joachim, C. (1988) Active molecules for electron transfer and switching, *J. Chim. Phys.*, **85**, 1133.

33. West, W.; Pearce, S. (1965) The dimeric state of cyanine dyes, *J. Phys. Chem.*, **69**, 1894.

34. Brooker, L. (1991) *In Recent Progress in Chemistry of Nature and Synthetic Colouring Materials*; Hillson, P.; Sutherns, E. *Theory of the Photographic Process*, 3rd ed., 204.

DISCUSSION OF THE TOLBERT LECTURE

VINCENZO BALZANI
Dipartimento di Chimica "G. Ciamician"
Università di Bologna
40126 Bologna
ITALY

1. Questions directed to Prof. Tolbert

EBBESEN (NEC): When you plotted the bandgap versus the chain length, did you use the cutoff value of the absorption band or the absorption peak energy? Furthermore, did you check for solvent effects? Finally, are you sure that the molecules remain in the linear conformation (linear chain) and do not form other isomers in solution?

TOLBERT: In answer to your first question, since these molecules are not fluorescent, the values shown are the peak values. We do not have the 0-0 values. The shift between peak value and 0-0 value is approximately constant among the various chains.

We have checked some solvent effects. Since all of these molecules are symmetrical, the solvent effect is minimal with the exception of the cyanine. This is another clue that we are getting symmetry collapse. If you look at the solvatochromic behavior of the cyanines, you see that the one with thirteen carbon atoms shows a very large hypsochromic shift, consistent with symmetry collapse since you now have a dipole moment.

As to the presence of other rotamers, this is a valid concern. For the ferrocenes we do not have coupling constant information, but for the other molecules we have the coupling constants and, again, the interaction varies because the extent of various rotamers in these linear systems is very low for the ions. What we have are the average values of coupling constants, which are consistent with largely trans coupling.

MÜLLEN (Mainz): When considering the role of solvent effects one should also consider that there is an ion-pairing effect. Therefore in discussing the effective conjugation length within negatively charged π chains, one should explicitly take into account the polarizing effect of the counter cation.

TOLBERT: Yes; in merocyanine, according to the work of Seth Marder, one can see bond alternation or not, depending on solvent and counterion. Calculations for our systems were gas-phase calculations,

J. Michl (ed.), Modular Chemistry, 135–139.

and measurements were carried out in DMSO as a solvent and K^+ as a counterion, i.e. under conditions, for which solvent-separated ions are known to exist, according to the work of Bordwell.

SEDDON (Belfast): In the slide of your cyclic voltammetric data for your bimetallic complexes, the separation of the two waves for the complex with the greatest distance between the two metal centers was less than 60 mV. Did you observe two discrete waves, or did you resort to deconvolution techniques?

TOLBERT: These were performed by my colleague, Larry Bottomley, and his coworker. For the long chains, separable waves were not observed, and the analysis of the data required deconvolution. Dr. Bottomley assures me that the numbers are accurate to within 20 mV.

WEGNER (Mainz): The question of electronic communication between two centers separated at distance of 10 Å or more and linked by a series of π-conjugated bonds is an old one and it is not surprising that you find that the two communicating centers lose information about each other as they move further and further apart. This was measured a long time ago very exactly for the case of oligomers of poly(diacetylenes) obtained by solid state polymerization of diacetylenes. A series of oligomers is obtained, in this case all contained within a single crystal of the monomer and terminated at both ends either by radical or carbene moieties. Each oligomer of DP = n + 2 could be addressed selectively by optical and ESR spectroscopies. For instance, the carbenes exhibit quintet excited states in equilibrium with singlet ground states. The larger the distance between the carbene pairs, the lower the exchange interaction, and the lower the energy separation between the singlet and quintet states. Eventually, at length in excess of 70 Å information at the one chain end about the other is lost. This has all been published long ago (*Note:* for a review, see [1]; for an original paper on probing oligomers with regard to end group interactions, see [2].)

TOLBERT: I am not trying to say that we have discovered something new, but a point that I am trying to make is that we have a new way of dealing with the conjugation length in a different way. Second, when you reach that limit, then you no longer can be talking about optical electron transfer, or about phenomena that take place on the time scale of optically dependent phenomena. Now you are going to have some time-dependent phenomena involving basically molecular structural changes and this is going to reduce the rate.

BALZANI (Bologna): If I may add a comment on this topic: even in the case of the so called mixed-valence compounds in coordination chemistry there is the problem of electronic communication (Hush, Robin and Day): when the two metal centers are close together, they interact, whereas when they are far apart, the interaction decreases down to the limit of zero interaction. The magnitude of the interaction, of course, also depends on the nature of the bridge linking the two metals.

TOLBERT: Right.

KUKI (Alanex): The extraction of an exponential fall-off constant from the DE of these conjugated bridge bis(ferrocenes) is quite dangerous. The formula $H_{AB}=H_0 \exp(-gd_{AB})$ is a WKB tunneling formula which assumes a fixed barrier height as d_{AB} is changed, whereas of course the electronic levels of a polyene bridge are changing in a definite way as the chain length is extended. Fitting the band gap *vs* length data in the bis(ferrocenyl)polyenes to this equation intrinsically assumes that g is a distance-independent constant. However, since the barrier height is in fact distance-dependent, the g attenuation factor thereby extracted is of very little meaning. Instead one may recognize that in the venerable WKB formula this g is proportional to the square root of the barrier height, which is the mismatch in the electronic energies of the relevant ferrocene and polyene bridge molecular orbitals. So the correct formula instead would be some variant of

$$H_{AB} = H_0 \exp(-c*\mathrm{sqrt}[E_{bridge}(d_{AB}) - E_{ferrocene}]*d_{AB})$$

where $\exp(-gd_{AB})$ is explicitly written in energy (and distance) dependent form, and c is a factor consisting of well known fundamental constants.

TOLBERT: I don't believe the WKB formulation is unique. The attenuation factor g is a general factor which describes exponential falloff of any coupling with distance. In fact, in the case presented here I don't believe the geometry remains constant, but incorporates relaxation phenomena in the mechanism. The equation presented does not take into account any particular coupling mechanism.

DIEDERICH (ETH Zürich): Have you checked Lambert-Beer's law for your extended unsaturated systems? I wonder whether you have aggregation and, as a result, mechanism for facile electron transfer from one chain to the other. This concerns me in particular for the α,ω-dipyridocyanines, since it is well known that the pyridinium ion and the dihydropyridine may undergo significant stacking interactions.

TOLBERT: These experiments were done in DMSO, for which there is apparently no aggregation. We see evidence for aggregation of the cyanines in aqueous solution. In the presence of cyclodextrins, the absorption increases dramatically, indicating a break-up of aggregation. In DMSO, the intensities are similar to those in water/cyclodextrin, and show no effect of added cyclodextrin. To be sure, we were not able to get good Beer-Lambert plots for the higher cyanines, because our method of generation does not produce quantitative deprotonation.

GRIMME (Cologne): You gave us a caveat about extrapolation of electronic properties for the cyanines. Don't you think that the same caveat should also be used for the properties of your polyenes?

TOLBERT: Right, there is no system which can maintain constant bond lengths over an extended number of double bonds. The origin of bond alternation in polyenes is exactly this effect. At the Hückel calculation level, this effect is not in evidence, but it shows up when the bond orders and resonance integrals are calculated it properly.

PRINZBACH (Freiburg): An experimental check would be the activation energy for cis-trans isomerization as a function of chain length, which should decrease with increasing chain length and reach some limiting value.

TOLBERT: With the short chains in DMSO we can see cis-trans isomerization; for C_5 approximately 95% is all trans; for the longer ones, we can no longer see other isomers. On the average, according to NMR coupling constants, they are all trans, and the calculation were done for all trans, and the calculations are not inconsistent with the experimental observations.

MICHL (Colorado): I agree with the comment of Prof. Wegner that there is no surprise in your results; what is more interesting to me is how does the molecular structure affect the speed of propagation of the perturbation. To say that it propagates with the frequency of vibration is a little too vague because it can be almost anything, it depends on the force constant of the vibration.

TOLBERT: It is the frequency of the vibrational mode associated with the electronic coupling.

MICHL: Yes, and it is difficult to say what it is, until we do some type of calculation or measurement of that mode.

TOLBERT: One of the experiments planned is to do the measurement. In order to do that we need this ferrocene complex, but with different end groups so that we can distinguish them using time-dependent spectroscopy. Our goal is to actually measure the rate of that process.

CHIDSEY (Stanford): What is the role of the counterion in determining the conduction? In simple theories, the counterions are considered to provide a uniform background charge. However, depending on the charge density I can imagine that they are not so innocuous. At low charge density and in low dielectric media, the carriers could be pinned to the counterions. A simple picture would be that they are localized around the counterions in hydrogenic orbitals with an effective Bohr radius. Is there evidence for any of this?

TOLBERT: Conduction in polyacetylene is very dependent on the identity and concentration of the counterion. This is ion "pinning", by analogy to ion pairing in solution. The optimum dopant concentration for potassium is approximately one metal ion for every six carbon atoms. At this concentration, the solitons can hop from site to site, and a smooth translation of solitons is unlikely.

2. Question directed to Prof. Balzani

PORT (Stuttgart): The $Ru(bpy)_3^{2+}$-anthracene-$Os(bpy)_3^{2+}$ system that you have illustrated is similar, but much worse, than a three component system that we have described a few years ago, where switching is completely reversible; your system is totally irreversible.

BALZANI: Yes, I know very well your paper published in *Chem. Phys. Lett.* [3] a couple of years ago. The system that I have briefly illustrated, and which has been published a few months ago in *Angew. Chem. Int. Ed.* [4], is different from your system in that we do not need light of two different colors to perform the experiment. Our system is, in fact, a "self-poisoning" system in the presence of oxygen.

3. References

1. Sixl, H. (1984) Spectroscopy of the intermediate states of the solid state polymerization reaction in diacetylene crystals, *Adv. Polymer. Sci.* **63**, 49-90.
2. Hartl, W. and Schwoerer, M. (1982) ENDOR of quintet-state bicarbenes in diacetylene single crystals, *Chem. Phys.*, **69**, 443-457.
3. Walz, J., Ulrich, K., Port, H., Wolf, H. C., Wonner, J., and Effenberger, F. (1993) Fulgides as switches for intramolecular energy transfer, *Chem. Phys. Lett.* **213**, 321-324.
4. Belser, P., Dux, R., Baak, M., DeCola, L., and Balzani, V. (1995) Electronic energy transfer in a supramolecular species containing the $[Ru(bpy)_3^{2+}]$, $[Os(bpy)_3^{2+}]$, and anthracene chromophoric units, *Angew. Chem. Int. Ed.* **34**, 595-598.

SUPRAMOLECULAR ARCHITECTURE IN LANGMUIR-BLODGETT FILMS

FRANCK ARMAND, JEAN-PHILIPPE BOURGOIN,
FRANÇOIS DOUBLET, SANDRINE ISZ,
MARIA-VICTORIA MARTINEZ, SERGE PALACIN,
HENRI PEREZ, ANNIE RUAUDEL-TEIXIER
Service de Chimie Moléculaire
Commissariat à l'Energie Atomique
Centre d'Etudes de Saclay
Gif sur Yvette
France

ABSTRACT. Two general strategies for building and organizing parallel conducting chains are developed. Both are based on the preorganization of precursors of a conjugated polymer by the Langmuir-Blodgett technique. Preliminary results are given.

1. Introduction

Although known for several decades, organic conductors have not yet fulfilled the technical requirements for massive industrial applications. The main reason for this failure probably lies in the low-dimensionality of these conductors: conjugated polymers, as charge-transfer salts, create linear conducting pathways. They are thus highly sensitive to defects (one single defect is able to block the conductivity of a whole conducting route) and their conductivity can be limited for intrinsic reasons, such as the Peierls transition. Answers to these weaknesses are to be found in lower defect densities and higher dimensionalities.

Numerous improvements in the stability and the conductivity of conducting polymers have already been achieved. One of the most impressive one was proposed by MacDiarmid, who showed that the conductivity of doped polyacetylene can be dramatically enhanced by making the conducting fibers parallel [1]. Making the conducting paths perfectly parallel and in close contact enables the charge carriers to jump easily from a chain to a neighbor, in order to avoid a defect. Unfortunately, polyacetylene is not stable enough to be used in practical applications. Stretching polymer fibers gives, moreover, a rough and limited improvement

J. Michl (ed.), Modular Chemistry, 141–152.

in the order of the conducting chains. Synthesizing well-ordered conducting polymers from scratch would be a major breakthrough in the field.

We are currently investigating different strategies to orient and organize one-dimensional organic conductors *by synthesizing them from preoriented precursors*. The Langmuir-Blodgett (LB) technique is used as an engineering tool to build highly ordered conducting films where the one-dimensional conducting pathways are parallel.

2. Strategy

2.1 CONDUCTING PATHWAYS PARALLEL TO THE SUBSTRATE

One-dimensional polymers [2] and mesogenic molecules [3] easily exhibit in-plane anisotropy in LB films. This phenomenon arises from the combination of long-range interactions between the amphiphilic molecules and the flow effect occurring during the transfer of the monolayer from the air-water interface to a solid substrate [4]. Polyphthalocyanines, polyglutamates, pyrylium salts or tetrafunctionalized monomeric phthalocyanines all form columns which are found more or less parallel to the direction of the transfer [2,3,5].

If a similar in-plane orientation could be obtained from molecules bearing precursors of a linear conjugated polymer on their periphery, a solid-state polymerization of the oriented precursors should lead to the formation of conjugated chains parallel to the main axis of the solid. (Figure 1).

This strategy requires: (i) a reliable self-organizing system A that forms in-plane oriented columns when deposited onto a solid substrate by the LB technique; this system will be used as a scaffold to orient the precursors of the future conjugated polymer; (ii) molecular precursors B of a conjugated polymer, grafted along the backbone of the columnar phase formed by the self-organizing system A; the precursors can be grafted by covalent or ionic bonds to the system A; (iii) and a solid state polymerization, performed after transfer of the complete system A+B onto a solid substrate; the precursors B are supposed to be oriented by the self-organizing system A, leading to an in-plane oriented conjugated polymer. This final polymer should exhibit parallel conducting pathways. The sections 3 and 4 are only devoted to the selection of the self-organizing system A. The grafting of the precursors B and their solid state polymerization will only be achieved when the best self-organizing system A is selected.

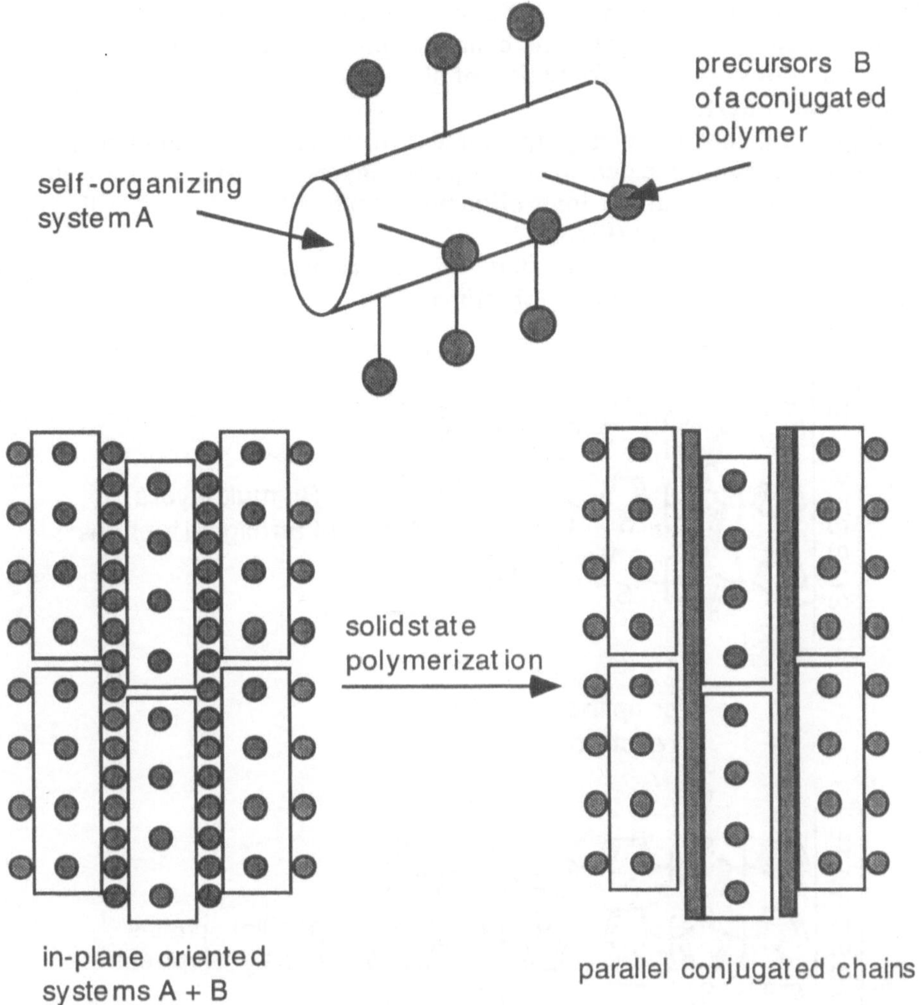

Figure 1: In situ polymerization of pre-oriented monomers (only one monolayer is represented, for clarity; the interdigitation of the precursors is hypothetical)

2.2 CONDUCTING PATHWAYS PERPENDICULAR TO THE SUBSTRATE

The LB technique is well suited for organizing amphiphilic molecules perpendicular to a solid substrate. The resulting film, analogous to the natural membranes, is stabilized by van der Waals interactions between adjacent aliphatic chains. Fully conjugated molecules, such as oligothiophenes, exhibit the same behavior, provided the conjugated chain is long enough to ensure the formation of stable two-dimensional aggregates at the air-water interface.

Oligothiophenes oriented perpendicular to the substrate could be coupled with their neighbors located in the adjacent layers in order to form longer oligomers. These resulting long oligomers would all be aligned in the same direction *by construction* (Figure 2).

Section 5 is devoted to the study of the orientation of oligothiophenes at the air-water interface and in LB films.

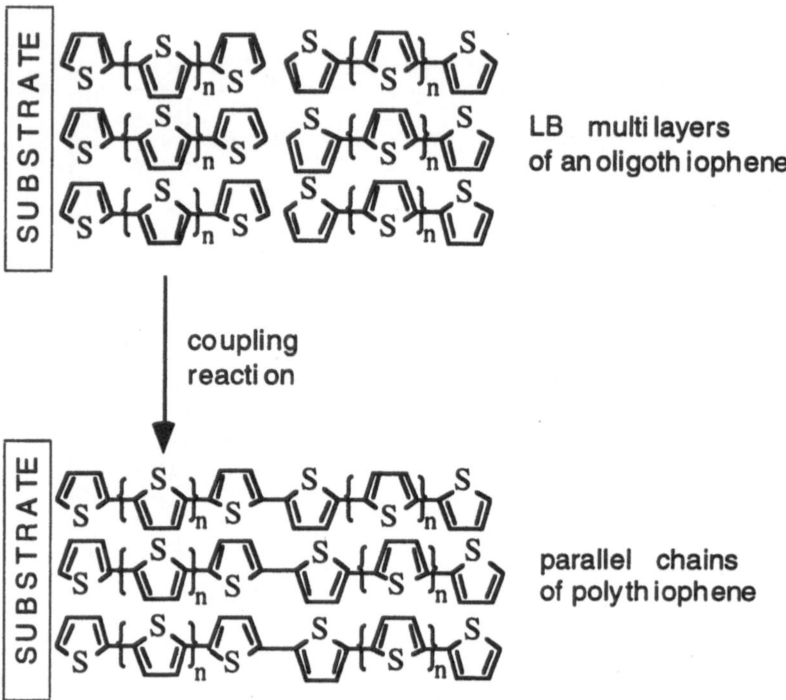

Figure 2: In situ polymerization of oriented oligothiophenes (only two layers are schematized for clarity)

3. Substituted polysiloxophthalocyanines do not exhibit a high in-plane anisotropy

Dichroic ratios up to 2.3 have been obtained with polysiloxophthalocyanines substituted with aliphatic chains on the periphery of the macrocycles [6].

We synthesized polysiloxophtalocyanines bearing functional groups such as pyridine rings [7]. These functional groups were intended to be used as linkers for the precursors of the final conducting polymer.

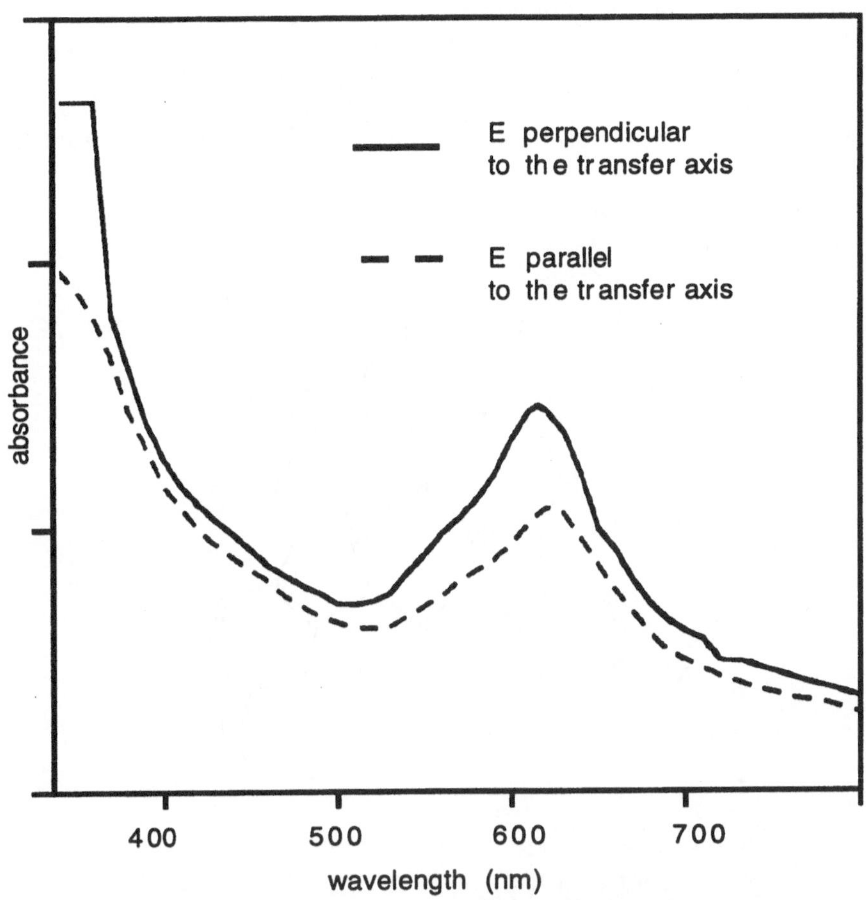

Figure 3: In-plane anisotropy in LB films of tetrakis(pyridyloxy)polysiloxophthalocyanine

Figure 3 exhibits the linear dichroism spectra in the UV-visible range of LB films of the polysiloxophthalocyanine A_1. The dichroic ratio was low (order parameter 0.33) and no improvement of the in-plane anisotropy was observed upon annealing.

This lack of high in-plane anisotropy may arise from the unsuitable symmetry of the two-dimensional phase formed at the air-water interface and the lack of mobility of the bulky oligomers. This system does not seem suitable as a scaffold to organize precursors of a conjugated polymer.

4. Mesogenic phtalocyanines exhibit high in-plane anisotropy

Among the molecular systems that exhibit high in-plane anisotropy in LB films, tetrasubstituted phthalocyanines [5] appear particularly attractive: the chain length, the linker between the macrocycle and the chains, the central metallic ion can all be easily modified.

We have systematically investigated the influence of these parameters on the in-plane orientation. Figure 4 gives examples of the molecules we have prepared and studied. The present paper will only focus on the best results, obtained with the copper tetrakis(*n*-hexylcarboxylate)phthalocyanine A_2.

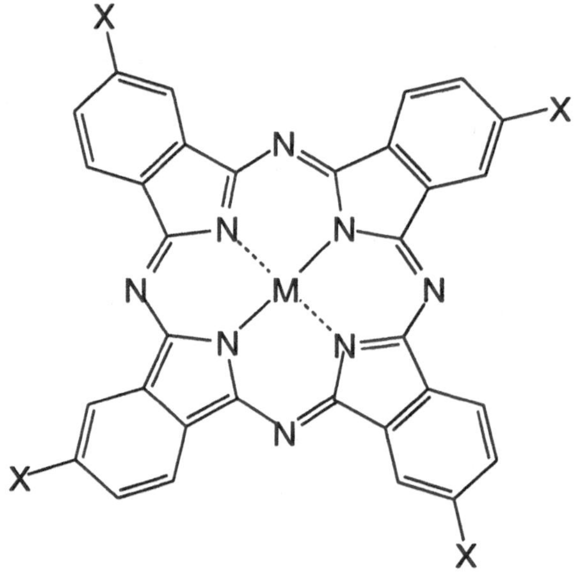

$X = COOC_6H_{13}$
$X = CON(C_8H_{17})_2$
$X = OC_5H_4N$

$M = Cu, Ni$

Figure 4: Tetrasubstituted phthalocyanines (only the symmetrical isomer is represented)

The compression isotherm of A_2 exhibited a plateau at 16 mN. m^{-1}. The molecular area was 80 Å^2 at the transfer pressure, just before the plateau. This area was compatible with a macrocycle standing on its edge at the air-water interface. The transfer onto solid substrates was regular and reproducible, with a transfer ratio close to 1.

Figure 5 gives several topographic AFM images recorded in the tapping mode on LB films made from A_2, varying the number of layers. The dark areas were regions where the last monolayer was missing, as shown by the depth profiles (not shown).

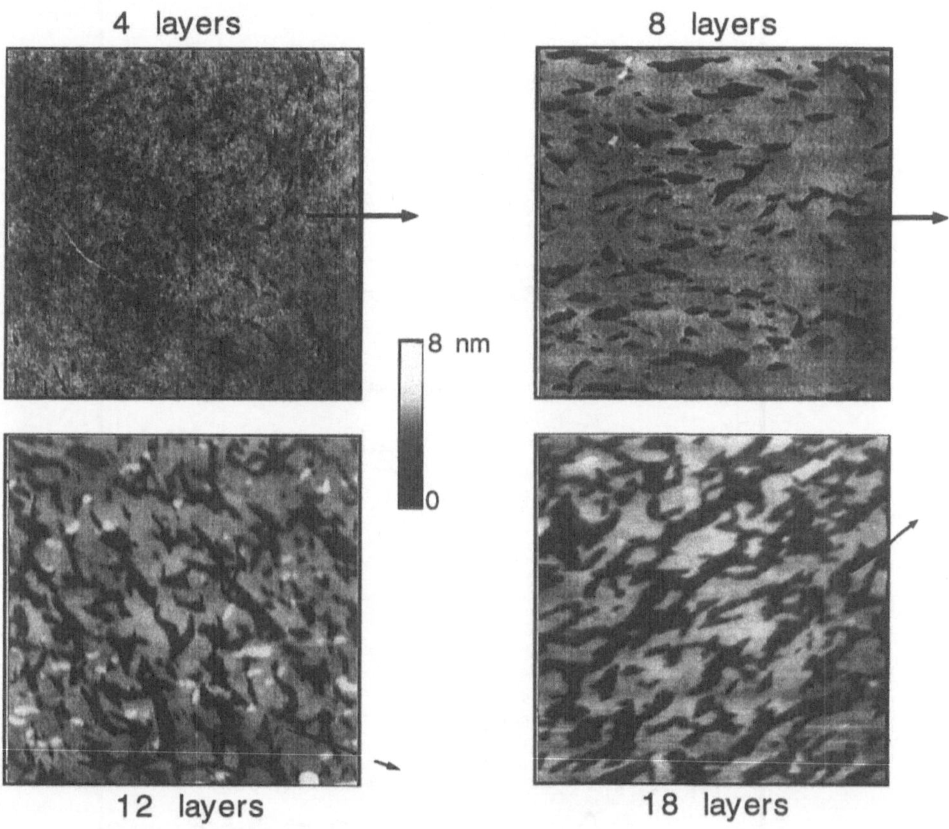

Figure 5: Topographic AFM imaging (2 X 2 μm) of the top layer of LB multilayers deposited on silicon. The imaging was made in the tapping mode on a Nanoscope III apparatus

The last layer imaged by AFM appeared less and less complete as the total number of layers increased. This behavior is consistent with a spontaneous solid state reorganization of the lamellar film that occurs just after the transfer: as the transfer ratio is lower than 1, each deposited layer is slightly incomplete. The next layer to be deposited fills the holes left by the unperfect deposition in the last deposited layer. If this process goes to completion, all the underlying layers should be complete while the top layer should exhibit numerous defects. The number of defects in the top layer

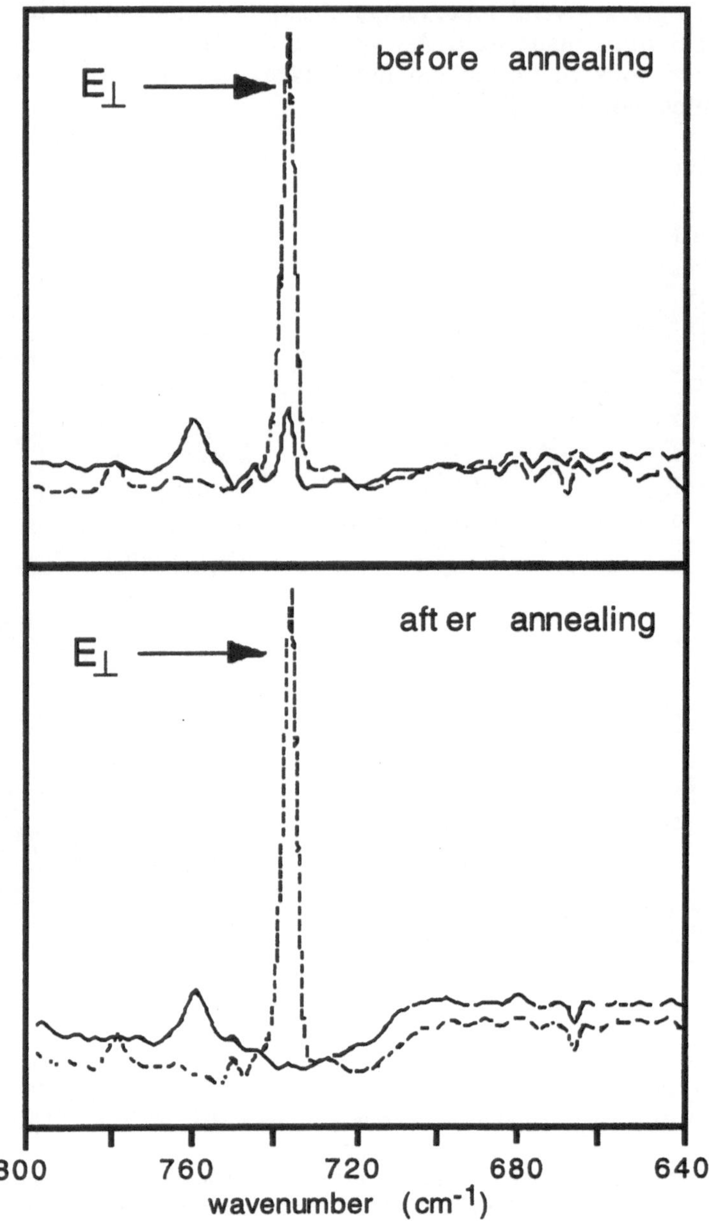

Figure 6: Linear dichroism spectra recorded in the infrared range on LB multilayers of copper tetrakis (*n*-hexylcarboxylate)phthalocyanine. (a) before annealing. (b) after annealing during 2 h at 200°C. The scale is the same for both spectra

should increase with the total number of layers, as observed by AFM. This result shows that the copper tetrakis(*n*-hexylcarboxylate)phthalocyanine is able to move easily within the lamellar structure. It is also noteworthy that the defects observed in the top layer by AFM exhibit a preferential orientation parallel to the transfer axis (given by the arrow).

Linear dichroism performed in the infrared range on LB films of A_2 confirmed that the macrocycles were standing on the edge, with their molecular plane perpendicular to the plane of the substrate. The in-plane anisotropy measurements were derived from the spectra given in Figure 6a. The order parameter was 0.94, which indicated that the macrocycles were well oriented perpendicular to the transfer axis. Annealing the sample for 2 hours at 200°C dramatically improved that anisotropy, as shown in Figure 6b. The order parameter after annealing was almost equal to 1. The macrocycles were then *perfectly* organized perpendicular to the transfer axis. An independent X-ray study has shown that the phthalocyanines form columnar mesophases crystallized in a square lattice [8]. AFM measurements on the top layer after annealing (not shown) confirmed that the defects were also oriented parallel to the transfer axis.

Copper tetrakis(*n*-hexylcarboxylate)phthalocyanine appear to be a suitable system A to organize precursors of a conjugate polymer. Work is now in progress to graft the precursors on the periphery of the macrocycles.

5. Thiophene-based oligomers can be organized by the Langmuir-Blodgett technique

We have synthesized several non-substituted and mono-substituted oligothiophenes. The longer ones (sexithiophene, quinquethiophene, 1-quaterthiophene carboxylic acid) were successfully transferred onto solid substrates, while the shorter ones (quaterthiophene, 1-terthiophenecarboxylic acid) did not give monolayers at the air-water interface.

The structure of the monolayer of sexithiophene (S_6) was extensively studied, in collaboration with Prof. Meir Lahav (Weizmann Institute of Science, Israel). Cryo Transmission Electron Microscopy [9] and AFM measurements performed on monolayers deposited either on a carbon-collodion grid or on mica indicated that the S_6 molecules were standing vertically with respect to the air-water interface and the substrate. The crystal structure of the monolayer was obtained from the electron diffraction patterns and fitted nicely with the 3-dimensional crystal structure of S_6 [10]. Topographic AFM imaging confirmed the thickness of the deposited monolayer (26Å). That figure is in perfect agreement with the length of the S_6 molecule. These results will be published in detail elsewhere.

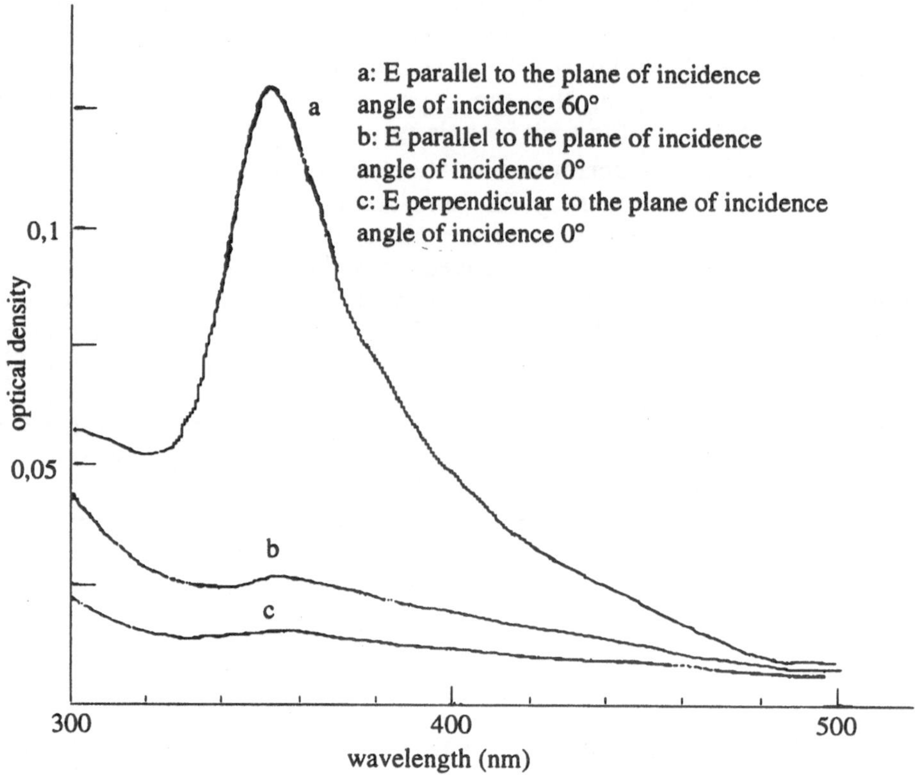

Figure 7: Linear dichroism spectra recorded in the UV-visible range on LB multilayers of sexithiophene. (E is the electric field of the incident light)

The orientation of the S6 molecules was confirmed by linear dichroism studies performed in the UV-visible range on LB multilayers of S6. Figure 7 gives three spectra recorded for different incidences and orientations of the electric field with respect to the normal to the substrate. No in-plane anisotropy was observed. The order parameter for the out-of-plane anisotropy was 0.95, indicating that the S6 molecules were almost perpendicular to the substrate, as already shown by electron diffraction and AFM measurements. The hypsochromic shift observed on the absorption band of the p-p* transition (l_{max} = 360 nm) with respect to the spectrum of the solution (l_{max} = 430 nm) confirmed the formation of highly ordered two-dimensional aggregates within the LB films. Similar results were obtained with quinquethiophene (order parameter 0.92) and with 1-quaterthiophene carboxylic acid (order parameter 0.82).

Preliminary attempts to link the oligothiophenes from one layer to the adjacent one in the solid lamellar structure were encouraging: significant amounts of S_{12} were observed in the mass spectrum of LB multilayers after performing cyclic voltammetry experiments using the LB multilayers as a modified electrode. It seemed that the radical cation S_6^+ formed during the first oxidation cycles was able to react with neighboring molecules to form the dimer S_{12}. As the oxidative potential was not kept constant, the final amount of dimer was low. These results will be detailed elsewhere.

6. Conclusion

We have shown that the LB technique allows the building of molecular systems that can serve as organizing scaffolds for the formation of well ordered conjugated polymers. Mesogenic tetrasubstituted phthalocyanines and oligothiophenes are good candidates for the direct formation of parallel conjugated pathways in the solid state. No polymerization or preorganized precursors have, however, yet been attempted. Thus, no comparison between the present strategy and other techniques (stretching of polymer fibers [1], chemical vapor deposition on stretched substrates [11]) can be made so far.

7. References

1. Theophilou, N., Swanson, D.B., MacDiarmid, A.G., Chakraborty, A., Javadi, R.P., McCall, R.P., Treat, S.P., Zuo, F. and Epstein, A.J. (1989) Highly conducting polyacetylene, *Synth. Met.* **28**, D35-D42.
2. Orthmann, E. and Wegner, G. (1986) Preparation of ultrathin layers of molecularly controlled architecture from polymeric phthalocyanines by the Langmuir-Blodgett technique, *Angew. Chem. Int. Ed. Engl.* **25**, 1105-1107.
3. Albouy, P.A., Vandevyver, M., Perez, X., Ecoffet, C., Markovitsi, D., Veber, M., Jallabert, C. and Strzelecka, H. Liquid crystalline order in Langmuir-Blodgett films of a disk-shaped heteroatomic salt as determined by X-ray diffraction, *Langmuir* **8**, 2262-2268.
4. Minari, N., Ikegami, K., Kuroda, S., Saito, K., Saito, M. and Sugi, M. (1988) Origin of the in-plane anisotropy in Langmuir-Blodgett films, *J. Phys. Soc. Jpn* **58**, 222-231.
5. Ogawa, K., Kinoshita, S., Yonehara, H., Nakahara, H. and Fukuda, K. (1989) Highly ordered monolayer assemblies of phthalocyanine derivatives, *J. Chem. Soc. Chem. Commun.* **8**,477-479.
6. Sauer, T., Arndt, T., Batchelder, D.N., Kalachev, A.A. and Wegner, G. (1990) The structure of Langmuir-Blodgett films from substituted phthalocyaninato-polysiloxanes, *Thin Solid Films* **187**, 357-374.
7. Martinez-Diaz, M.V., Torres, T. and Ruaudel-Teixier, A. (submitted) Functionnalized phthalocyaninato-polysiloxanes in a Langmuir-Blodgett supramolecular architecture, Seventh International Conference on Organized Molecular Films, Paper n° P-2.16.
8. Albouy, P.A. (1994) Structure of Langmuir-Blodgett films of copper phthalocyanine derivatives, *J. Phys. Chem.* **98**, 8543-8548.

9. Porzio, W., Destri, S., Mascherpa, M., Rossini, S.and Brückner, S. (1993) Structural aspects of oligothienyls from X-ray powder diffraction, *Synth. Met.* **55-57**, 408-413.
10. Weissbuch, I., Majewski, J., Margulis, L., Lahav, M. and Leiserowitz, L. (1993) Application of cryo-electron microscopy to self-assembled mixed monolayers for studying crystallinity and twinning, *J. Phys. Chem.* **97**, 8692-8695.
11. Lang, P., Hajlaoui, R., Garnier, F., Desbat, B., Buffeteau, T., Horowitz, G. and Yassar, A. (1995) IR spectroscopy evidence for a substrate-dependent organization of sexithiophene thin films vacuum evaporated onto SiH/Si and SiO_2/Si, *J. Phys. Chem.* **99**, 5492-5499.

CRYSTAL ENGINEERING OF IONIC SOLIDS

Polymorphism, Crystallography, Inter-Ring Interactions and C—H Hydrogen Bonding

CHRISTER B. AAKERÖY, TIMOTHY A. EVANS AND KENNETH R. SEDDON
School of Chemistry, The Queen's University of Belfast,
Stranmillis Road, BELFAST, BT9 5AG, Northern Ireland
WWW: http://www.ch.qub.ac.uk/krs/krs.html; e-mail: k.seddon@qub.ac.uk

ABSTRACT. This paper considers aspects of crystal engineering which are relevant to the design of salts as functional solids, with specific reference to the design of novel non-centrosymmetric materials for second harmonic generation. In addition, some aspects of polymorphism, current practice in the field of single-crystal X-ray diffraction, inter-ring interactions, and C—H hydrogen bonding in the salts of heterocyclic cations are discussed.

1. Introduction

In 1988, the editor of *Nature*, John Maddox, wrote:

> ***"One of the continuing scandals in the physical sciences is that it remains impossible to predict the structure of even the simplest crystalline solids from a knowledge of their chemical composition"***

This was true then, and is still largely true today, but this simple and accurate summary of the state of ignorance of the chemical community underlines the need for, and in essence defines, the field of crystal engineering [1,2]. This is a field in its formative years; it is at the interface between a number of crucial, but fundamentally different, areas, and has all the challenge and excitement expected of interdisciplinary research. We present, then, our thoughts on the crystal engineer's tool kit, and consider its effectiveness for granting topological control [3,4] over crystalline form, and hence control over such crucial physical phenomena as melting point, optical properties, thermal stability, solubility, colour, conductivity, crystal habit, and mechanical strength. This is an area of fundamental and pivotal interest to industry and academia, but it does not yet attract the full attention that it deserves.

The purpose of this paper is not to present a philosophy of crystal engineering - space does not permit this, and an up-to-date account of some recent developments can be found elsewhere [5]. Instead, the intention is to highlight, using examples taken from our

J. Michl (ed.), Modular Chemistry, 153–162.

own work, in particular, some of the applications of this new field, and indicate some of the possibilities which open up when chemists turns their attention from the intramolecular bond (with which they are familiar) to the world of intermolecular and interionic interactions and supramolecular chemistry [6], where the rule book remains to be written.

2. The Design of Novel Materials for Second Harmonic Generation

2.1 PRACTICAL CONSIDERATIONS

Recent years have seen considerable interest in nonlinear optical materials, including those capable of second harmonic generation (SHG)[¶] [7]. Such materials, and the basis of their properties, have attracted both academic and commercial investigators, and their applications include:

- Controlling the flow of optical data
- Advanced telecommunications
- Optical computing
- Improving densities in optical storage media
- Weapons guidance systems
- Laser defence systems

The existence of the phenomenon of SHG is determined by the crystal symmetry of the medium: SHG can only occur in a non-centrosymmetric medium [8]. However, this is only an on-off switch: the efficiency (*i.e.* intensity) of the SHG effect is governed, ***inter alia***, by the second-order coefficient of the polarizability of the medium, $\chi^{(2)}$, which (for molecular solids) is related to second-order coefficient of the polarizability of the molecule, β. Thus, provided the crystal is non-centrosymmetric (*i.e.* if $\chi^{(2)} \neq 0$), then the larger the value of the molecular coefficient, β, the more intense the emitted SHG radiation, and hence the more effective the material, provided that the molecules are organized within the crystal structure in such a way that the dipoles reinforce, rather than oppose, each other. Large values of β are related to the dipole moment of the molecule (or, to be more precise, to the difference in dipole moment between the ground state and the first excited electronic state), and hence efficient molecular SHG materials frequently usually have large ground state dipole moments. It is received wisdom in this field that molecules with a large dipole

---

¶ Second harmonic generation is the phenomenon of doubling the frequency of laser light, induced by passage of light through a non-centrosymmetric medium. Thus invisible infrared radiation of wavelength 1064 nm will be frequency-doubled to green 532 nm light.

moment tend to crystallize in centrosymmetric space groups,[¶] and therefore that attempts to create high-β materials are bedevilled by a tendency of these materials to be inactive in their normal crystalline form, because $\chi^{(2)}$ is zero. Irrespective of whether this folklore is correct or not [9], it is evident that merely preparing molecules with high β values does not guarantee that they will form SHG active crystals. Moreover, molecules with a high β value usually possess a small HOMO-LUMO gap, and hence are COLOURED. Indeed, large β values usually imply that the crystals are intensely coloured, and hence that they will absorb any visible light generated by the frequency doubling of infrared radiation. It follows that these materials are, for applications where the required radiation is visible, about as much use as a glass hammer! Despite these empirical observations, the open literature upon the synthesis of novel materials for SHG is dominated by papers for which the ***raison d'être*** is the synthesis of high β-value molecules, to the exclusion of all other practical considerations. In any practical device in a fully integrated optoelectronic circuit, the construction temperature is likely to approach 320 °C [10], and the properties of any useful materials **MUST** include:

- Thermal stability
- Involatility
- Optical clarity
- Lack of colour
- Mechanical strength
- Suitable crystal habit

and last, and **DEFINITELY** least:

- Large β values

Although for more specific uses (especially non-integrated applications), one or more of these restrictions may ne relaxed, the cynosure of designing new materials for SHG must be to prepare optically clear, colourless, thermally robust, mechanically strong, non-centrosymmetric crystals, with the value of $\chi^{(2)}$ (and hence β) being relatively unimportant, as long as it is not zero. Indeed, the current commercially important materials, potassium dihydrogen phosphate and lithium niobate(V), have risibly small values of $\chi^{(2)}$ compared with the least efficient of the organic materials available, and yet these are the industrial materials of choice.

¶ In a recent paper, Whitesell [9] has shown, statistically, using the structures in the Cambridge Structural Database, that there is an equal probability that molecules with large dipole moments will crystallize in polar (centrosymmetric) or non-polar (non-cetrosymmetric) space groups.

The question thus remains - why are so few synthetic papers concerned with designing materials with the **NECESSARY** properties for incorporating into practical devices? The answer is simple - because it is a much more difficult and diffuse problem to design materials which match the needs of industry than it is to maximize β. Nevertheless, crystal engineering, as will be demonstrated below, does offer solutions to these hitherto difficult problems. It is perhaps not as enticing to design colourless materials with high lattice energies as it is to be able to claim a molecule with the world's largest β value, but it is significantly more important.

2.2 THE CRYSTAL ENGINEER'S TOOL KIT

Before dealing with the problems raised in Section 2.1, it perhaps worth examining the tools available in the crystal engineer's tool kit. These include:

1. SYNTHETIC VECTORS
   - Hydrogen bonding
   - Inter-ring interactions (the misleadingly named π–π stacking)
   - Covalent bonds
   - Coulombic forces
2. COMPUTATIONAL CHEMISTRY
3. VISUALIZATION TOOLS
4. CAMBRIDGE STRUCTURAL DATABASE

These are, perhaps, worth examining in a little more detail.

2.2.1 *Synthetic Vectors*

Synthetic vectors are to the crystal engineer what covalent bonds are to the molecular chemist. They are the intermolecular or interionic cohesive forces which hold the crystalline structure together. These can include many varied forces, encompassing hydrogen bonds (see Section 5) [1], inter-ring interactions (see Section 4), conventional covalent σ-bonds,[¶] Coulombic forces and dipole-dipole interactions. The goal of the crystal engineer is to apply these synthetic vectors to the problems of the control of functional crystal structures with the same degree of success as the synthetic chemist designs and prepares funtionalized molecules. This goal lies a long time in the future, but the last decade (as so well illustrated by other papers in this book) has provided real and significant insights, and if we do not yet have rules, at least we now understand some of the guidelines.

¶ See the articles by Mike Zaworotko and by Jeffrey Moore, elsewhere in this volume.

2.2.2 *Computational Chemistry*

Just as computational chemistry is growing in importance in conventional chemical synthesis (especially drug design), then it has achieved a parallel, if not more significant, *rôle* in crystal engineering. The Cerius, CAChe and Spartan packages offer the ability to perform molecular mechanics, semi-empirical (especially AM1 and PM3), *ab initio* Hartree-Fock, and *ab initio* density functional calculations. Moreover, the results of these calculations can be visualized in three-dimensions - not only the structure of the molecules, ions or crystals, but also the molecular orbitals of the system (including the HOMO and the LUMO) and the total electron density and the charge and spin distribution.[¶] In addition, *inter alia*, lattice energy calculations are available, as are lattice optimization and crystal morphology packages. The use of computational techniques is central to the crystal engineer, and to ignore them would be as short-sighted as for the synthetic chemist to ignore NMR spectroscopy and X-ray crystallography!

2.2.3 *Visualization Tools*

Structural visualization is a rather different use of computer techniques than computational chemistry (although the uninitiated tyro, in our personal experience, regards the only worthwhile aim of computational chemistry to be the production of high quality graphics for publication!) - it enables the graduate and postdoctoral student not only to gain a quite remarkable insight into three-dimensional architecture and structure, in spatial terms, but also to enable structure-determining interactions to be identified and appreciated.

2.2.4 *Cambridge Structural Database*

The Cambridge Structural Database[§] offers a unique resource to the crystal engineer. Now available on CD-ROM, at trivial cost to academic institutions, it enables all organic and many inorganic (as long as they contain carbon) structures which are known to be rapidly scanned for structural or functional features, and to be imported in to, say, Cerius for visualization and detailed examination. Apart from the incalculable value of this resource for reevaluating known structures (see, for example, Section 5), it enables structures to be readily compared, using Etter's topological analysis [3,4], and many hypotheses to be tested using extant data, without the need for new structural determinations {*e.g.* [11]}.

2.3 CRYSTAL ENGINEERING AND SHG-ACTIVE MATERIALS

We have reported elsewhere on the development of several series of novel SHG-active salts with a wide range of desirable chemical and physical properties [12-15], using the strong and directional hydrogen-bonding between ions [16] as the principal synthetic vector, and

---

¶ See the article by Olivier Kahn, elsewhere in this volume.

§ Details available from:
Dr. Olga Kennard, Cambridge Structural Database, Cambridge Crystallographic Data Centre, CAMBRIDGE CB2 1EW, U.K.

it is not the purpose of a NATO ARW presentation to dwell on published material. However, it is worth, perhaps, expanding upon the crystal engineering principles which underpin this work, as they have a greater generality than the specific published targets. As early as 1989 [12], we proposed that it would be possible to use salts, in preference to neutral molecules, to enhance the thermal stability and eliminate the vapour pressure of novel materials for SHG applications. Of greater significance, however, we also proposed that it should be possible to use salts to "factorize" the properties of a material, so that the anion and the cation would have different, discrete functions. For example, in our own work on dihydrogen phosphates [12], hydrogen tartrates [13,14], hydrogen malates [17] and hydroxybenzoates [15], the anion is used to determine the crystal architecture, and impart thermal and structural stability, whilst the cation is used to "carry" the SHG activity.¶ In addition to the intellectual satisfaction of separating function and associating it with identifiable and discrete chemical units, the preparation of simple salts of this type is **MUCH** easier than the covalent synthesis of new materials in which it would be necessary to combine all the functions in one molecule. The use of salts, then, represents not just practical and synthetic convenience, but also represents significant economic, and hence commercial, advantage.

3. Polymorphism

3.1 INTRODUCTION

Polymorphism is a unique feature of the solid state, being a form of isomerism that is not conventionally considered by synthetic chemists. A recent court case in the USA has focussed attention on polymorphism, in a way that only multi-billion dollar litigation can! The true extent of polymorphism among crystalline materials is very difficult to assess (*vide infra*). However, as it has been more often encountered in areas of research where structure is of paramount importance, we are in strong agreement with McCrone [18] who stated (in 1965):

> ***"[...] every compound has different polymorphic forms and that, in general, the number of forms known for a given compound is proportional to the time and money spent in research on that compound."***

It is generally accepted that differences in lattice energies between different polymorphs are in the region 5-20 kJ mol^{-1}, and, especially for ionic compounds, these differences are

---

¶ In a spectacular and elegant independent demonstration of this principle, reported elsewhere in this volume, Olivier Kahn has factorized the properties of a salt such that it contains ferromagnetic anionic layers, and luminescent cationic layers: cooling the material below its Curie temperature results in a magneto-optical modification of its emission spectrum.

small compared to the total lattice energies of such materials. Thus, different methods of recrystallization are often sufficient to induce the formation of a new structural form (polymorph). However, as will be discussed in the following Section, the whole subject of crystal engineering in general, and polymorphism in particular, has been afflicted by the routine and normal practice of X-ray crystallographers (which is rather sad to have to report in 1995, which marks the centenary of the discovery of X-rays by Röntgen).

3.2 ARE CRYSTALLOGRAPHERS **BAD** SCIENTISTS?

The normal practice in determining a crystal or molecular structure is for the X-ray crystallographer to select one, or more, "perfect" crystals from the batch provided by the synthetic chemist. The process of selecting a single perfect crystal, and claiming (implicitly) it to be "characteristic" of the bulk material, is thus **scientifically indefensible**, as the selected single crystal is, **BY DEFINITION**, atypical of the bulk material from which it was removed. Thus the determined structure can only be ascribed to the single crystal for which it was determined, and **NOT** to the whole synthetic batch of material prepared. It may, or may not, represent the same compound as the bulk material. It may, for example, be a recrystallized minor impurity, or represent a different solvate, or a polymorph of the bulk material.

The above situation would not be so lamentable if there was not an easy remedy - but there is! It merely requires that the powder XRD pattern of the BULK sample should always be recorded, and compared with the simulated powder pattern from the single crystal data. This would provide absolute evidence as to whether or not the determined structure is that of the bulk material, and the whole procedure would take less than an hour. It is difficult to imagine another field of scientific endeavour where such appallingly lax procedures should be adopted as the norm, and be almost universally adopted, when such a simple remedy is available. It is quite incredible that the practice still persists, and we wish to suggest that it should, under normal circumstances, be an absolute prerequisite for publication in reputable journals that evidence be provided that the single crystal and the bulk material are crystallographically identical. The question of structural purity is, in many areas of chemistry, as important as that of chemical purity.

4. Inter-Ring Interactions

In Section 2.2.1, inter-ring interactions (frequently given the misleading term "π-π" interactions) were cited as synthetic vectors, alongside hydrogen bonds, which could lead to structural control in the solid state. The structural manifestation of inter-ring interactions is ring stacking, and this is an extremely common phenomenon found in the crystal structures of materials containing aromatic rings, particularly when these are functionalized. An important question which remains to be answered is: how strong are these interactions compared to, say, hydrogen bonds, and are they powerful enough to grant structural control? In previously unpublished work [19], we have been able do demonstrate computationally, using a simple electrostatic model based upon the Coulombic forces between localized computed atomic charges, that the inter-ring interactions are of the order of 20-30 kJ mol^{-1}, the same order of magnitude as a weak hydrogen bond. It is thus entirely reasonable to consider these interactions as synthetic vectors.

5. C—H Hydrogen Bonds Involving Heterocyclic Cations

Although there is now an increasing awareness of the existence of C—H···X (X = F, Cl, Br or I) hydrogen bonds in neutral organic molecules [20,21], there have been surprisingly few studies of C—H···X$^-$ interactions in organic salts. We have demonstrated the existence of extensive hydrogen bonding between aromatic C—H groups and halide ions, C—H···X, within a series of imidazolium salts [22-24], and in two polymorphs of pyridinium chloride [11]. At the time, we concluded [11]:

> ***"The existence of aromatic C—H···X hydrogen bonds in ionic compounds containing charged heterocyclic systems is very likely to be a regularly occurring phenomenon which needs to be further analyzed in a systematic and stringent fashion in order to rationalize the structural behaviour displayed by such ions."***

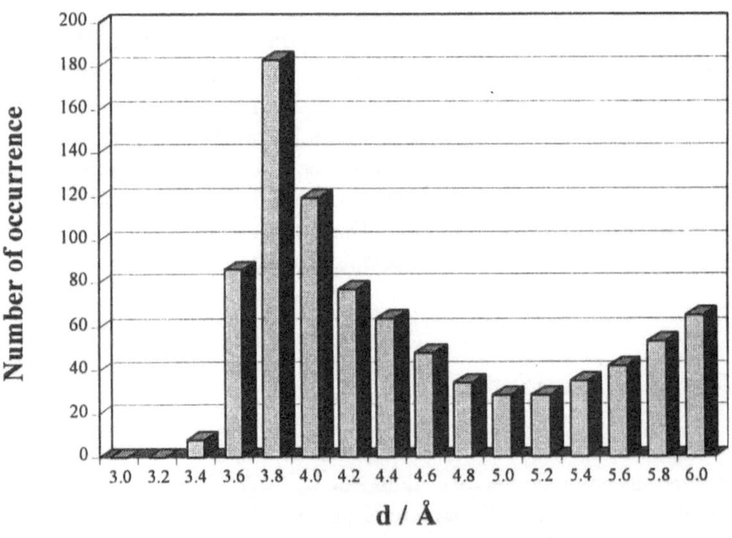

Figure 1: *The distribution of C-H...Cl contact distances between heterocyclic cations and chloride-containing anions, using data from the CSD*

By writing special code to supplement the normal search and display procedures which are supplied with the Cambridge Structural Database (CSD), we have now performed this further analysis, and demonstrated that C—H···X^- interactions in organic salts of charged heterocyclic systems are extremely common, but are precluded by adjacent bulky substituents (*e.g.* -CMe_3 or -$SiMe_3$) [25]. Figure 1 illustrates the distribution of C—H···X^- bond lengths for all well-defined structures in the CSD containing heterocyclic cations and chloride-containing anions. We have similar data for oxygen-, nitrogen-, sulfur-, fluoride-, bromide- and iodide-containing anions, and all show a strong propensity towards hydrogen-bonding. The tetramethylammonium cation also shows significant C—H···X hydrogen bonds, too. Thus, it is our contention that, although weak, C—H···X hydrogen bonds are present in many crystals, and that their presence has been largely unsuspected. As, say, the pyridinium cation is capable of forming hydrogen bonds with **ALL** its ring protons, then the summed effect of C—H···X hydrogen bonding is significant, despite each bond being individually weak. We cannot, as yet, associate an energy with these bonds, and this will be the subject of future research.

Acknowledgements
We are indebted to the BP Venture Research Unit, Unilever, and the Queen's University of Belfast for their generous funding of this project, to the EPSRC and Royal Academy of Engineering for the award of a Clean Technology Fellowship (to K.R.S.), and to Prof. J. Michl for his invitation allowing one of us (K.R.S.) to participate in this NATO ARW.

References

1. Aakeröy, C.B. and Seddon, K.R. (1993) The hydrogen bond and crystal engineering, *Chem. Soc. Rev.* **22**, 397-407.
2. Desiraju, G.R. (1989) *Crystal Engineering: The Design of Organic Solids*, Elsevier, Amsterdam.
3. Etter, M.C. (1990) Encoding and decoding hydrogen-bond patterns of organic compounds, *Acc. Chem. Res.* **23**, 120-126.
4. Etter, M.C., Macdonald, J.C. and Bernstein, J. (1990) Graph-set analysis of hydrogen-bond patterns in organic crystals, *Acta Crystallogr., Sect. B* **46**, 256-262.
5. Desiraju, G.R. (ed.) (1996) *The Crystal as a Supramolecular Entity*, Wiley, Chichester.
6. Lehn, J.-M. (1995) *Supramolecular Chemistry: Concepts and Perspectives*, VCH, Weinheim.
7. Chemla, D.S. and Zyss, J. (eds.) (1987) *Nonlinear Optical Properties of Organic Molecules and Crystals*, Vols. 1 and 2, Academic Press, New York.
8. Marder, S.R., Sohn, J.E. and Stucky, G.D. (eds.) (1991) *Materials for Nonlinear Optics, ACS Symp. Ser. 455*, 455, American Chemical Society, Washington D.C.
9. Whitesell, J.K., Davis, R.E., Saunders, L.L., Wilson, R.J. and Feagins, J.P. (1993) Do molecular dipole interactions influence solid-state organization?, *J. Phys. D* **26**, B56-B59.

10. Lytel, R., Lipscombe, G.F., Binkley, E.S., J.T., K. and Ticknor, A.J. (1991) Electrooptic polymer wave-guide devices - status and applications, in S.R. Marder, J.E. Sohn and G.D. Stucky (eds.), *Materials for Nonlinear Optics, ACS Symp. Ser. 455*, American Chemical Society, Washington D.C., pp. 103-112.
11. Aakeröy, C.B. and Seddon, K.R. (1993) The crystal-structures of pyridinium chloride revisited - evidence for extensive C-H...Cl hydrogen-bond interactions, *Z. Naturforsch., Teil B* **48**, 1023-1025.
12. Aakeröy, C.B., Hitchcock, P.B., Moyle, B.D. and Seddon, K.R. (1989) A novel class of salts for second harmonic generation, *J. Chem. Soc., Chem. Commun.*, 1856-1859.
13. Aakeröy, C.B., Hitchcock, P.B. and Seddon, K.R. (1992) Organic salts of L-tartaric acid - materials for second harmonic generation with a crystal-structure governed by an anionic hydrogen-bonded network, *J. Chem. Soc., Chem. Commun.*, 553-555.
14. Aakeröy, C.B. and Hitchcock, P.B. (1993) Hydrogen-bonded layers of hydrogentartrate anions - building-blocks for crystal engineering, *J. Mater. Chem.* **3**, 1129-1135.
15. Aakeröy, C.B., Bahra, G.S., Hitchcock, P.B., Patell, Y. and Seddon, K.R. (1993) Crystal engineering - hydrogen-bonded salts of hydroxybenzoic acids for second harmonic generation, *J. Chem. Soc., Chem. Commun.*, 152-156.
16. Aakeröy, C.B., Seddon, K.R. and Leslie, M. (1992) Hydrogen-bonding contributions to the lattice energy of salts for second harmonic generation, *Struct. Chem.* **3**, 63-65.
17. Aakeröy, C.B. and Nieuwenhuyzen, M. (1994) Hydrogen-bonded layers of hydrogen malate anions - a framework for crystal engineering, *J. Am. Chem. Soc.* **116**, 10983-10991.
18. McCrone, W.C. (1965) Polymorphism, in D. Fox, M.M. Labes and A. Weissberger (eds.), *Physics and Chemistry of the Organic Solid State*, Interscience, New York, pp. 725-767.
19. Aakeröy, C.B. and Seddon, K.R. (1992) unpublished results.
20. Desiraju, G.R. and Parthasarathy, R. (1989) The nature of halogen ... halogen interactions. Are short halogen contacts due to specific attractive forces or due to close packing of nonspherical atoms?, *J. Am. Chem. Soc.* **111**, 8725-8726.
21. Jeffrey, G.A. and Saenger, W. (1991) *Hydrogen Bonding in Biological Structures*, Springer-Verlag, Berlin.
22. Abdul-Sada, A.K., Greenway, A.M., Hitchcock, P.B., Mohammed, T.J., Seddon, K.R. and Zora, J.A. (1986) Upon the structure of room-temperature halogenoaluminate ionic liquids, *J. Chem. Soc., Chem. Commun.*, 1753-1754.
23. Abdul-Sada, A.K., Al-Juaid, S., Greenway, A.M., Hitchcock, P.B., Howells, M.J., Seddon, K.R. and Welton, T. (1990) Upon the hydrogen-bonding ability of the H(4) and H(5) protons of the imidazolium cation, *Struct. Chem.* **1**, 391-394.
24. Avent, A.G., Chaloner, P.A., Day, M.P., Seddon, K.R. and Welton, T. (1994) Evidence for hydrogen-bonding in solutions of 1-ethyl-3-methylimidazolium halides, and its implications for room-temperature halogenoaluminate(III) ionic liquids, *J. Chem. Soc., Dalton Trans.*, 3405-3413.
25. Aakeröy, C.B., Evans, T., Pálinkó, I. and Seddon, K.R. (1993-1995) unpublished results.

DISCUSSION OF THE PALACIN AND SEDDON LECTURES

W. E. BILLUPS
Department of Chemistry
Rice University
Houston, Texas 77251
U.S.A.

Session VI involved two unrelated lectures. The first was presented by Dr. Serge Palacin who addressed the topic of Supramolecular Architecture in Langmuir-Blodgett films. He is currently pursuing the building of a two-dimensional conducting film using tetrasubstituted monomers. The preorientation of the tetrasubstituted monomers showed a significant effect on the mechanism of the two-dimensional coupling. By combining one- and two-dimensional systems, he hopes to realize a three-dimensional network including two-dimensional conducting layers separated by linear conducting rods.

The discussion of this lecture was initiated by a series of questions from Professor STUPP (Illinois). He wanted to know whether the two dimensional polymerizations could be controlled to give polymers which deviate from linearly polymerized materials. In his answer to this question Dr. PALACIN stated that X-ray measurements indicate that the average extent of polymerization is over thirty molecules. Dr. STUPP commented that over a patch of 20 or 30 molecules there must be a very specific direction of polymerization in order to create or involve a two-dimensional structure. Dr. PALACIN stated that his goal was to make sure that at least two or more connections per molecule were formed. In this way more than a one-dimensional system would be formed. Professor WEGNER (Mainz) spoke in general about the difficulty in carrying out these experiments.

In response to a question from Professor MICHL (Colorado) regarding the thickness of the monolayer, Dr. PALACIN stated that he was not able to make this measurement, although he thought that he was dealing with a monolayer which was about 40 Å thick.

Professor DIEDERICH (ETH Zürich) discussed his collaborative work with Professor Ringsdorf involving the effect of pressure on the formation of monolayers. The discussion of this lecture ended with a comment from Professor OZIN (Toronto) on spatially controlled polymerizations and the dimensionality of polymerizations.

J. Michl (ed.), Modular Chemistry, 163–166.

The second lecture was presented by Professor SEDDON (Belfast) who stated that the purpose of his talk would be to present a philosophy of crystal engineering. He pointed out that much molecular chemistry, or supramolecular chemistry, relies upon the synthesis of large, elaborate, polyfunctional organic molecules, often representing a huge investment of labor, to produce low yields of the desired material - academically fascinating and challenging, but of little use for industrial applications. An alternative approach is to prepare simple salts, hence factorizing the desired properties into two parts. One desired property set is carried by the cation and one by the anion. He described novel SHG salts in which anions form stable, reproducible, hydrogen-bonded networks (thus granting the solid a high lattice energy), in which either the anions or the cations are chiral (thus creating non-centrosymmetric structures), and in which the cations carry the desired polarizability (thus controlling the intensity of the SHG produced).

The speaker considered the problems of polymorphism to compounded (and confounded) by the poor quality of the scientific method employed within the crystallographic community. It was argued that new criteria for the publication of crystal structures should be introduced, and that crystallographers should act like scientists, rather than technicians. Specifically, a powder pattern measurement should always accompany a single crystal structure determination in order to assure that the crystal selected for the latter is representative of the sample and not some rare polymorph. This lecture provoked extensive discussion and a disparity of opinion.

Professor MICHL contended that many investigators are only interested in the molecular structure of their compounds. He pointed out that 90%, or maybe 99%, of the organic chemists who omit their X-ray powder patterns simply do not care about crystal packing. They only want the molecular structure. An extensive discourse followed in which the speaker strongly disagreed.

Professor TOLBERT (Georgia Tech) suggested that most of us operate on the basis of probability and he asked what was the possibility of polymorphism occurring from recrystallization from the same solvent. After several exchanges, Professor SEDDON indicated that about 10 percent of the materials can come out in different forms in the same solvent by changing temperature gradients or concentration. He went on to suggest that most laboratory situations are such that polymorphism is near certainty. For example, one drop of water in acetone is not very, very uncommon, yet he knows of examples like that which cause the formation of polymorphs.

Professor WARD (Minnesota) offered salient comments on conformational polymorphism. He also thought the morphology indexing of crystals was lacking in the literature. He emphasized the importance of this information to the materials community.

As a representative of the materials community, Professor OZIN was in agreement with the speaker. He responded by saying that a lot of people in the audience and elsewhere are putting together "weird and wonderful systems" and that these materials are often not phase pure. He pointed out that we tend to pull out a crystal and make a big "hullabaloo" about it. He contends that this is causing a lot of trouble in the field. He pointed out that layered materials are particularly troublesome and with two and three dimensional materials it is even worse. Professor SEDDON agreed. Professor OZIN went on to say that as materials become "fancier" more possibilities for polymorphism would exist. He suggested that these materials are incredibly sensitive to the environment and often change in reversible manners. He even went on to suggest that the powder pattern might be dependent on the time of the year that the crystal is taken. He suggested that the materials community should be much more sensitive to this phenomenon. Professor SEDDON thought that this should apply to the organic community as well. He concluded that the materials community was the most aware of it.

Professor Jim WUEST (Montréal) asked if studies of polymorphism had been carried out in the presence of applied or selected fields. After some discussion it was decided that no such studies had been carried out.

Professor PRINZBACH (Freiburg) livened the discussion by asking the speaker, "Do you synthesize your crystals or do you prepare the crystals?" He (PRINZBACH) said that it was not just semantic. A satisfactory answer was not forthcoming.

Professor WHITESELL (UT Austin) described a three hour dialogue that he had attended on the structure of benzene. He concluded that the crystal structure of benzene was of no interest, but that no one had any idea why it crystallized, nor could anyone come up with a computational program to predict it.

Finally, Professor DIEDERICH commented that all of this pessimism might not be too important since in a previous NATO workshop he had attended several software packages were promised which would help with the polymorphism area. Nevertheless, he emphasized the problems that the pharmaceutical industry is experiencing as they are required to submit relevant data on all polymorphs.

The speaker presented results suggesting that the forces between stacked aromatic rings are comparable with the strengths of conventional hydrogen bonds. Thus, the interaction between aromatic rings (egregiously referred to, in many papers, as π-π stacking) was presented as a useful synthetic vector in the crystal engineer's tool-kit. Professor KAHN (Bordeaux) pointed out that he liked Seddon's strategy in designing his systems and Professor OZIN insisted on learning what sort of thought process was involved.

Evidence for the widespread hydrogen-bond donor ability of the C-H bonds of heterocyclic aromatic cations and a method of redefining the concept of van der Waals' radii in the solid state was presented. Professor SEEMAN wanted to know if Professor Seddon had ever done some thermodynamics or other experiments and demonstrated that these bonds actually are bonds. In his answer to this question Professor SEDDON insisted that they are not weak, as might have been expected. This subject came up several times during a series of questions by Professor OZIN who professed to having read Seddon's published work.

The speaker was asked by the discussion leader to comment on the controversial proposal in *Science* that a short (less than 2.5 Å), very strong, low-barrier hydrogen bond in the transition state, or in an enzyme-intermediate complex might supply as much as 80 $kJmol^{-1}$ of energy. He responded by saying that he did not find that huge. He went on to say that the whole field is short of quantitative measurements because they are extremely difficult to do. Professor DIEDERICH pointed out that there are gas phase values up to 90 $kJmol^{-1}$ for ionic hydrogen bonds.

THE BENZENE RING AS MODULUS: EXTENDED POLYBENZOID DISC STRUCTURES AND THEIR SUPRAMOLECULAR ORDERING

PETER HERWIG, MARKUS MÜLLER, AND KLAUS MÜLLEN*
Max-Planck-Institute for Polymer Research,
Ackermannweg 10, 55128 Mainz, Germany

ABSTRACT. Polyaromatic hydrocarbons (PAH) have lately become increasingly important due to their electronic and opto-electronic properties. Furthermore, two-dimensional PAHs serve as model compounds for graphite. Herein, we present a modular approach toward graphite segments in which soluble polyphenyl precursors are fused to new, well-defined disc structures. In order to achieve our goal, we have applied Kovacic's methods of intermolecular cyclodehydrogenation to intramolecular fusion processes. This concept has enabled us to prepare a wide variety of large polybenzoid compounds with different properties. The ability of these materials to assemble into well ordered systems has also been the subject of our studies. We have been able to prepare space filling, monomolecular films of unsubstituded PAHs, supported on MoS_2 surfaces, that have been monitored by scanning tunneling microscopy and low-energy electron diffraction. In case of alkyl substituted hydrocarbon derivatives liquid crystallinity has been observed via x-ray diffractometry and solid-state deuterium NMR-spectroscopy.

1. Introduction

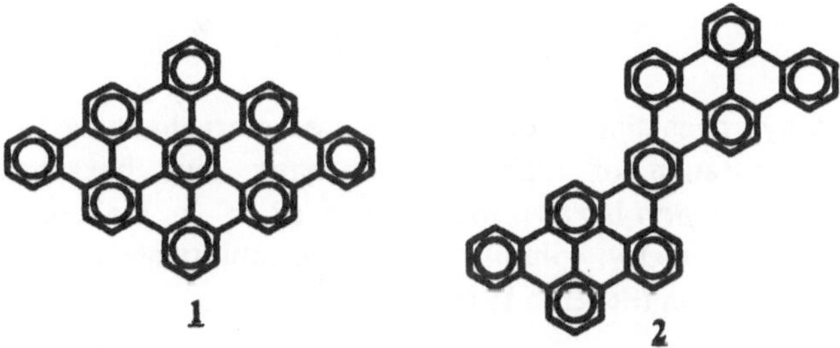

For many years chemists have been attempting to synthesize molecular objects of differing size and shape in order to achieve the phenomenon of supramolecular ordering under the influence of weak intermolecular forces. Building blocks for idealized structures are rods, loops and spheres [1]. The introduction of large discs with well defined size and shape into this field has so far been hampered with problems related to both synthesis and processing.

J. Michl (ed.), Modular Chemistry, 167–177.

Herein, we present a modular approach towards graphite subunits in which benzene rings are successively fused to form disc structures. Typical title structures are the polybenzoid hydrocarbons **1**, **2**, and **4**. The construction principles which we have adopted allow for a systematic variation of the size and shape of the molecules and can also be extended to polymeric analogues. Processing of the disc molecules has been achieved from solution and also by vacuum techniques in which the formation of monomolecular adsorbate layers or columnar structures are of primary concern.

2. Synthetic Methods toward Polybenzoid Hydrocarbons

Ribbon-type graphite subunits, such as [*n*]acenes or rylenes (e.g. **14**), are electronically attractive due to their low band gap. We have recently used pentacene as active component of field effect transistors with a high charge carrier mobility, whereby the mode of processing appears to be crucial [2]. Unfortunately, stability problems arise in low band-gap materials even in the case of the lower homologues [3, 4]. The present approach is therefore restricted to those polybenzoid hydrocarbons that are stable according to Clar's predictions [5].

The synthetic routes adopted for the fusion of benzene rings involve:

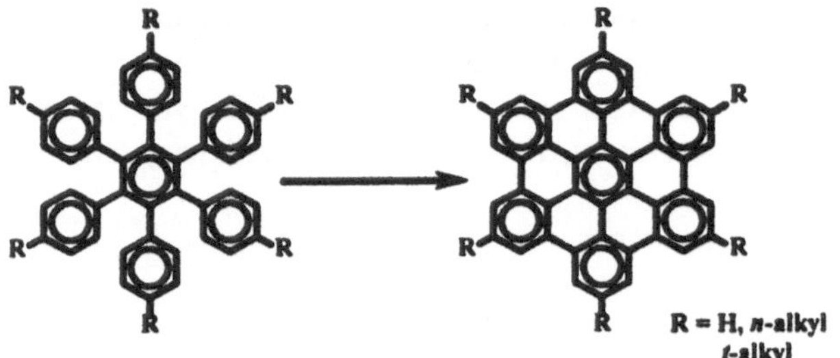

R = H, *n*-alkyl *t*-alkyl

3 4

Scheme 1

(i) cyclodehydrogenation of oligophenyl precursors that have been prepared by cyclotrimerisation of diphenyl acetylenes: thus hexaphenylbenzenes **3** produce hexa-*peri*-hexabenzocoronenes (R = H, *n*-alkyl, *t*-alkyl) **4** with hexagonal symmetry. Solubility and processability are achieved, thereby, via alkyl substitution (Scheme 1) [6];

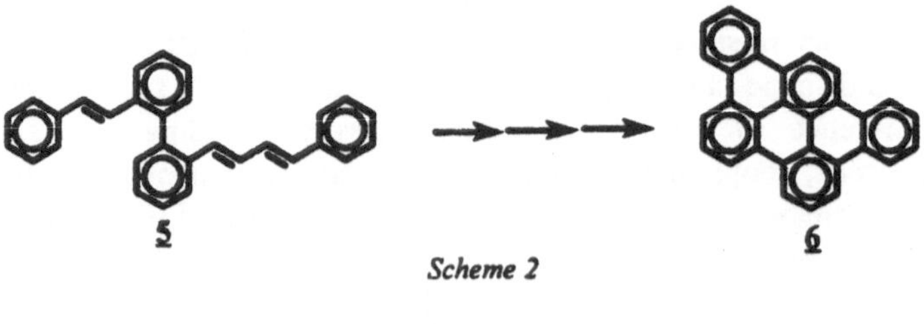

5 6

Scheme 2

(ii) intramolecular Diels-Alder cycloaddition of oligophenylene vinylene precursors: the simplest case is that of **5** which transforms into **6** via a sequence involving cycloaddition, dehydrogenation and electron-transfer induced cyclodehydrogenation (Scheme 2) [7, 8].

It is crucial that both routes, (i) and (ii), can be adapted to produce disc molecules possessing much larger areas: when a higher homologue of **5**, namely compound **7**, is subjected to a sequence according to (ii), one faces an ambiguity since the final cyclodehydrogenation can, in principle, lead to both **1** and **9** (Scheme 3).

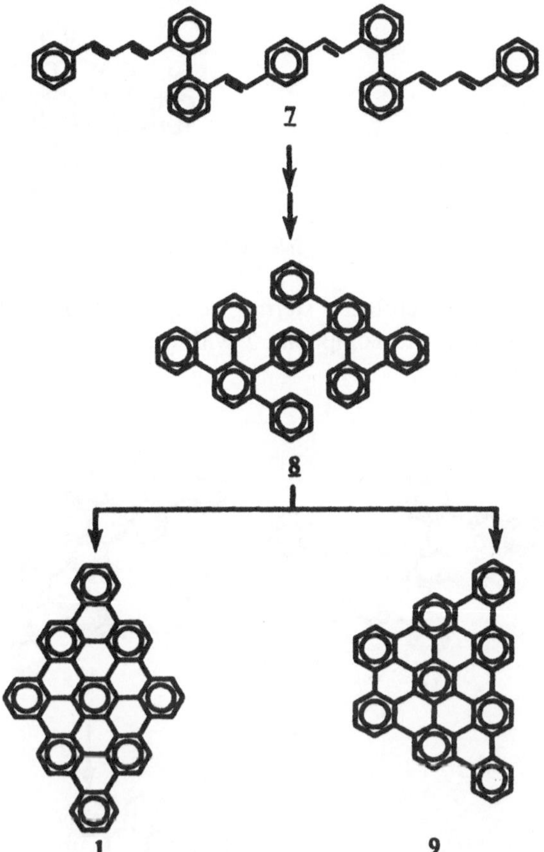

7

8

1

9

Scheme 3

This shortcoming can be avoided, however, when using a starting compound with a different mode of linking of the dienophile and diene components. The transformation of the terphenyl system **10** into **1** proceeds with high yield and, at the same time, provides convincing proof for the nature of the fusion product (Scheme 4).

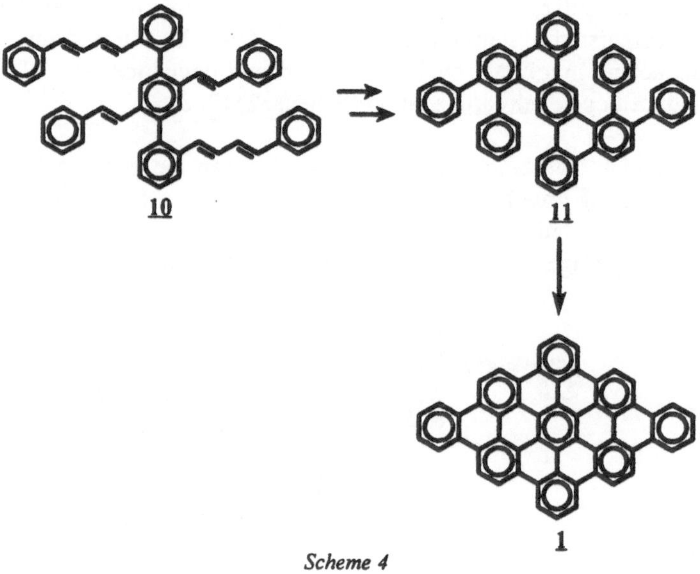

Scheme 4

A slight modification of the terphenyl starting compound, namely the synthesis of the analogue **12**, provides access to compound **2** with a completely different shape of the polybenzoid frame (Scheme 5).

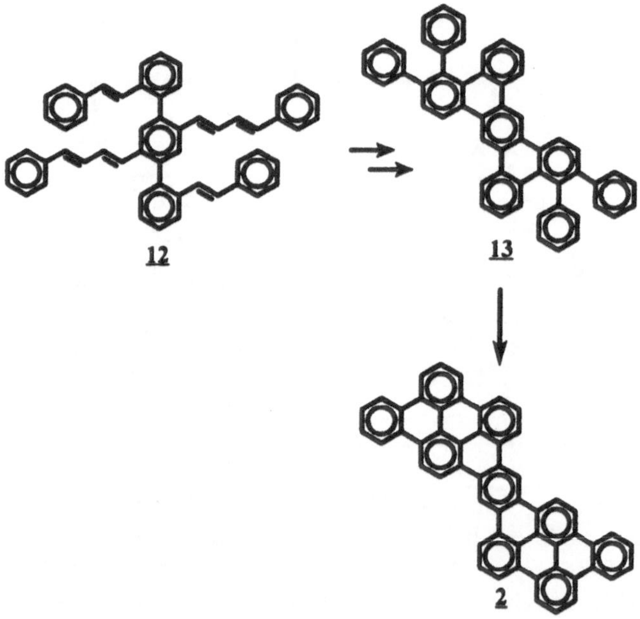

Scheme 5

The key design principles can thus be summarized as follows: Benzene is used as a rigid modulus whereby one crucial fusion process is cyclodehydrogenation with the transformation of *ortho*-terphenyl into triphenylene systems as a typical example. Accordingly, in the construction of the oligophenyl and oligophenylenevinylene precursors branching will be avoided; instead, the building blocks should always be able to converge to closed loops. Alkyl substitution will in many cases be unavoidable to achieve soluble and processable discs.

3. Supramolecular Ordering of Polybenzoid Hydrocarbons

Having available such extended polyaromatic hydrocarbons, two supramolecular motifs can be considered: the formation of columnar arrangements as a model of three-dimensional graphite, and the formation of monomolecular adsorbate layers as a model of two-dimensional graphite. With respect to the latter topic it should be mentioned that graphitic structures have recently been manipulated by using the tip of an atomic force microscope and "graphene" sheets been achieved [9].

A typical structural motif of polycyclic benzoid hydrocarbons is that of a column formed via π-stacking. Crystallization, electrocrystallization and charge-transfer complex formation of large discs are considered as possible approaches towards multi-layered solid-state structures. The high electrical conductivity observed for stack-type radical cation salts of small polybenzoid hydrocarbons would strongly suggest to subjekt larger disc molecules to related electrocrystallization experiments [10]. A possible consequence could be the minimization of on-site Coulomb repulsion in the charged layers. Desiraju and Gavezzotti [11, 12] have analyzed the packing modes occurring in single crystals of neutral polybenzoid aromatic hydrocarbons. It is clear that an inclusion of much larger discs is an important ingredient of such an approach. We have grown single crystals of e.g. hexa-*t*-butyl-hexa-*peri*-hexabenzo-coronene **4**, which shows a sandwich herringbone motif [13] in contrast to a simple herringbone motif of the unsubstituted hexa-*peri*-hexabenzo-coronene [14].

More interesting from a practical point of view - e.g. processability or fiber formation - is the occurrence of mesophases, possibly with columnar stuctures. It follows from an analysis by differential scanning calorimetry and X-ray diffractometry that, e.g., hexa-*peri*-hexabenzocoronene **4** with six dodecyl groups forms a very stable columnar mesophase with a hexagonal superstructure [15]. It should be noted that triphenylene, a related, but smaller disc molecule requires alkoxy or alkoxycarbonyl groups for the formation of discotic mesophases which in addition exist in a much smaller temperature range [16]. It is shown by solid-state deuterium NMR-spectroscopy that the large discs can maintain a stable columnar super structure, while at the same time having a lower degree of ordering than smaller discs [15]. Here again, transition to even larger discs is crucial in order to correlate molecular structure with the ability of forming stable discotic mesophases. It should be noted that photoconductivity measurements for discotic mesophases obtained from substituted triphenylenes have revealed a high charge carrier mobility [17]. The novel disc structures presented herein demand for related studies since increasing photoconductivity is relevant e.g. for electrophotography [18].

The very high isotropization temperatures of hexaalkyl-hexa-*peri*-hexabenzo-coronene **4** (often in excess of 400°C) suggest a comparison with the so-called carbomesophases formed above 400°C during the production of technical graphite from an isotropic coke [19]. In the latter process it is the formation of mesophases from the disc-type starting compounds which is responsible for the anisotropy of the final graphite structure. The novel alkyl substituted polybenzoid discs could allow a study of the graphitization while starting from well-defined mesophases.

Using graphite itself as a reference system for the analysis of large polybenzoid hydrocarbons there is still another challenge: the formation of highly ordered monomolecular layers on substrate surfaces. Such layers are relevant for many fields of research, e.g. as active components in sensors or, somewhat more remote, as elements of molecular electronics [20]. One anticipates such layers to result from a subtle balance of substrate-adsorbate and adsorbate-adsorbate interactions and to sensitively depend upon size and symmetry of the molecules. One method for the preparation of monomolecular adsorbate layers is the vacuum deposition of unsubstituted poly-aromatic hydrocarbons on to the 0001-cleavage planes of graphite or molybdenum sulfide. The resulting patterns can be monitored by low-energy electron diffraction [21], which is displaying the reciprocal lattice vectors to derive the unit mesh in direct space [8]. In view of the nessecary coincidence of the molecular and substrate symmetries and the formation of monomolecular superstructures, a comparison of disc molecules with different symmetries such as **1** and **4** is highly relevant. It appears that in both cases the molecules lie flat on the surface, forming a close packing arrangement.

Heating of the layers might induce a fusion of the single molecules by intermolecular dehydrogenation prior to their desorption. As a consequence, one would form polymeric monolayers similar to the above mentioned "graphene" sheets. A related experiment is the codeposition of the polybenzoid hydrocarbons with suitable donors or acceptors thus producing intermolecular charge transfer and probably in-plane conductivity.

An immobilization of a single molecule, e.g. as part of monomolecular adsorbate layer on highly ordered pyrolytic graphite, is also a precondition for its observation by scanning tunneling microscopy. This method has recently found a more frequent application [22]. However, there are extremely few successful studies of scanning tunneling spectroscopy in which due to a submolecular resulution one would achieve a chemical sensitivity of the STM technique. Prototype molecules would be alkyl substituted aromatic hydrocarbons, which possess spatially separated domains of significantly different electronic properties. Thereby the molecular electronic states of the alkyl units are energetically far away from the Fermi level of the electrode. Here again, the size of the graphite subunits appears to be crucial.

While a stable two-dimensional pattern is obtained for 3,10-didodecylperylene (**14**) [23], there is no unambigous local assignment of the current-potential curves to an aromatic or aliphatic domain. The situation is totally changed for hexadodecyl-hexa-*peri*-hexabenzocoronene. It follows from the lattice parameters of the unit cell that the sixfold symmetry axis of the molecule is reduced to a C_2 symmetry in the two-dimensional crystal. The current-potential curve taken over the aliphatic domain is symmetric and cannot be distinguished from the curve taken for graphite covered with solvent. The curve taken for the aromatic domain is asymmetric by one order of magnitude and possesses the shape of a diode function. This is the first example of scanning tunneling spectroscopy, a diode-type potential curve for a single molecule which at the same time is visualized by STM [6].

4. Toward Polymers

In order to learn about the electronic and opto-electronic properties of graphite it is essential to investigate larger polyaromatic hydrocarbons. Therefore it is challenging to apply our concept of cyclodehydrogenation to hexagonal graphite segments like **16**, which could be prepared easily via cyclotrimerisation of an alkylated ditriphenylenyl ethine. It is clear that a suitable choice of the building blocks will lead to even larger hexagonal disc structures.

R = *n*-alkyl

15 16

Scheme 6

It is obvious when facing building blocks such as **5**, **7**, **10**, or **12** that transition to higher homologues with phenylenevinylene subunits will also provide access to related polymers. The shape of the polybenzoid discs is thereby determined by the topology of the phenylenevinylene precursors. The ladder type polymer **18** may hold as an example for an one-dimensional carbon strip. The access to this polymer is provided by transition of the Diels-Alder reaction and following aromatization of oligophenylen-vinylenes to higher homologues like the poly-*para*-phenylene derivative **17**. This synthetic concept may open the door toward nanostructures with a well defined width and periphery.

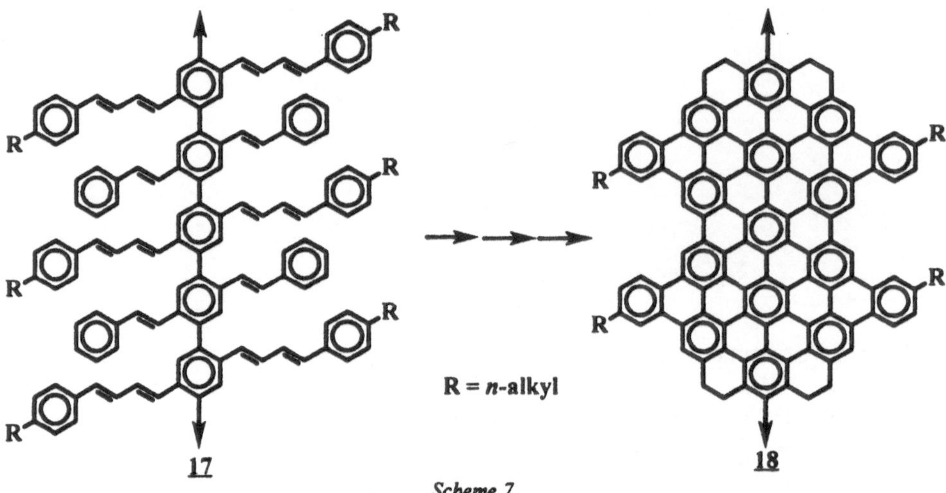

Scheme 7

On the other hand, the present design principles for polyaromatic hydrocarbons offer chances e.g. for the construction of non-planar polycycles like **20** with helical arrangements and for the incorporation of hetero atoms, which in turn allow complexation of metal centers. One of the six [5]helicene units in **20** disturbing the planarity of the aromatic system is marked.

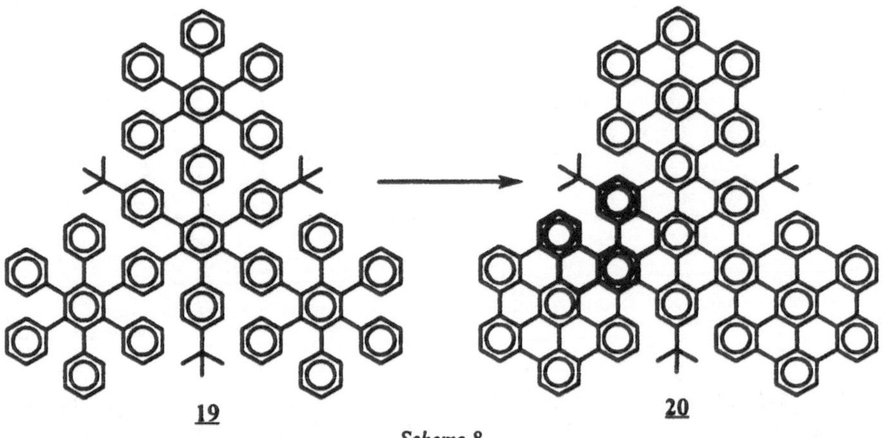

Scheme 8

Another possibility of curving hydrocarbons is the introduction of five membered rings into the framework of polyaromatic compounds. Beside their interesting electronic properties polybenzoid derivatives containing five membered rings even without any alkyl substitution show a remarkable solubility in common organic solvents. A simple model compound for five membered ring systems is presented by compound **23**.

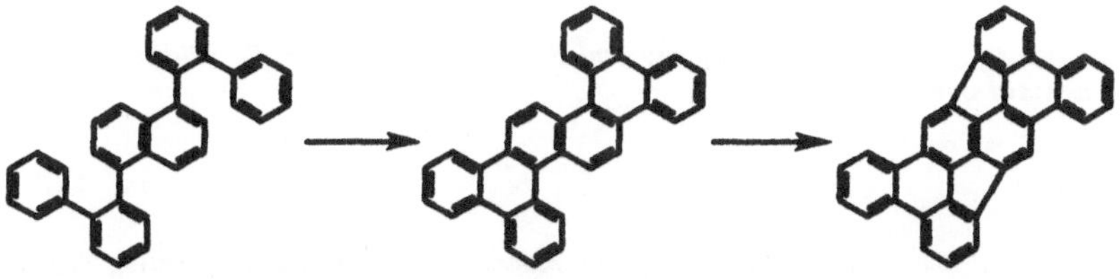

21 22 23

Scheme 9

To combine the attractive electronic properties of these aromatic compounds with the good processability of conventional polymers it is possible to incorporate the systems into polymers. The easiest approach for this purpose is to blend the graphite segments with linear polymers such as polystyrene, a more demanding method is to copolymerize soluble precursor molecules like **24** with styrene to yield the polymer **25**, which can be afterward cyclodehydrogenated under oxidative conditions.

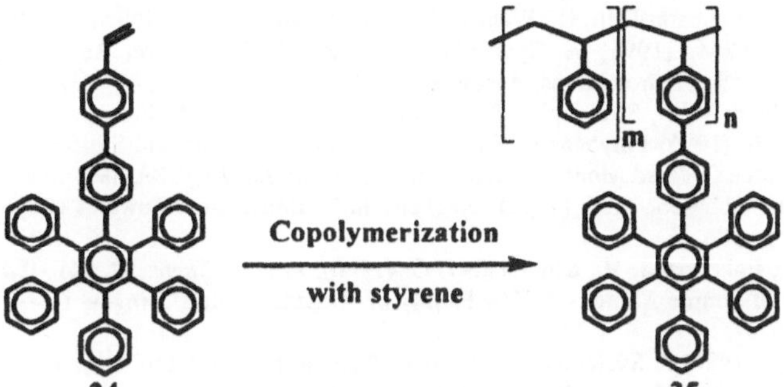

Copolymerization with styrene

24 25

Scheme 10

5. Conclusion

The present contribution has been focussing on a new synthetic approach toward extended polybenzoid aromatic hydrocarbons and the resulting opportunities for supramolecular ordering (i) in two-dimensional adsorbate layers on substrates and (ii) in three-dimensional columnar arrangements. Clearly, the scope of the present approach toward novel macromolecular and supramolecular architectures becomes even wider when activating other means of organization such as metal intercalation or hydrogen bonding.

A straightforward extension of our research concerns the role of the resulting molecular and supramolecular structures as electronic materials. Typical examples which have been already mentioned are the photoconductivity occurring in columnar meso-phases and an eventual increase of the charge carrier mobility [17] and the electrical conductivity seen in monomolecular adsorbate layers.

Financial Support by the Volkswagenstiftung and the Bundesministerium für Bildung und Forschung is gratefully acknowledged.

References

[1] Dietrich, B., Viout, P., and Lehn, J.-M. (1993) *Macrocyclic Chemistry*, VCH Verlag, Weinheim.

[2] Brown, A.R., Pomp, A., deLeeuw, D.M., Klaassen, D.B.M., Havinga, E.E., Herwig, P., and Müllen, K. (1996) Precursor-route Pentacene Metal-Insulator-Semiconductor Field-Effect Transistors, *J. Appl. Phys.* 79, 2136 - 2138.

[3] Horn, T., Wegener, S., and Müllen, K. (1995) Poly[n]acene precursors via repetitive Diels-Alder reactions with dehydrobenzenes, *Macromol. Chem. Phys.* **196**, 2463 - 2474.

[4] Scherf, U., and Müllen, K., (1995) *Ladder Polymers - An Outstanding Challenge for Polymer Synthesis*, Advances in Polymeric Science, Springer (review) **123**, 1 - 40.

[5] Clar, E. (1972) *The Aromatic Sextet*, John Wiley & Sons, London.

[6] Stabel, A., Herwig, P., Müllen, K., and Rabe, J.P. (1995), *Angew. Chem.* **107**, 1768 - 1770; (1995) Diodelike Current-Voltage Curves for a Single Molecule - Tunneling Spectroscopy with Submolecular Resolution of an Alkylated, peri-condensed Hexabenzocoronene, *Angew. Chem. Int. Ed. Engl.* **34**, 1609 - 1611.

[7] Müller, M., Mauermann-Düll, H., Wagner, M., Enkelmann, V., and Müllen, K. (1995), *Angew. Chem.* **107**, 1751 - 1754; (1995) A Cycloaddition-Cyclode-hydrogenation Route from Stilbenoids to Extended Aromatic Hydrocarbons, *Angew. Chem. Int. Ed. Engl.* **34**, 1583 - 1586.

[8] Müller, M., Petersen, J., Strohmaier, R., Günther, C., Karl, N., and Müllen, K. (1996), *Angew. Chem.* **108**, 947 - 950; (1996) Polybenzoid C_{54} Hydrocarbons: Synthesis and Structural Characterization in Vapor-Deposited Ordered Monolayers, *Angew. Chem. Int. Ed. Engl.* **35**, 886 - 888.

[9] Hiura, H. and Ebbesen, T.W. (1995) Graphene in 3-Dimensions: Towards Graphite Origami, *Adv. Mater.* **7**, 582 - 586.

[10] Kröhnke, C., Enkelmann, V., and Wegner, G. (1980), *Angew. Chem.* **92**, 941 - 942; (1980) Radical Cation Salts of Simple Arenes - A New Family of "Organic Metals", *Angew. Chem. Int. Ed. Engl.* **19**, 912 - 913.

[11] Gavezzoti, A. (1988) A Systematic Analysis of Packing Energies and Other Packing Parameters for Fused-Ring Aromatic Hydrocarbons, *Acta. Cryst.* **b44**, 427 - 434.

[12] Desiraju, G.R., and Gavezzoti, A. (1989) From Molecular to Crystal Structure; Polynuclear Aromatic Hydrocarbons, *J. Chem. Soc., Chem. Commun.*, 621 - 622.

[13] Herwig, P., Enkelmann, V. and Müllen, K., to be published.

[14] Goddard, R., Haenel, M.W., Herndon, W.C., Krüger, C., and Zander, M. (1995) Crystallisation of Large Planar Polycyclic Aromatic Hydrocarbons: The Molecular and Crystal Structures of Hexabenzo[bc,ef,hi,kl,no,qr]coronene and Benzo[1,2,3-bc:4,5,6-b'c']dicoronene, *J. Am. Chem. Soc.* **117**, 30 - 41.

[15] Herwig, P., Kayser, C.W., Müllen, K., and Spieß, H.W. (1996) Columnar Mesophases of Alkylated Hexa-peri-hexabezocoronenes with Remarkably Large Phase Widths, *Adv. Mat.* **8**, 510 - 513.

[16] Tinh, N.H., Gasparoux, H., and Destrade, C. (1981) Homologous Series of Disc-Like Mesogenes with Nematic and Columnar Polymorphism, *Mol. Cryst. Liq. Cryst.* **68**, 101 - 111.

[17] Bengs, H., Closs, F., Frey, T., Funhoff, D., Ringsdorf, H., and Siemensmeyer, K. (1993) Highly photoconductive discotic liquid crystals, Structure-property relations in the homologous series of hexa-alkoxytriphenylenes, *Liq. Crystals* **15**, 565 - 574.

[18] Adam, D., Schumacher, P., Simmerer, J., Häußling, L., Paulus, W., Siemensmeyer, K., Etzbach, K.-H., Ringsdorf, H., and Haarer, D. (1995) Photoconductivity in the Columnar Phases of a Glassy Discotic Twin, *Adv. Mater.* **7**, 276 - 280.

[19] Marsh, H. (Ed.). (1989) *Introduction to Carbon Science*, Butterworth, London.

[20] Swalen, J.D., Allara, D.L., Andrade, J.D., Chandross, E.A., Garoff, S., Israelachvili, j., McCarthy, T.J., Murray, R., Pease, R.F., Rabolt, J.F., Wynne, K.J., and Yu, H. (1987) Molecular Monolayers and Films, *Langmuir* **3**, 932 - 950.

[21] Zimmermann, U., and Karl, N. (1992) Epitaxial growth of coronene and hexa-peri-benzocoronene on MoS_2 (0001) and Graphite (0001): a LEED study of molecular size effects, *Surf. Sci.* **268**, 296 - 306.

[22] Pomerantz, M., Aviram, A., McCorcle, R.A., Li, L., and Schrott, A.G. (1992) Rectification of STM Current to Graphite Covered with Phtalocyanine Molecules, *Science* **255**, 1115 - 1118.

[23] Stabel, A. (1995) PhD thesis, University of Mainz, Mainz, Germany.

DISCUSSION OF THE MÜLLEN LECTURE

RALPH G. NUZZO
Department of Chemistry
University of Illinois at Urbana-Champaign
Urbana, Illinois 61801
U.S.A.

In addition to the lecture of Prof. Müllen, the participants discussed a lecture by Prof. Whitesell, who showed an intriguing example of how the distribution of molar mass in a step-growth polymerization could be narrowed by directed synthesis after organizing the chains by self-assembly at a surface. In this way, the large-scale structure could be easily refined (viz. the molecular weight of the chains) and the high-yield (but imprecise) chemistries of step-growth polymerization exploited for the rough construction of the molecular framework. Since Prof. Whitesell was unfortunately unable to provide a manuscript of his lecture, we do not render the discussion of his talk.

The theme that emerges most clearly from the two presentations (of Klaus Müllen and Jim Whitesell) is that there exist diverse pathways by which modules can be converted to structures of considerable molecular complexity. For Whitesell, the module of interest is a polymer (a polypeptide) tethered to a surface bound template which then in turn organizes via a self-assembly process. For Müllen, the molecular complexity is constructed via completely convergent sequences of molecular synthesis. Both works demonstrate the utility and current limitations of chemistries suited to the construction of high molecular weight materials of discrete structure.

Müllen described the syntheses of well-defined polycyclic benzenoid sheets. These precisely tailored pieces of graphite-like sheets are perhaps the most beautiful and complex hydrocarbons ever prepared by direct synthesis. The well defined nature of the structures enables their exploitation as modules in several generalizable ways (e.g. via the phase properties of these materials both in the solid-state and at a phase boundary, such as at a solid surface).

Considerable discussion followed. The synthesis of extremely large polycyclic hydrocarbons exhibiting, for example, ribbon or disc-like motifs in the solid state, was hailed as being a synthetic tour-de-force. The successful use of what were felt to be inherently low yield reactions to prepare them was most surprising. The Kovacic method of cyclodehydrogenation, which proceeds by a complex radical mechanism, was found to provide an extremely good entry

J. Michl (ed.), Modular Chemistry, 179–181.

into the complex cycles described by Müllen. Müllen noted that the intramolecular fusion reactions are sufficiently efficient for the all-hydrocarbon systems he describes to at least be considered synthetically useful at the research scale. The reactions do not tolerate functional substitution, however, and thus new methods for these cases are urgently needed.

Significant discussion attended the electronic structures evidenced by Müllen's complex conjugated systems. These molecules are poised between the limits offered by small molecules (e.g. benzene) and extended systems (e.g. graphite). Why, one wonders, is Müllen's large "aromatic" rhombus (C_{54}) orange? How do the electronic properties of this molecule help rationalize its enormous thermal stability (e.g. being able to survive heating to >450 °C without decomposition)?

Answers were sought based on the pioneering predictions of Clar. In simple terms, one needs to establish whether there are important insights to be gained by analyzing how and to what degree the orbital topology of the molecule influences the "compartmentalization" of the π-electron density. From crystal structures, one knows that there exist alternations of bond lengths in these materials which suggest an underlying inhomogenity in the π-system overlap manifold. The optical spectra clearly establish that a large band gap also exists. The systems are thus not truly delocalized in an extended way as might be inferred by making an explicit analogy to graphite. There does not appear to be a good consensus as to the underlying physics which leads to the observed corrugations of the electron potential surface in extended systems of this size and degree of crossed-conjugation. The effects of Coulomb localization are well appreciated. The necessary conceptual and theoretical bridges to truly extended systems are yet to be revealed, however.

MÜLLEN notes that pragmatic considerations also enter into the consideration of what limits attend the synthesis of truly extended systems. The limitations imposed by reactions occurring with a high, but less than quantitative, yield are well appreciated. Solubility also figures very centrally. This follows simply from the fact that syntheses are typically not unimolecular in nature and are thus best carried out in solution.

CHIDSEY (Stanford) suggested that more consideration be given to developing chemistries which can link smaller modules into extended structures via reactions occurring in the solid state.

SMALLEY (Rice) further elaborated on the utility of, and need for, modules which could serve as molecular precursors to a single graphite sheet.

MÜLLEN noted that this is a very important goal but one for which the synthetic chemistries are still too limited to effectively address.

CHIDSEY and SMALLEY raised the question with Müllen whether there exist possibilities for elaborating the structure of currently available large modules using low molecular weight synthons. For example, can we envision using a carbon nanotube (μm in length) as a starting material and use the reactions of Müllen to elaborate this into something of higher complexity and function? If one uses directed synthesis to make an object of comparable size, how could error corrections be carried out? More generally, how do we remove defects from large objects made by the sequential combination of chemical reactions? The participants strongly supported the notion that research conducted to address these questions would be both timely and critical for enabling further developments in the field.

MÜLLEN and many participants discussed specific aspects of his novel polybenzenoid materials. Can they be used to study aggregation thermodynamics? Are they potentially useful as photothermal or photoconductive materials? Can they be modified to enhance non-linear optical properties? Significant reasons exist to expect continued progress being made towards realizing such applications.

STUPP (Illinois) made the very interesting suggestion that the materials Müllen prepared might find their most promising applications as "fillers" in well-defined molecular composites. Stupp correctly noted that there exists less significant needs for wholly new classes of materials than for ones which can be used to improve the properties (e.g. mechanical) of commercial materials such as commodity plastics, etc. Stupp suggested that rather than “sticking” or “stitching” things together, one instead might imagine using the polybenzenoid architecture as a platform to orient and stiffen a low-molecular weight liquid crystal. The materials might be blended with linear polymers to give composites with advanced properties based on rationalized molecular design. The dominant and most provocative point Stupp raised here is that the field needs to look beyond the limitations inherent in the consideration of modules as being a materials class in itself. In Stupp's view, progress is more likely to result if the modules concept can be incorporated and synergistically applied to existing types of materials, especially polymers.

It is an intriguing notion indeed!

DENDRIMERS: NANOSCOPIC MODULES FOR THE CONSTRUCTION OF HIGHER ORDERED COMPLEXITY

DONALD A. TOMALIA
Michigan Molecular Institute
1910 W. St. Andrews Road
Midland, Michigan 48640, USA

ABSTRACT. An overview of "molecular modules" found within the hierarchies of biotic and abiotic molecular evolution strategies was described. Systematic advancement toward structural complexity in either case has required synthetic control over *critical molecular design parameters (CMDP's);* such as, (a) *size* (b) *shape* (c) *surface chemistry* (d) *rigidity/flexibility* and (e) *topology*. This control of CMDP's has been amply demonstrated with dendrimer substrates using either sub-nanoscopic (1-10Å) or nanoscopic (10-1000Å) reagents. These unique properties have validated the use of dendrimers as "structure controlled nanoscopic modules."

New combining rules have been defined for these dendrimer modules which have evolved from a packing/parking dilemma referred to as "sterically induced stoichiometry" (SIS). It was concluded that dendrimers may be used as nanoscopic modules to systematically produce higher structural complexity by strategies that parallel biotic systems, but require only abiotic building blocks.

Introduction

The organizational hierarchy of biotic matter, from the smallest of atoms to living organisms, is generally viewed as the advancement of structural complexity with increasing size of certain constitutive building blocks (modules)[1-3]. This is illustrated by hierarchy (A), module categories I-VI in Figure 1. Unique differentiated properties and behaviors are usually associated with each of these categories of complexity.

Abiotic chemistry may be viewed hierarchically in much the same way. Traditionally, synthetic chemistry has been described as the science of electron movement leading to either chemical bond formation, rearrangement or fragmentation. Although this describes some of chemistry's dynamic aspects, from a *dimensional* or *spacial* perspective, this science also involves the construction, assembly (coupling) and

J. Michl (ed.), Modular Chemistry, 183–191.

hierarchical evolution of "*modules*." This thought is consistent with basic definitions by Webster as described below:

> **mod·ule \ˈmäj-(ˌ)ü(ə)l\ *n* [L *modulus*] 1 : a standard or unit of measurement 2 : the size of some one part taken as a unit of measure by which the proportions of an architectural composition are regulated 3 a : any in a series of standardized units for use together**

Dimensional Hierarchy of Organic Matter

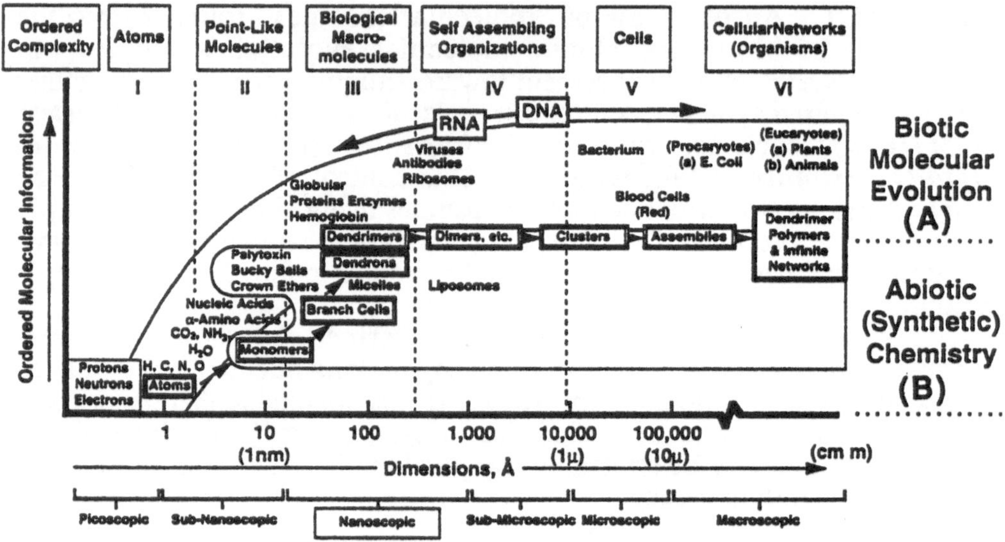

Figure 1. Dimensional hierarchy of organic matter: the advancement of structural complexity as a function of size.

This definition underpins the notion of fixed *atomic modules* that upon coupling may lead to larger more complex *molecular modules* as proposed by John Dalton in his "New System of Chemical Philosophy (1809)[4]. The 19th century was a rich period for the discovery of many fundamental modules and connectors (functional groups). Without understanding contemporary principles, these early chemists constructed sub-nanoscopic (5-10Å) modules based on all three hybridization states of carbon to initiate the areas of "aliphatic" and "aromatic" chemistry (see Figure 2)[4-5]. More recently Platonic hydrocarbons; such as, tetrahedrane[6], cubane[7], dodecahedrane[8], buckminsterfullerenes[9] as well as related architectural

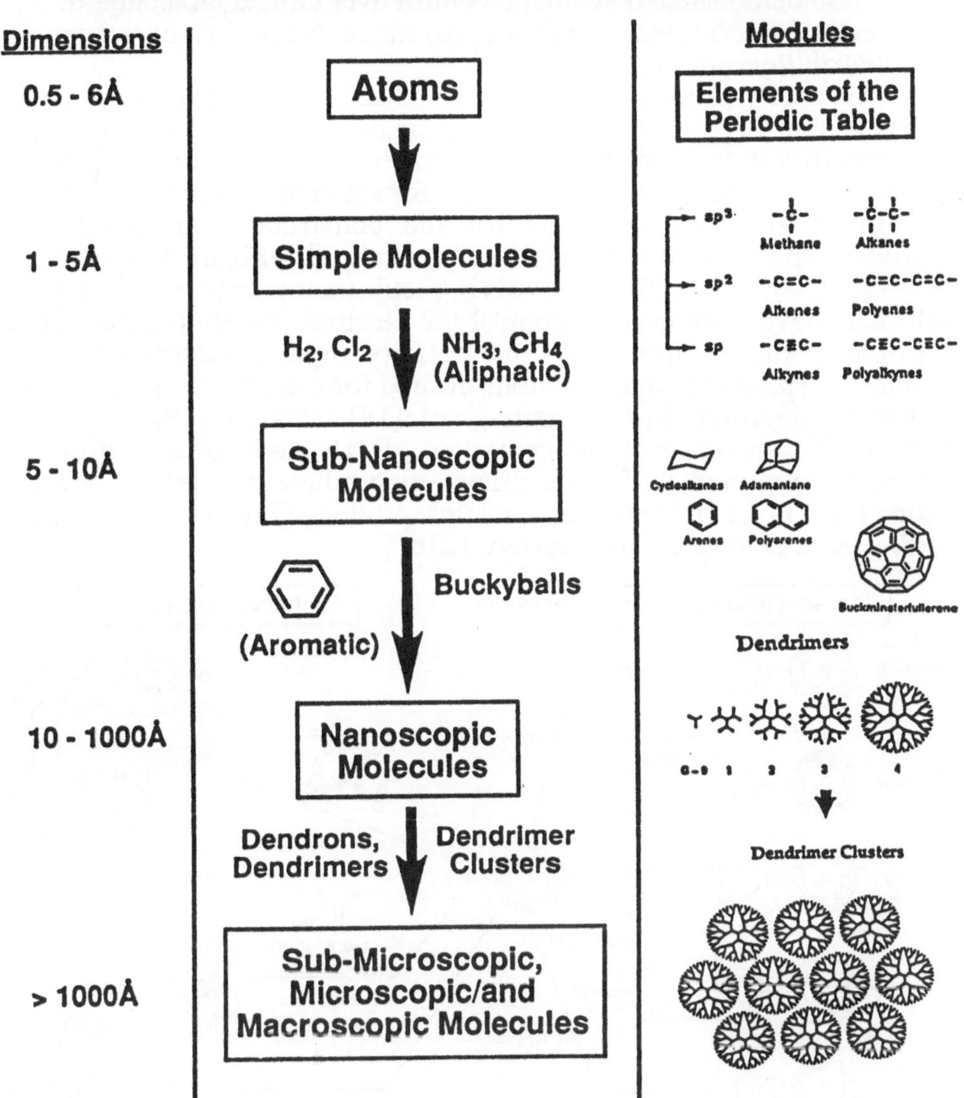

Modular Aspects of Synthetic Chemistry
Dimensions
0.5 - 6Å
1 - 5Å
5 - 10Å
10 - 1000Å
> 1000Å
Atoms
Simple Molecules
H2, Cl2
NH3, CH4
(Aliphatic)
Sub-Nanoscopic Molecules
(Aromatic)
Buckyballs
Nanoscopic Molecules
Dendrons, Dendrimers
Dendrimer Clusters
Sub-Microscopic, Microscopic/and Macroscopic Molecules
Modules
Elements of the Periodic Table
sp3
sp2
sp
Methane
Alkanes
Alkenes
Polyenes
Alkynes
Polyalkynes
Cycloalkanes
Adamantane
Arenes
Polyarenes
Buckminsterfullerene
Dendrimers
G-0
1
2
3
4
Dendrimer Clusters

Figure 2

structures; such as, rods[10-11], plates, rings and trees (dendrons/dendrimers)[12-13] have been reported. These constructions have not only advanced modular dimensions and complexity to nanoscopic proportions, but have also demonstrated synthetic control over critical molecular design parameters (CMDP's) such as; (a) *size*, (b) *shape*, (c) *surface chemistry*, (d) *rigidity/flexibility* and (e) *topology*.

The ability to precisely control the CMDP's of dendrimers by either "divergent, convergent synthesis"[14] or ligand-chelate assembly[15] has unequivocally validated their use as fundamental nanoscopic modules[16a-c]. Dendrimers may be used as substrates for sub-nanoscopic reagents (See (A), Figure 3) or as reactants for the construction of *nanoscopic compounds*, *clusters* and *polymers* (see hierarchy (B), Figure 1 and column (B), Figure 3). Reactions involving sub-nanoscopic reagents and dendrimers have been used to control the electron density (nucleophilic/electrophilic surface moieties), the polarity (hydrophilic/hydrophobic character) or specific design functions desired for a dendrimer surface (i.e., electron conducting[17a-b] paramagnetic[18], catalytic[19] or photon absorbing[20]. Nanoscopic reagents (e.g., DNA, antibodies or proteins) have been combined with dendrimers to produce new gene delivery vectors[12c,21a-b], immuno-diagnostics products[21c] or cell specific targeting devices for biological systems[21d].

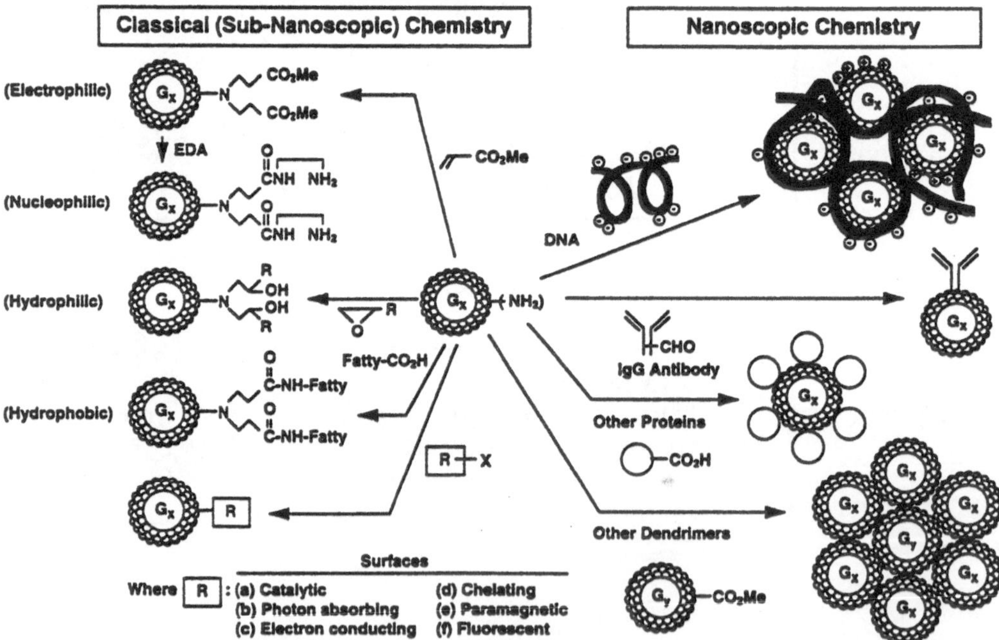

Figure 3. Sub-nanoscopic chemistry (A) and nanoscopic chemistry (B) on dendrimer surfaces (where: $G_x = G_y$ = generations 1 - 9).

Known chemistry associated with these nanoscopic constructions clearly differentiate dendrimers from sub-nanoscopic modules[3,5]. For example, the valency of a G=4, PAMAM dendrimer (NH_3 core) is 48 based on the number of primary amine surface (Z) groups available to react with sub-nanoscopic reagents (i.e., methyl acrylate or acetyl chloride <u>where</u>: dia. < 10Å) clearly demonstrates this stoichiometry. Reactions of this same dendrimer surface with nanoscopically sized entities such as IgG antibodies (dia.: ≃ 100Å) show that the stoichiometry is reduced to 2-4 under the most ideal reaction conditions. We have referred to these new combining rules as "sterically induced stoichiometry"[3]. They simply evolve from the "packing/parking dilemma"[27] created by covalently combining two nanoscopically sized reactants as illustrated in Figure 4. These new combining rules[3,5], unusual scaffolding effects[17a,18,19], unique guest-host relationships[22,23a-c] and novel supramolecular assemblies[24] appear to be directly related to the nanoscopic dimensions and complexity of dendrimers.

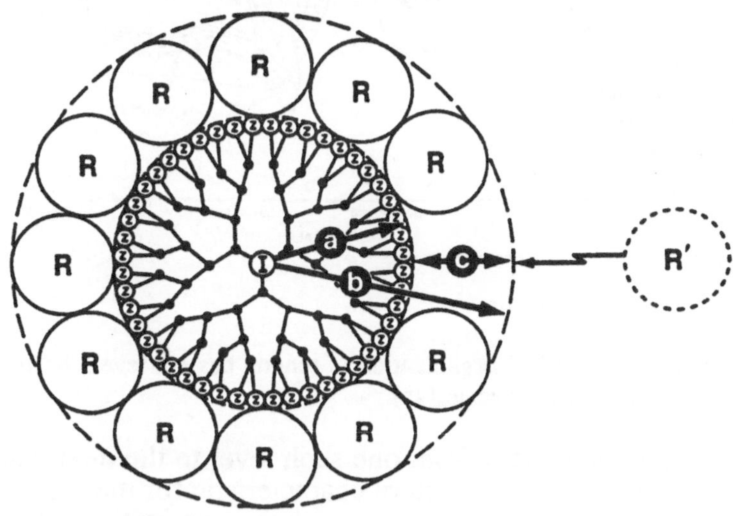

Molecular Space V_c Available for Attachment of Reagent (R')

$$= V_a - V_b = \frac{4}{3}\pi(r_a^3 - r_b^3)$$

Figure 4. Sterically induced stoichiometry (SIS resulting from the reaction of a nanoscopic dendrimer substrate possessing 48 surface groups (Z) with a nanoscopic reagent (R').

It should be noted that within the biological hierarchy of modules (Figure 1 (A)) there are certain well known pervasive patterns or properties which characterize the respective hierarchical levels of structural organization (i.e., I-IV). For example, progression through the dimensional hierarchies related to the construction of a biological tendon reveal that highly specified interfaces and architectural relationships are defined as these "dimensional modules" are organized and connected to produce higher complexity (see Figure 5)[5,25].

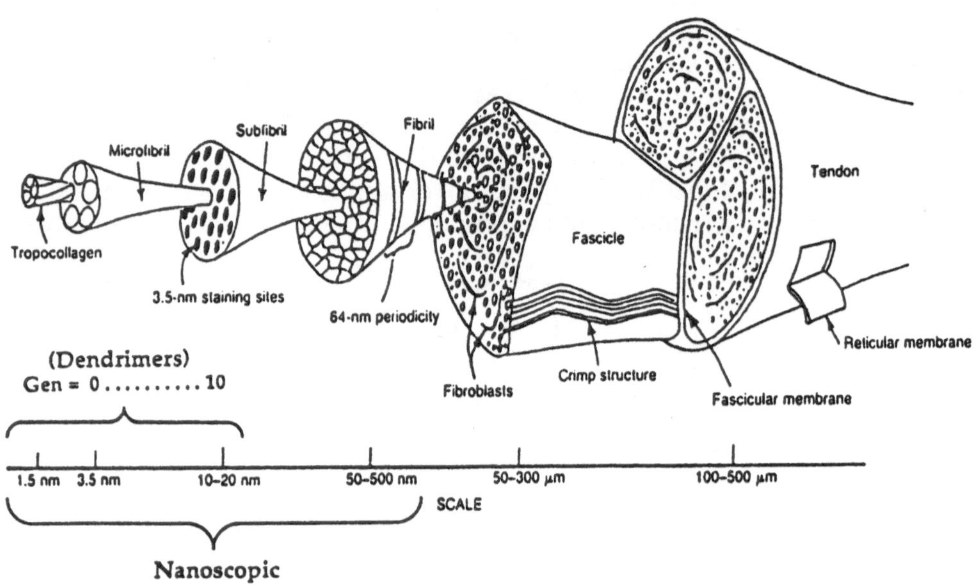

Figure 5. Hierarchical organization of tendon has six levels from the molecular to the macroscopic scales[25].

Advancement from one such level to the next usually involves the breaking of a property pattern characteristic for the preceding level and the establishment of a new order of behavior in the succeeding level. In essence, *"the whole becomes not only more than, but very different from the sum of its parts*[26]." At the same time, the emerging properties for a particular dimensional, or complexity level also provide the basis for the development of fundamental building blocks for the next higher level of complexity.

In summary, it will be interesting to determine if abiotic, dendrimer modules will exhibit similar hierarchical property differences as are observed for biological systems. If so, as these modules are assembled to higher structural complexity by design or coupled to produce dimensionally larger ordered arrays (e.g., Figure 1 (B)), one can only speculate on the possible new properties and characteristics that these megamolecules will exhibit.

Acknowledgements

We would like to thank the U.S. Army Research Office, U.S. Army Edgewood Research, Development and Engineering Center (ERDEC), Dade International and Dendritech Inc. for their support in funding this research.

References

1. Mason, S.F. (1991) *Chemical Evolution,* Clarendon Press, Oxford.
2. Joyce, G.F. (1992) Directed molecular evolution, *Scientific Am.* **267**, 90-97.
3. Tomalia, D.A., Naylor, A.M., and Goddard III, W.A. (1990) Starburst dendrimers: molecular-level control of size, shape, surface chemistry, topology, and flexibility from atoms to macroscopic matter, *Angew. Chem.* **102**, 119-157; *Angew. Chem., Int. Ed. Engl.* **29**, 138-175.
4. Heilbronner, E. and Dunitz, J.D. (1993) *Reflections on Symmetry,* VCH Press, Weinheim.
5. Tomalia, D.A., Durst, H.D. (1993) Genealogically directed synthesis: Starburst*/cascade dendrimers and hyper-branched structures, in E. Weber (ed.), *Topics in Current Chemistry Vol. 165: Supramolecular Chemistry I - Directed Synthesis and Molecular Recognition*, Springer-Verlag Berlin Heidelberg, pp. 193-313.
6. Maier, G., Pfriem, S., Schäfer, U., and Matusch, R. (1978) Tetra-tert-butyltetrahedron, *Angew. Chem.* **90**, 552-553, *Angew. Chem. Int. (ed) Engl.* **17**, 520.
7. Eaton, P.E. and Cole, T.W. (1964) Cubane, *J. Am. Chem. Soc.* **86**, 3157-3158.
8. a) Paquette, L.A. (1989) Dodecahedranes and allied spherical molecules, *Chem. Rev.* **89** (5), 1051-1065. b) Melder, J.P., Pinkos, R., Fritz, H., and Prinzbach, H. (1989) *Angew. Chem.* **101**, 314, The pagodane route to dodecahedranes: polyfunctionalized pentagonal dodecahedranes and dodecahedrenes, *Angew. Chem. Int. (ed) Engl.* **28** (3) 305-310.
9. Taylor, R. and Walton, D.R.M. (1993) The chemistry of fullerenes, *Nature* **363**, 685-693.
10. Müller, J., Base, K, Magnera, and T.F., Michl (1992) Rigid-rod oligo-p-carboranes from molecular tinkertoys. An inorganic Langmuir-Blodgett film with a functionalized outer surface, *J. Am. Chem. Soc.* **114**, 9721-9722.
11. Yang, X., Jiang, W., Knobler, C.B., and Hawthorne, M.F. (1992) Rigid-rod molecules: carborods. Synthesis of tetrameric p-carboranes and the crystal structure of bis(tri-n-butylsilyl) tetra-p-carborane, *J. Am. Chem. Soc.* **114**, 9719-9721.

12. a) Tomalia, D.A., Dewald, J.R., Hall, M.J., Martin, S.J., and Smith, P.B. (1984) *Preprints of the 1st SPSJ International Polymer Conference*, Soc. of Polym. Sci. Japan, Kyoto, p. 65. b) Tomalia, D.A., Baker, H., Dewald, J., Hall, M., Kallos, G., Martin, S., Roeck, J., Ryder, J., and Smith, P. (1985) A new class of polymers: Starburst-dendritic macromolecules, *Polymer J*. (Tokyo) **17**, 117. c) Tomalia, D.A. (1995) Dendrimer molecules, *Scientific American* **272**, 62-66.
13. Voit, B.I. (1995) Dendritic polymers: from aesthetic macromolecules to commercially interesting materials, *Acta Polymer*. **46**, 87-99. b) Fréchet, J.M.J. (1994) Functional polymers and dendrimers: reactivity, molecular architecture, and interfacial energy, *Science* **263**, 1710-1715. c) Newkome, G.R., and Moorefield, C.N. (1994) Metallo- and metalloido-micellane™ derivatives: incorporation of metals and nonmetals within unimolecular superstructures, *Macromol. Symp*. **77**, 63-71.
14. a) Fréchet, J.M.J., Jiang, Y., Hawker, C.J., Philippides, A.E. (1989) *Proc. IUPAC Int. Symp. Macromol.* Seoul, Korea, p. 19. b) Hawker, C.J. and Fréchet, J.M.J. (1990) Preparation of polymers with controlled molecular architecture. A new convergent approach to dendritic macromolecules, *J. Am. Chem. Soc.* **112**, 7638-7647.
15. Campagna, S., Denti, G., Serroni, S., Juris, A., Venturi, M., Ricevuto, V., and Balzani, V. (1995) Dendrimers of nanometer size based on metal complexes: luminescent and redox-active polynuclear metal complexes containing up to twenty-two metal centers, *Chem. Eur. J.* **1** (4), 211-221.
16. a) Tomalia, D.A. (1994) Starburst/cascade dendrimers: fundamental building blocks for a new nanoscopic chemistry set,*Advanced Materials* **6**, 7/8, 529-539. b) Tomalia, D.A. (1993) Starburst™/cascade dendrimers: fundamental building blocks for a new nanoscopic chemistry set, *Aldrichimica Acta* **26**, 91-100. c) Tomalia, D.A. (1995) Dendrimers - nanoscopic supermolecules according to dendritic rules and principles, in J.S. Siegel (ed.), *Supramolecular Stereochemistry*, NATO ASI Series, Kluwer Academic Publishers, Dordrecht **473**, pp. 21-26.
17. a) Miller, L.I., Hashimoto, T., Tabakovic, I., Swanson, D.R., and Tomalia, D.A. (1995) Delocalized π-stacks formed on dendrimers, *Chem. Mater*. **7**, 9-11. b) Duan, R.G., Miller, L.I., and Tomalia, D.A. (1995) An electrically conducting dendrimer, *J. Am. Chem. Soc.* **117**, 10783-10784.
18. Wiener, E.C., Brechbiel, M.W., Brothers, H., Magin, R.L., Gansow, O.A., Tomalia, D.A., and Lauterbur, P.C. (1994) Dendrimer-based metal chelates: a new class of magnetic resonance imaging contrast agents, *Magn. Res. Med.* **31**, 1-8.

19. Knapen, J.W.J., van der Made, A.W., de Wilde, J.C., van Leeuwen, P.W.N.M., Wijkens, P., Grove, D.M., and van Koten, G. (15 December 1994) Homogenous catalysts based on silane dendrimers functionalized with arylnickel(II) complexes, *Nature* **372**, 659-663.
20. Xu, Z. and Moore, J.S. (1994) Design and synthesis of a convergent and directional molecular antenna, *Acta Polymer.* **45**, 83-87.
21. a) Kukowska-Latallo, J.F., Bielinska, A., Johnson, J., Spindler, R., Tomalia, D.A., and Baker, Jr., J.R. (1995) Efficient transfer of genetic material into mammalian cells using Starburst dendrimer synthetic vectors, *Proc. Natl. Acad. Sci.* , in press. b) Tomalia, D.A. (September 28, 1995) Second Annual Artificial Self-Assembling Systems for Gene Transfer Conference, Wakefield, Massachusetts. c) Singh, P., Moll, F., Lin, S.H., Ferzli, C., Yu, K.S., Koski, R.K., Saul, R.G., and Cronin, P. (1994) Starburst™ dendrimers: enhanced performance and flexibility for immuno-assays, *Clin. Chem.* **40**, 1845-1849. d) Wu, C., Brechbiel, M.W., Kozak, R.W., and Gansow, O.A. (1994) Metal-chelate-dendrimer-antibody constructs for use in radio-immunotherapy and imaging, *Bioorganic & Medicinal Chemistry Letters,* **4** (3), 449-454.
22. Jansen, J.F.G.A.; de Brabander-van den Berg, E.M.M.; Meijer, E.W. (1994) Encapsulation of guest molecules into a dendritic box, *Science* **266**, 1226-1229.
23. a) Naylor, A.M., Goddard III, W.A., Keifer G.E., and Tomalia, D.A. (1989) Starburst dendrimers. 5. Molecular shape control, *J. Am. Chem. Soc.* **111**, 2339-2341. b) Ottaviani, M.F., Bossmann, S., Turro, N.J., and Tomalia, D.A. (1994) Characterization of Starburst dendrimers by the EPR technique. 1. Copper complexes in water solution, *J. Am. Chem. Soc.* **116**, 661-671. c) Ottaviani, M.F., Cossu, E., Turro, N.J., and Tomalia, D.A. (1995) Characterization of Starburst dendrimers by electron paramagnetic resonance. 2. Positively charged nitroxide radicals of variable chain length used as spin probes, *J. Am. Chem. Soc.* **117**, 4387-4398.
24. Newkome, G.R. et al., (1990) Synthesis and characterization of two-directional cascade molecules and formation of aqueous gels, *J. Am. Chem. Soc.* **112**, 8458-8465.
25. Baer, E., Hiltner, A., and Keith, H.D. (1987) Hierarchical structure in polymeric materials, *Science* **235**, 1015-1022.
26. Anderson, P.W. (1972) More is different, *Science*, **177**, 393-396.
27. Mansfield, M.L., Rakesh, L. and Tomalia, D.A. (22 Aug 1996) The random parking of spheres on spheres, *J. Chem. Phys.* **8**, 3245-3249.

ASSEMBLING TRIARYLMETHYLS INTO MESOSCOPIC-SIZE POLYRADICALS: HOW TO MAINTAIN STRONG INTERACTIONS BETWEEN MULTIPLE SITES IN A SINGLE MOLECULE?

ANDRZEJ RAJCA
Department of Chemistry
University of Nebraska
Lincoln, NE 68588

ABSTRACT. Design criteria and their implementation in synthesis of high-spin organic polyradicals, molecules with multiple interacting sites, are outlined. For high-spin polyradicals, where maintainance of through-bond interactions between many sites with "unpaired" electrons in a single molecule is essential, the following design criteria are identified: (1) defects, (2) through-space interactions, geometry (out-of-plane twisting), (3) connectivity between sites, dimensionality. Selected structures for polyethers, precursors for high-spin polyradicals, are discussed.

1. Introduction

Organic high-spin polyradicals with large number of interacting sites with "unpaired" electrons may serve as models for understanding of magnetic phenomena on the mesoscopic scale and, therefore, provide directions for search of novel materials [1]. In particular, through-bond interactions between many sites are especially important in providing strong and controllable coupling between the sites, possibly leading to interesting properties within a single molecule.

For mesoscopic polyradicals, there is a possilibity of observing magnetic phenomena associated with anisotropy barrier (E_A) in a single molecule. For example, superparamagnetism, which corresponds to a slow thermal relaxation of magnetization (high anisotropy barrier from spin-orbit coupling mechanism), was observed in large highly symmetric transition metal clusters [2-4]. Also, Iwamura reported an onset of superparamagnetism in a $S = 9$ nonacarbene, but the origin of the effect is not clear [5]. In polyarylmethyl polyradicals, based on three-coordinated carbon sites, the contribution of spin-orbit coupling effects to

J. Michl (ed.), Modular Chemistry, 193–200.

E_A is likely to be very small, leaving the possilibity of a major contribution to E_A from the classical magnetic dipole-dipole interactions; consequently, the elongated shape of polyradical could significantly increase E_A. Estimates of this effect in polyarylmethyl polyradicals can be found in the previous work [6]. If the dipole-dipole interactions could be make dominant, the calculation of the barrier is simple, making possible unambigous interpretation of new phenomena such as quantum tunneling of magnetization in ferromagnetic particles [4,7].

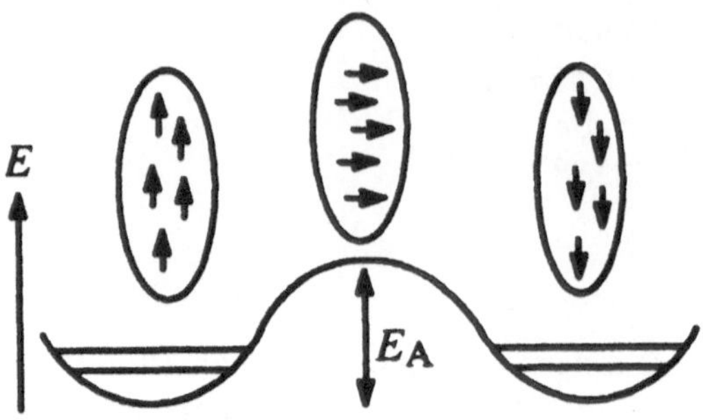

In this contribution, the design criteria and their implementation in the synthesis of high-spin polyradicals are outlined.

2. Spin Coupling

Molecules with two interacting sites, organic diradicals, are a convenient starting point for our discussion. For example, Schlenk and Thiele hydrocarbons [8,9] may conceptually be viewed as two spin sites interacting via a spin coupling unit;

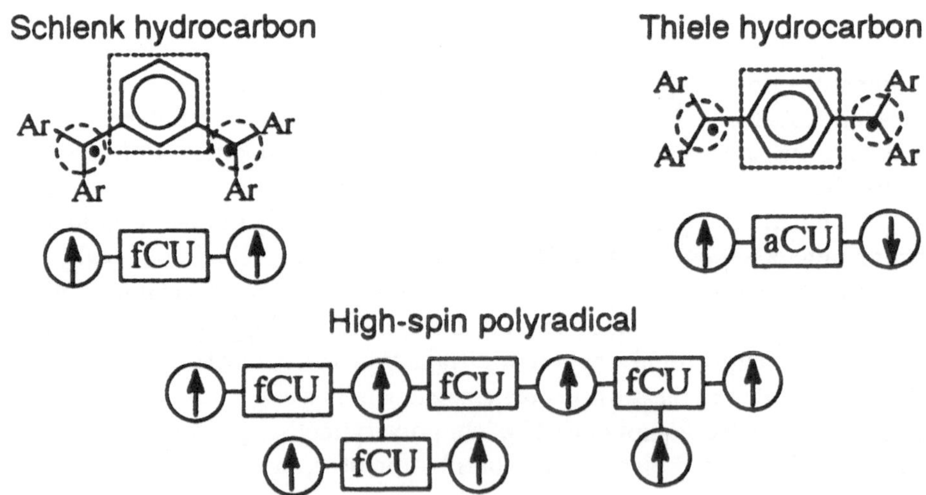

1,3-phenylene is a ferromagnetic coupling unit (fCU) and 1,4-phenylene is an antiferromagnetic coupling unit (aCU) [1]. In the case of strong spin coupling, as in Schlenk and Thiele hydrocarbons, connectivity within the coupling unit determines whether the spin coupling is ferromagnetic or antiferromagnetic. The concept of spin sites and coupling units is especially adequate for 1,3-phenylene-based systems because of their localized electronic structure [10].

Increase in number of interacting sites is straightforward in the case of spin coupled systems, i.e., organic polyradicals may be viewed as ensembles of alternating spin sites and coupling units. Connectivity between the coupling units determines the total spin of polyradical; the especially interesting high-spin polyradicals have fCU's as coupling units, implemented in numerous polyarylmethyl polyradicals [1].

3. Design of High-Spin Polyradicals

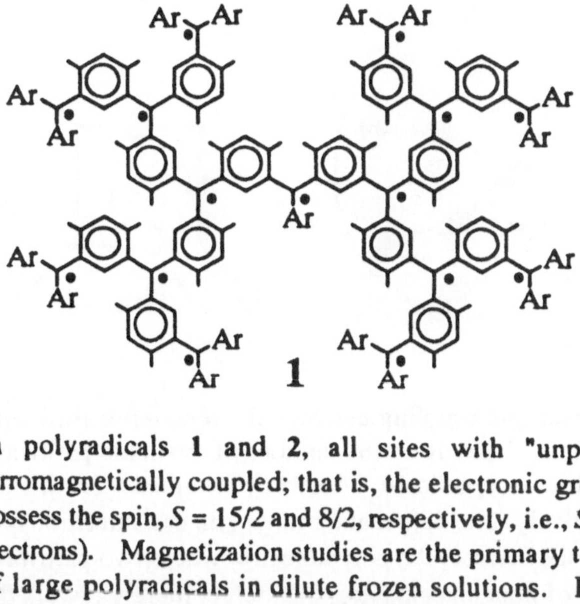

Ar = 4-*tert*-butylphenyl

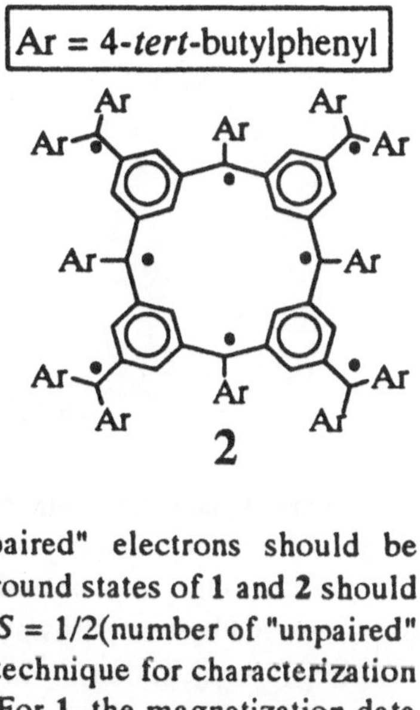

In polyradicals **1** and **2**, all sites with "unpaired" electrons should be ferromagnetically coupled; that is, the electronic ground states of **1** and **2** should possess the spin, $S = 15/2$ and $8/2$, respectively, i.e., $S = 1/2$(number of "unpaired" electrons). Magnetization studies are the primary technique for characterization of large polyradicals in dilute frozen solutions. For **1**, the magnetization data indicate a mixture of spin systems with a broad distribution of S "centered" in the $S = 5/2 - 7/2$ range, far from the expected value for S [6]. For **2**, spin systems with a very narrow distribution of S, centered at $S = 3.8$, near the expected value for S are found [11]. These results are better understood when the possible design aspects, which are associated with a maintainance of through-bond interactions between many sites in a single molecule, are identified: (1) defects, (2) through-space interactions, geometry (out-of-plane twisting), (3) connectivity between fCU's and sites, dimensionality.

Discussion of defects, which we define as failures to generate "unpaired" electron at the triarylmethyl site, must begin with chemistry associated with generation of polyradicals. Preparation of a polyradical starts with a convergent multi-step synthesis of a polyether, where Ar_3C-OMe sites correspond to Ar_3C sites in a polyradical. Polyether is isolated as a single pure compound; i.e., all Ar_3C-OMe moieties are intact within the limit of detection of typical characterization techniques for organic compounds. Subsequent conversion of polyether to polyradical is carried out concurrently at all triarylmethyl sites, according to eq. (1):

$$Ar_3C\text{-}OMe \xrightarrow[\text{THF}]{\text{Li or Na/K}} Ar_3\ddot{C}^{-} \xrightarrow[\text{180 K}]{I_2} Ar_3\dot{C} \qquad (1)$$

Because the reactions in eq. 1 are not quantitative, polyradicals will contain defects. For polyradicals **1** and **2**, the sites, the connecting pathways, and defects may be represented by dots, bars, and open circles, respectively [11].

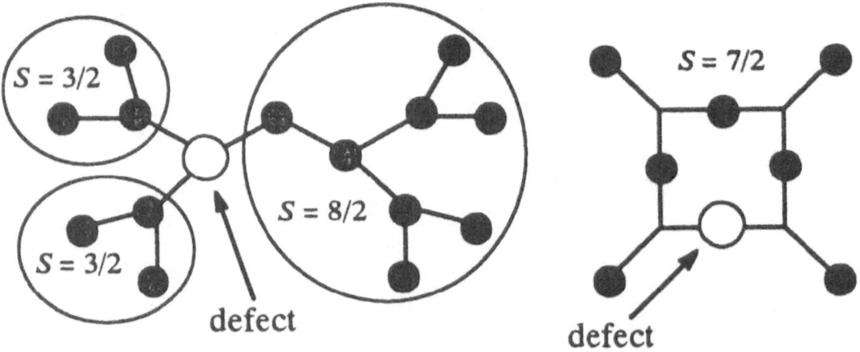

How the defect will affect spin coupling between the remaining sites with "unpaired" electrons in polyradical depends on the number of available pathways for spin coupling. In the dendritic polyradical **1**, where any two nonadjacent sites are connected by one pathway only, one defect at an inner site can interrupt spin coupling (π-conjugation). In the macrocyclic polyradical **2**, where two pathways are available between any pair of nonadjacent sites, one defect merely lowers the spin by 1/2 and all remaining sites with "unpaired" electrons are spin coupled [11]. We refer to such polyradicals as 0-proof and 1-proof, respectively. These differences between **1** and **2** would qualitatively account for their magnetization data, mentioned in the preceding paragraphs. Among many possible models, quantitatively accounting for magnetization data for **1**, the most straightforward approach is analogous to percolation on Bethe lattice with identical probabilities of defect at each site [6].

Through-space interaction and geometry (out-of-plane twisting) may also have a role in 1 and 2. In the dendritic polyradical 1, steric congestion may cause significant out-of-plane twisting for the π-conjugated system and through-space antiferromagnetic interactions between the denritic branches; both effects may weaken the through-bond ferromagnetic coupling. Moderate size macrocyclic rings, such as in 2, restrict the available conformations, limiting the impact of through-space interactions; furthermore, a moderate out-of-plane twisting of calix[4]arene ring provides steric shielding of the triarylmethyl sites, adding to the stability to the polyradical, and does not diminish the through-bond spin coupling.

Connectivity between fCU's and triarylmethyl sites in 1 and 2 is quite different; e.g., fCU's, 1,3-phenylene and 1,3,5-phenylyne, connect two and three nearest neighbor sites, respectively. The role of such connectivity is revealed when Heisenberg Hamiltonian (eq. 2) is applied to describe spin coupling in high-spin tri- and tetraradicals, assuming identical coupling constant, J, for the nearest neighbor ferromagnetic spin coupling. The energy gaps between the high-spin ground states and the lowest energy excited states clearly favor cyclic over branched connectivity and 1,3,5-phenylyne over 1,3-phenylene, as fCU [12].

$$H = -2J\sum_{i,j=nn} S_i S_j, \quad S_i = S_j = 1/2 \tag{2}$$

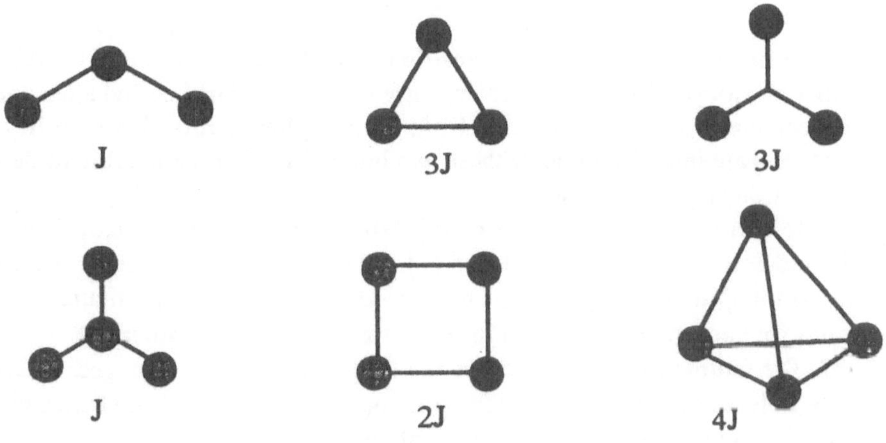

J 3J 3J

J 2J 4J

For molecules with very large number of interacting sites, which may be viewed as fragments of lattices, it may be instructive to consider the dimensionality of the lattice. In particular, magnetic phenomena are dependent on dimensionality, with two-dimensional lattices being the most sensitive to the details of the model Hamiltonians [13].

Implementation of the design aspects (1), (2), and (3) is illustrated by two polyethers, **3** and **4**, precursors for high-spin large polyradicals. Both polyethers are prepared according to multistep convergent synthetic routes; after each macrocyclic ring forming step, isomers are separated. This procedure allows for isolation of **3** and **4** as pure single isomers. Both polyethers are well soluble in organic solvents. ^{1}H NMR (500 MHz) spectra for all three isolated isomers of **4** in solution are broadened at ambient temperature but become sharp at elevated temperatures (75 °C).

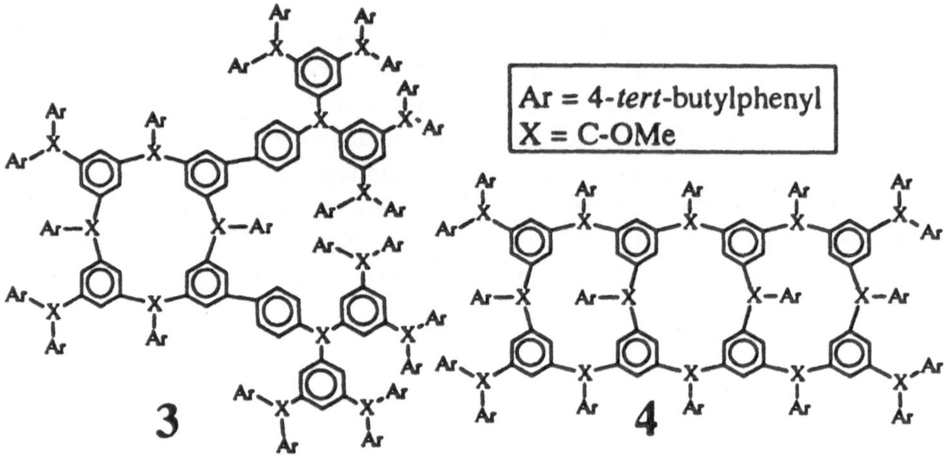

Ar = 4-*tert*-butylphenyl
X = C-OMe

In hexadecaether **3**, two dendritic pentaether branches are attached to a hexaether macrocyclic core. There are only two inner triarylmethyl sites, where one defect may interrupt spin coupling; their biphenyl moieties simplify the synthesis and are intended to make these two inner sites less vulnerable to defects in the reaction sequence in eq. 1.

In tetradecaether **4**, three calix[4]arene rings are annelated; **4** may be considered as a fragment of a 2-strand [1]. There are no sites where one defect can interrupt spin coupling. The elongated shape of **4** is of particular interest in search for superparamagnetism in the corresponding polyradical.

The future synthetic targets are the folded 2-strands, i.e., cages, where up to three defects do not interrupt the spin coupling. Elongated shapes can be implemented with biphenyl moieties, as shown for cage **5**.

While the complexity of synthesis of fragments of 2-strand is considerable, their adherence to all (1) - (3) design aspects should provide very-high-spin polyradicals. In particular, their superior resistance to defects compared to single macrocycles is evident, as illustrated by the parameter "Q", percentage of polyradicals with uninterrupted spin coupling (Table 1).

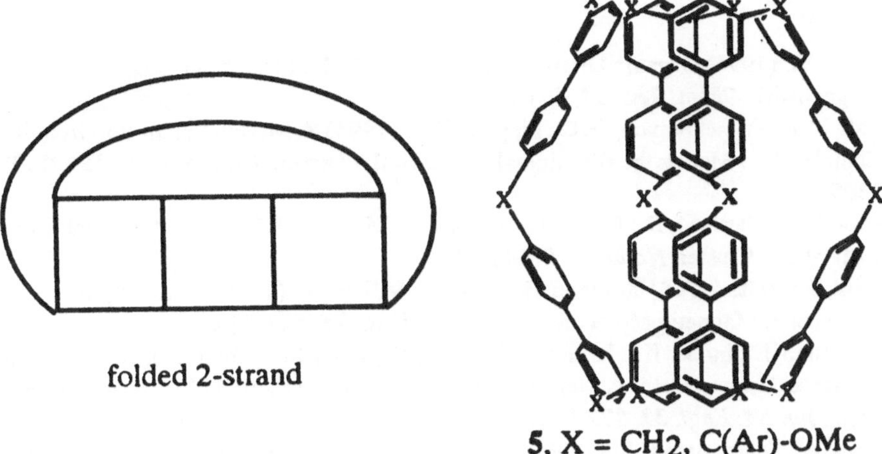

folded 2-strand

5, X = CH_2, C(Ar)-OMe

Table 1. Defect analysis of selected k-proof (k = 1 or >1) polyradicals. ("n" is the number of sites, "Q" is the percentage of polyradicals with all remaining sites coupled, and the 95 % yield per site is assumed.)

	n	Q
	8	98.6
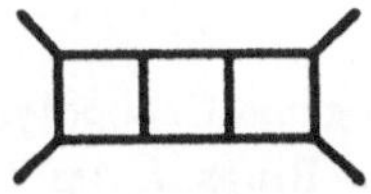	14	>98
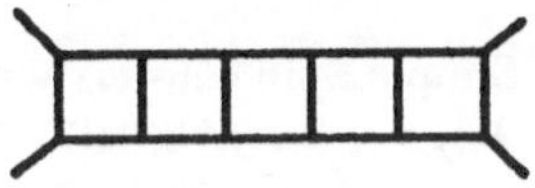	26	>97
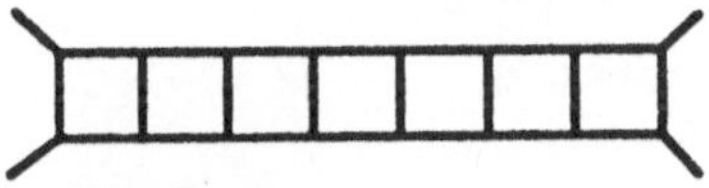	38	>96
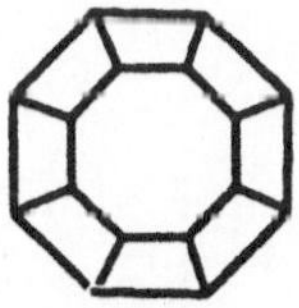	24	>99.3
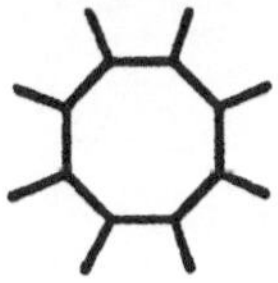	16	94.3

4. References

1. Rajca, A. (1994) Organic Diradicals and Polyradicals: From Spin Coupling to Magnetism? *Chem. Rev.* **94**, 871-893.
2. Taft, K. L.; Papaethymiou, G. C.; Lippard, S. J. (1993) A Mixed-Valent Polyiron Oxo Complex That Models the Biomineralization of the Ferritin Core, *Science* **259**, 1302-1305.
3. Sessoli, R.; Gatteshi, D.; Caneschi, A.; Novak, M. A. (1993) Magnetic Bistability in a Metal-Ion Cluster, *Nature* **365**, 141-143.
4. Politi, P.; Rettori, A.; Hartmann-Boutron, F.; Villain, J. (1995) Tunneling in Mesoscopic Magnetic Molecules, *Phys. Rev. Lett.* **75**, 537-540.
5. Nakamura, N.; Inoue, K.; Iwamura, H. (1993) A Branched-Chain Nonacarbene with Nonadecet Ground State: A Step Nearer to Superparamagnetic Polycarbenes, *Angew. Chem. Int. Ed. Engl.* **32**, 872-874.
6. Rajca, A.; Utamapanya, S. (1993) Toward Organic Synthesis of a Magnetic Particle: Dendritic Polyradicals with 15 and 31 Centers for Unpaired Electrons, *J. Am. Chem. Soc.* **115**, 10688-10694.
7. Awschalom, D. D.; DiVincenzo, D. P.; Smyth, J. F. (1992) Macroscopic Quantum Effects in Nanometer-Scale Magnets, *Science* **258**, 414-421.
8. Schlenk, W.; Brauns, M. (1915) Zur Frage der Metachinoide, *Chem. Ber.* **48**, 661-669.
9. Montgomery, L. K.; Huffman, J. C.; Jurczak, E. A.; Gendze, M. P. (1986) The Molecular Structures of Thiele's and Chichibabin's Hydrocarbons, *J. Am. Chem. Soc.* **108**, 6004-6011.
10. Rajca, S.; Rajca, A. (1995) Novel High-Spin Molecules: π-Conjugated Polyradical Polyanions. Ferromagnetic Spin Coupling and Electron Localization, *J. Am. Chem. Soc.* **117**, 9172-9179.
11. Rajca, A.; Rajca, S.; Desai, S. R. (1995) Macrocyclic π-Conjugated Cabopolyanions and Polyradicals Based upon Calix[4]arene and Calix[3]arene Rings, *J. Am. Chem. Soc.* **117**, 806-816.
12. Belorizky, E.; Fries, P. H. (1993) Exact Solutions for Simple Spin Clusters with Isotropic Heisenberg Exchange Interactions, *J. Chim. Phys. (Paris)* **90**, 1077-1100.
13. Mattis, D. C. (1985) *The Theory of Magnetism II*, Springer-Verlag, Heidelberg.

DISCUSSION OF THE TOMALIA AND RAJCA LECTURES

GEORGE R. NEWKOME
Department of Chemistry
South Florida University
Tampa, FL 33620
U.S.A.

Professor NEWKOME begins: I am going to go back and add a slight perspective to these two talks. It really goes back to 1978, with Professor Fritz Vögtle's paper in *Synthesis*. In that paper, he defined really two new types of organic construction: one which now has taken on the term "dendritic," and can be defined by a mathematical progression. In that paper he called it a "cascade synthesis." Also in Vögtle's 1978 paper in small print is a second mode of construction which he called a "tire tread synthesis," the concept was synthetically shown in the slide from Professor Rajca's presentation. It was then in 1985, when Don and I published in separate papers the thought process of utilizing a cascade methodology to construct polymers via a mathematical progression. Don used a 1:2 branching mode of construction, I used a 1:3 pattern for the basic construction of molecular trees.

As time has gone on, it has become obvious that as you grow these macromolecular spheres, there are really only several places where chemical reactions can actually occur. They can take place on the surface, but if they do, you then have a large number of loci for functionalization or substitution. And so early on, in fact in that 1985 paper, we coined the term "unimolecular micelle," and since we have been dealing with semantics at this meeting, I defined a unimolecular micelle as a single molecule that possessed micellular character. In 1991, we proved that things do go inside, and stay inside the macromolecule--in fact recently, Professor Meijer really proved that one can put in and secure things inside the macromolecule; he will be telling you about that later in this meeting. The point is as the chemistry of these cascade molecules unfolds, there are several very basic things that start to spin forth--if you take into account the "unimolecular micelle" concept.

J. Michl (ed.), Modular Chemistry, 201–218.

And from that, I would like to show you three or four (no more) slides depicting our current thought process concerning the creation and use of the resultant unimolecular micelles.

It works out, that the framework, and everything which goes with it, is really secondary to the basic concept, if you want to do chemistry inside the molecule; it can be demonstrated very clearly here.

We saw in Tomalia's presentation the creation of random networks as demonstrated in Slide 1. Multiple surface functionality, such as amines, react with carboxylic acids to form polyamides. The greater the size, greater the number of architectural possibilities. And so this example defines a

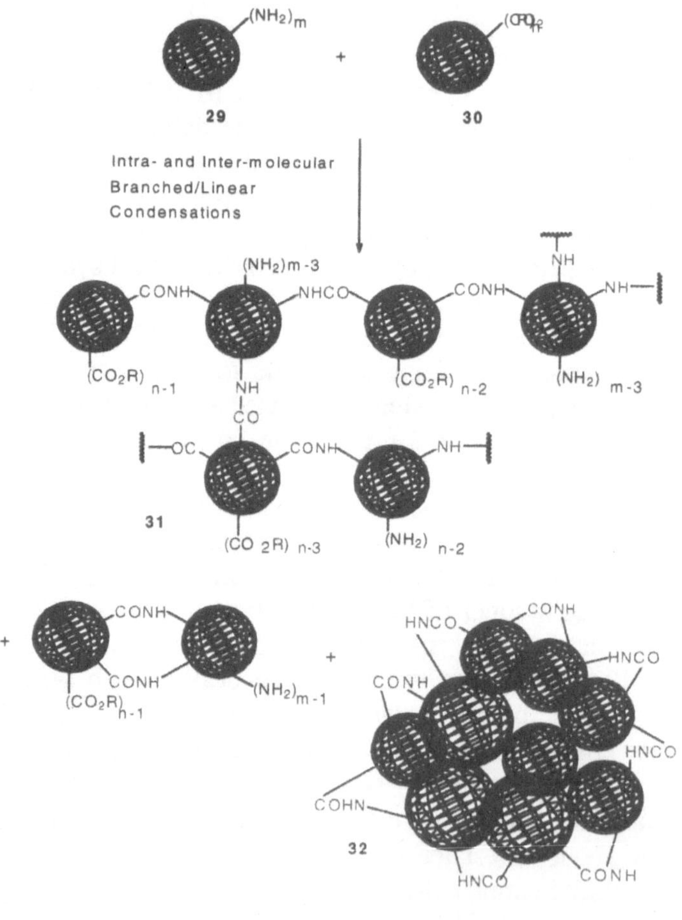

Slide 1

"random process" because there is no way of stopping or controlling the structure of the resultant macromolecule--it's just one out of many reactions which, in fact, are generated in this random procedure.

In an attempt to control the chemistry and thus the construction process, we can start to take advantage of the inside region of these macromolecules. What we have done is to utilize the synthetic methodology to create complementary functionality, which can reach out and attach itself inside another macromolecule. This is demonstrated in Slide 2. Assemblage of components by this mode of connectivity affords facile access to specific networks. You should easily envision the application of this procedure to form specific assemblages with a predetermined number of binding loci.

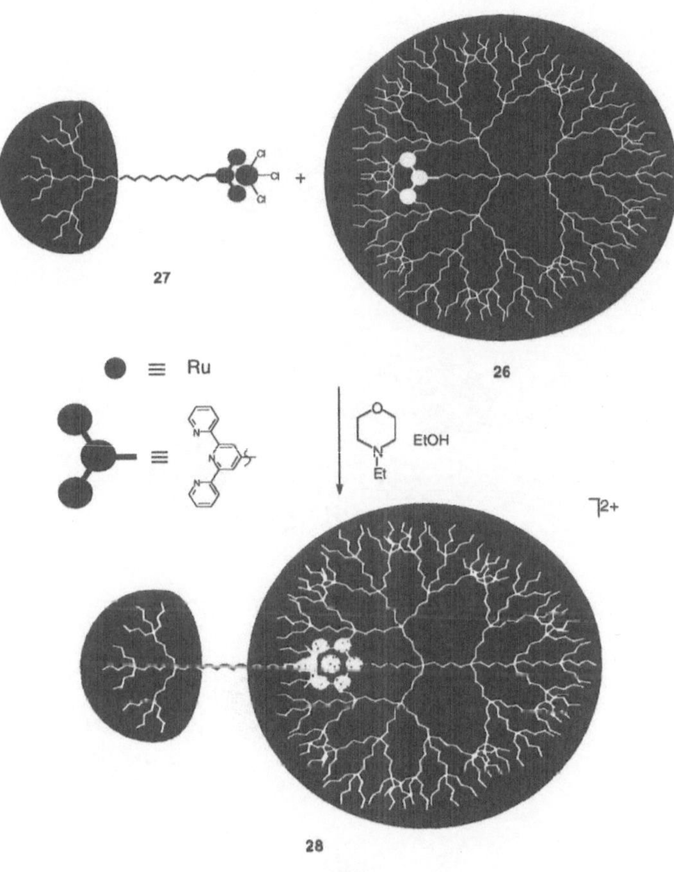

Slide 2

Conversely, one would generate a system where in fact you have specific sites of the molecule and you would in fact put them together in a very ordered process. The putting of things together in a dimeric species is, in fact, the first part of an entire assemblage, if you wish, of making large networks see Slide 3.

As you look at the assembly process, one really looks at the overall scope of current construction process, you have a choice of totally ordered networks to those that are totally random. And you are going to find multiple possible combinations from one extreme to the other.

The perfect network assembly goes back to an interesting analogy of yesterday, and that is a crystal. Because many times we isolate a single crystal and in fact it is one out of many we choose--whether that is in fact representative of the mixture may not be known, because most organic chemists extrapolate the data from that one crystal to the bulk. As we start to

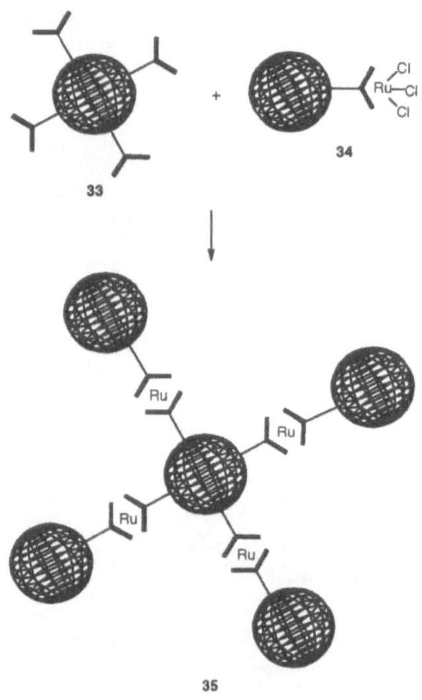

Slide 3

look at the specific effects, dendritic construction using dendrons are going to take place also in a very ordered and very specific way - the later example - to something which can be very, very random. And I think as we start to move forward to matrix assembly we will be making networks of larger and larger sizes, it is of absolute paramount importance that we define what is in fact put together by random process and what is put together by a specific process. And I might point out that it is really the difference between, if I may use the liberty in this room, the industrial thought process vs. the academic thought process where in fact we want something cheap vs. something expensive. We need to have always, in the construction process, an absolute predetermined molecular assembly - so that we can know what a perfect network is like. If you make something by a random process than there is not necessarily a correlation to a perfect model.

And with that I should also say that since I'm the guy who is going to transcribe this, I personally am not going to do it--I'll tell you ahead of time--sorry Josef. I have a secretary who is very good at this and so it reminds me of the story of when I gave a lecture in Japan. I was standing up talking, and of course there was an interpreter on the other side of the floor taking my words and putting them in Japanese for the audience and throughout this talk everyone is laughing and since it was a serious talk I couldn't understand.

So afterwards I said why is everybody laughing. They said, "Well we really couldn't get a chemical translator--we got somebody from Pharmacy." Since my secretary knows zero Chemistry, my comment would be that would you please give your name and, on top of that, be very articulate in what you say because she will translate everything exactly the way you say it!

And lastly, I would like to ask, since this is a very diverse group and we've been dealing with semantics, a challenge to Josef is that he should put together a glossary of terms so that this diverse group of people can in fact understand a common language.

Questions:

KAHN (Bordeaux): I would like to comment on Professor Rajca's presentation. You mentioned that one of your main goals was to observe a quantum tunneling effect, and for sure it's quite a challenging goal. However, I am a bit skeptical about the success, using organic radicals. The quantum tunneling effect requires a very strong magnetic anisotropy. In your case, the only anisotropy you can play with is the shape anisotropy, and I am not sure that such a shape anisotropy could be sufficiently pronounced to give rise to a large energy barrier. By the way, you are probably aware that in the field of Inorganic Chemistry, a magnificent case of quantum tunneling effect has been recently described. It concerns a Mn_{12} cluster with eight Mn (III) and four Mn (IV) ions and a S+10 ground state spin. Could you comment on my skepticism?

RAJCA: The possibility that the anisotropy is dominated by the shape anisotropy has both advantages and disadvantages. From the point of view of understanding, the dominance of the shape anisotropy would allow very simple calculation of the anisotropy barrier, using classical dipole-dipole interactions; thus, the complexity of the alternative mechanism for the anisotropy, based on spin-orbit coupling, would be avoided. The disadvantage is associated only, especially in spin diluted systems such as ours. This problem of the barrier height is addressed in a footnote in one our papers in 1993 [1]; the basic requirements for a significant height of the barrier are highly elongated molecular shape and the large number of unpaired electrons. An example of the system, that is the step toward that goal, is the polyradical based on the triple calix[4]arene **4**, which was just discussed. To my knowledge, only superparamagnetism was observed in the manganese cluster; whether the anisotropy barrier was predominantly traversed by the thermal motion or tunneling was not established unequivocally.

KAHN: It is not correct. In the Mn_{12} cluster I spoke about, the quantum tunneling effect has been characterized in an unambiguous way. What is very appealing with this system is that it exhibits the phenomenon of bistability at the purely molecular scale. I think that it is the first report of that kind. The theory of the phenomenon was recently developed by Vilain and co-workers from Grenoble in France.

TOMALIA: In response to a question, "How does one bind, bond, or attach cellulite IG1 to a dendrimer in a way that conserves its interesting properties?", the following answer was offered: A number of ways! One is by oxidizing the carbohydrate portion of the constant region to an aldehyde and then condensing it with amine functions on a dendrimer surface to give a Schiff's base and then doing a hydrogenation. And by the way, when you do that you actually conserve the recognition characteristics of the variable region. I might add that there is a product coming out from Baxter Healthcare International based on the nanoscopic connection of a dendrimer (G5) to a suitable IgG antibody exactly like I showed you. This antibody conjugate is to be used in a heart attack diagnostics kit. A CKMB kit, a creatinine kinase kit. Its performance benefit is that it will reduce the time for diagnosis from 40 minutes down to below 5 minutes. So there are some very distinct commercial advantages derived from these nanoscopic dendrimer modules by simply connecting them to Nature's nanoscopic modules such as an antibody.

SEEMAN (New York): I have two questions for Don. The first one is, you have 50 Å Fc immunoglobulin domains targeted to bind to a 50 Å particle, and you said, "Well how many should we get". And you said, "four" and I was thinking back to my freshman chemistry and I said to myself, "twelve". So my question is why is four the magic number and not twelve.

TOMALIA: It appears that the two 50 Å recognition sites (variable regions) that form the Y-like shape of the biped antibody, carve out substantially more space than just the space accounted for by a single 50 Å FC (constant region) domain. Even if you take an excess of the whole IgG and allow it to react with the dendrimer, the maximum number we can get on the surface is four. So this reduces the stoichiometry from 96 surface groups to four. It's definitely related to what George [Newkome] is talking about. Now we have a valency of 8, considering the two recognition groups per IgG antibody. I don't know if I have a slide to show that, but it's a very nice way to take the 96 dendrimer surface groups and reduce them down to a very controllable eight recognition site based on the dendrimer attached antibodies.

SEEMAN: The second part of my question is a technical question. How do you keep rogue strands from growing when you grow a dendrimer. Does this involve alternation of a generation or something like that?

TOMALIA: Yes, what happens, if it wasn't clear, it shuts itself off at each of these growth stages either when you add ethylene diamine or methyl acrylate. At these growth stages, you saturate to the valence limit of either the ester or amine precursor from the previous generation. Even if you add a thousand-fold excess it still shuts itself off. The amazing thing is that you don't need that much excess to actually see very mono-disperse material.

TOMALIA: Do we have an example of a dendrimer containing electrically conducting groups inside the core? We don't have those types of dendrimers, however as you heard earlier, Jeff Moore does. We have electron conducting groups only on the surface. It was one of the slides I slid through so quickly because I was running out of time. These conducting groups are TCNQ related moieties that we put on the surface. We find that conductance is generation specific. It gives a conductivity that is approximately in the same order of magnitude as polyacetylenes. It's electrically conducting, it's not a redox site or due to ionic conduction, but it is very generation dependent. We know it involves aromatic stacking and it's isotropic--it conducts in all three directions, it doesn't have to be doped and it's not sensitive to moisture or to air. This work has been accepted for publication in the *J. Am. Chem. Soc.* and will appear in the next few months. It is a surprising scaffolding effect offered by dendrimers which manifests isotropic electrical conductivity. What we have not done, however, is to demonstrate some of the beautiful redox type phenomena for dendrimers that have been reported here in Vincenzo Balzani's poster [at this meeting] as well as in the most recent issue of the *Chem. Eur. J.* If we have a chance, I would like to come back to that later if we need some provocative issues to talk about.

MORTIMER (UCLA): Can you assemble, order, or array dendrimers on a surface?

TOMALIA: Order and array yes. Let me give you an example: If you heard the talk earlier about constructing multilayers by assembling proteins (cationic) and alternating them with anionic polymers, we've done something similar to that with dendrimers. Is that what your are talking about? Basically, we took a negatively charged dendrimer followed by a positively charged dendrimer on top and simply alternated the charged layers of dendrimer. It saturates at each stage as you just dip in a beaker containing anionic dendrimer followed by dipping in a beaker containing cationic dendrimer and you obtain many, many layers in this fashion.

MORTIMER: I saw your rather disordered looking arrays of nucleophilic dendrimers reacted with electrophilic dendrimers. Where and how do obtain order?

TOMALIA: None of the slides I have shown demonstrate this order you asked about. There was no connection to this issue in these slides. These were just examples of covalent connectivity observed to give supermolecular organizations by covalency reacting electrophilic with nucleophilic dendrimers. These dendrimers will, however, self organize under other conditions to produce supramolecular arrays in very unusual ways, much like biological "actin". In the poster session [at this meeting], or if you are interested, I have one more quick slide to show you now how dendrimers can organize into "actin-like" (rod type) structures. Tonight I'll have it in more detail in the poster session.

CHIDSEY (Stanford): This is a question for Professor Rajca. I'd like to follow up on the issue of quantum tunneling. Perhaps it is overly practical of me, but I would like to suggest that one of the obvious reasons to study "modular chemistry" in the context of magnetic materials would be to get magnetic objects that are at room temperature with the minimum possible size. It seems to me that what you want to know is what is the minimal size object for which quantum tunneling doesn't occur in archival data storage applications. On that basis, I think it's valuable as we think about assembling objects with the possible application of magnetics, to get fairly serious. At least in your situation, you have a fairly simple dipolar coupling mechanism and the anisotropy could be characterized by some eccentricity. It should be possible to say how big something like that would have to be before it's archival. Is it even nanoscopic any more? Or are we back up to micron-sized objects? Can you comment on this? I think it actually bears on the issue of when does modular chemistry start to meet real materials science.

RAJCA: I would emphasize the importance of any new phenomenon or material in novel possible applications, that greatly expand our technology and knowledge; an example would be to use quantum tunneling in computing. Organic polyradicals may provide particularly simple models for quantum tunneling on the mesoscopic scale, providing better understanding for the phenomenon.

CHIDSEY: I'm sorry but I think that you have avoided my questions, not that I don't think there are new opportunities, it's just that I'm saying there is this paradigm which a graduate student listening to this talk might be tempted to think is accessible to one. My question is, it's not such a hard problem to calculate, how big does an object have to get for reasonable parameters before it has archival storage capabilities.

RAJCA: I commented on the barrier height in my answer to Professor Kahn.

CHIDSEY: No that was to be able to *observe* quantum tunneling, I'm asking to be able to shut down quantum tunneling. So I'm being very provocative by saying, in Chemistry Departments, especially my own, it is very tempting to look at these kinds of situations and make obvious analogies to existing storage devices. I suspect, I'll make the claim although I haven't done any calculations to back this up, that we're talking about objects which are now close to one micron in size before they become stable against quantum tunneling, and I submit that one ought to be at least open to saying that's not where the applications lie for this kind of modular chemistry. I think it's beautiful chemistry, but there is sort of this naive temptation among the less sophisticated in the community to make these direct connections, and I just want to vocally and openly say that I think that it's an important obligation to identify where we know the applications can't be as much as looking for the new opportunities.

KAHN: I would like to come back briefly to this point. Usually, the quantum tunneling effect is observed in superparamagnetic particles, containing typically 400-500 metal atoms. The investigation of this phenomenon is a very active area of research in the physics community. What I spoke about is something else. In the Mn_{12} cluster, I already mentioned, the quantum tunneling effect occurs in a molecular, perfectly defined object, with a small size. I really think that it would make sense to look for new molecular compounds of the same kind. The observation of bistability at the purely molecular scale, without cooperative interactions between the molecules within the lattice, is still a challenge.

WEGNER (Mainz): This is a comment to one of your [Tomalia] provocative statements when you compared to polydispersity of your dendrimers of different generations to that of ordinary linear polymers of similar molecular weight. This comparison is a bit unfair because the dendrimers have been growing in only up to ten generations. That is, their construction involves not more than ten reaction steps. Because each addition to a reactive chain end has a finite probability of success even under the most carefully controlled conditions of growth one ends up with a polydispersity of chain lengths and thus molecular mass distribution. The polydispersity is a consequence of the sequential repetition of the same chemical step for a thousand or more times. As the dendrimers have a history of only 4-10 steps, they must be "monodisperse" by the very nature of the statistics involved in their growth.

TOMALIA: That is a good comment and I wasn't trying to say that dendrimers are better, bigger, or more useful than classical polymers. I was simply trying to make a comparison by showing we're dealing with a continuum of molecular weights when we think of classical polymerization as opposed to well defined growth stages (molecular weights) when we speak of dendrimers.

WEGNER (Mainz): But you used the wrong comparison.

TOMALIA: The only way I have to normalize the two systems is to invoke molecular weights (MW) and that's the only direct comparison I can make. I agree with you, Gerhard, there are many events going on when you do a "one pot" classical, random polymerization. Of course, from biology we learn that the ability to control structure at pre-defined stages is what Nature has shown us is very important in building modules for evolutionary purposes.

When you have polydispersity of structure (molecular weight) you can't go on to the next step and design the next emerging set of properties. In fact, I have a couple of slides that are more provocative than that if I could just show one simple slide. That is the advancement from one complexity (dimensional) level to the next usually involves the breaking of a property pattern and the establishment of a new order of behavior in the succeeding level. Therefore as one progresses from one dimensional level to the next modular level a very prophetic thing happens. It was stated by Phil Anderson (see abstract) in 1972. He said that, "the whole becomes not only more than but very different from the sum of its parts." The only way you can sort out these subtle differences is to have control over structure. You lose a lot of those subtle differences when you have polydispersity. My comment is that we go to each of these complexity levels, we've been able to, as organic chemists, sort out optical activity or we can sort out geometric isomers and, for instance, we can sort out a variety of emerging properties only if we are able to control critical design parameters of structure such as size, shape, surface chemistry, topology, and flexibility. In polymers, we look at statistical properties--here you can look at articulate properties the way the organic chemist does. And Nature has chosen that strategy just based on hierarchies we've seen.

COLEMAN (Lyon): I would like to challenge you [Tomalia] totally on your concept that the dendrimers are proteins--they're not. The natural analogs of the dendrimers are the glyco forms which you find on protein surfaces. They're divergent but there is a big problem in these things that Nature doesn't actually care very much about polydispersity of these systems. They're highly disperse and they still function beautifully. So why are you talking about proteins when you should be talking about glyco forms?

TOMALIA: My answer to that is that proteins have pretty well defined and precise scaffolding.

COLEMAN: So do glyco forms.

TOMALIA: And that's largely what the uniqueness of dendrimer is all about. Dendrimers really offer pretty precise scaffolding and also very precise shapes. There are strong comparisons that you can make when we look at a generation 5 dendrimer (diameter = 53 Å) compared to hemoglobin (diameter = 55 Å). As you can see, it has the same size, same shape, however, different function. Furthermore, comparative electrophoresis experiments and size exclusion chromatography experiments show that they behave similarly because of their analogous sizes and shapes.

COLEMAN: It doesn't have the same size because most of these proteins have been glycosylated and are in fact twice as big as you're actually saying. The 1gg molecule has several points of glycosylation which makes it much larger than you state and larger than you calculate for covering the surface of your dendrimer. And that's why it doesn't fit 12, it fits four, because you've got these "bloody" glycoforms around blocking it.

TOMALIA: That may be part of it. But the real issue is, you could not up until this time, make such a size and shape comparison between a totally synthetic polymer and a natural macromolecule that I'm aware of. It is truly unique to see synthetic polymers behave like proteins and have protein chemists state on a regular basis, that these dendrimers are acting like proteins. They are precise bundles of mass with reproducible sizes and shapes and that's what we are focusing on. We're not focusing on the polydispersity or the richness of the surfaces that are found in the natural polymers. That is a more complex issue which will be dealt with later. Let me give you another example. In the case of gene transfection, we mimic histones which are a very discreet cluster of proteins, right? It has a specific size, when we mimic that size, that is when we get our best gene transfection. In this case, we didn't deal with the notion that histones may have a lot of glyco scenery on it.

We simply followed the notion that it had a particular size and a particular shape and it had a unique, conserved function in Nature as a scaffolding for compacting DNA. When we followed those simple rules, we were successful. When we violated those rules we failed.

COLEMAN: You [Tomalia] could do the same thing for histone with any latex having a high choice density.

TOMALIA: I would like to have you [Coleman] show us that because there are many, many companies out there looking for such a simple solution.

COLEMAN: It's a size choice ratio.

TOMALIA: It's not that simple. Then you need to talk to some important pharmaceutical companies because they're looking for those materials. I am chairing a gene transfection conference entitled "Artificial Self-Assembling Systems for Gene Transfer" next month in Boston. You should show up there and describe your approach because there are many pharmaceutical companies that would pay you millions of dollars if your idea would work. It has not been demonstrated for a latex--it is very unlikely to work. Latexes are cross-linked insoluble polymers to being with. They have to undergo "endocytosis" (penetration) through the cell membrane. You would have to take a cross-linked polymer inside the cell and have it do all of these--it just doesn't work!

PRINZBACH (Freiburg): Volume or size is not a very definite descriptor of cell activity. The question I have is for you, Prof. Tomalia. You had examples of where you had the participants similar in shape and size where cell activity broke down. The size can be small, and it can be large.

TOMALIA: And the answer to your question is we don't understand all the rules. These are the new still undefined, evolving rules of this new nanoscopic chemistry set. These rules may be as new and undefined as they were for Dalton nearly 200 years ago when he was describing the behavior of atoms as building blocks. When Dalton was looking at these atomic level building blocks (spheroids) he didn't initially know what those rules were. We are looking at nanoscopic level building blocks (spheroids) and we are trying to define and articulate rules that describe the behavior of these dendrimers while constructing more complex systems. I can honestly say that we still are quite naive in that quest. We do not know where these rules will take us.

NEWKOME: I would like to interject here if I may. If you [Tomalia] have thirty-two or sixty-four groups in the surface, and you pick out six of those, the number of isomers that you have, because of simply the mere mode of construction, are going to be overwhelming. I can't imagine that you're going to able pick out six specifically each time--and then go back and prove it. I don't see any way of proving, at least with our current technology.

TOMALIA: I agree with you [Newkome] on that point. If, however, you differentiate dendrimer sectors (dendrons) and basically it's a problem that has been solved, but it hasn't been completely shared in the literature, then it is possible to direct attachments to predefined regions of the dendrimer surface. There is a synthetic problem that has been solved. Basically it involves taking one dendron consisting of surface groups type A. You couple it to a core to make a single dendron. You then take a second dendron consisting of surface groups B to make a second differentiated dendron followed by a third differentiated dendron to give a dendrimer with differentiated dendrons containing surface groups A, B, and C. You then have differentiated reactivity at three different regiospecific sites on the dendrimer surface. For example, you may have ester groups here, you may have amino groups there and some other functionality in the third sector.

Now you have designed in differentiated reaction selectivity for subsequent reaction to create higher ordered complexity based on the regiospecifically place functionality. Now that is what we are presently doing in our Laboratory. Unfortunately, I'm not prepared to tell or talk about that today. It's been done at Fréchet's lab, it's been done in our laboratory, I think it's being done in Jeff Moore's laboratory, and this is the excitement of being able to selectively assemble shape designed constructs. Imagine having a "Velcro" in one dendrimer sector that sticks to "Velcro" on another--either a spheroid or a rod. You have the potential for self-assembly, you have the potential for discrete covalent bond formation, you have the opportunity to build nano-structures very predictably. That's what we're in he midst of. That's the excitement associated with these new nanoscopic building blocks.

NEWKOME: This is for Fréchet chemistry and that's absolutely totally correct. But this is a different story because you [Tomalia] were specifically synthesizing each of these dendron arms with thirty-two or sixty-four functional groups and you said that you can pick out six specific sites on that surface--I don't see how that can be done at this time. I think that you can synthesize the specificity onto or into the molecular structure but to selectively pick specific sites out of many similar sites is not possible.

TOMALIA: I can advance a dendrimer synthesis to a desired generational level to produce the valency that I want. If I want to have six amino groups, I can simply use a masked amine containing reagent that will sterically allow only six of those reagents to react on the dendrimer surface. This can be adjusted by the size of the reagent used.

NEWKOME: You're missing the point, I think it's the point they're getting at, and that is there's a difference if you construct it, you can put in whatever specific functionality you want wherever you want it, by either convergent or divergent route.

Once the molecule is made, and you have on it in hand, the surface will possess a surface with identical or nearly so functionality. I don't see how you can come along and pick out more than one pattern of identical groups in any repeat manner because of the number of similar groups and our lack of synthetic talents to accurately and repeatedly select a predetermined pattern on the surface.

TOMALIA: Oh, I misunderstood your question, yes that's correct. It is true, "*sterically induced stoichiometry*" (SIS) can reduce that multiplicity but it will not recognize a regiospecific site on a dendrimer surface. It will not pick out the exact spot--you're absolutely correct. The way to reach this objective is by this kind of differentiated dendrimer synthesis that I described above involving a differentiation of dendron sectors. Even then one has regiospecific control only to a desired sector, not necessarily to an individually targeted group within that sector.

MALLOUK (Penn State): I have a question for Don Tomalia. One of the things that determines the size of the dendrimer you get, when the outer branches ultimately bump into each other, is the size of the chemical bond in the primary building block. This seems to be an interesting opportunity for tinkertoy approach, where you have very long connectors made by coordinate covalent coupling or something like that. With large enough synthons you might get to micron scale objects after a few generations.

TOMALIA: That's absolutely correct! Many of the things on this slide have been demonstrated and unfortunately they're not in color to differentiate some of the possibilities. For example, one can select, as a core, an atom, one can select a small molecule or a functional group or a macromolecule. If you use a macromolecule core, you can attach dendrons (trees) along the core backbone. We have the electron micrographs which I didn't have time to show you. These constructions allow you to make rod shaped dendrimers such as shown here. You simply attach dendrons along the backbone of a macromolecule or you grow them and you get beautiful rod shaped cylinders that are 50 Å in diameter and as long (i.e., 4,000-5,000 Å) as the macromolecular core is.

It is an interesting exercise because you start out with a random molecular core, and you can see it uncoil and become a rod shaped entity by electron microscope. When you initiate your amplification chemistry here (referring to the macromolecular core), the steric congestion forces it into rod shapes that actually show up in the electron micrograph as clusters of rods all organized and parallel to each other. So shape design is possible--a variety of these lego-type toys are possible; wherein you can design shapes, you can have design shapes, you can have designed clefts by means of the procedures I showed previously where you have two sectors (dendrons) of the same size and you surround a smaller dendron to produce a little cleft. You can design and develop all sorts of combinations and permutations. I think George [Newkome] you've constructed barbell type dendrimers with hydrophobic alkanes in the interior and have shown some beautiful supramolecular characteristics and organizations. So to answer your question, there is almost an infinite number of possibilities.

MALLOUK: My question really is, here your [Tomalia's] synthon still has 1.5 Å carbon-carbon bonds as the primary building unit, and that's where the length scale of the largest possible dendrimer is determined. Regardless of the molecular architecture, the object becomes sterically crowded at a size of 70 Å or so. So the question I have is, is there an effort to use larger building blocks, for example long rigid rod connectors?

TOMALIA: Rigid rod connectors are a perfect example of reactive shapes we are using to connect dendrimers. I visualize with a lot of excitement the possibilities if you start connecting dendrimers in this fashion. Presently I'm working with Larry Miller (University of Minnesota). You may remember Larry Miller made so-called reactive rigid rod molecular level rulers. We are connecting dendrimers with these reactive spacer agents. As we speak, we putting them together with these spacers and we're designing certain defined spaces between the dendrimers. If that's what I understand your question to be, we get certain rigidity designed in these dendrimer matrices.

I have to say that because of the steric effects that can occur when you grow out too far, you can grow out far enough to for a mushroom-like cap that covers over a reactive dendron focal point. You can lose all reactivity at the focal point after a certain generation level is reached. It literally swallow up a reactive initiator core. If you amplify it too far, the mushroom cap has enough dynamic conformations to cover up that reactive site after a certain generational level. The way we determined these steric effects was to perform a reduction of a dendrimer possessing a disulfide initiator core. We let it come apart and then we tried to get them to re-oxidize as a function of generation. We saw varying degrees of resistance to reoxidation as a function of the generational level. This was all controlled by nanoscopic steric effects imposed on the reactive focal point of the dendron.

MEIJER (Eindhoven): On paper everything is possible, but by saying this I don't think you [Tomalia] give credit to the people that have made the few synthetic schemes available that work in a precise way. So there is a large difference between paper chemistry and laboratory experiments and the number of practical schemes is very limited and you have to work for years in order to add a new scheme to this series. And in that case you will end with a hundred milligrams. You know that the best; it's easy to go through the first generations, but going to the higher generations is the big deal. It's not without reason that all the examples that you show are based on the PAMAM's.

TOMALIA: Those are the rules that define the nanoscopic frontier. These are rules that seem kind of strange to most classical organic chemists since they expect certain well known things to happen (i.e., in the case of these disulfide core dendrimers and their steric effect that you might find associated with a classical neopentyl system. In this example, however, it is now nanoscopic in dimension.

MEIJER: Yeah, but what I mean is changing the core structure is easy to do, however, in going through the whole series of generations, you're talking about a lot of work. That's the point I'd like to make.

NEWKOME: I'm curious, what is the commercial purity of the PAMAM?

TOMALIA: According to electrospray mass spectroscopy we believe through generation seven (7), we see a substantial amount of ideal dendrimer structure. What is the purity of our pharmaceutical grade dendrimer? About 95%! The technical grade is not that pure. The technical grade is much less than that.

WUEST (Montréal): I have a question for a horticulturist like you [Tomalia]. Let's suppose that you have a tree with a beautiful, luxurious, spherical growth of interwoven branches, and by some means you are able to cut one of the branches near the trunk. That branch will not actually fall until there is a big windstorm. Otherwise, it will stay in the tree. Can you do an analogous experiment with dendrimers, and would you learn something interesting by doing it?

TOMALIA: That's a very interesting comment. As I understood the question, you [Wuest] asked if one could make a dendron connection to a core labile and have on of these wedges depart under certain prescribed reaction conditions. We have circumstantial evidence that we can. We don't have hard data. In fact, there are some reasons to want to do that. What we have done is design labile ester groups deep down in the dendrimer construct so that a proton could penetrate and hit the labile dendron connector and a piece would pop out; an exact piece on demand. It's a structure design, it's a strategy that I think is quite exciting.

NEWKOME: One of the things that is very interesting, is to listen to this conversation, when one starts to talk about dendrimers. It's hard to imagine that only 10 years ago, to an organic chemist, pure compounds stopped at about the mass of 2,000-3,000 a.m.u. Today, we can make precise, or nearly so, molecules up to a million in atomic weight. And in fact there is a paper which is coming out that suggests a mass weight of 4,000,000 is possible. The problem we have, of course, is how does one prove it. And so the limitation, which we are facing in this area, is how do you go about instrumentally proving what we have? The key fact is that from 1828, when Friedrich Wöhler converted ammonium cyanate to urea--it took until 1985 for us to prepare precise molecules possessing molecular weight greater than ca. 2,000 molecular weight. Millions of molecules were prepared from 1820-1985, almost all were constrained to this molecular weight ceiling. Imagine the number of possibilities that are now available by different combinations of molecular "bricks". Introducing these ideas and materials to material scientists and with new instrumentation of the analytical chemist, one can not imagine the breadth and diversity of macromolecules of the future--now that there is no molecular weight ceiling to limit our imaginations. You can now end up with an absolute totally perfect, macromolecule incorporating or coated with the desired functionality. I think we've really just started to look at the demise of the barrier between organic chemistry, inorganic chemistry, polymer chemistry, and material science.

I think this is one of the reasons that the semantics, as noted throughout this Workshop, is becoming so difficult. As we have to get together to discuss the issues, we will have to define what the key words are, what they mean to us, so that we can in fact talk in a common language.

Thank you all.

References:

1. Rajca, A. and Utamapanya, S. (1993) Toward Organic Synthesis of a Magnetic Particle: Dendritic Polyradicals with 15 and 31 Centers for Unpaired Electrons, *J. Am. Chem. Soc.*, **115**, 10688-10694.

SELF ASSEMBLY OF MOLECULAR MATERIALS

S. I. STUPP, K. E. HUGGINS, L. S. LI, L. H. RADZILOWSKI, M. KESER, V. LEBONHEUR, and S. SON
Departments of Materials Science and Engineering and Chemistry
Materials Research Laboratory
Beckman Institute for Advanced Science and Technology
University of Illinois at Urbana-Champaign
Urbana, IL 61801

ABSTRACT. Self assembled organic nanostructures with distinct shapes have great potential as the constituents of molecularly engineered materials. We have demonstrated in this manuscript that nanostructures of uniform shapes and dimensions can be obtained with a toolbox of self assembling block molecules and polymerizable chiral compounds. These nanostructures have supermolecular molar masses which extend beyond those of compounds accessible by conventional organic chemistry. In the solid state some of these systems have been found to organize into macroscopic structures such as films which develop useful properties as a result of the three dimensional or two dimensional ordering of the nanostructures. Some of the interesting properties observed are third order nonlinear optical properties combined with laser damage resistance, and also highly defined surface properties in macroscopic objects. Polar ordering of nanostructures has also been observed, suggesting that many properties such as second harmonic generation, piezoelectricity, and pyroelectricity might be accessible by molecular self assembly.

1. Introduction

Materials must be regarded as functional structures that take macroscopic form, and thus their molecular design is a lot more complex than that of functional chemical compounds. In fact, materials are rarely chemical compounds or single crystals, and are often mixtures of molecules or dispersions of materials in other materials (composites). Materials are rarely harvested directly from chemical reactors and thus their properties must be enhanced or at least preserved as they are processed into macroscopic structures. This also means their targeted properties must survive the presence of defects which are always present in macroscopic structures. One attractive methodology for future organic materials would be to use discrete molecular nanostructures as their building units and to rely on their self-assembly to generate macroscopic functional structures. If molecular nanostructures are to be considered as the *modular* units of materials, we must have the ability to

J. Michl (ed.), Modular Chemistry, 219–240.

generate molecular objects with dimensions ranging from 1 to 100 nanometers and molar masses that may range from 100,000 to 100,000,000 daltons. Such nanostructures cannot be synthesized presently with conventional organic chemistry, and thus our interest is to explore the formation of the nanostructures by bulk self-assembly. In a second hierarchical level we are interested in self-assembly of the nanostructures to create macroscopic systems.

Figure 1 illustrates schematically a collection of nanostructures that could be considered modular constituents of materials. Each of these nanostructures confines hundreds of thousands or millions of atoms into a volume of well defined shape. Nanostructures with simple global shapes are interesting as the building blocks of materials because they will self-assemble in three dimensions with predictable order. Predictable order can be expected from objects such as two-dimensional polymers shaped as molecular plates, nanotubes, parallelepipeds, and ellipsoids. For example, large two-dimensional (2D) polymers can only fill space efficiently by forming large domains with a single stacking direction. Tubes, on the other hand, must align uniaxially to fill space, and other nonspherical molecular objects with identical size and shape could self-organize into superlattices. Such distinct

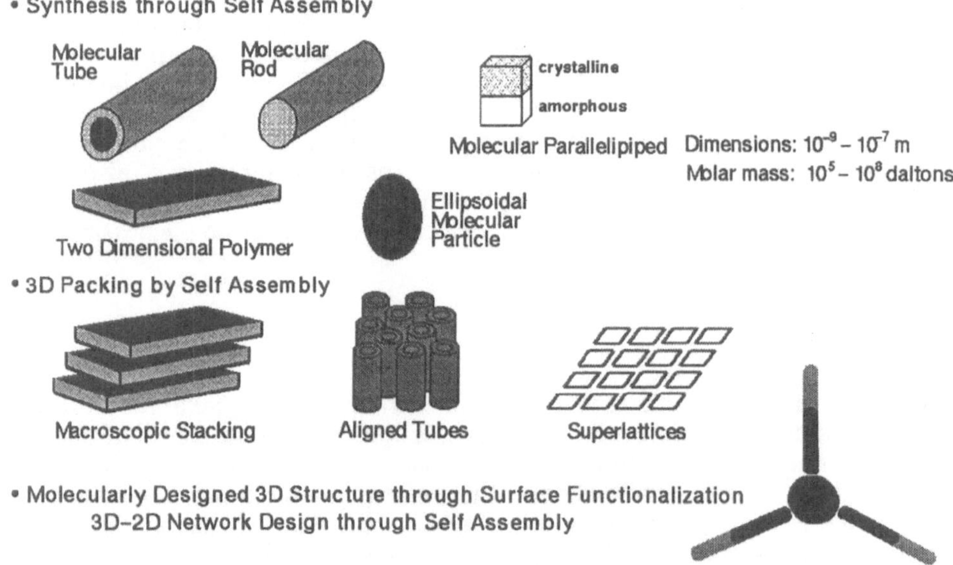

Figure 1. Schematic representation of nanostructures with distinct shape. The figure also shows examples of their expected stacking in three dimensions and the self assembly of dissimilar nanostructures.

types of three-dimensional order are extremely important in defining the properties of macroscopic structures. The stacking of 2D polymers can be important in defining surface properties in molecular materials, and commonly oriented identical cavities of tubular structures could sort, transform, and direct molecules in membranes and solid catalysts. Superlattices formed by nanostructures could be very interesting as sensor surfaces which present a specific arrangement of nano-scale sites to bind molecules, including proteins and other macromolecules. If the superlattice prevails throughout the material, the disapperance of one layer, by degradation or normal wear, would still present to the external environment the same sensing surface. If we learn to create nanostructures with complex shape that would mimic folded proteins, we can expect unique three dimensional arrangements which could become a rich source of new properties in materials. Another interesting aspect of using *shape invariant* molecular nanostructures to design materials is the possibility of forming 2D or 3D networks by self-assembly through the control of their surface chemistry. This concept is schematically illustrated in figure 1. Self assembled nanostructures could also be converted into macromolecules by confined polymerization. This conversion of supermolecular aggregates to shape-invariant polymers would contribute to the structural stability of nanostructure-based materials. Such polymers would retain their shapes during solid to liquid state transitions which is important in the processing of materials and their long-term stability.

One of the challenges in creating molecular nanostructures by self-assembly is to understand how one encodes information in chemical structure to create finite phases. The self-assembly of organic molecuels into 3D crystals and liquid crystals is not only common but extensively studied. Nanostructure formation with organic molecules requires a balance among entropic factors, nonspecific contact energies among structural units, and specific noncovalent interactions. Specific noncovalent interactions such as hydrogen bonds, coulombic forces, chiral interactions, and strong dipole-dipole forces are the ones commonly considered in synthetic chemistry because they are the most tangible, thus a great deal of learning remains in achieving the goal of nanostructure synthesis by self-assembly. For example, if one considers the formation of the mushroom nanostructure or the parallelipiped shown in figure 2, specific noncovalent interactions have to be balanced against entropic factors in order to frustrate the formation of an infinite crystal or mesophase. Also, self assembling methodologies for the nanostructures of materials should be preferably bulk processes so that scale up will remain a realistic goal.

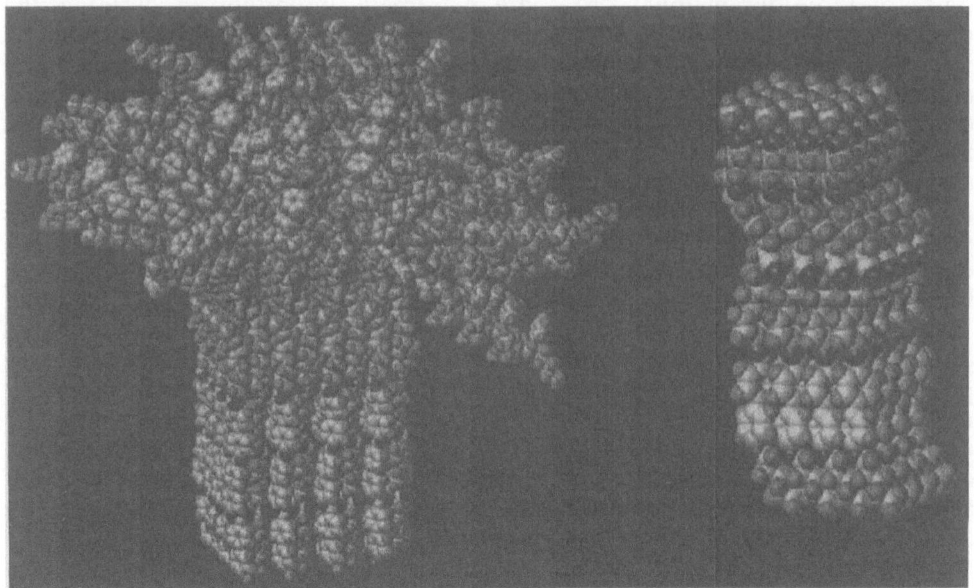

Figure 2. Molecular graphics rendition of a "mushroom" nanostructure formed by self assembly of rodcoil molecules, and of a small parallelipiped formed by aggregation of chiral molecules.

2. Two-Dimensional Polymers

The first objective pursued in our laboratory on molecular nanostructures was the self assembly of a 2D polymer using chiral molecules and molecular recognition events among them to form the plate-shaped structure. A molecular graphics rendition of this molecular object is shown in figure 3.

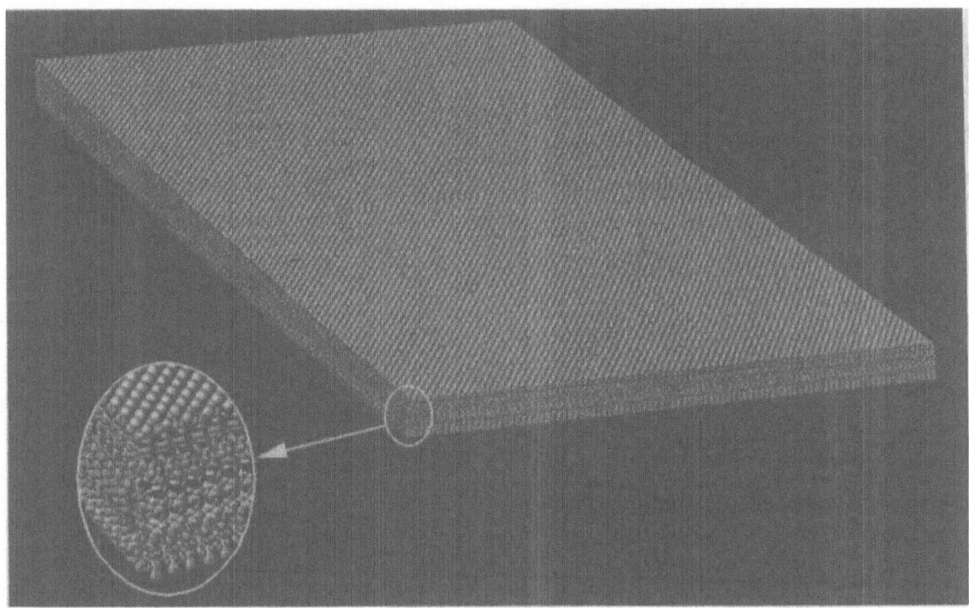

Figure 3. Molecular graphics rendition of a two-dimensional polymer.

We previously reported on the synthesis of the precursor molecule to form this 2D polymer [1,2]. The original molecule used for this purpose was a chiral oligomer which required 21 steps, and its structure is shown in figure 4. The molecule contained two functional groups with the capacity to react and form covalent bonds in order to stitch a planar assembly of molecules into a 2D polymer. As explained schematically in figure 4, precursor molecules with only one functional group would react to form one dimensional molecular objects such as comb polymers, whereas bifunctional ones with functional groups at opposite termini would react to form infinite network gels. These symmetric molecules that form gels are of course less challenging from a synthetic point of view, and the presence of functions at the termini would stabilize self assembly into layered structures. If both functional groups are positioned too close to each other the reaction among functional groups is then likely to form ladder polymers which are also one-dimensional. In order to stitch 2D polymers the two functions must be spaced by several bonds, as they are in our experimental system, so that the paths followed by polymerization of functions confined to different planes are not strongly correlated in space. The lack of spatial correlation should avoid formation of ladder polymers.

The functional groups used in the first experimental system that formed 2D polymers were acrylate and nitrile groups. Nitrile groups are known to react and form imine bonds [3-5] and thus can serve as stitching functions. The nitrile groups substitute the stereocenter, and the combination

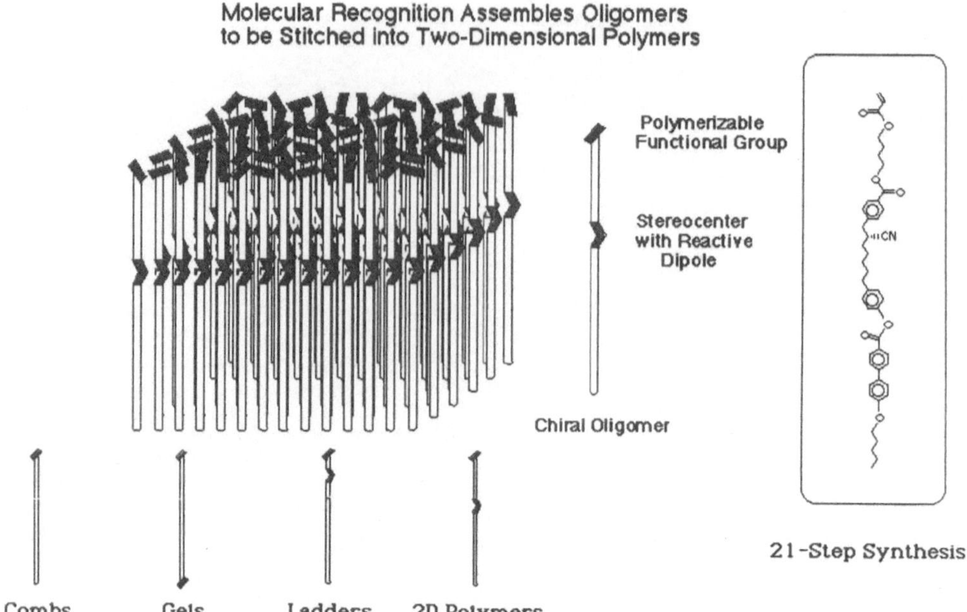

Figure 4. Schematic representation of the self assembly of a two-dimensional polymer through molecular recognition events among enantiomerically pure chiral molecules. The chemical structure of the precursor is shown on the right hand side. Chemical reactions among double bonds and nitrile groups stitch a two-dimensional polymer. Analogous molecules that would stitch one dimensional comb and ladder polymers or gels are shown schematically in the bottom of the figure.

of very strong dipole moment and common handedness as a result of enantiomeric enrichment in the system should promote ferroelectric-like dipole-dipole associations among precursor molecules. The importance of these homochiral interactions would be their ability to induce orientational ordering of layered molecules, thus confining the reactive functions to planes. We demonstrated the importance of plane confinement of functions for 2D polymer synthesis using a computer simulation. In this simulation stitching reactions are represented by random walks in function-containing planes. For bifunctional oligomers of the kind used in our experiments the simulation shows that orientational order within the layers is necessary in order for the system to react to a high enough extent so that high molar mass plates are formed. The plot in figure 5 shows the fraction of one million oligomers in a layered lattice which get connected by covalent bond forming reactions in the presence and absence of orientational order. In this system there would be three planes of reaction since the product of reaction in our system should be a bilayer plate (verified by x-ray diffraction). The central plane was assumed

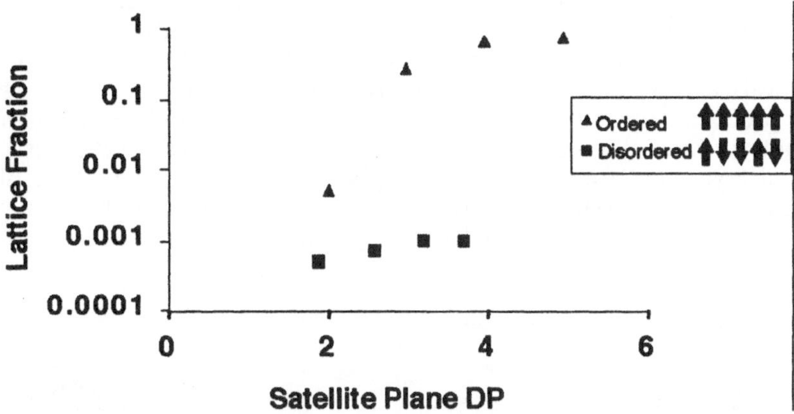

Figure 5. Plot from a computer simulation in which random walks in three separate planes connect units in a layered lattice (see text). The graph plots the fraction of precursors which become connected in a 1,000,000-unit lattice as a function of degree polymerization in two of the planes. One curve corresponds to orientationally ordered units and the second to a layer with disordered units.

to have 90% of its functional groups reacted with covalent backbones having an average degree of polymerization of 10. The plot then indicates the fraction of oligomers connected as a function of degree of polymerization in the two planes above and below the main one, each with 30% of its functions reacted. The conditions chosen for the simulation match our experimental findings [2]. The computer simulation emphasizes the importance of designing structures for molecular recognition events that order molecules in the proper orientation for extensive reaction. In this case the homochiral interactions among strong dipoles could have served this function, but other factors could contribute as well. As the plot in figure 5 indicates, 2D polymers are difficult to form with orientationally disordered layers. Also it is clear from the plot that in ordered layers the degrees of polymerization necessary to form 2D polymers are very small when several functional planes are present.

The compound shown in figure 4 was found to form layered structures at room temperature [2] and to react as the solid is heated through solid to liquid crystal transitions without the use of any special catalyst. The product obtained is a fusible solid which is also 90% soluble and has molar masses that can reach millions of daltons relative to polystyrene standards (based on gel permeation chromatography). It is interesting to explore structural differences among systems at intermediate levels of covalent stitching between self assembled layered oligomers and 2D polymers. In one system we polymerized the oligomer in dilute solution to create comb-like one dimensional polymers with degrees of polymerization of about 10. In another system these comb structures were heated in solution to a temperature of 110∞C, a process which should result in the partial reaction of nitrile groups. As shown in figure 6, we observe in the comb polymer layered structures in which molecules are tilted [6]. This is known based on the electron diffraction pattern obtained from this image. As the chemical reaction proceeds we envision the formation of "ladderoid" structures which should not yet have 2D architecture. In these structures the tilting is lost and orthogonal layers are formed. Interestingly, the electron diffraction obtained from these structures (shown in figure 6) exhibits hexagonal geometry but cannot be indexed with hexagonal lattice parameters.

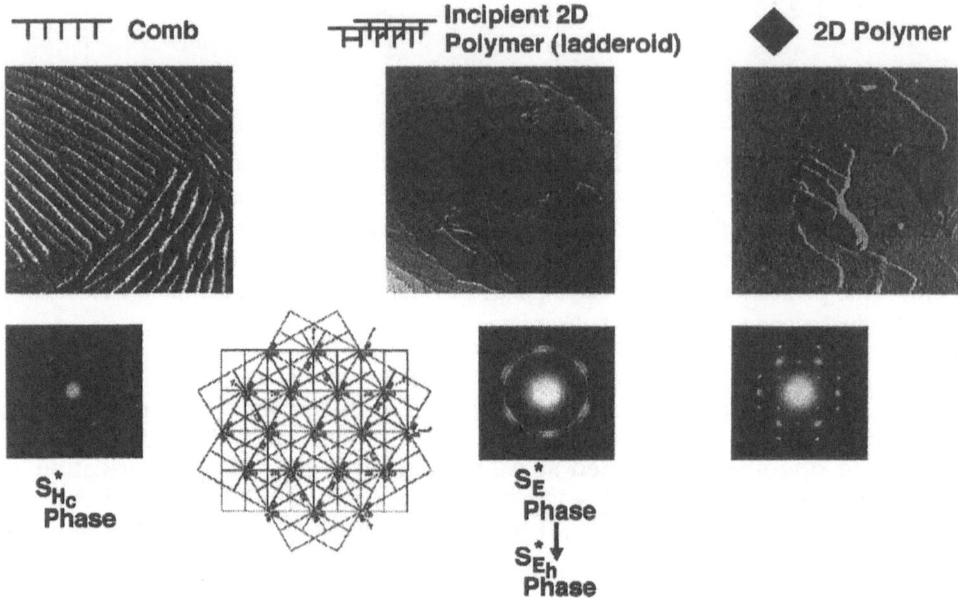

Figure 6. Transmission electron micrographs and elecron diffraction patterns of structures at different stages of evolution toward a 2D polymer. The structure at left corresponds to comb polymers formed by reaction among acrylate groups, the one in the middle is a structure formed with acrylate as well as some nitrile group reaction forming a "ladderoid". Shown at the right hand side is a micrograph of stacked 2D polymers and its corresponding diffraction pattern.

Instead this structure appears to be a dynamic one in which three orthorhombic lattices coexist related to each other by ±60∞. On the other hand, the transformation of the self assembled oligomers to 2D polymers results in the formation of single crystal stacks as clearly revealed by the electron diffraction pattern. Thus a high level of order is observed at room temperature when the product of the reactions are 2D polymers stacked to form thin films.

The concept of 2D polymers derived synthetically from layered assemblies suggests many concepts in advanced materials. As illustrated schematically in figure 7, 2D modular constituents of molecular materials could offer many useful properties as the structural chemistry in sublayers is varied.

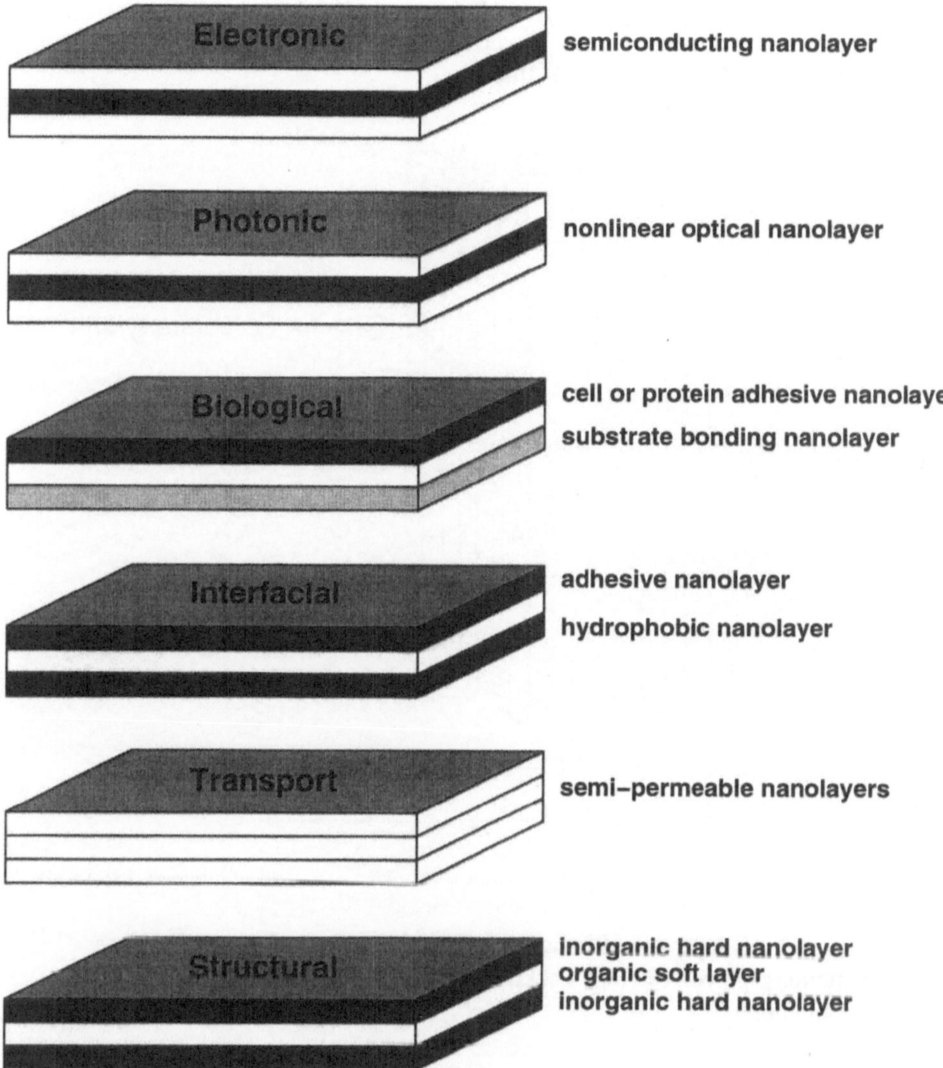

Figure 7. Schematic representation of nano-thickness 2D polymers derived form layered monomers conveying the concept that chemical variation in their sub-layers can lead to many useful structures for advanced materials.

For example, photonically active sublayers which exhibit nonlinear optical properties could offer extremely high temporal stability when 2D polymers stack to form films, specially those which form single crystal stacks as found in the system described above. This particular idea was previously demonstrated through experiment in our laboratory [1]. More recently, we have created 2D assemblies containing electronically conjugated sublayers based on diacetylene chemistry [7,8]. Figure 8 shows one of the structures synthesized, consisting of bilayer plates in which two planes contain polydiacetylene backbones and a middle layer connects comb polymers through polymeric hydrogen bonding.

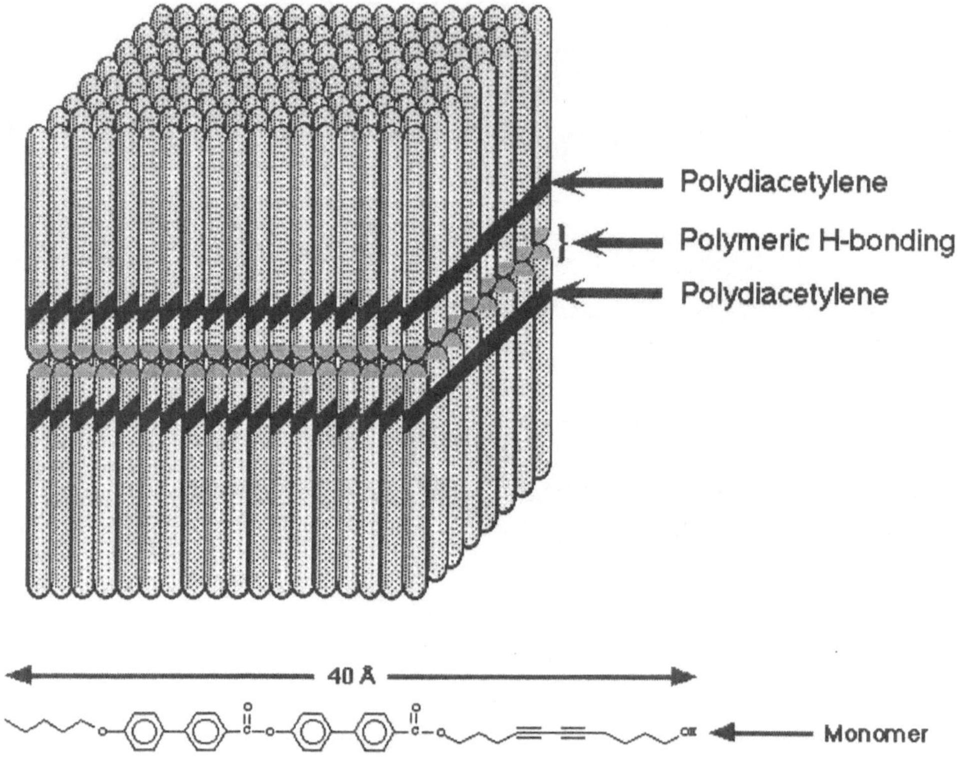

Figure 8. Bilayer two-dimensional assemblies formed by two planes containing polydiacetylene backbones and a hydrogen bonding layer in the middle.

Figure 9 shows an electron micrograph of the plate-like assemblies, and Table 1 shows their third order nonlinear optical susceptibilities.

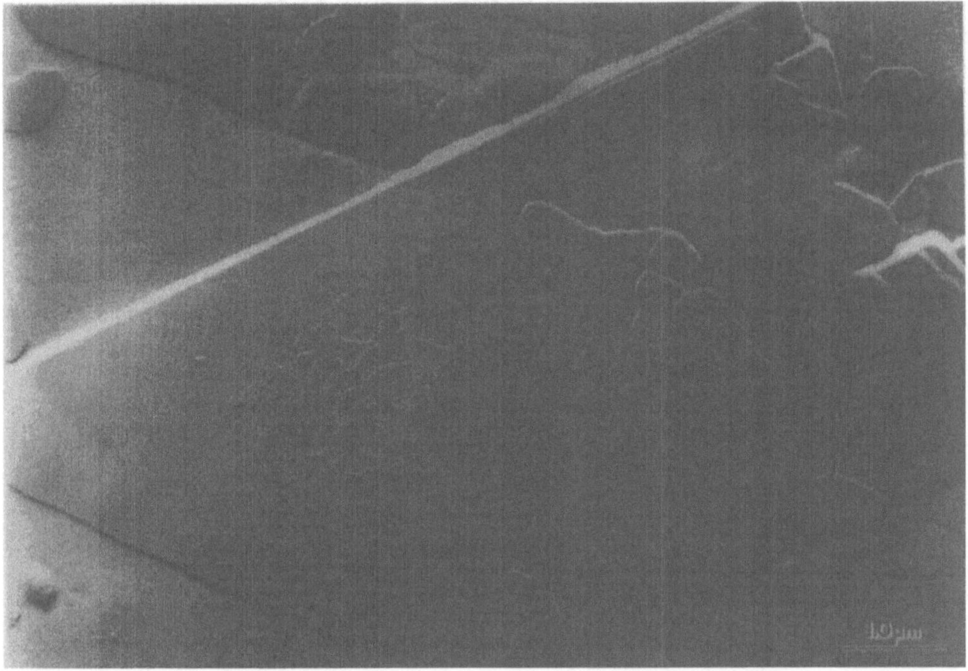

Figure 9. Transmission electron micrograph of two-dimensional assemblies formed by hydrogen bond forming polydiacetylenes.

The color differences among films are observed upon heating as expected from previous work on polydiacetylenes [9]. The transition from blue to red occurs upon heating and it is reversible, however, the thermochromic transition from red to orange upon further heating is irreversible. The third order susceptibilities of these films were obtained using sum frequency experiments and not four wave mixing experiments, and thus our films had to be exposed to high laser intensities. It is then clear that an important property of these films with 2D molecular architecture is a remarkable resistance to laser radiation. These films remained perfectly stable after 432,000 pulses of 20 ns at 20Hz, conditions that would have destroyed many conjugated polymers. This photochemical stability would be extremely important in developing photonic materials that exhibit third order effects. The synthesis of the precursor oligomer is shown in Scheme 1 and described in greater detail elsewhere [8,10]. Some of the other concepts illustrated in figure 7 are related to possibilities of creating sublayer chemistries of interest in biomaterials or adhesion, and then expressing the molecular properties of the modular constituents through their macroscopic stacking in head-to-tail fashion. Also, as illustrated in figure 7 one may design systems in which mineralization of sublayers could be induced based on sublayer chemistry, thus leading to the formation of organic-inorganic hybrid structures with 2D architecture. Such hybrids could have interesting optical and mechanical properties and would be analogous in nanostructure to materials that occur in nature such as nacreous aragonite.

We have developed two additional methodologies for the synthesis of 2D polymers, one involving molecules we describe as *rodcoils* [11] and another involving molecules having *hairpin* conformations [12]. All four methodologies are summarized in figure 10.

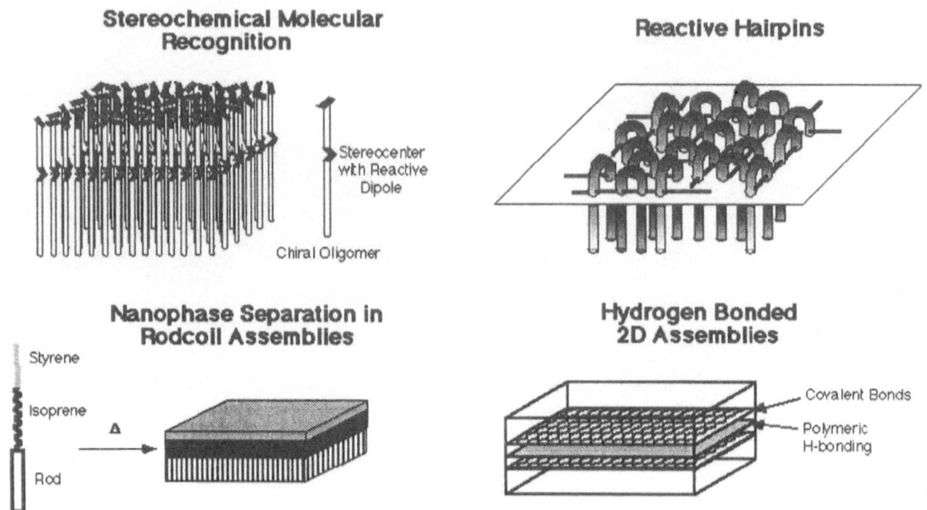

Figure 10. Schematic representation of four methodologies used to form 2D polymers, layered monomers orientationally ordered through homochiral molecular recognition, thermal crosslinking of layered triblock rodcoil molecules, layered reactive hairpin molecules, and topochemically polymerized monomers forming hydrogen bonded planes.

The rodcoil approach utilizes triblock molecules which may be regarded as miniature triblock polymers such as the one shown in figure 11.

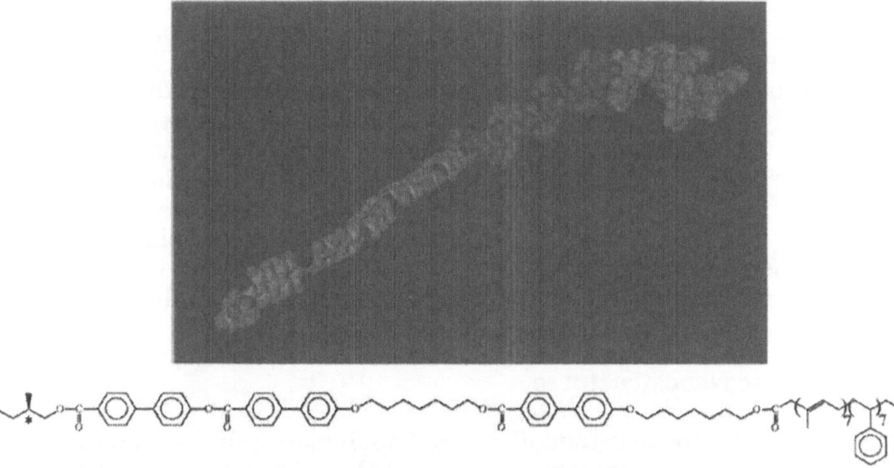

Figure 11. Molecular graphics and chemical structure of a triblock self assembling molecule that can form 2D polymers.

As illustrated in figure 12 these block molecules self assemble into nanophase separated structures with three distinct sublayers.

Figure 12. Transmission electron micrograph showing the nanophase separation of layered rodcoil triblock molecules forming three distinct nanolayers. The light regions contain oligostyrene segments, and the dark ones contain oligoisoprene segments stained with OsO_4 and rod segments ordered in a crystalline array. Crosslinking reactions in the middle isoprene nanolayer stitches two-dimensional polymers.

Thus, formation of 2D polymers can be achieved by simply heating the nanostructured layers to temperatures close to 200∞C in order to form crosslinks among isoprene structural units in the middle block. This results in the formation of 2D polymers with molar masses on the order of 10^7 daltons relative to polystyrene standards, and a thickness of approximately 8-10 nanometers. The hairpin approach takes advantage of the folding of precursor molecules to create a 2D polymer with only a single plane of reaction. Scheme 1 shows the synthesis of a precursor molecule used for this purpose which was found by x-ray diffraction to fold into hairpins, and reacts to form polymers through the reaction of diacetylene groups in the smectic state, that is while hairpins remain organized into layered assemblies.

3. Strip and Disc Nanostructures

We discovered recently that rodcoil block copolymers self assemble to form strip-like and disc-like nanostructures [13,14]. An important difference between the rodcoil molecules that formed layered phases and those that formed discrete strip or disc nanostructures is the volume fraction occupied by coil segments. It is also possible that differences in chemical sequence in rod segments also play a role in the self assembly of discrete vs. infinite layered phases. Figure 13 shows a family of rodcoil molecules investigated in our laboratory, those with the highest molecular weight coil segments (polyisoprene) formed discrete aggregates (strips and discs).

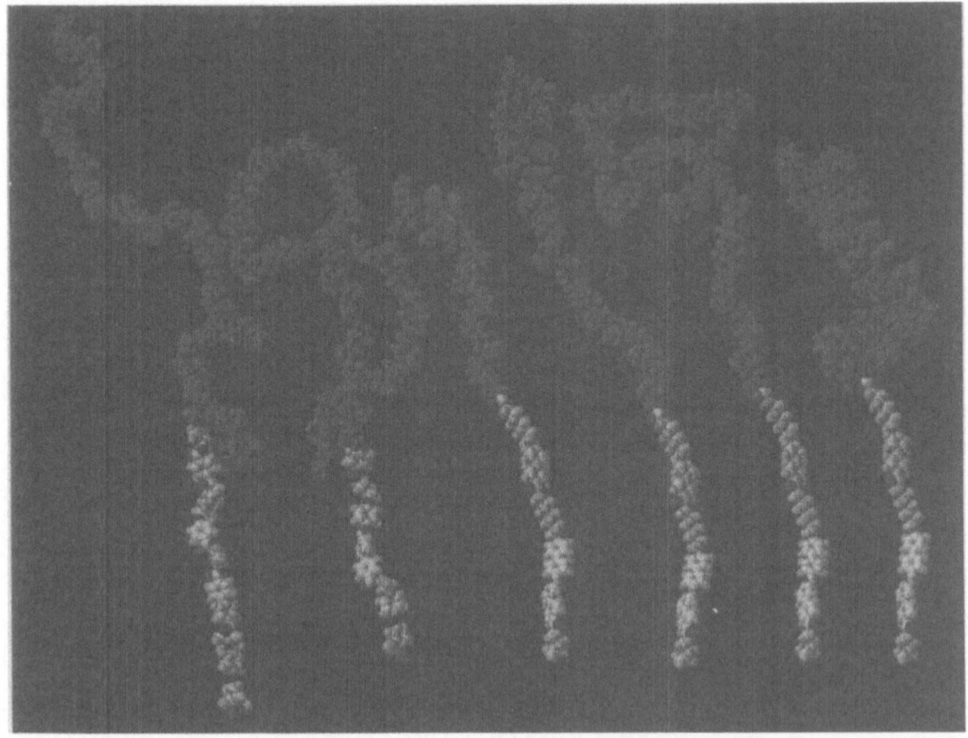

Figure 13. Molecular graphics of rodcoil molecules. The two rodcoil block copolymers shown on the right hand side were found to self assemble into nano-sized strips and discs, respectively.

Figures 14 and 15 show electron micrographs revealing strip and disc nanostructures, respectively, in which light regions represent the domains where rod segments are segregated and dark regions the volume occupied by coil segments.

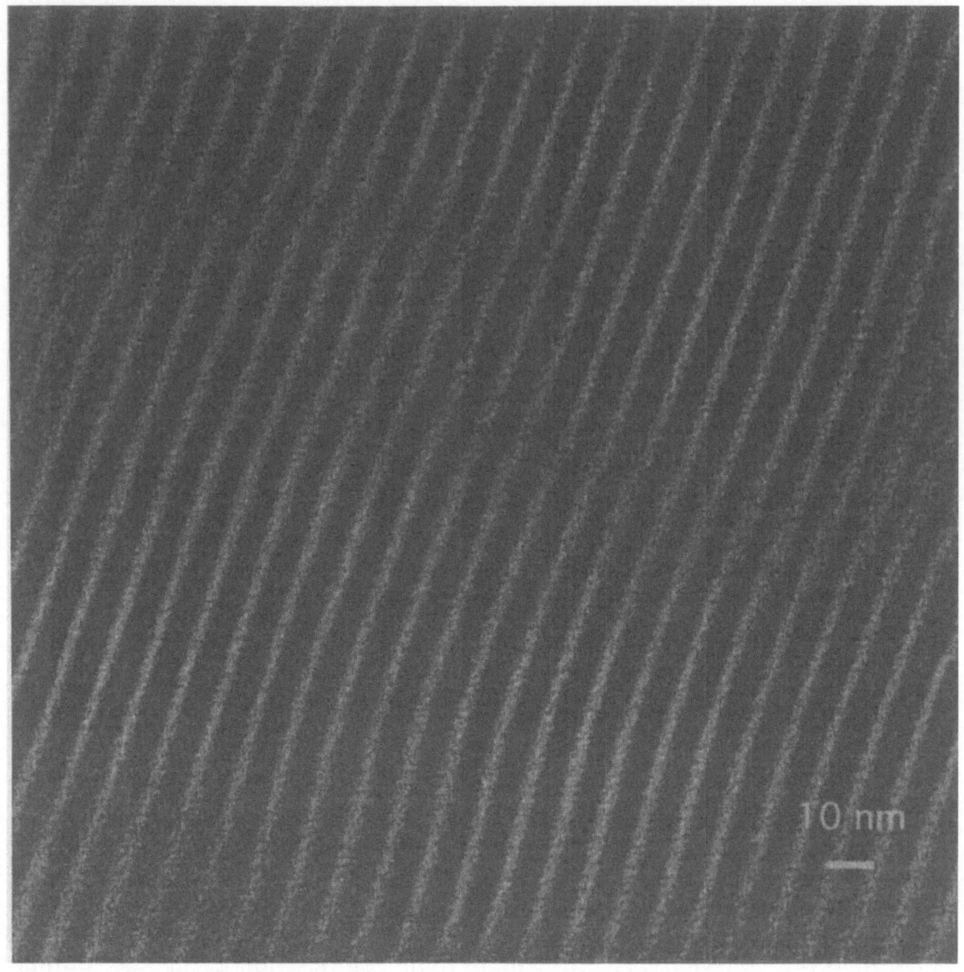

Figure 14. Fourier filtered transmission electron micrograph of strips formed by rodcoil polymers.

Figure 15. Fourier filtered transmission electron micrograph showing a hexagonal superlattice of self assembled rodcoil polymer nanostructures.

The origin of electron scattering contrast in this case is the reaction of double bonds in polyisoprene with OsO_4.

Rodcoil polymers with a rod volume fraction equal to 0.36 gave rise to strip-like nanostructures, whereas those with a rod volume fraction of 0.25 aggregated to form the disc-like structures shown in figure 15. The disc-like aggregates shown in figure 16 with dimensions on the order of 7-10 nanometers are very clearly organized into a hexagonal superlattice.

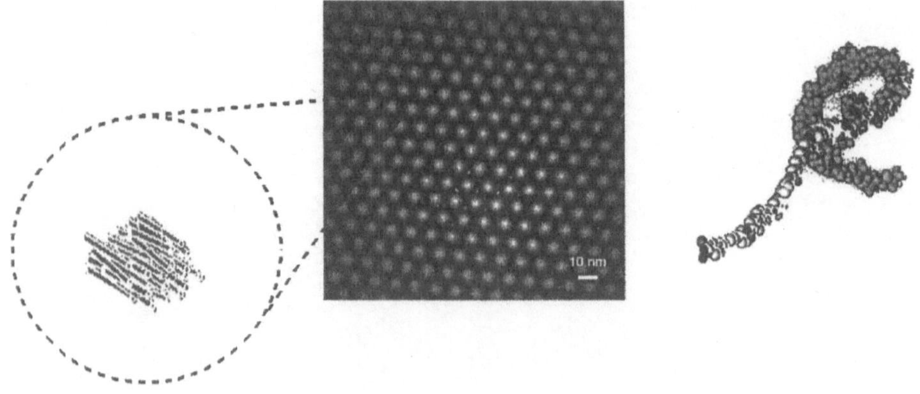

Figure 16. Transmission electron micrograph and schematic representation of the disc-like "nematoid" nanostructures formed by self assembling rodcoil polymers. The structure of the precursor polymer is shown on the right hand side.

In our opinion an important element of the polymer physics behind this self assembling behavior is preservation of entropy in coil segments. The formation of discrete nanostructures as opposed to infinite nematics or smectics allows a greater exploration of conformational space by coil segments. A limited aggregation of rod-like molecular segments allows the conformationally flexible coil segments to splay at the periphery of nanostructures. For high molecular weight rod-coil copolymers, the formation of discrete micellar phases in order to minimize stretching of coils was predicted theoretically by Raphael and DeGennes [15] and Williams and Fredrickson [16]. The chain stretching free energy for rod-coil copolymers was calculated by Williams and Fredrickson to be given by

$$F_{cs} = \frac{\pi\sigma^2\upsilon R_p^{\,3}}{2a^2}\left[f(\rho) + \frac{1}{4\rho} - \frac{1}{4}\left(\frac{2R_p/L}{3\sigma d^2\lambda}\right)^{1/3}\right]$$

$$F_\gamma + F_{ms} - F_{ls} \propto \frac{2^{1/2}(v\chi_s)^{1/2}}{x} + \frac{\lambda x}{4\kappa}\left(\beta - \frac{1}{4}\left(\frac{4x}{3\lambda}\right)^{1/3}\right) - \frac{\lambda 2}{16\kappa}$$

where ρ is the ratio of the hemisphere radius, R_s, to the aggregate radius, R_p, σ is described in this work as a grafting density (density of rod-coil junctions), d is the rod diameter, L is the rod length, a is proportional to the root-mean-square distance between neighboring Gaussian repeat units of volume υ, λ is the ratio of coil to rod volume fractions, $\kappa = Na^2/L$, $\nu = \kappa/\lambda$, and $x = R_p/L$.

Based on diffraction, electron tomography, and spectroscopic analysis (polarized infrared) [17,18], our current model of the disc-like nanostructure is that shown in figure 16. The strip-like object is simply the one dimensional elongated version of the disc since the nano-discs which form the hexagonal superlattice derive from the "fracturing" of strips occurring as rod volume fraction decreases [14]. The molecular architecture in the interior of the disc-like nanostructures at room temperature may resemble a vitrified nematic phase over a nanoscale volume. This is based on the fact that these aggregates do not show any evidence of internal crystallinity, and the way they pack in three dimensions suggests that coils emanate from both sides of the rod aggregate. This would imply nematic order in the interior of the nanostructure (n=-n, where n denotes the director), and coil splaying from both ends of the aggregate would create the disc-like shape. A disc-like shape would explain the formation of a hexagonal superlattice in these systems. We refer to this structure as a "nematoid", since it would be a nano-liquid crystal in analogy to the concept of a nanocrystal.

4. Mushroom Nanostructures

A more anisometric nanostructure with less symmetry than a disc would be the one shown in figure 2, which is shaped as a mushroom. A mushroom shaped nanostructure could pack in three dimensions with less symmetry than spheroidal molecular nanostructures (micelles, fullerenes, spherical dendrimers), cylinders, or discs. Furthermore, contrasting chemistry (e.g, hydrophobic vs. hydrophilic) in cap and stem regions of the mushroom and polar stacking of the nanostructures in three dimensions could generate macroscopic films with great chemical contrast on opposite surfaces. In targeting such structures, we have used the rodcoil structures shown below to create anisometric nanostructures [19], where n + m = $\overline{9}$.

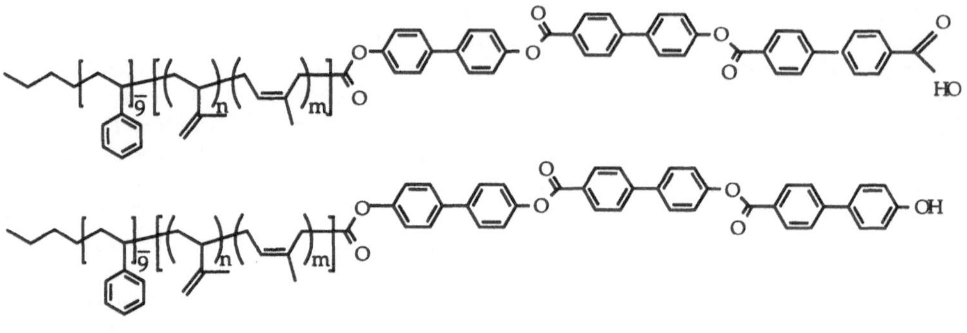

These triblock molecules have at opposite termini methyl and phenolic or methyl and carboxyl groups. These molecules also have a chemically periodic rod segment with a strong tendency to aggregate in parallel fashion, and a terminal block segment of atactic oligostyrene. The nonstereoregular block should not only frustrate crystallization but also limit molecular aggregation as a result of steric crowding, and this could lead to the formation of discrete nanostructures.

We have observed contrasting surface properties in macroscopic solvent cast films of the triblock molecules. These films generate instantly one hydrophobic surface and one hydrophilic surface in the plane of the film [19]. We also have evidence for the formation of finite mushroom nanostructures about three times larger than that shown in figure 2. Evidence for their formation by self assembly was obtained from electron diffraction and high resolution electron microscopy [20]. Most importantly, x-ray diffraction and electron microscopy reveal that at a second hierarchical level, a periodic structure self assembles with up to 100 layers of the mushroom nanostructures. This layered structure is very clearly revealed in ultramicrotomed sections of a film which is one micron in thickness. The periodicity of these films has a thickness comparable to one triblock molecule, suggesting that nanostructures stack with polar order. Polar order is of course an extremely important structural feature in materials since it is connected with properties such as piezoelectricity, pyroelectricity, ferroelectricity, and second order nonlinear optical susceptibility. Having the ability of producing macroscopic structures with dissymmetric surface chemistry opens the door to self assembling coupling agents for composite materials, self assembling solid lubricants, catalyst supports, self assembling membranes, and possibly molecularly designed and self assembling electrochemical cells. Because of the connection between polar order and so many properties of materials, this type of anisometric nanostructure is an excellent model for the concept of function integration in molecular materials.

5. Conclusions

We have demonstrated in this manuscript that self assembly of uniformly shaped and sized molecular nanostructures can be obtained with a toolbox of self assembling block molecules and polymerizable chiral compounds. These nanostructures have supermolecular molar masses which extend beyond those accessible in compounds by conventional organic chemistry. In the solid state some of these systems have been found to organize into macroscopic structures such as films which develop useful properties as a result of the three dimensional or two dimensional ordering of the nanostructures.

6. Acknowledgements

The work described here was supported by the National Science Foundation Grants DMR 93-12601 and DMR 89-20538, Office of Naval Research Grant N00014-93-1-0534, and the Department of Energy Contract DEFG02-91-ER45439. The authors are grateful to Aaron Amstutz for his assistance with molecular graphics.

7. References

1. Stupp, S. I., Son, S., Lin, H. C., and Li, L. S. (1993) Synthesis of Two-Dimensional Polymers, *Science*, **259**, 59.

2. Stupp, S. I., Son, S., Hong, X., Li, L. S., Lin, H. C., and Keser, M. (1995) Bulk Synthesis of Two-Dimensional Polymers. The Molecular Recognition Approach, *J. Am. Chem. Soc.*, **117** (19), 5212.

3. Grassie, N. and McNeil, I. C. (1956) The Thermal Degradation of Polymethacrylonitrile. Part I. Separation of Coloration and Depolymerization Reactions, *J. Chem. Soc.*, **1956**, 3929.

4. Renschler, C. L., Sylvester, A. P., Salgado, L. V. (1989) Carbon Films from Polyacrylonitrile, *J. Mater. Res.*, **4** 452.

5. Usami, T., Itoh, T., Ohtani, H., and Tsuge, S. (1990) Structural Study of Polyacrylonitrile Fibers during Oxidative Thermal Degradation by Pyrolysis-Gas Chromatography, Solid State ^{13}C Nuclear Magnetic Resonance, and Fourier Transform Infrared Spectroscopy, *Macromolecules*, **23** 2460.

6. Li, L. S., Hong, X. J., and Stupp, S. I., Novel Liquid Crystalline Structures of a Chiral Side Chain Polymer and its Phase Transitions, *Liquid Crystals*, in press.

7. Huggins, K. E., Stork, K. F., Son, S., Bohn, P. W., and Stupp, S. I. (1995) Nonlinear Optical Characterization of Two-Dimensional Polydiacetylenes, *APS Bulletin*, **40** (1), 158.

8. Huggins, K. E., Son, S., and Stupp, S. I., Two-Dimensional Supermolecular Assemblies: Structure and Third Order Photonic Properties, submitted for publication.

9. Wenz, G., Müller, M. A., Schmidt, M., and Wegner, G. (1984) Structure of Poly(diacetylenes) in Solution, *Macromolecules*, 17, 837.

10. U.S. Patent No. 5,412,144 (1995) Organic Materials with Nonlinear Optical Properties.

11. Stupp, S. I., Lee, M. S., Son, S., Li, L. S., and Keser, M. (1993) Test Tube Synthesis of Two Dimensional Polymers and Other Molecular Objects, *ACS Polymer Preprints*, **34** (1), 184.

12. Son, S. and Stupp, S.I., unpublished results.

13. Radzilowski, L. H., Wu, J. L., and Stupp, S. I. (1993) Monodisperse Rodcoil Polymers, *Macromolecules*, **26**, 879.

14. Radzilowski, L. H. and Stupp, S. I. (1994) Nanophase Separation in Monodisperse Rodcoil Polymers, *Macromolecules*, **27** (26), 7747.

15. Raphael, E. and de Gennes, P.-G. (1992) Aggregation of Flexible-Rigid-Flexible Triblock Copolymers, *Makromol. Chem., Macromol. Symp.*, **62**, 1.

16. Williams, D. R. M. and Fredrickson, G. H. (1992) Cylindrical Micelles in Rigid-Flexible Diblock Copolymers, *Macromolecules*, **25**, 3561.

17. Radzilowski, L. H., Carragher, B. O., and Stupp, S. I., Three Dimensional Self Assembly of Rodcoil CoPolymer Nanostructures, submitted for publication.

18. Radzilowski, L. H. and Stupp, S. I., Self Assembly of Rodcoil Polymer Molecules into a Nanostructure, submitted for publication.

19. Stupp, S. I., LeBonheur, V., and Walker, K. (1995) Materials with Self Organized Surfaces: 2D Polymer Assemblies, *ACS Polymer Preprints*, **36** (1), 562.

20. Stupp, S. I., LeBonheur, V., Walker, K., Li, L. S., Huggins, K., and Amstutz, A., Synthesis of Materials by Self Assembly of Molecular Nanostructures, submitted for publication.

TOWARDS OLIGOPHENYLENE CYCLES AND RELATED STRUCTURES: A REPETITIVE APPROACH

A.-D. SCHLÜTER, V. HENSEL, P. LIESS AND K. LÜTZOW
Freie Universität Berlin
Institut für Organische Chemie
Takustr. 3, 14195 Berlin, Germany

ABSTRACT. The synthesis of linear and kinked oligophenylene telechelics which are being used as modules for the construction of phenylene rods and cycles is reported. All compounds, though conformationally rigid, are kept soluble through the attachment of flexible alkyl chains.

1. Introduction

Cyclic oligophenylenes[1] and related conformationally rigid structures[2] have been attracting the interest of chemists for more than five decades. Issues which have been investigated with these compounds include aromaticity, host/guest chemistry, aggregation behavior, and molecular constructions. The reaction most commonly used for cyclization is the Kharasch coupling, which involves the oxidative CC-coupling of the Grignard derivatives of halogenated aromatics with stoichiometric amounts of copper (II) chloride. The yields of cyclic product strongly depend on the number of coupling steps and are generally very low. This coupling is (reasonably) limited to the preparation of symmetrical products. Recently, Moore synthesized a large hexagonal ring consisting of 12 phenylacetylene units.[3] The precursor required for ring closure was obtained in a repetitive synthesis involving protective group chemistry. In spite of the dimensions of the targeted ring (diameter 2.2 nm) the final coupling went very well and furnished the macrocycle in a yield of 70%. Moore's strategy allows in principle the construction of all kinds of polygons as long as the geometrical factors let the termini get close enough together for ring closure. Stimulated by this elegant study and given the experience in our group with Suzuki and Stille-type polycondensation[4], we decided to develop a repetitive synthesis on the basis of oligophenylenes which could then be used for the synthesis of, e.g., large, hexagonal cycles and monodisperse rods with functionalized termini (oligophenylene telomers). An objective of this work in the distant future is to try to provide synthetic access to fragments of honeycomb-like macromolecules on oligophenylene basis. This paper describes the characteristics of the strategy and first results of its application.

J. Michl (ed.), Modular Chemistry, 241–250.

2. Strategy

The strategy is repetitive.[5] It uses the same building blocks, coupling reactions, and functional groups over and over again in order to minimize the enormous effort otherwise involved in the synthesis of large molecular objects. The strategy also rests upon the use of place-holders (as few as possible) which may be activated into functional groups (for coupling) whenever necessary. This way an oligomerization which otherwise would compete can be prevented and the products obtained are monodisperse. In order to keep the conformationally rigid targets soluble, all building blocks involved are decorated with flexible alkyl chains. In the terminology of "Modular Chemistry"[6] in which *modules* are used (like the sticks and connectors in a set of childrens tinkertoys) to construct *objects*, the building blocks correspond to the module *stick* and the coupling chemistry to the module *connector*. The coupling results in an *object* which may be used to construct larger objects (e.g., an oligophenylene cycle), in which case it acts as a *module*. Figure 1 shows two important growth strategies, involving symmetrical and unsymmetrical modules. The Xs and Ys symbolize functional groups through which two modules can be connected. The Ps indicate "protected" functional group whereby the term protected is somewhat misleading, at least in the sense in which it is used in protective group chemistry. X^P and Y^P are substituents which can be converted into X and Y, and should therefore rather be referred to as place-holders.

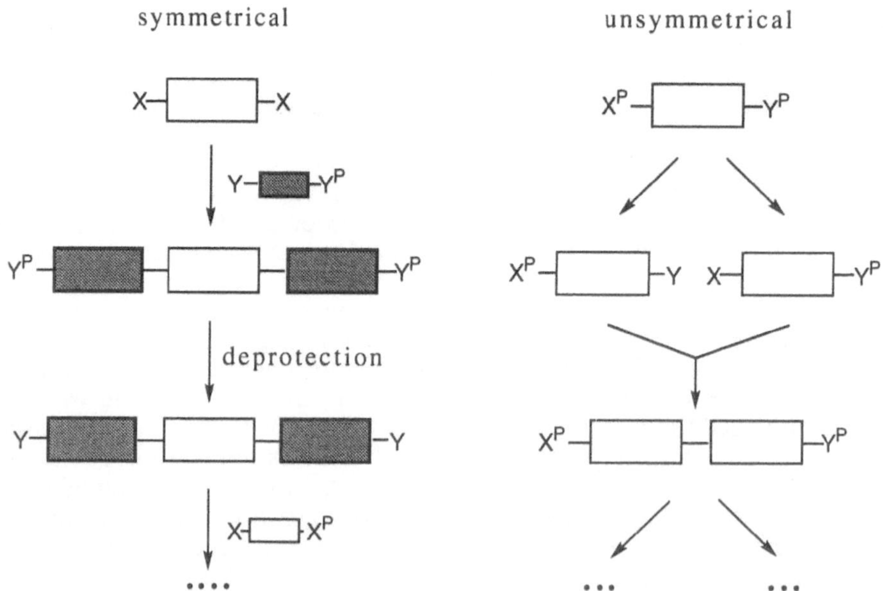

Figure 1. Growth strategies in repetitive synthesis starting from symmetrical and unsymmetrical building blocks (modules).

The conversion of X^P into X and of Y^P into Y in the presence of the other respective place-holder Y^P and X^P proceeds without any mutual interference. Trimethylsilyl (TMS) and bromo groups are used as place-holders. TMS at the terminus of an oligophenylene module is especially valuable because it can easily and cleanly be converted into an iodo, bromo, and boronic acid function (Fig. 2). The bromo place-holder is convertible into boronic acid or TMS. These few manipulations are already sufficient to do all necessary couplings of modules. Everything is designed so that only one reaction is needed to bring about the coupling, the Suzuki cross-coupling of boronic acids and aromatic halides under Pd(0)-catalysis.[7] This easily performed reaction is known to proceed regiospecifically and with high yields. Additionally, it is highly sensitive to the nature of the halogen involved. Iodoaromatics couple significantly faster than bromoaromatics, which are faster than their chloro analogs.[8] As a result, modules containing both bromo and iodo sites undergo coupling first at the iodine. This characteristic of the Suzuki reaction has been incorporated into the strategy as one of its main features. The specificity is observed even when the iodo site is more sterically hindered than the bromo site (Fig. 3). Suzuki coupling of compounds **A** and **B** gives product **C** in an isolated yield of 96% which is indicative of an exclusive formation. The regiospecificity of this reaction was further established by a careful mass-spectroscopical investigation of the raw coupling product which did not give any indication for the formation of the other potential coupling product.

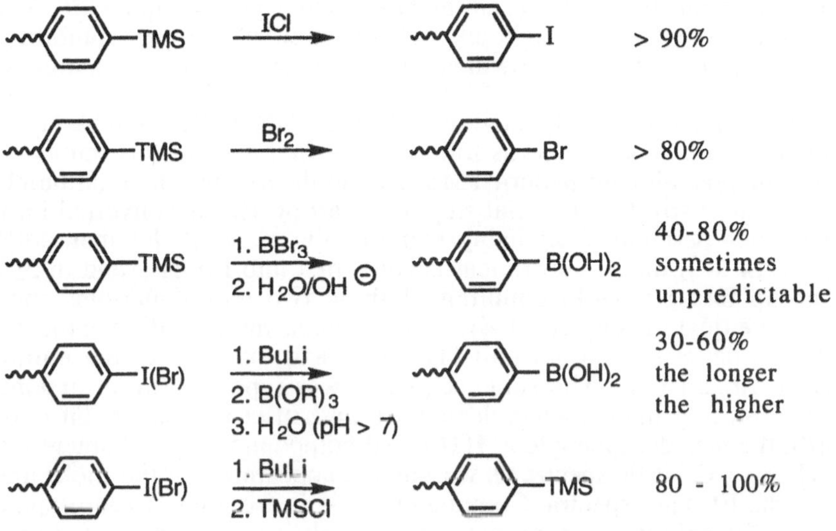

Figure 2. The limited set of place-holder and functional group manipulations required for the repetitive synthesis of oligophenylene rods and cycles.

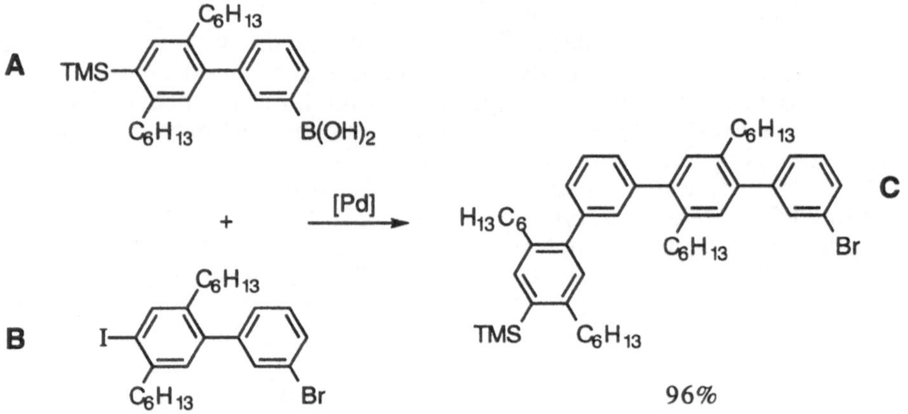

A TMS C_6H_{13} C_6H_{13} $B(OH)_2$

\+ [Pd]

B I C_6H_{13} C_6H_{13} Br

C $H_{13}C_6$ TMS C_6H_{13} C_6H_{13} C_6H_{13} Br

96%

Figure 3. The model coupling of **A** and **B** gives **C** as the only product, thus proving the regiospecificity of the Suzuki coupling at the iodo rather than bromo sites.

3. Results

Table 1 contains a selection of modules with differently and identically functionalized termini for the synthesis of oligophenylene rods and cycles. They were prepared on the 5-30 g scale as analytically pure compounds. Most of them are substituted with straight alkyl chains in order to impart solubility. The successful application of the growth strategy is illustrated by two examples.

The first example uses the biphenyl modul e**1(2)** (Fig. 4). The numbering of this compound is to be understood as follows: The first digit of this number (one) indicates the place-holder pattern TMS/Br, and the second, the total number of benzene rings (two). In the initial step, one part of **1(2)** is converted into **2(2)** and the other part into **3(2)** using some of the place-holder manipulations described. In **2(2)** the Br substituent is converted into $B(OH)_2$ and in **3(2)** the TMS group into I. Suzuki coupling of these two monofunctional modules furnishes the quaterphenylene **1(4)**, where, again, the first digit indicates the substitution pattern at the termini (TMS/Br) and the second, the number of phenylene rings (four). The same sequence was repeated (in an exponential manner) affording the octaphenylene **1(8)**, and, after its conversion into **2(8)** and **3(8)**, the hexadecaphenylene **1(16)**. All compounds were characterized by their ^{1}H and ^{13}C NMR spectra as well as the appearance of the molecular ion peaks in the EI mass spectra. Compound **1(16)** is the longest rod prepared so far. It is still soluble in common organic solvents, which is an important prerequisite to the continuation of the growth procedure. The next object would be **1(32)** which contains 32 benzene rings in a strictly linear sequence and is 13.5 nm long.

TABLE 1. Selection of modules for the synthesis of oligophenylene rods and cycles. Different termini: X/Y = TMS/Br, TMS/$B(OH)_2$, I/Br; Identical termini: X = Br, Y = TMS. In all cases: R = hexyl.

Different termini 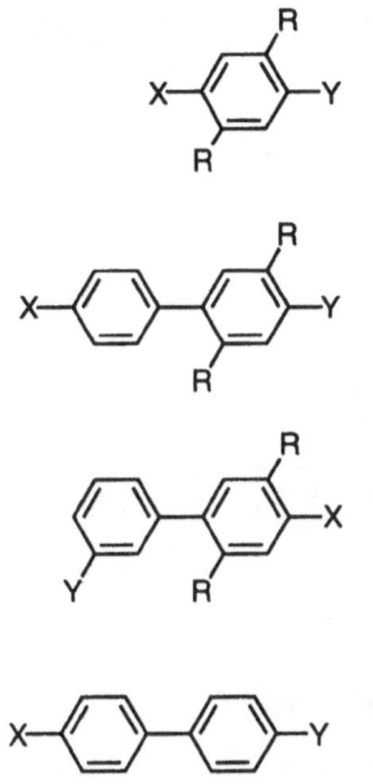	Identical termini 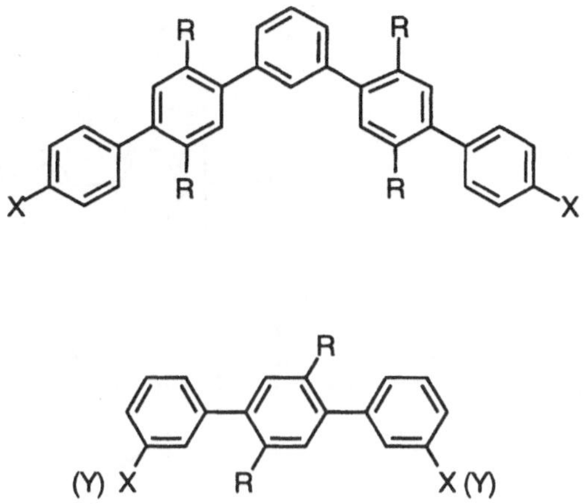
R, X, Y, R	R, R, X, R, R, X
R, X, Y, R	R, (Y) X, R, X (Y)
R, X, Y, R	
X, Y	

The rods can be used not only as modules for further constructions but also to establish the long sought reference kit for GPC measurements of rigid-rod polymers. Even though it would be necessary to have the longer compounds **1(32)** and perhaps even **1(64)** available for this purpose, the first four members of this series were already used in this sense. Figure 5 shows the elution curve obtained after a co-injection of compounds **1(2)**, **1(4)**, **1(8)**, and **1(16)**. The molecular weights obtained from referencing the retention times to polystyrene standard are higher than the actual ones (from mass spectrometry), by a factor of approximately 1.6. This finding is in qualitative agreement to the general view that rigid-rod macromolecules have a smaller hydrodynamic volume than flexible ones of the same molecular weight. More research will certainly be necessary before a reliable standard system is developed.

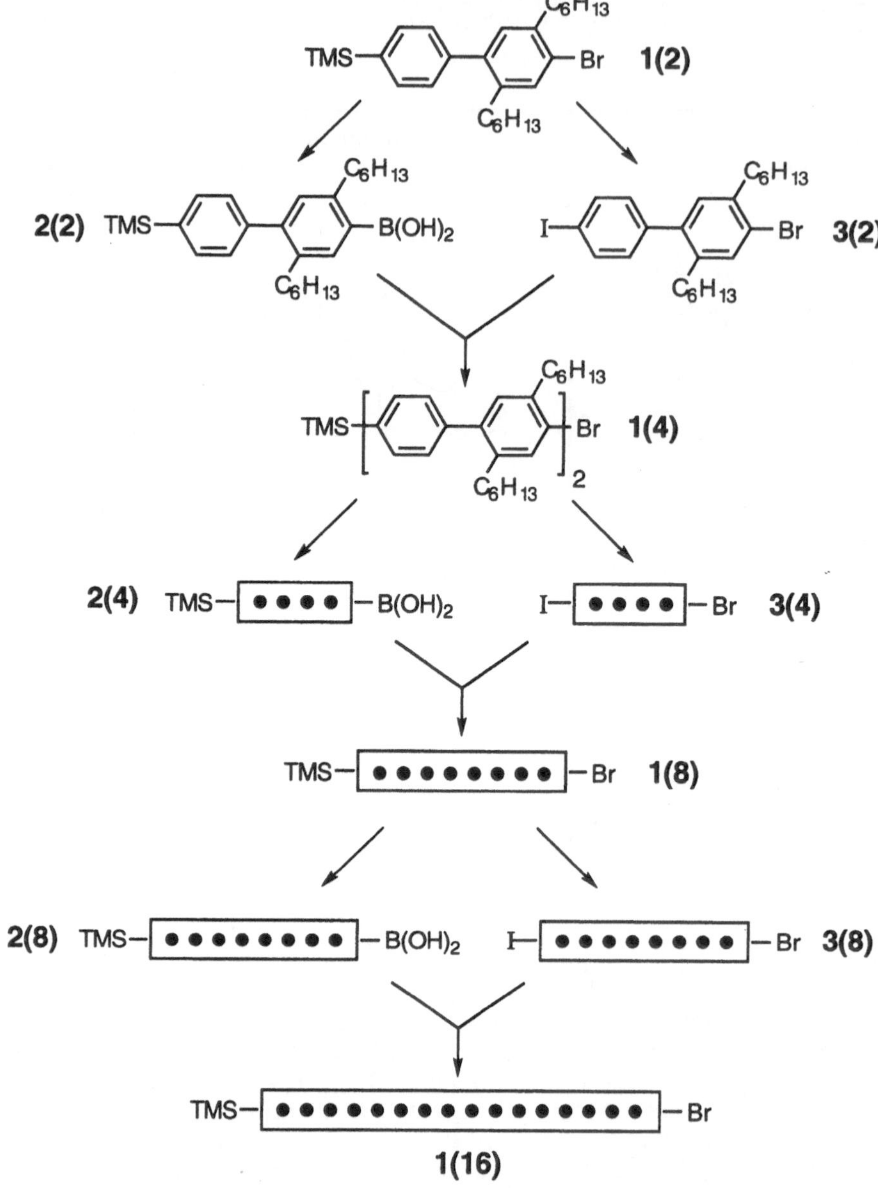

Figure 4. Repetitive synthesis of oligophenylene rods (telechelics) using an exponential growth strategy.

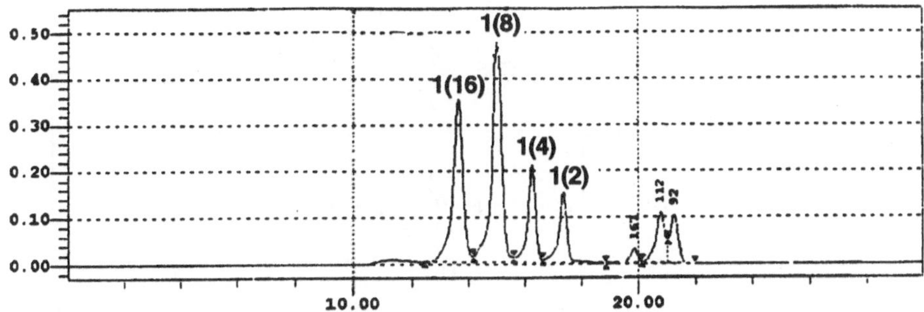

Figure 5. GPC elution curves of oligophenylene rods versus polystyrene standard. The molecular weights versus polystyrenes and, in parentheses, the actual m.w.'s are: **1(2)**: 785 (472); **1(4)**: 1280 (791); **1(18)**: 2230 (1419) and **1(16)**: 4485 (2711).

The second example illustrating the applicability of the repetitive strategy is the synthesis of the cyclic oligophenylene **8.** It starts from module **4(2)** (Fig. 6). By activation of one of its place-holders at the time, the monofunctional compounds **5(2)** and **6(2)** were obtained. Their Suzuki-type coupling gave module **4(4)** which afforded after a few steps compound **4(6)** in good yields and on the g-scale. Replacement of the TMS group in **4(6)** by $B(OH)_2$ afforded **7.** Compound **7** represents half of the targeted ring **8** and has the proper substitution pattern for cyclization. The cyclization could actually be brought about by the addition of 0.5 mol-% of tetrakis(triphenylphosphine)palladium(0) to a heterogenous system of a 10 mM solution of **5(6)** in toluene and 1 M sodium carbonate which was then refluxed for 24 h. Since the experiments have not been completed yet, an accurate number for the yield cannot be given. Besides NMR evidence the structure of **8** rests upon the EI mass spectrum which shows the molecular ion peak as basis peak (!) at the expected mass m/z = 1922 and with the calculated isotope pattern.

Macrocycle **8** is the largest oligophenylene cycle known today. It is highly soluble in standard organic solvents like chloroform, methylene chloride, tetrahydrofuran, and toluene. Due to hindered rotation about the CC single bonds in the ring, it may be obtained as a mixture of the eight possible atropic isomers. The side-to-side distance in **8** is 1.48 nm.

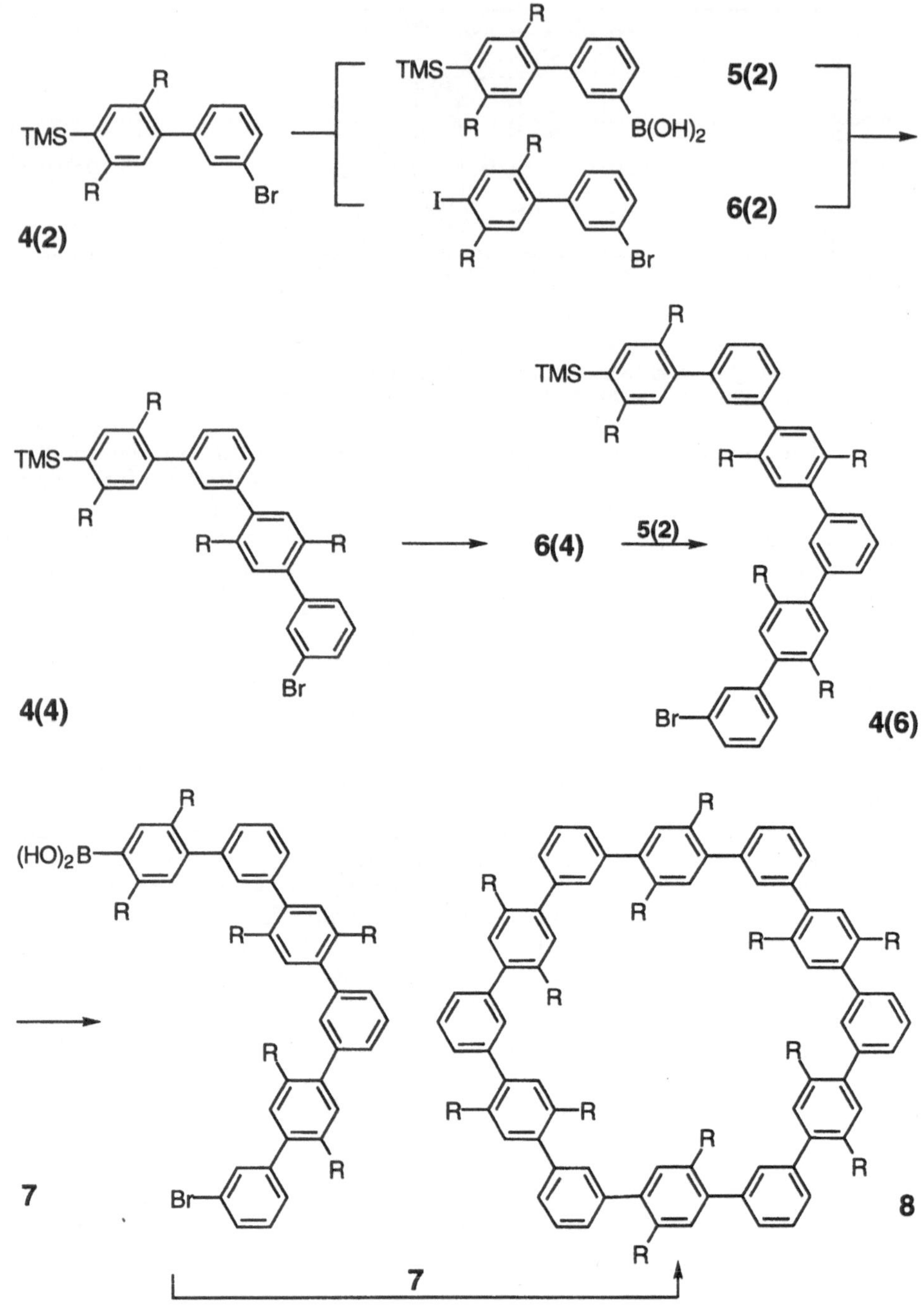

$R = C_6H_{13}$

Figure 6. Repetitive synthesis of an oligophenylene cycle using a linear growth strategy.

4. Summary

The concept of repetitive synthesis has been successfully applied to oligophenylenes. A series of rigid-rod telechelics and the first representative of a series of cycles has been prepared. Figure 7 shows a box of the few tools that a synthetic chemist needs in order to do repetitive synthesis with oligophenylene modules. The box contains a set of building blocks, a few place-holder activation reactions, and the Suzuki reaction as the key coupling reaction. The existing building blocks of this box should still be complemented by ones with three place-holders and/or functional groups as well as by heteroaromatic ones. This way future research would not only be restricted to larger cycles than **8** and longer rods than **1(16)**. It may then even be possible to approach far more complex systems like certain parts of honeycomb-like oligophenylenic networks with and without heteroaromatics.

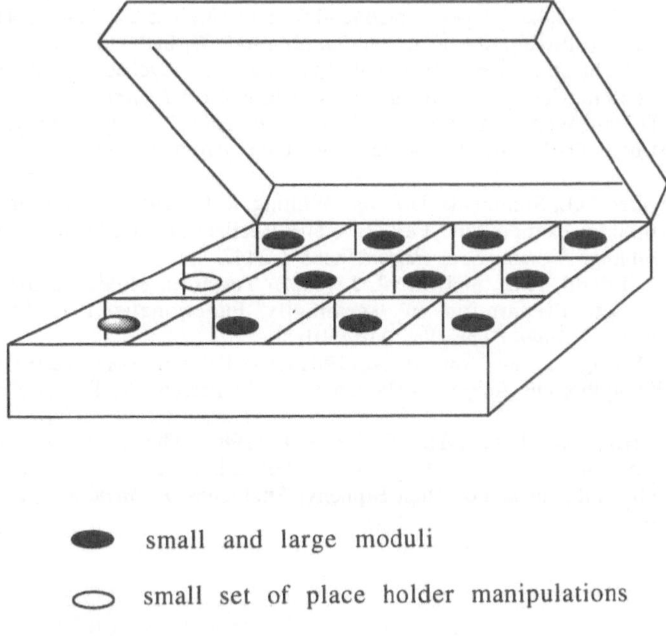

Figure 7. Magic box for minimium effort repetitive synthesis of oligophenylene objects of various kinds.

5. Acknowledgement

We thank Dr. W. Lamer and T. Kolrep for the GPC measurements and Dr. Lehmann, BAM, Berlin, for recording the mass spectrum of the cycle. Financial support by the Fonds der Chemischen Industrie is gratefully acknowledged.

6. References

1. Staab, H.A. and Binnig, W. (1967) Synthese und Eigenschaften von Hexa-*m*-phenylen und Octa-*m*-phenylen, *Chem. Ber.* **100,** 293; Fujioka, Y. (1984) Syntheses and Physical Properties of Several Polyphenylenes Containing Mixed Linkages, *Bull. Chem. Soc. Jpn.* **57,** 3494; Cram, D.J., Kaneda, T., Helgeson, R.C., Brown, S.B., Knobler, C.B., Maverick, E., and Trueblood, K.N. (1985) Host-Guest Complexation. 35. Spherands, the First Completely Preorganized Ligand Systems, *J. Am. Chem. Soc.* **107,** 3645; Percec, V., Okita, S. (1993) Synthesis and Ni(0)-Catalyzed Oligomerization of Isomeric 4,4"'-Dichloroquaterphenyls, *J. Polym. Sci.: Part A: Polym. Chem.* **31,** 877.
2. Vögtle, F., and Kadei, K. (1991) Großflächige Makrocyclen aus *p*-Quaterphenyl-Bauteilen, *Chem. Ber.* **124,** 903; Scott, L.T., Cooney, M.J., and Johnels, D. (1990) Homoconjugated Cyclic Polydiacetylenes, *J. Amer. Chem. Soc.* **112,** 4054; de Meijere, A., Jaekel, F., Simon, A., Borrmann, H., Köhler, J., Johnels, D., and Scott, L.T. (1991) Regioselective Coupling of Ethynylcyclopropane Units: Hexaspiro[2.0.2.4.2.0.2.4.2.0.2.4]triaconta-7,9,17,19,27,29-hexayne, *J. Amer. Chem. Soc.* **113,** 3935; deMeijere, A., Kozhushkov, S., Puls, C., Haumann, T., Boese, R., Cooney, M.J., and Scott, L.T. (1994) Hexaspiro[2.4.2.4.2.4.2.4.2.4.2.4]dotetraconta-4,6,11,13,18,20,25,27,32,34,39,41-dodecain-ein explodierendes [6]Rotan, *Angew. Chem. Int. Ed. Engl.* **33,** 869.
3. Moore, J.S. and Zhang, J. (1992) Effiziente Synthese makrocyclischer Kohlenwasserstoffe mit Durchmessern im Nanometerbereich, *Angew. Chem. Int. Ed. Engl.* **31,** 922.
4. Schlüter, A.-D. and Wegner, G. (1993) Palladium and Nickel Catalyzed Polycondensation - The Key to Structurally Defined Polyarylenes and Other Aromatic Polymers, *Acta Polym.* **59,** 44.
5. Igner, E., Paynter, O.I., Simmonds, D.J., and Whiting, M.C. (1987) Studies on the Synthesis of Linear Aliphatic Compounds. Part 2. - The Realization of a Strategy for Repeated Molecular Doubling. *J. Chem. Soc., Perkin Trans. I* 2447.
6. Kaszynski, P., Friedli, A.C., and Michl, J. (1992) Toward a Molecular-Size "Tinkertoy" Construction Set. Preparation of Terminally Functionalized [*n*]Staffanes from [1.1.1]Propellane", *J. Amer. Chem. Soc.* **114,** 601.
7. Miyaura, N., Yanagi, T., and Suzuki, A. (1981) The Palladium-catalyzed Cross-coupling Reaction of Phenylboronic Acid with Haloarenes in the Presence of Bases. *Synth. Commun.* **11,** 513.
8. Gray, G.W., Hird, M., Lacey, D., Toyne, K.J. (1989) The Synthesis and Transition Temperatures of Some 4,4"-Dialkyl- and 4,4"-Alkoxyalkyl-1,1':4',1"-terphenyls with 2,3-or 2',3'-Difluoro Substituents and of Their Biphenyl Analogues. *J. Chem. Soc., Perkin Trans. II* 2041.

DISCUSSION OF THE STUPP AND SCHLÜTER LECTURES

GERHARD WEGNER
Max-Planck-Institut für Polymerforschung
Postfach 31 48
D-55021 Mainz
Germany

The discussion following the lectures of Professors Stupp and Schlüter addressed a number of features specific to the material presented by these speakers, but also questions of more general nature. Many questions specifically addressed to Prof. Stupp dealt with the problem of the relation between the symmetry of packing of his rod-coil polymers and the molecular design of these objects. Dr. BUNZ (Mainz) asked for the origin of the shape in the rod-coil objects and its relation to the molar masses of the components. Professors NEWKOME (South Florida) and MEIJER (Eindhoven) compared the rod-coil polymers to mushrooms or dumbbells and asked for the packing principles of such objects, and in particular asked for the reason why they form a polar structure as reported and do not pack in a symmetric fashion as one would ordinarily expect. Prof. WEGNER hinted that the polar structure might be induced by the surface to which the films of these materials are cast. Prof. OZIN (Toronto) wanted to know whether the ordered phases of the rod-coil polymers show a domain structure similar to polycrystalline aggregates of ordinary materials, and what the origin of the monoclinic symmetry was, which was reported.

In his answers to these questions Prof. STUPP insisted that the major driving force to structure formation was a thermodynamic incompatibility between rod- and coil-segments in the molecules. The packing requirements of the rods and the tendency of the system to minimize free volume between the rod segments then drives the system to adopt particular symmetries. Shape complementarity was an important design principle following from such considerations. He reiterated that in his laboratory there was no evidence for any other packing modes than the one he had reported, and he agreed that the interaction with substrate surfaces in the stage when the superlattices are obtained as films may play an important role. The term "superlattices", he

J. Michl (ed.), Modular Chemistry, 251–254.

said he had chosen simply because the motifs of these lattices are very large. The use of this term in the present context was objected to by Prof. WEGNER who pointed out that the term "superlattice" is frequently used to describe periodic features superimposed onto a lattice of normal motifs, for instance a "superlattice of crystal defects" such as stacking faults, etc. No agreement was reached concerning nomenclature and semantics in this point. However, replying to Prof. OZIN's question, Prof. STUPP said that there was very clear evidence of domain structures and grain boundaries in the ordered films, also the domains were generally very large and of the size or larger than the field of view in TEM. He pointed out that the rod-coil bundles are giant smectogens and consequently all features seen with low molecular smectogens and their liquid crystal phases are or ought to be seen here as well.

Further questions to Prof. Stupp addressed the formation and properties of the 2-dimensional objects obtained by solid-state polymerization of layered assemblies. Prof. SEDDON (Belfast) asked for the control of molar mass and size distribution of the 2-D-systems. Prof. CHIDSEY (Stanford) wanted to know how frail the 2-D-objects are and how their structural integrity is preserved, and Dr. PALACIN (Harvard) inquired about the degree of conversion in the topochemical polymerization and whether residual monomer was left inside or was extracted. In his answers Prof. STUPP pointed out that there was no control possible so far considering the size and shape of the 2-D-objects, but that a regular shape is not necessarily a goal. For instance, in the formation of membranes from these materials an odd shape is advantageous and gives the reason for formation of pores. The 2-D-objects are quite stable and can be handled as "solutions". At this point the definition of what was meant by "solution" remained open, in other words whether it was a colloidal suspension of anisometric objects of small size or something else. In any case, residual monomer was not extracted although this was possible. Prof. STUPP said that the degree of conversion vs. polymerization of the acrylic double bonds was generally nearly quantitative whereas the degree of conversion vs. addition across the nitrile groups was generally low and not more that 50 percent as determined spectroscopically.

Questions to Prof. SCHLÜTER addressed a number of issues concerning the strategy and scope of the synthesis of large rings having some degree of shape persistence. Prof. MÜLLEN (Mainz) asked for effects of components such as thiophene, pyridyl, etc., substructures on the efficiency of ring formation and also on rotational isomers. This question was also addressed by Prof. STODDART (Birmingham). Prof. BALZANI (Bologna) wanted to know whether such large rings could serve as spacers or components in studies relating to energy or charge transfer between molecular entities, and Prof. WUEST (Montréal) raised a discussion of how "simple" or "complex" a strategy to the synthesis of large cycles needs to be, in other words whether "one-pot" synthesis of large rings are feasible. He insisted that this is a more or less necessary requirement if this field should ever be considered of practical use. In his answers Prof. SCHLÜTER pointed to the restricted mobility of the segments of the large rings as the main reason for the high yield in ring synthesis. Thus, any component may be included as long as a conformational analysis would not point to increased mobility of the constituents to be converted to the desired ring structure. Atropisomers may be possible products, but the barriers to rotation are not high enough to prevent isomerization although this may be--of course--a question of substituent size and other factors controlling the height of the barrier of rotation of moieties contained in the ring segments. The strategy of the ring synthesis so far followed a stepwise and thus traditional approach. Even if this is a somewhat tedious undertaking and needs many steps it nevertheless delivers the desired structures in amounts which allow a sufficient characterization of the properties. Once this is done, other approaches and even "one-pot" syntheses could be tried which generally gives mixtures with the usual separation problems.

A more general discussion was induced by a remark of Prof. WEGNER who wanted to know what "modules" are in the context of synthetic chemistry. Would one talk about modules and modular chemistry as soon as a level of size and organization was reached where properties of macroscopic nature appear? Such properties would need cooperative interactions between many components of the structure, typical examples for such properties would be electrical conductivity, mechanical elasticity or hardness, etc. Or was the term "modular chemistry" merely coined to address structures of a certain complexity and hierarchical nature? Prof. OZIN warned that the discussion should not be too narrowly focused on organic chemistry, but the wealth of inorganic and biological structures which fall under the term "modular" should not be neglected. He also disliked the term "hierarchy" because it was currently used in other contexts. Prof. WEGNER offered to change "level of complexity" to "hierarchy" if that clarified the issue. But would one consider the synthesis of benzene from 3 molecules of acetylene with a cobalt catalyst as "modular"? Prof. MICHL (Colorado) stated at this point that in his opinion most definitions in organic chemistry are somewhat fuzzy and one should not induce unnecessary complications by being too rigorous. Thus, he would not use the term "modular" in the context of the benzene synthesis, but he would definitely use it in the context of synthesis of large rings such as the ones seen in Prof. Schlüter's presentation. Size of the structure and complexity of the composition are the criteria to be used. As an example, he would not refer to the formation of carbon nanotubes as described by

Prof. Smalley as modular chemistry, however, the nanotubes could very well be used as components in modular chemistry.

GRAPHENE IN 2 & 3 DIMENSIONS

THOMAS W. EBBESEN
NEC Research Institute
4 Independence Way
Princeton, NJ 08540, U.S.A.

1. Introduction

The contents of this presentation are probably at the very edge of the theme of this meeting. However I hope it will contribute to the discussion precisely because it brings a different point of view from an area of science which has changed rapidly during the last decade. It is quite possible that some of the new concepts that have developed can be used in modular chemistry, especially in the design of modules from simple building blocks and assembly of modules into larger units.

The discovery of C_{60} and C_{70} ten years ago [1] has led to a complete new understanding of pure carbon materials [2,3]. Among the new concepts that have evolved is that one can lift a graphitic sheet, graphene, out of its 2 dimensions and give it a variety of 3 dimensional forms which had seemed unthinkable until then. Secondly, the properties of these structures can, in principle, be controlled simply by controlling their geometry.

In graphene, the basic building unit is a hexagon of sp^2 carbons. These hexagons form larger structures and building blocks through covalent bonding. The resulting structures can be flat sheets (honeycomb structure), ribbons, bent and folded open networks, cylinders (open nanotubes), etc.. An infinite variety of closed and open structures can also be formed by the introduction of rings other than hexagons (Euler's theorem). For instance, one pentagon introduces a 60 degree disclination, so 12 pentagons (4π) are necessary to close any structure independent of the number of hexagons present. C_{60} is one of the smallest of such closed structures.

The formation of 3 dimensional graphene structures also involves a change in the bonding character, i.e. sp^2 - sp^3 rehybridization. Rehybridization is the source of the out-of-plane flexibility of graphene and results in well defined folding and cutting patterns along the symmetry axes of the material [2, 3]. Rehybridization lines (sp^3 defect lines) could be used to delimit sp^2 areas for various purposes such as defining conductive zones.

Next more detailed explanation is given both with regards to Euler's theorem and rehybridization before their potential applications to modular chemistry are considered.

2. Euler's Theorem

J. Michl (ed.), Modular Chemistry, 255–260.

Euler's theorem relates the number of vertices (V), edges (E) and faces (F) of an object as follows:

$$V - E + F = X \qquad (1)$$

with $X = 2(1\text{-}g)$

X is Euler's characteristic and is related to the number of holes g. For a sphere like C_{60}, $g = 0$ while for a taurus $g = 1$.

For graphitic hexagonal networks, the sp^2 carbons imply that $3V = 2E$ and a more practical expression of Euler's theorem can be derived [4]:

$$3n_3 + 2n_4 + n_5 - n_7 - 2n_8 - 3n_9 = 12\ (1\text{-}g) \qquad (2)$$

where n_x is the number of polygons having x sides (and x vertices).

Equation (2) is very useful in that it tells how the presence of rings other than hexagons deform the flat hexagonal network and how many are necessary to balance the number of each type in order to obtain a closed structure, i.e. having a total disclination of 4π. For example: $g = 0$, $n_5 = 12$ i.e., 12 pentagons are necessary to close the hexagonal network whatever its size. Furthermore, n_5, a pentagon, creates a positive $4\pi/12 = 60°$ disclination while a heptagon, n_7, generates a negative 60° disclination as illustrated in Fig.1.

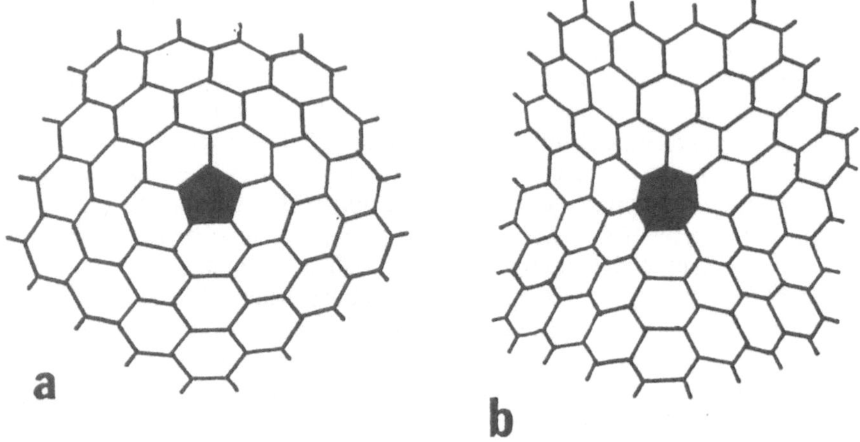

Figure 1. Schematic view of the opposite 60° disclinations induced by a pentagon (a) and a heptagon (b) in a hexagonal network (reproduced from Advance Materials, ref. 3).

By combining molecular polygons having different number of sides, novel 3 dimensional structure, can be formed. This approach might be very practical for modular chemistry as will be discussed in part 4.

3. Rehybridization

When a graphene sheet is bent away from its preferred flat 2 dimensional structure, it looses some of its sp^2 character and gains some sp^3 character in the bend. This ability to rehybridize is what gives graphene its out-of-plane flexibility. It has been shown elsewhere that the bending, folding and tearing of graphene occurs preferentially in certain directions of the graphene sheet which can be explained as follows [2,3].

In the bending or folding of the graphene sheet, defect lines having stronger sp^3 character are formed. However this sp^2-sp^3 rehybridization always involves the 2 carbon atoms of the bond that is bent. The sp^2 hexagonal network of graphene have four distinct pairs of carbon atoms along which the defect line with sp^3 character can form as shown in Fig. 2. Among these pairs, 2 are along the [100] symmetry axis and the other 2 along the [210] direction. Since there are equivalent symmetry axes every 60° and offset by 30° for the two principal axes, there is an axis every 30° as can be seen in Fig. 2. So one would expect that graphene folds and tears along directions which are multiples of 30°. This is indeed observed as shown elsewhere and illustrated in Fig. 3 [2,3]. Except for the defect line of type IIa, all the others have 2 conformations (boat and chair) [2].

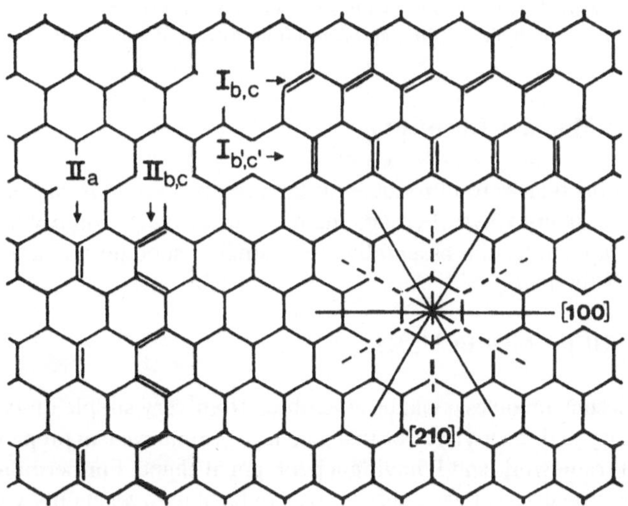

Figure 2. Hexagonal network of graphene, its symmetry axes and the 4 different pairs of carbon atoms across which defect lines having sp^3 character will preferentially form (reproduced from Advanced Materials, ref. 3).

It must be remembered that these folding and tearing directions are those that are the most favorable energetically because the lines remain uninterrupted. With enough energy, other directions may also occur but the defect line will be jagged on a microscopic scale. In any case, once folded onto itself the sheet will be stabilized by the inter-layer van der Walls forces. If the sheet is completely folded over with a sharp bend, the defect line will be essentially sp^3 with the formation of dangling bonds along the edge. This configuration could then be stabilized, or fixed, by chemical passivation (reaction with the dangling bonds).

Molecular materials could also be folded and patterned using similar principles as discussed below.

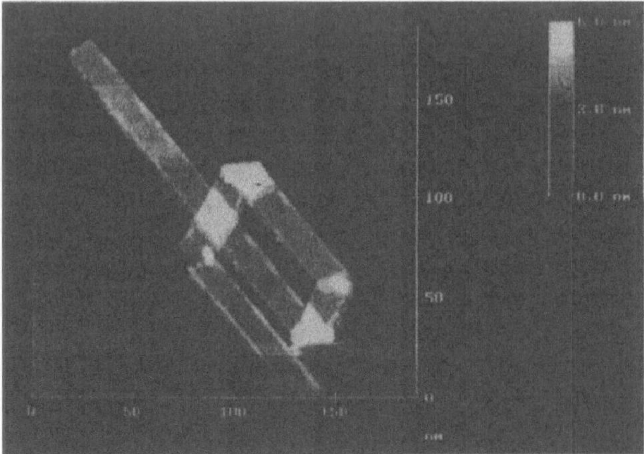

Figure 3. AFM picture of a graphitic ribbon folded 4 times on the surface of HOPG. Notice that the folding has occurred along angles that are multiples of 30° (reproduced from Advanced Material, ref.3)

4. Potential Uses in Modular Chemistry

The concepts that were introduced in the previous sections can be put to use in modular chemistry in many ways. Perhaps the most attractive possibility is the design of new molecules and supramolecular assemblies from smaller subunits using the geometrical principles of Euler's theorem.

4.1 DESIGNING EULER MOLECULES

Large molecular modules could be assembled from very simple units having fairly well defined geometry and reactivity. For instance lets assume we can prepare molecule H having a hexagonal framework and P having a pentagonal shape. Furthermore lets assume that H and P are functionalized at each corner so as to be able to react only with each other (and not with themselves) as shown in Fig. 4. Then by mixing H and P in proportions of 10:6, large spheres might result in one or more reaction steps. Half spheres could be isolated and then used to prepare tubes in a secondary step. Such schemes could be extended to include other polygons (heptagonal, square, etc., molecules) which with the right reactivity could serves as the basic units to assemble a variety of larger structures through covalent bonding, coulombic interactions, hydrogen bonding, etc.. In other words, from a relatively small set of subunits, or tectons [5], having different geometries, a great variety of 3 dimensional structures could be assembled.

From these considerations, we can propose a new class of molecules, eulerenes, which would be defined as follows: molecules that are assembled by using the principles of Euler's theorem to achieve a desired geometry. Fullerenes is a sub class of Eulerenes.

Supramolecular assemblies of the Euler type are found routinely in nature in order to achieve closed structures from hexagonal packing of subunits, e.g. protein coating of viruses. However, to my knowledge, there are no eulerenes, or Euler assemblies, that have

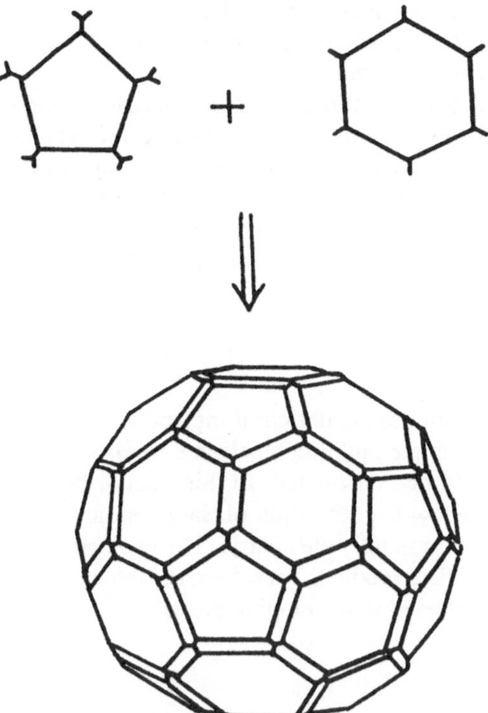

Figure 4. Example of an Eulerene formed from hexagonal and pentagonal subunits and whose final 3 dimensional structure is determined by the principles of Euler's theorem.

been prepared in the laboratory. Attempts have been made to make fullerenes by wet synthesis but without success so far. Eulerenes might be much simpler to assemble because the bonds between the cycles should be more apt at bending compared to the rigid sp^2 network of the fullerenes. Eulerenes could be very useful as encapsulants and as units for the construction of larger molecular skeletons.

Using the principles of Euler's theorem from the smallest building unit to the larger supramolecular assemblies might also favor the formation of hierarchical structures.

4.2 REHYBRIDIZATION AND OTHER GEOMETRICAL FACTORS

As with the case of graphene, rehybridization properties of a molecular material could be used to create folds, patterns, zones and linkages. In other words, one could design molecular structures that have a line of atoms or single points which easily rehybridize so that the structure can be made to fold or link to another object in a well defined manner. Such a scheme could also be applied to encapsulate other materials.

Rehybridization can be used to delimit conducting zones in a continuous network of a given material. A continuous network from microscopic to macroscopic scales has many advantages. For instance in the case of graphene, it should allow one to access molecular size small conductive wires and devices from larger macroscopic areas without any interruption in the connectivity. Therefore it has the potential to overcome the problem of addressability of molecular scale modules.

There are other geometrical notions that are important for modular chemistry. For a given material, tubes are stronger than the open flat sheet. It might also allow fine tuning of the properties of the material. For example, the electronic properties of carbon nanotubes depend on their diameter and their helicity [6,7]. The tubular structures also appear to avoid the low dimensional instabilities, such as Peierls distortion, associated with molecular "wires" [6] of similar dimensions and gives the material strength and elasticity [7].

5. Conclusion

Modular chemistry implies the notion of molecular architecture and therefore of geometry. For this purpose, simple building blocks are needed from which a variety of 3 dimensional assemblies could be constructed. In this regard, building units, or tectons, having features that would allow the formation of larger structures according to Euler's theorem would be promising for modular chemistry. The geometry will also determine the properties and functions of the final product. Therefore, the issues of property, function and architecture must all be considered in developing the science and technology of modular chemistry.

6. References

1. Kroto, H. W. , Heath, J. R., O'Brian, S. C., Curl, R.F. and Smalley, R. E. (1985) C_{60}: Buckminsterfullerene, *Nature* **318**, 162-163.
2. Hiura, H., Ebbesen, T. W., Fujita, J., Tanigaki, K. and Takada, T. (1994) Role of sp^3 defect structures in graphite and carbon nanotubes, *Nature* **367**, 148-151.
3. Ebbesen, T. W. and Hiura, H. (1995) Graphene in 3-dimensions: Towards graphite origami, *Adv. Mater.* **7**, 582-586.
4. Terrones, H. and Mackay, A.L.(1992) The geometry of hypothetical curved graphite structures, *Carbon* **30**, 1251-1260.
5. Wuest, J.D., in Mendenhall, G.D., Greenberg, A., Liebman, J.F. (Eds.) (1988) *Mesomolecules: From Molecules to Materials*, Chapman & Hall, New York.
6. Mintmire, J.W., Dunlap, B.I. and White, C.T. (1992) Are fullerene tubules metallic?, *Phys. Rev. Lett.* **68**, 631-634.
7. Ebbesen, T.W. (1994) Carbon nanotubes, *Annu. Rev. Mater. Sci.* **24**, 235- 264.

FULLERENE FOOTPRINTS: CYCLOADDUCTS OF CARBON RINGS

DOUGLAS L. STROUT AND GUSTAVO E. SCUSERIA
Center for Nanoscale Science and Technology,
Rice Quantum Institute and Department of Chemistry,
Rice University, MS 60
Houston, Texas 77005-1892, USA

ABSTRACT. Since their discovery, the fullerenes have challenged scientists with a wide array of problems concerning their properties, behavior and potential applications. One such fundamental question is that of the assembly of these carbon cages. Much effort has been directed at obtaining an understanding of the process whereby graphite is transformed into fullerenes. In this work, a theoretical study is carried out to explore the nature of carbon clusters, including the fullerenes themselves as well as other structures which may be intermediates along the path to fullerenes. The experiments which generate fullerenes are interpreted by the use of theoretical calculations, and this interpretation is used to develop a model for a fullerene assembly pathway.

1. Introduction

A fundamental question about C_{60} and the other fullerenes is the mechanism of their formation. What happens to graphite after it is vaporized which leads to the appearance of carbon cages? Various models have been advanced. The "pentagon road" model [1] suggests that fullerenes form from the successive addition of small carbon species to the edges of curved graphitic precursors until cage closure results. The "fullerene road" [2] advances the idea that carbon cages of all sizes are formed in the experiments but then add or lose C_2 radicals until a 60-atom cage is formed. Also, "ring stacking" models [3] exist which propose that C_{60} and the larger fullerenes are formed by the sequential addition of specific carbon rings. Several fullerene formation mechanisms have been discussed by Curl [4].

Recently, a powerful tool has been introduced for examining the details of graphite vaporization experiments. An experimental technique, known as "ion chromatography," [5] has revealed new insights into the fullerene formation experiments. With this technique, the cluster ions produced by graphite vaporization are mass-selected and injected into a tube that is filled with an inert gas, usually helium. The ions move through the length of the tube, and collisions between the clusters and the inert gas cause the clusters to be separated by cross-section, as the more compact clusters exit the tube at earlier times. The resulting plot of cluster intensity versus time is called an "arrival time distribution" (or ATD).

von Helden *et al.* [6] performed ion chromatography on the graphite vaporization products and reported several peaks in the ATD. For carbon clusters with 10-20 atoms, a single peak appeared. Beginning with C_{21}, two peaks were observed, and for 30 or more atoms, three more peaks were seen, for a total of five. These

J. Michl (ed.), Modular Chemistry, 261–274.

authors also developed a method for assigning the arrival time peaks to structural cluster isomers. For each proposed structural isomer, they optimized the geometry using the semiempirical PM3 method. Each optimized geometry was then used for a Monte Carlo simulation of the collisions between the proposed cluster and the helium atoms. This procedure yields a theoretical cross-section for the cluster, which is then converted to a convenient scale called the "mobility" scale. The experimental arrival times are also converted to mobilities, and if a match is obtained, the proposed cluster is assigned to the experimental peak.

According to the von Helden *et al.* model, the peak that is seen beginning at C_{10} corresponds to a monocyclic carbon ring and is called the "Ring I" peak. The second peak observed at C_{21} is denoted the "Ring II" peak and is matched to planar bicyclic rings. One of the three peaks appearing at C_{30} is named "Ring III" and is interpreted as planar tricyclic structures. Another peak is of very low intensity and has been dubbed "3-D ring." The final peak, corresponding to the most compact structures observed in the experiment, is assigned to fullerenes.

Following the method of von Helden *et al.*, Book *et al.* [7] developed a mobility code and used this program to illustrate two important points about the theoretical mobilities. First, calculated mobilities were shown to be insensitive to molecular vibration. This validates the use of static, equilibrium geometries for the mobility calculations. Also, these mobilities were shown to be rather insensitive to the choice of the level of theory employed when obtaining optimized geometries. Therefore, fast, semiempirical methods can be used to obtain valid geometries quickly.

Using the Book *et al.* mobility code, Strout *et al.* [8] proposed an alternative to the von Helden *et al.* model. This alternative advanced the idea that bimolecular cycloaddition processes between monocyclic carbon rings were sufficient to explain the "Ring" isomers in the ion chromatography experiments. In this "cycloaddition model," monocyclic structures are retained as Ring I, and Ring II is interpreted as 2+2 cycloadducts of carbon rings, following the previous concept of planar bicyclic rings. However, it is shown that 2+4 cycloadducts match the Ring III mobilities. Those cycloadducts are nonplanar and therefore more compact than the 2+2 cycloadducts. Since the original publication of these cycloaddition ideas, we have further demonstrated [9] that 4+6 cycloadducts also have mobilities which match Ring III experimental values. For all mobility calculations in this work, geometries have been optimized with a semiempirical tight-binding (TB) potential [10]. Figure 1 shows 36-atom examples of 2+2, 2+4, and 4+6 cycloadducts, and Table 1 demonstrates that 2+4 and 4+6 cycloadducts have mobilities consistent with Ring III.

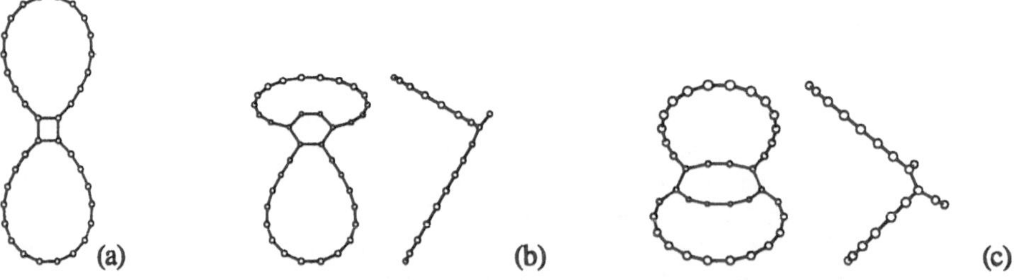

Figure 1. C_{36} Cycloadducts: (a) 2+2 (b) 2+4 (c) 4+6

TABLE 1. Comparison of experimental Ring III mobilities to those calculated for 2+4 and 4+6 cycloadducts (mobilities in cm^2/volt second)

Cluster	Ring III	Adduct	4+6	2+4
C_{36}	3.70	18+18	3.75	3.64
C_{40}	3.26	22+18	3.31	3.23
C_{44}	2.95	22+22	2.97	2.87
C_{48}	2.7	26+22	2.65	2.71

The ion chromatography method of interpreting graphite vaporization experiments has power because it reveals information about clusters that are produced in the process that ultimately leads to fullerenes. In fact, some of the other structures which appear in this experiment may be intermediates in the formation of fullerenes, and understanding the nature of those intermediates sheds light on the fullerene formation mechanism itself. However, mobility calculations can only demonstrate that a proposed cluster is consistent with experimental observation. Neither the von Helden *et al.* model nor the cycloaddition model can be proven using mobility data alone.

Using high level ab initio calculations, it will be shown that the cycloaddition model is consistent with the experimental observations about Ring II and Ring III. The 2+4 and 4+6 cycloadducts not only satisfy the mobility data but also explain the appearance threshold of 30 atoms for the Ring III peak. These cycloadducts will then be used to develop a model of fullerene assembly based on reactions that are shown to yield reasonable theoretical energetics.

2. The Case for 2+4 and 4+6 Cycloadducts

The mobility results establish the cycloadducts as **possibilities** for the clusters generated by graphite vaporization, but it remains to be shown why 2+2 cycloadducts would appear at C_{20} whereas 2+4 and 4+6 cycloadducts should not be seen until C_{30}. The question of whether or not a given cluster appears involves the energetics of the reaction that produces it. Detailed, accurate energies of reactants, products and transition states are needed, and for that high-level ab initio calculations are required. Geometries for the monocyclic rings, cycloadducts and transition structures are optimized with the Hartree-Fock (HF) method using a double zeta (DZ), or 4s2p, basis set formed by contracting the Huzinaga-Dunning (9s5p) primitive Gaussian set [11].

Energies at the DZ HF geometries are computed with density functional theory (DFT) using a double zeta plus polarization (DZP) basis, consisting of the DZ basis plus polarization function α_d=0.75. This DFT method employs the nonlocal (gradient-corrected) Becke [12] exchange functional and the nonlocal Perdew [13] correlation functional. This Becke-Perdew method is denoted BP. Hartree-Fock calculations are carried out by using the TURBOMOLE quantum chemistry package, [14] and DFT is introduced through a separate code [15] written in our research group which interfaces with TURBOMOLE.

Density functional theory is implemented as a hybrid with Hartree-Fock, meaning that the exchange and correlation energies are computed using the Hartree-Fock electron density rather than a DFT density. The DFT total energies are computed by subtracting the HF exchange from the HF total energy and adding the DFT exchange and correlation energies. That is,

$$E^{DFT} = E^{HF} - E_x^{HF} + E_x^{DFT} + E_c^{DFT} \tag{1}$$

for a general hybrid DFT method, where E_X and E_C represent exchange and correlation energies, respectively.

The particular clusters studied include monocyclic rings C_{10}, C_{14}, and C_{18}. These rings are chosen because they are Hückel 4n+2 rings. The cycloadducts selected are C_{20}, C_{24}, C_{28}, C_{32}, and C_{36}. This range of cycloadducts includes clusters on both sides of the appearance threshold for Ring III and should allow for a thorough exploration of the cycloaddition processes. 2+2, 2+4, and 4+6 cycloaddition energetics are quantified using the theoretical methods outlined above. The DZ HF ground states for 2+4 and 4+6 adducts and their transition states are electronic singlets. Transition states for 2+2 adducts have been optimized with triplet electronic configurations, because of the Woodward-Hoffmann rules for cycloadditions. The 2+2 adducts themselves are predicted by DZ HF to have triplet ground states. However, DZP BP predicts singlet states for the 2+2 adducts, and the results reported in this work are for singlet 2+2 cycloadducts.

The first question to be answered by these calculations is whether the transition barriers for 2+2, 2+4, and 4+6 cycloaddition are a function of cluster size over the range of 20-36 atoms. If the barriers to 2+4 and 4+6 are appreciably higher for sizes below C_{30} than above C_{30}, then the appearance threshold for Ring III can be explained simply as a result of prohibitively high reaction barriers below the threshold.

DZP BP results are shown in Table 2 and do not show the radical changes in energy required to explain the appearance thresholds. The 2+2 barriers are within a few tenths of an eV of each other over the range 20-36 atoms (a tenth of an eV is 2.3 kcal/mol), which is consistent with experiment because 2+2 adducts are observed throughout the size regime considered. However, 2+4 and 4+6 barriers across the C_{30} threshold also do not vary by more than a few tenths of an eV. The DZP BP barriers do not support any kind of assertion that 2+4 of 4+6 cycloadducts should form selectively at sizes above C_{30}.

TABLE 2. DZP BP barriers for 2+2, 2+4, and 4+6 cycloaddition (energies in eV)

Process	C_{20} 10+10	C_{24} 10+14	C_{28} 14+14	C_{32} 14+18	C_{36} 18+18
2+2	2.38	2.66	2.06	2.69	1.91
2+4	1.61	1.27	1.05	0.90	0.93
4+6		1.80	1.38	1.33	0.91

However, a cycloaddition may occur and be immediately followed by the corresponding retrocycloaddition. This makes it important to determine ΔE for the cycloaddition processes because the reverse barrier equals the forward barrier minus ΔE. The DZP BP ΔE data for the cycloadducts are shown in Table 3, and these data provide a clear explanation for the observed appearance thresholds. The 2+2 cycloadditions are exothermic by more than 2 eV (nearly 50 kcal/mol) over the entire range from 20-36 atoms. Once formed, a 2+2 cycloadduct would be very stable and tend not to dissociate. For larger clusters, such as C_{32} and C_{36}, the 2+4 and 4+6 cycloadducts have stability similar to the 2+2 adducts. However, for the smaller clusters, the 2+4 and 4+6 cycloadducts are much less stable and may dissociate back to their constituent monocyclic rings. This is the explanation that demonstrates that 2+4 and 4+6 cycloaddition are consistent with experimental observations about Ring III.

TABLE 3. DZP BP ΔE for 2+2, 2+4, and 4+6 cycloaddition (energies in eV)

Process	C_{20}	C_{24}	C_{28}	C_{32}	C_{36}
	10+10	10+14	14+14	14+18	18+18
2+2	-2.35	-2.46	-2.51	-2.46	-2.42
2+4	-0.24	-1.26	-1.73	-2.10	-2.20
4+6	-0.59	-1.29	-1.81	-2.21	-2.25

If resistance to retrocycloaddition is the key to whether a given cycloadduct will be experimentally observed, then the DZP BP level of theory can be used to establish an empirical rule that a cycloaddition process must be exothermic by more than 2 eV to produce an adduct that resists dissociation. This rule would predict 2+2 adducts at all sizes from 20-36 atoms and 2+4 and 4+6 adducts at sizes with more than 30 atoms, which is consistent with the experimental observations. This rule is qualitative in nature and only applies to the DZP BP level of theory, and there certainly exists no physical reason why 2 eV should be the threshold. Higher levels of theory, such as coupled-cluster theory (as CCSD or CCSD(T)), could reflect the same behavioral trends as DZP BP and perhaps show a much different threshold.

The cycloaddition model, therefore, is consistent with two important experimental observations. First, the cycloadducts have been shown to have cluster mobilities that match the mobilities observed in the ion chromatography experiments. Also, the energetics of the cycloaddition reactions provide a clear and simple explanation for the observed cluster appearance thresholds. The cycloaddition model provides a complete and plausible interpretation of the graphite ion chromatography experiments.

3. From Cycloadducts to Rings and "Hoops"

What happens to the cycloadducts after they are formed? It has been suggested previously that 2+2 cycloadducts can transform to fullerenes, [16] but this seems unlikely because 2+2 adducts with 20-30 carbon atoms fail to produce fullerenes despite strong theoretical evidence [17] that fullerenes are much more stable than carbon rings that are seen in this size range. However, a 2+2 cycloadduct can undergo a 2+2 retrocycloaddition process which breaks two bonds to form a single large monocyclic ring. The newly formed large ring can react to form still larger cycloadducts. Larger 2+2 cycloadducts can react to form larger rings, and so on. Rings with as many as 60 atoms have been detected by ion chromatography techniques [16].

2+4 and 4+6 cycloadducts have no such one-step reaction that leads to a larger ring. However, these clusters can, with a modest barrier, undergo cyclization reactions around the site of the initial cycloaddition. In the present model, the cycloadduct undergoes a sequence of cyclizations that results in a cylindrical carbon species that is denoted here as a carbon "hoop." For a C_{36} 2+4 cycloadduct of two C_{18} carbon rings, the transformation to a hoop has been mapped out using the semiempirical TB method. TB optimized structures for the minima along the path are shown in Figure 2, and the TB energetics are presented in Table 4. At the TB level of theory, each step in the hoop formation is exothermic and has a modest barrier. The overall process is exothermic by more than 2.5 eV. A similar hoop assembly exists for the C_{36} 4+6 cycloadduct. The resultant hoop is identical to the one formed from the 2+4 adduct. Both cycloadducts are viable starting materials for hoop formation.

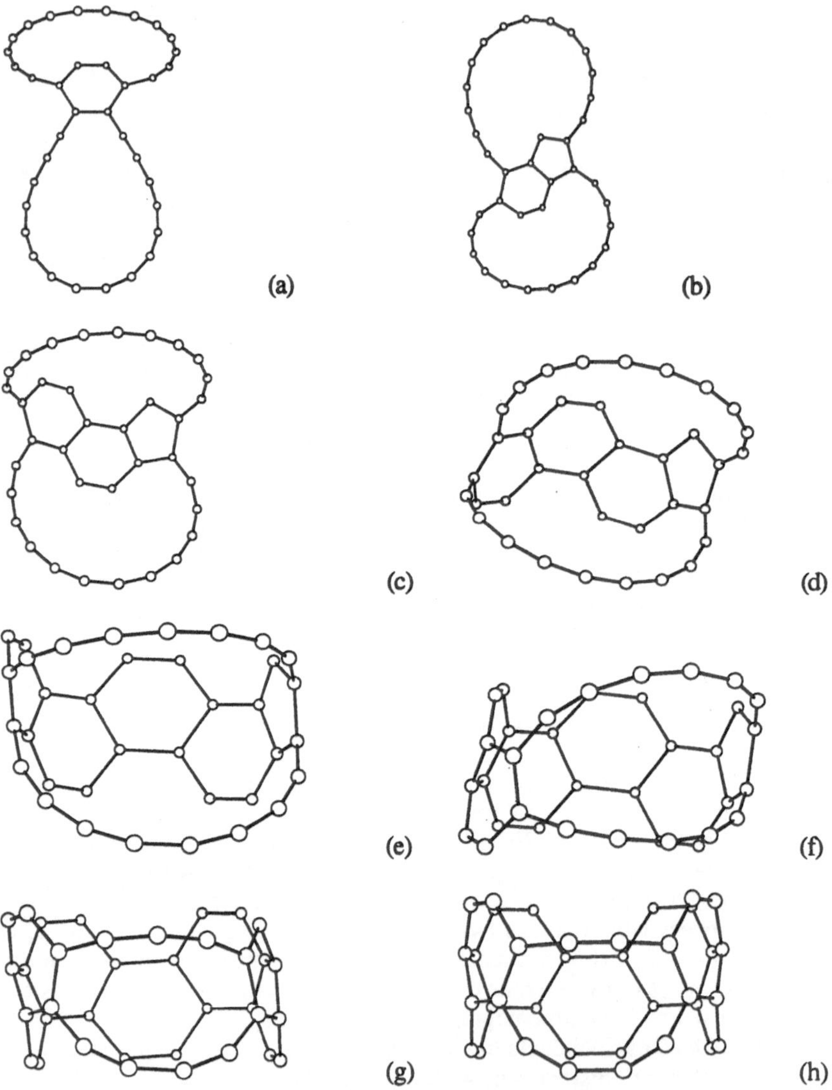

Figure 2. TB pathway from C_{36} 2+4 cycloadduct to carbon "hoop"

TABLE 4. TB energetics for hoop assembly shown in Figure 2 (energies in eV).

Step	E(minima)	Barrier (eV)
(a, adduct)	0.00	0.87
(b)	-0.29	1.16
(c)	-0.51	1.12
(d)	-0.85	1.01
(e)	-1.09	0.95
(f)	-1.47	1.00
(g)	-1.53	0.99
(h, hoop)	-2.66	

Do these results hold up at higher levels of theory? DZ HF optimizations have been performed out on a sequence of the hoops themselves. Using these DZ HF geometries, DZP BP energies have been computed for hoops with sizes ranging from 28 to 60 atoms. The DZP BP energetics are detailed in Table 5. This data illustrates that, as cluster size increases, hoops become increasingly stable with respect to monocyclic precursors.

TABLE 5. DZP BP energies of hoops with respect to monocyclic rings (in eV)

Hoop size	Rings	Hoop
C_{28}	14+14	-2.03
C_{32}	14+18	-2.45
C_{36}	18+18	-3.28
C_{40}	18+22	-3.80
C_{44}	22+22	-4.05
C_{48}	22+26	-4.21
C_{52}	26+26	-4.70
C_{56}	26+30	-5.24
C_{60}	30+30	-5.32

DZP BP energies at DZ HF geometries predict that the transformation from cycloadducts to hoops is energetically favorable for all cluster sizes for which the Ring III species are observed experimentally. The pathway from cycloadducts to hoops introduces a significant amount of graphitic (sp^2) character to the carbon clusters. Can a hoop become a fullerene, which is an all-sp^2 molecule?

4. From Hoops to Cages

Carbon hoops have a large number of triple bonds between atoms whose bond angles are not 180°. The bond angle strain should make those sites very susceptible to reactions which rearrange the bonds. One such reaction is the 1,2-carbon shift. In a 1,2-carbon shift, a hexagon becomes a pentagon with a dangling atom attached, as illustrated in Figure 3. After two 1,2-shifts have occurred on neighboring triple bonds, the two dangling atoms can bond to each other to form a new pentagon or hexagon. In this way, a hoop can close its top and bottom faces with new polygons and thereby form a spherical cage.

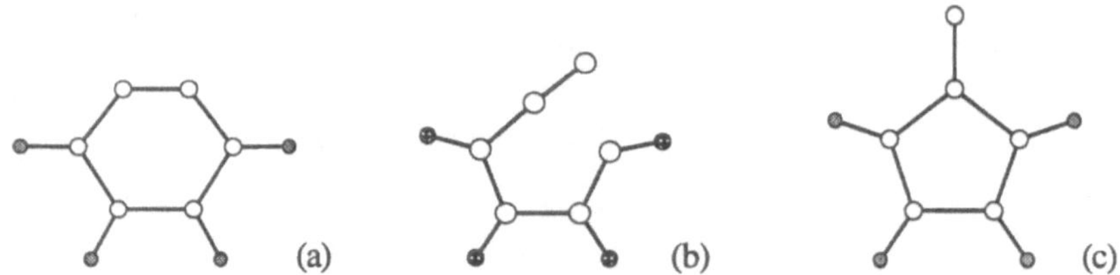

Figure 3. 1,2-carbon shift: (a) reactant (b) transition state (c) product

Energetics of 1,2-carbon shifts have been explored with the DZP BP method on DZ HF optimized geometries. Calculations are first performed on small model systems which simulate a section of a hoop, as shown in Figure 4. For these systems, which have hydrogen atoms replacing the connections to the rest of the hoop, minima and transition states have been optimized. The DZP BP energetics are presented in Table 6. For actual hoops, minima have been optimized for C_{32}, C_{36}, and C_{40}, and the energetics are also shown in Table 6.

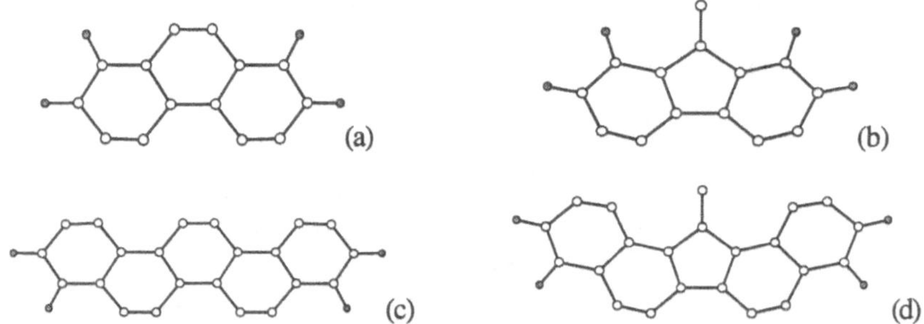

Figure 4. Model systems for 1,2-carbon shifts: (a) $C_{14}H_4$ reactant (b) $C_{14}H_4$ product (c) $C_{22}H_4$ reactant (d) $C_{22}H_4$ product

TABLE 6. Energetics of 1,2-carbon shifts (energies in eV)

Molecule	Barrier	ΔE
$C_{14}H_4$	+1.67	+1.15
$C_{22}H_4$	+1.92	+1.28
C_{32} hoop		+1.91
C_{36} hoop		+1.33
C_{40} hoop		+1.69

Transition barriers of 2 eV and ΔE around 1.3 eV require that the hoop put approximately 3.3 eV into the 1,2-shift process before any energy can be recovered.

That is the energy to perform two 1-2-carbon shifts. After the first two shifts, the first new polygon forms, and the electronic energy of the cluster drops significantly, which fuels further carbon shifts. These new carbon shifts form more polygons, which liberates more energy for more reactions. The process continues until cage closure occurs, with a very large total drop in the cluster's electronic energy.

For the C_{36} hoop discussed in the previous chapter, the path from hoop to initial cage closure has been mapped using the semiempirical TB method. Minima and transition states are optimized, and the TB energetic data are shown in Table 7. Minima along the path are illustrated in Figure 5. The total process lowers the electronic energy of the cluster by 10.6 eV. However, it is interesting to note that the closed cage is initially not the D_{2d} isomer which is known to be most stable. The cage still must find a way to its most stable form.

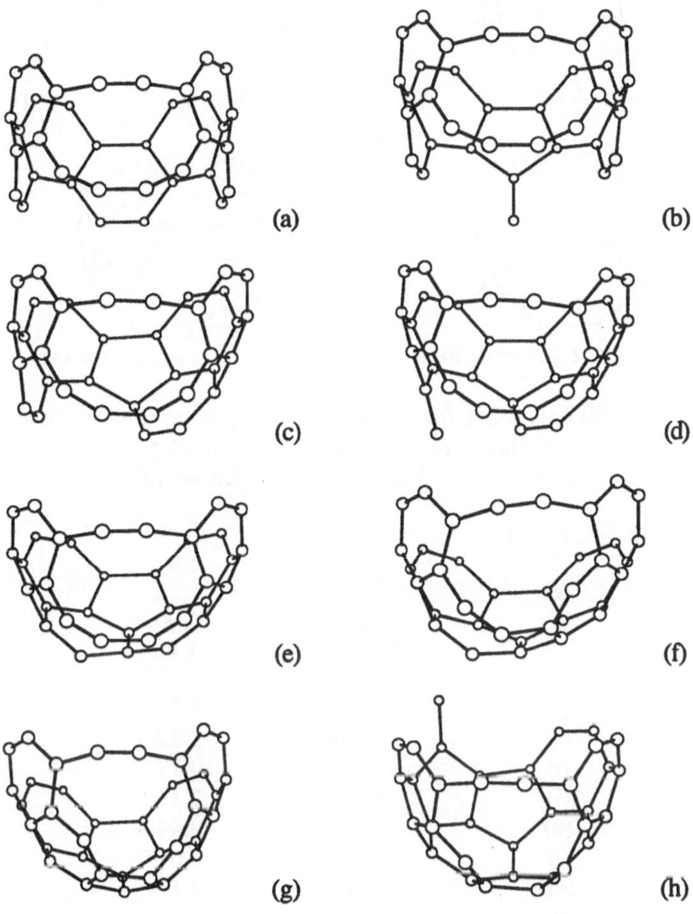

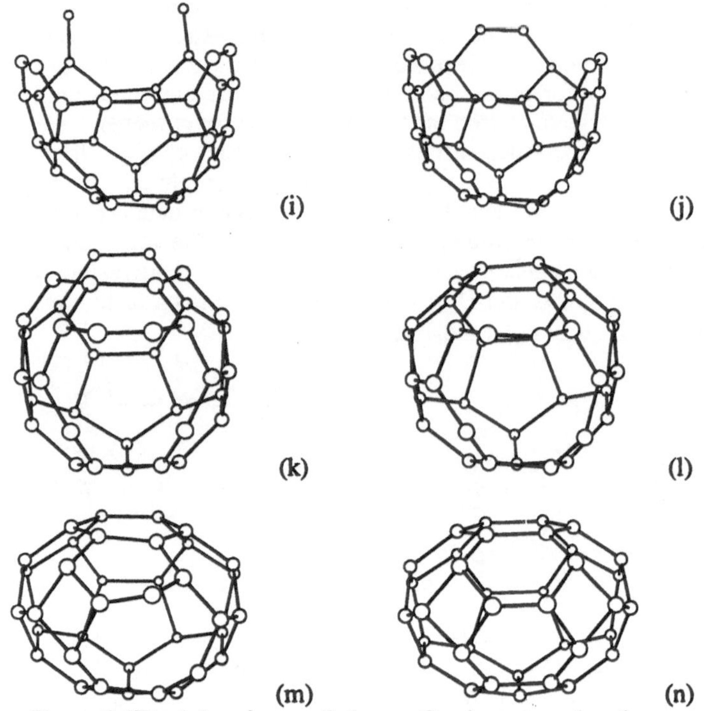

Figure 5. TB minima for a path from a C_{36} hoop to a closed cage

TABLE 7. TB energetics for a path from a C_{36} hoop to a closed cage, as shown in Figure 5 (energies in eV)

Step	E(minima)	Barrier (eV)
(a, hoop)	0.00	2.42
(b)	+1.21	2.62
(c)	-1.53	2.22
(d)	-0.56	0.00
(e)	-2.32	0.16
(f)	-4.12	0.00
(g)	-6.39	3.39
(h)	-4.14	3.22
(i)	-2.11	0.19
(j)	-5.77	2.18
(k)	-3.88	0.44
(l)	-9.54	0.66
(m)	-8.90	0.10
(n)	-10.57	

5. Annealing

Once a carbon cage is closed, there is no guarantee that the cage will be the most stable isomer for the particular fullerene. In fact, it is highly improbable that the fullerene generation experiments would yield the most stable isomer of any fullerene upon initial cage closure. However, the cages can isomerize by undergoing processes that rearrange the polygons composing the cage. Such reactions are collectively referred to as annealing processes.

The best known annealing process is the Stone-Wales, [18] or pyracylene, rearrangement. Two pentagons and two hexagons exchange places in a section of a fullerene surface. The Stone-Wales reaction on the buckminsterfullerene molecule has been characterized by theoretical means [19]. A more general class of Stone-Wales type reactions, in which a C_2 bond rotates 90° irrespective of the number of atoms in the surrounding polygons, has also been studied [20]. Both Stone-Wales studies employed the semiempirical MNDO method, as well as HF and DFT methods [19,20].

Results from these studies predict that the barrier to Stone-Wales type rearrangements is, in general, about 6-7 eV, which is the highest barrier of any reaction in the present fullerene assembly model. However, by the time cage closure occurs, an enormous lowering of the electronic energy of the cluster has taken place, with a corresponding conversion of that energy to vibration. This vibrational energy is the means whereby the annealing barriers can be overcome. To illustrate the magnitude of the lowering of the electronic energy, DZP BP calculations have been performed on several fullerenes, with comparison to the energies of the original monocyclic rings, and the results are shown in Table 8. Since fullerenes are more stable than monocyclic rings by 20 eV and more, overcoming the barriers to Stone-Wales type rearrangements is not a problem. TB energetics for a complete path from two C_{30} rings to buckminsterfullerene are shown in Figure 6.

TABLE 8. DZP BP energies of fullerenes relative to monocyclic precursors (in eV)

Fullerene	Rings	Rel. energy
C_{32}	14+18	-18.86
C_{36}	18+18	-21.86
C_{40}	18+22	-25.88
C_{44}	22+22	-30.29
:		
:		
C_{60}	30+30	-50.13

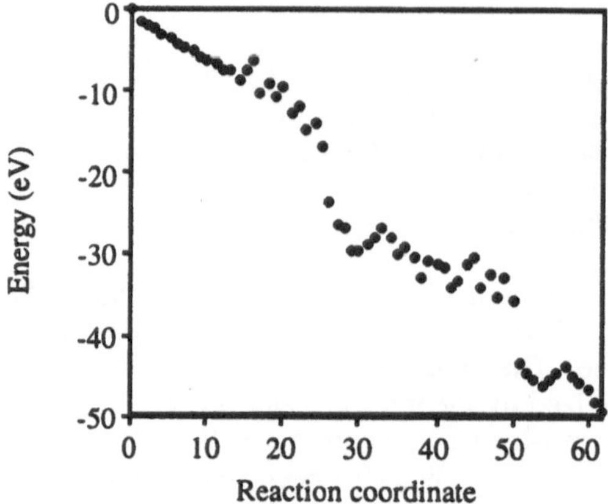

Figure 6. TB energetics for a path from two C_{30} rings to buckminsterfullerene

6. Conclusion

The cycloaddition model provides a complete, explicit pathway from monocyclic carbon rings to fullerenes. This model is consistent with current experimental results and elucidates the road to fullerenes with energetic information obtained at high levels of theory. Also, this model is simply a path to closure of a carbon cage, any cage, rather than relying on contrived mechanisms which are directed toward specific cages. These are the advantages of the model.

In the cycloaddition model, experimentally observed monocyclic rings collide to create a variety of cycloadducts. The appearance of 2+2, 2+4, and 4+6 cycloadducts can be supported by experimental data. 2+2 adducts open via a retro-2+2 process to form larger monocyclic rings, which could be called "second generation" rings. Second generation rings collide with other rings to form large cycloadducts. Large 2+2 cycloadducts can open to form third generation rings, and so on.

2+4 and 4+6 cycloadducts do not have a low-energy path to second generation monocyclic rings. Rather, they undergo a series of cyclizations to form carbon hoops. Carbon hoops undergo 1,2-carbon shifts and further cyclizations to generate closed cages. These cages anneal to find the most stable isomer. The elements of the cycloaddition model are represented graphically in Figure 7.

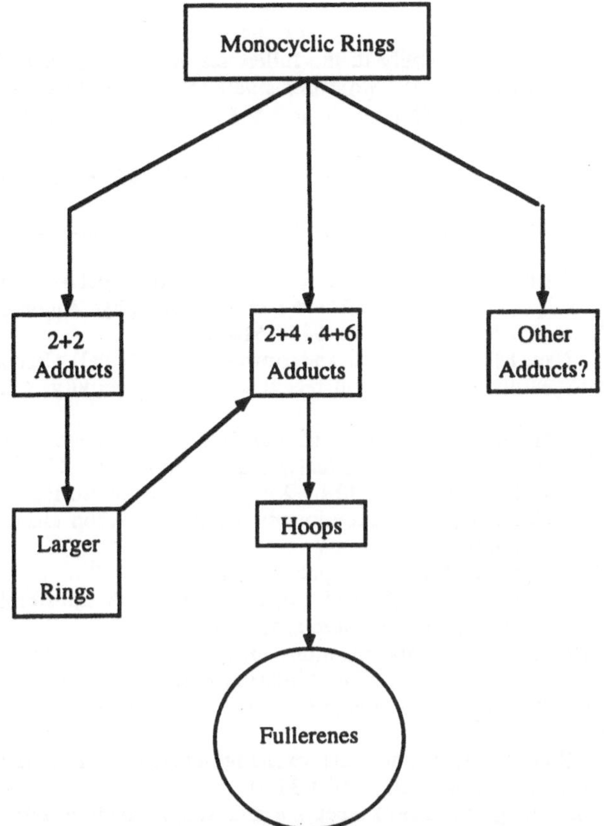

Figure 7. The cycloaddition model for fullerene formation

The cycloaddition model is a useful tool for understanding fullerene formation. Reaction pathways in this model are both consistent with experimental observations and plausible from a theoretical point of view. A new interpretation of the graphite vaporization ion chromatography results is developed into a fullerene assembly model based on reactions which are shown by theoretical calculations to have reasonable energetics. We believe that this cycloaddition model is a novel, successful approach toward the elucidation of the fullerene formation process.

Acknowledgments

This work was supported by the National Science Foundation (CHE-9321297) and the Welch Foundation. Acknowledgment is made to the donors of the Petroleum Research Fund, administered by the American Chemical Society, for partial support of this research.

References

1. Smalley, R.E. (1992) Self-assembly of the fullerenes, *Acc. Chem. Res.*, **25**, 98-105.
2. Heath, J.R. (1992) Synthesis of C_{60} from small carbon clusters: A model based on experiment and theory, in G.S. Hammond and V.J. Kuck (eds.), *Fullerenes*, American Chemical Society, Washington, DC, pp. 1-23.
3. Wakabayashi, T. and Achiba, Y. (1992) A model for the C_{60} and C_{70} growth mechanism, *Chem. Phys. Lett.*, **190**, 465-468.
4. Curl, R.F. (1993) On the formation of the fullerenes, *Phil. Trans. R. Soc. Lond. A*, **343**, 19-32.
5. Kemper, P.R. and Bowers, M.T. (1990), *J. Am. Soc. Mass Spectrom.*, **1**, 197. Kemper, P.R. and Bowers, M.T. (1991) Electronic state chromatography: Application to first-row transition metal ions, *J. Phys. Chem.*, **95**, 5134-5146.
6. von Helden, G., Hsu, M.-T., Gotts, N. and Bowers, M.T. (1993) Carbon cluster cations with up to 84 atoms: structures, formation mechanism, and reactivity, *J. Phys. Chem.*, **97**, 8182-8192.
7. Book, L.D., Xu, C.H. and Scuseria, G.E. (1994) Carbon cluster ion drift mobilities. The importance of geometry and vibrational effects, *Chem. Phys. Lett.*, **222**, 281-286.
8. Strout, D.L., Book, L.D., Millam, J.M., Xu, C.H. and Scuseria, G.E. (1994) How unequivocally do ion chromatography experiments determine carbon cluster geometries?, *J. Phys. Chem.*, **98**, 8622-8626.
9. Strout, D.L. and Scuseria, G.E. (1995) unpublished results.
10. Xu, C.H., Wang, C.Z., Chan, C.T. and Ho, K.M. (1992) A transferable tight-binding potential for carbon, *J. Phys. Condens. Matter*, **4**, 6047-6054.
11. Huzinaga, S. (1965) Gaussian-type functions for polyatomic systems. I, *J. Chem. Phys.*, **42**, 1293-1302. Dunning, Jr., T.H. (1970) Gaussian basis functions for use in molecular calculations. I. Contraction of (9s5p) atomic basis sets for first row atoms, *J. Chem. Phys.*, **53**, 2823-2833.
12. Becke, A.D. (1988) Density-functional exchange-energy approximation with correct asymptotic behavior, *Phys. Rev A*, **38**, 3098-3100.
13. Perdew, J.P. and Wang, Y. (1992) Accurate and simple analytic representation of the electron-gas correlation energy, *Phys. Rev. B*, **45**, 13244-13249. Perdew, J.P., Chevary, J.A., Vosko, S.H., Jackson, K.A., Singh, D.J. and Fiolhais, C. (1992) Atoms, molecules, solids, and surfaces: Applications of the generalized gradient approximation for exchange and correlation, *Phys. Rev. B*, **46**, 6671.
14. Ahlrichs, R., Bär, M., Häser, M., Horn, H. and Kölmel, C. (1989) Electronic structure calculations on workstation computers: The program system TURBOMOLE, *Chem. Phys. Lett.*, **162**, 165.
15. Odom, G.K. and Scuseria, G.E. (1995) unpublished results.
16. Hunter, J.M., Fye, J.L., Roskamp, E.J. and Jarrold, M.F. (1994) Annealing carbon cluster ions: A mechanism for fullerene synthesis, *J. Phys. Chem.*, **98**, 1810-1818.
17. Taylor, P.R., Bylaska, E., Weare, J.H. and Kawai, R. (1995) C_{20}: fullerene, bowl, or ring? New results from coupled-cluster calculations, *Chem. Phys. Lett.*, **235**, 558-563.
18. Stone, A.J. and Wales, D.J. (1986) Theoretical studies of icosahedral C_{60} and some related species, *Chem. Phys. Lett.*, **128**, 501-503.
19. Murry, R.L., Strout, D.L., Odom, G.K. and Scuseria, G.E. (1993) Role of sp^3 carbon and 7-membered rings in fullerene annealing and fragmentation, *Nature*, **366**, 665-667.
20. Murry, R.L., Strout, D.L. and Scuseria, G.E. (1994) Theoretical studies of fullerene annealing and fragmentation, *Int. J. Mass Spectrom. Ion Proc.*, **138**, 113-131.

DISCUSSION OF THE EBBESEN AND SCUSERIA LECTURES

RICHARD E. SMALLEY
Department of Chemistry
Rice University
Houston, TX 77251
U.S.A.

MICHL (Colorado): To Scuseria: I have a question concerning the data that you gave us for the mobility of ions. Were these ions produced by post-ionization? Or were they produced directly in the initial laser vaporization?

SCUSERIA: In Bower's group they used what they call "residual ions." These are the ions that are made in the laser vaporization in the nozzle. In Jarrold's group they used a laser to ionize the neutrals after they come out of the nozzle - it's a post-ionization.

MICHL: So the evidence then pertains to the neutral chemistry, not just to the ions.

SCUSERIA: Yes, that's right.

MICHL: The other thing I wanted to ask had to do with the last transparency you showed us which had all these intermediates. But I saw nowhere on there the rings that you call 3D. Do you have any ideas about what they are?

SCUSERIA: As I showed you, there are certain graphene sheets, some of them with pentagons, that will match the mobility. So, it is not clear to us what are the 3D rings. Now, I believe that these experiments are a snapshot of the process where you are making fullerenes. So the fact that the 3D rings are very low in intensity means that they play no important role.

On the contrary you can take the argument that by the time you take the snapshot everything important has already happened - so actually the 3D rings are low abundance because they have converted themselves into fullerenes very efficiently. And we have argued with Smalley about this for quite a long time.

J. Michl (ed.), Modular Chemistry, 275–286.

The snapshot is taken only a microsecond or two after you go back to all carbon atoms. Where as you saw in the data, the fullerenes, the cages are both even and odd. Which means there had been no time to anneal them. Because if you were to anneal them you would have got rid of the cages with an odd number of carbons. The fullerenes can only have an even number of carbon atoms. So all this to me is evidence that this is a snapshot and that some of the reactions have been completed and have gone all the way to fullerenes, but that what you are looking at is a snapshot of the intermediates.

MICHL: If we were to throw out from chemistry all intermediates that don't accumulate, we would not have very many left. Not seeing a lot of intensity here in your snapshot doesn't mean that they don't play an important role. Maybe they are just very reactive.

SCUSERIA: That's true too. They could be very, very reactive. That's true. They don't have a clear role in this model.

PRINZBACH (Freiburg): First a question for Ebbesen. What are the energetic costs for the rehybridization per carbon atom?

EBBESEN (NEC): That is a very good question. When you look at small molecules, as you know, the rehybridization energy is very large. We've done some *ab initio* calculations on sheets, and it is very difficult to do the calculations very accurately. You can't fold molecules but you can see the sheets fold very easily. It is reversible. The issue that you are probably asking is what happens when you take the sheets and you put a defect line in it. When you do calculations, and we've tried very many conformations, most of the time it's much higher energy with a kink in it. But in some cases it is almost isoenergetic. What is interesting is that the pi energy is always stabilized by introducing a kink. The nuclear repulsion is what determines the balance. And so if we could continue these calculations, the best indication is that we would come close to isoenergetic. When you think of a molecule you cannot extrapolate because of the extent of the delocalization. You wouldn't see a benzene ring fold over an edge, but we see graphene sheets do that. There is plenty of evidence besides ours to show that.

PRINZBACH: I was asking this, because I am one of these organic chemists who would like to synthesize these materials. After all we have synthesized dodecahedrane which is partially dehydrogenated. And I've been looking for a calculation for C_{20}. And I've seen at least 4 or 5 calculations, the last one by Pople and others. All disagree. What is your prediction?

SCUSERIA: Yes, I know. I'm one of the "others" (laughter). Well, C_{20} is very complicated. It is a very difficult molecule for theory because it has all these dangling bonds. The best calculation that you mentioned is already a couple of years old. There are other people who have recently done very high level calculation which show that the C_{20} fullerene is not the lowest in energy, but rather the open bowl and the monocyclic ring are very close in energy and lower than the fullerene.

PRINZBACH: But to a chemist the conclusion was not quite correct, because a practicing chemist knows about kinetic stabilization. So when you start with $C_{20}H_8$ it might well be that you can isolate C_{20} and yet these people tell you that they have calculated C_{20} and that it will not be stable. You cannot isolate it.

SCUSERIA: But $C_{20}H_8$ and C_{20} are very different molecules.

PRINZBACH: Yes, but chemists know that they can isolate molecules that are high in energy, far from the minimum, if they are kinetically stabilized. So I understand that if you start out from a carbon sheet, you probably will not be able to come to the C_{20} fullerene, but if you start out from a molecule which has already the skeleton, with all the carbon-carbon single bonds, why shouldn't you arrive, not at the absolute total minimum, but instead at a somewhat higher minimum, why not? Why don't you calculate the mechanism? How a C_{20} fullerene would, you know, disintegrate into two C_{10} molecules.

SCUSERIA: C_{20} fullerene is a minimum on the potential energy surface. You might be able to synthesize it. I'm not saying the molecule is not there. In this paper, the question we were addressing is "what is the lowest energy structure for C_{20}?" And there is some debate about that. There is no trace whatsoever in any experiment of C_{20} fullerene. What that means, simply, is that you may be missing it. As these things start reacting you can easily go to C_{20}, and you won't come back, okay? And that's what we believe happens in fullerene formation. By the way, we believe that you get C_{70} and C_{76}, C_{78} when for some reason you miss the huge C_{60} thermodynamic well. If you miss it, it goes to C_{70}. The next well, the next target is then C_{70}, C_{76}, etc.

MÜLLEN (Mainz): To Ebbesen: In the origami part of your work, when you use the AFM tip as a cutting tool, could you describe the type of graphene you are actually studying? Is it prepared with this method?

EBBESEN: This was HOPG - Highly Oriented Pyrolytic Graphite. The best graphite you can buy. It has domains which are crystalline in the *c* direction and which are very large, several μm to hundreds of μm before you encounter some domain edge. When you do the AFM, even the STM, and fold the sheets, you find that the folds are just one or two atomic layers.

MÜLLEN: How thick is this? Is it one sheet, or multilayer?

EBBESEN: It is multilayer, but we just lift off the one layer, and we can see layers underneath. It is like many sheets on top of each other. It's not quite what you have discussed in your talk, which is very nice, it's like a dream to me. I'm glad to see that people are doing this. I didn't know about your work.

KAHN (Bordeaux): I have a question for Scuseria:. At the beginning of your talk you mentioned that you used all calculation techniques, some of them arising from the Hartree-Fock scheme and some of them arising from density functional theory. Could you tell me which one appears to be the most reliable, say in terms of matching between experimental and calculated data, for instance for mobility?

SCUSERIA: The reason we use all of these different tools is because these molecules are large for he size of systems that we can do very accurate calculations. So we use the empirical and semi-empirical methods to explore ideas. We come up with crazy ideas all of the time, so it's just an exercise in playing with a computer and seeing what comes out. When we find something interesting, we come back to the problem and redo it with higher levels of theory: Hartree-Fock and density functional theory, and density functional theory comes in many different flavors these days: local density approximation, gradient corrected functionals, etc. But the best possible scenario is that all of these methods pretty much give the same picture - which is never the case. When you're lucky, they give the same qualitative feature. In this particular case most of them give the same qualitative picture, but to all of this energetics I show you, you can easily add 0.2 - 0.3 eV error bars, coming from the different methods. In the case of C_{20}, what is the lowest energy structure for C_{20}? All of these methods predict completely different things, so I won't go on record saying the lowest energy form for C_{20}, is this, that or the other. But if you look at the activation barriers for these 2+2 and 2+4 cycloadditions, all these methods predict more or less the same barriers, the same exothermicities within 0.1 to 0.2 eV.

KAHN: When you say elaborated Hartree-Fock method, what do you mean? Do you mean that you perform calculations with a large CI?

SCUSERIA: No, we can't do CI on these things. I would love to do CI, but we don't do CI. We do Hartree-Fock at the SCF level, OK, and we use a decent basis set that has polarization functions, and then we do density functional theory too. I just didn't want to focus on those aspects because I gathered that most of this audience wouldn't be interested. I had one transparency over which I went very fast and which had all the details of the calculations.

TOLBERT (Georgia Tech): Thomas, I guess now I'm a little confused about what you're saying about the nature of these folds. In my imagination you have a radius of curvature associated with these folds which is going to be approximately that of a nanotube, but what you were just describing a little while ago sounded like you actually meant that these things folded flat.

EBBESEN: Eventually. They start off as a curved sheet, and the tip comes along, and rips an edge, and folds it over. So as we scan, of course there is a force from the scan, and we actually depress this hump in the process.

TOLBERT: OK, but I would argue that at that point you have something very different. In other words, if it actually bends double -- because you are talking about a tremendous amount of strain. The strain involved is say the radius of curvature in a nanotube is not so great. I would like to know what it is per bond, but...

EBBESEN: A nanotube has strain, of course. Calculations that you (Smalley) were involved in show that if you take a ribbon and roll it up (along the long axis) at a certain temperature, the dangling bond energy exceeds the strain of curvature and it stays closed. Now what stabilizes the fold is the contact between the sheet, the van der Waals forces. Now you can move the AFM tip back and the sheet unfolds. Sometimes you can go back and forth repeatedly.

TOLBERT: I guess what I would want to argue and I think is not available from the experiment, and maybe you're neglecting in the calculation, is that in fact the radius of curvature goes below that level where in fact you're breaking bonds, and now you're getting free valences, or sp valence going to be...

EBBESEN: I agree with you. That's a good point. But that's why I showed you a picture of progressive shots where the sheets are folded in accordion style, and of course the bending at the edges must be closer to sp^3...

TOLBERT: But my point is not that they will be close to sp^3, but that they don't even exist as such any more -- that they're broken.

EBBESEN: No, no, they're not. They're not broken, because in that case you cannot fold them back and forth.

TOLBERT: At the point at which they are really folded flat back.

EBBESEN: Yes. That's the point. And that is what is extraordinary. There was this famous experiment by Bacon in 1960 where he was looking at fibers which were rolled-up graphite sheets. He wanted to show they were like a scroll, and since they were very good conducting fibers, he ran an intense electric current through them and showed that the sheets became unrolled from the scroll. So these folds are reversible. There's no question about that. Of course, in the presence of chemical impurities and so forth, they may not reverse.

MICHL: I think you mentioned this briefly: but it seems that you could fix this formation, this sort of folded structure, by introducing something that will saturate dangling bonds -- atomic hydrogen or something like that. Has that sort of thing been done?

EBBESEN: No, we have not done it. Of course, we would like to do this in a systematic way. Take a sheet, fold it over, then expose it to a gas and fix that structure. Now it holds down because of the van der Waals forces. When we do this, it is exposed to the air, but it is still reversible.

MICHL: If you have enough impurities around while you're doing this to fix all of the dangling bonds, then why can you fold it back?

EBBESEN: Because the fold is not sharp enough to generate dangling bonds in these examples.

MÜLLEN: Looking at the folding process, as if it were an old fashioned model of surfaces. So we talk about yield. We talk about the available quantities. We talk about reproducibility. Say you made a folded sheet and you find tomorrow or the next day, it is not sufficient -- I need some more. Can you make this same specimen again?

EBBESEN: That is a good question, and we wish we could do it today. What we do know is that if you have a light enough touch with the AFM and so forth, if you have the right conditions, you always get the same folding angles. I didn't bring this with me, but now we can cut single layers, for example. We can make a huge hexagon -- not single benzene ring -- but a huge hexagon out of a single sheet. The first people who contacted me when we published that paper in Nature, were all diamond people. The diamond people were very interested in these ripples, because that ripple structure (they sometimes call it waffle graphite) may be an intermediate to diamond. Some of them wrote me and said if you can make a lot of this I can make seeds for diamond growth. Now we can't do that. We don't have enough to make seeds.

JANATA (PNL): Rick, can you reduce the nanotubes to make them into polyanions?

SMALLEY (Rice): We haven't tried that. One suspects you can.

EBBESEN: I can answer that...

JANATA: Would you estimate how much charge you could put on a nanotube?

EBBESEN: With our best oxidized nanotubes, the whole surface is covered with oxides, not just the tip, which is eaten opened. You can't see it by TEM. But we know they are on the surface from other types of analysis. Most of these are -COOH, carboxylic groups, and so when you put those oxidized nanotubes in aqueous solution the pH drops and becomes acidic. So you have lots of these COO^- groups that are charged and they help to form a dense suspension. Now with the single-shell nanotubes, the problem is that the nanotubes come apart.

JANATA: My question is what happens when you reduce them.

EBBESEN: Oh. Reduce? There is some work we have done but never published. Ball Labs did some work on reduction with alkali metals. In that case they fall apart.

SMALLEY: If you put these as a slurry into a hydrocarbon solvent and you put electrodes on them, they all go to the positive electrode. The electric field just collects them, just sweeps them out of the solution. So they certainly pick up negative charge, and go to the positive electrode. Now how much they pick up per nanometer of length, we don't know.

JANATA: You could do it by dispersion of sodium in THF.

EBBESEN: I think they would fall apart. That's exactly what I was trying to say.

JANATA: So they behave differently than fullerenes?

EBBESEN: Yes, very differently. I had long discussions with the people at Bell Labs before they published the paper about this. And we had the same experimental results as they.

JANATA: What about those that are capped -- the ones which are closed nanotubes?

EBBESEN: The ones we tried were all closed. This work has not been done on single-shell nanotubes. It would be interesting to see if single-shell nanotubes have the same behavior. The multi-shell ones fall apart. And there's a reason for that I won't get into. The alkali won't get in. Somehow it has to open a hole and gets underneath and they rip the whole nanotube open, leaving intercalated graphite.

JANATA: If you could put only one electron, the energy of the capacitor tube without a counterion would certainly be sufficient to rip it apart.

EBBESEN: Maybe.

JANATA: You could do a gedanken experiment: and say if I put 2 electrons or 3 electrons what would be the density that...

SMALLEY: The fact that it falls apart means that you do see intercalated graphite after you do this. There has not been any demonstration that I know of that a nanotube which has complete atomic, molecular integrity of its outer sheet, that it's really closed and isolated, that it would be eaten apart by alkali metals. Do you know of one?

EBBESEN: No. If you could anneal the nanotubes - there are defects in the nanotubes. Perhaps you could demonstrate it.

SMALLEY: Well, let's presume that you have one that is free of defects. Presume you had one that is perfect, like for example, a single-walled nanotube.

EBBESEN: They might have 5 and 7 membered rings too.

SMALLEY: Let's presume that they're perfect. Do you have any evidence that it would be attacked by alkali metals? I would guess that they would survive and behave like other fullerenes. The single-walled nanotubes have less curvature to them than the ordinary fullerenes like C_{60} and C_{70} by a substantial amount.

JANATA: Well, in the experiment they used, they have a tremendous current density.

SMALLEY: You're talking about the arc.

JANATA: Exactly.

SMALLEY: Yes.

JANATA: So that means there's nothing fundamentally wrong with the high charge density being on the nanotube, and jeopardizing the stability.

SMALLEY: Right. Even a single-walled nanotube is a lot bigger than C_{60} per nanometer, so just from the form and the physical capacitance and coulomb repulsion, it would be no problem. But a chemical attack...

EBBESEN: The multi-shell nanotubes, when you grow them in the arc, you can calculate that there is an average of five charges per tube at any given time. But when you do the field emission, there might be much more charge on the end.

MALLOUK (Penn State): A comment about this last point. I will read this *Science* article about the alkali metal reaction of the nanotubes. But of the alkali metals, sodium does not react with graphite to form intercalation compounds. The others do. I was just wondering if you recall which metal they used. They were more interested in the heavier metals, weren't they?

EBBESEN: Yes, that's a good point. They tried rubidium, of course, and potassium, so it may be an issue of intercalation.

WUEST (Montréal): To get synthetic chemists to think seriously about devising ways to extend nanotubes and related structures under mild conditions, you have to provide them with a fairly clear picture of what the edges, tips, and boundaries actually look like, because that is where further reactions will presumably take place. This must be cleared up if further chemistry is to be done in a rational way. Without more information, it is hard to even begin thinking about the problem.

SMALLEY: We've spent quite a number of years in my group taking multi-walled nanotubes out of the arc, mounting individual ones on electrodes, opening up the tip to continue their growth. One thing we've learned is that the open tip of a multi-walled nanotube is certainly not a simple thing that you imagine in that pretty picture from the computer image where all of the atoms are either on one graphene sheet or another. It's quite clear that there is extensive bridging between the open edges and there are hundreds of different ways that it can be done. With a single-walled nanotube, of course, there is just one layer and no bridging, but there's all sorts of ways of closing with a fullerene cap.

I assume that one of the problems of taking traditional organic synthesis to these materials is that they're not soluble and so you have trouble finding out whether you've done anything. What we can offer you is a starting material. We can offer you, for example, single-walled nanotubes. It will be your responsibility to find something that will open them up and maintain them in this active open form as they grow.

TOLBERT: Assuming we could do so, how much would you offer?

SMALLEY: In nanotubes or money? Well, millions of nanotubes (laughter). These days we're shipping out for free some tens of milligrams of samples, where 50% or greater are single-walled nanotubes with overlying junk. But then the question of whether it has grown in length, you could tell later by microscopy of these individual objects, or if it has actually connected to an electrode, you may be able to electrochemically see the effects of growing. There are people who have been stimulated into thinking about doing this. The potential impact of this area is huge. These tubes, or longer more perfect versions of them, ought to be the strongest tensile strength fibers ever grown. They should be electrically conductive; it's quite feasible that you could replace all the copper wire in the world, if you can learn to grow them efficiently.

EBBESEN: There are people who have grown them electrochemically. Harry Kroto has an experiment based on a high temperature melt (lithium chloride) where they electrochemically grow nanotubes. There is also a paper out in *Advanced Materials* in which they grow fibers, carbon fibers, in another high temperature electrochemical experiment. And so I think that might also be a solution, not a synthetic solution reaction, but they combine electrochemically in this experiment because they are conducting, and it might be that they grow along the edge, just like you've proposed in the arc.

CHIDSEY (Stanford): I suggest another model substrate: the edge plane of graphite. There you have a least 10^{14} carbon atoms cm^{-2}, and it might be better than the millions of nanotubes as a starting place for synthetic chemistry. Obviously, you're now limited to surface analysis, but at least there are surface analytical techniques that are well calibrated for doing that, I think, although not with the resolution and beauty of NMR. And the question that provokes here is -- and I would love to hear any of the organic chemists who might have thought about this -- if you get something like what Professor Smalley talked about, burning the nanotubes back, if you burn the edge plane of graphite back you might have carboxylic acids there, maybe ketones. Do people that have a familiarity with oxygenated aromatics have a good idea how you might get to something like a chlorinated edge plane that would then be susceptible to something like the coupling chemistry, palladium catalyzed or otherwise that we heard about today? Because that is a strategy that I can imagine doing in my lab without any idea of aromatic synthetic chemistry, at least I think I know how to begin.

TOLBERT: That's been done actually -- not on HOPG -- but certainly on graphite fibers as well as some fairly crude graphite samples to get covalent bonds to graphite. It's a standard technique.

SMALLEY: Now the question is can you go and extend the graphene planes.

CHIDSEY: Yes, I know you can modify graphite... something that could be more like the kinds of reactions we heard about today, in which we could use something, let's say the palladium catalyst coupling reaction. Certainly there are lots of things like the amidation reactions, and all sorts of things you could do on that kind of surface, and that's what you could study, but nothing as elegant as extending the aromaticity.

SMALLEY: The edge plane of graphite you can guarantee does not look like a bunch of edges of graphene sticking out with their bonds dangling. They will all look like this picture we put up, with the edge folding over, with bridges.

CHIDSEY: But after oxidation, I suspect after oxidation they have -- I'm just guessing, but I bet they have oxygen functionality.

EBBESEN: They do. That's what I just said in my talk. If you oxidize nanotubes, they are covered with carboxylic groups, ketones, and phenol groups, and the ratios are 4:2:1, roughly. And this we found independently, but if you look in the literature of the edge-on carbon chemistry, all this chemistry done 20 - 30 years ago, we have just exactly the same proportions they determined directly by specific synthetic reactions that would only occur on these groups and then seeing what products they would get.

BRICKS AND OPEN-SHELL BUILDINGS IN MOLECULAR MAGNETISM

Olivier Kahn
Laboratoire des Sciences Moléculaires, Institut de Chimie de la Matière Condensée de Bordeaux
33608 Pessac, France

ABSTRACT. Molecular magnetism may be considered as the facet of supramolecular chemistry dealing with open-shell units. Along this line, we will first introduce the concept of magnetic brick, or module, and then we will present some examples of design of one-, two- and three-dimensional magnetic compounds from such bricks. Some of these compounds are molecular-based magnets. They exhibit a spontaneous magnetization below a critical temperature.

1. Introduction

Molecular magnetism is a new field of research dealing with the synthesis and the study of the physical properties of molecular assemblies involving open-shell units [1]. It is essentially interdisciplinary, bringing together organic, organometallic and inorganic chemists as well as theoreticians, physicists, and material science people.

The heart of the discipline concerns the design and the synthesis of new molecular assemblies exhibiting bulk properties such as long-range magnetic ordering or bistability with hysteresis effect, which confers a memory effect on the system. In a certain sense, molecular magnetism may be considered as the facet of modular chemistry dealing with open-shell modules. Actually, instead of modules, we usually speak of bricks, probably because we view ourselves as masons. First, we synthesize magnetic bricks, then we try to assemble them in a controlled fashion to build an edifice exhibiting interesting physical properties

J. Michl (ed.), Modular Chemistry, 287–302.

[2,3]. In this short paper, we would like to illustrate such an approach on some examples arising from the work recently performed in our research group [4,5].

2. Magnetic bricks (or modules)

The bricks we play with to design our open-shell molecular assemblies are characterized by three factors, namely the shape, the chemical functionality, and the spin distribution. The first two factors are common to all facets of supramolecular chemistry. The third one is specific to molecular magnetism. Let us illustrate these ideas on an example, namely the $[Cu(opba)]^{2-}$ brick shown in Figure 1, opba standing for *ortho*-phenylenebis(oxamato).

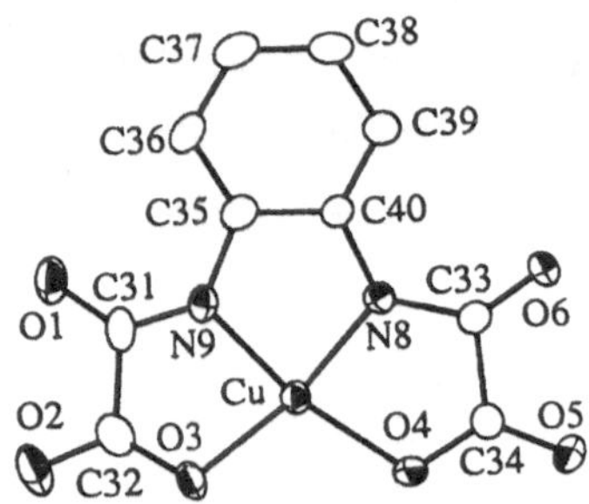

Figure 1. Structure of $[Cu(opba)]^{2-}$.

The shape of this brick is almost planar and is appropriate for the design of one- or two-dimensional networks. On the other hand, it cannot be the structuring motif of a three-dimensional network. The chemical functionality is characterized by the presence of four oxygen atom donors, two on the left-hand and two on the right-hand side of the brick, so that $[Cu(opba)]^{2-}$ can play the role of a bisbidentate ligand, and can bridge two metal ions. The spin distribution factor deserves to be discussed in a more thorough manner. Experimentally, the spin distribution can be deduced from polarized neutron diffraction data [6-11]. Theoretically, it can be calculated by different methods. The most appropriate seems to be the Density Functional Theory (DFT) [12]. In the case of $[Cu(opba)]^{2-}$, no polarized neutron diffraction data are yet available. The calculations indicate that the spin density arising from the Cu(II) ion (d^9 configuration) is strongly delocalized, not only toward the oxygen and nitrogen atoms bound to the copper atom, but also toward the peripheral oxygen atoms. This pronounced spin delocalization, due to the conjugated character of the C-N and C-O bonds, will favor strong antiferromagnetic interactions between the Cu(II) ion of $[Cu(opba)]^{2-}$ and the ions coordinated through the oxamato oxygen atoms. In addition to the positive spin densities on the copper, nitrogen, and oxygen atoms, a weak negative spin density is found on the carbon atoms of the oxamato groups, on either side of the molecular plane (π symmetry). These negative spin densities arise from spin polarization [13,14].

The spin density distribution as obtained in a DFT calculation for another Cu(II) brick, Cu(oxpn) with oxpn = N,N'-bis(3-aminopropyl)oxamido is depicted in Figure 2 [15]. Cu(oxpn) is a bidentate ligand which can be viewed as a half of $[Cu(opba)]^{2-}$ [16].

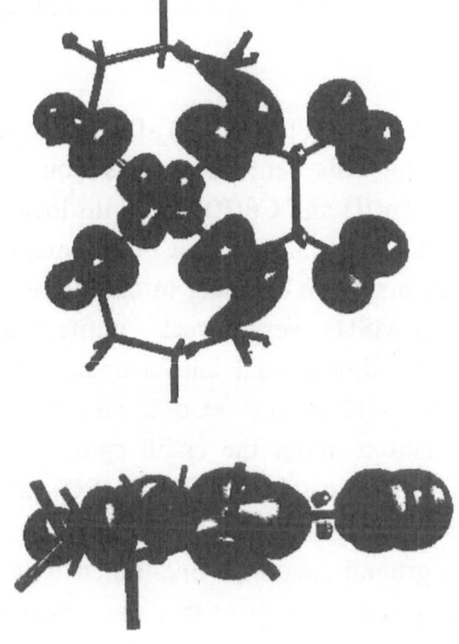

Figure 2. Spin density distribution in Cu(oxpn) as obtained in a DFT calculation; the spin density is positive everywhere except on the carbon atoms of the oxamato bridges, where it is negative.

3. Design of One-Dimensional Ferrimagnets

Let us come back to $[Cu(opba)]^{2-}$. The reaction of this brick with the Mn(II) ion in DMSO, in the presence of a small amount of water, affords a linear chain compound of formula $MnCu(opba)(H_2O)_2$•DMSO, whose structure is represented in Figure 3 [17].

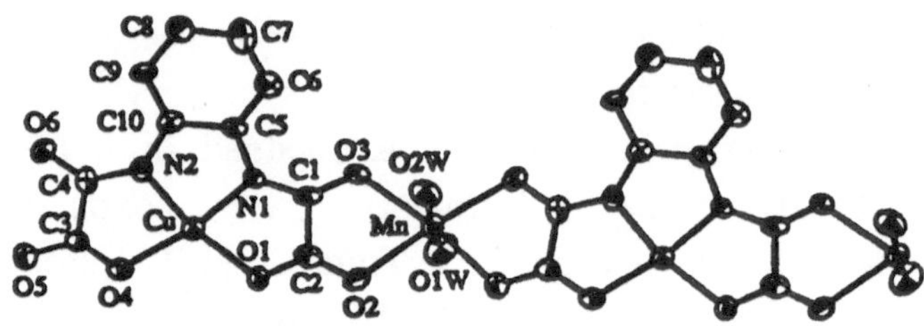

Figure 3. Structure of the linear chain compound $MnCu(opba)(H_2O)_2$•DMSO (from ref. 17).

The magnetic behavior of $MnCu(opba)(H_2O)_2$•DMSO is characteristic of one-dimensional ferrimagnetism. This behavior results from the antiferromagnetic coupling of the adjacent Mn(II) and Cu(II) ions with local spins $S_{Mn} = 5/2$ and $S_{Cu} = 1/2$, respectively, and noncompensation of the magnetic moments in the ground state. There is no magnetic ordering in one dimension, and in the case of $MnCu(opba)(H_2O)_2$•DMSO very weak antiferromagnetic interchain interactions lead to a three-dimensional antiferromagnetic ordering at $T_c = 5$ K. An external magnetic field of 5.0 kOe is sufficient to overcome these interchain interactions, and to align the chain spins in a ferromagnetic-like fashion. It follows that $MnCu(opba)(H_2O)_2$•DMSO may be defined as a metamagnet built from ferrimagnetic chains.

The nature of the ground state of a one-dimensional ferrimagnet may be characterized by the spin density map in this state. Such a spin density map has been determined for a very similar chain compound, of formula $MnCu(pba)(H_2O)_3$•$2H_2O$ with pba = 1,3-propylenebis(oxamato), from both polarized neutron diffraction data and a DFT calculation. The agreement between the experimental and calculated maps is excellent. The results deduced from polarized neutron diffraction are shown in Figure 4. The map presents a large positive and almost spherical spin density around the Mn(II) ion ($S_{Mn} = 5/2$) and a weak negative spin density around the Cu(II) ion (S_{Cu} =

1/2) localized in the plane of the Cu(pba) fragment. It is also remarkable that despite its weakness the negative spin density arising from Cu(II) is more delocalized toward the bridging network than the positive spin density arising from Mn(II). This situation must be related to the fact that the Cu-N and Cu-N bonds are shorter and more covalent than the Mn-O bonds of the bridging network..

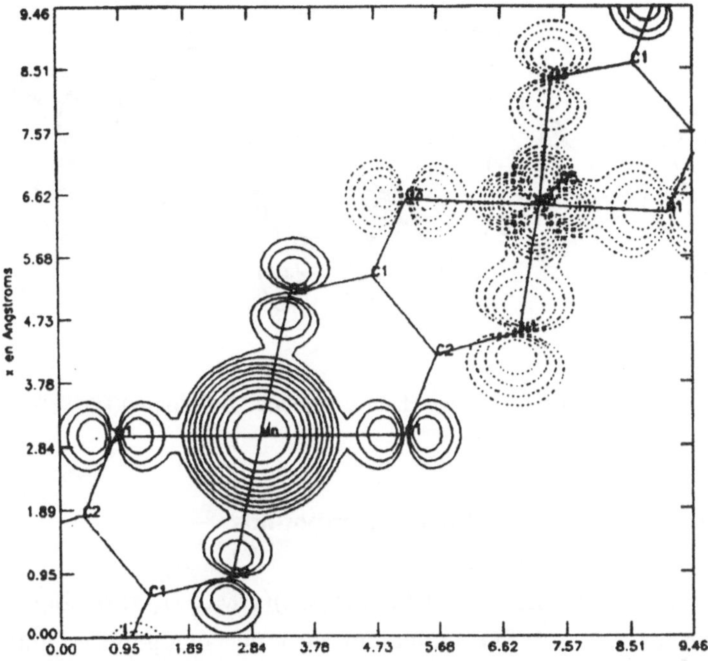

Figure 4. Spin density map in the ground state of the one-dimensional ferrimagnet $MnCu(pba)(H_2O)_3{\bullet}2H_2O$; the positive spin density around Mn(II) is represented by full lines, and the negative spin density around Cu(II) is represented by dotted lines.

If the same reaction between $[Cu(opba)]^{2-}$ and Mn(II) is carried out in the total absence of water, another compound is obtained, of formula $MnCu(opba)(DMSO)_3$ [18]. This compound lies the zig-zag chain structure represented in Figure 5. Two DMSO molecules are bound to the manganese atom in a *cis* position, and the third DMSO molecule occupies the apical position in the copper coordination sphere. The magnetic properties of $MnCu(opba)(DMSO)_3$ are very similar to those of $MnCu(opba)(H_2O)_2{\bullet}DMSO$, except that owing to the bulkiness of the DMSO molecules in the manganese and copper coordination spheres, the Mn(II)Cu(II) chains are almost perfectly isolated within the crystal lattice. MnCu(opba)$(DMSO)_3$ exhibits a one-dimensional magnetic behavior down to 1.7 K.

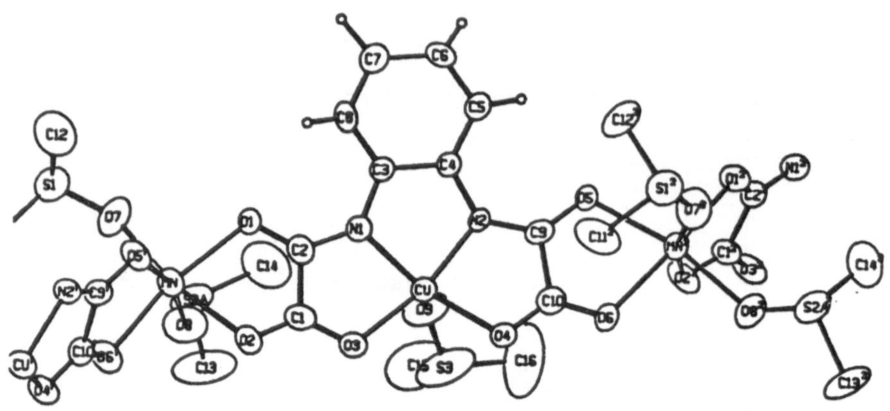

Figure 5. Structure of the zig-zag chain compound MnCu(opba)$(DMSO)_3$ (from ref. 18).

4. Two-Dimensional Mn(II)Cu(II) Compounds

The zig-zag chain structure of MnCu(opba)$(DMSO)_3$ resulting from the *cis* coordination of the DMSO molecules around manganese may suggest a strategy to increase the dimensionality of the system. This strategy consists in replacing the DMSO molecules bound to the manganese atom by the bisbidentate ligands $[Cu(opba)]^{2-}$, which results in a crosslinking of the chains, as schematized in Figure 6. The reaction of a Mn(II) salt with $(NBu_4)_2$[Cu(opba)] in 2 : 3 stoichiometry affords a compound of formula $(NBu_4)_2Mn_2[Cu(opba)]_3$•6DMSO•$H_2O$ [18]. Up to now, it has not been possible to solve the crystal structure. However, it is most likely that this structure consists of layers with Mn_6Cu_6 edge-sharing hexagons. The manganese atom would be surrounded by three Cu(opba) groups. Such an arrangement shown in Figure 7 has actually been found in a related compound, in which the tetra-*n*-butylammonium cation is replaced by a radical cation [19,20]. Probably the NBu_4^+ cations and noncoordinated DMSO molecules are located between the honeycomb layers.

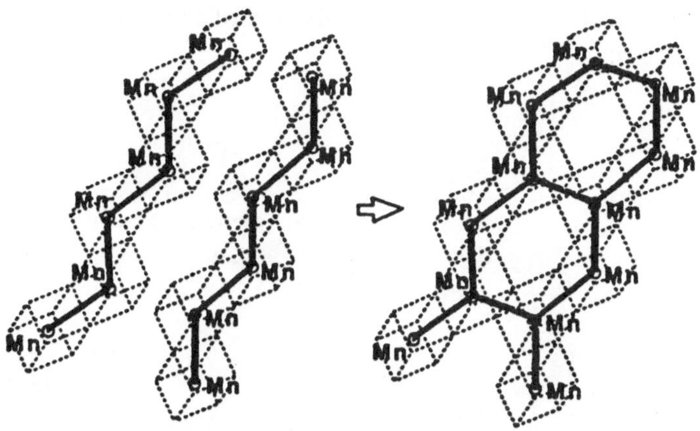

Figure 6. Formation of a honeycomb two-dimensional lattice through crosslinking of zig-zag chains.

The magnetic behavior of $(NBu_4)_2Mn_2[Cu(opba)]_3{\cdot}6DMSO{\cdot}H_2O$ is characteristic of a ferrimagnetic situation with Mn-Cu antiferromagnetic interactions and noncompensation of the local spins in the ground state. At T_c = 15 K, a three-dimensional magnetic transition is observed, with a spontaneous magnetization below this temperature. The field dependence of the magnetization at 5 K reveals a saturation magnetization of 7 Nβ mol^{-1}, which confirms that in the magnetically ordered state all the S_{Mn} = 5/2 local spins are aligned along the field direction and the S_{Cu} = 1/2 local spins along the opposite direction.

The anionic copper(II) brick $[Cu(opba)]^{2-}$ can be prepared with almost all kinds of countercations, including for instance the photophysically active chromophore $[Ru(bipy)_3]^{2+}$, with bipy = 2,2'-bipyridine. From $[Ru(bipy)_3][Cu(opba)]$ we have synthesized the compound of formula $[Ru(bipy)_3]Mn_2[Cu(opba)]_3{\cdot}13H_2O$ in which magnetic and photophysical properties coexist [21]. As a matter of fact, this material shows a long-range magnetic ordering at T_c = 12 K along with a luminescence whose characteristics depend on the magnetic phase. We do not intend to develop here this preliminary result. However, we would like to stress that the design of molecular-based materials exhibiting two types of physical properties, if possible in a synergistic fashion, is probably one of the main issues in this area of research [15,22].

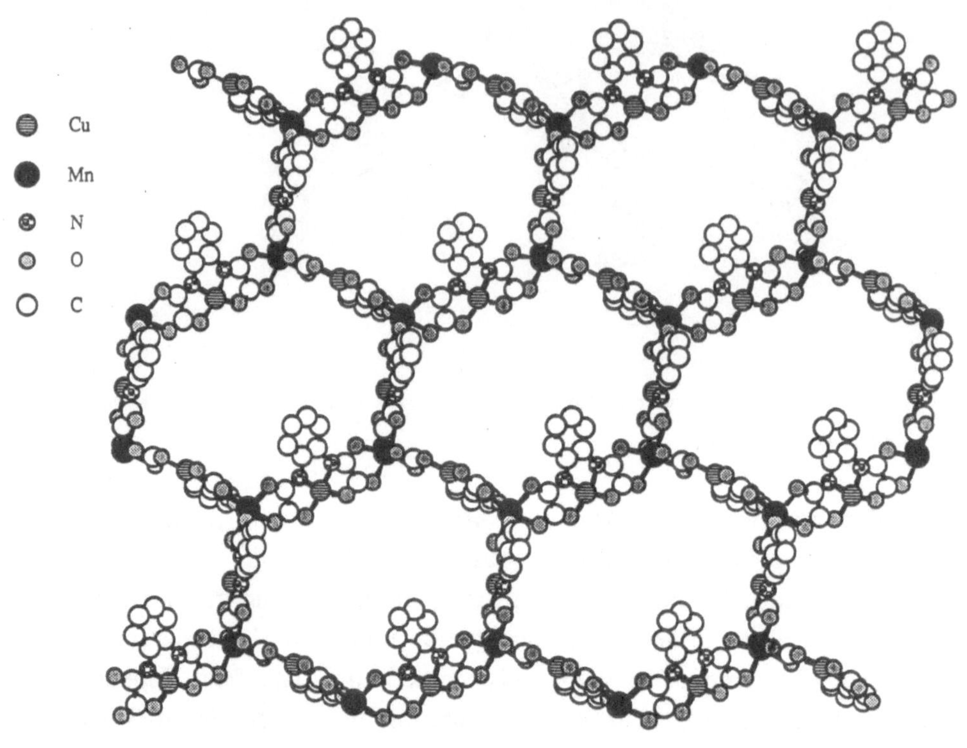

Figure 7. Structure of a layer in $(rad)_2Mn_2[Cu(opba)]_3(DMSO)_2{\bullet}2H_2O$, with rad^+ = 2-(1-methylpyridinium-4-yl)-4,4,5,5-tetramethylimidazoline-1-oxyl-3-oxide (from ref. 19).

5. A Molecular-Based Magnet with Three Kinds of Spin Carriers, and a Three-Dimensional Fully Interlocked Structure

The anionic copper(II) precursor $[Cu(opba)]^{2-}$ can also be prepared with radical cations. One of them, noted rad^+, is 2-(1-methylpyridinium-4-yl)-4,4,5,5,-tetramethylimidazoline-1-oxyl-3-oxide :

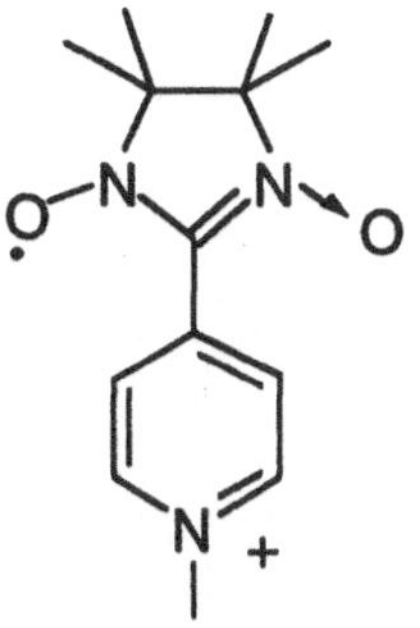

where the unpaired electron is equally shared between the two N-O groups. The reaction of Mn(II) chloride with a large excess of $(rad)_2[Cu(opba)]$ in DMSO affords well-shaped single crystals of a compound of formula $(rad)_2Mn_2[Cu(opba)]_3(DMSO)_2{\bullet}2H_2O$ [19,20]. The structure of this compound is quite amazing. There are two equivalent two-dimensional networks, noted A and B. Each network consists of honeycomb layers as shown in Figure 7. These layers stack above each other in a graphite-like fashion. The A and B networks are quasi perpendicular, and interpenetrate each other with a full interlocking of the Mn_6Cu_6 hexagons as shown in Figure 8. The topology is that of a wire netting. The hexagons are connected further though the radical cations which bridge two copper atoms belonging to networks A and B, respectively. This affords Cu_A-rad-Cu_B-rad chains. Another peculiarity of this structure concerns the chirality of the manganese sites. A manganese atom surrounded by three bidentate ligands is obviously chiral. In the structure there is a perfect alternation of Λ and Δ sites occupying the adjacent corners of a hexagon; two such sites are linked by a Cu(opba) group.

$(rad)_2Mn_2[Cu(opba)]_3(DMSO)_2{\bullet}2H_2O$ exhibits a long-range magnetic ordering with the appearence of a spontaneous magnetizätion at $T_c = 22.5$ K as shown in Figure 9. The field dependence of the magnetization reveals a peculiar behavior. The radical cations interact ferromagnetically with the Cu(II) ions along the Cu_A-rad-Cu_B-rad chains [23]. It turns out that below T_c and in zero field the radical spins are aligned in the direction opposite to that of the Mn(II) spins. When the field increases, a progressive spin decoupling between the radicals and the anionic skeleton $Mn_2[Cu(opba)]_3$ occurs and the radical spins tend to orient along the field direction.

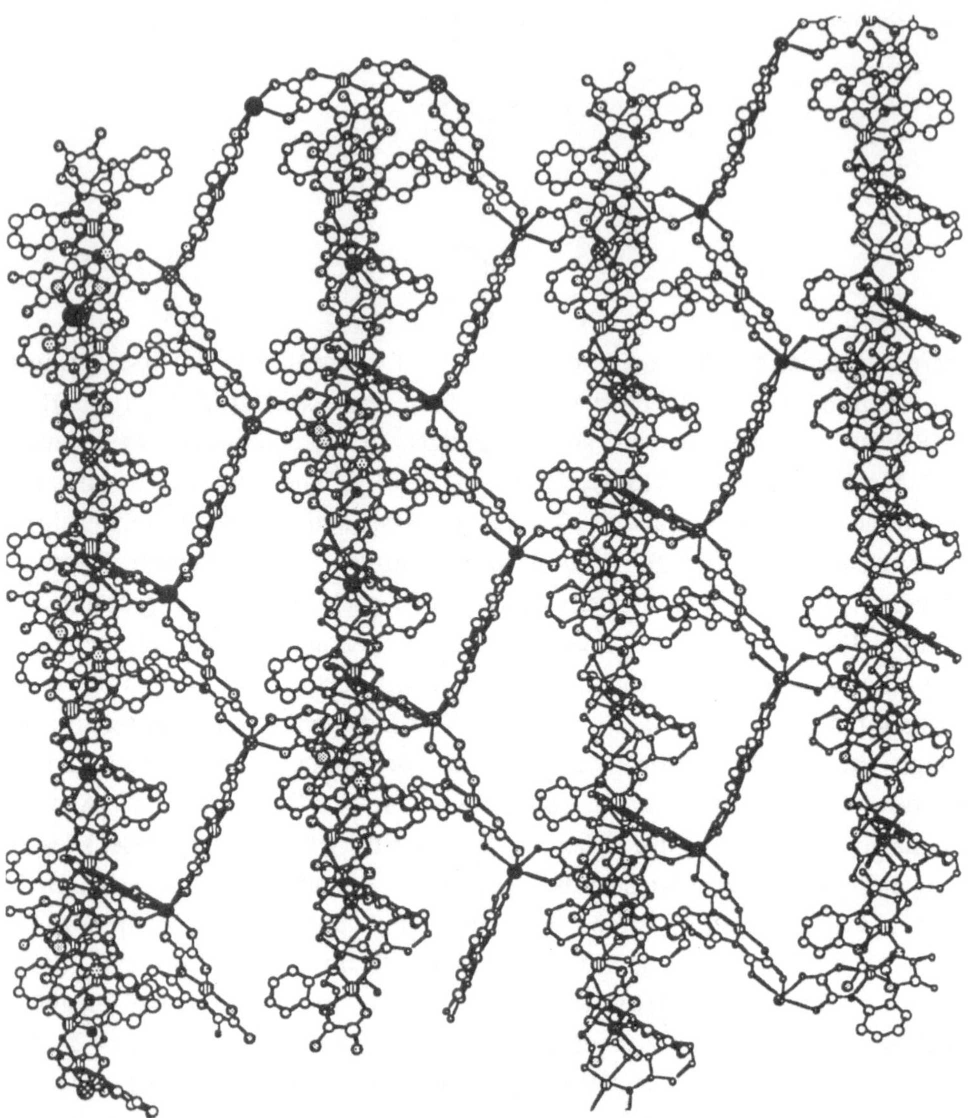

Figure 8. Interpenetration of the two quasi perpendicular networks in $(rad)_2Mn_2[Cu(opba)]_3(DMSO)_2 \cdot 2H_2O$ (from ref. 20).

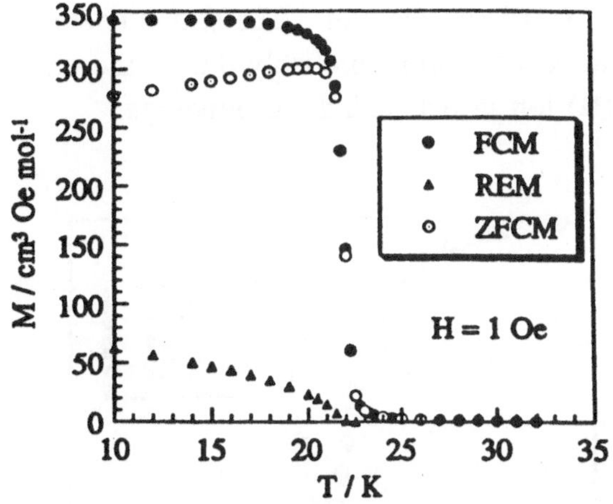

Figure 9. Magnetization versus temperature curves for $(rad)_2Mn_2[Cu(opba)]_3(DMSO)_2{\bullet}2H_2O$; FCM = field-cooled magnetization, REM = remnant magnetization, ZFC = zero-field-cooled magnetization (from ref. 19).

6. Co(II)Cu(II) Two-Dimensional Molecular-Based Magnets with Large Coercive Fields

The Mn(II)Cu(II) molecular-based magnets described above exhibit a magnetic hysteresis loop with a weak coercive field, of a few tens of Oe at most. When manganese is replaced by cobalt in the two-dimensional compound $(cat)_2Mn_2[Cu(opba)]_3{\bullet}S$, where cat^+ stands for a cation and S for solvent molecules, a large coercive field is obtained [24]. Let us describe the magnetic properties of one of these compounds, of formula $(rad)_2Co_2[Cu(opba)]_3{\bullet}0.5DMSO{\bullet}3H_2O$. The field-cooled-magnetization (FCM), zero-field-cooled magnetization (ZFCM), and remnant magnetization (REM) curves are displayed in Figure 10. The FCM shows a break around T_c = 34 K. The REM obtained in switching off the field at 2 K, and warming in zero field, is strictly equal to the FCM below T_c. All the information induced by the field is retained. As for the ZFCM, it is negligibly small up to 15 K. The magnetic domains are randomly oriented, and the domain walls do not move. The ZFCM increases slightly as T increases above 15 K, due to the thermal agitation which displaces the domain walls. FCM and ZFCM merge at T_c. The field dependence of the magnetization at 1.7 K shown in Figure 11 reveals a magnetization of about $7x10^3$ cm^3 Oe mol^{-1}. The strong coercivity observed in these Co(II)Cu(II) compounds is most probably due to the unquenched orbital momentum of the Co(II) ion in octahedral surroundings.

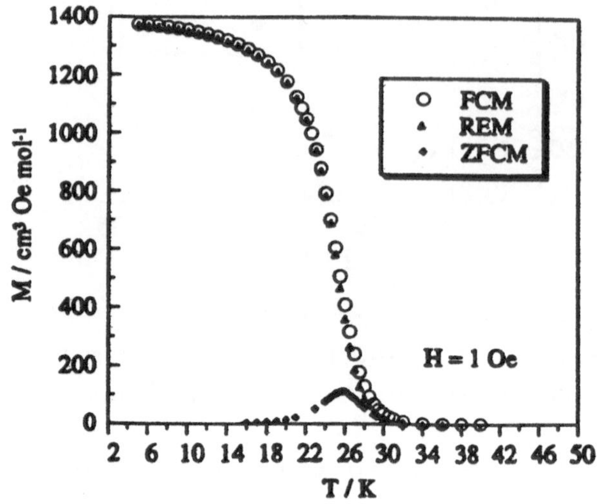

Figure10. Magnetization versus temperature curves for $(rad)_2Co_2[Cu(opba)]_3 \cdot 0.5DMSO \cdot 3H_2O$ (from ref. 24).

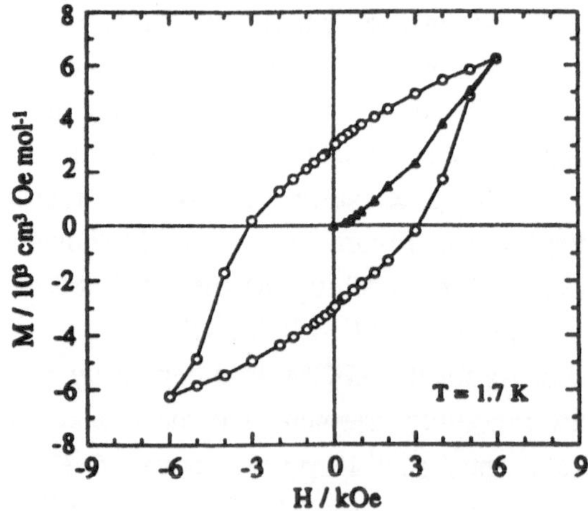

Figure 11 : Magnetic hysteresis loop at 1.7 K for $(rad)_2Co_2[Cu(opba)]_3 \cdot 0.5DMSO \cdot 3H_2O$ (from ref. 24).

7. 4f - 3d Molecular-Based Materials

The $[Cu(opba)]^{2-}$ brick, or module, can also be utilized to design lanthanide(III) - copper(II) networks. The stoichiometry is usually $Ln_2[Cu(opba)]_3 \cdot S$, where Ln stands for a lanthanide(III) ion and S for solvent molecules. Let us restrict ourselves to one series of compounds of this kind, of formula $Ln_2[Cu(opba)]_3 \cdot 3H_2O \cdot xDMF$, Ln being a rare earth atom from Sm to Yb, and Y. The structure consists of infinite ladders as shown in Figure 12. The sidepieces of the ladders are made of alternating LnCu(opba) motifs, and the rungs are made of Cu(opba) motifs joining two Ln(III) ions. Along the *a* direction, the ladders stack above each other. We do not intend to discuss here the physical properties of these compounds. Let us mention, however, that the magnetic properties are quite original, although the Ln(III)-Cu(II) interactions are much weaker than the Mn(II)-Cu(II) interactions [25]. For Ln = Gd, the compound exhibits a long-range ferromagnetic ordering at T_c = 1.1 K.

8. Conclusion

In this paper, we only presented a very small part of our work in the field of molecular magnetism. What we attempted to do is to follow the chemical and physical adventures of a unique brick, $[Cu(opba)]^{2-}$. From this brick, we succeeded to synthesize linear and zig-zag ferrimagnetic chains, as well as two- and three-dimensioanal compounds exhibiting a spontaneous magnetization below a certain critical temperature. Some molecular-based magnets showing a strong coercivity have also been synthesized. Furthermore, we showed that it was possible to design open-shell molecular assemblies presenting both magnetic and photophysical properties. Our goal in this short presentation was to point out that molecular magnetism combines the beauty and the aesthetic appeal of supramolecular chemistry, and the excitement of new physics.

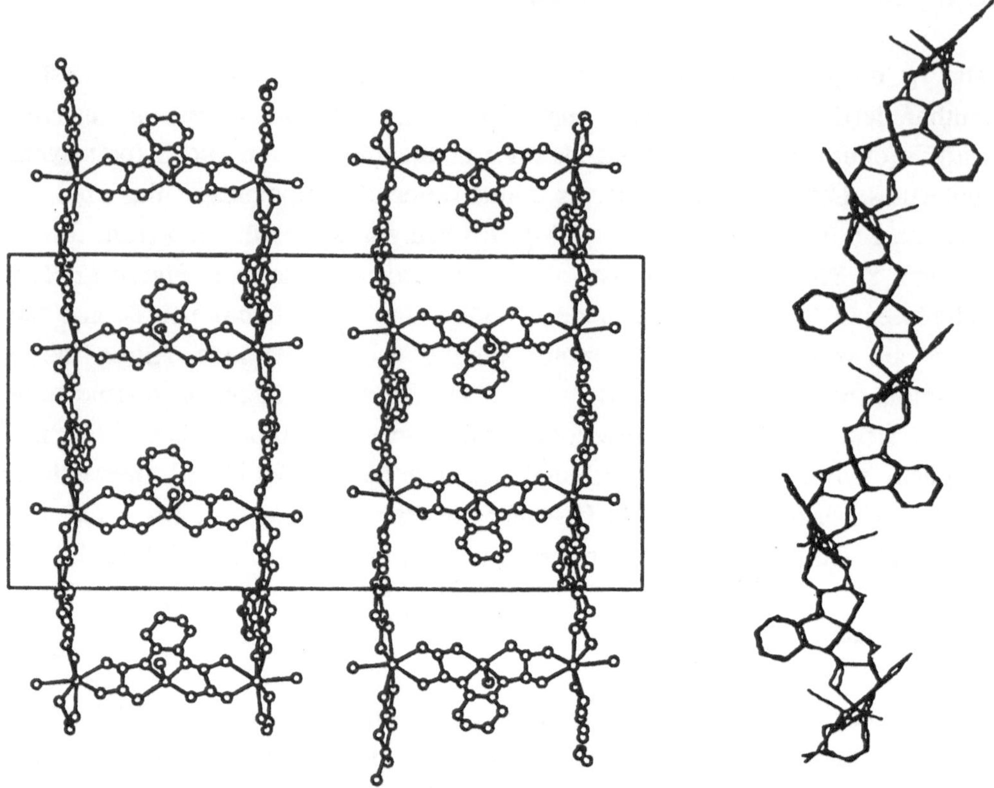

Figure 12. View along the *a* direction of two adjacent ladders in $Ln_2[Cu(opba)]_3 \cdot 3H_2O \cdot xDMF$, and view along the *c* direction of one of the bars of the ladders.

References

1. Kahn, O. (1993) *Molecular Magnetism*, VCH, New York.
2. Kahn, O. (1985) Dinuclear complexes with predictable magnetic properties, *Angew. Chem. Int. Ed. Engl.* **24**, 834.
3. Willett, R. D., Gatteschi, O., and Kahn, O. Eds. (1985) *Magneto-Structural Correlations in Exchange Coupled Systems*, NATO ASI Series C, vol. 140, Reidel, Dordrecht.
4. Kahn, O. (1987) Magnetism of the heterobimetallic systems, *Struct. Bonding (Berlin)* **68**, 89.
5. Kahn, O. (1995) Magnetism of heterobimetallics: Towards molecular-based magnets, *Adv. Inorg. Chem.* **43**, 179.
6. Brown, P. J., Capiomont, A., Gillon, B., and Schweizer, J. (1979) Spin densities in free radicals, *J. Mag. Mag. Mat.* **14**, 289.
7. Figgis, B. N., Mason, R., Smith, A. R. P., Varghese, J. N., and Williams, G. A. (1983) Spin density and structure of aquabis(2,2'-bipyridine)-di-μ-hydroxo-sulphatodicopper(II) tetrahydrate at 4.2 K, *J. Chem. Soc. Dalton Trans.* 703.
8. Zheludev, A., Barone, V., Bonnet, M., Delley, B., Grand, A., Ressouche, E., Rey, P., Subra, R. and Schweizer, J. (1994) Spin density in nitronyl nitroxide free radical. Polarized neutron diffraction investigation and ab initio calculation, *J. Am. Chem. Soc.* **116**, 2019.
9. Zheludev, A., Grand, A., Ressouche, E., Schweizer, J., Morin, B. G., Epstein, A. J., Dixon, D. A., and Miller, J. S. (1994) Experimental determination of the spin density in the tetracyanoethenide free radical, $[TCNE]^{\cdot -}$, by single-crystal polarized neutron diffraction. A view of a π^* orbital, *J. Am. Chem. Soc.* **116**, 7243.
10. Ressouche, E., Boucherle, J. X., Gillon, B., Rey, P., and Schweizer, J. (1993) Spin density maps in nitroxide-copper(II) complexes. A polarized neutron diffraction determination, *J. Am. Chem. Soc.* **115**, 3610.
11. Gillon, B., Cavata, C., Schweiss, P., Journaux, Y., Kahn, O., and Schneider, D. (1989) Spin density in the heterodinuclear compound $Cu(salen)Ni(hfa)_2$: A polarized neutron diffraction study, *J. Am. Chem. Soc.* **111**, 7124.
12 Becke, A. D. (1989) *Density Functional Theories in Quantum Chemistry: Beyond the Local Density Approximation*, Becke, A. D. Ed., A.C.S. Symposium Series, Washington D.C., vol. 394, pp. 166 and references therein.
13. Kollmar, C., and Kahn, O. (1993) Ferromagnetic spin alignment in molecular systems: An orbital approach, *Acc. Chem. Res.* **26**, 259.
14. Kahn, O. (1994) Molecular magnetism: New language and new objects, *Comments Cond. Mat. Phys.* **17**, 39.

15. Baron, V., Gillon, B., Plantevin, O., Gousson, A., Mathonière, C., Kahn, O., Grand, A., Öhrström, L., Delley, B. (1996) Spin dernsity maps for an oxamido-bridged Mn(II)Cu(II) binuclear compound; polarized neutron diffraction and theoretical studies, *J. Am. Chem. Soc.* in press..

16. Mathonière, C., Kahn, O., Daran, J. C., Hilbig, H.; and Köhler, F. H. (1994) Complementarity and internal consistency between magnetic and optical properties for the $Mn^{II}Cu^{II}$ heterodinuclear compound $[Mn(Me_6\text{-}[14]\text{-}N_4)Cu(oxpn)](CF_3SO_3)_2$ (Me_6-[14]-N_6 = (±)-5,7,7,12,14,14-hexamethyl-1,4,8,11-tetraazacyclotetradecane; oxpn = N,N'-bis(3-aminopropyl)oxamide), *Inorg. Chem.* **32**, 4057.

17. Stumpf, H. O., Pei, Y., Ouahab, L., Le Berre, F., Codjovi, E., and Kahn, O. (1993) Crystal structure and magnetic behavior of the ferrimagnetic chain compound $MnCu(opba)(H_2O_2\text{•}DMSO$ (opba = o-phenylenebis(oxamato) and DMSO = dimethyl sulfoxide), *Inorg. Chem.* **32**, 5687.

18. Stumpf, H. O., Pei, Y., Kahn, O., Sletten, J., and Renard, J.P. (1993) Dimensionality of $Mn^{II}Cu^{II}$ bimetallic compounds and design of molecular-based magnets, *J. Am. Chem. Soc.* **115**, 6738.

19. Stumpf, H. O., Ouahab, L., Pei, Y., Grandjean, D., and Kahn, O. (1993) A molecular-based magnet with a fully interlocked three-dimensional structure, *Science* **261**, 447.

20. Stumpf, H. O., Ouahab, L., Pei, Y., Bergerat, P., and Kahn, O. (1994) Chemistry and physics of a molecular-based magnet containing three spin carriers, with a fully interlocked structure, *J. Am. Chem. Soc.* **116**, 3866.

21. Turner, S. S., Michaut, C., Kahn, O., Ouahab, L., Lecas, A., and Amouyal, E. (1995) A molecular-based magnet incorporating the $[Ru(bpy)_3]^{2+}$ chromophore, New. J. Chem. **19**, 773.

22. Day, P. (1993) The chemistry of magnets, Science **261**, 431.

23. Caneschi, A., Gatteschi, D., Sessoli, R., and Rey, P. (1989) Toward molecular magnets: The metal-adical approach, *Acc. Chem. Res.* **22**, 392.

24. Stumpf, H. O., Pei, Y., Michaut, C., Kahn, O., Renard, J. P., and Ouahab, L. (1994), Bimetallic molecular-based magnets with large coercive fields, *Chem. Mater.* **6**, 257.

25. Guillou, O., Kahn, O., Oushoorn, R. L., Boubekeur, K., and Batail, P. (1992) One- and two-dimensional rare earth - copper molecular materials, *Inorg. Chim. Acta* **198-200**, 119.

SEMICONDUCTOR NANOCRYSTALS AS MOLECULES AND BUILDING BLOCKS

LOUIS BRUS
AT&T Bell Labs
Murray Hill, NJ, 07974, USA

ABSTRACT. In this short contribution I discuss some molecular properties of nanocrystals that are relevant to their use as components in complex materials and devices.

1. Introduction

Semiconductor nanocrystals made by organometallic synthesis are essentially new classes of large inorganic molecules. Some aspects of their properties can be best understood using solid state ideas, and some aspects using molecular ideas. In the progression from small clusters to large nanocrystals, ones sees the evolution from principally molecular to principally solid state properties. Perhaps the most obvious and easily observed size dependent property is the slow, asymptotic evolution of the band gap with increasing size. In 1983 my colleagues and I accidentally observed an increased band gap in 4.5 nm CdS colloidal nanocrystals [1]. We worked out the simple theory of quantum size effects, and discrete excited states, for the electronic spectra, redox potentials, and ionization potentials of nanocrystals.. Successively better syntheses of crystalline and nearly monodisperse crystallites have been found in the decade since then, so that now excellent size control and well resolved room-temperature spectra are possible, as shown in figure 1 [2]. Very detailed optical characterization studies have been done. Due to the high quality of the organometallic synthesis, CdSe has become the prototypical direct gap nanocrystal system.

2. Size Regimes

Figure 2 shows a schematic of size regimes in both spectroscopic and kinetic properties [3]. The smallest clusters tend to adopt unique bonding patterns not observed in the solid state unit cell, in order to satisfy broken surface bonds. As a cluster grows, at some point the structure begins to closely approximate an excised fragment of the bulk crystal. Yet, such nanocrystals

J. Michl (ed.), Modular Chemistry, 303–308.

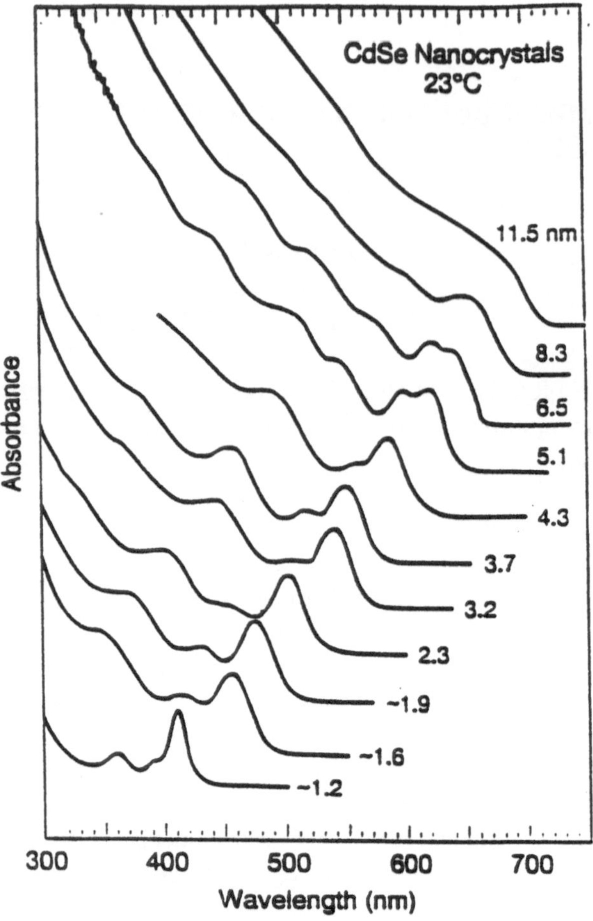

Figure 1. Size dependent CdSe spectra, adapted from ref. 2.

have discrete excited states and a larger band gap than the bulk solid. As size increases, the excited state spectrum becomes more dense, and the bulk band gap slowly appears.

Even as the band gap forms and the spectra appear continuous, the interaction with the radiation field is weak. Electric dipole transitions are strongest, as in molecular spectroscopy. As size increases still more, the interaction with the field becomes strong and must be included on an equal basis in Schrodinger's equation. Polariton and cavity resonance effects begin to occur, and the spectroscopy becomes solid-state-like.

In large (e. g., 50 nm) nanocrystals, the binding energy of an electron-hole pair is less than kT at 23 C, and pairs dissociate into individual carriers. Under normal excitation conditions, several dissociated pairs exist simultaneously, and so their collective decay kinetics is a many body scattering problem. In small nanocrystals, only one pair exists under normal excitation conditions. This pair can not dissociate as both carriers are confined by the surfaces. This confined pair decays unimolecularly and acts very much like a molecular excited state.

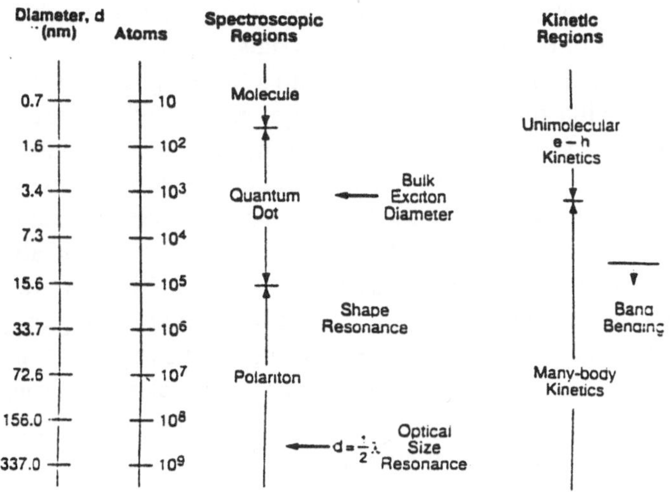

Figure 2. Schematic Size Regimes, adapted from ref. 3

3. Direct and Indirect Gap Quantum Size Effects

The direct or indirect nature of a bulk semiconductor is a consequence of a selection rule derived from translational symmetry. In nanocrystals, translational symmetry is absent, and the selection rule is expected to be weakly broken. What happens experimentally? In direct gap CdSe the band gap transition is strongly dipole allowed for all sizes. The optical spectrum is discrete, and actually somewhat simplified as only dipole allowed transitions appear in the spectrum.

In indirect gap Si, the indirect gap property is present in the smallest (ca. 1.2-1.5 nm) nanocrystals yet examined [3,4]. The band gap transition is vibronically induced by TO (transverse optic) and TA (transverse accoustic) modes, and the unimolecular radiative rate is slow, on the order of 10^4 s^1. The optical spectrum appears continuous above the band gap, apparently because all possible transitions are present with roughly equal intensity due to vibronic interaction. Despite the fact that the radiative rate is slow, the 23 C quantum yield of emission is much higher than in bulk crystalline Si wafers. This happens because the non-radiative processes which dominate recombination in bulk Si - three carrier Auger recombination and defect catalyzed recombination - are effectively quenched in nanocrystal Si.

4. Nanocrystal Interaction with the Local Environment

In excited electronic states with one electron-hole pair, and in ionized nanocrystals, an electric field due to the charged carriers exits the nanocrystal and interacts with the local environment, over a range of a few nm. This is not true in bulk semiconductors and devices. This local field has consequences in kinetics that are much like solvation effects in Marcus electron transfer theory.

For example, in electroluminescnce of materials in which electrons and holes are transported through touching nanocrystals, a rate limiting step is thought to be electron injection from one nanocrystal into a touching nanocrystal containing a hole [5]. This injection process releases a large amount of eletrostatic energy, as two charged nanocrystals convert into two neutral nanocrystals(one of which is electronically excited). For touching 2 nm Si nanocrystals, the free energy release ΔG is about 1.0 eV in vacuum. In the Marcus theory of electron transfer at 23 C, a fast activationless transfer occurs only if ΔG equals the reorganization energy λ. The internal reorganization energy of Si vibrations is very small, on the order of 0.01 eV. Thus the injection process is in the extreme "Marcus inverted region", and is very slow in vacuum. In a water environment, where the outer (water polarization) reorganization energy is 0.4 eV, the injection rate is many orders of magnitude faster. This result may offer one explination why wet porous silicon thin films show more efficient electroluminescence than dry films [6].

5. Materials containing Nanocrystals

I mentioned above the remarkable case of porous Si thin films, made by electrochemical etching of wafer Si in alcoholic HF solutions. Microscopically, porous Si is composed of partially fused nanocrystals that resemble irregular Si wires. The color of the electroluminescence, from about 550 nm to 900 nm, is determined by the size of the nanocrystals. The film is epitaxial on the Si wafer, which directly forms one electrical contact.

CdSe direct gap nanocrystals, with colors tunable across the visible, are better luminescent centers than indirect gap Si. Luminescent diodes incorporatingCdSe nanocrystals and conducting organic polymers such as PVK (polyvinylcarbazole) appear to offer more design flexibility and color tunability than do diodes made of just conductive polymers alone [7]. For example, emission comes from either the organic polymer or the nanocrystals as a function of applied voltage. Both porous Si and the organic diodes appear to contain novel transport physics that is not well understood presently.

Monodisperse nanocrystals capped with organic ligands, such as CdSe capped with trioctylphosphine oxide, can be crystallized into opal-like three dimensional supercrystals [8]. These macroscopic supercrystals are expected to be insulators because of the passivating organic surface layer. Supercrystals should become more semiconducting with conducting organic surface layers. They might be doped by interstitial species, much as C_{60} crystals are doped. One could imagine that the transport kinetics, from nanocrystal to nanocrystal, could be adjusted by the polarity of the interstitial material, as in the electron-hole injection case above.

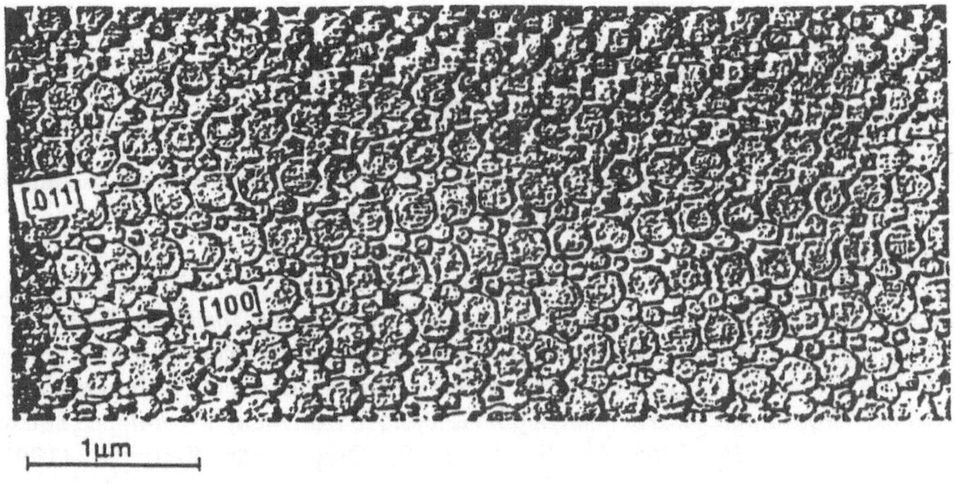

Figure 3. Natural superlattice opal, adapted from ref. 9

Naturally occurring gem opals are supercrystals made of monodisperse silica spheres, typically several hundred nm in size. Figure 3 shows the remarkable example of a natural opal made from two different sphere sizes, forming a superlattice layered structure of the AlB_2 structural type [9]. In analogy, and as an example, one might imagine nanocrystal opals in which planes of small spheres of SiO_2 would separate planes of CdSe nanocrystals. In such structures the dimensionality of charge transport could be varied. Such considerations should lead to a rich new area in electronic materials.

References

1) (a) Rossetti, R., Nakahara, S., Brus, L. *(1983)* Quantum size effects in the redox potentials, resonance Raman spectra, and electronic spectra of CdS crystallites in aqueous solution, *J. Chem. Phys.* **79,** *1086* (b) Brus, L, (1984) Electron-electron and electron-hole interactions in small semiconductor crystallites: the size dependence of the lowest excited electronic state, *J. Chem. Phys.* **80**, 4403.

2) Murray, C. B., Norris, D. J., Bawendi, M. G. (1993) Synthesis and characterization of nearly monodisperse CdE (E=S, Se, Te) semiconductor nanocrystallites, *J. Am. Chem. Soc.* **115,** 8076

3) Brus, L., Szajowski, P., Wilson, W., Harris, T., Schuppler, S., Citrin, P. (1995) Electronic spectroscopy and photophysics of Si nanocrystals: relationship to bulk c-Si and porous Si, *J. Am. Chem. Soc.* **117,** 2915.

4) (a) Littau, K., Szajowski, P., Muller, A., Kortan, A., Brus, L. (1993) A luminescent Si nanocrystal colloid via a high-temperature aerosol reaction, *J. Phys. Chem.* **97,**

1224 (b) Wilson, W., Szajowki, P., Brus, L. (1993) Quantum confinement in size-selected , surface oxidized silicon nanocrystals, *Science* **262**, 1242 (d) Schuppler, S. etal (1995) Size, shape, and composition of luminescent species in oxidized Si nanocrystals and H-passivated porous Si, *Phys.Rev. B52*, 4910.

5) Brus, L. (1996) Model for Carrier Dynamics and Photoluminescence Quenching in Wet and Dry Porous Silicon Thin Films, *Phys. Rev.* **B53**, 4649

6) (a) Halimaoui, A. etal, (1991) Electroluminescence in the visible range during anodic oxidation of porous silicon films, *Appl. Phys. Lett.* **59**, 304 (b) A. Bsiesy etal (1993) Voltage-controlled spectral shift of porous silicon elecroluminescence, *Phys. Rev. Lett.* **71**, 637. (c) Ligeon, M. etal (1993) Analysis of the electroluminescence observed during the anodic oxidation of porous layers formed on lightly p-doped silicon *J. Appl. Phys.* **74**, 1265 (d) E. Kooij, R. Despo, and Kelly, J. (1995) Electroluminescence from porous silicon due to electron injection from solution, *Appl. Phys. Lett.* **66**, 2552

7) (a) Dabbousi, B., Bawendi, M., Onitsuka, O., and Rubner, M. (1995) Electroluminescnece from CdSe quantum-dot/polymer composites, *Appl. Phys. Lett.* **66**, 1316 (b) Colvin, V., Schlamp. M., Alivisatos, A. P. (1994) Light-emitting diodes made with cadmium selenide nanocrystals, *Nature* **370**, 6488

8) Murray, C., Kagan, C., Bawendi, M. (1995) Self-organization of CdSe nanocrystallites into three-dimensional quantum dot superlattices , *Science* **270**, 1335.

9) Sanders, J. (1980) Close-packed structures of spheres of two different sizes. I Observations on a natural opal, *Phil. Mag.* **A42**, 705.

DISCUSSION OF THE KAHN AND BRUS LECTURES

Discussion Leader: A. W. Coleman

JOSEF MICHL
Department of Chemistry and Biochemistry
University of Colorado
Boulder, CO 80309-0215
U.S.A.

1. Questions directed to Prof. Kahn

KASZYNSKI (Vanderbilt): Olivier, could you comment on spin polarization in the σ-electron fragments and spin delocalization in the π-electron fragments? How does one obtain efficient spin-spin exchange between organic species, or between organometallics?

KAHN: Two factors determine the spin density map. The first is spin delocalization. This is easy to understand by considering the distribution of the half-occupied molecular orbital. The second factor, much more difficult to understand, is spin polarization, which may provide some negative spin density. This cannot be understood at the self-consistent field approximation level, and one definitely needs to go beyond. For example, one way to understand why you have some negative spin density zones is to perform a large CI. But, that doesn't exactly answer your question.

Briefly, I can show you the structure of a system that has been designed and synthesized in my group, and in which all the interactions occur through space and not through bonds (Figure 1). It has a two-dimensional honeycomb-like network, with six kinds of hexagons, and is built from two kinds of bricks: a radical cation, in which the pyridinium group is in the meta position, and a hexacyanometallate.

There are four kinds of interaction between the modules. For the radical cation, you find a large positive spin density on the two terminal oxygen atoms and the two nitrogen atoms, and quite significant negative spin density on the sp^2 carbon atom. You also have an interaction between the negative spin density

J. Michl (ed.), Modular Chemistry, 309–321.

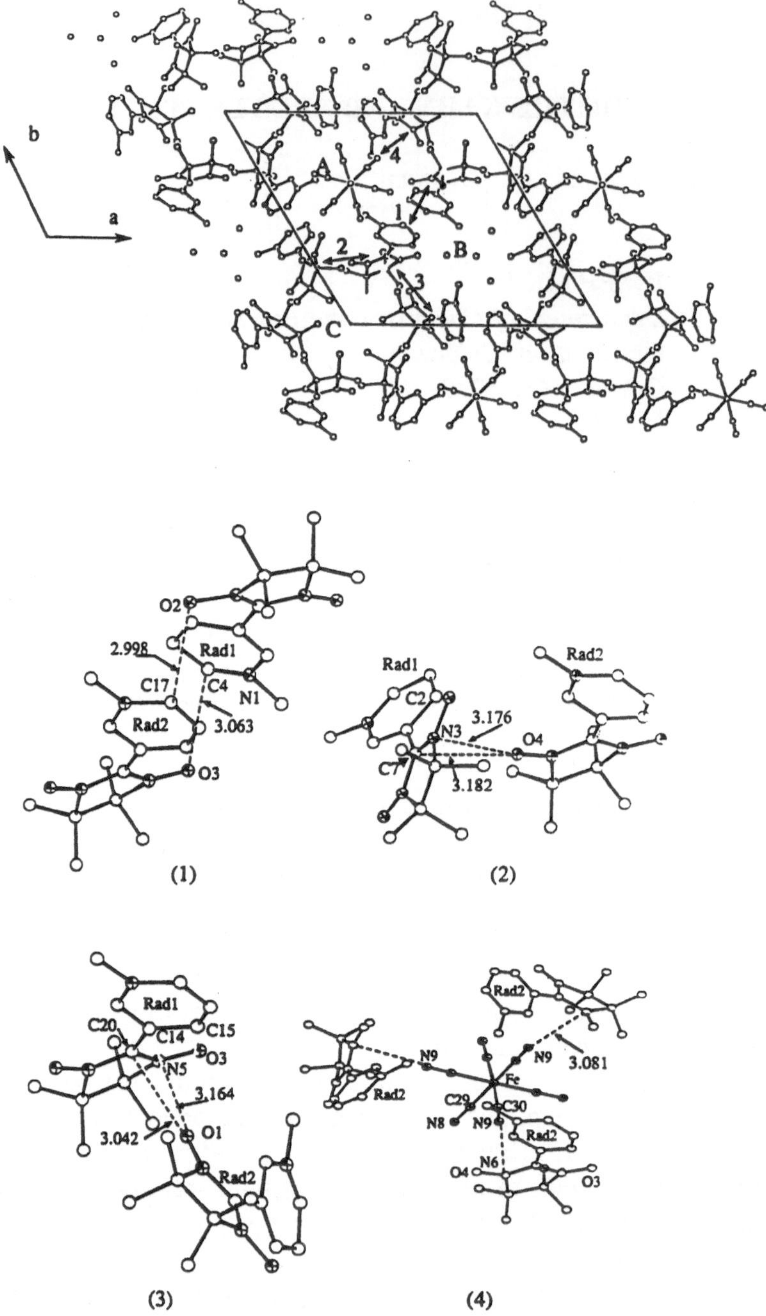

Figure 1. X-ray structure of a magnetic material from O. Kahn's laboratory.

on the module number one and the positive spin density on the module number two. That is a very powerful tool for achieving a ferromagnetic interaction. And when you are able to design some spin carrier in such a way that the small spin density of one unit preferably interacts with the large spin density of another unit, you achieve a ferromagnetic interaction, and thus attain your goal.

CHIDSEY (Stanford): I have a question for Olivier but it is actually motivated by what Lou Brus said at the end. What are the design elements that make for a specific function? And, in that regard, I'd like to return to the issue of storage and I'd like to know what the design elements for coercivity are? What is it that pins the magnetization? I'm not going to worry about this, but suppose some top graduate student comes to my office and says that he is thinking about starting up a research program in magnetization and modular design as a junior faculty member, so what do I look for? It seems to me that we haven't yet articulated the most important thing if you want to make permanent magnets, which is how you lock the magnetization in.

KAHN: Coercivity is governed by two factors. The chemical factor, that is, the chemical composition, and the physical factor. As far as the chemical factor is concerned, the crucial role is played by the magnetic anisotropy of the spin carriers. If you use manganese 2+, which is very isotropic, you don't get any coercivity or just very tiny coercivity. If you replace manganese 2+ with cobalt 2+, which is well known to be very strongly anisotropic, you end up with a system that exhibits a huge coercivity. As far as the physical factors are concerned, the key role is played by the grain shapes and sizes, and similar factors, which are not easy to control in molecular crystals. So, I think that it's an almost hopeless task to design a purely organic magnet which would exhibit a large coercivity. Possibly, if the magnetic interactions are large enough, you could get magnetic ordering around room temperature, but even so, the coercivity will be close to zero. What you have to keep in mind is that you need strongly anisotropic spin carriers.

CHIDSEY: But then you need an anisotropic shape. Is that what you are telling us?

KAHN: No, not necessarily. Magnetic anisotropy arises from the fact that you have a large orbital contribution in the ground state.

MICHL (Colorado): I've been told over the years that one cannot make an organic ferromagnet in one or two dimensions and that three dimensions are required. You talked briefly about systems that you called magnetic that were one-dimensional or two-dimensional. Can you elaborate on that, please?

KAHN: Yes, it is correct to say that there is definitely no magnetic ordering in one dimension. It is absolutely not possible to design a purely one-dimensional magnet that exhibits long-range order. It is possible to get a two-dimensional magnet if you have some magnetic anisotropies. It is not easy to do, but it is possible. Usually, the magnetic ordering is a three-dimensional process. In the case of our graphite-like compounds, it is not yet clear whether the magnetic ordering is two-dimensional, in which case it would arise from the fact that there is some magnetic anisotropy, or whether three-dimensional interactions are important. This is an important issue in my work.

MICHL: And the magnetic anisotropy you are talking about is just the difference between $g_{\perp}$ and $g_{\parallel}$?

KAHN: It is the difference between $g_{\perp}$ and $g_{\parallel}$ as well as orbital content within the ground state.

WARD (Minnesota): You showed pictures that had chains of different colors. Were you implying that they were crystallographically different?

KAHN: No. Colors were used to enhance the appeal of the transparency. The two networks which interpenetrate each other are crystallographically identical.

WARD: How much of a problem is paramagnetism in materials like this?

KAHN: The magnetic properties of those molecule-based magnets are very complicated. There is nothing similar in the field of solid state magnetism. First of all, you have antiferromagnetic interaction between the large spins and the small spins, which give rise to a ferrimagnetic network. In addition, there are the antiferromagnetic interactions between the copper-manganese skeleton and the radicals. And this leads to very peculiar magnetization curves. But it was not the objective of my lecture to discuss this.

WARD: You don't refer to that as ferrimagnetism?

KAHN: It's even more complicated. There is a superposition of antiferromagnetic interaction between two non-equivalent spin lattices which typically produces ferrimagnetism, and ferromagnetic interaction between the organic part and the manganese-copper part. Almost everything is new in that field of molecular assemblies.

You are familiar with the magnetization curve for a normal magnet. When you apply an extremely small field such as the magnetic field of the Earth, you get a large magnetization and when you increase further the magnetic field, you do not increase the magnetization further but you reach saturation. Our material doesn't behave that way. It is a magnet in the magnetic field of the Earth, but when we increase further the magnitude of the magnetic field, the magnetization increases smoothly. This is due to the fact that you progressively turn the radical spins and align them along the direction of the manganese spins. It's a very peculiar behavior.

SEDDON (Belfast): You showed some beautiful results from theory. How do they correlate with the experiment?

KAHN: We attempted to use several techniques and local density theory provides the most valuable results by far. But, don't ask me why this is so. Actually, it's something that I cannot understand from the purely theoretical point of view, why local density theory is so appropriate for providing accurate spin density maps. But, the fact is that it requires much less time and money in comparison with the Hartree-Fock scheme followed by an extended CI, by a factor of 100 in terms of computing time.

KUKI (Alanex): I was really thrilled by your description of these luminescent magnets. Ideas just start popping into one's head, which I am sure you have thought about a good bit. There are two very specific things. One is to be able to use the luminescence as an optical readout of the magnetic state, of the bit stored there in magnetic memory. That sounds obviously fantastic, it's possible to look at a special location and see if it's lit up in the right color. The other, and this may be much harder, would be to apply magnetic field to an entire ensemble of these particles, and then have the magnet take that magnetization only if illuminated by a dye laser. Have you tried it?

KAHN: Not yet. It's a brand new result, and our system only works below 70 K, I must confess. We have still many steps to do before we are able to use this.

STUPP (Illinois): Can you tell us about the physical form and morphology of your networked compounds? Also, I am interested in knowing if the size of the cations can affect the magnetization.

KAHN: The first question deals with the shape of the crystals and morphology of the compounds. Most of the one-dimensional compounds crystallize in an orthorhombic lattice, and the crystal has orthorhombic shape. Which means that it is very easy to find the chain direction and to do anisotropic measurements. Three of the compounds crystallize in the $P2_12_12_1$ space group, some in the triclinic space group. And, to be quite frank, I don't remember exactly the shape of the crystals. What is clear is that those crystals are extremely robust. At the very beginning when we obtained the first crystals we realized that we are now at something new, because the crystals were much more robust than usual in this kind of chemistry.

As for the second question, T_c doesn't depend much on the size of the cation. This strongly suggests that the main mechanism of magnetic ordering is two-dimensional.

2. Questions directed to Prof. Brus

MALLOUK (Penn State): I have a comment about the last part of Lou Brus's talk which maybe he would have addressed had we had a little more time. The Graetzel cell has an Achilles heel at present, namely the mechanism of transport from one electrode to the other. If you look at the energetic situation, at the maximum power point there is about 0.9 V of free energy wasted getting all the charge separation to occur. If you could cut that down to about 0.3 V, which would be reasonable given the potentials of the two electrodes, you could double the efficiency.

The main part of the problem is the fact that as far as the redox potential is concerned, the I^-/I_3^- couple is in a bad place compared to the ruthenium II/III potential, which is the one to which it is interfaced. The authors have recognized this, and they have looked at a number of one-electron outer-sphere redox couples. And what always happens is that the open-cell photovoltage goes up, but the quantum yield for charge separation goes down. So the iodide appears to be a special case. It makes iodine, which is why you have to operate at such a high potential. Then, that cascades down to the $I_2^{\cdot}$ radical and then within a few microseconds this forms triiodide. Because this is an inner sphere couple now, this inhibits the background. So, it is a current challenge in molecular design and maybe in the design of the surface, to see whether one could come up with a one-electron couple at the right potential and prevent that back reaction from happening. Then you could get the cell from about 10% efficiency to 20%.

BRUS: I think you are exactly right. The whole process with the cell advantage certainly is just based on trial and error over a period of five years. It's only after the fact that the design analysis was formulated. In the end, it may well work, after five more years of hard work.

OZIN (Toronto): I can't remember exactly when this was done, but Texas Instrument used to have a silicon particle based HBr/Br_2 solar cell. What happened to that and what are the lessons we learned?

BRUS: I was never able to figure it out because it was basically entirely proprietary. They pursued this inside the company and I have never talked to any of the guys who have actually done it. So I can't really comment on what you have just said.

OZIN: All I remember is that the silicon particles were prepared in a very clever way. They would start with poor quality silicon, and the growth process of these spheres produced a higher quality low-defect product.

MALLOUK: Yes, they used metallurgically-based silicon and millimeter or submillimeter sized particles. In the process the impurities diffused to the outside and they could be etched off. So, they made onion type p-n junctions, made a contact with the inner junction, and then they had the opposite type p-n-n-p junctions connected together. So they had basically two solid state photodiodes in series. And then, this thing made hydrogen from the HBr. It was 10% efficient or so, and it was cheap. But not cheap enough to market, so it died.

TILLEY (Berkeley): I am interested in superlattices and I'm interested in spaces for making small particles. I was wondering if you could just elaborate on what's been done? You alluded to it in your talk.

BRUS: Well, the CdSe system was found by accident.

STUPP: You showed us a three dimensional array of nanocrystals. What was in those nanocrystals?

BRUS: It's not known what's in the interstitial spaces, but I would presume that it is the solvent, which is a mixture of toluene and methanol.

There are a number of ways you might use the interstitial space to influence the kinetics and the recombination. So, right now you have an insulator on the surface and an insulator in the interstitial spaces, and that is the least useful situation. You might think of actually running the synthesis to begin with with a conducting organic molecule on the surface so that you might increase the conductivity. You have a band structure from one sphere to the next with the basis of the band structure being the eigenstates that are on the individual spheres. It's exactly like a C_{60} crystal in this regard. You could increase the coupling from one sphere to the next by this conductive organic material, and then the bandwidth would go up. You can decrease the coupling by putting polar material into the interstitial spaces. That will tend to localize the charge right on one sphere again and introduce electron-phonon coupling in this system. The whole crystal should be a semiconductor as it stands right now. It's not doped in any sense but you could dope it as you dope C_{60}, by putting in things interstitially. If you put conductive organic material into the interstitial spaces, you might go to a system that would be analogous to a type II superlattice in solid state. Maybe the electrons would move through the spheres and the holes would move through the conductive organic. It depends on the relative offsets of the bands, and so forth. So, that's formally analogous to the Graetzel system that I was talking about except that in a Graetzel system it is a redox carrier which is carrying the second charge through the electrolyte.

TOLBERT (Georgia Tech): Lou, I am wondering about the electronic structure of your emitting species. Are you really talking about a spheroid in which the electron density is larger on the outside than the inside? Do you really have a hole on the inside and a charge on the outside?

BRUS: No, it is the opposite of that, actually. You can treat these things two ways, theoretically. You can just do a straight tight-binding calculation on the structure and find the most stable state of the electron, and it is actually a one-electron wave function which has its density in the center of the sphere and a node on the surface. And so, in the simplest approximation both the electron and the hole are in the center of the particle.

TOLBERT: So there is no charge migration of the hole?

BRUS: That's right. It's just there, it's in the center of the particle. And a consequence is that the absorption spectrum is insensitive to solvation. The kinetics are sensitive to solvation because of the Marcus argument, but the absorption spectrum is insensitive to solvation. The system is molecular in the sense that the spectrum depends upon particle size and that solvation affects the carriers kinetics. And it's also molecular in the sense that for small particles there is just one electron-hole pair inside. And, as I was saying in the talk, the electron and the hole can't separate because of the small size of the particle. So, they are forced by the geometry of the lattice to be on top of each other. That's exactly the same as in a molecule, where they are on top of each other in an excited state because they are both on the same molecule. So, they interact with each other directly. There is the Coulomb and the exchange interaction between the electron and the hole, and there is electron-phonon coupling, electronic-vibrational (vibronic) coupling with the lattice, which introduces the shift of the vibrational coordinates, and so forth. All of that follows through exactly the same as in a molecule.

It's different only when you get to the spectral electron-hole pair inside the particle. Then it becomes much more complicated, and that's a function of size. Under normal excitation conditions, the absorption cross-section of a sphere scales with its volume. So, if you double the diameter, you have a factor of ten larger chance of getting two electron-hole pairs in. So, under constant light flux, or under constant electron irradiation, in situations where you are creating some excited state or photoconductivity, just increasing the size in all these systems you will go into a regime with multicarrier kinetics rather than single electron-hole pair (single excited state) kinetics.

ZAWOROTKO (Halifax): I have a question about the synthesis of the cadmium selenide clusters or nanostructures. I'm familiar with a paper from two years ago by Fenske on copper selenide clusters. It attracted my attention because at that time, it was the largest isolated inorganic complex. And, it was, I believe, $Cu_{73}Se_{146}$, phosphate about 70 or 80. Anyway, the basic point I would like you to address is, that in this synthesis, the simple strategy was to mix the terminating ligand in the right ratio with the copper and the selenium, so you have a sort of dendrimeric situation where there are orders of size, and there are just discrete clusters throughout a continuum. Is that what happens in the cadmium case, too? Is it just a continuum of molecular weight and sizes of the nanostructures? Or, are there discrete dendrimeric orders as you build up the lattice?

BRUS: It's a different strategy than the one used in Fenske's synthesis. Ian Dance, in Australia, has been synthesizing for several decades small fragments of cadmium sulfide with phenyl groups on the surface. His work is exactly analogous to Fenske's, with the example that you mentioned. He has crystallized out and isolated a number of these successively larger clusters. I think the largest one that's been obtained in cadmium selenide is something like cadmium 17, sulfur 32 or so forth. That line has been followed by the workers at DuPont, Ian Wang and Norman Heren and also by Weller's group. And they have been changing the ligands and trying to go to larger size. But the synthesis in the pure triethylphosphine oxide basically has to do with the idea of separating nucleation from growth. And so, if you just have so many nuclei and you feed in feedstock in such a way that there are no more nuclei formed, the smaller ones just grow. I think the dispersity is equal to $\Delta n/n$ as n grows. So, if you let it grow fairly large, the dispersity gets narrower. It never goes exactly to what Fenske has, namely, perfect molecules. But the standard deviation in the diameter can be something like 2 or 3%. That's enough to produce the crystallization of spheres that I described, but there is still some rotational disorder. And each cadmium selenide particle probably has a little bit different surface structure. In fact, if I remember Fenske's article correctly, in his crystal structure of the largest particle he says that there is a lot of disorder on the surfaces.

ZAWOROTKO: It's also very interesting because that one has 20 Å channels running through the crystals. I assume the cadmium ones are tetrahedral clusters?

BRUS: Yes, this is all tetrahedral. It's wurtzite basically, so it's uniaxial. Actually the particles are a little bit elliptically shaped, and correlated with the c axis of the wurtzite lattice. If you look carefully, the symmetry is spherical only in zero order, and in fact C_{3v} or D_{3h} is a closer point group. But the numerical effect of going to the lower symmetry is not very large. It's a slight deviation from the spherical model.

PORT (Stuttgart): Do you see already a way to use porous silicon in the sense of modular chemistry? If I understood correctly does it require a control of the defects in this system?

BRUS: The use everyone has thought of so far has simply been to make a light emitting diode out of it. Basically, to put contacts on it and run holes in from one side and electrons in from the other side and to get light out. And that's the reason there have been 500 papers. Because in the silicon industry people would like to put little light emitting diodes on the edges of silicon chips. They would like to grow this material for the diode directly out of the silicon and not have to introduce some other material. This works very poorly if the silicon is dry, and it works really well if the silicon is wet. In fact, the only device which works well is a liquid junction light emitting diode. And I think that this is so because of the Marcus theory argument that I gave. If you have a line of crystallites and the hole is coming from one end and an electron is coming from the other end, they can hop. Then they get next to each other, and in that last hop the hole jumps into a crystallite that contains an extra electron to make a neutral excited state. And even if the two particles are exactly the same size, that hopping process releases a tremendous amount of energy because you have two charged particles going into two neutral particles. So, when you release all this energy, you have to dissipate it somehow, and that's very hard to do in vacuum, but easy to do if you have a polar solvent around. So, I think for that reason it works better when it is a liquid junction. But, the fact that it is a liquid junction is a tremendous psychological barrier to overcome for the device people who are interested in electronics. And also, the lifetime is too long. So, even if you get excited state formed in high yield it still fluorescences at a very slow rate, and you could not hope to modulate this slow rate. About 50 μs at room temperature is the typical lifetime of the excited state. The whole thing is efficient as a CW light bulb, but it can't be modulated very well. It can only be modulated by making it into, say, a CW diode and then modulating externally or something like that. So, when all these things were realized the interest has turned off. The value of it is that this layer is epitaxial to the silicon lattice. So, in fact it's set up to connect molecularly into silicon. If you had some kind of sensor arrangements, some organic polymer that was measuring something and you wanted to make good electrical chemical contact to the silicon wafer, you might do it through the porous silicon layer, since the layer is in fact epitaxial to the silicon it's grown on. There is one good junction there. It's just very hard to make another good junction.

MICHL: Lou, I am curious about your answer to Laren Tolbert. That had to do with the nature of the distribution of the hole and of the excited electron in the particles. You said they were both centered in the middle.

BRUS: There are two separate bands involved here. There is the valence band and the conduction band. You're exciting from the extreme top of the valence band into the bottom of the conduction band. In cadmium selenide, the conduction band is basically an s-type orbital on cadmium. And so it's the lowest orbital. In a bulk crystal it is just a running wave, and the factor inside the unit cell is just this s orbital on the cadmium atom. So, the orbital is a totally symmetric combination. In the spherical nanoparticle now, it's a totally symmetric combination of those s orbitals on the cadmiums, with a node on the surface all the way around. The valence band basically is the 3p orbitals on seleniums. It is a totally symmetric combination of these which is at the top of the valence band.

MICHL: So, it's really a selenide to cadmium charge transfer transition?

BRUS: That's right.

SITA (Chicago): I'm curious about the nature of the porous silicon samples. Are they hydrogen-atom terminated on the surfaces? What's the polydispersity that you get from the process?

BRUS: The porous silicon etching makes a porous silicon that has almost perfect hydrogen termination. There are two types of termination which are known for silicon lattice to be perfect electrically. One is hydrogen termination or alkane termination, but principally hydrogen termination, and the other is oxide termination. And this etching process makes directly the small size particle and it also makes directly the hydrogen termination. The oxide termination is not very stable with time so the choice I made in the gas-phase process was to use the oxide termination. I didn't have time to talk about it, but there is a second little oven there. We first make faceted little silicon particles in the first oven. They flow into the second oven. There is about 30 ms of exposure to high temperature oxygen and a self-limiting growth of the oxide. That puts about 8 Å of fairly good quality oxide on the particles. That has two advantages. One is that it terminates them electrically, but the second is that you have now an oxide surface which you can handle. So, you can use all the chemistry which is known for silica particles to deal with it. We make an ethylene glycol colloid and we separate the particles by size-selective precipitation and by chromatography. The initial size distribution is quite wide. We can separate the particles into four or five batches by chromatography. But still the size distribution within each batch is limiting the spectra. In the spectra that I didn't show because we ran out of time, you get a luminescence spectrum that looks Gaussian-like. What's actually happening is that on the high-energy side, we see the smaller particles emitting, and on the low-energy side, the bigger particles. We can use size-selective luminescence or size-selective excitation to work out the band shapes of individual particles. It's basically like hole-burning an inhomogeneous distribution.

We haven't found any way to dissolve, or to handle as a solution, large hydrogen-terminated silicon particles. You immediately get a scum, which goes to the surface, and things of that sort. So, there is no good way to work with that. If you had particles supported on some substrate, you could do this and go back and forth and make them smaller and so forth. And in porous silicon you can do that. A number of papers have been published describing etching back and forth and watching the luminescence shift to the blue as the average size got smaller with time.

MALLOUK: Lou, we were taught in school that the reason silicon has a low absorption coefficient and a long radiative lifetime is because it's an indirect gap semiconductor. And the expectation was when you make it little, $\boldsymbol{k}$ is no longer a good quantum number, and it becomes a direct gap semiconductor. And then you should have a higher extinction coefficient and a faster radiative rate. So is there some other symmetry reason in silicon, or something like that, that causes the radiative rate to remain pretty much the same?

BRUS: Well, you are exactly right. As you know, this thinking dominated the early stages of all of this work that it would become partially direct gap-like, and that would increase the luminescence. But, it's just like in molecular spectroscopy. You can break symmetry and the breaking can either be very weak or it can be quite strong. So, you have a system that has a zero order symmetry and then you apply some external field that has its own symmetry and breaks it. It can be a very weak field, because a small splitting of the higher order levels, or a very large field. Now, size turns out to be a weak perturbation. What finite size does is that it mixes the indirect gap with the nearest direct gap. But in silicon that nearest direct gap is 2 eV higher and it's all the way across the Brillouin zone. So, for that reason, it's a weak effect. And in fact at the same time in silicon you have a vibronic interaction, since the fluorescence is dipole forbidden, but it's vibronically allowed in the bulk material. If you look at the fluorescence in the infrared it's just like a molecular fluorescence that has a false origin and is vibrationally allowed. That vibronic interaction is also size-dependent, and as it turns out, it is the stronger effect. There is an effect on the radiative rate, but it's principally due to the vibronic interaction. The lifetimes that I have measured are in fact somewhat longer that you would expect. So, something must be going on with the wave function additionally to what I have just said. It can't be simply a particle in the box to give these results. There is electronic fine structure in silicon because of the high spatial degeneracy and maybe it has something to do with that. But that's an unresolved question.

HIERARCHICAL INORGANIC MATERIALS: STEALING NATURE'S BEST SECRETS

Modular Chemistry Over Three Length Scales

GEOFFREY A. OZIN*, DEEPA KHUSHALANI AND SCOTT OLIVER
Materials Chemistry Research Group,
Department of Chemistry,
University of Toronto,
80 St. George Street,
Toronto, Ontario,
Canada M5S 1A1

ABSTRACT. This paper describes the stages that led up to our discovery of the use of vesicle-type templates in the synthesis of hierarchical inorganic materials that resemble biomineralized ultrastructures. The formation of the ultrastructure begins with the self-assembly of a mixed surfactant-cosurfactant, patterned vesicle template with a mesolamellar architecture. Vesicle patterning arises from its phase-separation into domains. The transport of inorganic precursors through the cosurfactant domains, and the nucleation and growth of the inorganic phase in the surfactant regions of the vesicle, lead to the remarkable ultrastructured products described in this paper. This accomplishment shows that vesicles can sculpt impressive macroscopic forms and surface ornamentations. A "cellular" model that is inspired by biological design principles for synthesizing hierarchical materials, with inorganic and organic components connected at interfaces, accounts for the observations. Based on this work, the prospects for the synthesis of biologically inspired, self-assembling, inorganic macrostructures look most promising.

1. Introduction

In this lecture, the various research phases that led to our discovery of the synthesis of hierarchical inorganic materials will be presented. Structural control of these materials spans the nanoscopic to macroscopic length scales [1-7]. The "molecular beaker synthesis" strategy that we have adopted to create inorganic materials with elaborate ultrastructures, is inspired by Nature's principles of biomineralization [8]. The inorganic fabric in our materials is sculpted into a variety of extraordinary patterns and

J. Michl (ed.), Modular Chemistry, 323–333.

forms, that bear a striking resemblance to some of Nature's inorganic structures [9].

Some examples of hierarchical inorganic materials afforded by Nature include the curved costal silica rods of choanoflagellates, solid silica microspheres in biogenic iridescent opal, silica skeletons of diatoms and radiolarians, macroporous shells and egg shell-like objects with intricate surface patterning. Interestingly, morphological control in these biominerals is aided by the highly adaptable structure of amorphous biogenic silica [10]. However, a detailed understanding of the chemical processes involved in the formation of these ultrastructures is still lacking [8-10].

To date, materials chemistry has not been much of a match for Nature in the synthesis of materials where form controls function over all constructional length scales. This is because a new synthesis principle is required that has the capacity to mimic Nature's higher level of cellular dexterity and templating skills for sculpting elaborate biomineral ultrastructures. In the course of our work, we have been able to synthesize morphologies that bear a resemblance to Nature's biominerals mentioned above [1-7]. We propose a model involving a "cellular" synthesis paradigm, in which surfactant-cosurfactant vesicles facilitate the assembly, define the mesoscopic structure and sculpt the macroscopic form.

2. Results and Discussion

The thought process that led to our discovery of a synthetic route to hierarchical inorganic materials began with developments in the field of molecular recognition, self-assembly and replication [11]. This involved thinking about organics recognizing organics through molecular replication. In particular, base-pairing in DNA and protein synthesis using RNA. This molecular templating function proceeds by the recognition and assembly of complementary building-blocks of organic molecules. This results in the formation of a supramolecular organic replica. Consequently, an intruiging question is raised to whether organics can recognize and organize inorganics to produce replicas, or what we like to visualize as inorganic lithomorphs.

We had previously spent a considerable amount of time using small organic molecules for organizing inorganics into oxide and non-oxide-based open-framework materials [12]. These self-assembling frameworks have exquisite arrays of crystallographically defined nanopores. The methodology to generate such ordering emphasizes the power of small organic molecules to template inorganic replicas with nanoscopic structural features. They have structure-property attributes that are ideal for molecular recognition applications, including size and shape selective catalysis, membranes for molecule specific separations and purifications, hosts for nanostructures in quantum electronics, non-linear optics, information storage and processing, and materials for molecule discriminating chemical sensing [12, 13].

A natural progression of ideas and literature clues led us to begin thinking about higher level templating of inorganic ultrastructures. One could not help but be impressed by the trend of materials scientists using Nature's best ideas for creating advanced composite materials that meet a complex set of performance criteria [14].

Equally inspiring were the attempts to understand biomineralization processes that create Nature's elaborate inorganic constructions [8-10]. Nature does not waste energy manipulating materials and structures without function. It eliminates those that do not perform adequately and economically. It has refined the ultrastructure and form of materials over evolutionary time scales. Simply stated, biological structures work. Recognition of the above was a turning point in our approach to materials chemistry. Our way of thinking about materials research began to be motivated by Nature.

A basic construction paradigm can be envisaged involving the selection, localization, concentration and nucleation of elements, control of nanostructure, definition of spatial segregation, command of interfaces and form, with culmination of a macrostructured architecture. This processing strategy, which may be repetitive, can result in smart machines, devices and systems, having efficient functional components, intelligent responses to external signals, and controlled interaction with other materials and components.

It took the seminal discovery of mesoporous silicas by Kresge *et al.* [15] to help us realize the enormous potential of templating inorganic replicas using supramolecular organic assemblies. This yielded materials with ordered structural features beyond the nanoscopic size regime. Consequently, we had ideas of simultaneously building-in structural features over nanoscopic (<10 Å), mesoscopic (10-10^3 Å), microscopic (10^3-10^4 Å) and macroscopic (> 10^4 Å) length scales, in much the same way that Nature creates biomineral ultrastructures. However, to template beyond the mesoscopic size regime using amphiphile assemblies would require higher order assemblies and a new way of thinking about the nucleation, growth and shaping of inorganic components. It was inevitable that a supramolecular, vesicle-based templating strategy would provide us with a solution for the synthesis of hierarchical materials with higher order structural complexity and fluidity of form over increasing length scales. For us, this was a revelation and a challenge.

The literature was replete with information on water-surfactant binary and oil-water-surfactant ternary phase diagrams, as well as the equilibrium fluid mesostructures of amphiphile assemblies that are believed to dominate in different regions of composition-phase space [16]. The surfactant mesostructures that have been identified include spherical and cylindrical micelles, their swollen and inverted analogues, lyotropic liquid crystalline phases with hexagonal, cubic and lamellar shapes, discoid structures, and macroscopic unilamellar and multilamellar vesicles. Also, a vast literature was available concerning the thermodynamic and structural properties of mixtures of surfactants, as well as the effects on these properties of solvent, temperature, pH and electrolyte [17].

Initial efforts toward the synthesis of hierarchical inorganic structures focused on learning how to make inorganic replicas of supramolecular mesoscopic templates. At the time that we started in 1992, the surfactant templating mechanism of hexagonal mesoporous silicas was not fully understood. An appealing proposal involved silica polymerization on a pre-assembled hexagonal liquid crystal mesophase [15]. Much work has been subsequently reported to try to clarify the templating mechanism of the mesoporous silicas. All models that have been proposed offer some variation on the surfactant templating of mesoporous silicas. What phenomenon was responsible for the origin of the spectacular biomineral-like worm-shaped morphology? The resemblance in size and form to the siliceous curved costal rods of the choanoflagellates and the spicules of marine sponges was remarkable and thought provoking [7-10]. The PXRD patterns and TEM images of our worm-shaped forms show that they contained well-ordered mesopores with extensive regions of channel curvature.

These observations led us to propose a model that could account for the constructional stages of the worm-shaped morphology mesoporous silicas [7]. It involved the novel concept of the curved forms being related to the topology of colloidal silica-surfactant membrane-bound reaction spaces. They facilitate the assembly of the mesoporous structure from silica-coated micelles. In essence, a colloidal silica network in a synthesis gel is coated with a surfactant bilayer [19]. This construct is entirely analogous to the "silicalemma" membrane which plays a vital role in the formation and sculpting of biosilicified ultrastructures (i.e., biosilicification requires membrane-bounded vesicles for precise structural organization, spatial constraints, ordered silica particle aggregation and chemical regulation of the silica deposition process) [8-10]. Under acidic or basic conditions, the silica is transported across the bilayer membrane to polymerize on the micellar templates. During this process, "vectorial coalescence" of the silica-surfactant micelles occurs through necking at high curvature contact points, thermodynamically driven by the minimization of surface free energy [19]. This creates the hexagonally close-packed channel architecture of the mesoporous silicas. Their overall morphology is constrained to the shape of the silica-surfactant membrane-spaces present in the synthesis gels (i.e., as a cast or mold). This model encouraged us to intensify our search for vesicle templating of macroscopic forms and patterns in inorganic materials.

The next big breakthrough in our laboratory came when Scott Oliver obtained an SEM image of an aluminophosphate-based macroscopic honeycomb that he had synthesized using amphiphilic alkylamines in glycol based solvents [1]. These surfactant-glycol systems are known to be microstructured in much the same way as their aqueous counterparts [20]. The resemblance of the honeycomb morphology to those found in diatoms and radiolarians was, to us, amazing [8-10]. We realized that we were on the verge of elucidating the "science and art" of vesicle templating of inorganic ultrastructures. The honeycomb macrostructure was established to be a well-ordered mesolamellar aluminophosphate with an Al : P ratio of 1 : 2 and an organic content of 40-50 wt.%. The inorganic and organic lamellae themselves lacked long-range order and in this respect paralleled that found in silicified structures in biology [1].

Scott Oliver succeeded in growing single crystals of a surfactant-phosphate assembly that we believe to be a precursor to the species responsible for templating this spectacular ultrastructured honeycomb, see below. The single crystal XRD structure of this assembly was subsequently solved [1], Figure 2. It contains an intricate hydrogen-bonded network of interdigitated decylammonium cations arranged orthogonally to a layer of dihydrogen-phosphate counteranions.

We realized that we had unravelled a key piece of the vesicle templating theme of micelle and liquid crystal templating [18].

Deepa Khushalani, Alex Kuperman and myself had a different templating mechanism for the above mesoporous silicas [7]. Our experiments led us to realize the importance of the morphologies of these structures. Elaborate curved worm-shaped morphologies containing well-ordered mesopores were predominantly obtained, Figure 1. This recognition turned out to be a turning point in our way of thinking about the

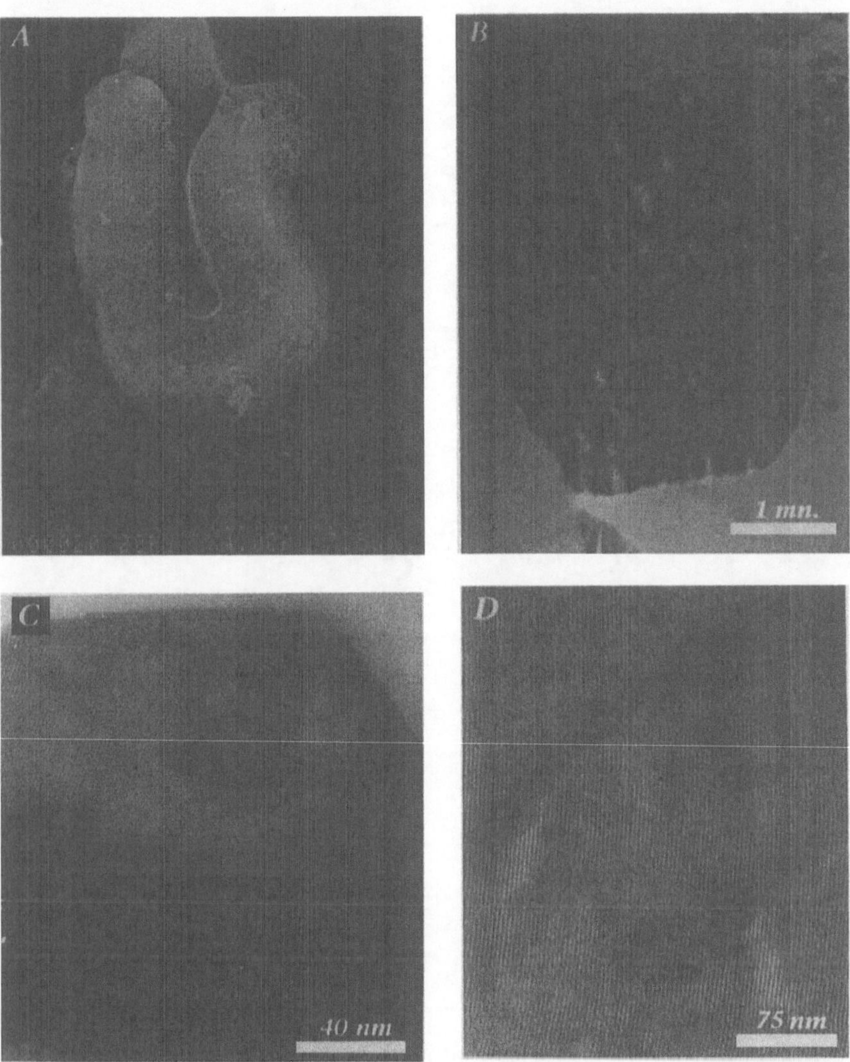

Figure 1. Micrographs of the worm-shaped morphology a) Overall SEM image. b) SEM orthogonal cross-sectional image. c) TEM orthogonal cross-sectional image. d) TEM transverse cross-sectional image. The above show that aqueous silica-surfactant-based chemistry can be orchestrated to produce sculptured forms of silica having a morphology similar to those of the curved silica rods found in the choanoflagellates but with an internal structure based on an ordered honeycomb arrangement of mesoporous channels [7].

puzzle. A preliminary combinatorial approach was designed to establish the synthetic parameters required for this type of templating process. Initial targets included effects of surfactant head group, length of the alkylammonium chain, degree of oligomerization of the glycol, amount of water, pH, reaction temperature and time. It became clear that the patterning and overall morphology of the inorganic ultrastructures

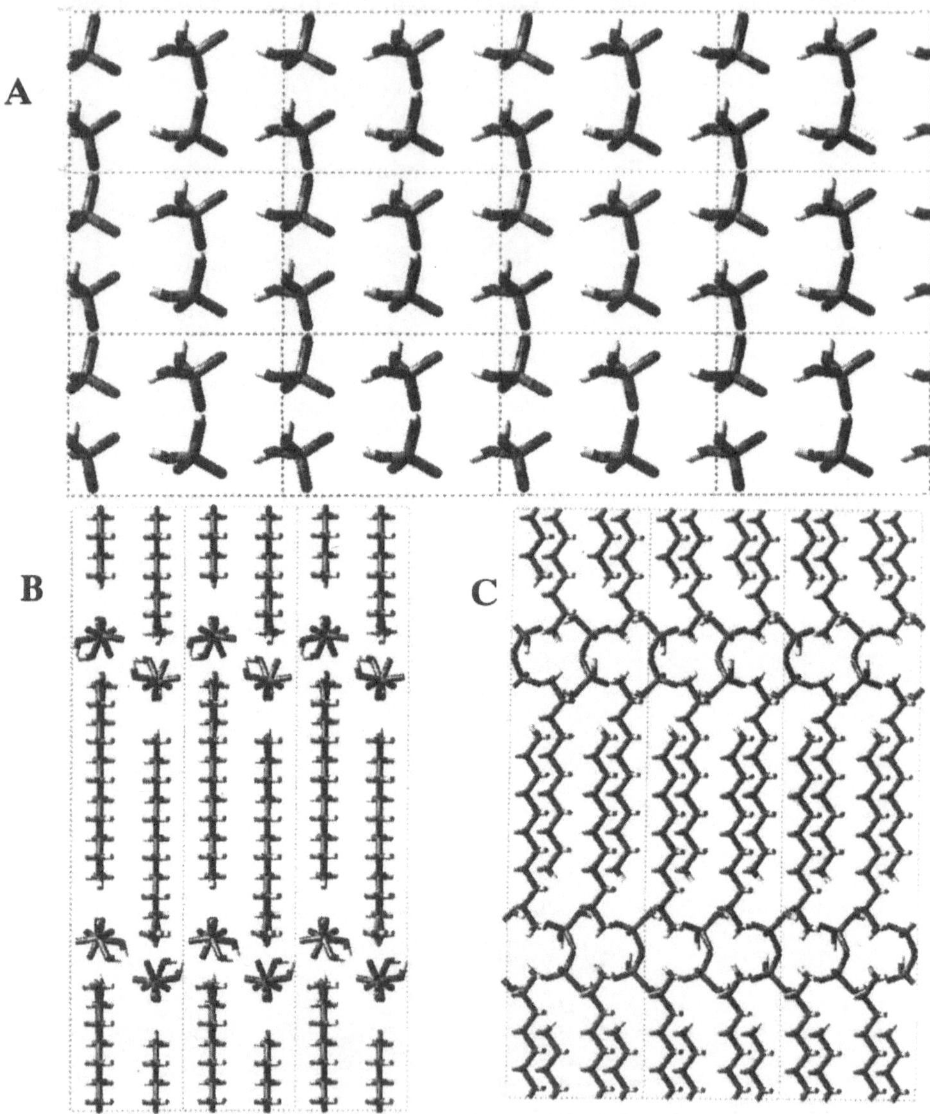

Figure 2. Cerius™ projections of the crystal structure of $(C_{10}H_{21}NH_3^+)(H_2PO_4^-)$ [1]. a) [010]-projection of one phosphate layer. Each dihydrogen phosphate forms four hydrogen bonds to its nearest neighbours. b) [001]-projection, showing the interdigitated, lamellar nature of the structure. c) [100]-projection. The phosphate layers are connected through hydrogen-bonded alkylammonium cations, creating an extended three-dimensional framework.

only occurred for a specific choice of reagents and reaction conditions. These observations implied we were dealing with control of structure and templating behavior of surfactant and glycol(i.e., cosurfactant) based supramolecular assemblies in different regions of composition-phase space.

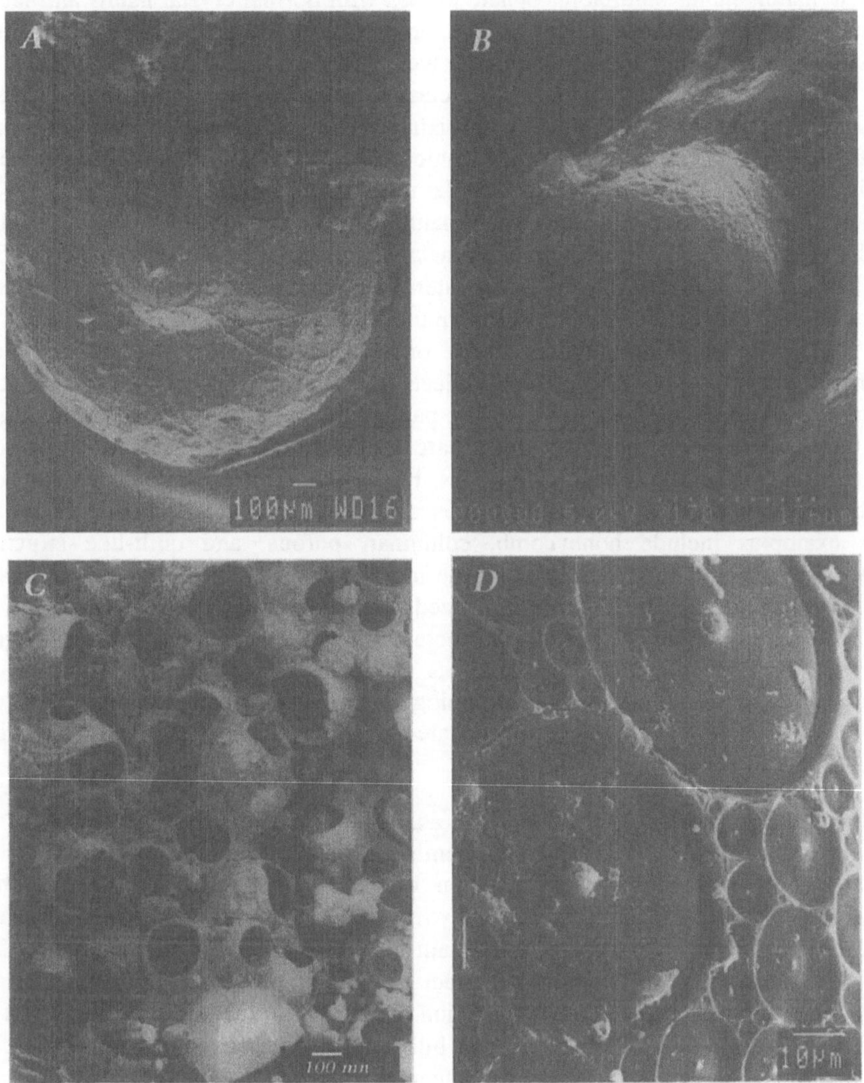

Figure 3. A collection of SEM images of the morphologies and surface patterns of mesolamellar aluminophosphate synthesis products formed in the TEG-alkylammonium dihydrogen phosphate-alumina system [1-5]. a) Surface patterned solid spheroid.[u] b) Surface patterned shell-like structure attached to a spheroid.[d] c) Separate clumps of "egg shell" structures.[u] d) Collections of surface bowls on a spheroid surface.[d] [d] = decylamine-, [u] = undecylamine based synthesis products.

In the work described in this paper, tetraethylene glycol (TEG) was used for manipulating decylamine and undecylamine surfactants [1-5]. Through a series of experiments performed, it was determined that an alkylamine containing a minimum ofeight carbons was a prerequisite for vesicle templating. Shorter amines were found to template nanolamellar aluminophosphate phases with normal crystal habits and facets, devoid of surface patterning [21]. The organic and inorganic lamellae of these nanostructures are both crystallographically well-ordered.

We then began to categorize the classes of macroscopic morphologies, pores and surface patterns that we were synthesizing. The product yields were essentially quantitative and phase pure. The inorganic fabric was exclusively a mesolamellar aluminophosphate, "molded" into spectacular macroscopic forms and fine patterns. The syntheses reproducibly yielded three dominant morphologies. Two of them comprised of solid spheroids and hollow shells with macroscopic pores and surface patterns. The third had the form of flakes without any interesting surface features. A representative collection of ultrastructured objects is shown in Figures 3 and 4.

The major morphological forms of interest are comprised of millimeter dimension solid and hollow spheroidal objects which display a range of micron scale protrusions and surface patterns. A surface patterned solid spheroid is shown in Figure 3a. Barnacle and egg shell-like objects are found attached to the surface of solid spheroids, Figure 3b, or in separate clumps, Figure 3c. The solid spheroids display a variety of micron scale surface patterns, such as collections of surface bowls, Figure 3d. Other examples include honeycomb, columnar, porous, and quilt-like structural features, Figure 4. The resemblance of the morphologies of these artificial inorganic assemblies to some of Nature's biomineralized structures has been noted [1-5].

The formation of these ultrastructures initially seemed to be an enigma. However, the observations and concepts outlined above provide us with an understanding of the origin of the morphologies and surface patterns observed in the TEG-alkylammonium dihydrogen phosphate-alumina system [4]. This takes into account the following:

(a) the multifunctional role of TEG: (i) as a solvent to enable the self-assembly of the vesicle template; (ii) for viscosity and chemical control of the rates of mineralization, transport, polymerization and deposition of the mesolamellar aluminophosphate regions of the ultrastructure; (iii) as a polydentate ligand to form "controlled-release" (TEG)Al(III) complexes; (iv) as cosurfactant for the control of bilayer curvature [22]; (v) as a demixing agent to promote surfactant-TEG phase separation (domains) and patterning of vesicle bilayers; (vi) as ion channels to facilitate the transport of (TEG)Al(III) complexes through vesicle bilayers, to access reactive phosphate sites and permit mesolamellar aluminophosphate nucleation and growth [23].

(b) "cellular" processing events that ensue during the adhesion, fusion, fission, reshaping and collapse of vesicles [24]. These events facilitate the shaping of the macroscopic morphology and patterning of the mesolamellar aluminophosphate material. The approach of vesicles can lead to adhesion and aggregation. This can also result in vesicles in tension which may distort into truncated spherical shapes. The equilibrium deformation that minimizes the surface free energy involves the balance of adhesion (i.e., hydrophobic, electrostatic, hydration, steric) and elastic (i.e., stretching, bending) forces between the amphiphilic surfaces of the vesicle bilayers.

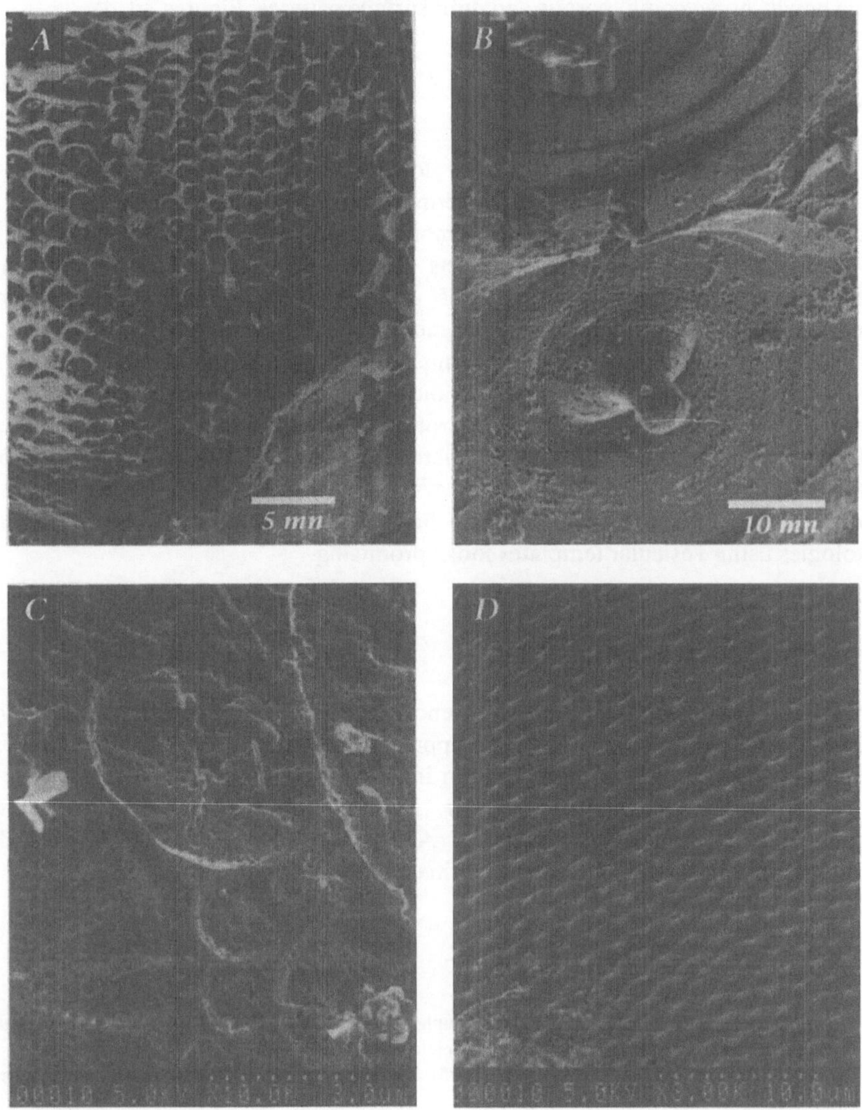

Figure 4. Further examples of SEM images of the aluminophosphate morphologies and surface [1-5]. a) Surface honeycomb.[d] b) Perforated columnar protrusion.[u] c) Sub-micron surface pores.[d] d) Surface "eiderdown" morphology.[d] [d] = decylamine-, [u] = undecylamine based synthesis products.

The vesicles may become internally stressed by alterations in the environment, temperature or packing, which may lead to rupture or local breakthrough followed by fusion or fission (e.g., the fused, shell-like structures, Figure 3c). This could be followed by vesicle reshaping (e.g., columnar shapes, Figure 4b) and collapse (e.g., surface bowls, honeycomb, porous and quilt surface patterns, Figures 3d, 4).

3. Conclusions

The "cellular" processing model serves to explain the origin of the worm-like morphologies of the mesoporous silicas formed in the surfactant-silica system. It also accounts for the seemingly unrelated morphologies and surface patterns of the mesolamellar aluminophosphate products formed in the TEG-alkylammonium dihydrogen phosphate-alumina system. Just as micelles can facilitate the self-assembly of inorganic-organic composites with mesoscopic patterning, our research has demonstrated that vesicles can sculpt impressive macroscopic forms and surface ornamentations. Our model is inspired by biological design principles for synthesizing hierarchical materials with inorganic and organic modular components connected at interfaces. This paradigm for creating hierarchical materials with self-selected-form could lead to complete structural systems with multi-functional properties. Based on this development, the prospects for new biologically inspired "synthetic" materials technologies using vesicular templates looks promising.

4. Acknowledgements

Financial assistance from the Natural Sciences and Engineering Research Council of Canada in support of this work is deeply appreciated. SO expresses his appreciation to the Ontario Graduate Scholarship program in partial support of his research. We are grateful for the expert assistance of Dr. Neil Coombs (Imagetek, Toronto) with Scanning Electron Microscopy and Dr. Alan Lough for the single crystal data. Insightful discussions with Dr. Alex Kuperman are also acknowledged.

5. References

1. Oliver, S., Coombs, N., Kuperman, A., Lough, A. and Ozin, G.A. (1995) Artificial inorganic assemblies that mimic radiolaria and diatom skeletons, submitted.
2. Oliver, S., Coombs, N. and Ozin, G.A. (1995) Synthetic hollow aluminophosphate microspheres, submitted.
3. Oliver, S., Coombs, N. and Ozin, G.A. (1995) Hierarchical inorganic-organic composite mesh-like ultrastructures, submitted.
4. Oliver, S. and Ozin, G.A. (1995) Skeletons in the beaker: synthetic hierarchical inorganic materials: A "cellular" model, submitted.
5. Oliver, S. and Ozin, G.A. (1995) Macroporous inorganic shells, submitted.
6. Yang, H., Coombs, N., Kuperman, A., Mamiche-Afara, S. and Ozin, G.A. (1995) Hierarchical materials chemistry: Synthesis of oriented mesoporous silica thin films on mica, submitted.

7. Khushalani, D., Kuperman, A., Ozin, G.A. and Coombs, N. (1995) Natural and synthetic mesoporous silicas, submitted.
8. Mann, S. (1995) Biomineralization and biomimetic materials chemistry, *J. Mater. Chem.* **5**, 935-946.
9. Richard, M. (1990) *Ouvrage Dedie à la Memoire du Professor Henry Germain*, Koeltz Scientific Books, Konigstein; Simkiss, K. and Wilbur, K.M. (1989) *Biomineralization: Cell Biology and Mineral Deposition*, Academic Press, San Diego; Mann, S., Webb, J. and Williams, R. (1989) *Biomineralization: Chemical and Biochemical Perspectives*, VCH Publishers, New York; Takahashi, K. (1991) *Radiolaria: Flux, Ecology and Taxonomy in the Pacific and Atlantic*, Ocean Biocoenisis Series, **3**, Woods Hole, Mass.; Anderson, O.R. (1983) *Radiolaria*, Springer-Verlag, New York; Lowenstam, H.A. (1989) *On Biomineralization*, Oxford University Press, Oxford; Mann, S. (1983) *Mineralization in Biological Systems: Structure and Bonding*, Springer-Verlag, Berlin.
10. Mann, S. and Perry, C.C. (1986) Structural aspects of biogenic silica, in D. Evered and M. O'Connor (eds.), *Silicon Biochemistry: Proc. of CIBA Foundation Symp. Ser.* **121**, Wiley, New York, pp. 40-58.
11. Ball, P. (1994) *Designing the Molecular World: Chemistry at the Frontier*, Princeton University Press, Princeton.
12. Bowes, C.L. and Ozin, G.A. (1995) Self-Assembling frameworks, *Adv. Mater.*, in press; Kuperman, A., Nadimi, S., Oliver, S., Ozin, G.A., Garcès, J.M. and Olken, M.M. (1993) Non-aqueous synthesis of giant crystals of zeolites and molecular sieves, *Nature* **365**, 239-242.
13. Ozin, G.A. (1992) Nanochemistry: Synthesis in diminishing dimensions, *Adv. Mater.* **4**, 612-649.
14. Vincent, J. (1990) *Structural Biomaterials*, Princeton University Press, Princeton; National Materials Advisory Board (1994) *Hierarchical Structures in Biology as a Guide for New Materials Technology*, National Academy Press, Washington D.C.
15. Kresge, C.T., Leonowicz, M.E., Roth, W.J., Vartuli, J.C. and Beck, J.S. (1991) Ordered mesoporous molecular sieves synthesized by a liquid crystal templating mechanism, *Nature* **359**, 710-712; Beck, J.S., Vartuli, J.C., Roth, W.J., Leonowicz, M.E., Kresge, C.T., Schmitt, K.D., Chu, C.T-W., Olson, D.H., Sheppard, E.W., McCullen, S.B., Higgins, J.B. and Schlenker, J.L. (1992) A new family of mesoporous molecular sieves prepared with liquid crystal templates, *J. Am. Chem. Soc.* **114**, 10834-10843.
16. Ogino, K. and Abe, M. (1993) *Mixed Surfactant Systems*, Dekker, New York.
17. Hoffmann, H. and Pössnecker, G. (1994) The mixing behavior of surfactants, *Langmuir* **10**, 381-389; Pilsl, H., Hoffmann, H., Hoffmann, S., Kalus, J., Kencono, A.A., Lindner, P. and Ulbricht, W. (1993) Shape investigation of mixed micelles by small angle neutron scattering, *J. Phys. Chem.* **97**, 2745-2754; Treiner, C. and Makayssi, A. (1992) Structural micellar transition for dilute solution of long chain binary cationic surfactant systems: A conductance investigation, *Langmuir* **8**, 794-800.
18. Firouzi, A., Kumar, D., Bull, L.M., Besier, T., Sieger, P., Huo, Q., Walker, S.A., Zasadzinski, J.A., Glinka, C., Nicol, J., Margolese, D., Stucky, G.D. and Chmelka, B.F. (1995) Cooperative organization of inorganic-surfactant and biomimetic assemblies, *Science* **267**, 1138-1143; Huo, Q., Margolese, D., Ciesla, U., Demuth, D., Feng, P., Gier, T.E., Sieger, P., Firouzi, A., Chmelka, B.F., Schuth, F. and Stucky, G.D. (1994) Organization of organic molecules with inorganic molecular species into nanocomposite biphase arrays, *Chem. Mater.* **6**, 1176-1191; Chen, C-Y., Burkett, S.L., Li, H-X. and Davis, M.E. (1993) Studies on mesoporous materials: II Synthesis Mechanism, *Micropor. Mater.* 2, 27-34.
19. Iler, R. K. (1979) *The Chemistry of Silica*, Wiley, New York.
20. Martino, A. and Kaler, E.W. (1995) Phase behavior and microstructure of nonaqueous microemulsions *Langmuir* **11**, 779-784; Thomas, B.N., Safinya, C.R., Plano, R.J. and Clark, R.N. (1995) Lipid tubule self-assembly: Length dependence on cooling rate through first order phase transition, *Science* **267**, 1635-1638; Haslop, W.P. Allonby, J.M., Akred, B.J. and Messenger, E.T., (1986) U.S. Patent Number 4,618,446; Akred, B.J., Messenger, E.T. and Nicholson, W.J. (1987) U.S. Patent Number 4,659,497.
21. Oliver, S., Kuperman, A., Lough, A., Ozin, G.A., Garcès, J.M., Olken, M.M. and Rudolf, P. (1994) New insights into the mode of formation of $AlPO_4$-n molecular sieves, *Stud. Surf. Sci. Catal.* **84**, 219-225.
22. Hoffmann, H. (1994) Fascinating phenomena in surfactant chemistry, *Adv. Mater.* **6**, 116-129.
23. Fendler, J.H. (1982) *Membrane Mimetic Chemistry*, John Wiley & Sons, New York.
24. Israelachvili, J.N. (1991) *Intermolecular and Surface Forces*, Academic Press, New York.

ASSEMBLY OF ORIENTED NANOMETER CHANNELS ON ORGANIC LAYERS

SUE FENG AND THOMAS BEIN*
Department of Chemistry, Purdue University, West Lafayette, IN 47907-1393, USA

ABSTRACT. Strategies for the assembly of molecular sieve crystals on gold substrates are discussed. The article focuses on the growth of oriented nanometer structures on organic layers. Three different models of self-assembled multilayer mixed organophosphonate films are prepared on gold surfaces. The first model uses spacer molecules having a chain length similar to the phosphonate chains but containing methyl head groups. In model II, the phosphonate surface acts as a base layer, and a longer alkyl spacer molecule protrudes from the surface. The third model system contains a mixture of end groups such as bromide and methyl with different chain length and ratios. The bromide end groups in the mixed layers were converted into organic templating agents similar to those used in the preparation of the bulk synthesis gel. The first two types of mixed functional films show profound effects on the surface crystallization of zincophosphate molecular sieves, with a nonlinear dependence of both surface crystal density and orientation on phosphonate surface group density.

1. Introduction

Many different strategies are being developed to create chemical structures of ever increasing complexity. In looking beyond single molecules, the power of assembling structures from preformed modules is being recognized. In particular, self-assembly using strong or weak binding forces between a *template* and the chemical modules of interest can lead to systems vastly different from individual molecules, including monolayer assemblies on metal surfaces [1], nano- and mesoporous metal oxide lattices [2], and biomimetic crystallization of oriented simple solids on organic layers [3].

Of particular interest are assemblies that are designed with a certain physical function in mind, such as catalysis, molecular recognition in sensors, or optical nonlinearity. Our interest in nanoporous assemblies on surfaces originates from the desire to construct highly selective chemical sensors. Consider the great variety of zeolite molecular sieve structures for this purpose. Zeolites are typically defined as crystalline, porous tectoaluminosilicates (or other metal oxides), where oxygen bridges two tetrahedrally coordinated framework atoms [4,5]. Close to one hundred different framework topologies [6] and many more compositions are known, resulting in a vast range of different sorption properties for small and medium size molecules, with pore sizes ranging from about 0.3 to 3 nm, and internal surface properties ranging from hydrophilic to hydrophobic and from basic to acidic. If a zeolite microcrystal is considered to function as a module with selective sorption properties, it will be desirable

J. Michl (ed.), Modular Chemistry, 335–344.

to assemble arrays of such microcrystals on appropriate sensor substrates such as piezoelectric oscillators [7]. The resulting device is expected to show an enormous increase in surface area compared to the bare substrate, and a steep sorption isotherm at low partial pressure *(amplification)*, as well as the molecular sieve selectivity and specific surface interactions of the zeolite *(selectivity)*. Furthermore, many zeolite materials are chemically and thermally stable, which is of great importance in practical applications.

Previous work in this context has attacked the attachment problem with several strategies. Zeolite/sol-gel composite films were deposited by dip-coating a suspension of the zeolite crystals in a tetraethylorthosilicate-derived sol on piezoelectric oscillators [8,9]. These films are chemically and mechanically stable, and offer significant microporosity and molecular selectivity. Potential limitations include additional porosity introduced by the sol-gel glass film, and clogging of zeolite pores by the silicate monomers or polymers of the sol-gel matrix. A second approach is based on the bonding of zeolite crystals to a molecular coupling layer on the gold electrode of a piezoelectric oscillator (such as a quartz crystal microbalance, QCM). Thus, a layer of (3-mercaptopropyl)-triethoxysilane was first attached to the gold, followed by exposure to a suspension of the zeolite crystals. The advantage of the resulting films is the absence of a binding matrix, and therefore uninhibited access to the zeolite "modules". Several studies have demonstrated the high molecular selectivity of chemical sensors based on this concept [10,11,12].

Much greater demands on the modular assembly process arise if *oriented* molecular sieve channels are desired. Oriented nanometer channel structures are of substantial interest for size-selective chemical sensors, separation membranes, and other novel devices. The third approach, displayed in Figure 1, is based on the idea that certain functional organic layers should be capable of assisting in the surface-nucleation and growth of molecular sieves. The surface is expected to act as a template for crystal growth. In our recent studies we have discovered that single component organo-phosphonate films can promote the growth of oriented molecular sieves, specifically zinco-phosphate and aluminophosphate crystals [13,14]. We found that several factors are important for the surface nucleation process, including the presence of the phosphonate surface, a specific solution templating agent, substrate orientation, gel concentration, and synthesis temperature. The crystals are attached to the surface with one of their triangular faces in the case of zinco-phosphate, and with oriented vertical channels in the case of aluminophosphate molecular sieves.

In order to understand more about the chemistry at the organic-inorganic interfaces we have designed three models with a variety of mixed phosphonate layers on gold surfaces. The first model uses spacer molecules that have a chain length similar to the phosphonate chains but contain "inert" methyl head groups. In model II, the phosphonate surface acts as a base layer, and a longer alkyl spacer molecule is embedded in the surface. The third model system contains a mixture of end groups such as -Br, and $-CH_3$, with different chain length and ratios. The -Br end groups in the mixed layers were converted into organic templating agents which are similar to those used in the preparation of the bulk synthesis gel.

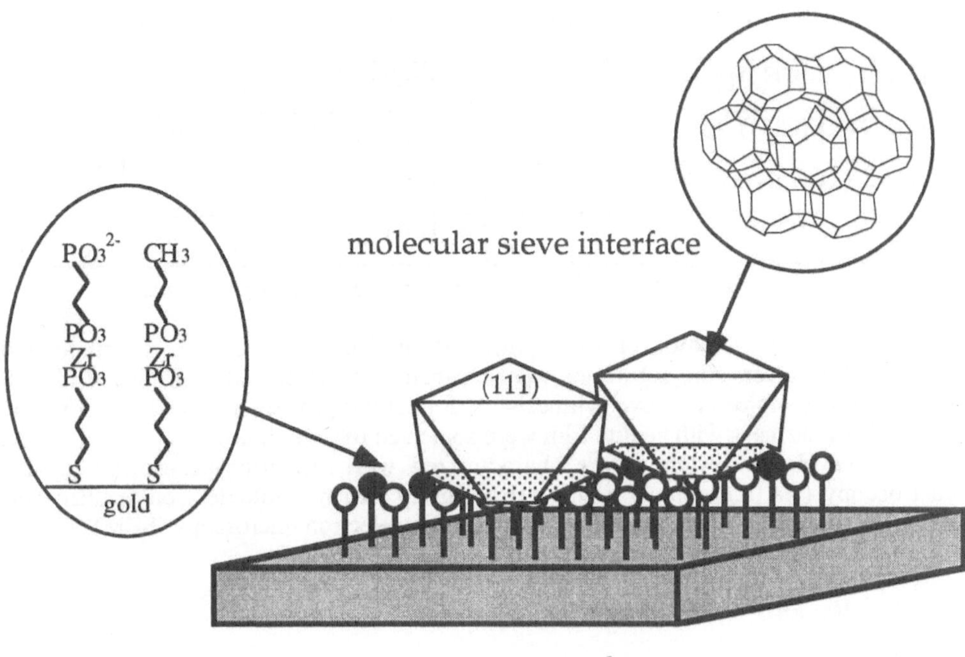

Figure 1. Growth of zincophosphate molecular sieves on mixed alkyl/organophosphonate layers on gold (model I) (schematic).

2. Experimental Section

Trilayers consisting of a phosphorylated alkylthiol, zirconium, and mixed phosphonic acids where prepared as follows. A self-assembled monolayer was first formed on the gold substrate (on Cr/Si) via adsorption (for 48 h) of 11-mercapto-1-undecanol (MUD) from a 1.0 mM solution in ethanol. The MUD monolayers were phosphorylated for 1 h in a solution of 0.2 M phosphorus oxychloride ($POCl_3$) (Aldrich, 99%) and 0.2 M 2,4,6-collidine (Aldrich, 99%) in dry acetonitrile in a nitrogen filled glovebox, and then rinsed thoroughly with acetonitrile. The trilayers were formed by exposure at room temperature to a 5 mM solution of $ZrOCl_2$ and then to a 1.0 mM (total molar concentration) solution of the pure or mixed phosphate/phosphonate molecules (shown in Table 1) in ethanol (#6 in ether, 5+6 in 1:1 v/v ethanol/ether) for 12-48 h, followed by a rinse in the respective solvent.

Model III contains mixed organic films of compounds 1 and 3, and also 1 and 2 as the outermost layer with molar ratios 1/1, 1/2, and 1/5 in solution. The mixed films were further reacted to convert the bromide end group to quaternary ammonium salts. The bromide mixed films were treated with a 0.2 M solution of dabco (1,4-diazabicyclo[2.2.2]octane) in anhydrous ether at ambient temperature for 1 hour, then rinsed thoroughly with ether and dried with a nitrogen stream.

TABLE 1. The molecules studied as components of the third layers[a]

1. H_2PO_3-O-$(CH_2)_{11}$-Br	**2.** H_2PO_3-O-$(CH_2)_5$-CH_3
3. H_2PO_3-O-$(CH_2)_6$-Br	**4.** H_2PO_3-$(CH_2)_4$-PO_3H_2
5. H_2PO_3-$(CH_2)_{10}$-PO_3H_2	**6.** H_2PO_3-O-$(CH_2)_{17}$-CH_3
7. H_2PO_3-O-$(CH_2)_{11}$-CH_3	**8.** H_2PO_3-O-$(CH_2)_{10}$-OH

[a] Compound **4** formed disordered films and **8** presented solubility problems.

For the preparation of zinco-phosphate molecular sieve films on model I-III mixed organic layers, the substrates were cleaned with water and then placed into the zincophosphate molecular sieve synthesis gel for crystal growth, as described in ref. 13. The resulting surfaces with zeolite film were sonicated for two minutes.

The above systems were characterized with reflection-absorption infrared spectroscopy (RAIR), contact angle, ellipsometry, X-ray photoelectron spectroscopy (XPS), grazing-angle X-ray diffraction, and scanning electron microscopy (SEM).

3. Results and Discussion

3.1. MODEL I MIXED FILMS

In model I, we study a system composed of two types of molecules with similar chain length, i.e., compound **7** [H_2PO_3-O-$(CH_2)_{11}$-CH_3] and **5** [H_2PO_3-$(CH_2)_{10}$-PO_3H_2], but with different end groups to investigate the possible domain formation and dilution effects of the phosphonate groups.

We explore three issues: First, the distribution of the PO_3H_2 groups in mixed monolayers as a function of the molar fraction of the phosphonate molecules in solution. Second, whether the mixed tri-layers form large single component domains or not. Last, the effects of the mixed films on the growth of zeolite molecular sieve crystals, including density and orientation.

3.1.1. *Composition and structural information*

In the RAIR spectrum of the pure H_2PO_3-$(CH_2)_{10}$-PO_3H_2 (DBPA) (**5**) as the outer layer (not shown), only CH_2 vibrations, 2921 cm^{-1} (υ_a) and 2851 cm^{-1} (υ_s) are present, similar to previous observations with related systems [15]. With increasing molar concentration of the alkyl-terminated compound **7**, the spectra show the expected increase of the CH_3 intensity, υ_a 2964 cm^{-1} and υ_s 2878 cm^{-1}, while the methylene stretch frequencies do not change much. The frequencies of the CH_2 stretching modes have been used to differentiate between crystalline-like (ordered) and liquid-like (disordered) states of the alkyl layers. A general trend is that the frequency of the CH_2 (υ_a) mode increases from about 2918 to 2927 cm^{-1}, when going from a crystalline-like to a disordered alkyl layer. We conclude that the degree of order in model I mixed films with compounds **5** and **7** as top layers does not change much when changing the top layer composition.

The cosine of the advancing water contact angle on model I mixed layers with compounds **5** [H_2PO_3-$(CH_2)_{10}$-PO_3H_2] and **7** [H_2PO_3-O-$(CH_2)_{11}$-CH_3] shows a linear relationship with the phosphonate molar ratios in the adsorption solution, suggesting (i) that there is no strong preference for adsorption of compound **5** vs. **7** in the outer third layers, and (ii) that Cassie's Law stating $\cos \theta = \chi_1 \cos \theta_1 + \chi_2 \cos \theta_2$ [16], appears to hold reasonably well (χ_1 and χ_2 are the mole fractions of the two components in the mixed layer, and θ_1 and θ_2 are the contact angles on pure layers of the two components). Because the observed relationship is linear, it is reasonable to assume that the solution phosphonate ratio is reproduced on the surface (XPS data show a monotonous increase of P/S and P/Zr atomic ratios with solution concentration of P, but it is difficult to extract precise film compositions from the data because the effect of the film structure on the photoelectron mean free paths is not always known [17]).

We are not aware of studies of similar mixed multilayer films but there is extensive literature on single-layer mixed thiol films. For example, Bain et al. [18,19] concluded from XPS and wettability experiments of coadsorbed OH and CH_3 terminated thiols on gold that molar solution ratios are not generally reproduced on the gold surface, and that a component with longer alkyl chains introduces a disordered oleophilic phase (see model II below).

3.1.2. *Effects of mixed film composition on crystal growth*

The mixed alkyl/organo-phosphonate films with compounds **7** and **5** were used for crystal growth studies of zinco-phosphate molecular sieves (see Figure 1). We observe three significant effects:

(i) Scanning electron micrographs (SEMs) reveal that the zinco-phosphate crystal density increases as the mole fraction of the phosphonate functional groups increases (Figure 2).

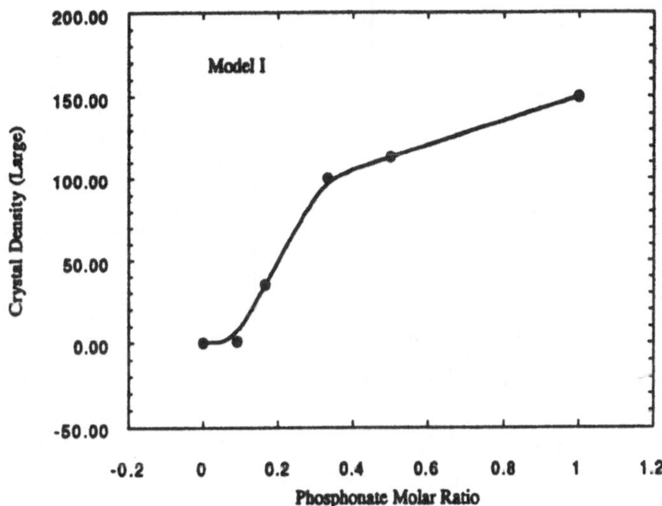

Figure 2. Model I crystal density (large crystals, > 1 μm) vs. the phosphonate molar ratio in the adsoption solution. Crystal density is the number of crystals (> 1 μm) per 50 X 50 μm^2.

There appears to be a minimum required molar ratio of the phosphonate groups which must be present in order to promote significant formation of a layer of crystals. This threshold molar ratio is about 0.2 in the case of model I. (ii) Two different sizes of crystals are found on the surface. The large crystals are larger than 1 μm, and smaller crystals are smaller than 0.5 μm. Finally, scanning electron micrographs show that the crystals are less oriented on mixed organic films than on the 100 % phosphonate films. Grazing-angle X-ray diffraction was used to determine the crystal orientations.

Figure 3A shows normalized (vs. intensity of the strong (111) peak at 5.9° 2θ, 26,000 cps) grazing-angle X-ray diffraction patterns of zinco-phosphate molecular sieve crystals on model I films where the phosphonate molar ratio (χ_P) ranges from 1.00 to 0.091. For example, the intensity of the (311) peak at 11.6° 2θ increases from (a) to (e); this indicates an increasing disorder in the system.

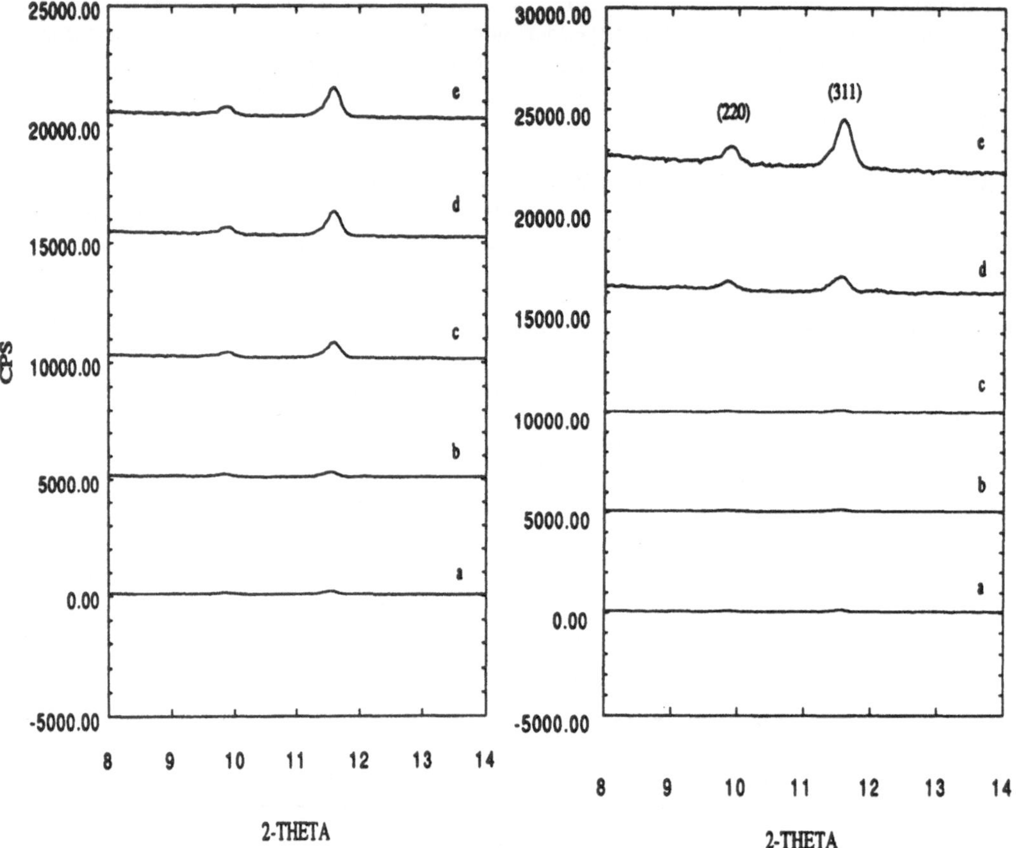

Figure 3. A (left). Grazing-angle X-ray diffraction patterns normalized to the (111) peak of zinco-phosphate molecular sieve crystals on model I mixed organo-phosphonate films where the phosphonate molar ratios are: (a) 1.00, (b) 0.50, (c) 0.333, (d) 0.167, and (e) 0.091. 2θ = 8-14°. B (right). Patterns for model II films where the phosphonate molar ratios are: (a) 1.00, (b) 0.938, (c) 0.909, (d) 0.833, and (e) 0.50. 2θ = 8-14°.

3.2. MODEL II MIXED FILMS

3.2.1. *Composition and structural information*

Model II uses the phosphonate film **5**, [H_2PO_3-$(CH_2)_{10}$-PO_3H_2] as a base and adds in a longer spacer group, compound **6** [H_2PO_3-O-$(CH_2)_{17}$-CH_3] at different mixing ratios. The cosine of the water contact angle of the mixed films of model II shows an interesting behavior — at low phosphonate concentrations it changes much less than linear with phosphonate content. We propose that the longer component **6** [H_2PO_3-O-$(CH_2)_{17}$-CH_3] bends towards the surface and partially blocks component **5** [H_2PO_3-$(CH_2)_{10}$-PO_3H_2] with its methylene chain on the surface, such as the proposed bending of $HS(CH_2)_{21}CH_3$ on a base of dodecanethiol on gold [19]. However, the ellipsometry data of model II mixed films (assuming $n = 1.54$ for the organic film [20]) show a fairly linear relationship between the mixed film thickness and the phosphonate molar ratio in the solutions containing compounds **6** and **5**. Because ellipsometry data reflect molecular density on the substrate, the results are not strongly affected by the local surface structure (which is probed by contact angle). Hence, these results establish that the solution phosphonate molar ratio is approximately reproduced on the substrate.

3.2.2. *Effects on crystal growth*

The substrates with different ratios of [H_2PO_3-O-$(CH_2)_{17}$-CH_3] (OPA): [H_2PO_3-$(CH_2)_{10}$-PO_3H_2] (DBPA) were placed into the zinco-phosphate molecular sieve synthesis gel to investigate the effect of the mixed organo-phosphonate film on crystal growth. No crystal growth is observed on the surface when the ratio of [H_2PO_3-O-$(CH_2)_{17}$-CH_3] (OPA): [H_2PO_3-$(CH_2)_{10}$-PO_3H_2] (DBPA) is 1:0, which might be expected because only methyl groups are exposed on the surface. When the ratio of OPA:DBPA is 1:1, electron microscopy still indicates an almost crystal-free surface. Only at high phosphonate surface density, significant crystal density was observed (Figure 4).

If the resulting crystal density is plotted as a function of the phosphonate molar ratio (not shown), the data indicate that the minimum phosphonate molar ratio for significant growth is about 0.8, which is much higher than in model I (threshold ratio: ca. 0.2). Furthermore, the crystals are only highly oriented at high phosphonate concentrations on the surface. Grazing angle X-ray diffraction data displayed in Figure 3B show that the X-ray intensity of non –(111) peaks such as (311) decreases from e to a in the Figure, when the phosphonate molar ratio increases from 0.5 to 1.00.

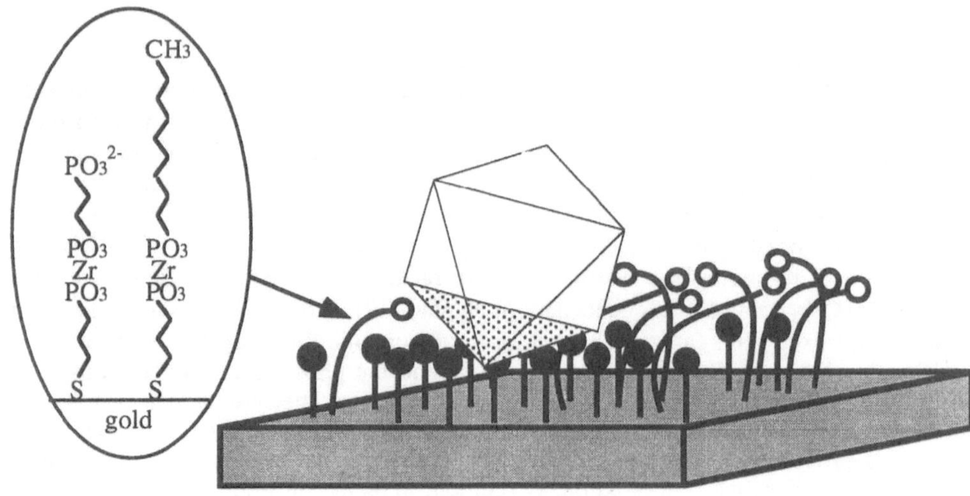

Figure 4. Growth of zincophosphate molecular sieves on mixed alkyl/organophosphonate layers on gold (model II) (schematic).

3.3. MODEL III MIXED FILMS

Some preliminary results regarding the mixed films of model III are shown in Table 2. We observe that the contact angles converge at about 70° after the quaternization. At this point, the yield of the surface quaternization reactions is unknown, but the similar contact angles suggest that a chemically similar surface has been formed in each case. After the conversion of the alkyl bromide termini in the films to quaternary ammonium salts, the films were exposed to the zincophosphate molecular sieve synthesis gel. Scanning electron micrographs show that the model III systems studied so far did not promote surface nucleation and growth of molecular sieves.

TABLE 2. Advancing water contact angles of model III films with compounds **1** [H_2PO_3-O-$(CH_2)_{11}$-Br], **2** [H_2PO_3-O-$(CH_2)_5$-CH_3], and **3** [H_2PO_3-O-$(CH_2)_6$-Br] before and after reaction with dabco.

Ratio of compounds **1**:**3**	contact angles before reaction	contact angles after reaction with dabco	Ratios of compounds **1**:**2**	contact angles before reaction	contact angles after reaction with dabco
1:1	65	63	1:1	74	71
1:2	75	69	1:2	76	71
1:5	77	70	1:5	82	71

The objective of using model III systems is to establish whether the presence of "templating molecules" as surface head groups can enhance or change the nucleation process of the molecular sieve crystals. The absence of surface crystallization could be related to (i) the absence of a template pattern commensurate with the zeolite crystal structure, or (ii) to the confinement of organic templating molecule head groups on the surface, which could lead to a loss of templating effect, compared to similar molecules in bulk synthesis.

4. Conclusions

In summary, the above studies elucidate the influence of phosphonate head group surface assemblies on molecular sieve growth on organic layers. In mixed layers composed of "active" phosphonate groups and alkyl chains of about equal length, a relatively small fraction of the active groups is sufficient for effective surface growth, but the molecular sieve crystals are only oriented at high phosphonate coverage in the organic layer. In model II with long "spacer" molecules, we find that even when the phosphonate-to-spacer ratio is as large as 1:1, almost no crystals are formed on the surface. In this case, the space between the longer, bent-down methylene chains is probably too small for nuclei to penetrate into the voids and reach the "buried" phosphonate head groups to initiate the crystal growth. However, as the phosphonate functional group density is increased beyond 80% in the mixture, apparently there are enough phosphonate sites exposed to induce effective surface growth.

The studies provide useful information for extended control of molecular sieve crystal growth at the inorganic-organic interface. The oriented zeolite structures are promising candidates for applications such as catalytic membranes with true molecular selectivity, or controlled access of molecules of pre-selected size to a sensor surface.

5. Acknowledgments

The authors greatly appreciate funding from the U.S. National Science Foundation, the Exxon Education Foundation, and from the Purdue Research Foundation.

6. References

1 Ulman, A. (1991) *An Introduction to Ultrathin Organic Films: From Langmuir-Blodgett to Self-Assembly*; Academic Press, New York.

2 Huo, Q., Margolese, D. I., Ciesla, U., Demuth, D. G., Feng, P., Gier, T. E., Sieger, P., Firouzi, A., Chmelka, B. F., Schüth, F., and Stucky, G. D. (1994) Organization of Organic Molecules with Inorganic Molecular Species into Nanocomposite Biphase Arrays, *Chem. Mater.* **6**, 1176-1191.

3 Fendler, J. H. and Meldrum, F. C. (1995) The Colloid Chemical Approach to Nanostructured Materials, *Adv. Mater.* **7**, 607-632.

4 Breck, D. W. (1974; 1984) *Zeolite Molecular Sieves* Krieger, Malabar, FL.

5 Davis, M. E. and Lobo, R. F. (1992) Zeolite and Molecular Sieve Synthesis, *Chem. Mater.* **4**, 756-768.

6 Meier, W. M. and Olson, D. H. (1992) *Atlas of Zeolite Structure Types*, 3rd Edition, Butterworth-Heinemann, London.

7 Ward, M. D. and Buttry, D. A. (1990) In Situ Interfacial Mass Detection with Piezoelectric Transducers, *Science*, **249**, 1000-1007.

8 Bein, T., Brown, K., Frye, G. C. and Brinker, C. J. (1989) Molecular Sieve Sensors for Selective Detection at the Nanogram Level, *J. Am. Chem. Soc.* **111**, 7640-7641.

9 Yan, Y., Bein, T., Brown, K. D., Forrister, R. and Brinker, C. J. (1992) Molecular Recognition on Acoustic Wave Devices: Modified Zeolite-Silica Thin Films with Tailored Adsorption Properties, *Mat. Res. Soc. Symp. Proc.* Vol. **271**, 435-441.

10 Yan, Y. and Bein, T. (1992) Molecular Sieve Sensors for Selective Ethanol Detection, *Chem. Mater.* **4**, 975-977.

11 Yan, Y. and Bein, T. (1992) Molecular Recognition on Acoustic Wave Devices: Sorption in Chemically Anchored Zeolite Monolayers, *J. Phys. Chem.* **96**, 9387-9393.

12 Yan, Y. and Bein, T. (1995) Zeolite Thin Films with Tunable Molecular Sieve Function, *J. Am. Chem. Soc.* **117**, 9990-9994.

13 Feng, S. and Bein, T. (1994) Growth of oriented molecular sieve crystals on organophosphonate films, *Nature* **368**, 834-836.

14 Feng, S. and Bein, T. (1994) Vertical Aluminophosphate Molecular Sieve Crystals Grown at Inorganic-Organic Interfaces, *Science* **265**, 1839-1841.

15 Bent, S. F., Schilling, M. L., Wilson, W. L., Katz, H. E. and Harris, A. L. (1994) Structural Characterization of Self-Assembled Multilayers by FTIR, *Chem. Mater.* **6**, 122-126.

16 Adamson, A. W. (1990) *Physical Chemistry of Surfaces*, Fifth Ed., Ch. X, Wiley Interscience, New York. See also: Cassie, A. B. D. (1948) *Discuss. Faraday Soc.* **3**, 11.

17 Akhter, S., Lee, H., Hong, H.-G., Mallouk, T. E. and White, J. M. (1989) Structural characterization of multilayer metal phosphonate film on silicon using angular-dependent x-ray photoelectron spectroscopy, *J. Vac. Sci. Technol.*, **A 7**, 1608-1613.

18 Bain, C. D. and Whitesides, G. M. (1989) Formation of Monolayers by the Coadsorption of Thiols on Gold: Variation in the Head Group, Tail Group, and Solvent, *J. Am. Chem. Soc.* **111**, 7155-7164.

19 Bain, C. D., Evall, J. and Whitesides, G. M. (1989) Formation of Monolayers by the Coadsorption of Thiols on Gold: Variation in the Length of the Alkyl Chain, *J. Am. Chem. Soc.* **111**, 7164-7175.

20 Lee, H., Kepley, L. J., Hong, H.-G., Akhter, S. and Mallouk, T. E. (1988) Adsorption of Ordered Zirconium Phosphonate Multilayer Films on Silicon and Gold Surfaces, *J. Phys. Chem.*, **92**, 2597-2601.

DISCUSSION OF THE OZIN AND BEIN LECTURES

Session XII: Oriented Nanometer Channels in Zeolites and Hierarchical Inorganic Materials

CLIFFORD P. KUBIAK
Department of Chemistry
Purdue University
West Lafayette, IN 47907-1393
U.S.A.

1. Introduction

The twelfth panel discussion dealt with synthesis, oriented single crystal growth, and the chemistry of zeolites. A spirited and colorful hour of discussion followed the first thirty minute lecture by Geoff Ozin entitled: "Hierarchical Inorganic Materials" and the second thirty minute lecture by Thomas Bein entitled: "Assembly of Oriented Nanometer Channels on Organic Layers."

1.1 FUNDAMENTAL CONCEPTS

1.1.1 Controlling the Chemistry at Interfaces

At last we consider real zeolites. Perhaps no other area of modular chemistry can be cited to better illustrate the importance of controlling the chemistry at interfaces. This figures importantly in the preparation of surfaces appropriately chemically modified for the oriented growth of single crystal zeolite materials, as illustrated in Tom Bein's work. Templates can be viewed as a workbench for manipulating other large modular components. The term "work bench" is not inappropriate since it is often a surface that is being used as a platform for the synthesis of more complex modular structures. Here the size of the modular components makes the challenges of achieving chemical control large. The second area where control of the interface is important is in making more complex modular structures by piecing components together. How, for example, will we be able to manipulate graphene tubules to make connections to other molecular or supramolecular assemblies?

J. Michl (ed.), Modular Chemistry, 345–359.

1.1.2 Stealing Nature's Secrets: Biomineralization

Geoff Ozin showed us that there is much to be learned by stealing nature's secrets. Biomineralization occurs by processes that can involve the formation of highly complex and organized inorganic structures by synthesis on protein beta sheet templates. Consider the extracellular lorica formed by a unicellular flagellate (Stephanoeca diplocostata Ellis)[1,2]. The organism is surrounded by a basket of silica plates. It finds a way to synthesize the silica plates inside the cell, and then transfer them to the outside of the cell. During cell division, the organism can in fact divide the inorganic structures between the parent and the daughter cell. The magnetic sensors of magnetotactic bacteria contain single crystals of magnetite, synthesized within vesicles of the organism. The nucleation and growth of crystalline iron oxide in ferritin is an amazing example of organically templated inorganic synthesis. The aragonite crystals organized into lamellar sheets in cuttlefish bones is one example of nature's many uses of calcium carbonate. Another example of biomineralization is the lowly trilobite. These insect-like creatures existed on earth hundreds of millions of years ago. The trilobite figured out how to mineralize its own exoskeleton for strength; and subsequently, over hundreds of millions of years, became itself totally mineralized, so that today it is available for $ 6.95 in a local rock shop.

1.1.3 Templates for Oriented Growth

I have chosen to emphasize one of the important lessons that can be learned from biomineralization and used by both of our speakers, Geoff Ozin and Thomas Bein, and that is the idea of constructing appropriate templates for the oriented growth of inorganic materials. Bein showed us how an anionic phosphate patch can be prepared by attachment of thiol functionalized phosphonates to gold surfaces. These can then be layered with zirconium ions, and organic diphosphonates to prepare a phosphate functionalized substrate for the oriented growth of crystalline aluminophosphates. My own research group has used this approach to modify gold surfaces with "stand up" monolayers of dithiols for the subsequent attachment of crystalline gold clusters to the exposed thiol surface. Yet another area where template synthesis has been cleverly used in the synthesis of zeolites involves micellar templates. It is possible to prepare rod-like micelles where the tail groups point to the center of the structures. These micelles then present a cationic or anionic surface which is mineralized by different soluble inorganic ions. This is an approach to the synthesis of MCM materials. Mesoporous materials with parallel channel type structures can be imagined to result from the rod like micelles aligning with one another as mineralization is occurring. So, in this way we can use soluble molecular components to prepare tubular micelle assemblies. The micelle assemblies are mineralized to prepare zeolitic materials. There is an interesting symmetry to the fact that once the organic template synthesis has been used to achieve the synthesis of a mesoporous solid state inorganic material, the solid state inorganic materials may themselves be used as rigid templates for the synthesis of molecules and polymers.

2. Discussion

We begin with a question posed to Geoff Ozin.

2.1 SYNTHESIS

2.1.1 Reproducibility

BANASZAK HOLL (Michigan): You neglected to mention the reproducibility or yields of your syntheses. Are they reproducible?

OZIN: It's highly reproducible. By the way the one student who did all the work was Scott Oliver. There are only three forms. I didn't get a chance to show you the third form. They are plates which are not patterned. For good reasons, there is no surface curvature. There are spheres which are surface patterned and have a lamellar bulk structure. They are completely reproducible. They are so large that you can pick them up and put them onto the stage of an optical microscope, and you can study them by polarized optical microscopy techniques. The third morphology is egg-shaped. I didn't show you our eggs. Basically, we've made complete eggs, and we can control the holes within those shells. It is a reproducible system. I don't see any limitations at this stage. I know I didn't give you the details on the compositions. But the template is special in many ways. You really just have to design the right type of inorganic deposition chemistry. I know the word deposition is vague, because it can lead to an ionic type of material, or it can be a more covalent type of material. I was very worried at the beginning about the reproducibility. The challenge is now control. I mean it's not under control. Maybe nature's biomineralization wasn't under control at the beginning, and we have evolved from this to give you something like that green quilt. It's incredible. It's an inorganic material. It's not a polymer. So you can imagine what you would achieve if you can start controlling that. These are also layer by layer processes. It's repetitive processing the way that you create the system. Just think of the computer chip, and the way you can fabricate that. I think that with repetitive processing at the macro, meso, microscopic scale you have an opportunity to do incredible things. I think that the quick answer to your question is that we have reproducibility, and not a random process. We have a process that seems to involve a very small number of template objects.

2.1.2 Reproducibility vs. Control

WEGNER (Mainz): Along the same lines. You showed just for a very short time the very complex phase diagram you that have in these effective water systems. Now if you run chemistry, in this system you are going to change the phase state throughout the whole reaction. That brings back the question: what is the difference between reproducibility and control?

OZIN: The question on reproducibility is: can I make the same egg shapes with the same dimension and the same holes. I'm talking about 100,000 Å features in some of these patterns, so it's not like the honey combs I was showing you where I can talk in terms of Å. It's not like a normal diffraction pattern where I can talk with crystallographic precision. So "reproducible" in terms of macroscopic form, reproducibility in terms of mesoscopic form, which we have. These chemically sculpted forms all diffract equally well. They are only made, the ones I showed you, of a single phase. It's a mesolamellar inorganic material. Now "control", for example, comes into play if I wanted to have monodispersed vesicles. On the other hand, if you look at biogenic opal, you have a gradient of sub-micron dimension spheres. You don't have to end up with a close-packed uniform structure like sodium chloride. You end up with a functional gradient. So control in our system means being able to make only the quilt, or make only the eggs filled with holes.

WEGNER: Where are you on the phase diagram?

OZIN: That's a good question. In this whole area of supramolecular templating using the Ringsdorf type paradigm you can see that the synthesis system is not under control. Since Charles Kresge discovered "micelle templating of inorganic mesostructures", the whole world has been trying to figure out how that works. No one has convinced me yet how it works. So where are we? We're way behind Charles Kresge. We are now on a scale that is taking us to millimeters, hundreds of thousands of Å, and all the way down. So, I'm at the flight of the first airline, writing the first Ph.D. dissertation, trying to get a paper into *Nature* (and it looks like it will go in). So that's where I am. I'm at the first flight. It's not under control. But it is reproducible.

2.1.3 Biological Templates

SEEMAN (New York): This seems to be a terrific extension of the work originally started by Harting about 120 years ago.

OZIN: I don't know that work. What's it about?

SEEMAN: He was trying to do much the same thing, generate a synthetic part of the morphology of a biological system. Have you tried to put in small proteins or other things of a well defined nature to really get control.

OZIN: There are lots of proposals out there on biological templates. The natural objects I'm showing you are not templated by the synthetic amphiphiles that I use, but the message here is so clear in terms of the types of biological templates to use. At the University of Toronto we have some Centers of Excellence, one of which is protein engineering and crystallography. A program on biological templates for inorganic materials synthesis would be very interesting. It's definitely the future.

2.1.4 Bigger Templates

CHIDSEY (Stanford): I know within the context of self assembly and modular chemistry there is a lot of interest in using Å level objects to control even macroscopic shapes. Biology actually goes through the steps of building cells that have certain sizes, shapes, and orientations with respect to each other. Cells have special connectivities to make an organism. As a short cut to that in the near term, materials scientists might want to take advantage of a kind of top down strategy by combining larger scale templating with chemical mineralizations and the themes that are coming up here. I wonder if harder materials, still soft materials but not as soft as detergents, things like latex spheres, strands, sheets, might not be good templates for building other more complex structures. Has that field gone beyond just the simple composites like cement and polymer/ceramics?

OZIN: I think that is my plaster cast analogy. You break your arm. You get a plaster cast. You take it off and you have a replica of your arm.

CHIDSEY: I'm not talking about things on the micron scale that might have interesting mechanical properties by virtue of implanting mechanics.

OZIN: A latex sphere is fine. That sort of technology is under control. Surface functionalization of latex spheres, self assembly, and inorganic deposition. People have already been using arrays of latex spheres on planar substrates as lithographic masks. You just organize latex spheres and then you deposit semiconductors through them and you get a lovely pattern underneath which is a replica of that. I think that the concept of organizing modular objects on different length scales and creating topological replicas is so appealing. It's just a matter of controlling it.

2.1.5 Combinatorial Synthesis

MORTIMER (UCLA): I was wondering how sensitive your creations are to the exact conditions under which you make them, and whether you could use a combinatorial approach the prepare new structures.

OZIN: We are in the design stage. I didn't get the exact details in, but the choice of the cosurfactant, the nature of head groups, and the compatibility with the surface template are critical. If I'm on track with the concept of phase separated domains in the surface of the vesicle template, then it is behaving like an ion-channel for the transfer of inorganic precursors to the site of deposition. It turns out that I have to use a specific cosurfactant otherwise you don't get the structures that I showed. It is very sensitive to that choice. Consider the point about shifting the phase diagram, for example. Galen Stucky, who has done some wonderful work, following on from Kresge's discovery, is creating new phase diagrams. So, in the silica system, he uses words like silicatropic, and then you look at that and ask: what's the difference? We've gone from chloride as the counterion to slightly more exotic polyanionic inorganics. Once you put in something with, say, an 8^- charge, and it's a silicate cluster, then the game has changed. So what's templating what? There are a lot of words out there that people are using like "facilitated assembly", "prescribed assembly", "contingent assembly". I know they're all words, but it is likely a co-assembly or cooperative assembly process. There's a big jump from a Ringsdorf diagram to putting in an inorganic that is undergoing a deposition process at the interface between water and the organic template. There is a world of difference when trying to make some bone analogs. I'm part of the bone mineral group at the University of Toronto. Extending this to Ca^{2+} and various other interesting anions the phase diagram is completely different. The construction you are trying to make is ionically bound. So the sort of things Thomas Bein was talking about come into play. The organization of cations and anions, lattice, charge and stereochemical matching considerations are most important. That's going to be completely different compared to the shaping of silica.

2.1.6 What Have You Made?

YAGHI (Arizona State): Regarding the composition of your materials, to what extent can you figure out their composition? Can you write a balanced chemical equation for your materials?

OZIN: We know the stoichiometry of the material. They are stoichiometric.

YAGHI: Do you have byproducts?

OZIN: The byproducts are in trace amounts. The diffraction pattern I showed you was in fact atypical. They are normally cleaner than that. The materials are phase pure. We have analyzed the inorganic and organic components of these materials. I know the thermal stability of the sytems. The thermal analysis has been done.

2.1.7 What Are You Really Stealing From Nature?

PRINZBACH (Freiburg): I have a problem with the idea. Did nature have an idea when it produced it?

OZIN: Oh I'd love you to be with me next week in Spain as some of the guru's in this field get together to talk about the origin of life. Somebody else asked me that type of question in terms of was it accidental that organism created a hard shell?

PRINZBACH: What are you stealing (from nature)? What are you really stealing?

OZIN: What am I stealing? I just saw a great quote which is: "good scientists borrow ideas, great ones steal them". What am I stealing? I'm not sure. I'm trying to figure out what I've stolen. But I think that you can see how mistakes are made, how little bits of luck come along, and how suddenly you get a hollow shell filled with holes, suddenly that thing is breathing, or suddenly it's feeding, and surviving. I hope that I'm lucky enough to be part of working on that question before I retire. I really hope I'm involved with that.

2.2 SURFACE CHEMISTRY

2.2.1 Oriented Growth of Zeolite Crystals

COLEMAN (Lyon): This is a question for Tom Bein. How flat are your QCM surfaces, in reality. Gold on the QCM, how flat is it? Do your crystals all tend to go all over the place?

BEIN: Most of our gold substrates are made by vapor deposition, and have roughness at the 10 - 100 nm scale. The astounding thing is that these crystals that I talked about, the aluminophosphates, are so oriented. Maybe you didn't see the scale. The crystals are about 50 μm in diameter. These are gigantic crystals. You can see them with your eye as fine columns coming out. So it seems as if the growth eventually orients the crystals, with respect to the surface, to that scale. We are now looking at single crystal gold surfaces, single crystal planes.

COLEMAN: Can you actually flatten out the gold on the QCM, will it expose the quartz crystal?

BEIN: We have used diamond cut and also polished flat surfaces, but even if you deposit some gold through evaporation onto this polished surface, which looks like a mirror, it has about a 30 nm undulation of height, pretty severe. So, the crystallization doesn't seem to be very sensitive to that for gold. On the other hand, our surface studies with these mixed layers indicate that the growth process is very sensitive to surface structure on a nanometer or Å scale at the start. Then it becomes less and less dependent on the overall structure as the crystals grow.

WARD (Minnesota): One of the pictures you showed of lateral orientation on the surface is really intriguing. I have two questions. First, can you identify a reasonable heteroepitaxy between that face on the zeolite and the gold surface?

BEIN: Not at this point. Because this was just Friday before I left. We haven't really analyzed this system yet, but as we speak I think it is under the AFM.

WARD: The second question concerns the fact that thiols will etch gold during deposition. There is a method called grapho-epitaxy that is used to grow oriented patterns on micromachined silicon. When you have grooves or features "scratched" into the silicon surface, crystals will grow inside those features with their low energy faces contacting the walls of the grooves. I was wondering if you could possibly have pits in the surface and you were seeing growth inside the pits and getting an orientation which is a replica of the gold etch pits.

BEIN: We do not have evidence for gold etching, and our data suggest that we have formed dense layers of thiol phosphonic acids after deposition. The etch growth idea would be interesting, but I don't see that these pits would all be aligned with a certain orientation.

WARD: The gold is somewhat oriented, right? It's probably mostly [111].

BEIN: Yes, it's [111].

CHIDSEY: I want to comment on exactly this point. I can understand that there is a [111] surface that promotes a certain growth. But in Bein's picture, over many microns, you had the walls of the columns or the triangular crystals all oriented along the same direction, and unless that was a single crystal of gold underneath there is really no reason why over a micron length scale gold should have the same lateral orientation. The grains are rarely even a few thousand Å, let alone a few μm.

BEIN: You are absolutely right, but this was a single crystal. It was on mica. It was annealed gold deposited on mica that you can scan first, and it has large terraces on it, single crystal [111].

WARD: So the steps should be oriented.

BEIN: Yes, those are things that now obviously we can look at. We tried to set the system up such that we can scan both the surface at a molecular resolution and these crystals as they grow up. It turns out that there are many constraints to consider, including gel formation, the viscosity of the liquid, all these kinds of things.

COLEMAN: If you're actually on mica, could it be a mica edge effect? They go for microns. Maybe it could just be mica that you are picking up there. That will orient [111].

BEIN: Yes. Let's not worry too much about this one picture. I wouldn't want to put my head on the block for it. I just found it very intriguing. (To Chris Chidsey): Chris, regarding your comment on the gold, if we have polycrystalline gold, like the very first view graph that I showed, there we don't have this level of lateral crystal orientation. So, there seems to be a correlation. It's too early to say so much.

2.2.2 Sensors

TOLBERT (Georgia Tech): Since we're supposed to speculate, have you speculated about the sensors? If you are going to design chemical sensors that presumably rely on different zeolites on the same surface, and for which you will use microfabrication to address the various sensors present on the surface, the question is: what about the selectivity of different surfaces for various kinds of zeolites, and how you would achieve that goal?

BEIN: That's a very nice point. You don't have to get to that stage of control to actually have a sensor array, because, as I said, we have done most of our sensor studies on the molecularly attached, zeolite films, and they really work beautifully. We have measured response ratios above 100, sometimes even 500:1 of molecules that only adsorb on the zeolite surface and that go into the volume of the crystal. To create a selective sensor this goal of selective surface growth is not really necessary. What you can do is basically deposit certain little patches on the appropriate sensors, such as arrays of surfactants, to make chemically sensitive surfaces regions between the transducer pads. That would be one way. An even more primitive way would be to just buy ten QCMs and build them into a little tube, which really would not be a nanotube, but a macrotube, and then flow your gases through that. The logic would be the same, it's just less elegant and less fast. But, it works. We have already nanostructures for many hydrocarbons and gases. Slow equilibration can be an issue for very low partial pressures.

2.2.3 Epitaxy or Concentration Gradient?

WEGNER: Do you have any information on the chemistry of the interface between the growing crystal and the substrate? This question touches on the question of epitaxy versus a concentration gradient in a nucleation center.

BEIN: Well, that's a nice point that we are also addressing. Initially we have looked at phosphate crystals, zinc phosphate and aluminum phosphate crystals. So, at one step, there's always a phosphate terminating face in this growth process. Now if you present a phosphate terminated face on this organic layer and also allow for the slight flexibility of these chains (although AFM seems to indicate really nice order) then you have the connection. What you assume is that the next layer on this phosphate would be the metal, or the positively charged template (for Coulomb attractive reasons), and then the attachment of the next phosphate layer would occur. To test that we have looked at the bulk synthesis, normal bulk hydrothermal synthesis, and injected small alkyl phosphonic acids into the solution, and what we found, to our delight, was complete quenching of further crystal growth. The X-ray diffraction pattern stops at that level of peak height and peak width, and amorphous content. You've basically blocked crystal growth. What that means is that these phosphonic acids have a very strong adhesion, chemical bonding, apparently to all of the crystal faces, that contribute to further crystal growth. Because the phosphonic acids in this experiment are obviously molecular, they are very flexible to adapt to different local geometries. So there should also be a chemical bond between molecular phosphonic acid (on the surface) and the zeolite nucleus and growing crystal on the surface.

WEGNER: If I understand this correctly, you say that the nuclei which are formed in solution attract at the surface. It's not as the surface induces crystallization by surface nucleation.

BEIN: Yes, I think so. I mean the more attractive picture would be a higher level of epitaxial control. Like if you had a grid that looked like one of the terminating faces of the zeolite crystal, and then the fluid system, being as smart as we think they are, would just know that this is a crystal face and build on it. However, we don't even know about the mechanism of bulk zeolite crystallization, and that's actually something we hope to address in this *in situ* study. What are the species coming out of solution and building up the zeolite crystal? That is not known. That is very difficult to study, because of the corrosive high temperature conditions for the typical hydrothermal, high pressure synthesis in a Teflon autoclave. I think that this room temperature synthesis on a surface might be the first chance to study this.

2.3 DESIGN/APPLICATION: WHAT DO YOU WANT?

2.3.1 Strong or Weak Bonds?

KUKI (Alanex): We've had a lot of discussions about forming two-dimensional arrangements, advantages or disadvantages of strong covalent bonds, or weaker interactions. The alkane thiol approach that they use, or Tom Mallouk's ideas, seems to bring up a general point. Maybe it is obvious in retrospect, that you have this sort of dichotomy, if I understand what Chris Chidsey said. The weak interactions allow annealing, but they are not as stable. Well, here is the case where you have the weak interaction of the alkane thiol on gold, and then the zirconium crosslinks. The crosslinking can be done after some annealing. If that is the case, it seems to me you have the best of both worlds. Maybe what one is seeing, case by case, different kinds of chemistry, totally different approaches, making nets, is a rediscovery of this idea, the pedestal idea. Use weak interactions to allow annealing of two-dimensional arrangements, first, and then toggle in stronger bonding.

BEIN: Yes, I think so. Are you talking to me?

KUKI: Well it was your example that brought it up.

2.3.2 Really Weak Bonds

OZIN: Can I just throw something in? This goes back to what would seem to be an even weaker, more complex situation where you'd love to have some control, and that is simply creating a Langmuir-Blodgett film on a water surface. So, again by compressing and creating an ordered structure, you have a certain distance between the head groups, and certain charge, and you can inject inorganics into that aqueous solution, like Thomas did on a solid surface. You can control nucleation and growth, and only certain faces grow. Fendler, Mann, loads of people have studied this; only those crystals grow whose constituents, which might be cationic or anionic, interact with the film. So if you put down a cationic Langmuir-Blodgett film, the anions go in first, and the way the anions order, and the way the cations follow, and the way the crystal grows, it's all part of this same game. It's what I'm doing as well. We're starting off with, quite often, a two-dimensional template. It can be curved, obviously. This is important to understand. It doesn't have to be a rigid geometrically well-defined solid surface. It can be a more dynamic and flexible liquid surface.

KUKI: Right. It might seem to be a natural next step for people building large aromatic, very beautiful, covalent structures to extend that concept. The next level of organization might be best done with something softer, which could then be toggled into covalent bonding.

OZIN: The people in the biomineralization area have been talking about this since the beginning of the field.

2.3.3 Porosity, But Accessible Porosity?

PORT (Stuttgart): (To Thomas Bein) Thomas, what is the present status of getting molecules inside your large structures.

BEIN: I could say two things. You can get anything into the channels that can diffuse in, and has the driving force to do so. We are working a lot with organometallics as catalysts. We try to anchor them at specific sites. Geoff (Ozin) has also worked a lot in this area. So, metal containing molecules can diffuse in. You can react them inside, transform them to others. If your question was referring to the "wires", was that the point?

PORT: It was just a general question, and how you keep controlling getting things in each of those tubes.

BEIN: The thermodynamic description of this process is the adsorption isotherm, and the adsorption isotherm shows you that eventually you fill up all of the pores. You have a saturation situation when all of the pores are filled, and then you have communication with the surface. So that is not a problem if you fill up all of the pores.

PORT: Can you really check on that? Because I remember a lot of work on zeolites, and there people could not really get control of what was finally in the pores.

BEIN: I will tell you about one example, and that is infrared spectroscopy using a proton that is in the zeolite as the probe. If you look at the protons that are structurally associated with the zeolite, Bronsted acidity, and you take a basic molecule, like an amine, you can see that all of these protons react with the amine. The original vibration totally disappears, and is replaced by a shifted one. So you know that you have reached the interior of the zeolite crystal. That's one example. I am really not worried about this issue. Diffusion is possible and it is obviously necessary for catalysis, for example.

OZIN: Can I make a quick comment on that? I agree with Thomas. Getting things inside the zeolite, that's not the problem anymore. The concept is great, of creating these perfectly ordered structures, and you can see that we are going to go way past 100 Å in length scales. We are going to be in a really exciting regime, in terms of size and properties. That's not the problem. We know how to fill these porous structures. The problem is making thin films and growing large high quality crystals. That's the problem with the field, and what you are seeing there with Thomas and others, is that this is going to happen.

2.3.4 Strong or Weak Bonds? Depends on What You Want

CHIDSEY: I'd like to comment. The point that I want to make I think is related to what Atsuo (Kuki) was saying. That is that in the zirconium system, it's not as if it makes these nice hard bonds, which are a regular lattice. I think Tom (Mallouk), if he still were here would support that. It's more a glassy kind of situation. It's more like the plaster cast with silicate, than it is like making epitaxial zeolites, or calcium carbonates or phosphates, and it raises a really interesting point, which is: what do you want? Do you want a kind of glassy, continuous film, that may have better properties because it doesn't have any well defined grain boundaries in it; or do you want nice separate single crystal columns which are going to have all of the advantages of crystalline zeolites, but then you have to deal with how you stitch them together. I don't think it matters whether we are talking about zeolites or calcium carbonate, probably calcium carbonate materials are less mechanically robust in some ways than the silica materials {hoots of disapproval}. Well, I don't know for sure, because they don't have all the grain boundaries. It's application specific obviously, but I don't think that it's right to think that you can have both easily. Maybe you can grow single crystals that cover a macroscopic domain, but I think that it is much more likely that we are going to see one or the other of these two paradigms.

2.3.5 Nature's Way

OZIN: I hate to sound like Nature Boy here. I keep going back to nature, but nature knows how to take care of all of those problems. Many of nature's materials are combinations of amorphous regions and crystalline regions, combinations of organics and inorganics, and nature's structures work. She throws away the ones that don't work. If I am going to control the way in which shell cracks, and the way the cracks propagate, I know exactly what I have to do. So, I grow calcium carbonate crystals in a particular orientation. I alternate with a little bit of protein, just the right amount, the protein and the inorganic. I do it in a repetitive fashion, and I create one of the strongest shells around. You are absolutely right. Nature's interfaces are nature's way of doing epitaxy, and nature's way of doing epitaxy is charge matching, stereochemical and geometrical matching. Remember nature is not interested in optics and electronics.

CHIDSEY: That's just my point. It all depends on what you want to get.

OZIN: If nature were interested in creating a semiconductor superlattice, I'll bet you she'd do it as well. She's not interested in that.

2.3.6 Calcium Phosphates

COLEMAN: Just a point of clarification there. Have you actually got a calcium phosphate system that works? Because they are a pain in the butt normally because they are not very soluble.

OZIN: Calcium phosphate is much more of a challenge than any of the silica based systems, and there are many more phases.

COLEMAN: Well, hydroxyapatite.

OZIN: You don't just have to go after calcium hydroxyapatite. There are many materials that will form nice interfaces. I think that you have got to get control over the different phases. You are dealing with an electrostatic epitaxy problem in that case. It's more challenging, but it can be done.

COLEMAN: Have you done it?

OZIN: It's not under control in the way that you would like to see it done. But, yes, it's a great opportunity, definitely. We've got a paper coming out in *Supramolecular Chemistry*, which shows how to do mesoscopic templating of semiconductors. In that paper is the first example of a synthetic mesoscopic semiconductor. So, if you can do that you can do calcium phosphate and silicas. The whole field is open. Nature uses around 60 biominerals. I love this statement: Nature uses the most boring of inorganic materials to make the most exquisite structures. Chemists tend to use very complex starting materials, to make very boring structures, relative to nature. There is a difference in the whole way of thinking and I think it is coming together now, as we mix all this stuff up.

STUPP (Illinois): I just wanted to comment on the hydroxyapatite, because that's a field that I worked on. There are examples that we have found, for example poly(lysine) or poly(glutamic acid), where it's possible to nucleate large single crystals all of the apatite with the lysine and then you go to a more mesoscopic structure when you change the base to glutamic or poly(glutamic acid). So it is possible to manipulate the hydroxyapatite system, even though I agree that it's much more complex. Again, the interaction between the organic chain and the precursor ions controls the growth rates on different faces and you can definitely get mesoscopic or single crystal structures over large distances.

OZIN: That's a big field. There is even a Gordon Conference called "Calcium Phosphates". When you first got the zeolites, you find there are thousands of patents. The calcium phosphate field also has thousands of papers and patents. People have been working on this for a long time. The challenge is that it is hard to make a contribution to the bone biomineral field using our synthetic approach.

3. Conclusions

While there is much to be learned from an appreciation of nature's elegant simplicity for achieving the highest levels of form and function from the simplest materials, our discussion underscored the idea that the way we make things should depend on the end-use application. The chemical modification of surfaces for the oriented growth of crystalline mesoporous materials is promising. However, in terms of finding the best templates to achieve specific solid state syntheses, in general, nature is still very hard to beat. The discussion also echoed a theme heard throughout the conference: "soft" chemical interactions are useful for achieving order and annealing out mistakes; "hard" chemical interactions are useful for achieving permanence. The synthesis of complex modular solid state structures will likely require a graduated scale up of "harder" chemical interactions as assembly progresses.

4. References

1. Mann, S. (1993) Molecular tectonics in biomineralization and biomimetic materials chemistry, *Nature*, **365**, 499-505.
2. Mann, S., Archibald, D. D., Didymus, J. M., Douglas, T., Heywood, B. R., Meldrum, F.C., Reeves, N. J. (1993) Crystallization at inorganic-organic interfaces: biominerals and biomimetic synthesis, *Science*, **261**, 1286-1292.

DESIGN AND SYNTHESIS OF MACROMOLECULAR SYSTEMS CONSISTING OF CYCLODEXTRINS AND POLYMERS

AKIRA HARADA
Department of Macromolecular Science, Faculty of Science, Osaka University, Toyonaka, Osaka 560, Japan

ABSTRACT. Cyclodextrins (CDs) were found to form inclusion complexes with various polymers with high specificity to give stoichiometric compounds. In these complexes, polymer chains were threaded into cyclodextrins. The structures and properties of the complexes have been studied by spectroscopic methods. Polyrotaxanes were prepared by capping the chain ends with bulky groups.

1. Introduction

In recent years, much attention has been focused on molecular recognition of low molecular weight compounds in the field of biomimetic and supramolecular chemistry. Crown ethers, cryptands, and cyclophanes have been extensively used as host molecules. However, their guests have been limited to simple ions and small molecules. Host molecules which can recognize and respond to larger and more complicated compounds and even polymers are required.

In biological systems such as enzyme-substrates, enzyme-inhibitors, antigen-antibodies, DNA, RNA, and cell adhesion systems, "macromolecular recognition", recognition of macromolecules by macromolecules, plays an important role in constructing supramolecular structures and maintaining life. However, there have been no approaches toward macromolecular recognition by artificial host-guest systems.

Since cyclodextrins were discovered, a large number of inclusion complexes of cyclodextrins with low molecular weight compounds have been prepared and characterized [1]. However, there were no reports on the formation of inclusion complexes of cyclodextrins with polymers when we started our project. We found that cyclodextrins form complexes with various polymers with high selectivities in the crystalline state [2-4]. The design and synthesis of supramolecular architectures through macromolecular recognition by cyclodextrins are described in this article.

2. Complexes of Cyclodextrins with Hydrophilic Polymers

J. Michl (ed.), Modular Chemistry, 361–370.

Aqueous solutions of some nonionic polymers were added to saturated aqueous solutions of cyclodextrins to see whether the cyclodextrins would form complexes with these polymers. Table 1 shows the results of complex formation between cyclodextrins and some nonionic water-soluble polymers.

We found that cyclodextrins did not form complexes with some nonionic water-soluble polymers such as poly(vinyl alcohol) (PVA) and poly(acrylamide) (PAAm) by the same procedure as that for low molecular weight compounds. However, we found that although β-CD does not form complexes with PEG of any molecular weight, α-CD forms complexes with poly(ethylene glycol)(PEG) of various molecular weights to give crystalline compounds in high yields [5]. However, β-CD formed complexes with poly(propylene glycol)(PPG) to give stoichiometric compounds in high yields [6], although α-CD did not form complexes with PPG of any molecular weights. γ-CD formed complexes with poly(methyl vinyl ether) (PMVE), though α- and β-CD did not form complexes with PMVE [7].

TABLE 1. Complex formation of cyclodextrins with hydrophilic polymers

Polymer	Structure	Mn	Yield(%)		
			α-CD	β-CD	γ-CD
PVA	$-(CH_2CH(OH))-$	22 000	0	0	0
PAAm	$-(CH_2CH(CONH_2))-$	10 000	0	0	0
PEG	$-(CH_2CH_2O)-$	1 000	92	0	trace
PPG	$-(CH_2CH(CH_3)O)-$	1 000	0	96	80
PMVE	$-(CH_2CH(OCH_3))-$	2 000	0	0	84

2.1. COMPLEX FORMATION BETWEEN α-CD AND POLY(ETHYLENE GLYCOL)

When PEGs were added to a saturated aqueous solution of α-CD at room temperature, the solution became turbid and the complexes were formed as crystalline precipitates when the number average molecular weight (Mn) of PEG was higher than 200.

The rate of complex formation depends on the molecular weight of PEG. PEG of molecular weight 1000 forms complexes most rapidly. This observation may be due to the fact that the number of end groups decreases as the molecular weight increases. Addition of the PEG to a saturated aqueous solution of β-CD did not cause any changes in solution.

The complexes were isolated by filtration and centrifugation. Figure 1 shows the yields of the complexes of α-CD with PEG as a function of the molecular weight of PEG. The yields are determined on the basis of 2:1 (ethylene glycol unit : α-CD) stoichiometry, as described in the following section. α-CD did not form complexes with the low molecular weight analogs, ethylene glycol, diethylene glycol, and triethylene glycol. α-CD formed complexes with PEG of molecular weight higher than 200. The yields increased with increasing molecular weight and reached saturation at a molecular weight of about 800.

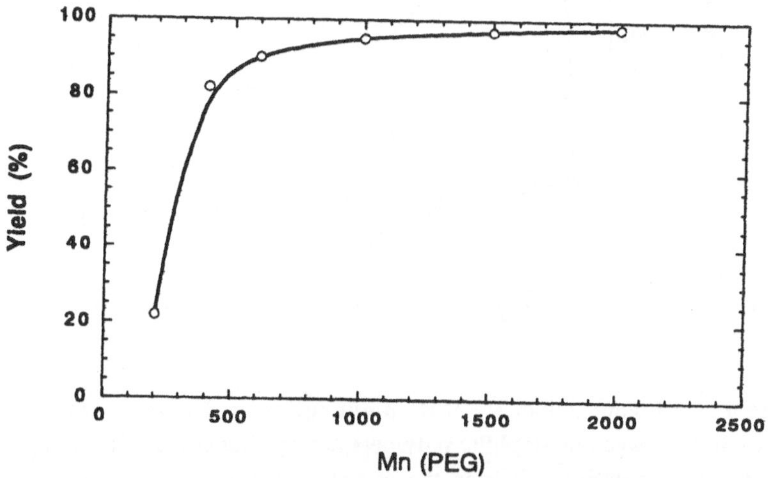

Figure 1. Yields of complexes of α-CD with PEG as a function of the MW of PEG.

2.2. COMPLEX FORMATION BETWEEN CD AND POLY(PROPYLENE GLYCOL)

β-CD did not form complexes with PEG of any molecular weight. However, β-CD formed complexes with PPG with molecular weights higher than 400. Figure 2 shows the yields of the complexes of β-CD with PPG as a function of the MW of PPG. β-CD did not form complexes with the low molecular weight analogs, such as propylene glycol, dipropylene glycol, and tripropylene glycol. β-CD formed complexes with PPG of molecular weight higher than 400. Yields of the complexes increase with increasing molecular weight. The complexes were obtained almost quantitatively between β-CD and PPG with molecular weight 1000. Then, the yields decrease with increase in molecular weight. This result is different from that of the complex formation between α-CD and PEG. This observation may be due to the fact that PPG is more hydrophobic than PEG owing to the methyl group of the main chain. γ-CD also formed complexes with PPG, even with PPG of low molecular weight. α-CD did not form complexes with PPG of any molecular weight. An α-CD cavity is too small for PPG to penetrate on account of the steric hindrance by methyl groups on the main chain.

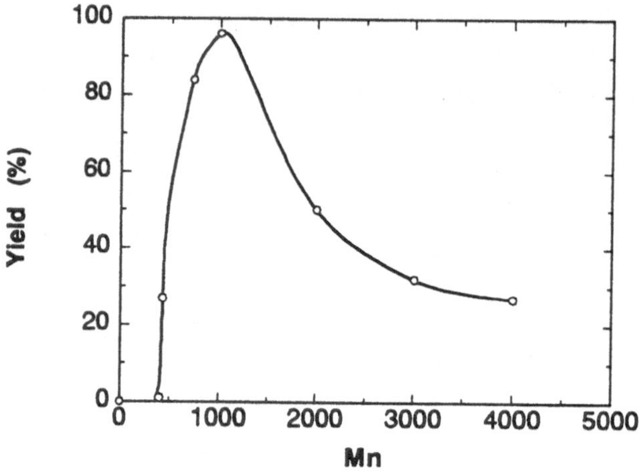

Figure 2. Yields of the complex of β-CD with PPG as a function of the MW of PPG.

Figure 1 and Figure 2 indicate that minimum chain length is required for the complex formation between hydrophilic polymers and cyclodextrins. These results show the importance of cooperative effects in the complex formation. The cooperativity is thought to result from the fact that a single polymer chain has many binding sites which are included by cyclodextrin molecules.

The neighboring CDs bound on a polymer chain interact with each other by hydrogen bonding. This view is consistent with the fact that PPG does not form crystalline complexes with 2,6-di-*O*-methyl-β-CD, 2,3,6-tri-*O*-methyl-β-CD, and the water-soluble β-CD polymer. These compounds are thought to be unable to include a PPG chain to form crystalline complexes, since they cannot form hydrogen bonds because of the lack of hydroxyl groups.

2.3. STOICHIOMETRIES OF THE COMPLEXES

The continuous variation plots for the complex formation between α-CD and PEG show that the stoichiometries of the complexes are 2:1 (ethylene glycol unit : CD). The stoichiometries were confirmed by ^{1}H NMR spectroscopy. The comparison of the integrals for the peaks of α-CD with those of PEG shows that a single α-CD binds two ethylene glycol units of PEG. It is notable that the stoichiometries of the complexes are always 2:1, even if α-CD and PEG are mixed in another ratio. The length of two ethylene glycol units corresponds to the depth of the cavity of α-CD (Figure 3).

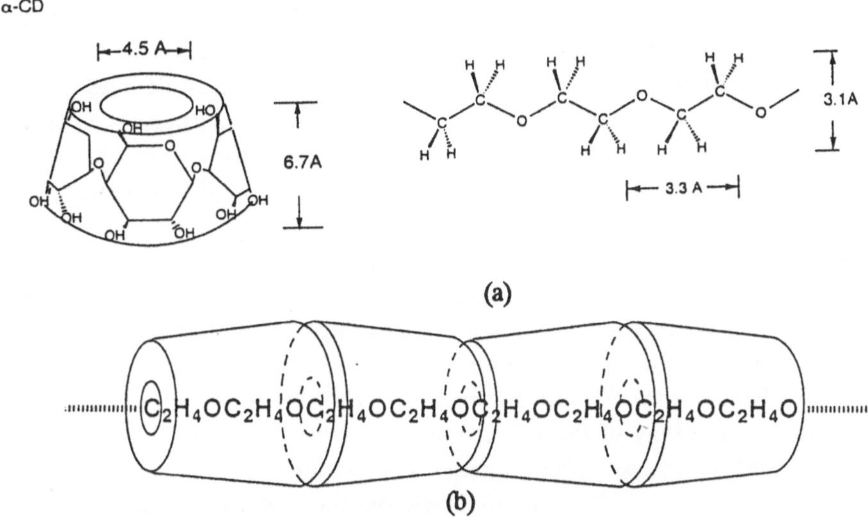

Figure 3. Dimensions of α-CD and PEG (a) and the proposed structure of the complex of α-CD with PEG (b)

2.4. PROPERTIES OF THE COMPLEXES

The complexes of α-CD with PEG of low molecular weight are soluble in water. The complexes of PEG of higher molecular weight can be dissolved in water by heating. The addition of low molecular weight guests, such as benzoic acid and propanol, to the suspension of the complex resulted in solubilization of the complex when the molecular weight of PEG was low. The complex formation is reversible. Although addition of salts, such as NaCl and KCl, did not cause any change in the solubility of the complexes, the addition of urea results in solubilization of the complexes. These results indicate that although ionic interactions between α-CD and the polymer are not important, hydrogen bonding formation plays an important role in forming the complexes between PEG and α-CD.

The decomposition points of the complexes are a little higher than that of the cyclodextrin. The complex of α-CD with PEG-1000 decomposes above 300 °C, whereas α-CD melts and decomposes below 300 °C. The PEG chain stabilizes α-CD.

2.5. STRUCTURE OF THE COMPLEXES

The X-ray powder patterns of the α-CD-PEG complexes show that the complexes are crystalline and the patterns are very similar to those of the complexes of α-CD with valeric acid or octanol, which have been reported to have an extended columnas structure, and are totally different from those of the complexes with small molecules such as acetic acid and propanol, which have a cage-type structure. These results indicate that the complexes of α-CD and PEG are isomorphous with those of the channel-type structure rather than the cage-type-structure.

Molecular model studies show that a PEG chain is able to penetrate α-CD cavities, while a PPG chain cannot pass through the α-CD cavity. These views are in accordance with our observation that α-CD forms complexes with PEG but not with PPG. β-CD did not form complexes with PEG. A PEG chain is too slim to fit in the β-CD cavity. However, β-CD forms complexes with PPG. Model studies indicate further that the single cavity of the CD accommodates two ethylene glycol units when ethylene glycol chain assumes a planar zigzag conformation.

Figure 4 shows the ^{13}C CP/MAS NMR spectra of α-CD and the α-CD-PEG complex. α-CD assumes a less symmetrical conformation in the crystal when it does not include a guest in the cavity. In this case, the spectrum shows resolved C-1 and C-4 resonances from each of the six α-1,4-linked glucose residues. In particular, C-1 and C-4 adjacent to a conformationally strained glycosidic linkage are observed at 80 and 98 ppm, respectively. In the spectrum of the α-CD-PEG complex, however, the peaks at 80 and 98 ppm disappeared. Each carbon of glucose can be observed as a single peak. These results indicate that α-CD adopts a symmetrical conformation and each glucose unit of the CD is in a similar environment. The X-ray studies of single crystals showed that although α-CD assumes a less symmetrical conformation when it does not include guests in the cavity it adopts a symmetrical conformation when it includes a guest in its cavities. The CP/MAS NMR spectra of complexes and uncomplexes CDs are consistent with the results on X-ray structural studies.

2.6. COMPLEX FORMATION BETWEEN CD AND PEG DERIVATIVES

PEGs with small end groups, such as methyl and amino groups, form complexes with α-CD. However, PEGs having bulky groups at either end of the PEG, which do not fit or pass through the α-CD cavity, such as 3,5-dinitrobenzoyl and 2,4-dinitrophenyl groups, did not form any complexes with α-CD. PEG having a large group at one end but a small group at the other end formed a complex with α-CD. These results indicate that α-CD includes a PEG chain from a small end group. Figure 3 (b) shows a proposed structure of the complex of PEG with α-CD. The inclusion complex formation of PEG in the α-CD channel is entropically unfavorable, However, formation of the complexes is thought to be promoted by hydrogen bond formation between neighboring cyclodextrins. Therefore, head-to-head and tail-to-tail arrangements are thought to be the most probable structures.

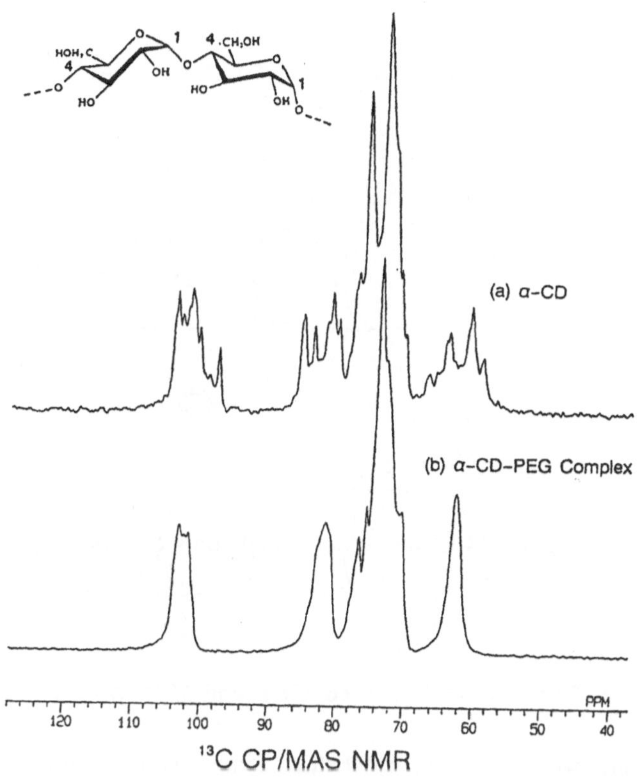

Figure 4. ^{13}C CP/MAS NMR spectra of α-CD (a: α-CD assumes a less symmetrical conformation] and the α-CD-PEG complex (b: α-CD adopts a symmetrical structure.)

3. Polyrotaxanes

We found that although α-CD formed complexes with PEG having small end groups, such as methyl and amino groups, it did not form complexes with PEG carrying large end groups, such as 2,4-dinitrophenyl groups and 3,5-dinitrobenzoyl groups. Therefore, we designed and synthesized polyrotaxanes in which many cyclodextrins are threaded on to a polymer chain by capping the polymer chain ends with bulky substituents as shown in Scheme 1. The inclusion complex of α-cyclodextrin with PEG-bisamine was treated with an excess of 2,4-dinitrofluorobenzene, which is bulky enough to prevent dethreading, in dimethylformamide [8-12] We obtained polyrotaxanes containing more than 100 cyclodextrins.

Scheme 1.

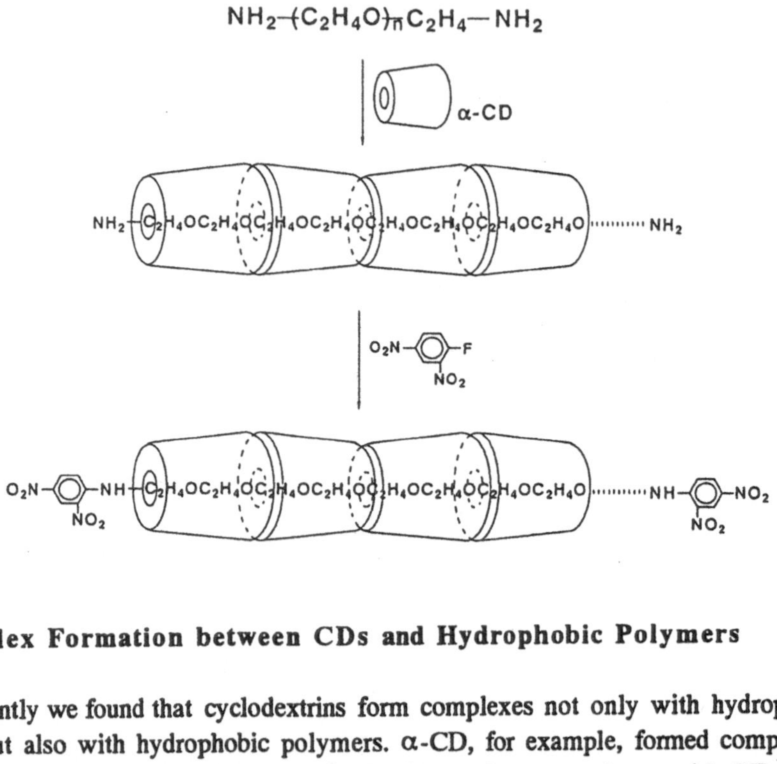

$$NH_2\text{-}(C_2H_4O)_n C_2H_4\text{-}NH_2$$

α-CD

NH_2–$C_2H_4OC_2H_4OC_2H_4OC_2H_4OC_2H_4OC_2H_4OC_2H_4OC_2H_4O$ ⋯⋯ NH_2

O_2N–C₆H₃(NO_2)–F

O_2N–C₆H₃(NO_2)–NH–$C_2H_4OC_2H_4OC_2H_4OC_2H_4OC_2H_4OC_2H_4OC_2H_4OC_2H_4O$ ⋯⋯ NH–C₆H₃(NO_2)–NO_2

4. Complex Formation between CDs and Hydrophobic Polymers

Recently we found that cyclodextrins form complexes not only with hydrophilic polymers but also with hydrophobic polymers. α-CD, for example, formed complexes with oligoethylenes (OE) efficiently, but β-CD did not form complexes with OE.[12] In contrast, although α-CD did not form complexes with PIB [13] (Figure 5), γ-CD formed complexes with poly(isobutylene)(PIB) efficiently.

OE was found to form inclusion complexes with α-CD not only from aqueous solutions of α-CD but also from DMF solutions of α-CD to give stoichiometric compounds in a crystalline state. The stoichiometries of the complexes are 3:1 (ethylene unit : α-CD). X-Ray diffraction studies and ^{13}C CP/MAS NMR spectra suggest that the OE chain is included in α-CD and the OE backbone in the complexes is more flexible than that in uncomplexed state.

α-CD did not form complexes with PIB of any molecular weight. β-CD and γ-CD formed complexes with PIB. The yields of the complexes with β-CD decreased with increase in the molecular weight of PIB. In contrast, the yields of the complexes with γ-CD increased with increase in the molecular weight and the complexes were obtained almost quantitatively with PIB of molecular weight of 1000. The chain length selectivity is totally reversed between β-CD and γ-CD.

$$\left(-CH_2-C(CH_3)_2- \right)_n$$

Polyisobutylene(PIB)

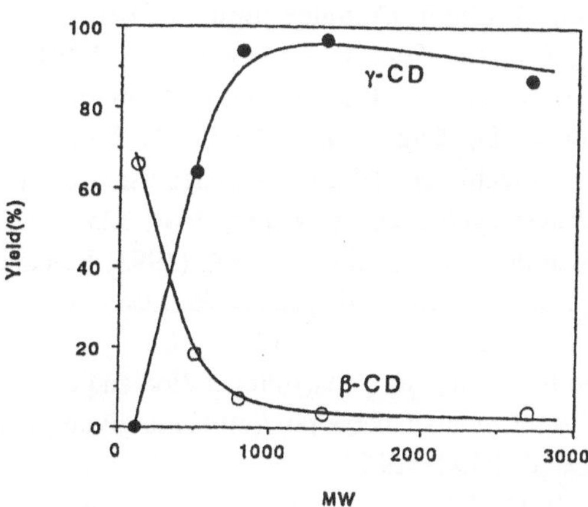

Figure 5. Yields of complexes of PIB with β-CD and γ-CD as a function of MW of PIB

5. Conclusion

Cyclodextrins were found to form inclusion complexes not only with low molecular weight compounds but with hydrophilic and hydrophobic polymers to give stoichiometric compounds. The selectivities shown by cyclodextrins toward polymers are much higher than for low molecular weight compounds. This is due to the fact that the guest polymers have a lot of recognition sites and is one of the reasons why living systems are composed of may kinds of macromolecules. This kind of complex formation can be utilized to create new supramolecular architectures and functions [14,15].

6. References

1. Bender, M. L. and Komiyama, M. (1978) *Cyclodextrin Chemistry*, Springer, Berlin.
2. Harada, A. (1993) Macromolecular recognition: inclusion complexes of polymers with cyclodextrins and preparation of polyrotaxanes, *Polymer News*, **18**, 358-363.
3. Harada, A., Li, J., and Kamachi, M. (1993) Macromolecular recognition, Formation of inclusion complexes of polymers with cyclodextrins, *Proc. Jpn. Acad.*, **69** *B*, 39.
4. Harada, A., and Kamachi, M. (1990) Complex formation between poly(ethylene glycol) and α-cyclodextrin, *Macromolecules*, **23**, 2821-2823.
5. Harada, A., Li, J., and Kamachi, M. (1993) Preparation and properties of inclusion complexes of poly(ethylene glycol) with α-cyclodextrin, *Macromolecules*, **26**, 5698-5703.
6. Harada, A. and Kamachi, M. (1990) Complex formation between cyclodextrin and poly(propylene glycol), *J. Chem. Soc., Chem. Commun.*, **1990**, 1322-1323.
7. Harada, A., Li, J., and Kamachi, M. (1993) Complex formation between poly(methyl vinyl ether) and γ-cyclodextrin, *Chem. Lett.* **1993**, 237-240.
8. Harada, A., Li, J., and Kamachi, M. (1992) The molecular necklace: a rotaxane containing many threaded α-cyclodextrins, *Nature*, **356**, 325-327.
9. Harada, A., Li, J., Nakamitsu, T., and Kamachi, M. (1993) Preparation and characterization of polyrotaxanes containing many threaded α-cyclodextrins, *J. Org. Chem.*, **58**, 7524-7528.
10. Harada, A., Li, J., and Kamachi, M. (1994) Preparation and characterization of a polyrotaxane consisting of monodisperse poly(ethylene glycol) and α-cyclodextrins, *J. Am. Chem. Soc.*, **116**, 3192-3196.
11. Wenz, G., and Keller, B. (1992) Threading cyclodextrin rings on polymer chains, *Angew. Chem., Int. Ed., Engl.* **31**, 197-199.
12. Wenz, G. [1994] Cyclodextrins as building blocks for supramolecular structures and functional units, *Angew. Chem., Int. Ed., Engl.* **33**, 803-822.
13. Harada, A., Li, J., Suzuki, S., and Kamachi, M. (1993) Complex formation between polyisobutylene and cyclodextrins: inversion of chain-length selectivity between β-cyclodextrin and γ-cyclodextrin, *Macromolecules*, **26**, 5267-5268.
14. Harada, A., Li, J., and Kamachi, M. (1993) Synthesis of a tubular polymer from threaded cyclodextrins, *Nature*, **364**, 516-518.
15. Born, M., and Ritter, H. [1996] Topologically unique side-chain polyrotaxanes based on triacetyl-β-cyclodextrin and a poly(ether sulfone) main chain, *Macromol. Rapid Commun.*, **17**, 197-202.

TOWARD MODULAR CHEMISTRY WITH THE DENDRITIC BOX AS MODULE

J.F.G.A. JANSEN, H.W.I. PEERLINGS, J.C.M. VAN HEST, E.W. MEIJER*
Laboratory of Organic Chemistry, Eindhoven University of Technology, P.O. Box 513, 5600 MB Eindhoven, The Netherlands

E.M.M. DE BRABANDER - VAN DEN BERG
DSM Research, P.O. Box 18, 6160 MD Geleen, The Netherlands

ABSTRACT. In this paper we like to present the dendritic box as a new tool toward modular chemistry. A dendritic structure with a densely-packed shell is prepared via modification of the fifth generation poly(propylene imine) dendrimers with N-BOC-L-phenylalanine groups. This so-called dendritic box has a diameter of approximately 4.5 nm, as determined by DLS and SAXS, and has internal cavities which can be used for the encapsulation of guest molecules. Supramolecular encapsulation is achieved by performing the construction of the box in the presence of the guest molecules, followed by dialysis to remove excess and adhered guest. A number of properties of encapsulated guests is critically influenced by the host, like the observation of induced optically activity and a solvent-independent fluorescence of dyes encapsulated as well as the formation of a triplet radical pair of 3-carboxy-proxyl. A shape-selective liberation of encapsulated guests is accomplished by a two-step hydrolysis of the shell. The potentials of these boxes as new supramolecular architectures within the nanometer regime are discussed and the physical locking of guest molecules is proposed as a new binding principle useful for modular chemistry.

J. Michl (ed.), Modular Chemistry, 371–384.

1. Introduction

The interest in dendrimeric macromolecules arises from the unique properties of these highly branched structures that have a defined number of generations and functional end groups [1,2]. The high degree of control over molecular weight and shape has led to the synthesis of unimolecular micelles [3], spherical and cone-shape mesostructures [4], as well as stratified dendrimers possessing generations of different structure [5]. Diameters of the spherical dendrimers range from 3-10 nanometers, enabling these structures to be building blocks of a new chemistry set [6].

After the initial reports on dendritic molecules [7], proposals have been made for the construction and applications of guest-host systems made out of dendrimers [1,2,8]. The concept of topological trapping by core-shell molecules is based on the fact that, at some stage in the synthesis of dendrimers, the space available for the new generation or end-group modification is not sufficient to accommodate all the atoms required for complete conversion (the so-called sterically induced stoichiometry) [1]. We will discuss here the synthesis of a dendritic box consisting of a flexible core with a rigid shell [9]. These boxes have internal cavities available in which guest molecules can be physically entrapped due to the rigid shell.

2. The dendritic box

The flexible core of our dendritic box is based on poly(propylene imine) dendrimers which were synthesized by the divergent approach [10]. A repetitive reaction sequence using the double Michael addition of a primary amine to acrylonitrile followed by the heterogeneously catalyzed hydrogenation of the nitriles to primary amines, yields diaminobutane-based poly(propylene imine) dendrimers DAB-dendr-$(NH_2)_n$ with n= 4, 8, 16, 32, 64, and 128 primary amine end groups. The unmodified dendrimers are very flexible and possess glass transition temperatures of approximately -40 °C and -65 °C for the CN- and NH_2-terminated dendrimers, respectively [10]. More recently a variety of end-group modifications have been reported [11]. For the construction of the rigid shell of the dendritic box a critical end-group modification of the cascade polyamine with an appropriate bulky group is performed. For instance, the N-hydroxy-succinimide ester of a tert-butyloxycarbonyl (t-BOC)-protected L-phenylalanine is allowed to react with the fifth generation poly(propylene imine) dendrimer in a CH_2Cl_2-triethylamine mixture (figure 1). Extended washing procedures were used to obtain pure dendritic box with a molecular weight of almost 24,000 [9].

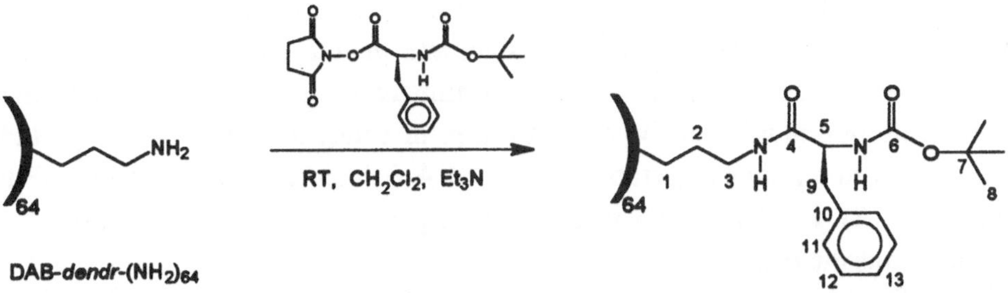

Figure 1. Schematic presentation of the synthesis of the amino acid-terminated poly(propylene imine) dendrimers, including an atomic numbering of the shell of the dendritic box.

Structure elucidation of the dendritic box was performed with a variety of characterization techniques; IR, UV, ^{1}H and ^{13}C NMR spectroscopy data are all in agreement with the structure assigned. However, the resonances in the ^{13}C NMR spectra showed a significant line broadening for the higher generations. Spin-lattice (T_1) and spin-spin (T_2) relaxation measurements were performed and the results for the shell atoms were compared with the corresponding data of the other lower generations (figure 2). The observed increase of T_1 relaxation times after the third generation is indicative of a decrease in molecular motion for the higher generations; an almost solid-phase behavior of the shell in solution is proposed. Further evidence for this close packing of the shell is found from chiroptical studies (see later). Presumably, intramolecular hydrogen bonding between several L-Phe residues in the shell is contributing to this solid-phase character. Unfortunately, MALDI-TOF and electrospray mass spectrometry studies have not been successful yet.

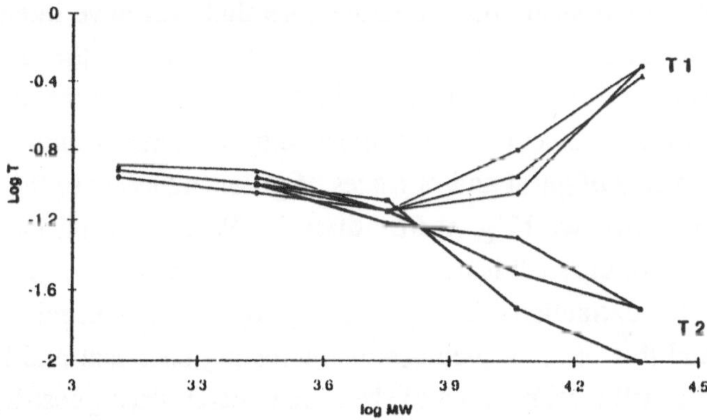

Figure 2. Double logarithmic graph of relaxation data (carbon T_1 and T_2) versus molecular weight (generation) for the carbon atoms 1, 2, and 3 (see figure 1) as recorded at 75MHz in chloroform.

CHARMm molecular mechanics calculations of the DAB-dendr-(N-t-BOC-L-Phe)$_{64}$ are performed to get insight into the three-dimensional structure. A globular architecture is found with an estimated radius of 23 ± 3 Å. Dynamic light scattering studies of the dendritic box in solution showed single particle behavior with a radius of gyration of 17 ± 4 Å (which resembles a radius of the box of 22 Å). Finally, SAXS measurements gave a radius of gyration of 18 Å for the dendritic box.

The choice of L-Phe as the amino acid component of the shell has been made from a study in which we compared a variety of amino acids of different size. By using larger amino acids like L-Trp it is not possible anymore to modify all the end groups due to the restricted space available verifying the sterically induced stoichiometry principle [1,6]. On the other hand by performing the modification reaction with smaller amino acids, like L-Ala and L-Leu such a dense packing is not achieved, as concluded from NMR and modeling studies, as well as the encapsulation experiments described in the next paragraph. L-Tyr in the t-BOC protected form is comparable with L-Phe as shell component with respect to dense packing, but lacks the good solubility in most organic solvents. Therefore, we selected the DAB-dendr-(NH-t-BOC-L-Phe)$_{64}$ as the dendritic box, being a nanometer-sized host system for a variety of guest molecules [9].

3. Encapsulation of guest molecules into the dendritic box

The experimental and modeling results prompted us to propose that we prepared molecules with a solid shell and a flexible core that will have internal cavities available for guest molecules. As the shell is constructed in the last step, it is possible to perform this coupling reaction in the presence of guest molecules. In fact, we encapsulated molecules with some affinity for tertiary amines within the dendritic box. Excess of guest and/or traces of guests adhering to the surface are removed by extensive washing and/or dialysis. When a dendrimer of lower generation was used, the shell is not dense enough to capture the guests and they were removed by extraction. A large variety of guest molecules have been encapsulated and this opens a plethora of interesting chemical and biochemical applications. We will discuss some of these nanometer-sized guest-host systems here as well as the properties of the guest molecules that are so critically influenced by the dendritic box.

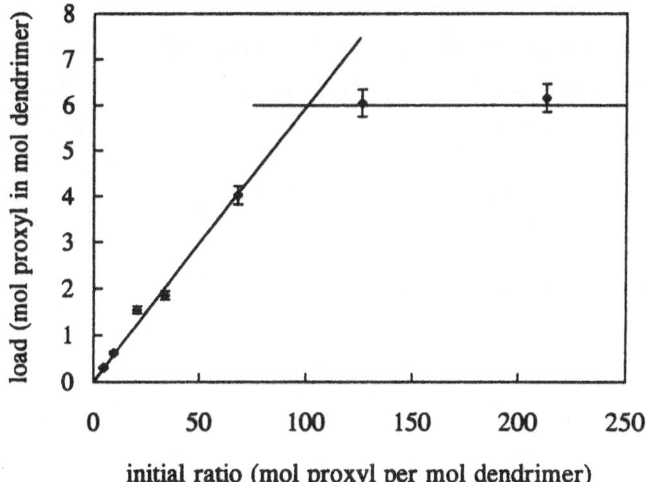

Figure 3. Number of 3-carboxy-proxyl radicals trapped in the dendritic box as determined by ESR spectroscopy versus the molar ratio of radical and dendrimer in the initial solution prior to the encapsulation reaction.

By carring out the encapsulation reaction in the presence of a varying concentration of 3-carboxy-proxyl, the number of entrapped radicals could be varied from 0.3 to 6.0 molecules per dendritic box as determined by ESR spectroscopy [12]. The number of 3-carboxy-proxyl radicals in the dendritic box does not increase above six (figure 3), clearly demonstrating that the maximum attainable number of radicals is restricted by the shape of the cavities in the box. The ESR spectra of 3-carboxy-proxyl@DAB-dendr-(NH-t-BOC-LPhe)$_{64}$ dissolved in 2-methyltetrahydrofuran are strongly temperature dependent. At 305 K an essentially isotropic ^{14}N-coupled ESR spectrum is observed, characteristic for a rapid rotational diffusion of the radical spin probes. Lowering the temperature results in a decreasing intensity of the isotropic spectrum and the appearance of an anisotropic ESR spectrum, consistent with a more restricted motion of the spin probe. In the temperature range from 150 to 250 K a superposition of the motionally narrowed (isotropic with A_{iso}(N)=1.40-1.42 mT) and the slow-motion (anisotropic with A_{zz} (N)=3.38 mT) spectrum is observed. This superposition indicates that the micro-environment of the encapsulated 3-carboxyl-proxyl molecules is not uniform over the interior of the dendritic box. A solid sample of 3-carboxy-proxyl@DAB-dendr-(NH-t-BOC-LPhe)$_{64}$ with more than 1.6 molecules per box shows at lower temperatures a (partial) ferromagnetic alignment of the radicals. The observation of a $\Delta_{ms} = 2$ ESR transition exhibiting a partially resolved 1:2:3:2:1 hyperfine coupling pattern due to two ^{14}N nuclei with A(pair) =1/2A(3-carboxy-proxyl) showed unambiguous spectral evidence of the presence of a triplet-state radical pair (figure 4). The intensity of the Δ_{ms} = 2 signal follows Curie law (I =C/T) between 4.2 and 100 K, consistent with a triplet ground state. To the best of our knowledge this is the first observation of an intermolecular ferromagnetic exchange interaction in a non-crystalline guest-host assembly. Since these types of interactions are often observed intramolecularly or in organic crystals, we are prompted to conclude that the dendritic box possesses some peculiar ordering properties apparently dictated by the architecture of the dendritic skeleton.

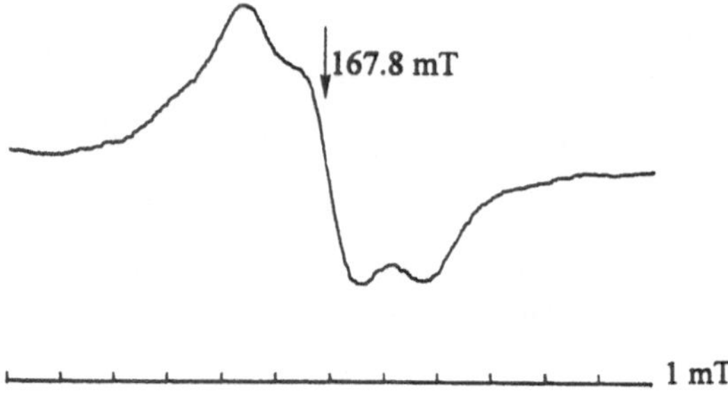

Figure 4. Δ_{ms} = 2 ESR spectrum of a solid sample of 3-carboxy-proxyl@DAB-dendr-(NH-t-BOC-LPhe)$_{64}$ at 4.2 K.

As another example we have encapsulated a variety of organic dye molecules into the dendritic box [9]. Rose Bengal is encapsulated in a similar fashion as the spin probe described above. The number of Rose Bengal molecules encapsulated could be estimated after prolonged dialysis by comparison of the UV spectra of guests that are inside or outside of the box. The relation between the number of encapsulated molecules of Rose Bengal as a function of the concentration of Rose Bengal used in the shell-forming reaction is depicted in figure 5. Also in this case the maximum number of guest molecules attainable is limited, in this case to four. It is tempting to propose that each of the four guest molecules is occupying one large cavity present in the dendritic box. Although the absorption spectra of Rose Bengal and Rose Bengal@DAB-dendr-(NH-t-BOC-LPhe)$_{64}$ are identical, there is a large difference in the fluorescence spectra as recorded in $CHCl_3$.The strong fluorescence at λ_{max} = 600 nm for Rose Bengal@DAB-dendr-(NH-t-BOC-LPhe)$_{64}$ is completely absent in the case of the supramolecular isomer of Rose Bengal outside the box. In the latter the fluorescence is quenched effectively. The emission of the guest-host system is relatively insensitive to solvent effects, hence, we believe that we have prepared a fluorescent sphere with an environmental-independent emission profile.

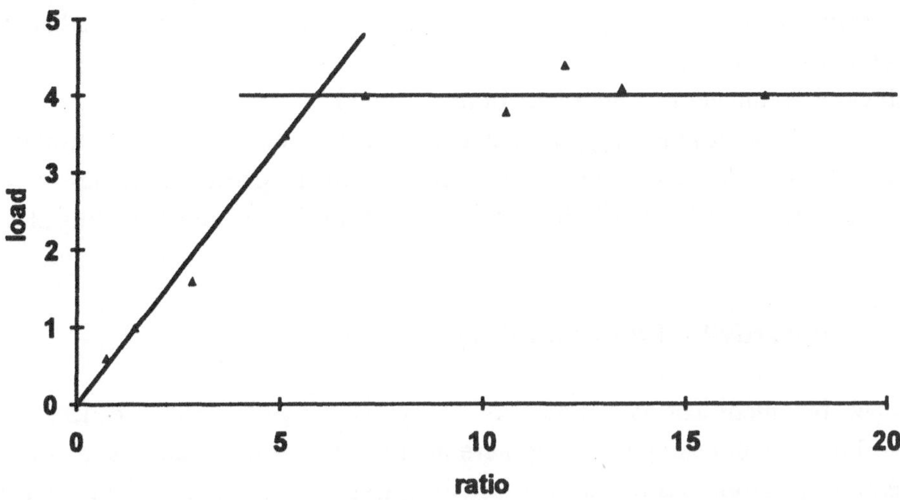

Figure 5. Number of Rose Bengal molecules encapsulated (load) in one dendritic box as determined by UV-vis spectroscopy versus the molar ratio of Rose Bengal and dendrimer during the encapsulation reaction.

Eriochrome Black T is a pH-dependent dye that is very soluble in polar solvents and can be encapsulated in the box. Due to the many (62) tertiary amines present in the interior of the box, Eriochrome Black T shifts its absorption spectrum from λ_{max} = 280 nm for free dye in CH_2Cl_2 to λ_{max} = 360 and 570 nm for dye in the box and in CH_2Cl_2. As soon as the absorption spectrum of free dye and encapsulated dye are different it is not possible to determine accurately the number of molecules encapsulated in a simple way. Since Eriochrome Black T is very soluble in water or acetonitrile, while the dye@box is insoluble in these solvents, we used this system to study the diffusion of the dye out of the box. Even after prolonged heating, dialysis or sonification the aqueous phase of the dispersion did not become colored due to diffusion. Therefore, it was concluded that the diffusion of dye out of the box is unmeasurably slow.

By comparing the encapsulation results of a large variety of dye molecules, it became apparent that many coplanar dye molecules with an ionic group can be encapsulated into the dendritic box. For concentrated solutions of large dyes in the encapsulation reaction the maximum number of dye molecules entrapped is four, which is related to the architecture of the dendritic box. Large three-dimensional dyes or coplanar dyes without ionic or polar groups are hard to encapsulate and only small numbers of the ratio guest per host are observed, typically around 0.1 as the average number of guests per box. Smaller polar guests can be encapsulated with maximum numbers beyond four, but in almost all cases an integer number of six or ten is found.

These results suggest that the procedure employed here produces a unimolecular compartmented structure in which guest molecules can be encapsulated and for which the diffusion out of the box is unmeasurably slow.

4. Optical activity of the dendritic box

Since the modification reaction of the poly(propylene imine) dendrimers is performed with enantiomerically pure amino acids, it is of interest to study the chiroptical properties of the box and the guest@box systems [13,14]. Much to our surprise, we noticed that the optical activity of the DAB-dendr-(NH-t-BOC-L-Phe)n decreases drastically on going from dendrimers of the first generation with four end groups ($[\alpha]_D$ = -11; c=1, $CHCl_3$) to the dendritic box of the fifth generation ($[\alpha]_D$ = -0.1; c=1, $CHCl_3$) with 64 end groups. This decrease in optical activity is not due to (partial) racemization of the amino acids employed, as was demonstrated with HPLC using a chiral stationary phase. The specific optical rotations as a function of generation are given in figure 6. A more thorough investigation employing a variety of different amino acid derivatives revealed that this decrease of optical rotation with increasing generation is a general phenomenon for all of the t-BOC-protected amino acids used [13]. Circular dichroism and optical rotatory dispersion measurements confirmed the results of the specific optical rotations. Using model systems we investigated the solvent dependence of the lower generations and found that for the L-Phe derivative the optical rotation is strongly influenced by the solvent. The optical rotations varied from $[\alpha]_D$ = 7.3 (c=1, toluene) to $[\alpha]_D$ = -6.4 (c=1, acetonitrile).

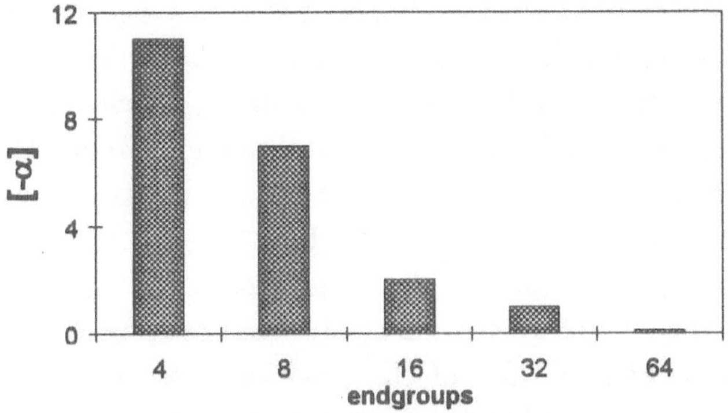

Figure 6. Specific optical rotations at the sodium D line for the t-BOC-L-Phe modified poly(propylene imine) dendrimers as measured in $CHCl_3$.

In order to investigate the importance of both the amide and the carbamate functionality in the end groups, a system is synthesized in which the carbamate group is replaced by an acetal moiety, while the shape is almost constant (figure 7). When the model compound of the propylamine and the end group was submitted to optical rotation measurements in various solvents, it was shown that this model compound showed only a marginal solvent dependence as the $[\alpha]_D$ value varied at values between 40 and 60. The specific optical rotations of the various generations of acetal-modified poly(propylene imine) dendrimers were measured and proved to be independent of the generation [15].

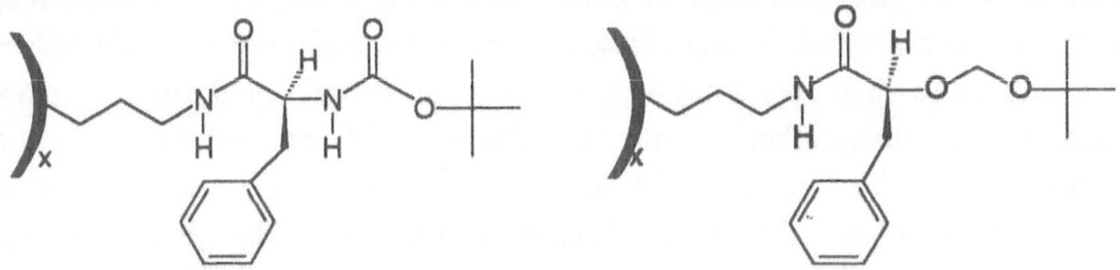

Figure 7. The dendritic box versus a modification in which the carbamate group is replaced by an acetal moiety.

In order to explain the peculiar optical behavior of the dendritic box, with its solid-like shell, it is first necessary to explain the difference in solvent dependence of the model compounds on the optical activity. A strong solvent, concentration, or temperature dependence of the optical rotation of organic compounds has been observed already about a century ago [16], but a detailed rational explanation is still lacking [17]. It is assumed that different solvents give rise to different distributions of conformations, which sometimes (but not necessarily) leads to large differences in optical rotations. If in the dendritic shell of the box several conformations are frozen in, an average optical activity will be observed. In the case of the dendritic box this will apparently tend to a vanishing optical rotation and for the acetal-modified dendrimers to a nearly constant value of about 42. Hence, the highly dense packing of end groups in the multiple-hydrogen bonded shell of the dendritic box gives rise to different frozen-in conformations, leading to an internal compensation of optical activity.

Stimulated by the observation of induced chirality of dyes dissolved into chiral bilayers and micelles [18], the circular dichroism (CD) spectra of a variety of dyes encapsulated in the dendritic box have been recorded. Induced circular dichroism spectroscopy is based on the transfer of chirality from the environment to an achiral dye and could therefore be applicable to these boxes. The vanishing optical activity is caused by a compensation effect and local optical activity is still thought to be present. In figure 8 the results are given for two samples of Rose Bengal@DAB-dendr-(NH-t-BOC-L-Phe)$_{64}$ with one and with four molecules of Rose Bengal per dendritic box on the average. Although both samples show identical UV spectra, a dramatic difference is observed in their induced CD spectra. The dendritic box with one molecule of Rose Bengal encapsulated exhibits an induced CD spectrum related to the UV spectrum, in which all bands possess a negative Cotton effect. However, an exciton-coupled spectrum is observed when four molecules of Bengal Rose are encapsulated on the average in a single dendritic box. This exciton coupling indicates the close proximity of chromophores with a certain fixed orientation [14]. All explanations for the induced CD observed are speculative, however, it is reasonable to assume that some chirality is present in the cavities of the dendritic box, despite the vanishing optical activity of the shell.

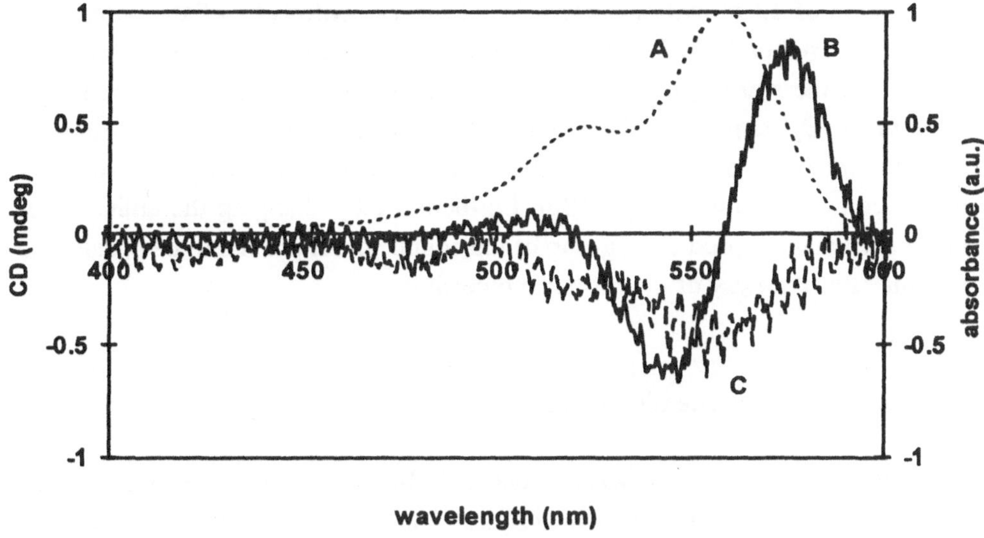

Figure 8. UV (A) and CD spectra of Rose Bengal@DAB-dendr-(NH-t-BOC-L-Phe)$_{64}$ containing one (C) and four (B) molecules of Rose Bengal.

5. Shape-selective liberation of encapsulated guests

So far, the dendritic box has been used to encapsulate guest molecules into the internal cavities present. The rigid, densely packed shell of the DAB-dendr-(NH-t-BOC-L-Phe)$_{64}$ limits the diffusion out of the box of almost all guest molecules studied up to now. Obviously, it is difficult to determine the diffusion of solvent molecules accurately, but all experimental data available so far show that small molecules like CH_2Cl_2 can penetrate through the rigid shell. If a dendritic box is made from the t-BOC protected glycine amino acid, a semi-permeable box is made [19]. This idea of tuning the density of the shell by decreasing the size of the end groups has been used to obtain a shape-selective liberation of guests form the dendritic box made from L-Phe [20].

After encapsulation of four molecules of Rose Bengal and 8-10 molecules of para-nitrobenzoic acid together in a dendritic box, hydrolysis of the t-BOC groups with formic acid (95% HCOOH, 16h) was performed. Subsequent dialysis of the reaction mixture (5% water in acetone) yielded a perforated dendritic box in which only the four molecules of Rose Bengal are entrapped, whereas all para-nitrobenzoic acid was dissolved in the acetone/water mixture. Rose Bengal cannot be liberated from the perforated box, not even after the addition of 12 N hydrochloric acid. However, hydrolysis of the outer shell using 12 N HCl under reflux for 2 h liberated Rose Bengal after dialysis (100% water) and the starting poly(propylene imine) dendrimer was recovered in 50-70% yield. By applying this two-step hydrolysis procedure to a variety of different mixtures of guest molecules it was shown that this shape-selective liberation is a general principle [20]. Furthermore by changing the amino acids in the shell and the protecting group of the amino acid it proved to be possible to fine-tune this pathway of liberation completely.

6. Toward modular chemistry and conclusions

We have discussed the synthesis of dendritic boxes possessing a unimolecular compartmented structure in which guest molecules are physically locked. Evidence is presented that the encapsulation is dominated by the architecture of the dendrimer and that some supramolecular ordering is present. Furthermore, a shape-selective liberation of guests can be accomplished by a two-step process.

It is envisaged that the binding between guests and dendritic box can be used as a new tool in modular chemistry. Therefore, we are working on large guest molecules that are partly inside and partly outside the dendritic box. The inside part is fictionalized with an anchoring group and a spacer is used to make these large guests compatible with the shell [21]. When the large part that is outside the box becomes functional or possesses functional groups, it should be possible to build larger structures, comparable with the key-and-lock principle, recently published by Newkome *et al.* [22]. It is very appealing to us to exploit the functionality of four as is found for the maximum number of many encapsulated guests. The approach described here based on the dendritic box will lead to modular chemistry at the nanoscopic level and by using mechanical bonding between modules.

7. Acknowledgement

The authors like to thank many of their collegues at the Eindhoven University and DSM Research for stimulating discussions, experimental assistance and enthusiasm to enter the field of dendrimers. Many of the names are found in the list of references to the original literature.

8. References

1. Tomalia, D.A., Naylor, A.M., Goddard ,W.A. III, Starburst-Dendrimere: Kontrolle von Grösse, Gestalt, Oberflächenchemie, Topologie und Flexibilität beim Übergang von Atomen zu makroscopischer Materie, (1990) *Angew. Chem.*, **102**, 119.
2. Fréchet , J.M.J., Functional Polymers and Dendrimers: Reactivity, Molecular Architecture, and Interfacial Energy, (1994) *Science*, **263**, 1710.
3. Newkome, G.R., Moorefield, C.N., Baker, G.R., Saunders, M.J., Grossman, S.H., Unimoleculare Micellen, (1991) *Angew. Chem.*, **103**, 1207.
4. Tomalia, D.A., Baker, H., Dewald, J., Hall, M., Kallos, G., Martin, S., Roeck, J., Ryder, J., Smith, P., A New Class of Polymers: Starburst-Dendritic Macromolecules, (1985) *Polymer Journal*, **17**, 117.
5. Wooley, K.L., Hawker, G.J., Fréchet, J.M.J., Unsymmetrical Three-Dimensional Macromolecules: Preparation and Characterization of Strongly Dipolar Dendritic Macromolecules, (1993) *J. Am. Chem. Soc.*, **115**, 11496.
6. Tomalia, D.A., Durst, D. (1993) *Topics in Current Chemistry*, Weber E. (ed), Springer-Verlag, Berlin, **165**, 93.
7. Buhleier, E., Wehner, W., Vögtle, F., "Cascade"- and "Nonskid-Chain-like" Syntheses of Molecular Cavity Topologies, (1978) *Synthesis*, **155**.
8. Maciejewski, M., Concepts of Trapping Topologically by Shell Molecules, (1982) *J. Macromol. Sci. - Chem.*, **A17**, 689.
9. Jansen, J.F.G.A., de Brabander - van den Berg, E.M.M., Meijer, E.W., Guest molecules encapsulated into a Dendritic Box, (1994) *Science*, **266**, 1226.
10. De Brabander - van den Berg, E.M.M., Meijer, E.W., Poly(propylenimin)-Dendrimere: Synthese in grösserem Massstab durch heterogen katalysierte Hiedrierungen, (1993) *Angew. Chem.*, **105**, 1370.
11. Jansen, J.F.G.A., Meijer, E.W., manuscript in preparation.
12. Jansen, J.F.G.A., Janssen, R.A.J., de Brabander - van den Berg, E.M.M., Meijer, E.W., Triplet Radical Pairs of 3-Carboxyproxyl Encapsulated in a Dendritic Box, (1995) *Adv. Mat.*, **7**, 561.
13. Jansen, J.F.G.A., Peerlings, H.W.I., de Brabander - van den Berg, E.M.M., Meijer, E.W., Optische Aktivität chiraler dendritischer Oberflächen, (1995) *Angew. Chem.*, **107**, 1321.
14. Jansen, J.F.G.A., de Brabander - van den Berg, E.M.M., Meijer, E.W., Induced Chirality of Guest Molecules encapsulated into a Dendritic Box, (1995) *Recl. Trav. Chim. Pays-Bas*, **114**, 225.
15. Peerlings, H.W.I., Jansen, J.F.G.A., de Brabander - van den Berg, E.M.M., Meijer, E.W., Optical activity of dendrimers with chiral end groups, (1995) *ACS, Polym. Sci. & Eng.*, **73**, 342.
16. Winther, C., Zur Theorie der optischen Drehung III, (1907) *Z. Phys. Chem.*, **60**, 590.
17. Eliel, E.L., Wilen, S.H. (1994) *Stereochemistry of Organic Compounds*, J. Wiley & Sons Inc., New York, **1076** (1994).
18. Kurzitake, T., Nakashima, N., Moritsu, K., Enhanced circular dichroism and fluidity of disk-like aggregates of a chiral, single-chain amphiphile, (1980) *Chem. Lett.*, **1347**.
19. Bosman, A.W., Janssen, R.A.J., Meijer, E.W., unpublished results.
20. Jansen, J.F.G.A., de Brabander - van den Berg, E.M.M., Meijer, E.W., The Dendritic Box: Shape-Selective Liberation of Encapsulated Guests, (1995) *J. Am. Chem. Soc.*, **117**, 4417.

21. Van Hest, J.C.M., Jansen, J.F.G.A., Meijer, E.W., unpublished results.

22. Newkome, G.R., Güther, R., Moorefield, C.N., Cardullo, F., Echegoyen, L., Pérez-Cordero, E., Luftmann, H., Wege zu dendritischen Netzwerken: Bis-Dendrimere durch Verknüpfung von Kaskadenmolekülen über Metallzentren (1995) *Angew. Chem.*, **107**, 2159.

DISCUSSION OF THE HARADA AND MEIJER LECTURES

J. FRASER STODDART
School of Chemistry
University of Birmingham
Edgbaston
Birmingham B15 2TT
United Kingdom

1. Summary by Prof. Stoddart

In summing up the talks by Harada and Meijer reflected on both mechanically-interlocked molecular compounds and dendritic polymers prepared recently in his own laboratories. He discussed how rotaxanes – and in principle polyrotaxanes – can be most easily self-assembled by a so-called slippage procedure, wherein a delicate steric match is achieved between the size of the stoppers on the dumbbell component and the ring component that is slipped over them with the help of some thermal energy. The process is a thermodynamically driven one on account of the matching recognition site present on the dumbbell and ring components of the rotaxane. Turning to the catenated variety of interlocked molecular compounds, the discussion leader described the first catenated cyclodextrins reported from his own laboratories in a full paper recently [1]. He pointed out that the first attempts to catenate cyclodextrins had been described back in the late 1950's by Prinzbach and his collaborators [2]. He also referred to an X-ray crystal structure of a [2]catenane involving dimethyl-β-cyclodextrin as one of the ring components. He emphasised that the crystallography, involving very highly hydrated crystals, was more akin to the practice of protein crystallography than to single crystal studies carried out on more conventional small molecular compounds. He pointed out that the existence of catenated cyclodextrins gives a unique insight into the nature of the noncovalent bonding interactions that cyclodextrins employ in binding substrates. The discussion leader then referred to some wholly synthetic cyclic oligosaccharides that have recently been described in detail in a full paper [3]. He described an intriguing crystal structure of a cyclic octasaccharide where approximately 1 nm diameter nanotubes are formed along one of the crystallographic axes. Again, the crystals are very highly hydrated. Finally, the discussion leader mentioned dendritic polymers, pointing out that he had investigated the structure of a naturally occurring

J. Michl (ed.), Modular Chemistry, 385–396.

carbohydrate dendrimer in the form of gum arabic back in the 60s as a Ph.D student. He described some very recent results [4] from his own laboratories on a convergent approach to the synthesis of carbohydrate dendrimers, before inviting questions from the audience for Professors Harada and Meijer.

2. Questions directed to Prof. Harada

PALACIN (Harvard): I have a question for Dr Harada. About this linear oligomer of cyclodextrin you build around the polymer chain and then remove the polymer. Did you try to put the polymer once again inside the rod after the polymerization? And, did you see any difference in the thermodynamics and in the kinetics of insertion?

HARADA: We have not tried to put the polyethylene glycol back into the cyclodextrin tube. We have tried some dyes, fluorescent probes, and iodine. Polyethylene glycols come out of the tubes rather easily because the complexes are stabilized by hydrogen bonding with the cyclodextrins. We can remove the polyethylene glycol chain by extraction with organic solvents.

MICHL (Colorado): Professor Harada – when you take out the polymer from the large tube that you synthesized, can you then put monomers back and polymerize them? And, is there anything special about the tacticity that you see in such polymers?

HARADA: Can I ask – what kind of monomer?

MICHL: Could you fill your tube with the monomer and would the chirality of each cyclodextrin affect the tacticity that you see in the tube you form?

HARADA: I have not tried this experiment. I will try it in the future, indeed, in the near future. There is a report that cyclodextrin and a monomer can complex and then the authors tried to effect a polymerization with a co-monomer. But, in this case, they obtained a polymer but not one incorporating cyclodextrins.

WEGNER (Mainz): I was interested in your report on the regular sliding of the cyclodextrins on the backbone. In other words, you demonstrated structures in which you have a head-to-head and tail-to-tail arrangement. Now, in the context of your other statement that it is a crystalline complex, which precipitates directly from solution, the question is – is this regular arrangement a consequence of simultaneous crystallization and complex formation? Do your complexes exist in solution at all? Frankly, your competitor in Germany, Gerhard Wenz, has no indication whatsoever for a regular arrangement when the complex exists in solution.

HARADA: Yes, that's a very important point. We didn't observe the existence of the complex in aqueous solution. We measured the viscosity and used some optical and spectroscopic methods. In aqueous solution, each component is almost dissociated. We see a type of complexation that is a kind of crystallization. And so, the polymer chain is some kind of pseudo-crystallization and it determines the orientation of the crystallization of the cyclodextrin. The cyclodextrins are able to form tunnels by using hydrogen bonding between the cyclodextrins. In the case of the single crystal of the complex of cyclodextrins, we think they could be ordered. In this case, the cyclodextrins are arranged head-to-head and tail-to-tail perfectly. And then we did another single crystal X-ray measurement and this experiment proved that the cyclodextrins alternate: they are head-to-head and tail-to-tail. We think that there are two possibilities to form such a complex. Cyclodextrins tend to form dimers in water. Of course, in water the cyclodextrins are isolated from each other. Once the guest molecule is in solution, the cyclodextrins form pairs. An alternative explanation can be that the equilibrium is very fast. One cyclodextrin threads on to the polymer chain, the next one threads on the other way round. If it is the best arrangement, that is the head-to-head, it is stabilized, and if it is the other arrangement, it will come off. I think the head-to-head and tail-to-tail arrangement is the most stable. However, I'm not sure that, in the case of the polymer, it is the perfect arrangement or not.

WEGNER: I'm sure that in Gerhard Wenz's case, where he has the freely floating chain in solution, which assembles the rings, it's a random arrangement.

HARADA: We have made measurements with STM. In the polyrotaxane, it differentiates between the head-to-head or tail-to-tail arrangements in some parts. And, of course, they are not perfect head-to-head or tail-to-tail. There are some parts with head-to-tail. Right now, we don't know which is the particular arrangement.

KUKI (Alanex): My question concerns the modular assembly of these very nice tubes. Perhaps you mentioned this point and I might have missed it. But consider the idea of crosslinking the cyclodextrins and then dethreading. Can you do that?

HARADA: Yes.

KUKI: You can crosslink them.

HARADA: You can.

KUKI: How long?

HARADA: On the average, fourteen cyclodextrins.

KUKI: Then the second question is – if it is possible to have cyclodextrins that are, say, acylated or otherwise functionalized to be more hydrophobic, then perhaps you can make a long tube which for half of its length is hydrophilic on the outside and for the other half is hydrophobic on the outside. Such a tube would orient say at an oil-water interface. There seem to be many possibilities for making very interesting tubes.

HARADA: Yes, I understand what you are saying. We are trying to make hydrophobic tubes using hydrophobic crosslinking agents. We would like to make ion channels so we are now preparing hydrophobic tubes. We are also interested in making amphiphilic tubes.

KUKI: How would you do that?

HARADA: It's more difficult to make the amphiphilic tubes. We have to make tubes that are symmetric and protected. It is rather difficult. If you have any ideas, I would like to hear them.

3. Questions directed to Prof. Meijer

EBBESEN (NEC): I have a question for Professor Meijer. You talked about your dendrimers and told us that you can locate dyes inside, like Rose Bengal. And you also talked about solvents. Solvents are probably not in there. I don't know what you can conclude on the basis of the changes you observe in the spectra. You need to have more solvent sensitive dyes inside. There are dyes that exist that can be used for that purpose. If there are a few solvent molecules in the first shell, their fluorescence will be changed dramatically. Rose Bengal is not in that category. And you mentioned excitons. There wouldn't be excitons inside the dendrimers unless the dye molecules are very close to each other. You are not going to get excitons over the distances that are present inside your dendrimers.

MEIJER: Firstly, we make these molecules with the dyes inside in methylene chloride and then we dry the material, dialyse it, and dry it again and then we put in other solvents. We have not only used Rose Bengal but now 50 different dyes, of which three are well known solvatochromic dyes. One was introduced by Effenberger – it is a nitromethoxybithiophene – and one was introduced by Verhoeven – it is a fluorescent probe which is luminescence sensitive for solvents. And what we see is a little bit remarkable. Some of these probes are insensitive for the solvent you add to the box. So it seems to be a phase-separated system, the dye is only in the environment of the interior of the box and it doesn't matter what the solvent is. We also have dyes, for instance the fluorescent dyes of Verhoeven, in which the fluorescence that you see mimics exactly what you see if it is free in solution. And we have the in-between situation. We are 99.9% sure that all the dyes are really encapsulated. Therefore, it looks as if some of these dyes have a sensitivity to solvents up to a couple of Å while other dyes have an influence which goes beyond a couple of Å. That's our interpretation. There are not many studies done on systems which are only a few Å separated from solvent, although there have been many experiments done on bilayers. But there, the flexibility is much larger than in this rigid phase-separated molecular box. It comes to the second question – that is induced CD. If we have one molecule encapsulated, the CD spectrum shows a Cotton effect that follows the UV. In the case of four molecules, we have a molecular exciton. There are two absorption bands, one has a positive CD and the other one is negative. (Another) Harada from Japan has elaborated this in great detail and there are many examples. It only says that the guests have one kind of fixed orientation.

MÜLLEN (Mainz). The photoisomerization of anthracene is a nice way of changing sites in a molecule when it is attached to a polymer chain to couple two polymers. Can you comment on the photochemistry going on in the dendritic box?

MEIJER: We have not had time yet to study in detail which type of molecules go into the molecular box and which ones don't. If you use a pure hydrocarbon, there is not really a driving force. We now use encapsulated photosensitive dyes in order to do a kind of photo-switching. We still have to go into the details of the photochemistry, but I think these cavities have some space in which you have to deal with similar rules as doing photochemistry in the solid state. But that is something we have still to study in detail.

MILLER (IBM): This question is not concerned with the solvation inside the dendritic box. It's really concerned with mobility. If you were to take something like pyrenecarboxylic acid, would you be able to see excimer formation?

MEIJER: We haven't put it in the box. Indeed, we haven't looked in detail at any kind of mobility of the guests. At this point, it's still difficult to go into these details, because we don't exactly know what the real structure is of the dendritic box. It's hard to come up with a model. We have avoided details before we study the photophysics of the system.

TOMALIA (MMI): Have you taken your system to a high enough limit in terms of generations that it will self-seal?

MEIJER: The highest we've made is the sixth generation with 128 residues and then we are running already into trouble with the heterogeneous catalyzed hydrogenation. Your idea is similar to this one, except that we make a shell in one step. It's a single step synthesis. While in the case of going to the largest generation, you're dealing with a lot of chemistry you have to do to make the largest generation before it self seals. You can't play with guests during that chemistry. Obviously, you also can use lower generations in which you do dynamic encapsulation and play around with physical means of collapsing chains in on themselves. But our method is a very robust method.

COLEMAN (Lyon): I have a tentative suggestion. I'm not sure it will work but I believe you can use xenon NMR to map out available spaces. It's been used in cyclodextrins and in calixarenes. You should get xenon inside. And then you should be able to map out the surfaces within the dendrimer. And then you should be able to map out the surfaces when you have got other systems inside the box. That might give you some information about what is going on inside the dendritic box.

MEIJER: A very nice suggestion, that we will follow.

STUPP (Illinois): When the optical rotation drops with the generation, is this caused by a change of shape or is it caused by different aggregation modes as you change the number of generations. I guess both change the environment. You've matched that experiment which described how the optical rotation of phenylalanine depends on solvents. Both of these change the environment so which one do you think is the main mechanism by which the $[\alpha]_D$ results.

MEIJER: When I saw the $[\alpha]_D$ of the box went back to zero, we returned to the whole series of different generations. I expected to see something similar to what you see with intrinsic viscosity – that is, you go through a maximum at which a complex organic molecule changes to a dendrimer. But that's not what we see here. It's more of a gradual decrease. And I think that has to be because of the hydrogen bonding which is possible in the shell. It already leads, at lower generations, to a freezing out of conformations. In the Poster, I have some very preliminary results on where we have played with different structures. And I have now the idea that it is not the contributions from a whole assembly of different conformations frozen in, but only of two. This situation is well known for amino acids. Sometimes they crystallize in a pseudoracemic arrangement because then the density is much lower than if it is a conglomerate. We think it is like this from the following experiment. If you do an experiment with the fifth generation, having all D-phenylalanine, $[\alpha]_D$ is zero. If you have L-tyrosine, it is almost the same – $[\alpha]_D$ is zero. But, in certain solutions, L-tyrosine and D-phenylalanine have opposite optical rotation. If you now make a dendrimer having 50% of the groups D-phenylalanine, the other 50% L-tyrosine, then we end up with a dendrimer box, which is still a box, which has an optical rotation of 18°. So, I think there is more than just freezing out of conformations. And that is the point I want to make here. When we started with dendrimers, I just thought of something which does not have much order. And, I am still thinking that is the case. But, many of the results we get point to something that is going into some kind of order, which is much more than I had expected. I never thought of getting four guests in a box. We came up with that figure after doing 20 or 30 different experiments and always we found four. So, what's special about four? And, also the encapsulation and the liberation. It's so strange, it's hard to believe it's correct. But we did it so often, that it has to be correct.

LINDSEY (N. Carolina State): I was going to suggest an experiment. You know Rose Bengal is an excellent sensitzer of singlet oxygen. A study of singlet oxygen formation might give you a nice probe of the microstructure in the dendrimers.

MEIJER: We're working on that in collaboration with a group at the University of Berlin using water soluble boxes. If the boxes are water soluble, then they could be used for some kind of anti-cancer therapy.

BALZANI (Bologna): To obtain some type of information about the position of the encapsulated molecule in the structure of the dendrimer, you can use a long-lived excited state inside and look for quenchers coming from outside. Quenchers of different size maybe can penetrate at different rates and can tell you something about the nature of the holes reaching into the interior of the dendrimer.

MEIJER: I think that is a good point to study. And, maybe we can have some discussion about what would be the best molecules. Although you should keep in mind that the interior of the dendrimer is very electron rich. For instance, if you add TCNQ to the box, it directly takes two electrons - one per TCNQ - from the box, resulting in a very stable charge-transfer complex. On the other hand, if you add C_{60} to it, irradiation brings it to the triplet state of C_{60}, and then it takes an electron from the box and you have the anion of the C_{60}. The dication of the box is a very stable species. However, many of these quenching experiments you suggest don't like to have tertiary amines involved. That is a difficulty.

KAHN (Bordeaux): I was wondering whether it would be possible to do some chemistry within the box or dendrimer in a selective fashion.

MEIJER: Yes. That's our main topic besides the modular chemistry.

KAHN: Could you comment a little bit more?

MEIJER: We had done one experiment. We call it "the ship in the bottle experiment". We have a molecular box made out of glycine. Paranitrobenzaldehyde and acetone can pass this shell and the tertiary amines in the interior are basic enough to do a classical aldol condensation. The product that you make can't pass the barrier again. That's one experiment we did but what we are actually aiming at is using that shape selectivity in order to have a catalyst inside. Then the substrate would come in and react, and depart again. The shell should be capable of selecting small substrates over large substrates.

KAHN: That seems to me to be a very appealing line of research.

MEIJER: Yes. I think it is possible. We have already attempted some experiments with large benzaldehydes and acetone and they don't react.

KAHN: Do you think you could carry out a reaction with a salt – the cation going into one box, and the anion in another box? Of the same dendrimeric unit?

MEIJER: Never thought of it. It is a little like what Don Tomalia has done with his polyamines and polyacids and then bringing them together.

KAHN: Yes, in a sense.

MEIJER: I don't want to compete with Don.

MICHL: Professor Meijer – have you looked at fluorescence depolarization, by any chance, to see if mobility might characterize the prisoner in the box?

MEIJER: Yes – in collaboration with Professor De Schryver. The difficulty is that the difference between a free and encapsulated fluoresceine was completely different from that of free and encapsulated Rose Bengal. But both were much less mobile inside the box. We did the experiment with structures in which we had four guests encapsulated, and then you have also to take into account that the energy transfers within the box. So, we have made molecules with only one dye encapsulated to avoid this internal energy transfer. These studies are in progress.

MICHL: I was struck by the fact that you saw induced circular dichroism with one molecule inside, although the source of chirality is relatively far away. Can you suggest that you had indeed some well-defined chiral structure in the center of your box? And perhaps you could use theory to figure out from the sign of the circular dichroism something about the conformation this forces on the dye molecule – and from that, perhaps you could conclude something about the shape of the cavity.

MEIJER: That's good. We haven't thought about the idea of using the theory and the sign of the CD. More strikingly, what you see in the CD of the box is that the optical activity of the shell is vanishing to almost zero and hence the induced CD of the guests is unexpected. However, the effects are very small.

GRIMME (Cologne): I think there must be strong host-guest interactions present in your system. Can you comment on these host-guest interactions?

MEIJER: Indeed, there is a very strong guest-host interaction. In many cases not only based on acid-base reactions, but they very easily form clathrate-type structures. We think that, in equilibrium where you have an excess of the guest, all the four cavities are always filled with guests independent of what you do with the dendrimer. I have to add one thing. You could ask whether these cavities are already present or not when the box is empty? I think this relates to the same topic as the induced fit versus the lock and key principle with enzymes. Dendrimers are confined in space with respect to their conformation. George Newkome showed this beautifully by adding salts to an amine or base to an acid – I don't remember exactly which one – that you can really go to the extended conformations of all the units and thereby induce the 'breathing' of the dendrimer. As soon as you have an excess of guests, they are in. What the time scale of exchange is, I do not know.

GRIMME: So you don't need a large excess?

MEIJER: No. And this is not a highly concentrated solution.

PRINZBACH (Freiburg): How stable thermally is your dendritic box? When does it lose its order? At what temperature? How much heat can you apply in order to do some chemistry?

MEIJER: The stability of these molecules is high until the point of about 200 degrees when the t-BOC groups start to eliminate from the structure. But, below 200°C, they are more or less stable, as determined by TGA.

PRINZBACH: But at 200°C, you can only introduce four guest molecules.

MEIJER: Oh, you mean doing the encapsulation at very high temperatures?

PRINZBACH: Yes. At 200°C.

MEIJER: Oh no, we never did it. We do the encapsulation in methylene chloride and that solvent has a boiling point of only 40°C.

PRINZBACH: What about dichlorobenzene?

MEIJER: What should that prove?

PRINZBACH: No, my question was whether you could increase the volume by increasing temperature. Can you?

MEIJER: No. I don't think so.

NEWKOME (South Florida): You have done the nitrogen NMR on your compounds. Both protonated and free? If you have the nitrogen NMR and you add to it four protons, in the form of these acids, is there a preferred position or location for those protons? And, would that give you any information as to the juxtaposition of the guest?

MEIJER: Good point. We did the natural abundance ^{15}N NMR in water where the primary amines are the most basic. And, if you add protons, they first go to the primary amines. If you then add more, they go to the two most inner amines, and then the following generations are protonated. This mechanism follows the theory of protonation of polyelectrolytes. In methylene chloride, probably the situation will be the other way around. The tertiary amines will be more basic compared to the primary amines. And then, if the same would hold for what we saw in water, that the inner are the most favourable ones, then that would be good for the encapsulation bringing the acid guests inside. But, it's hard to say.

NEWKOME: But you don't have primary amines, though, in your actual box?

MEIJER: No, but you start with primary amines. So, if you add the acid to it, you have an acid-base reaction and then the most basic amines could be protonated.

NEWKOME: No, I'm sorry. What I'm getting at though is – when you add your four guests, each will bring one proton along with it.

MEIJER: Yes.

NEWKOME: You then cap the whole thing off. Those four protons are still bound inside.

MEIJER: Yes, that's true.

NEWKOME: And the question is, from your NMR data, could you ascertain where they would be by looking at the chemical shift.

MEIJER: The molecular weight is then 24,000 with 4 guests, which is also 4,000 makes it 28,000, with almost 200 nitrogens. And then only four are protonated.

NEWKOME: Use deuterium.

MEIJER: You could be right.

OZIN (Toronto): Does the rule that you gave on protonation sites follow alkylations generally? That's one question. The second one, I think, is stupid but I'm going to ask it. I'm not a polymer chemist. But you can buy polytriethylenamine and polyethylenediamine. And you do the same types of experiments that you are doing. Sort of as controls. You mentioned things like hydrodynamic radii and things like that, using different techniques and you tried doing some of this host-guest stuff. How different are they?

MEIJER: There are no stupid questions, only stupid answers, I'm afraid. With respect to the alkylation, we haven't followed that with time. What we did is an alkylation with methyl iodide, and all the amines are alkylated. The primaries and all the secondaries. That means you have 126 positive charges in your dendrimer with the same number of iodides, too. And, that goes to full completion. If you do it with large iodides, like hexyl iodide, then it seems to be that you can only alkylate the periphery. It doesn't go into the interior. Going to the comparison with the linear chain, we do not always compare everything we do with the linear chain. But many of the reactions are done as well on linear chains. If you want to do a hydrogenation of polyacrylonitrile, it's impossible to do. You can't get all these nitriles to the surface of your catalyst. The same holds for benzyl ethers: you can't reduce them if they are in linear chains, as shown by Fréchet. Secondly, if you compare the viscosity of the linear polymer with that of the dendritic polymer, there is a distinct difference when you go beyond that point at which the dendrimer becomes a globular structure. The encapsulation was also tried on the lower generation. And then, it doesn't work. But, actually, we should check that again with, for instance, poly-(allyl amine) or something like that. It's a little bit like an experiment in which you are 99.9% sure it doesn't work. And, that's not the kind of experiments we like the most, but we should do it.

PORT (Stuttgart): Can you get ordered structures of dendrimers in the solid phase?

MEIJER: All the dendrimers we have studied so far are amorphous, except a few. In the case where we react the polypropylene-imine dendrimer with long-chain acid chlorides, we have a molecule which is very hydrophilic in the interior and hydrophobic on the exterior. There, we have X-ray data in which there is some crystallization of the side chains and so it's like a hairy ball. These ones seem to be ordered. Furthermore, we have some with pentafluorobenzene groups at the periphery and there seems to be some kind of stacking, giving rise to a liquid crystalline type ordering, but if you examine it with X-rays, then you don't see any ordering.

EBBESEN: I want to raise the issue of the space in the dendrimers. Is there no solvent inside the dendrimer at all?

MEIJER: If you have a dendrimer in solution, there are many solvent molecules somewhere around in that 'tree'. And depending on whether it is a good solvent or a poor solvent, the chains are extended or they are collapsed into the backbone. As soon as you have some 'glue' in the form of hydrogen bonding, then there is some distance between the core and the shell and it will be possible that you have some empty space with solvent. TGA of these dendrimers show the presence of solvents at the higher generations.

4. References

1. Armspach, D., Ashton, P.R., Ballardini, R., Balzani, V., Godi, A., Moore, C.P., Prodi, L., Spencer, N., Stoddart, J.F., Tolley, M.S., Wear, T.J., and Williams, D.J. (1995) Catenated cyclodextrins, *Chem. Eur. J.* **1**, 33-55.
2. Lüttringhaus, A., Cramer, F., Prinzbach, H., Henglein, F.M. (1958) Cyclisation von langkettigen dithiolen, *Liebigs Ann. Chem.* **613**, 185-198.
3. Ashton, P.R., Brown, C.L., Menzer, S., Nepogodiev, S.A., Stoddart, J.F., and Williams, D.J. (1996) Synthetic cyclic oligosaccharides. Syntheses and structural properties of a cyclo[(1→4)-a-rhamnopyranosyl-(1→4)-a-D-mannopyranosyl]-trioside and -tetraoside, *Chem. Eur. J.* **2**, 580-591.
4. Ashton, P.R., Brown, C.L., Boyd, S.E., Jayaraman, N., Nepogodiev, S.A., and Stoddart, J.F. (1996) A convergent synthesis of carbohydrate-containing dendrimers, *Chem. Eur. J.* **2**, 1115-1128.

MOLECULAR SELF-ASSEMBLY OF HYDROGEN-BONDED CRYSTALLINE NETWORKS

M. D. WARD and V. A. RUSSELL
Department of Chemical Engineering and Materials Science
University of Minnesota
Amundson Hall, 421 Washington Ave. SE
Minneapolis, MN 55455

ABSTRACT. Guanidinium and organosulfonate ions self-assemble into two-dimensional hydrogen bonding networks with the general formula $[C(NH_2)_3]^+ RSO_3^-$ having pseudo-hexagonal symmetry due to hydrogen bonding between six guanidinium proton donors and six sulfonate electron lone pair acceptors. These hydrogen bonded layers assemble in the third dimension in a manner which maximizes van der Waals interactions between R groups. The steric requirements of the R groups dictate whether this assembly results in "classical" bilayer motifs in which all the R groups are oriented to one side of a given sheet or "single layer" motifs in which R groups are oriented to both sides of a given hydrogen-bonded sheet. Synthesis and characterization of over 30 crystalline compounds reveal that this hydrogen-bonded network is robust, which aids significantly in materials design as the crystal engineering problem is reduced to one dimension. The pervasiveness of this network can be attributed to the "softness" of hydrogen-bonding which allows the sheets to pucker slightly in order to accomodate steric strain between R groups within the layers. Identical networks have also been synthesized from α,ω-alkanedisulfonates and aromatic disulfonates in which the hydrogen-bonded sheets are pillared by organic spacers and the hydrocarbon density between layers is reduced by a factor of two compared to the monosulfonate analogs. This leads to nanoporous "molecular sandwiches" in which the void space can be controlled simply by adjusting the structure of the organic pillar, suggesting rational design of host-guest compounds.

1. Hydrogen-bonding Modules for Network Structures

Organic molecular materials often possess interesting electronic properties, including nonlinear optical behavior, electrical conductivity, superconductivity, and ferromagnetism. Crucial to the design and synthesis of these materials is a thorough understanding, and ultimately control of, the assembly of constituent molecules into the supramolecular motif that defines the solid state structure. It is constructive to consider the constituent molecules as the fundamental building blocks of the solid state, or in the spirit of this symposium, as "modules." The formation of ordered solid state networks with a desired arrangement and dimensionality relies on an appropriate "topological director," that is, a module having a well defined functional group that can recognize complementary functional groups on other like molecules (homomeric assembly) or

J. Michl (ed.), Modular Chemistry, 397–407.

different molecules (heteromeric assembly). A crucial property of directors is bonding which is strong and highly directional relative to competing interactions. Formation of extended networks also requires "polyvalent" modules, that is, molecules having more than one bonding functionality. These capabilities are provided by molecules containing hydrogen bonding functionalities.

Several examples of ordered, extended hydrogen bonding networks have been reported that illustrate the important influence of this interaction on directing the organization of molecules in the crystallization of solid state materials. These reports have demonstrated that the local supramolecular organization about each module can be predicted with reasonable confidence based on molecular topology. Flat molecules having one-dimensional hydrogen-bonding topologies form "ribbon" or "tape networks,"[1,2,3,4] while tetrahedral-like topologies have afforded diamond-like networks [5,6]. However, control of three-dimensional packing generally is elusive owing to the contribution of weak dispersive interactions which hold individual networks together in the crystal. These interactions are numerous, weak, and relatively isotropic, which makes control of their assembly difficult.

One strategy for overcoming this obstacle involves robust two-dimensional hydrogen bonding networks whose assembly in the third dimension is governed simply by functional groups extending from the networks. The use of a two-dimensional network whose integrity is preserved during crystallization reduces the degree of freedom available in assembly, thereby reducing the design of the three-dimensional packing to a single dimension. The purpose of these functional groups is to "glue" the layers together by strong, directional interactions such as hydrogen bonding, metal-ligand coordination, or covalent bonding. Even weak, comparatively non-directional dispersive interaction can direct the assembly in the third dimension in a predictable manner if the two-dimensional network is sufficiently robust. In this sense, the two-dimensional network becomes the fundamental "module" that is assembled in the remaining dimension.

This strategy is illustated schematically in Figure 1, in which generic molecules assembled into a two-dimensional network form layers, with functional groups, attached to the molecules in the layer, which are capable of directing assembly in the third dimension. In the absence of these functional groups the hydrogen bonded layers will simply pack in a manner to minimize void space by meshing of protrusions of one layer into hollows of another [7,8]. The directors may be based on hydrogen bonds, metal-ligand bonding, hydrogen bonding, or simple van der Waals bonding. In the latter, the assembly would be directed by the tendency to maximize packing density by interdigitation of the functional groups. Such motifs offer unique opportunities to engineer interesting optical, magnetic, or conducting properties, depending upon the choice of molecules in the region spanning the layers. Furthermore, the formation of open networks should be possible by judicious control of the size and number of groups spanning the layers.

2. Two-Dimensional Hydrogen Bonding Networks: Guanidinium Alkyl- and Arylsulfonate Salts

The formation of three-dimensional solids from two-dimensional hydrogen-bonded networks therefore requires molecules with hydrogen bonding functionalities constrained to a single plane and polyvalent topologies allowing for assembly of continuous extended networks. The molecules must have functional groups extending normal to the plane which can direct assembly in the third dimension and provide desirable properties.

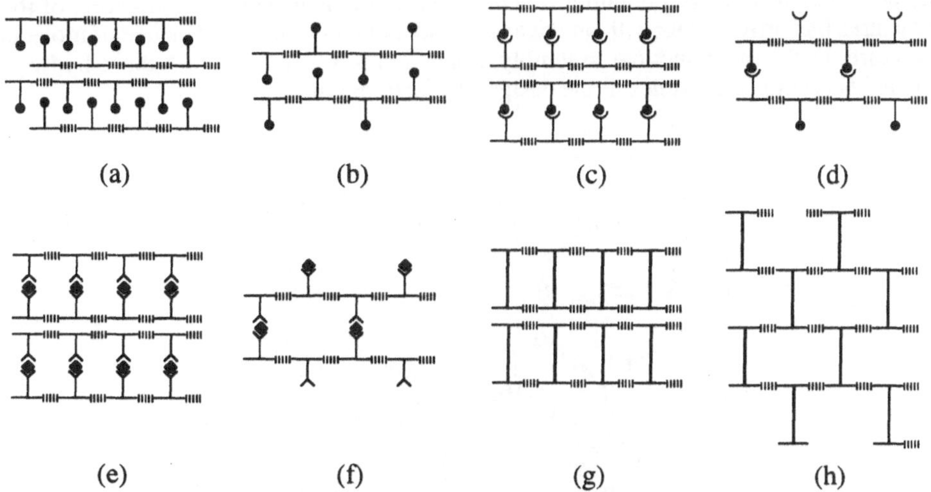

Figure 1. Schematic representation of possible assembly motifs for a generic two-dimensional hydrogen bonding network, depicted here as molecules held together by hydrogen bonds (IIIII), containing different types of functional groups that can direct assembly in the third dimension. The interactions between layers are depicted schematically as: (a),(b) = dispersive interactions; (c),(d) = hydrogen bonding; (e),(f) = metal-ligand binding; (g),(h) = covalent bonding. The first example in each pair depicts bilayer assemblies, while the second example in each pair depicts a alternating motif in each sheet which leads to packing in the third dimension described by translationally related single layers. The detailed molecular structures, additional ligands on metals sites, and solvent or guest incorporation into the lattice will influence the packing.

Recently we have synthesized networks based on guanidinium alkane- and arenesulfonate salts which possess the aforementioned structural elements (Figure 2) [9,10]. Single crystal x-ray diffraction of over 40 structures reveals that the pervasive network structure consists of two-dimensional quasihexagonal sheets with alkyl or aryl residues extending from the sheets (although the motif of these R groups may vary, as discussed below). This network is a consequence of the polyvalency of the guanidinium cations and sulfonate residues and their topological equivalence; that is, the cations have six protons and the sulfonates have six lone-pair proton acceptors. Accordingly, all the hydrogen bonding capacity is fulfilled within this network, which is important in forming robust networks [11].

The assembly of the guanidinium-sulfonate network can be viewed as the formation of a guanidinium-sulfonate hydrogen-bonded dimer which forms an $R_2^2(8)$ ring [12]. These dimers then assemble into one-dimensional "ribbons" or "tapes" via identical $R_2^2(8)$ interactions. The formation of these ribbons consumes four donor and four acceptor sites on the respective ions. The remaining hydrogen bonding sites connect adjacent ribbons via the formation of $R_2^2(8)$ and $R_3^6(12)$ rings. Packing in the third dimension is then realized by stacking of the sheets. In the case of alkane- and arylsulfonates the sheets are held together by van der Waals interactions between R groups. Our experience indicates that this network is fairly robust for many different R groups, but the stacking motif depends upon the steric requirements of the R group and the presence of hydrogen bonding functionality on the R group. Examination of these structures has provided us with an advanced understanding of the influence of molecular structure on the inter-network assembly, a key step toward exploiting these hydrogen-bonded networks as modules in the design of functional materials.

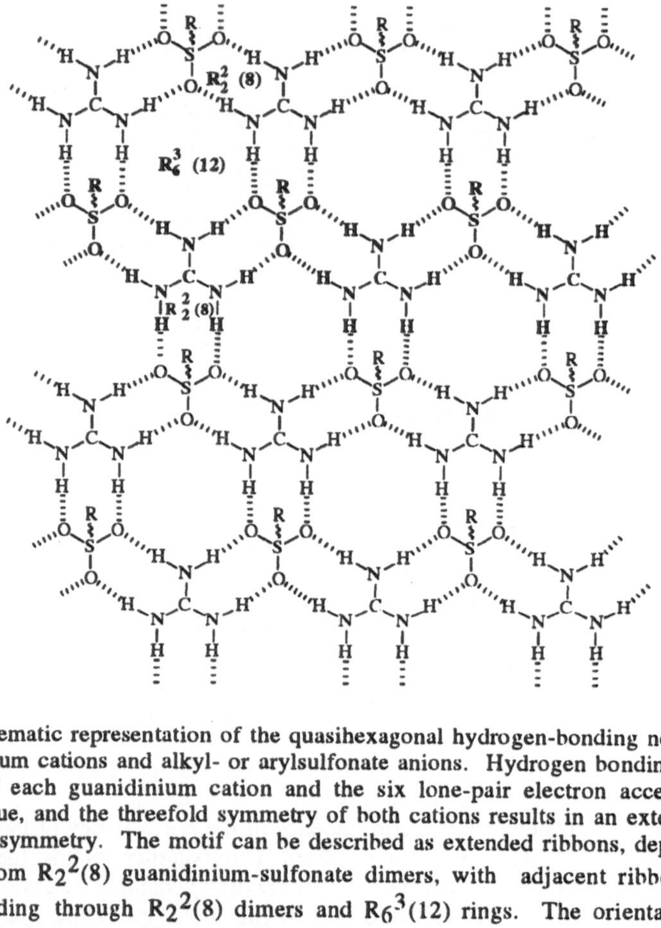

Figure 2. Schematic representation of the quasihexagonal hydrogen-bonding network formed from guanidinium cations and alkyl- or arylsulfonate anions. Hydrogen bonding between the six protons of each guanidinium cation and the six lone-pair electron acceptors of each sulfonate residue, and the threefold symmetry of both cations results in an extended network with threefold symmetry. The motif can be described as extended ribbons, depicted in bold, which form from $R_2^2(8)$ guanidinium-sulfonate dimers, with adjacent ribbons joined by hydrogen bonding through $R_2^2(8)$ dimers and $R_6^3(12)$ rings. The orientation of the R groups extending from the sulfonate residues is not specified.

2.1. ASSEMBLY OF NETWORKS BY VAN DER WAALS INTERACTIONS

The layering motifs observed for the guanidinium-sulfonate salts can be illustrated by the salts containing methanesulfonate, (1s)-(+)-10-camphorsulfonate, 1-naphthalenesulfonate, and 2-naphthalenesulfonate (**I** - **IV**). Guanidinium methanesulfonate crystallizes as large sheets having centimeter dimensions. Single crystal x-ray structural analysis reveals the presence of two-dimensional quasihexagonal sheets as depicted in Figure 2, which are parallel to the *ab* plane. These sheets form bilayers which stack normal to the *ab* planes (Figure 3). Consequently, the stacking sequence of the sheets comprises a non-polar region in which the methyl groups hold the sheets together by van der Waals interactions, and a polar region between the sheets in which Coulombic and van der Waals interactions are operative. The salt formed from 2-naphthylsulfonate crystallizes in the same motif, with a larger bilayer "thickness" compared to the methanesulfonate compound (10.82 vs. 5.96), as expected for the larger R group. This bilayer structure is evident from the macroscopic properties of these crystals, as crystals can be cleaved with a razor along the sheet direction, or by pulling the sheets apart with adhesive tape applied to opposite sides of a crystal. Atomic force microscopy of freshly cleaved crystals of guanidinium methanesulfonate in a saturated solution of the salt also reveals large molecularly flat *ab* faces with molecular-scale contrast that has hexagonal symmetry.

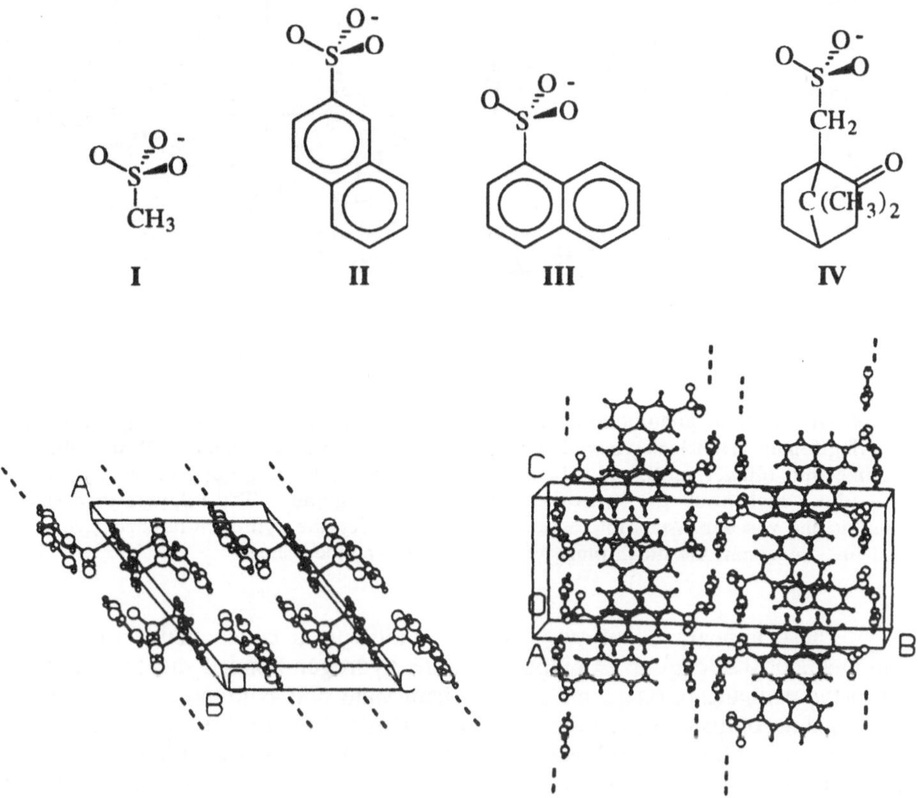

Figure 3. Unit cells of guanindinium methanesulfonate (left) and guanidinium 2-naphthylsulfonate (right) as viewed along the hydrogen-bonded sheets. Both compounds crystallize into bilayer motifs in which interdigitation of the R groups provide a "van der Waals" glue which holds the hydrogen bonded sheets together. The bilayers then stack by Coulombic and van der Waals interactions directly between the hydrogen-bonded layers. The dashed lines represent the mean plane of the hydrogen-bonded sheet.

The assembly of guanidinium salts of (1s)-(+)-10-camphorsulfonate and 1-naphthalenesulfonate differ dramatically from the aforementioned salts with respect to two principal features (Figure 4). First, these salts do not form bilayers; rather, the hydrogen-bonded sheets stack by interdigitation of translationally related single layers, in which the R groups on adjacent ribbons within a sheet orient to opposite sides of the sheet (Figure 5). Second, the hydrogen-bonded sheets in these compounds are puckered so that they more closely resemble "pleated" or "accordion" sheets. This puckering results from bending of adjacent ribbons, as if the hydrogen bonds between the ribbons act as a hinge. The magnitude of the puckering can be defined by the dihedral angle, θ_{IR}, between the adjacent ribbons.

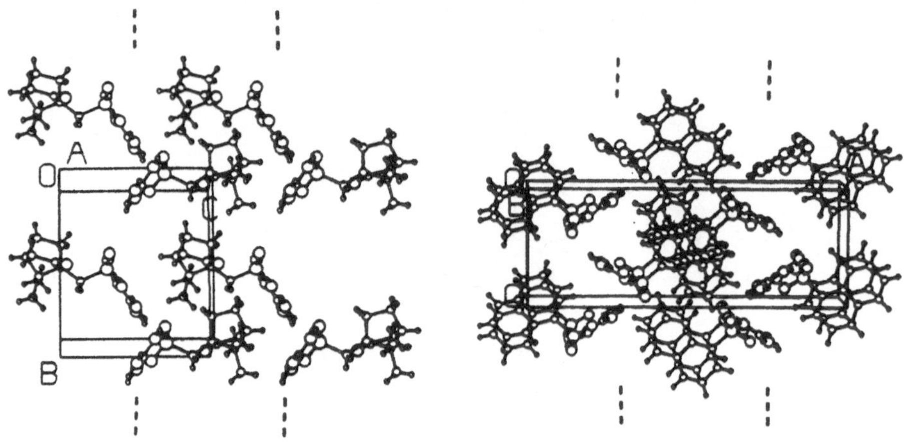

Figure 4. Unit cells of guanindinium (1s)-(+)-10-camphorsulfonate (left) and guanidinium 1-naphthylsulfonate (right) as viewed along the hydrogen-bonded sheets. Both compounds crystallize into bilayer motifs in which interdigitation of the R groups provide a "van der Waals" glue which holds the hydrogen bonded sheets together. The bilayers then stack by Coulombic and van der Waals interactions directly between the hydrogen-bonded layers. The dashed lines represent the mean plane of the hydrogen-bonded sheet.

These appearance of two different motifs is a consequence of the steric requirements of the R group extending from the hydrogen-bonded sheets. The bilayer motif of the methanesulfonate and 2-napthylsulfonate salts is favored by high packing density (and accompanying van der Waals interactions) that results from interdigitation of the R groups within the bilayer. If the R group is described by either spheres or cylinders it can be shown that close packing in the bilayer motif is possible only if the diameter of the R, as viewed normal to the hydrogen bonded sheet, is less than $d_{S-S}/\sqrt{3}$, where d_{S-S} is the center-to-center distance between nearest sulfonate residues (Figure 6). If the diameter exceeds this value interdigitation is not possible and the salts resort to the single layer motif in order to achieve dense packing. The R groups also interdigitate in this motif, albeit in a different manner than in the bilayer structure. The naphthalene sulfononates are particularly illustrative; 1-naphthylsulfonate has a larger projected area and therefore forms the single layer rather than the bilayer motif.

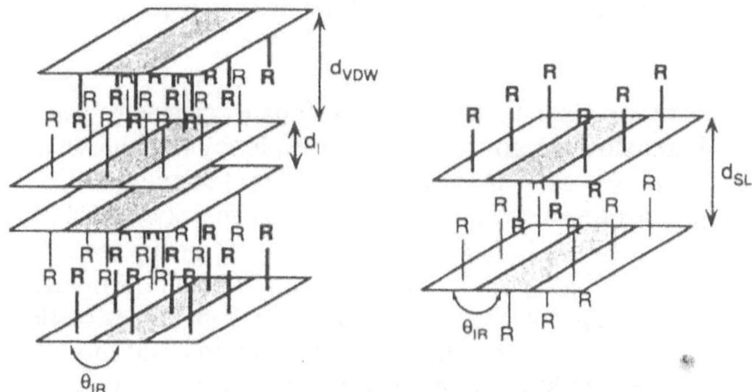

Figure 5. Schematic representation of the bilayer and single layer motifs in guanindinium-sulfonate salts. In the bilayer motif all the R groups of a sheet are oriented to the same side, whereas in the single layer motif the orientation of the R groups on adjacent ribbons alternative above and below the sheet. The puckering of the sheet is defined by the dihedral angle, θ_{IR}, between ribbons.

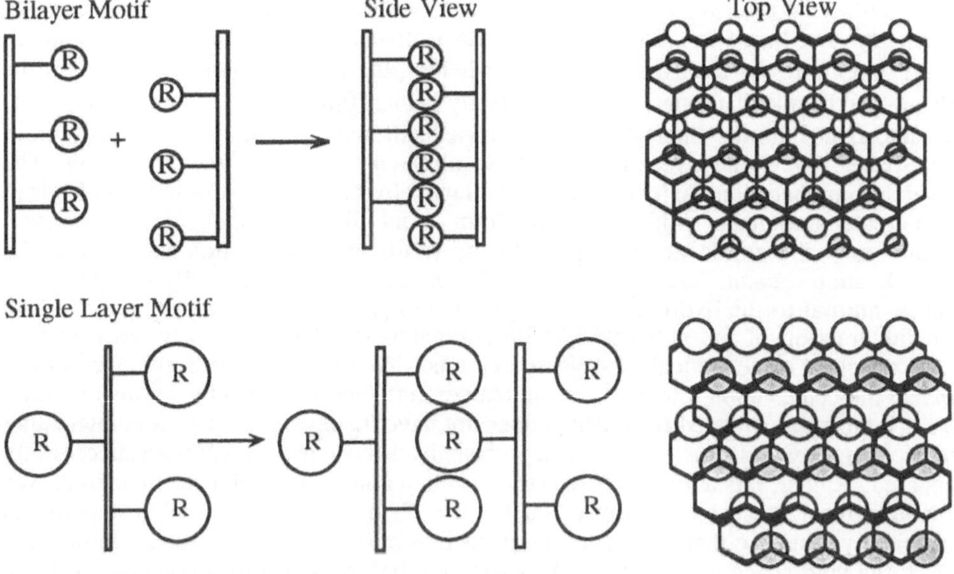

Figure 6. Schematic represention of the steric influence of the R group on the layering motif in guanindinium-sulfonate salts. Interdigitation of R groups all arranged on the same side of the hydrogen-bonded sheet is possible if the R group projected diameter < $d_{S\text{-}S}/\sqrt{3}$, where $d_{S\text{-}S}$ is the distance between nearest sulfonate groups. This results in the bilayer structure (top). If the diameter > $d_{S\text{-}S}/\sqrt{3}$, interdigitation is not possible and the single layer motif, in which R groups on adjacent ribbons are oriented to opposite sides, is formed as this allows interdigitation and efficient packing between the hydrogen-bonded sheets.

2.2 COVALENTLY LINKED HYDROGEN-BONDED SHEETS: ENGINEERING OF NANOPOROUS LATTICES

The crystal structures of the alkane- and arenedisulfonates clearly reveal that the layering motifs in the third dimension are governed by the tendency to achieve efficient packing within the bilayer. In these cases, there is one R group for every sulfonate ion in each layer. However, if disulfonates were used to build these networks so that they would "crosslink" the hydrogen bonded sheets the density of R groups would effectively be reduced by a factor of two (Figure 1g,h). This seemed to be a reasonable design strategy for the construction of nanoporous layers in which the robustness of the guanidinium-sulfonate network allows it to be considered as the fundamental module. In principle, the porosity of the layers can then be controlled simply by varying the length of the disulfonate spacer. The simplest spacers are the alkanedisulfonates $^{-}O_3S\text{-}(CH_2)_n\text{-}SO_3^{-}$. Molecular models reveal that for n = 0 (dithionite or $S_2O_6^{2-}$) and n ≥ 2, covalently linked bilayer motifs are possible. For n = 1, the tetrahedral geometry of the spacer prevents the formation of parallel sheets.

We have found that for n = 0, 2, and 4, that covalently linked bilayers are indeeed formed. In the case of n = 0, S-S linked bilayers crystallized in the hexagonal space group $P6_3mc$ are observed (Figure 7). When n = 2 and 4 the space group symmetry is reduced to $C2/m$ and $P2_1/n$, respectively, because the alkane chains in bilayer region lower the symmetry. Nevertheless, the hydrogen-bonded sheets in the ethane- and butanedisulfonate salts are perfectly planar (θ_{IR} = 180°) and exhibit the ideal quasihexagonal motif. Interestingly, the ethanedisulfonate has a considerable amount of void space between the layers, the crystal overall having a packing fraction of 0.59. The larger spacing between the hydrogen bonded layers in the butanedisulfonate salt leads to even greater amounts of void space in the form of one-dimensional channels. However, in this case the voids are occupied by acetonitrile solvent molecules from the crystallization solvent. The acetonitrile molecules are oriented vertically in the bilayer, that is, normal to the hydrogen bonded sheets and parallel to the butane chains. The packing fraction of the network without solvent is 0.51, whereas with solvent it is 0.68, which is more typical of organic crystals. These crystals lose the acetonitrile solvent molecules upon standing at room temperature, accompanied by a transformation to a crystalline unsolvated form which does not have the layered motif. It is interesting to note that a layered structure based on propanedisulfonate has not yet been successfully prepared. Rather, this anion leads to crystals with a nonlayered solid state packing. We conclude from these observations that for n = 2, the hydrogen bonded network is sufficiently robust so that the low packing fraction of 0.59 can be tolerated. In the case n = 3, the void space is too large to be sustained, but too small to accomodate solvent to stabilize the bilayer motif. However, for n = 4 the void space is large enough to accomodate solvent molecules which act as "supports" for the bilayer; without the solvent the void space cannot be sustained.

These results suggest that more extensive pillaring of the hydrogen-bonded sheets can be realized by using longer disulfonates coupled with solvent or guest molecules having lengths comparable to the expected separation between the sheets. The two-dimensional hydrogen bonded sheet is robust enough to be considered as a reliable "module," and consequently, the design of nanoporous bilayers simply relies on

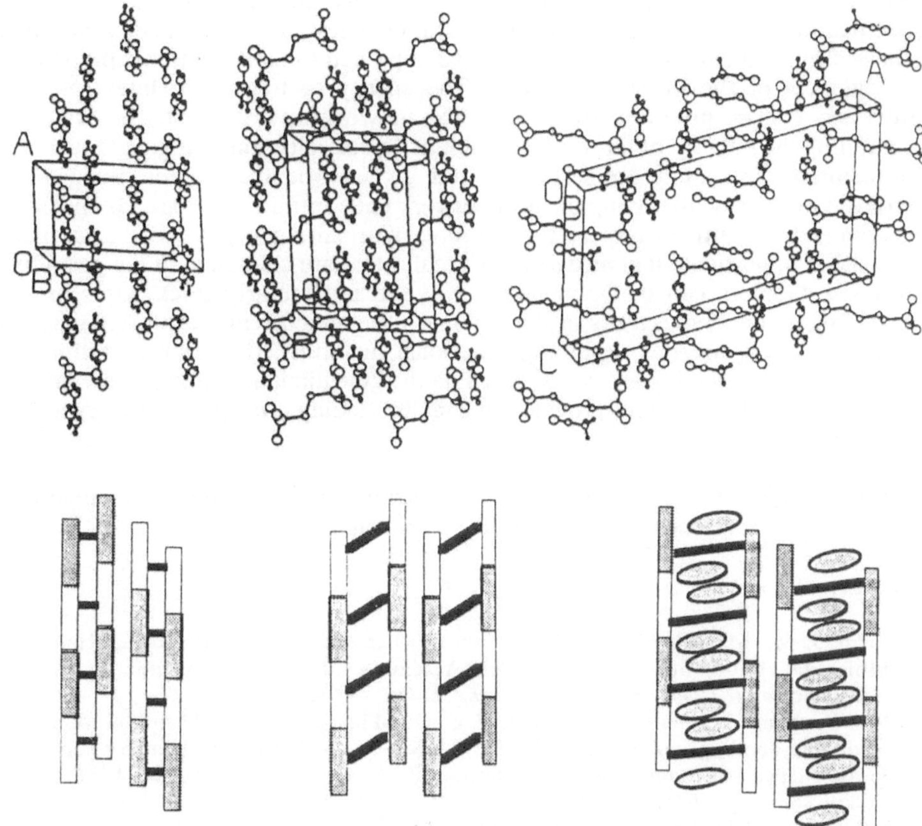

Figure 7. Unit cells of the guanidinium salts of dithionite (left), ethanedisulfonate (middle), and butanedisulfonate (right) illustrating the covalently linked bilayers. In the butanedisulfonate salt large channels that result from low density packing of the butane chains in the bilayer are filled with acetonitrile solvent molecules. Schematic representations of the bilayer motifs are shown below each example.

changing the length of one dimension. While it is tempting to compare such nanoporous materials with the well-studied inorganic aluminosilicate zeolites, the collapse of the organic networks upon removal of solvent makes such a comparison contentious. Rather, it is probably more appropriate to consider these materials as complementary, and to focus on the properties that distinguish the molecular networks from their inorganic counterparts.

Inorganic zeolites have proven to be quite utilitarian in the synthesis of novel host-guest materials for with novel optical properties, as well as catalytic environments for a variety of reactions. While the molecular networks cannot approach the thermal and structural stability of the inorganic zeolites, their molecular nature provides unique opportunities with respect to tailoring the host environment to more precise specifications. That is, the environment about the guest molecule can be systematically adjusted by molecular design, which can be exploited to tune electronic properties and arrangement of the guest molecules. The energy of transition states for reactive processes can be modified by varying the network environment, for example, by controlling the polarity of the voids through judicious selection of organic functionality. Furthermore, if reactions are performed in these networks the reaction products can be retrieved by simply dissolving the host network under mild conditions, compared to the rather harsh conditions required to dissolve inorganic zeolites. For example, it is reasonable to postulate that monomer molecules that are clathrated in the guanidinium-sulfonate bilayers can be polymerized to yield molecularly thick wires or sheets, depending on the connectivity of the polymer. We are currently exploring reactions in these bilayers as well as designing functional materials, including two-dimensional magnetic and conducting networks, based on the guanidinium-sulfonate networks. The studies described above have provided us with the guiding principles necessary to achieve these goals.

Acknowledgments. The authors gratefully acknowledge the financial support of the National Science Foundation and the Office of Naval Research.

References

1. Lehn, J. M., Mascal, M., DeCian, A., and Fisher, J. J. (1992) Molecular Ribbons from Molecular Recognition Directed Self-Assembly of Self-Complementary Molecular Components, *J. Chem. Soc. Perkin Trans.* **2**, 461 - 463.
2. Fan, E., Yang, L., Geib, S. J., Stoner, T. C., Hopkins, M. D., and Hamilton, A. D. (1995) Hydrogen-bonding control of molecular aggregation - Self-complementary subunits lead to rod-shaped structures in the solid state *J. Chem. Soc. Chem. Commun.* 1251-1252.
3. Zerkowski, J. A., MacDonald, J. C., Seto, C. T., Wierda, D. A., and Whitesides, G. M. (1994) New Varieties of Crystalline Architecture Produced by Small Changes in Molecular Structure in Tape Complexes of Melamines and Barbiturates, *J. Am. Chem. Soc.* **116**, 4305 - 4315.
4. Lehn, J.-M., Mascal, M., DeCian, A., and Fisher, J. J. (1990) Molecular Recognition Directed Self-Assembly of Ordered Supramolecular Strands by Cocrystallization of Complementary Molecular Components, *J. Chem. Soc. Chem. Commun.* 479-481.
5. Simard, M., Su, D., and Wuest, J. D. (1991) Use of Hydrogen Bonds to Control Molecular Aggregation. Self-Assembly of Three-Dimensional Networks with Large Chambers, *J. Am. Chem. Soc.* **113**, 4696.
6. Ermer, O. (1988) Fivefold-Diamond Structure of Adamantane-1,3,5,7-tetracarboxylic Acid, *J. Am. Chem. Soc.* **111**, 3747 - 3754.
7. Kitaigorodskii, A. I. (1973) *Molecular Crystals and Molecules*, Academic Press, New York.
8. Perlstein, J. (1994) Molecular Self-Assemblies. 2. A Computational Method for the Prediction of the Structure of One-Dimensional Screw, Glide, and Inversion Molecular Aggregates and Implications for the Packing of Molecules in Monolayers and Crystals, *J. Am. Chem. Soc.* **116**, 455 - 470.
9. Russell, V. A., Etter, M. C., and Ward, M. D. (1994) Layered Materials by Molecular Design: Structural Enforcement by Hydrogen Bonding in Guanidinium Alkane- and Arenesulfonates, *J. Am. Chem. Soc.* **116**, 1941 - 1952.

10. Russell, V. A., Etter, M. C., and Ward, M. D. (1994) Guanidinium Para-Substituted Benzenesulfonates: Competitive Hydrogen Bonding in Layered Structures and the Design of Nonlinear Optical Materials, *Chem. Mater.* **6**, 1206 - 1217.
11. Etter, M. C. (1991) Hydrogen Bonds as Design Elements in Organic Chemistry, *J. Phys. Chem.* **95**, 4601 - 4610.
12. Etter, M. C., MacDonald, J. C., and Bernstein, J. (1990) Graph Set Analysis of Hydrogen-Bond Patterns in Organic Crystals, *Acta Crystallogr.* **B46**, 256 - 262.

FROM MOLECULES TO CRYSTALS

MICHAEL J. ZAWOROTKO
Department of Chemistry, Saint Mary's University, Halifax, Nova Scotia, B3H 3C3, Canada

ABSTRACT. This paper addresses strategies for propagation of molecular symmetry into space group symmetry by invoking supramolecular chemistry and the concept of modular self-assembly. Three classes of compound are particularly amenable to modular self-assembly: cocrystals, salts and coordination polymers. A representative compound from each class is presented. A cocrystal based upon $[Mn(CO)_3(\mu_3\text{-}OH)]_4$, a T_d cubane cluster, crystallizes in the high symmetry cubic space group *P*n-3m. A salt based upon trimesic acid, 1,3,5-benzenetricarboxylic acid, generates a 2-D framework with 3-fold symmetry in a trigonal space group. The Zn sustained coordination polymer has D_{4h} point group symmetry at the metal and crystallizes in *P4/mmm* with Z=1. Synthetic and structural details concerning these compounds are presented along with a discussion of the implications of these results in the context of crystal engineering of porous solids.

1. Introduction

Are crystal structures predictable? This question was provocatively addressed by Maddox in 1988 [1] and more recently in an article by Gavezzotti [2]. The answers were definite and qualified negatives, respectively. The focus of this contribution will be an attempt to answer this question with a qualified yes, the qualification being that one must carefully select residues for symmetry (i.e. point group) and functionality (noncovalent molecular recognition sites) in order to impart control over molecular recognition, self-assembly and, ultimately, crystal packing, crystal system and space group. Since the vast majority of crystal structures were completed to analyze the molecular structure rather than the crystal structure (i.e. the CSD is not a truly representative or random sample) it is not appropriate or sometimes even realistic to address this question by using the existing literature. A series of recent results from our laboratory supports this assertion, placing particular emphasis upon self-assembly of complementary molecules or "modular self-assembly" (Figure 1).

J. Michl (ed.), Modular Chemistry, 409–421.

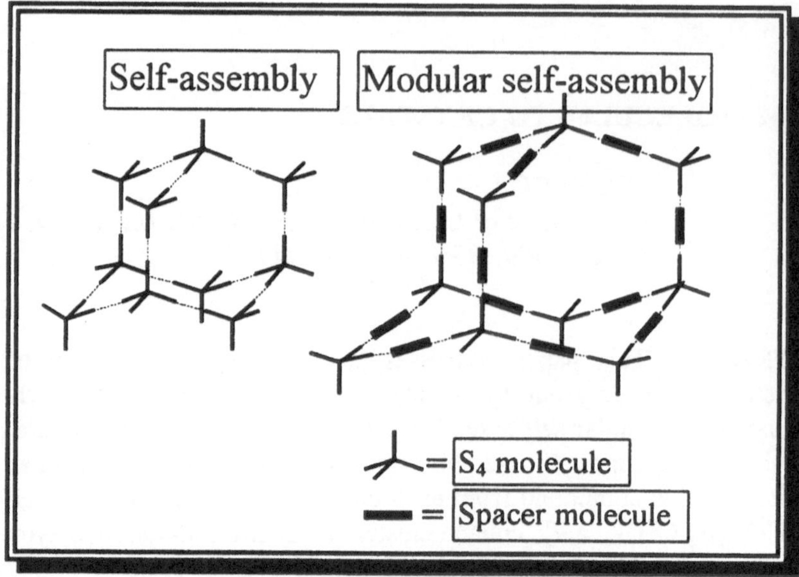

Figure 1. A comparison of the self-assembly and modular or multi-component self-assembly approaches to construction of infinite frameworks. A diamondoid network is illustrated.

2. Results and Discussion

The residues we have utilized for this study contain **previously known** organic, organometallic and inorganic moieties and represent three classes of compound that are by their nature modular: cocrystals, salts and coordination polymers. We have chosen to present the results of three crystal structures that are representative of these three classes of compound but are chemically and topologically unrelated apart from one aspect: they possess space group symmetry and therefore structural features that are clearly dictated by the molecular symmetry of the residues present in the compound. The three crystal structures help answer the question of whether crystal structures can be predictable, however, in many ways we feel that they afford more questions than answers since each structure, being inherently modular, is prototypal for a possibly large range of new solids. The structures are addressed individually and then summarized as a group.

2.1 $[Mn(CO)_3(\mu_3\text{-}OH)]_4 \cdot 2C_6H_6$, $\mathbf{1} \cdot 2C_6H_6$ - A DIAMONDOID COMPOUND IN A DIAMONDOID SPACE GROUP

As revealed in Figure 1, a modular approach to construction of diamondoid networks would involve self-assembly of the following pleromers [3]: an S_4 hydrogen bond donor moiety (a "director") with a complementary linear hydrogen bond acceptor or spacer (a "propagator"). In the case of a T_d moiety with linear spacers that can sustain 3-fold symmetry, the resulting structure would be expected to crystallize in a space group with -43m (i.e. T_d) symmetry. Such is indeed the case for the title compound, which crystallizes in the cubic space group Pn-3m. An adamantoid portion of the crystal packing in $\mathbf{1} \cdot 2C_6H_6$ is illustrated in Figure 2 and reveals how molecules of **1** are linked by benzene molecules in such a manner that a cubic diamondoid network is generated. The full T_d symmetry of **1** is propagated in the crystal since the benzene molecules engage in symmetrical bridge π-hydrogen bonds between the μ_3-OH moieties of **1**. Ironically, when **1** is crystallized in its pure form it crystallizes in the much lower symmetry space group R-3 and the only crystallographic symmetry retained is a 3-fold axis [4]. The generation of a large void is apparent and is filled by interpenetration of a second independent diamondoid framework. The crystal structure of $\mathbf{1} \cdot 2C_6H_6$ is therefore topologically analogous to that of cuprite (Cu_2O), which involves tetrahedral oxygen atoms that are linked by linear Cu atoms. Full details of the crystal structure of $\mathbf{1} \cdot 2C_6H_6$ were recently published along with those of a number of closely related structures generated from other arenes or more conventional linear spacer or propagator molecules such as 4,4'-dipyridyl, 1,4-diaminobenzene and 1,2-diaminoethane [5].

Diamondoid structures have also been prepared for tetrahedral organic compounds which self-assemble [6,7] and for tetrahedral transition metals such as Cu^I and Ag^I and appropriate linear bifunctional ligands [8]. The latter also represent an inherently modular situation but interpenetration of up to nine independent networks has thus far mitigated against realization of the full potential for porosity.

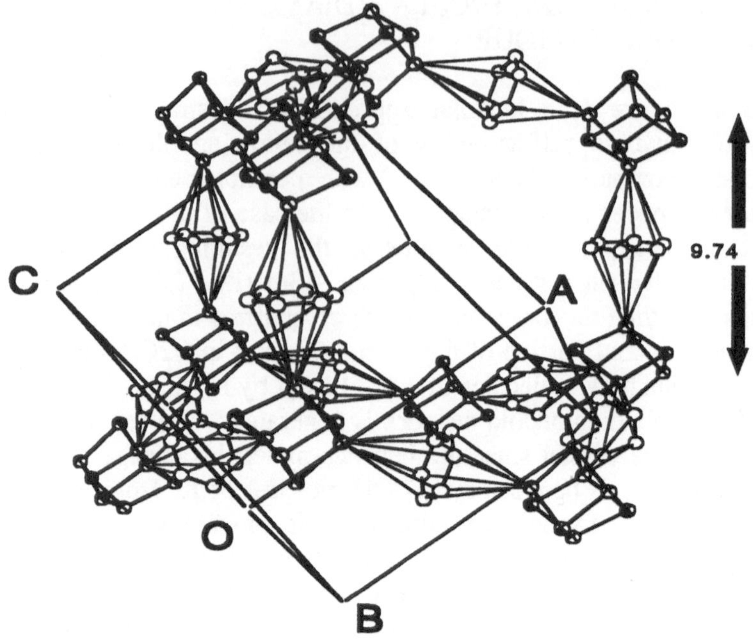

Figure 2. A perspective ORTEP view of anadamantoid portion of one of the two interpenetrated diamondoid networks that exist in 1·2benzene. The T_d point group of 1 is propagated into -43m crystallographic site symmetry because the benzene molecules sustain 3-fold symmetry.

2.2 $[NMe_2H_2]_{12}[(H_3TMA)(H_2TMA^-)_3(HTMA^{2-})_3(TMA^{3-})]$, TMA = 1,3,5-$(CO_2)_3C_6H_3$, A HONEYCOMB GRID SUSTAINED BY HYDROGENBIS(CARBOXYLATE) HYDROGEN BONDS.

Trimesic acid (1,3,5-benzenetricarboxylic acid), H_3TMA, represents a prototypal example of a "tecton" that self-assembles to form a honeycomb grid, in this case with cavities that are *predictably* ca. 14Å diameter. Unfortunately, although H_3TMA acts as a host matrix for a number of small guest molecules [9], it does not typically realize its full potential as a porous solid because of interpenetration. Indeed, the cavity in pure H_3TMA is filled by three independent grids [10] and we know of only one report in which H_3TMA eschews interpenetration [11]. Furthermore, H_3TMA does not possess a modular structure and is therefore not fine-tuneable. We have therefore investigated several deprotonated forms of H_3TMA that necessarily involve a countercation. As would be expected, the number of protons removed from H_3TMA and the nature of the counterion both influence the nature of the resulting network. In order for a honeycomb or chicken-wire grid to be generated the average charge on the anion would have to be 1.5. Such a grid would be sustained via hydrogen(biscarboxylate) hydrogen bonds. These hydrogen bonds occur in Speakman salts [12], are amongst the strongest hydrogen bonds known and would require either disproportionation and/or crystallographic disorder for generation of a honeycomb grid. Compound **2** represents our first crystallographically characterized H_3TMA. As Figure 3 reveals, a honeycomb grid is indeed generated. There are several remarkable features about the structure:

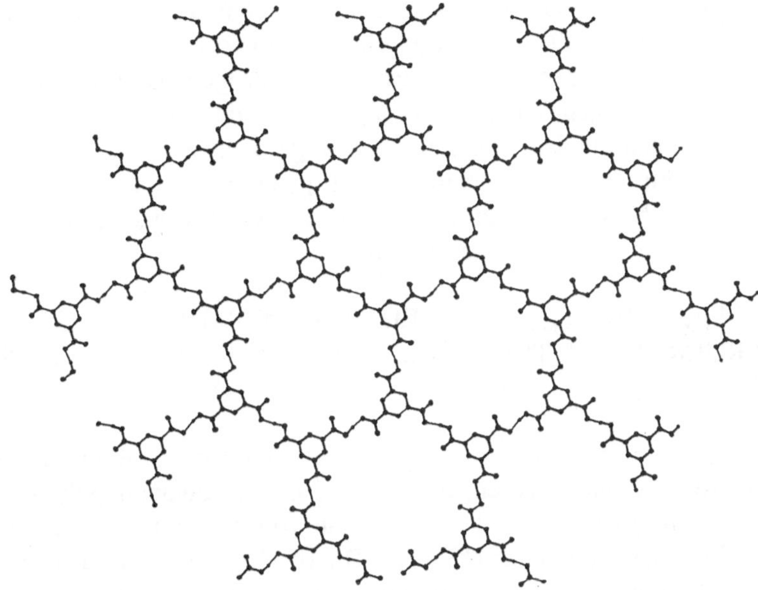

Figure 3. One of the anionic honeycomb grids observed in **2**. Cations are omitted for the sake of clarity.

(i) there is no interpenetration of grids despite the fact that the cavity of the honeycomb is only slightly smaller than that of H_3TMA itself; (ii) although the overall stoichiometry is $[NMe_2H_2]_{1.5}[H_{1.5}TMA]$, hydrogen atoms are located so that the crystal is effectively a cocrystal involving all four of the possible TMA residues, H_3TMA, H_2TMA^-, $HTMA^{2-}$ and TMA^{3-}; (iii) the 3-fold symmetry of the resulting network manifests itself crystallographically by the choice of crystal system (trigonal) and space group (R-3). (i)-(iii) can be attributed to the influence of the cations, in particular cation-anion hydrogen bonding, on the crystal packing. That the cations do not interfere with self-assembly of anions is expected since the hydrogen(biscarboxylate) hydrogen bond is one of the strongest hydrogen bonds known. Attachment of cations to carboxylate moieties occurs in a manner that generates two crystallographically independent honeycomb grids, both of which have crystallographic 3-fold symmetry. Grid A, illustrated in Figure 3, consists of one mole of H_3TMA to three moles of $HTMA^{2-}$. Grid B is built from one mole of TMA^{3-} to three moles of H_2TMA^-. The O····O distances in the hydrogenbiscarboxylate hydrogen bonds range from 2.523(5) to 2.551(5)Å. Hydrogen atoms were located and refined and their placement was also supported by C-O distances, which were observed to average 1.202 and 1.321Å in the C-OH and C=O moieties of the carboxylic acid, respectively, and 1.258Å in the carboxylate groups. An examination of the environment of the cations revealed that they are positioned so that they always donate a hydrogen bond to a carboxylate group rather than a carboxylic acid moiety. Once again, this is expected based upon the expected hierarchy of hydrogen bond strengths which dictates that ionic hydrogen bonds are stronger than and will occur in preference to molecular hydrogen bonds. The cations arrange themselves so as to cross-link adjacent honeycomb grids in a manner that generates a center of inversion between the grids. N···O hydrogen bond contacts, which range from 2.711(6) to 2.808(10) Å, are within the expected range.

2.3 $[Zn(4,4'\text{-dipyridyl})_2(SiF_6)]_\infty$ - A NEUTRAL OCTAHEDRAL GRID WITH PORES LARGER THAN THOSE FOUND IN NATURALLY OCCURING ZEOLITES.

Interest in coordination polymers has expanded dramatically in the past two years with several groups having presented examples of coordination polymers based upon modular self-assembly of known metal coordination environments and multifunctional exodentate ligands [13-17]. The driving force for such research is the correlation between mineralomimetic architectures and bulk physical properties such as conductivity, polarity and porosity. In such a context, there already exist examples of 2-D square and hexagonal/honeycomb grids generated from square planar or trigonal metals, respectively, coordinated to linear bifunctional ligands such as *4,4'*-bipyridine or pyrazines. Furthermore, as detailed earlier, 3-D diamondoid coordination polymers based upon tetrahedral or S_4 coordination environments have been generated by either coordinate covalent bonds or hydrogen bonding. However, given the ubiquity of octahedral metal environments, it is somewhat surprising that simple 3-D octahedral polymers remain largely unexplored. For the purpose of this study, we define an octahedral polymer as an infinite 3-D framework that is sustained by octahedral metal centres that are cross-linked by linear bifunctional ligands. We exclude infinite frameworks which consist of both octahedral metal centres and those with a different coordination geometry. In such a context, to our knowledge μ-cyano polymers such as $Fe^{III}{}_4[Fe^{II}(CN)_6]_3$, Prussian Blue [18], the poorly characterized $[Cd(CN)_2(quinoxaline)]_n$ [13c] and triply interpenetrated $[Cd(pyrazine)\{Ag_2(CN)_3\}\{Ag(CN)_2\}]$ [17c] represent the only examples of such structures.

Figure 4 illustrates how cross-linking of cationic square grids with suitable linearbifunctional anionicligands could in principle afford neutral

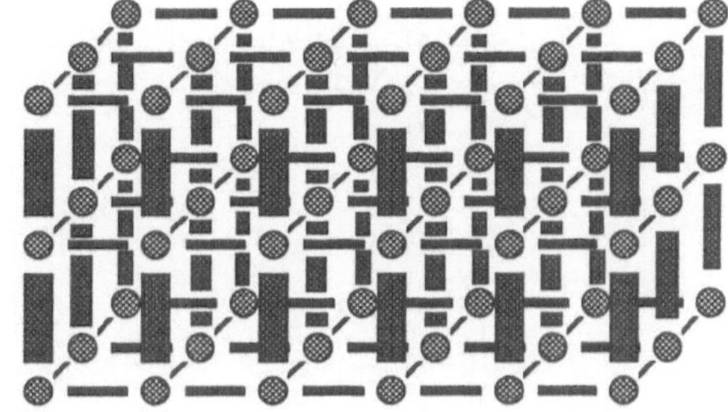

Figure 4. A schematic representation of how neutral octahedral frameworks might be generated. In principle, any octahedral transition metal would sustain the framework in the presence of appropriate bifunctional ligands.

octahedral polymers with porosity controlled by the length, volume and chemical type of the spacer ligands. $[Zn(4,4'\text{-bipyridine})_2(H_2O)_2]SiF_6$, **3**, polymerizes to afford square grids with sides of length 11.43Å [13a]. Unfortunately, unlike $[Cd(4,4'\text{-bipyridine})_2](NO_3)_2$, which eschews interpenetration [14a], the large cavity thus generated is self-included by hydrogen bonded hexafluorosilicate counterions and a second independent square grid. It occured to us that the anhydrous form of **3** could be expected to fit the hypothetical network proposed in Figure 4 since SiF_6^{2-} anions are capable of acting as a linear bridge between transition metal moieties such as $[Co(N\text{-vinylimidazole})_4]^{2+}$ [19] and $[Cu(5\text{-phenylpyrazole})_4]^{2+}$ [20]. We therefore synthesized the anhydrous form of **3**, $[Zn(4,4'\text{-bipyridine})_2]SiF_6$, **4**.

4·*x* dimethylformamide crystallizes in the tetragonal space group *P4/mmm* with Z = 1 and therefore sits around a crystallographic 4/mmm position. This alone is quite a remarkable feature since the space group symmetry manifests the maximum D_{4h} point group symmetry that can be sustained by an MX_4Y_2 species. The space group symmetry and cell dimensions are therefore entirely rational or indeed predictable based upon the chemical composition of **4**. However, an even more remarkable feature of **4** is that it eschews interpenetration and therefore, as Figure 5 reveals, large square channels are generated parallel to the crystallographic *c*-axis. As would be anticipated these channels have essentially

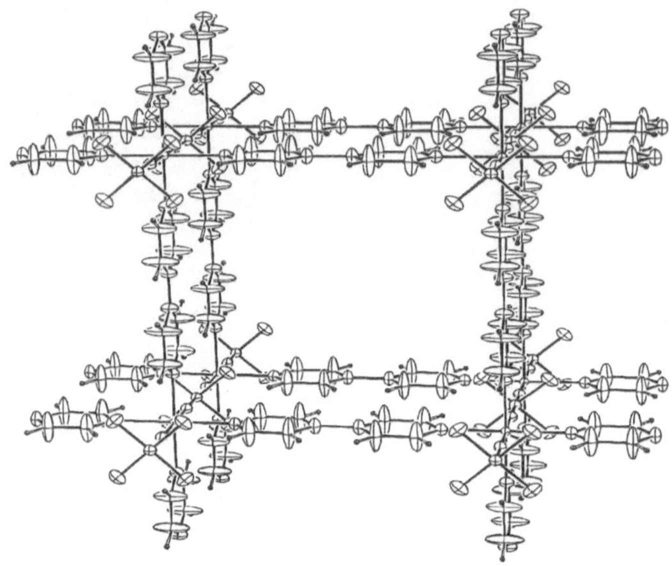

Figure 5. An ORTEP view of one of the square channels that run parallel to the crystallographic c-axis in **4**.

the same dimensions as those seen in the interpenetrated dihydrate, 11.3959(11) x 11.3959(11)Å, and, with effective pore sizes of ca. 8 x 8Å, are larger in cross-dimensional area than any that we are aware of in natural zeolitic solids [21]. The channels represent ca. 50% of the volume of the crystal since the density of **4** in the absence of solvent is just 0.87 compared to 1.856 in the corresponding dihydrate, **3**. The μ-SiF_6^{2-} moieties bridge in a perfectly linear fashion with Zn-F distances of 2.082(10)Å, shorter than the M-F distances encountered in previous studies concerning μ-SiF_6^{2-} units. The Zn-N distances, 2.157(8)Å, are also within the expected range. That there is no interpenetration in **4** can possibly be attributed to the orientation of the *4,4'*-bipyridine ligands. These ligands are disposed such that the pyridine moieties are almost coplanar with each other and the crystallographic *ac* and *bc* planes, respectively. The former is quite unusual since such a conformation for *4,4'*-bipyridine is sterically disfavoured and the latter is important since it precludes porosity and interpenetration along either the crystallographic *a* or *b* directions. We attribute this conformation and orientation to the existence of C-H···F hydrogen bonds from α-C-H hydrogen atoms to fluorine atoms of the SiF_6^{2-} anions. C-H···F and C···F contacts, 2.342 and 3.429(9)Å, respectively, are well within ranges expected for significant C-H···F interactions [22].

Although **4** does not compare to other zeolitic solids in terms of stability in aqueous environments or potential catalytic activity, we consider the following features to be particularly salient: (a) **4** represents, to our knowledge, the first example of a neutral octahedral coordination polymer and, therefore, there are no counter ions occupying the microchannels (compounds with comparable or larger pores have recently been reported but either cations or anions are present in the pores [13b,23]); (b) **4** is inherently modular and all three components are in principle interchangeable; (c) the dimensions of the channels are precisely what one would expect based upon the geometric features of the three components; (d) the channels in **4** are hydrophobic; (e) the space group of **4** can be regarded as a direct manifestation of the point group symmetry (i.e. the structure of **4** can be regarded as having been "crystal engineered" even from the perspective of space group symmetry). We consider **4** to be the prototype of a potentially wide range of porous octahedral structures and are actively investigating the generality of this class of solid and its ability to incorporate medium sized hydrophobic molecules of cross-sectional area consistent with the dimensions observed in **4**. Further details concerning compound 4 will soon be published elsewhere [24].

3. Summary

In summary, the X-ray crystal structures of compounds **1**, **2** and **4** illustrate that mineralomimetic architectures can be generated from even the weakest of noncovalent interactions. For example, compound 1 demonstrates that π hydrogen bonding can be inherently directional and strong enough to control self-assembly, crystal structure and space group. These results therefore support the recent work of other groups which is suggesting that the key to designing solid state supramolecular architecture lies with controlling symmetry and functionality at the molecular level. In other words, it is reasonable to assert that crystal packing is a *de facto* manifestation of even subtle molecular properties rather than *vici-versa.* Indeed, even in our group, such a strategy has thus far afforded predictable 1-D (strands), 2-D (sheets or carpets) or 3-D (superdiamondoid, octahedral) networks. Furthermore, when coupled with modular approaches of the type described herein, there is inherently a "fine-tuneability" in the resulting architecture and all three classes of compound, cocrystals, salts and coordination polymers, are relatively unexplored in the context of systematic examination.

Our particular interest at present concerns designing solids with pores that are fine-tunable but it should be clear that that the principles of crystal engineering should be applicable to control of other bulk properties of interest to the materials scientist and that they should be applicable to films as well as to crystals. However, on the down side, Maddox's comments concerning our inability to predict crystalline architecture from molecular structure [1] are still almost as generally true as they were in 1988.

4. Experimental

4.1 **1**·2C_6H_6.

1·2C_6H_6 was prepared by recrystallizing **1**·2C_6H_5Me [25] from distilled benzene. X-ray structure of **1**·2C_6H_6: $C_{24}H_{16}Mn_4O_{16}$, orange cubes (0.35 x 0.40 x 0.40mm), cubic, space group *Pn-3m*, a = 11.2472(4)Å, V = 1422.77(5)Å^3, Z = 2, ρ = 1.82 Mgm^{-3}. Data were collected on an Enraf-Nonius CAD-4 diffractometer with $Mo_{k\alpha}$ radiation (λ = 0.71073Å). Non-hydrogen atoms were refined with anisotropic thermal parameters. Values of R = 0.042 and R_w = 0.052 were obtained for 213 out of 264 reflections with $I > 2.5\sigma(I)$ and 28 parameters.

4.2 Compound **2**.

2 was prepared by recrystallizing trimesic acid from dimethylformamide, dmf, over a period of several weeks. Dimethylamine is a thermal decomposition product of dmf. X-ray structure of **2**: $C_{32}H_{42}MN_4O_{16}$, colorless (0.20 x 0.50 x 0.60mm), trigonal, space group *R-3*, a = 33.412(5), c = 17.465(3)Å, V = 16885(4)Å^3, Z = 18, ρ = 1.31 Mgm^{-3}. Values of R = 0.053 and R_w = 0.049 were obtained for 2766 out of 4889 reflections with $I > 2.5\sigma(I)$ and 481 parameters.

4.3 Compound **4**.

0.31 g (1.0 mmol) of $[Zn(OH_2)_6]SiF_6$ was suspended in 25 ml of 1-4dioxane and 25 ml of benzene and the mixture was refluxed in a Dean-Stark apparatus to azeotropically remove the water. The powder of $ZnSiF_6$ thus formed was dissolved by adding 25 ml of dimethylformamide (DMF) to the above mixture. Most of the benzene and 1,4-dioxane were removed by rotoevaporation at this point. The mixture was cooled and a solution of 0.31 g (2.0 mmol) of *4,4'*-bipyridine in 10 ml of 1,4-dioxane was added. The solution was refluxed for 30 minutes. The resulting pale gold/yellow solution was cooled and allowed to stand at room temperature for 12 hours. 0.48 g of colorless crystals of **2** were afforded. X-ray structure of **4**·*x*DMF: As for **1**·2C_6H_6 except $C_{20}H_{16}F_6N_4SiZn{\cdot}x$DMF , colorless cubes (0.30 x 0.30 x 0.40mm), tetragonal, space group *P4/mmm*, with a = 11.3959(11), c = 7.6775(9) Å, V = 997.05(15)Å^3, Z = 1, ρ = 0.87 Mgm^{-3} (octahedral polymer only) or 1.27 Mgm^{-3} (including disordered solvent). The solvent molecules were observed to be disordered in a manner which could not be readily resolved. Solvent atoms were therefore treated as carbon atoms and refined with fixed isotropic thermal parameters and variable site occupancy. Values of R = 0.063 and R_w = 0.061 were obtained for 426 out of 569 reflections with $I > 2.5\sigma(I)$ and 48 parameters. Hydrogen atoms of the *4,4'*-bipyridine moieties were placed in calculated positions with D_{C-H} = 1.00Å.

5. References

1. Maddox, J.S. (1988) Crystals from first principles, *Nature* **335**, 201.
2. Gavezzotti, A. (1994) Are crystal-structures predictable? *Acc. Chem. Res.* **27**, 309.
3. Lehn, J.-M. *Supramolecular Chemistry: Concepts and Perspectives* (1995) VCH, Weinheim, p. 12.
4. Holman, K.T. and Zaworotko, M.J. (1995) Crystal and molecular structure of $[Mn(CO)_3(\mu_3\text{-}OH)]_4$, *J. Chem. Crystallogr.* **25**, 93-95.
5. Copp, S.B., Holman, K.T., Sangster, J.O.S., Subramanian, S. and Zaworotko, M.J. (1995) Supramolecular chemistry of $[M(CO)_3(\mu_3\text{-}OH)]_4$: A modular approach to crystal engineering of superdiamondoid networks, *J. Chem. Soc., Dalton Trans.* 2233.
6. Ermer, O. (1988) Fivefold-diamond structure of adamantane-1,3,5,7-tetracarboxylic acid, *J. Am. Chem. Soc.* **110**, 3747.
7. Simard, M.; Su, D.and Wuest, J.D. (1991) Use of hydrogen-bonds to control molecular aggregation -self-assembly of 3-dimensional networks with large chambers, *J. Am. Chem. Soc.* **113**, 4696.
8. Zaworotko, M.J. (1994) Crystal engineering of diamondoid networks, *Chem. Soc. Rev.* 283, and references therein.
9. Herbstein, F.H. (1987) Structural parsimony and structural variety among inclusion complexes (with particular reference to the inclusion compounds of trimesic acid, N-(p-tolyl(-tetrachlorophthalimide, and the Heilbron "complexes"), *Topics in Current Chemistry* **140**, 108.
10. Duchamp, D.J. and Marsh, R.E. (1969) Crystal structure of trimesic acid (benzene-1,3,5-tricarboxylic acid), *Acta Crystallogr.* **B25**, 5.
11. Herbstein, F.H., Kapon, M. and Reisner, G.M. (1987) Catenated and noncatenated inclusion complexes oftrimesic acid, *J. Incl. Phenom.* **5**, 211.
12. Speakman, J.C. (1971) Acid salts of carboxylic acids, crystals with some "very short" hydrogen bonds, *Structure and Bonding* **12**, 141.

13. (a) Gable, R.W., Hoskins, B.F. and Robson, R. (1990) A new type of interpenetration involving enmeshed independent square grid sheets - the structure of diaquabis-(4,4'-bipyridine)zinchexafluorosilicate, *J. Chem. Soc. Chem. Commun.* 1677. (b) Abrahams, B.F., Hoskins, B.F., Michail, D.M. and Robson, R. (1994) Assembly of porphyrin building-blocks into network structures with large channels, *Nature*, **369**, 727. (c) Abrahams, B.F., Hardie, M.J., Hoskins, B.F., Robson, R. and Sutherland, E.E. (1994) Infinite square-grid ($Cd(CN)_2(N)$) sheets linked together by either pyrazine bridges or polymerizable 1,4-bis(4-Pyridyl)butadiyne bridges arranged in an unusual crisscross fashion, *J. Chem. Soc., Chem. Commun.* 1049. (d) Hoskins, B.F., Robson, R. and Scarlett, N.V.Y. (1995) Six interpenetrating quartz-like nets in the structure of $ZnAu_2(CN)_4$, *Angew. Chem., Int. Ed. Engl.* **34**, 1203. (e) Batten, S.R., Hoskins, B.F. and Robson, R. (1995) 2,4,6-tri(4-pyridyl)-1,3,5-triazine as a 5-connecting building-block for infinite nets,*Angew. Chem., Int. Ed. Engl.* **34**, 820.
14. (a) Fujita, M., Kwon, Y.J., Washizu, S. and Ogura, K. (1994) Preparation, clathration ability, and catalysis of a 2-dimensional square network material composed of Cadmium (II) and 4,4'-bipyridine, *J. Am. Chem. Soc.* **116**, 1151. (b) Fujita, M., Kwon, Y.J., Sasaki, O., Yamaguchi, K. and Ogura, K. (1995) Interpenetrating molecular ladders and bricks,*J. Am. Chem. Soc* **117**,7287.
15. (a) Carlucci, L., Ciani, G., Proserpio, D.M. and Sironi, A. (1994) Interpenetrating diamondoid frameworks of silver(I) cations linked by N,N'-bidentate molecular rods, *J. Chem. Soc., Chem. Commun.* 2755; (b) Carlucci, L., Ciani, G., Proserpio, D.M. and Sironi, A. (1995) 1-dimensional, 2-dimensional, and 3-dimensional polymeric frames in the coordination chemistry of $AgBF_4$ with Pyrazine - the first example of 3 interpenetrating 3-dimensional triconnected nets, *J. Am. Chem. Soc.* **117**, 4562. (c) Carlucci, L., Ciani, G., Proserpio, D.M. and Sironi, A. (1995) Novel networks of unusually coordinated silver(I) cations - the wafer-like structure of $(Ag(Pyz)_2)(Ag_2(Pyz)_5)(PF_6)_3$ and the simple cubic frame of $(Ag(Pyz)_3)(SbF_6)$, *Angew Chem. Int. Ed. Engl.* **34**, 1895.
16. Gardner, G.B., Venkataraman, D., Moore, J.S. and Lee, S. (1995) Spontaneous assembly of a hinged coordination network,*Nature* **374**, 792.
17. (a) Kitazawa, T., Nishikiori, S., Kurodo, R. and Iwamoto, T. (1988) Novel clathrate compound of cadmium cyanide host with an adamantane-like cavity. Dadmium cyanide-carbon tetrachloride (1/1), *Chem. Lett.* 1729. (b) Kitazawa, T., Nishikiori, S., Yamagishi, A., Kuroda, R. and Iwamoto, T. (1992) Tetrahedral guest in a tetrahedral cavity - a neopentane molecule encaged in a 3-dimensional cadmium cyanide framework, *J. Chem. Soc., Chem. Commun.* 413. (c) Soma, T.,Yuge, H. and Iwamoto, T. (1994) 3-dimensional interpenetrating double and triple famework structures in $(Cd(bpy)_2(Ag(CN)_2)_2)$ and $(Cd(Pyrz)(Ag_2(CN)_3)(Ag(CN)_2))$, *Angew. Chem., Int. Ed. Engl.* **33**, 1665.
18. Buser, H.J., Schwarzenbach, D., Petter, W. and Ludi, A. (1977) The crystal structure of Prussian Blue: $Fe_4(Fe(CN)_6)_3$ x H_2O, *Inorg. Chem.* **16**, 2704, and references therein
19. Driessen, R.A.J., Hulsbergen, F.B., Vermin, W.J. and Reedijk, J. (1982) Synthesis, structure, spectroscopy, and magnetism of transition-metal compounds with bridging hexafluorosilicate groups. Crystal and molecular structure of catena(-μ-

hexafluorosilicato)tetrakis(N-vinylimidazole)cobalt(II), *Inorg. Chem.* **21**, 3594.

20. Keij, F.S., de Graaff, R.A.G., Haasnoot, J.H., Reedijk, J. and Pedersen, E. (1989) An unusual hexafluorosilicato-bridged chain compound - crystal structure of caten- (μ-Hexafluorosilicato)tetrakis(5-phenylpyrazole)Copper(II), *Inorg. Chim. Acta* **156**, 65.
21. Meier, W.M. and Olson, D.H. *Atlas of Zeolite Structure Types*, 3rd revised edition, Butterworth-Heinemann, 1992.
22. Shimoni, L., Carrell, H.L., Glusker, J.P. and Coombs, M.M. (1994) Intermolecular effects in crystals of 11-(Trifluoromethyl)-15,16-dihydrocyclopentaphenanthren-17-one, *J. Am. Chem. Soc.* **116**, 8162.
23. (a) McCarthy, T.J., Tanzer, T.A. and Kanatzidis, M.G. (1995) A new metastable 3-dimensional bismuth sulfide with large tunnels-synthesis, structural characterization, ion-exchange properties, and reactivity of KBi_3S_5, *J. Am. Chem. Soc.* **117**, 1294. (b) Jiang, T.; Lough, A.J.; Ozin, G.A.; Young, D.; Bedard, R.L. (1995) Synthesis and structure of the novel nanoporous tin(IV) sulfide material TPA-SNS-3, *Chem. Mater.* **7**, 245.
24. Subramanian, S. and Zaworotko, M.J. (1995) Porous solids by design: $[Zn(4,4'\text{-}bpy)_2(SiF_6)]_n \bullet x$ DMF, a single framework octahedral coordination polymer with large square channels, *Angew. Chem. Int. Ed. Engl* **34**, 2127.
25. Clerk, M.D. and Zaworotko, M.J. (1991) High-nuclearity manganese carbonyl complexes: structures of $[\{Mn(\mu_3\text{-}OH)(CO)_3\}_4]$ and $[Mn_7(\mu_3\text{-}OH)_8(CO)_{18}]$, *J.Chem.Soc., Chem. Comm.* 1607-1608.

DISCUSSION OF THE WARD AND ZAWOROTKO LECTURES

JAMES D. WUEST
Département de Chimie
Université de Montréal
Montréal, Québec H3C 3J7
Canada

1. Introduction

During this workshop, we have seen how it is becoming possible to build predictably ordered solids from "sticky" molecular modules, which I call *tectons*. This subject has been one of the principal themes of the lectures in Session 14, it has been the focus of several poster presentations, and it is of great interest to many participants at this meeting. In my brief introductory remarks, I want to talk about two issues related to the problem of designing tectons suitable for the assembly of ordered molecular solids.

1.1. NATURE OF THE INTERMOLECULAR INTERACTION

A chemist interested in controlling the assembly of ordered molecular solids must select an intermolecular interaction that will make the molecules associate in a particular way. Ideally, this interaction should be strong and directional, so that the degree of association will be high and the geometry of association will be easily predicted. Two obvious choices are hydrogen bonds or coordination to metals. Coordination to metals is attractive because it is typically strong and can be expected to produce robust networks. In addition, syntheses of coordination complexes can be extremely simple, as the lecture of Prof. Zaworotko has nicely illustrated, and metals offer a number of distinct coordination geometries, which can lead to the construction of networks with a variety of different architectures. These features seem so attractive that it is reasonable to ask why a chemist might prefer to use hydrogen bonds or other weak interactions in order to control the assembly of ordered solids. Because I have made extensive use of hydrogen bonding in my own work, I feel that I must try to answer this question. One advantage is that ordered

J. Michl (ed.), Modular Chemistry, 423–432.

hydrogen-bonded networks can be constructed from a single discrete module, whereas networks held together by coordinative interactions require at least two different components, a metal and a suitable ligand. My work, as well as that of Jeff Moore and others, has shown that it is possible to assemble architectures with interesting features from a single molecular component. Such contructions have an elegant simplicity. In addition, hydrogen bonds are typically weaker and more easily bent than coordination bonds, so hydrogen-bonded networks should be more deformable and may therefore be more resistant to mechanical stress and fracture. For similar reasons, ordered hydrogen-bonded networks may be less entropically unfavorable. Another advantage of hydrogen-bonded networks is that editing of defects may occur more readily, and networks held together by hydrogen bonds can in principle be disassembled more easily than those joined by coordinative interactions. This is important because it may facilitate the recovery of molecules trapped inside microporous hydrogen-bonded networks. Furthermore, the largely organic interiors of hydrogen-bonded networks might be expected to have a particularly strong and useful affinity for organic guests. It may also prove to be easier to devise ways to crosslink the primarily organic tectons that make up hydrogen-bonded networks, thereby converting architectures maintained by weak noncovalent interactions into those permanently joined by covalent bonds. A final reason to try to use weak interactions such as hydrogen bonds to control order in solids is that it is more difficult, making it a particularly challenging and interesting problem to attack. Networks maintained by coordinative interactions and by weaker interactions such as hydrogen bonds each have characteristic features that make them suitable for certain purposes and unsuitable for others, so it is likely that chemists will continue to need to use both types of interactions in order to develop the full potential of a modular tectonic strategy for making predictably ordered structures.

1.2. PREDICTABILITY OF CRYSTAL STRUCTURES

Crystal structures cannot yet be predicted in detail, and they are likely to remain unpredictable for many years to come. However, for many types of work, a detailed prediction is not actually necessary. As we have seen in our own work, for example, rigid tectons that incorporate four self-complementary hydrogen-bonding sites in a nominally tetrahedral orientation have a strong tendency to form interpenetrating diamondoid networks. The distance from the center of one tecton to its neighbor can be estimated, and a simple calculation can provide an approximate degree of interpenetration. Such estimates can be used to assess the expected openness of a particular diamondoid network and can lead the chemist to identify and synthesize tectons that should produce particularly open architectures. Detailed knowledge of the actual molecular symmetry of the tecton or the exact extent of deformation of the diamondoid network is not essential. What is important, at least in initial studies of such systems, is to be able to anticipate the connectivity of the tectonic modules, which in this case results with a high degree of predictability from their rigid, nominally tetrahedral geometry. Much progress in this area has been made without needing to predict three-dimensional structures in detail.

Many participants at this meeting have encountered the phenomenon of polymorphism in their work, and polymorphism has been a principal subject of the lecture of Prof. Ward. However, no one has clearly stated that the structures of many polymorphs are only trivially different from one another. It is important to keep this in mind. I have chosen one simple example, that of anthranilic acid, because it was studied in detail by Prof. Ward. Anthranilic acid has three known polymorphs. One is zwitterionic, and it is distinctly different from the other two. The other two are quite similar; in both, the carboxyl groups form cyclic hydrogen-bonded pairs, and layered structures are produced. For many purposes, the differences between these two non-zwitterionic polymorphs can be considered minor. I would suggest that one of the chief advantages of building ordered solids from structurally well-defined tectonic modules with strong, specific sites of intermolecular attraction is that the number of polymorphs with non-trivial differences should be minimized.

In his lecture, Prof. Ward has described beautiful work that shows how surfaces can direct the formation of specific polymorphs. However, it is not clear that surfaces can have such pronounced directing effects when the individual molecules have strong intrinsic preferences for certain types of contact. It is true that in the hydrogen-bonded networks studied by my group, not more than about half of the lattice energy can be attributed to hydrogen bonding of the tectonic modules. Nevertheless, hydrogen bonding makes an extremely important contribution in our systems, and structures deprived of extensive hydrogen bonding are surely much less stable. We have built into these tectonic modules very specific instructions for the formation of certain types of structures. While polymorphs with minor structural variations are still conceivable, it is unlikely that external effects, including interactions with surfaces, will change fundamental preferences dictated by the number and orientation of sticky sites of intermolecular contact.

2. Comments from Prof. Zaworotko and Prof. Ward

I would now like to begin the general discussion by inviting the two speakers, Prof. Ward and Prof. Zaworotko, to respond to any of the comments I have made in my introduction. Then we can proceed to questions from the audience.

ZAWOROTKO: In general, I agree with you about polymorphism. Some types of compounds are prone to polymorphism and others are not. If you choose the right system, polymorphism is not likely to be a problem. For example, monocarboxylic acids can associate to form hydrogen-bonded chains or cyclic dimers. The possibility of polymorphism is clear. However, if you use the association of carboxylic acids to generate a three-dimensional architecture, where interactions are controlled in all three dimensions, then polymorphism is almost precluded; at least, there are fewer opportunities for polymorphism.

WARD: I also agree. In part of the work I talked about today, we examined molecules with largely isotropic interactions, and extensive polymorphism is observed. In our work with guanidinium sulfonates, however, the interactions are highly directional. We have now studied the structures of about fifty different compounds in this series, and we have seen a few examples of polymorphism in which the normally characteristic hydrogen-bonded sheet structure is not favored. In general, however, it is a very robust network, and polymorphism is seldom observed, although it is not entirely absent. It is important to remember that in many materials, such as pharmaceuticals, polymorphism will remain an important problem because the individual molecules do not have rigid structures and strong directional intermolecular interactions.

3. Questions directed to Prof. Wuest

CHIDSEY (Stanford): Would it be possible to control the interpenetration of hydrogen-bonded networks by carrying out crystallizations of tectons in the presence of solvents or additives that themselves have secondary interactions with the tectons? If these interactions are important but weaker than the interactions of the tectons with themselves, then the tectons may associate normally by hydrogen bonding, but tectons that form interpenetrating networks may be prevented from getting too close.

WUEST: This is an interesting idea, but I'm not sure how to use it. There are many things that can be done to affect the crystallization of sticky tectonic modules. For example, we can introduce additives that hydrogen bond to the sticky ends of a growing crystal. In fact, we normally use carboxylic acids as solvents for crystallization, so the tectons are strongly solvated by hydrogen bonding to the carboxylic acids. However, this alone does not prevent interpenetration.

4. Questions directed to Prof. Ward

MÜLLEN (Mainz): You have emphasized the commensurate match between the growing crystal layer and the underlying substrate. It seems to me that organization in the initial layer and packing in the bulk crystal are subject to completely different control mechanisms. In between there must be stresses. How can crystallization on surfaces be used to create perfect single crystals?

WARD: We are talking about heteroepitaxy in the first layer and homoepitaxy, or normal crystal growth, in subsequent layers. In cases where there is heteroepitaxy, there is no perturbation of the original layered structure; the second and succeeding layers can then grow without stress, and the crystal can in principle be free of defects. However, if the first layer reconstructs on the lattice of the substrate because it has a lower free energy when it becomes commensurate, then you are right. Subequent layers will be strained and stressed. I would suggest that either of two things can happen. One possibility is that the crystal will grow with a large number of defects in that region. The other possibility is that during crystal growth, the initial layer will reconstruct back to its native form if interaction with the upper layer is stronger than that with the underlying substrate. I don't know if that actually happens or not. We are trying to image crystallization layer by layer in order to follow the strain and see how it evolves.

KAHN (Bordeaux): Could you explain what you mean by *pseudopolymorphism*? In addition, I wonder if anyone is aware of studies of crystallization in the presence of very large magnetic fields. This would be interesting if the species being crystallized were paramagnetic, because the local spins might be aligned, thereby producing a novel polymorph in which the magnetic field stabilizes a metastable phase that would not normally be available.

WARD: Pseudopolymorphism is commonly used to describe solvates of a given compound. The term probably should not be used, because solvates are not polymorphs.

WEGNER (Mainz): When you look at the AFM data presented on the first slide, you see that nucleation of the first layer occurs at certain points. What are these points? Are they defects on the HOPG?

WARD: I would like to say that nucleation occurs on perfect areas, but I can't say that unequivocally because I can't exclude the possibility that nucleation begins at point defects. We see no evidence of growth at steps. We have recently done work related to earlier studies of liquid crystals in molecular corrals on graphite. You can take a piece of HOPG and make single-layer holes or pits in the surface. When you then grow monolayers on this modified HOPG, growth occurs at the terraces around the holes, and the holes do not fill in until later. This may be because they are too small, and so if nucleation occurs near the edges of the holes, their small size may prevent critical nuclei from being formed. Alternatively, the orientation of the edges may be unfavorable for the epitaxially oriented nuclei. The first nucleation appears to be a random process on the surface; whether a point defect is there or not I can't say because I can't see that by AFM very easily.

In the case of crystallization on succinic acid, initial growth always occurs at a step site, because it presents an angle that matches the dihedral angle of two closely-packed planes in the crystal. Again, whether a defect is present along the step site and plays a role in nucleation is difficult to say.

SEDDON (Belfast): The experiments in which crystallization occurs on succinic acid are very elegant. Can you devise a system that induces the crystallization of a polymorph not previously known to exist?

WARD: This is an interesting question. We have not done this by direct means, but I have mentioned that a combinatorial approach might be appropriate. The dicarboxylic acids provide an interesting series of substrates for such an experiment, because the dihedral angle can be tuned from 90° to 45° just by the choice of substrate. In principle, then, this provides one library of substrates that can be used. Other libraries certainly exist that would provide similar structural variability. To answer your question, I would have to say that I don't know how to *design* systems for inducing the formation of a new polymorph, but I think they can be uncovered by a combinatorial strategy.

WUEST: I think that a critical experiment in this area is to devise a substrate that induces the formation of a new polymorph, particularly of a molecule that has already been very well studied. By doing that, you show that you really understand how crystallization occurs on a surface, and the practical implications would be extremely important.

PALACIN (Harvard): Can your approach be used to modify the size and shape of the crystals of a given polymorph?

WARD: In the system I described that produces the red polymorph, the morphology is in fact modified somewhat, and the long and short axes have been interchanged. In other systems that we have looked at, however, it appears that morphology is ultimately determined by fundamental preferences of the crystal; once you grow away from the substrate on a flat surface, the bulk properties of the crystal take over. What we can do is modify polymorph selectivity and induce rapid nucleation, but we do not see evidence for significant modifications of morphology, at least not yet.

CHIDSEY: In many models of fast crystal growth, dislocations are considered to play an important promoting role. Is this true in your systems?

WARD: In order for growth to take place near equilibrium, it must occur by dislocations. In fact, the monolayer that I described grows to a large crystal by dislocations. Dislocations generate the steps that create the second, third, fourth, amd succeeding layers, and growth occurs layer by layer.

STUPP (Illinois): Could you say something about the nature of defects in molecular crystals? What is equivalent to a point defect in atomic solids?

WARD: I think that the most common defect in organic crystals is the inclusion of solvent. This means that a solvent molecule is trapped in the solid; solvent of crystallization, which is included periodically in the solid, is a different matter. Little is known about defects in organic crystals.

TOLBERT (Georgia Tech): It seems to me that defects might be considered assets rather than problems. For example, have you tried to incorporate in a growing crystal other molecules that might be similar but that might help accommodate a dislocation?

WARD: Defects can of course be assets. For example, a seed crystal can be considered to represent a controlled defect. Work of the type you suggest has been done, and molecules have been designed so that they resemble a parent molecule and can bind to its lattice, but they contain functionality that serves as a blocking group and makes subsequent attachment of the parent molecules to a particular crystal face more difficult. This results in a modification of the crystal morphology. In this sense, designed defects have been introduced to change morphology. This can have significant practical importance. Similarly, dye molecules have been incorporated as defects in inorganic crystals to making lasing materials. So defects can definitely be useful.

5. Questions directed to Prof. Zaworotko

SEEMAN (New York): You did not say much about how your crystallizations depend on solvent. Are your results solvent-independent? If they are not, what features of the solvent play a role?

ZAWOROTKO: Solvent is present and certainly plays a role. Several generalizations guide us in our choice of an appropriate solvent. For example, acetonitrile is useful when transition metals are involved because it is a weakly coordinating ligand that can enter networks and bind metals when space is available and needs to be occupied. For work with trimesic acid, which is difficult to dissolve, methanol is the solvent of choice. It does not appear to interfere with ionic hydrogen bonds, but it is a bad solvent for crystallizations that would be expected to produce hydrogen bonds between neutral molecules. This is presumably because ionic hydrogen bonds are strong enough to dominate association. In such ionic systems, the choice of solvent is less important.

OZIN (Toronto): In the structure with the Ag-Ag bonds, the silvers are d^{10}. What is the bond length, and is the interaction of the silvers passive or active?

ZAWOROTKO: The bond length is 2.97 Å, which is just slightly longer than the distance in silver metal. There are also other examples of constrained silver dimers with T-shaped geometries. Ours might be the only example of a case where the silvers are not constrained to point toward one another. However, there are examples where macrocyclic chelating ligands force the silvers together, and then the same distance is observed. The distance is well within the sum of the van der Waals radii, so it is a real interaction.

WARD: How important is solvent in the formation of structures derived from your cubic metal hydroxide?

ZAWOROTKO: Solvent is critical in the formation of molecular networks. In the case you are referring to, we use benzene; if we use solvents that can accept hydrogen bonds more strongly than an arene, then they will bind to the cube instead. Obviously, the solvents must be kept dry. That is the only precaution we must take. When strong hydrogen bonds are involved, we should normally get four-fold network structures with holes. With methanol as a solvent, we get a three-component, two-dimensional, interwoven system. In this case, the solvent participates, presumably because a simple diamondoid structure would not be close-packed. Solvent is especially important in such systems.

WARD: In many cases, removal of solvent from crystalline solvates causes the space group to change to one of lower symmetry. Is this related to what you observe?

ZAWOROTKO: In our case, the solvent can propagate the tetrahedral symmetry of the metal hydroxide module. This is what happens when the solvent is benzene. It allows the tetrahedron to become part of the space group. Our metal hydroxide module has such a strong affinity for other molecules that we had a great deal of difficulty crystallizing it free of solvents. After much effort, however, we found that it could be crystallized from chloroform. Under these conditions, there is no 4 bar symmetry, and the molecules are close-packed when they crystallize on their own. When they are crystallized with something else, the solvent tends to propagate the symmetry. The tetrahydrate of the metal hydroxide module is a discrete entity that hydrogen bonds to adjacent tetrahydrates to give an infinite three-dimensional array. Crystals of the type we tend to see could be called solvates, but I prefer to call them cocrystals because I feel that they involve two complementary components, whereas in a solvate the included molecules can be considered to play a more passive role by filling space.

WARD: There is an interesting analogy between the phase diagrams of block copolymers and what you see in your cocrystals. In your systems, you are introducing a second component; in addition, you are changing the length scale.

ZAWOROTKO: I think cocrystals are unfairly maligned. All you have to do is choose the components correctly and then mix them together. If they are complementary, the symmetry at the molecular level will be propagated. With cocrystals, there is essentially an infinite number of permutations.

Occasionally one hears speculation that if a more stable polymorph of a particular compound comes into existence, then eventually it will no longer be possible to obtain the less-stable forms. Is there any scientific basis for this?

WEGNER: It is simply not true. It is contradicted by experience. In the course of my earlier work on solid-state polymerizations, my group acquired a great deal of familiarity with polymorphs. We could reproducibly form up to five polymorphs of the same compound in the same laboratory.

WARD: On the other hand, I am aware of problems experienced by pharmaceutical companies when an undesired polymorph is difficult to avoid.

CHIDSEY: There is much interest in the formation of particular crystal morphologies by electrodeposition and other forms of deposition. Organic modular chemistry might make a contribution in this area because there is much empirical information about additives that may bind to inorganic and metallic defects and thereby control morphology in electroplating. However, there is no rational molecular or modular understanding of these phenomena.

SEDDON: Could you tell us more about how your coordination networks can be assembled and disassembled?

ZAWOROTKO: Hydrogen-bonded systems are ideal because they can be destroyed by warming in an appropriate solvent or reformed by cooling. The coordination networks tend to be much less soluble, and their formation is less easily reversed. In some cases, crystallization must be carried out at 90 °C.

MICHL (Colorado): I would like to know more about how your metal hydroxide cubane-type molecules might be modified structurally to produce new modules.

ZAWOROTKO: These compounds are well known, and studies of the manganese derivatives date from the 1970s. In particular, conversion of the hydroxy groups into alkoxy groups should be a simple substitution reaction. The cubane motif is more robust and common than most chemists realize. In fact, I suppose that every metal has corresponding tetrahedral cubane-type clusters.

WUEST: Have you attempted to study the porosity of your square channel compounds?

ZAWOROTKO: Our principal goal is to generate analogous systems with other metals. We are interested in porosity, and we would like to try to include larger molecules in the pores, but that is not our primary focus. There is much that needs to be done, and I would prefer to begin by first mapping out the basic rules of crystal engineering. This will not be quick, nor will it be the work of a single group, but there is an explosion of interest in this field, and many exciting developments have occurred recently. For example, we now have examples of paramagnetic solids with infinite frameworks, organic solids with pore openings larger than those of naturally-occurring zeolites, and structures with large cavities that are fully closed. If we can put these elements together, and create materials with useful properties on demand, then we are in business. But there is still much to do with the simplest of systems.

BOTTOM UP CONSTRUCTION OF PHOTOCHEMICAL MOLECULAR DEVICES BY MODULAR CHEMISTRY

VINCENZO BALZANI
Dipartimento di Chimica "G. Ciamician" Università di Bologna
via Selmi 2, 40126 Bologna, Italy

ABSTRACT. Photochemical molecular devices (PMDs) are structurally organized and functionally integrated molecular systems capable to perform useful light-induced functions. The "bottom up" construction of PMDs is an important field of application of modular chemistry. The first step is the design and synthesis of components (modules) having suitable light-related properties (absorption spectrum, excited state lifetime, energy, and redox potentials, etc.). Such modules have then to be assembled in ordered arrays. Ordering does not simply concern the achievement of the desired structure, but also the properties related to the function that the device has to perform In this paper we illustrate the modular aspect of some PMDs which are currently the object of our investigations, namely (i) PMDs based on energy-transfer processes, including dendrimers for light harvesting, and (ii) PMDs based on structural changes (light driven mechanical molecular machines).

1. Introduction

1.1. PHOTOCHEMICAL MOLECULAR DEVICES (PMDs) AND MODULAR CHEMISTRY

A most important driving force of chemical research is currently the extension of the concept of device (or machine) to the molecular level [1]. A *molecular device* is an

J. Michl (ed.), Modular Chemistry, 433–449.

assembly of molecular modules (i.e., a supramolecular species) designed to achieve a specific function. Each module performs one or more single acts, while the entire device performs one or more complex functions, characteristic of the assembly. The extension of the concept of "device" to the molecular level is of interest not only for basic studies, but also for the growth of nanoscience and the development of nanotechnology.

Molecular devices, like macroscopic devices, need energy to operate. For several reasons, the most convenient form of energy to make molecular devices work is light. We are interested to design and construct *photochemical molecular devices* (PMDs) [2], i.e. structurally organized and functionally integrated systems capable to perform useful light-induced functions. In a PMD at least one of the modules must play the role of interface towards light. This module may be connected, either directly or through other components, to a module that plays the role of interface towards use. Photochemical molecular devices can be based on processes involving energy transfer, electron transfer, or structural changes.

The "bottom up" construction of PMDs is an important field of application of modular chemistry. The first step is the design and synthesis of components (modules) having suitable light-related properties (absorption spectrum, excited state lifetime, energy, and redox potentials, etc.). Such modules have then to be assembled in ordered arrays. Ordering does not simply concern the achievement of the desired structure, but also the properties related to the function that the device has to perform [2]. For example, if we are interested to construct a light harvesting antenna (*vide infra*), the molecular modules must be ordered in the spacial structure to allow efficient energy transfer along the desired direction.

In the last few years a great variety of PMDs have been designed and constructed using organic or inorganic components [2,3]. In this paper we will illustrate the modular aspect of some PMDs which are currently the object of our investigations, namely (i) PMDs based on energy-transfer processes, and (ii) PMDs based on structural changes (mechanical molecular machines).

1. 2. FUNCTIONS OF PMDs

In nature photons are exploited by living organisms as energy in photosynthetic processes and as information in visual processes (Figure 1). Taking inspiration from Nature, scientists are trying to construct PMDs, much simpler than the natural ones, that can use light as an energy supply to perform energy-expensive functions (e.g., conversion of sunlight into chemical or electrical energy), or as an input signal for sensors or information processing (Figure 2).

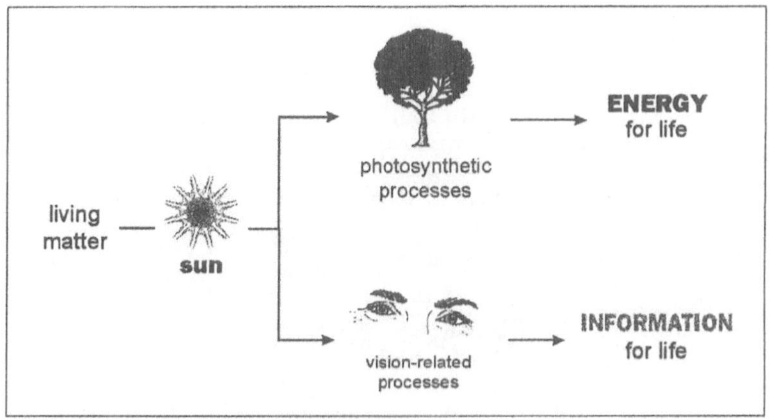

Figure 1. In nature light is used for energy and information purposes.

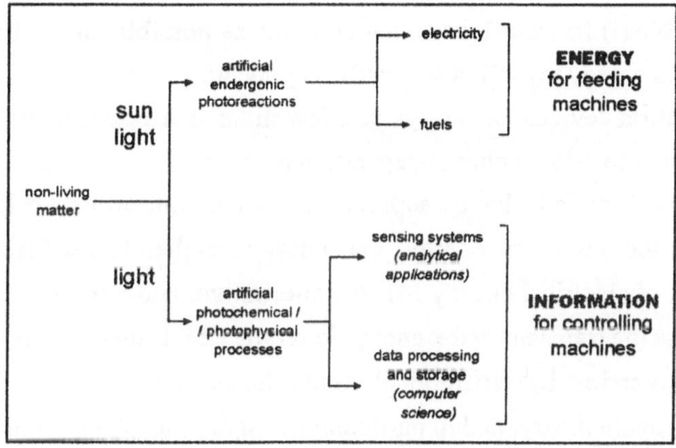

Figure 2. Energy- and information-related artificial photochemical processes.

1.3. PMDs FOR ENERGY CONVERSION

The structure of natural photosynthetic systems suggests that artificial systems for solar energy conversion should be based on two fundamental types of devices (Figure 3):

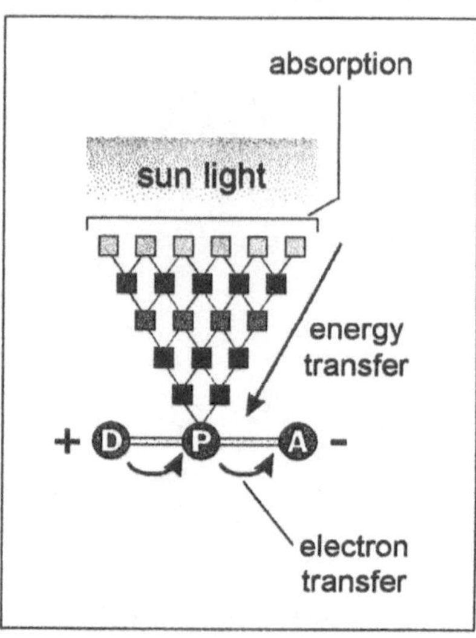

Figure 3. Schematic representation of an artificial device capable to harvest sunlight for energy-conversion purposes.

- light harvesting devices (called also antenna devices), i.e. large arrays of molecular components capable (i) to absorb as much sunlight as possible and (ii) to channel the resulting excitation energy towards a selected component;

- charge separation devices, i.e. arrays of a few molecular components capable to use the excitation energy to cause a charge-separation process.

Following photoinduced charge separation, part of the absorbed light energy is transiently converted into redox energy. This energy can then be used in various ways. One is generation of electrical energy [4]. An alternative, and perhaps more important, possibility is to use the transient redox energy to produce high energy chemicals ("fuels"). Solar energy conversion by artificial photosynthesis is certainly one of the most challenging goals in chemistry, and in particular in supramolecular chemistry. A possible scheme of artificial photosynthesis and some of the basic principles involved in this process have been discussed elsewhere [3e].

1.4. PMDs FOR INFORMATION PURPOSES

Exploitation of photons for information purposes can be performed by two different routes (Figure 4) [3e]:

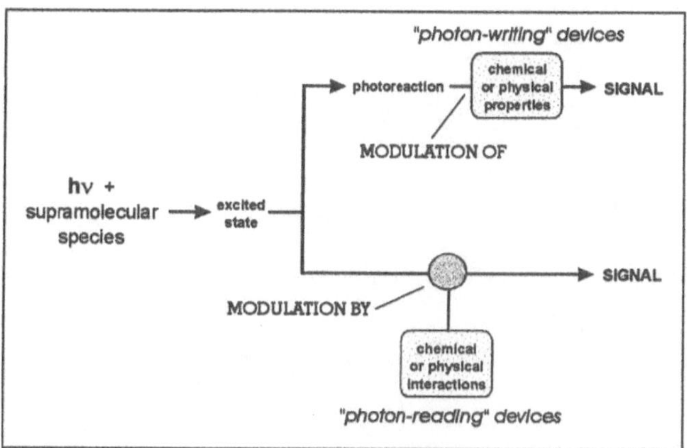

Figure 4. Photochemical reactions can be used to obtain information either "writing" on matter or "reading" the state of matter.

The first one ("photon writing") involves the occurrence of a photoreaction that causes ("writes") some changes in the properties of the device, reflected in a monitorable signal. The second route ("photon reading") is based on some kind of interaction between the device and external species which affects the response of the device to light excitation. Some excited state manifestation (most usually, luminescence) can therefore be used "to read" the interaction. A great variety of photon-writing and photon-reading PMDs have been designed in the last few years [3].

2. PMDs based on energy transfer

2.1. INTRODUCTION

Light absorption by a molecule generates electronic energy. For several practical purposes an important function is represented by the possibility of transmitting electronic energy to another molecule over a more or less long distance, where energy can be used for chemical purposes or reconverted into light. This function can only be obtained by an appropriate sequence of suitable modular components organized in the dimension of space, energy, and time [2].

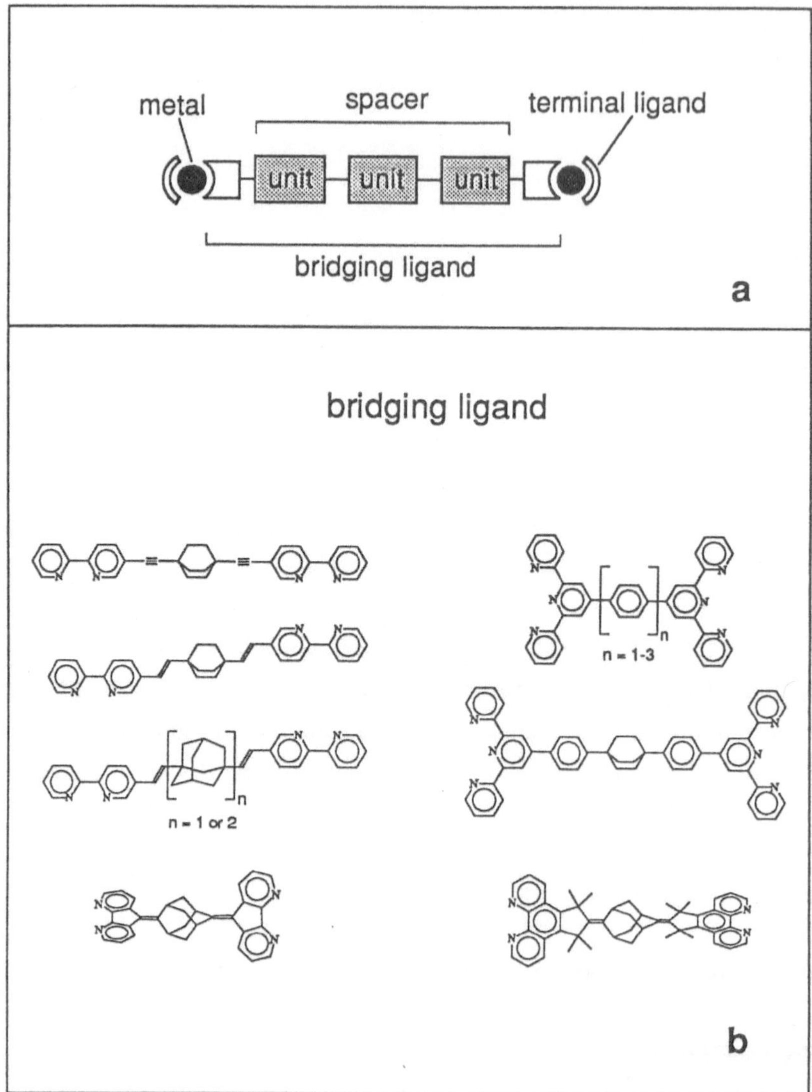

Figure 5. (a) Schematic representation of a rod-like dinuclear complex; (b) bridging ligands used to construct the dinuclear complexes

2.2. ROD-LIKE WIRES FOR ENERGY TRANSFER

In collaboration with the groups of P. Belser and A. von Zelewsky (Fribourg, CH), F. Vögtle (Bonn, D), J.-P. Sauvage (Strasbourg, F), and E. C. Constable (Basel, CH) we have investigated energy-transfer processes in a number of hetero-dinuclear Ru/Os complexes of a variety of polypyridine bridging ligands, with metal-to-metal distance up to 2.4 nm (Figure 5) [5,6]. As a consequence of the transfer of energy, the Ru-based luminescence is quenched and the Os-based luminescence is sensitized. The lifetime of the excited states and the rate of the energy-transfer process are measured by pulsed laser excitation.

The observed energy-transfer rate constants are in the range 10^6-10^{12} s^{-1}, depending on the length and chemical nature of the spacer. In the complex shown in Figure 6a, the rate constant for energy transfer from the Ru-based to the Os-based moiety is 4.4×10^6 s^{-1} [5d]. When the bicyclo[2.2.2]octane module is removed (Figure 6b), the rate of energy transfer increases by more than four orders of magnitude [5c].

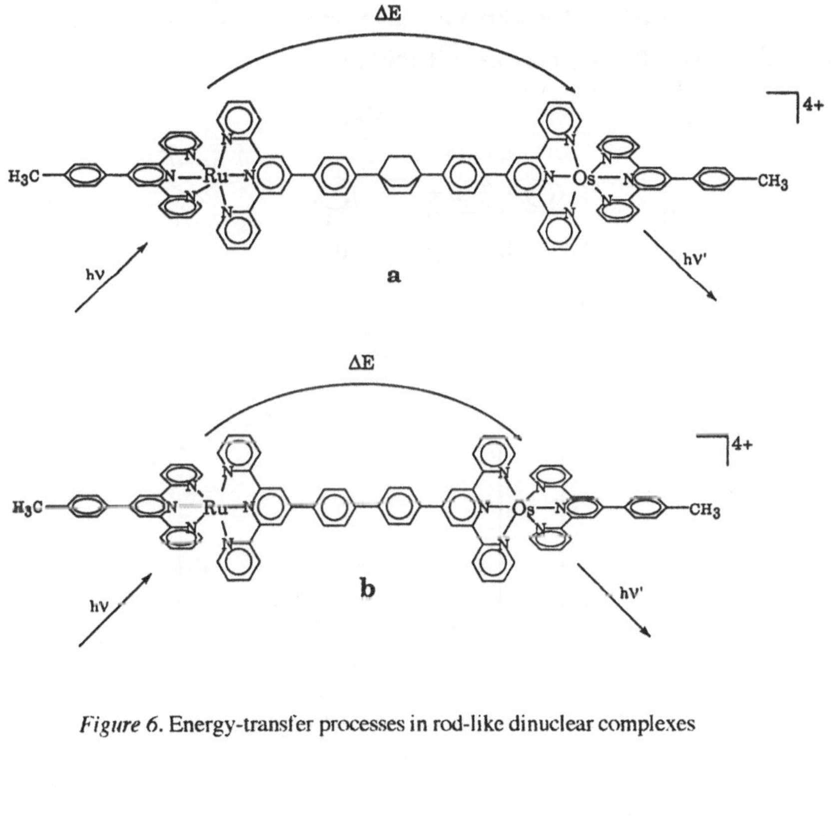

Figure 6. Energy-transfer processes in rod-like dinuclear complexes

Current developments concern the synthesis of (i) longer modular bridging ligands, (ii) compounds containing a greater number and a larger variety of metal centers, and (iii) compounds where energy transfer can be switched on/off by a chemical, photochemical, or electrochemical stimulus.

2.3. WIRES CONTAINING AN "ACTIVE" SPACER

An attempt to switch electronic energy transfer in a molecular wire has led to the synthesis of the compound shown in Figure 7, which contains anthracene as a spacer. In deaerated solution, excitation of the Ru-based moiety is followed by fast and complete energy transfer to anthracene which, in its turn, transfers energy to the Os-based moiety, as expected because of the relative energy of the lowest excited states in the three components. In aerated solution, however, the energy flow stops because excited anthracene reacts with oxygen giving rise to an anthracene endoperoxide and other species whose lowest excited state lies higher in energy than the donor excited state [7]. Unfortunately the system does not show a reversible behavior.

Other systems based on photochromic active spacers, where energy or electron transfer can be switched on/off, have been described[8].

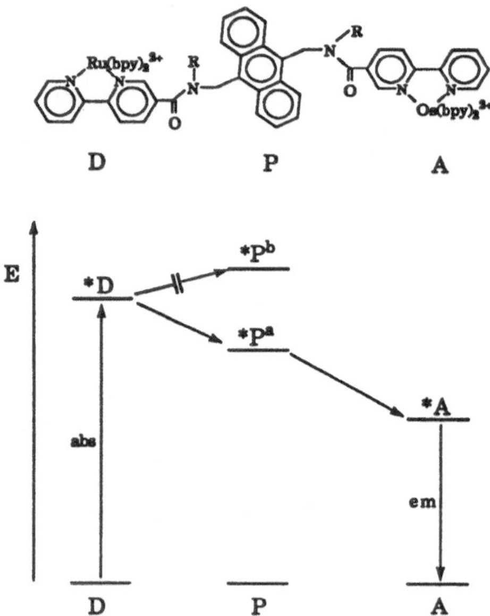

Figure 7. Energy transfer through an active spacer.

2.4. LIGHT HARVESTING ANTENNAS

In collaboration with the groups of S. Campagna and S. Serroni (University of Messina) and G. Denti (University of Pisa) we have synthesized tree-like (dendritic) multicenter transition-metal complexes based on Ru and Os as metals, 2,3-dpp and 2,5-dpp as bridging ligands, and bpy or biq as terminal ligands [9]. This has been made possible by the design of a divergent iterative procedure based on the "complexes as metals and complexes as ligands" synthetic strategy (Figure 8).

A fundamental building block for the iterative synthesis is the "protected" (methylated) complex shown in Figure 8b. Such a complex can first be used as a "complex metal" because of the two labile Cl^- ligands. After deprotection, it can play the role of "complex ligand" because of the presence of the two chelating sites. The iterative synthetic strategy shown in Figure 8 is characterized by a full, step-by-step control of the growing process. Therefore, different building blocks containing different metals and/or ligands can be introduced at each step. Moreover, each deprotected compound can be used as a ligand core in convergent synthetic processes with "complex metals" carrying terminal ligands, to yield sterile dendrimers of higher generation [9f].

More than 60 compounds having 2, 3, 4, 6, 7, 10, 13, or 22 metal centers have been obtained. The largest compound so far prepared contains 22 metal atoms, 21 bridging ligands (2,3-dpp), and 24 terminal ligands (bpy). Altogether, it is made of 1090 atoms, with a molecular weight of 10890 daltons and an estimated size of 5 nm [9bf].

Besides exhibiting endo- and exo-receptor properties, the transition-metal based dendrimers show some unique properties: (i) strong absorption of visible (solar) radiation (ε=202000 $M^{-1}cm^{-1}$ at 542 nm for the species containing 22 metal centers [9f]); (ii) luminescence both in rigid matrix at 77 K and in solution at room temperature [9]; (iii) a great number of reversible reduction (ligand-based) and oxidation (metal-based) processes (for tetranuclear compounds, as many as 18 reduction and 4 oxidation steps can be found between +2.0 and -3.0 V vs SCE [9ag]); (iv) a non negligible electronic interaction allows fast intercomponent energy and electron transfer [9]. Since the properties of the modular components are known and different modules can be located in the desired positions of the dendrimer array, synthetic control of the various properties can be obtained. It is therefore possible to construct arrays where the electronic energy migration pattern can be predetermined, so as to channel the energy created by light absorption on the various components towards a selected module. Future developments will be in the direction of

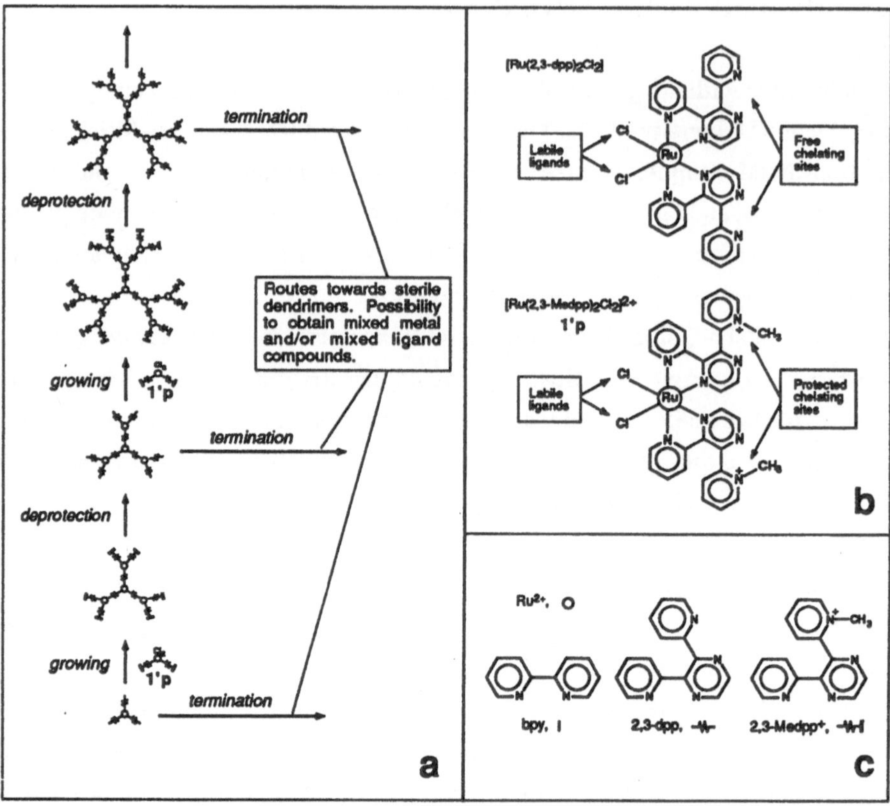

Figure 8. (a) Iterative synthetic procedure for the synthesis of dendrimers based on transition-metal complexes; (b) a fudamental building block; (c) metal and ligands used.

preparing (i) larger and larger compounds by using iterative synthetic strategies, and (ii) compounds with a larger variety of metal centers, bridging ligands and terminal ligands to further increase the number or "pieces of information" incorporated in the supramolecular array.

3. PMDs based on structural changes

3.1. INTRODUCTION

As we have discussed in detail elsewhere [2], photoinduced structural changes can constitute the basis of a number of interesting devices, particularly for switch on/off applications. Photoinduced switching of receptor (coordination) ability and modification of cavity size have since long been known [10] and an example of photoinduced activation of coreceptor catalysis has recently been reported [11]. Such devices are based on the structural changes caused in a multicomponent system by the photoinduced isomerization of a suitable component, e.g. azobenzene. A dendrimer which contains six photoisomerizable azobenzene units and 25 benzene rings has also been prepared [12]. Upon light excitation the shape and size of its void regions are forced to change, which in principle can lead to changes in its host-type properties. Azobenzene units have also been inserted in a ring of a [2]catenane, yielding a photoswitchable catenane [13].

In collaboration with the group of J. Fraser Stoddart (University of Birmingham, U.K.), we are currently engaged in design and development of PMDs based on supramolecular species like pseudorotaxanes, rotaxanes and catenanes in which changes in the relative positions of the components are caused by photoinduced electron transfer processes [14]. Such devices, as well as those mentioned above, can be considered as prototypes of simple mechanical molecular machines.

3.2. MECHANICAL MOLECULAR MACHINES

A photochemically driven mechanical molecular machine [14b] is shown schematically in Figure 9. In the dark the wire containing the electron-donating naphtho unit threads spontaneously the macrocycle containing the two electron-accepting viologen-type units. The driving force for this self-threading process is a charge-transfer interaction which takes place when the wire threads the macrocycle. Light excitation of a suitable photosensitizer (P) causes reduction of the viologen macrocycle and thereby destroyes the electron donor-acceptor interaction. As a consequence, the macrocycle and the wire separate (unthreading process). When oxygen is allowed to enter the solution, the reduced viologen unit of the macrocycle is reoxidized and the naphtho unit threads again the macrocycle.

Figure 10 shows schematically other types of photochemical molecular machines that are currently under investigation in our laboratories. Molecular machines based on electrochemical [14d, 15-17] or chemical [14e, 15] stimuli have also been reported.

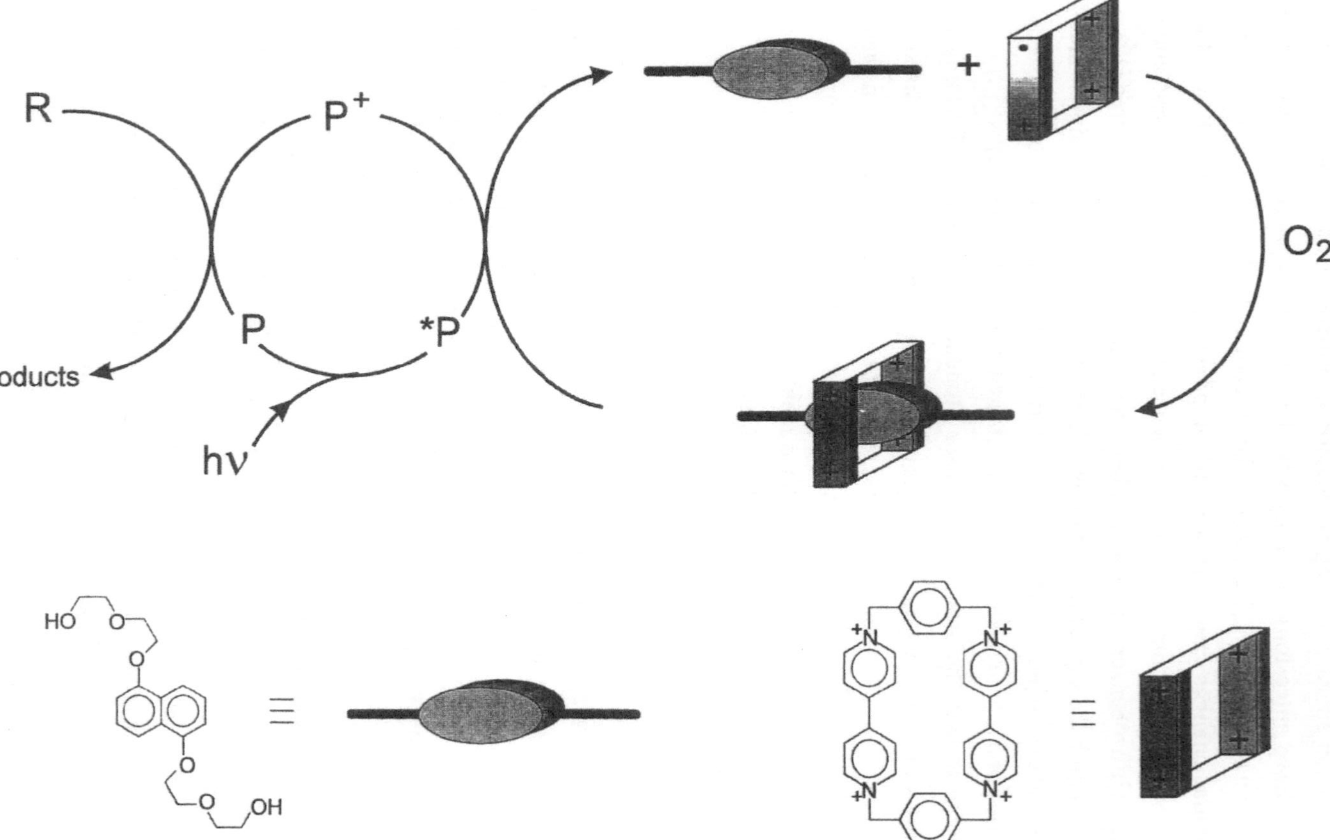

Figure 9. A photochemically driven molecular machine; P is an electron-transfer photosensitizer (e.g., $Ru(bpy)_3^{2+}$); R is a sacrificial electron donor (e.g., triethanolamine).

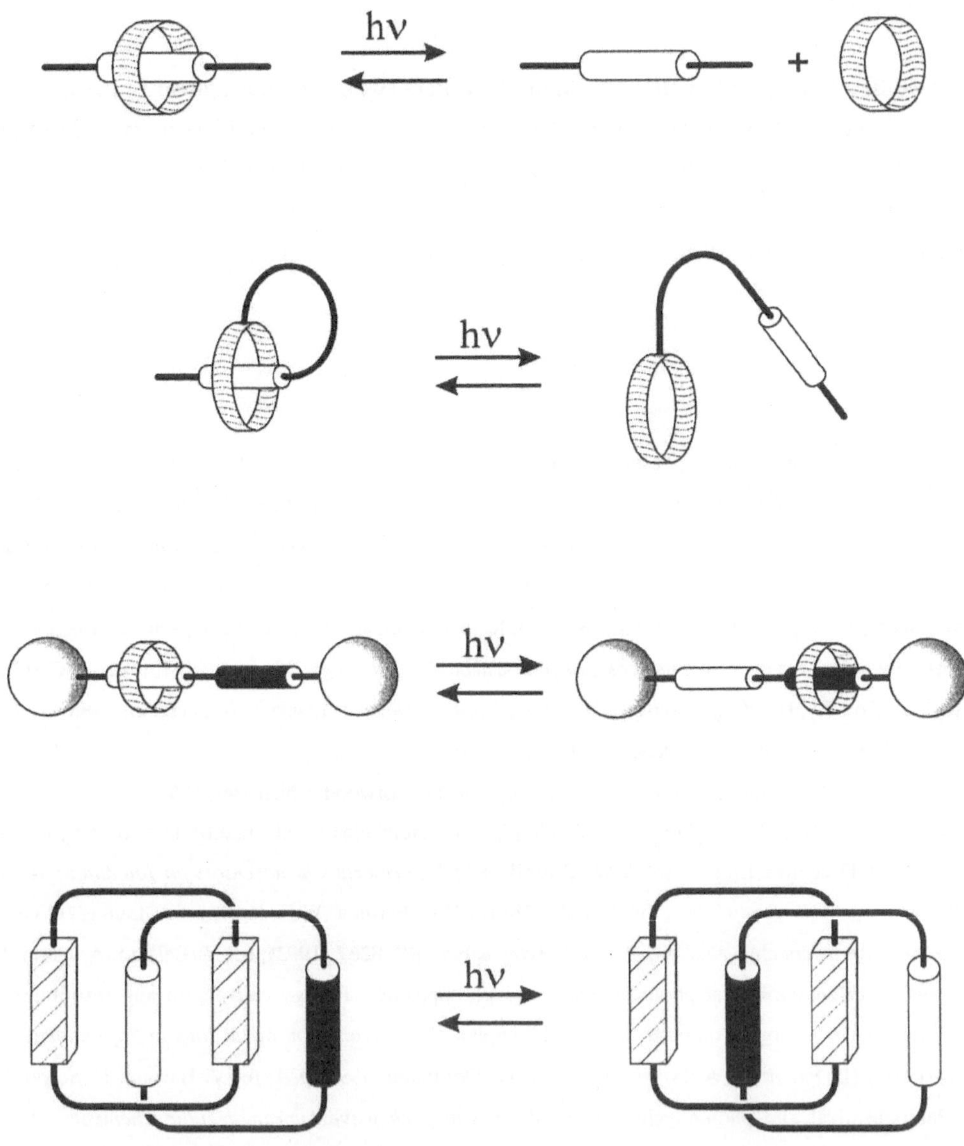

Figure 10. Photochemical molecular machines based on rotaxanes and catenanes.

4. Acknowledgments

I would like to thank my colleagues and coworkers (whose names appear in the quoted references) of the " G. Ciamician" Department of Chemistry of the University and of the FRAE-CNR Institute for their fundamental contribution to the studies described in this paper. This work was supported by the Ministero della Università e della Ricerca Scientifica and by the Consiglio Nazionale delle Ricerche (Progetto Strategico Tecnologie Chimiche Innovative).

5. References

1. (a) A. Aviram, M.A. Ratner, "Molecular rectifiers", *Chem. Phys. Lett.*, **29**, 277 (1974); (b) F.L. Carter (ed.) *Molecular Electronic Devices II*, Dekker, 1987; (c) V. Balzani, L. Moggi, F. Scandola, "Towards a supramolecular photochemistry. Assembly of molecular components to obtain photochemical molecular devices", in *Supramolecular Photochemistry* (V. Balzani ed.), Reidel, Dordrecht, 1987, p. 1; (d) J.-M. Lehn, "Supramolecular chemistry: scope and perspectives. Molecules, supermolecules, and molecular devices", *Angew. Chem. Int. Ed. Engl.* **27**, 90 (1988); (e) F.L. Carter, R.E. Siatkowsky, H. Woltjien (eds.) *Molecular Electronic Devices*, Elsevier, Amsterdam, 1988; (f) J.-M. Lehn, *Supramolecular Chemistry*, VCH, Basel, 1995.
2. V. Balzani, F. Scandola, *Supramolecular Photochemistry*, Horwood, Chichester, 1991.
3. (a) A.P. de Silva, C.P. McCoy, "Switchable photonic molecules in information technology", *Chem. Ind.*, 1994, December 19, 992; (b) A.W. Czarnik (ed.) *Fluorescent Chemosensors for Ion and Molecule Recognition, ACS Symp. Ser.*, 1993, Vol. 538; (c) B.L. Feringa, W.F. Jager, B. de Lange, "Organic materials for reversible optical data storage", *Tetrahedron*, **49**, 8267 (1993); (d) V. Balzani, A. Credi, F. Scandola, "Supramolecular photochemistry and photophysics. Energy conversion and information-processing devices based on transition metal complexes", in *Transition Metal Ions in Supramolecular Chemistry* (L. Fabbrizzi, A. Poggi eds.), Kluwer, Dordrecht, 1994, p. 1; (e) V. Balzani, F. Scandola, "Photochemical and photophysical devices", in *Comprehensive Supramolecular Chemistry* (D.N. Reihnhoudt ed.), Pergamon Press, Oxford, Vol. 10, in press.
4. B. O'Regan, M. Graetzel, "A low-cost, high-efficiency solar cell based on dye-sensitized colloidal TiO_2 films", *Nature*, **353**, 737 (1991).
5. (a) L. De Cola, V. Balzani, F. Barigelletti, L. Flamigni, P. Belser, A. von Zelewsky, M. Frank, F. Vögtle, "Photoinduced energy and electron transfer processes in supramolecular species. Tris(bipyridine) complexes of Ru(II)/Os(II), Ru(II)/Ru(III), Os(II)/Os(III), and Ru(II)/Os(III) separated by a rigid spacer",

Inorg. Chem., **32**, 5228 (1993); (b) F. Vögtle, M. Frank, M. Nieger, P. Belser, A. von Zelewsky, V. Balzani, F. Barigelletti, L. De Cola, L. Flamigni, "Rigid rod-like metal complexes of nanometric dimension. Synthesis, luminescence properties, long-range energy transfer", *Angew. Chem. Int. Ed. Engl.*, **32**, 1643 (1993); (c) F. Barigelletti, L. Flamigni, V. Balzani, J.-P. Collin, J.-P. Sauvage, A. Sour, E.C. Constable, A.M.W. Cargill Thompson, "Rigid rod-like dinuclear Ru(II)/Os(II) terpyridine-type complexes. Electrochemical behavior, absorption spectra, luminescence properties, and electronic energy transfer through phenylene bridges", *J. Am. Chem. Soc.*, **116**, 7692 (1994); (d) F. Barigelletti, L. Flamigni, V. Balzani, J.-P. Collin, J.-P. Sauvage, A. Sour, "Luminescence properties of rigid rod-like dichromophoric species. Dinuclear Ru/Os terpyridine-type complexes with 2.4 nm metal-to-metal distance", *New. J. Chem.*, **19**, 793 (1995); (e) M. Frank, M. Nieger, F. Vögtle, P. Belser, L. De Cola, V. Balzani, F. Barigelletti, L. Flamigni, "Dinuclear Ru(II) and/or Os(II) complexes of bis-bipyridine bridging ligands containing adamantane spacers. Synthesis, luminescence properties, intercomponent energy and electron transfer", *Inorg. Chim. Acta*, in press; (f) L. De Cola, V. Balzani, F. Barigelletti, L. Flamigni, P. Belser, S. Bernhard, "Photoinduced energy and electron transfer processes in dinuclear Ru(II) and/or Os(II) complexes connected by a linear rigid bis-chelating bridge", *Rec. Trav. Chim. Pays Bas*, **114**, 534 (1995).

6. For a review, see: V. Balzani, A. Juris, M. Venturi, S. Campagna, S. Serroni, "Luminescent and redox-active polynuclear transition-metal complexes", *Chem. Rev.*, in press.
7. (a) P. Belser, R. Dux, M. Baak, L. De Cola, V. Balzani, "Electronic energy transfer in a supramolecular species containing the $Ru(bpy)_3^{2+}$, $Os(bpy)_3^{2+}$, and anthracene chromophoric units", *Angew. Chem. Int. Ed. Engl.*, **34**, 595 (1995); (b) L. De Cola, V. Balzani, P. Belser, R. Dux, M. Baak, "Electronic energy transfer in supramolecular systems. Self-poisoning and self-educating systems", *Supramol. Chem.*, **5**, 297 (1995).
8. (a) J. Walz, K. Ulrich, H. Port, H.C. Wolf, J. Wonner, F. Effenberger, "Fulgides as switches for intramolecular energy transfer", *Chem. Phys. Lett.*, **213**, 321 (1993); (b) V. Goulle, A. Harriman, J.-M. Lehn, "An electro-photoswitch: redox switching of the luminescence of a bipyridine metal complex", *J. Chem. Soc., Chem. Commun.*, 1034 (1993); (c) S.L. Gilat, S.H. Kawai, J.-M. Lehn, "Light-triggered electrical and optical switching devices", *J. Chem. Soc., Chem. Commun.*, 1439 (1993).
9. (a) G. Denti, S. Campagna, S. Serroni, M. Ciano, V. Balzani, "Decanuclear homo- and heterometallic polypyridine complexes. Synthesis, absorption spectra, luminescence, electrochemical oxidation, intercomponent energy transfer", *J. Am. Chem. Soc.*, **114**, 2944 (1992), and references therein; (b) S. Serroni, G. Denti, S. Campagna, A. Juris, M. Ciano, V. Balzani, "Arborols based on luminescent and redox-active transition metal complexes", *Angew. Chem. Int. Ed. Engl.*, **31**, 1493 (1992); (c) S.

Roffia, M. Marcaccio, C. Paradisi, F. Paolucci, V. Balzani, G. Denti, S. Serroni, S. Campagna, "Electrochemical reduction of 2,2-bipyridine and bis(2-pyridyl)pyrazine ruthenium(II) complexes used as building blocks for supramolecular species. Redox series made of eight, ten and twelve redox steps", *Inorg. Chem.*, **32**, 3003 (1993); (d) A. Juris, V. Balzani, S. Campagna, G. Denti, S. Serroni, G. Frei, H.U. Güdel, "Near infrared luminescence of supramolecular species made of Os(II) and/or Ru(II)-polypyridine components", *Inorg. Chem.*, **33**, 1491 (1994); (e) S. Serroni, A. Juris, S. Campagna, M. Venturi, G. Denti, V. Balzani, "Tetranuclear bimetallic complexes of Ruthenium, Osmium, Rhodium and Iridium. Synthesis, absorption spectra, luminescence, and electrochemical properties", *J. Am. Chem. Soc.*, **116**, 9086 (1994); (f) S. Campagna, G. Denti, S. Serroni, A. Juris, M. Venturi, V. Ricevuto, V. Balzani, "Dendrimers of nanometer size based on metal complexes. Luminescent and redox-active polynuclear metal complexes containing up to 22 metal centers", *Chem. Eur. J.*, **1**, 211 (1995); (g) Roffia, *et al.*, work in progress.

10. S. Shinkai, T. Nakaji, T. Ogawa, K. Shigematsu, O. Manabe, "Photoresponsive crown ethers 2. Photocontrol of ion extraction and ion transport by a bis(crown ether) with a butterfly-like motion", *J. Am. Chem. Soc.*, **103**, 111 (1981).
11. F. Würthner, J. Rebek, Jr., "Light-switchable catalysis in synthetic receptors", *Angew. Chem. Int. Ed. Engl.*, **34**, 446 (1995).
12. H.B. Mekelburger, K. Rissanen, F. Vögtle, "Repetitive synthesis of bulky dendrimers. A reversibly photoactive dendrimer with six azobenzene side chains", *Chem. Ber.*, **126**, 1161 (1993).
13. F. Vögtle, W.M. Müller, U. Müller, M. Bauer, K. Rissanen, "Photoswitchable catenanes", *Angew. Chem. Int. Ed. Engl.*, **32**, 1295 (1993).
14. (a) P.L. Anelli, P.R. Ashton, R. Ballardini, V. Balzani, M. Delgado, M.T. Gandolfi, T.T. Goodnow, A.E. Kaifer, D. Philp, M. Pietraszkiewicz, L. Prodi, M.V. Reddington, A.M.Z. Slawin, N. Spencer, J.F. Stoddart, C. Vicent, D.J. Williams, "Molecular meccano. 1. [2]rotaxanes and a [2]catenane made to order", *J. Am. Chem. Soc.*, **114**, 193 (1992); (b) R. Ballardini, V. Balzani, M.T. Gandolfi, L. Prodi, M. Venturi, D. Philp, H.G. Ricketts, J.F. Stoddart, "A photochemically-driven molecular machine", *Angew. Chem. Int. Ed. Engl.*, **32**, 1301 (1993); (c) P.R. Ashton, R. Ballardini, V. Balzani, M.T. Gandolfi, D.J.-F. Marquis, L. Pérez-García, L. Prodi, J.F. Stoddart, M. Venturi, "The self-assembly of controllable [2]catenanes", *J. Chem. Soc., Chem. Commun.*, 177 (1994); (d) P.R. Ashton, R. Ballardini, V. Balzani, A. Credi, M.T. Gandolfi, S. Menzer, L. Pérez-García, L. Prodi, J.F. Stoddart, M. Venturi, A.J.P. White, D.J. Williams, "Molecular meccano. 4. The self-assembly of [2]catenanes incorporating photoactive and electroactive π-extended systems", *J. Am. Chem. Soc.*, in press; (e) R. Ballardini, V. Balzani, A. Credi, M.T. Gandolfi, S.J. Langford, S. Menzer, L. Prodi, J.F. Stoddart, M.

Venturi, D.J. Williams, "Simple molecular machines. Chemically-driven unthreading and rethreading of a [2]pseudorotaxane", *Angew. Chem. Int. Ed. Engl.*, in press.

15. R.A. Bissell, E. Cordova, A.E. Kaifer, J.F. Stoddart, "A chemically and electrochemically switchable molecular shuttle", *Nature*, **369**, 133 (1994).
16. A. Livoreil, C.O. Dietrich-Buchecker, J.-P. Sauvage, "Electrochemical triggered swinging of a [2]catenate", *J. Am. Chem. Soc.*, **116**, 9399 (1994).
17. L. Zelikovich, J. Libman, A. Shauzer, "Molecular redox switches based on chemical triggering of iron translocation in triple-stranded helical complexes", *Nature*, **374**, 790 (1955).

CONSTRUCTION OF SOLID/SOLID INTERFACE MODELS USING MODULAR CHEMISTRY: THE Si/SiO_2 INTERFACE

SUNGHEE LEE
Brown University
Department of Chemistry, Providence, RI 02912

MARK M. BANASZAK HOLL
The University of Michigan
Department of Chemistry, Ann Arbor, MI 48109

WEI HSUI HUNG AND F. R. MCFEELY
IBM T.J. Watson Laboratory
P.O. Box 218, Yorktown, NY 10598

ABSTRACT. Elucidating the structure and reactivity of solid-solid interfaces remains a conundrum for scientists. The interfaces, exhibiting both dramatic and subtle structural differences from the bulk solids, frequently exist over a very small region which leaves little sample available to analyze. Modular chemistry approaches have an important role to play in gaining an understanding of solid-solid interfaces. By using modular pieces to construct one or both sides of the interface, a spectroscopically amenable, well-defined interface can be generated. A series of experiments using this method will be presented for the case of the Si/SiO_2 interface, a critical component of MOSFET devices (Metal Oxide Semiconductor Field Effect Transistor). Specifically, it will be shown how model interfaces constructed from spherosiloxane clusters such as $H_8Si_8O_{12}$ and Si(100)-2x1 have facilitated the interpretation of core-level photoemission studies of the interface, furthered understanding of the structural implications of the photoemission experiments, and provided insight into the reactions of hydrogen, a major contributor to device failure at the interface.

1. Introduction

Study of the structure and reactivity of solid/solid interfaces is complicated by several factors.[1] The interface region is difficult to study by many spectroscopic techniques because it contains a small total number of atoms to be detected and is difficult to resolve because of the large background generated by the nearby bulk solids. In addition, the interface region is often disordered, reducing the effectiveness of diffraction techniques. Interfaces are also particular sensitive to

J. Michl (ed.), Modular Chemistry, 451–460.

synthesis conditions and can change content and/or structure depending on how the solid/solid juncture is made.

The general synthetic methods for forming solid/solid interfaces have been high temperature reactions in which one solid diffuses into another or gas/solid reactions. In either case, a thermodynamic product is obtained and the arduous task of determining the interface structure ensues. A particularly important solid/solid interface from the industrial standpoint, Si/SiO_2, is prepared by the high temperature reaction of silicon and O_2 gas. This interface plays an extremely important role in microelectronics because it exhibits a remarkably high degree of electrical perfection. Naively, one might expect that the geometrical constraints placed upon the interface to obtain the remarkable electrical properties would necessarily enforce the presence of a simple, easy to elucidate structure. Despite intense study since the advent of Si/SiO_2 based transistors, the structure of the interface region, and the nature of chemical moieties present, remains hotly debated. The debate is both spirited and timely because of the current effort in the microelectronics industry to make devices with SiO_2 thickness of ~30-50 Å. Assuming the stoichiometric interface is 10 Å thick, this means the interface region is a full 20-30% of the dielectric layer, as compared to 1% or less for most current devices.[2]

2. Modular Formation of Si/SiO_2 Interface Models

Modular chemistry offers a more spectroscopically amenable method for the construction of solid/solid interfaces. Well-characterized surfaces and modular components take the place of the surfaces and gases of the thermal oxidation process. In the case of the Si/SiO_2 interface, we have utilized the spherosiloxane clusters $H_8Si_8O_{12}$, $H_{12}Si_{12}O_{18}$, and $H_{14}Si_{14}O_{21}$ to generate model interfaces (Fig. 1).[3]

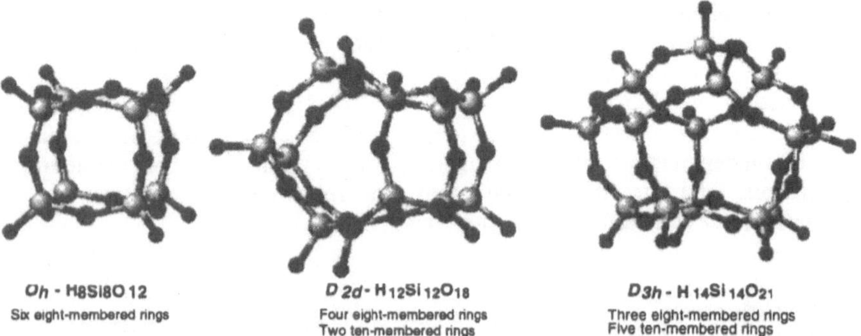

Figure 1. The Spherosiloxane clusters used for model interface construction.

The spherosiloxane clusters have been a favorite among chemists seeking modular approaches to controlling solid state structure[4] and those seeking to mimic surface binding sites on silica.[5] The clusters also provide a reagent uniquely suited for constructing a model of the Si/SiO_2 interface.

One of the primary reactions observed on the Si(100)-2x1 surface, the orientation of silicon used for device manufacture, is the addition of a bond across the dimer of silicon atoms present on the surface.[6] In particular, Si-H bonds have been observed to add across the dimer sites. Thus, the termination of the Si-O cluster by hydrogen provides a site for reaction with the surface. The cubic symmetry of $H_8Si_8O_{12}$ means that regardless which Si-H bond reacts, the same product will be formed. $H_8Si_8O_{12}$ is composed entirely of eight-membered rings, so if the cluster reacts with the surface by the simple addition of an Si-H bond across a surface dimer, a model interface can be formed which is controlled in terms of stoichiometry and structure. By making use of the three different spherosiloxane clusters the number of eight and ten membered rings can even be controlled. While the Si-H bonds provide a convenient reaction pathway, they do introduce more hydrogen than is present in a Si/SiO_2 interface generated thermally from dry O_2 gas. Hydrogen does play a major role in device failure via reactions in the interface region however, so the presence of hydrogen in well-defined sites is useful for the study of device failure mechanisms.[7] To summarize, the spherosiloxanes provide a readily synthesized modular unit which have a useful vapor pressure for gas phase dosing of single crystalline silicon surfaces. All the elements present in the unit are known to play an important role in the structure and reactivity of the Si/SiO_2 interface. By constructing a model interface in a modular fashion, we are able to test spectroscopic assignments made for the device interfaces and the resulting model provides a test bed for exploring chemical reactions in the interface region.

The three clusters react with the surface dimers of the Si(100)-2x1 surface via a Si-H bond to yield a cluster attached via one vertex to the surface.[8] In this manner, three different model interfaces can be generated. The Si $2p_{3/2}$ core-level photoemission spectra of the interfaces generated are shown in Figure 2. In each case, the large peak defined at 0.0 eV in binding energy is the feature due to the bulk silicon substrate. The remaining four peaks are assigned, in order of increasing binding energy, to surface Si-H (0.4 eV), surface Si bound to the cluster vertex (1.0 eV), the cluster vertex Si bound to the surface (2.1 eV), and the $HSiO_3$ groups of the cluster respectively (3.7 eV).[9] A comparison of these model interfaces to the thermal Si/SiO_2 interface used in devices is instructive. The diameters of the clusters are 10 Å, 12 Å, and 14 Å for $H_8Si_8O_{12}$, $H_{12}Si_{12}O_{18}$, and $H_{14}Si_{14}O_{21}$, respectively.[10] The Si/SiO_2 interface region, at least as defined by compositional change, is believed to be about 10 Å thick. Thus, the dimensions of

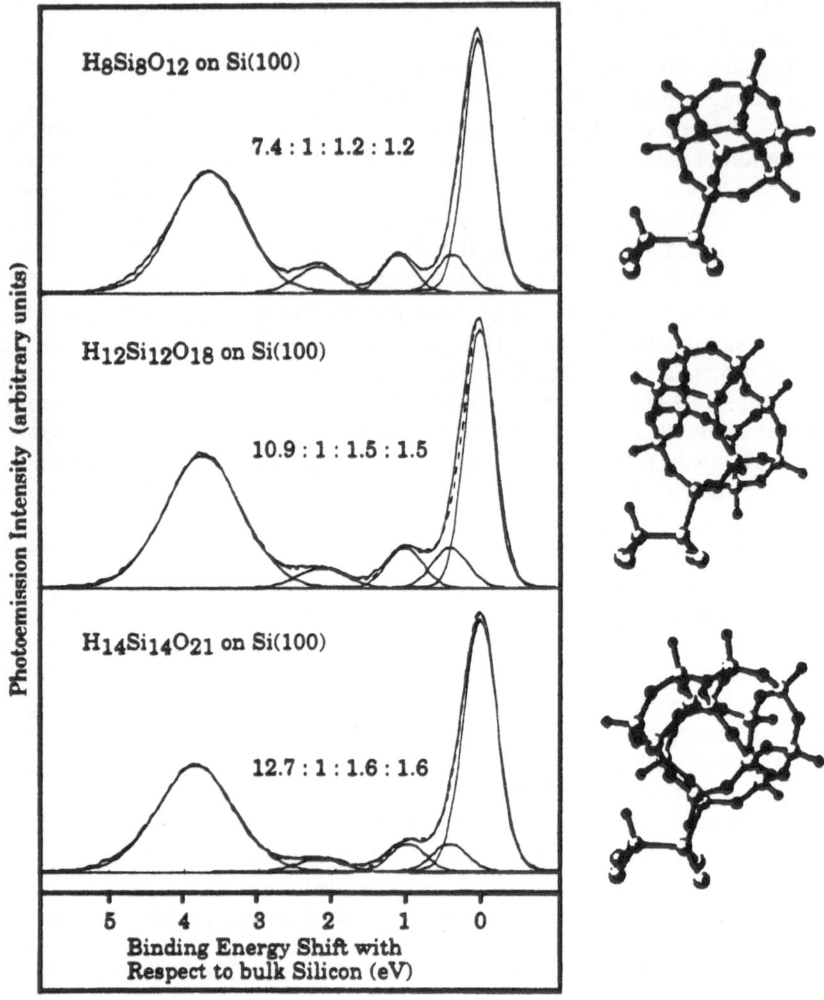

Figure 2. Si $2p_{3/2}$ core-level photoelectron spectra of the $H_8Si_8O_{12}$, $H_{12}Si_{12}O_{18}$, and $H_{14}Si_{14}O_{21}$ derived model interfaces. The incident photon energy used was 170 eV. The clusters are shown bound to a surface dimer by one vertex, the four silicon atoms below each surface dimer are also shown.

the model are about right. Although interfaces made with dry O_2 contain much less hydrogen, the hydrogen content is on the same order of magnitude as interfaces grown using wet oxidation processes.[11] Amorphous SiO_2 has a distribution of Si-O ring sizes, by analogy one expects the interface to have a similar distribution. The model interfaces constructed have a pre-determined distribution of ring sizes.

Finally, the exact structural moieties making up the interface region are not known, and a considerable disagreement exists in the literature regarding the best hypotheses for structural features making up the interface region.[12] By using modular techniques, we have been able to construct a model interface which provides a starting point for the empirical determination of binding energy shifts at the Si/SiO_2 interface. The key experimental model system prior to these results consisted of the reaction of water with the Si(100)-2x1 interface.[6b] Dosing with water adds one-half a monolayer of hydroxide bound to the silicon surface. The model interface made by water dosing can be considered to have one layer of silicon (the surface layer) and one layer of oxygen.[13] In contrast, the spherosiloxane clusters imitate a multilayer silicon oxide interface region. The symmetry of the $H_8Si_8O_{12}$ cluster decreases from O_h to C_{3v} upon binding to the surface. Including the initial surface layer of silicon atoms, the model interface derived from $H_8Si_8O_{12}$ has five layers of silicon and three layers of oxygen. A similar analysis of the other two model interfaces shows that $H_{12}Si_{12}O_{18}$ and $H_{14}Si_{14}O_{21}$ both give 5 silicon and 4 oxygen layers.[8b] The structure of Si/SiO_2 interface derived from dry O_2 is a matter of debate, however we can estimate the number of layers present based on the total width of approximately 10 Å. Our most recent interface model consists of 4 layers of silicon atoms and only two layers of oxygen atoms.[12b] Models proposed by other workers contain a varying number of layers ranging from completely abrupt, 1 layer each of silicon and oxygen, to more extended interfaces containing 5 layers of silicon and 4 layers of oxygen. Since the completely abrupt interface can be ruled out by numerous groups photoemission experiments, we should focus on the more extended 10 Å thick interface models as being the best candidates to provide an accurate picture of the thermal Si/SiO_2 interface. It is gratifying that the spherosiloxane based interface models have an appropriate number of silicon and oxygen layers.

Often when modular chemistry is discussed, it is in the context of linking up the modules to form a unique type of extended structure. While we created a two-dimensional array of modules on a surface in the reactions discussed above, we have not interlinked the modules themselves. Unfortunately, there is an intrinsic difficulty to characterizing the products formed by Si-O-Si linking reactions using core-level photoemission spectroscopy. An SiO_4 fragment in the vicinity of a silicon surface has the same binding energy shift with respect to bulk silicon as an $HSiO_3$ fragment, about 3.7 eV. Thus, for the linking reactions we are most interested in, those which may lead us to unique phases of SiO_2 which we could segregate at the interface, the ability to characterize the products becomes a significant problem. However, valence band spectroscopy, combined with mass spectroscopy, does provide some ability to observe cluster to cluster linking occurring. Upon heating a cluster derived model interface to 320 °C, we begin to see evidence of structural changes in both the core-level Si $2p_{3/2}$ spectra[8] and the valence band spectra shown in Figure 3. In particular, the valence band spectra indicate a further drop in symmetry and by 540 °C the valence band spectrum is indistinguishable from that of the thermal Si/SiO_2 interface. Mass spectroscopy indicates production of H_2 upon heating. The combination of the two results, and the formation of a new peak at 2.8 eV in the Si 2p core-level spectra, suggests that we may be linking clusters via the Si vertices. However, intensity increases which occur in the 1-2 eV region preclude a simple vertex linking picture from being proposed as the only reaction occurring, and indeed suggest that some degree of cluster decomposition is also occurring.[4]

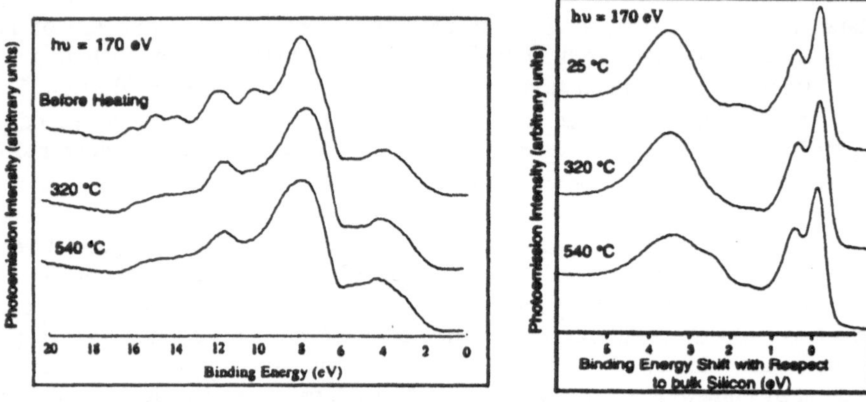

Figure 3. Panel A: Valence Band spectra showing the thermal reaction of the model interface derived from $H_8Si_8O_{12}$. Panel B: Si 2p core-level spectrum of the same thermal reaction. Note that the Si $2p_{1/2}$ core-levels have not been stripped in this spectrum.

3. Reactions of Atomic Hydrogen at the Si/SiO_2 Interface

The new model interfaces have proven themselves useful with respect to understanding reactions which occur at thermal Si/SiO_2 interfaces. In a recent experiment we exposed a Si/SiO_2 interface grown on Si(100) using dry O_2 to a flux of hydrogen radicals.[14] The most prominent change in spectral features is an apparent increase in region D at ~3.6 eV, traditionally assigned exclusively to

"Si^{+4}" or SiO_2. At the same time we see a decrease in intensity in regions C, B, and A. The decrease in region C is particularly remarkable as this is the region which had been generally assigned to silicon oxide fragments containing hydrogen.[15] Fortunately, examination of binding energy shifts versus fragment types provided by the model interfaces allows a possible explanation of the observations. The model interfaces demonstrate that an $HSiO_3$ fragment in an oxide matrix has a binding energy shift of 3.6 eV with a full width at half maximum (FWHM) of 1.2 eV. Upon exposure of the Si/SiO_2 interface to hydrogen radicals we see a new peak grow in at 3.6 eV with FWHM of 1.2 eV. Thus, the model interfaces provide a plausible assignment for the type of species formed during the reaction.

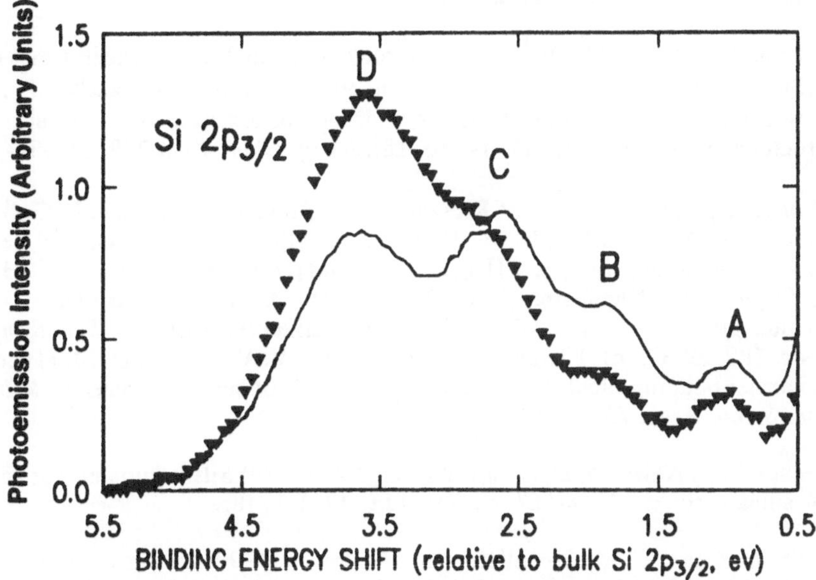

Figure 4. The Si $2p_{3/2}$ core level spectrum of the Si/SiO_2 oxide interface region before and after exposure to a flux of hydrogen radicals

4. Summary

We have used a modular unit, the spherosiloxane family of clusters, to synthesize the first set of model systems for the Si/SiO_2 interface. These model systems have provided insight into the spectroscopic assignments, structure, and reactivity of the thermally grown Si/SiO_2 interface.

References

[1] For the purposes of this paper, we are considering covalently bound solids such as ceramics, metals, and semiconductors.

[2] Many excellent reviews of the Si/SiO_2 interface have been published. a) *The Physics and Technology of Amorphous SiO_2* Ed. R.A.B. Devine Plenum Press, 1988. b) *SiO_2 and Its Interfaces* Ed. Pantelides, S.T., Lucovsky, G. *Mat. Res. Soc. Symp. Proc.* 1990. c) *The Physics and Technology of Amorphous Silicon Dioxide*, Ed. Helms, C. R., Deal, B. E., Plenum: New York, 1988. d) *The Physics and Chemistry of SiO_2 and the Si-SiO_2 Interface 2*, Ed. Helms, C. R., Deal, B. E. Plenum: New York, 1993.

[3] a) Agaskar, P.A. (1991) New synthetic route to the hydridospherosiloxanes O_h-$H_8Si_8O_{12}$ and D_{5h}-$H_{10}Si_{10}O_{15}$, *Inorg. Chem.* **30**, 2707-2708. b) Agaskar, P. A. and Klemperer, W. G. (1995) The higher hydridospherosiloxanes: synthesis and structures of $H_nSi_nO_{1.5n}$ (n=12, 14, 16, 18), *Inorg. Chim. Acta* **229**, 355-364.

[4] For example, see the work of Klemperer: a) Brevett, C. S., Cage, P. C., Klemperer, W. G., Millar, D. M., Ruben, G. C. (1991) Synthesis and sol-gel polymerization of $[Si_8O_{12}](OCH_3)_8$, *J. Inorg. and Organomet. Polymers 1*, 335-342. b) Cagle, P. C., Klemperer, W. G., Simmons, C. A. (1990) Molecular architecture and its role in silica sol-gel polymerization, *Mat. Res. Soc. Symp. Proc.* **180**, 29-37. c) Klemperer, W. G., Mainz, V. V., Millar, D. M. (1986) A molecular building-block approach to the synthesis of ceramic materials, *Mat. Res. Soc. Symp. Proc.* **73**, 3-13.

[5] Feher, F. J., Newman, D. A. and Walzer, J. F. (1989) Silsesquioxanes as models for silica surfaces, *J. Am. Chem. Soc.* **111**, 1741-1748.

[6] a) S. M. Gates, C. M. Greenlief, and D. B. Beach (1990) Decomposition mechanisms of SiH_x species on Si(100)-(2x1) for x=2, 3, and 4, *J. Chem. Phys.* **93**, 7493-7503. b) Himpsel, F. J., McFeely, F. R., Morar, J. F., Taleb-Ibrahimi, A., Yarmoff, J. A. (1988) *Proceedings of the 1988 Enrico Fermi school on photoemission and absorption spectroscopy of solids and interfaces with synchrotron radiation*, North Holland, Varenna.

[7] a) Stathis, J. H. and Cartier, E. (1994) Atomic hydrogen reactions with P_b centers at the (100) Si/SiO_2 interface, *Phys. Rev. Lett* **72**, 2745-2748. b) Cartier, E., Stathis, J. H. and Buchanan, D. A. (1993) Passivation and depassivation of silicon dangling bonds at the Si/SiO_2 interface by atomic hydrogen, *Appl. Phys. Lett.* **63**, 1511-1512. c) Nijs, J. M. M., Druijf, K. G., Afanas'ev, V. V., van der Drift, E. and Balk, P. (1994) Hydrogen induced donor-type Si/SiO_2 interface states, *Appl.*

Phys. Lett. **65**, 2428-2430. d) Saks, N. S. and Brown, D. B. (1990) Observation of H^+ motion during interface trap formation, *IEEE Trans. Nucl. Sci.* **37**, 1624.

[8] a) Banaszak Holl, M. M. and McFeely, F. R. (1993) Si/SiO_2 interface: new structures and well-defined model systems, *Phys. Rev. Lett.* **71**, 2441-2444. b) Lee, S.; Makan, S., Banaszak Holl, M. M., McFeely, F. R. (1994) Synthetic control of solid/solid interfaces: analysis of three new silicon/silicon oxide interfaces by soft X-ray photoemission, *J. Am. Chem. Soc.* **116**, 11819-11826.

[9] Average binding energies from all three clusters are given to help locate the appropriate peaks for each feature. For additional detail regarding these assignments, see reference [8].

[10] Bürgy, H. and Calzaferri, G. (1990) Separation of the oligomeric silsesquioxanes $(HSiO_{3/2})_{8\text{-}18}$ by size-exclusion chromatography, *J. Chromatography* **507**, 481-486.

[11] Grunthaner, F.J. and Grunthaner, P.J. (1986) Chemical and electronic structure of the SiO_2/Si interface, *Mater. Sci. Reports* **1**, 65-160.

[12] a) Himpsel, F. J., McFeely, F. R., Taleb-Ibrahimi, A., Yarmoff, J. A. and Hollinger, G. (1988) Microscopic structure of the SiO_2/Si interface, *Phys. Rev. B.* **38,** 6084-6096. b) Banaszak Holl, M. M., Lee, S. and McFeely, F. R. (1994) Core-level photoemission and the structure of the Si/SiO_2 interface: a reappraisal, *Appl. Phys. Lett.* **65**, 1097-1099. c) Pasquarello, A., Hybertson, M. S., and Car, R. (1995) Si2*p* core-level shifts at the Si(001)-SiO_2 interface: a first-principles study, *Phys. Rev. Lett.* **74**, 1024-1027.

[13] We have chosen the ability to experimentally detect a distinct peak in the photoemission spectra as the measure of the number of layers which are counted as being part of the interface region. This should be a good indicator of compositional changes, but will not give a good indication of the distance the interface causes effects like strain.

[14] Hydrogen radicals were generated by exposing to 1.4×10^{-5} torr H_2 in the presence of a hot tungsten filament for 20 minutes.

[15] a) Ogawa, H., Terada, N., Sugiyama, K., Moriki, K., Miyata, N., Aoyama, T., Sugino, R., Ito, T. and Hattori, T. (1992) Silicon-hydrogen bonds in silicon oxide near the SiO_2/Si interface, *Appl. Surf. Sci.* **56-58**, 836. b) Hattori, T. and Ogawa, H. (1992) Detection of Si-H bonds in silicon oxide by X-ray photoelectron spectrum difference, *Appl. Phys. Lett.* **61**, 577. c) Hattori, T. (1991) Chemical

structure of ultrathin silicon oxide films and the oxide-silicon interface, *Thin Solid Films* **206**, 1. d) Grunthaner, P. J., Hecht, M. M., Grunthaner, F. J. and Johnson, N. M. (1987) The localization and crystallographic dependence of Si suboxide species at the SiO_2/Si interface, *J. Appl. Phys.* **61**, 629.

SYNTHESIS OF NEW MOLECULAR SYSTEMS

W. E. BILLUPS, WEIMEI LUO, DIANNE MCCORD,
AND ROBERT WAGNER
Department of Chemistry
Rice University
Houston, Texas 77251

ABSTRACT. The versatile cyclopropene synthon 1-bromo-2-chlorocyclopropene has been used to synthesize new di- and tricycloproparenes. Treatment of these cycloproparenes with silver ion in chloroform leads to rapid oligomerization via carbocation intermediates to give materials in which the aromatic systems are connected by six-membered rings. The ladder polymer formed by treating the linearly fused dicyclopropanaphthalene with silver ion in chloroform has been characterized by solid state nuclear magnetic resonance spectroscopy.

1. Introduction

The manufacture and manipulation of structures at a molecular level have been recognized as an important goal for more than thirty years [1]. The synthesis of linear rigid molecules composed primarily of six-membered rings (molecular lines) is of interest in this regard. Miller and his co-workers have prepared short molecular lines composed of polyacenequinone units via repetitive Diels-Alder reactions [2,3]. The Diels-Alder reaction has also been identified as a route to other novel belts or collars [4-9]. In this manuscript we summarize our recent work on the synthesis of six-membered rings using cycloproparenes [10] as starting materials.

The reaction is exemplified below using the simplest cycloproparene, benzocyclopropene **1**. In a typical experiment, the benzocyclopropene is dissolved in

$$\mathbf{1} \xrightarrow[CHCl_3]{Ag^+} \mathbf{2}$$

anhydrous chloroform and added dropwise to a stirred suspension of $AgBF_4$ (≈ 1 mole %) in anhydrous chloroform at 0 °C. The reaction is usually complete after a few minutes. Other examples in which benzocyclopropenes have been dimerized by silver ion in

J. Michl (ed.), Modular Chemistry, 461–471.

chloroform to yield dimers connected by six-membered rings are presented below in Table 1.

TABLE 1. DIMERIZATION OF BENZOCYCLOPROPENES BY SILVER ION

Cycloproparene	Product(s) Ratio	Yield[a] %
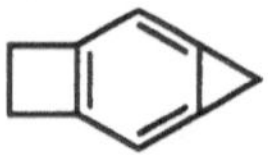	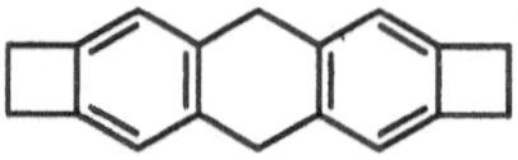	95
	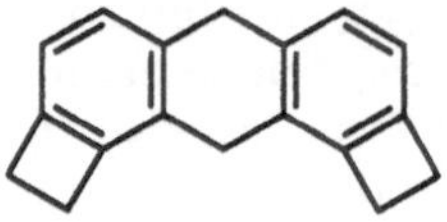 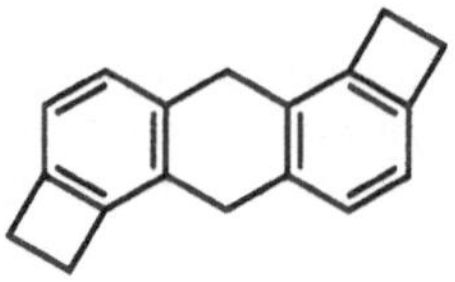 61:39	70
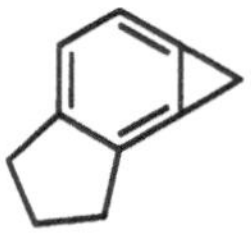	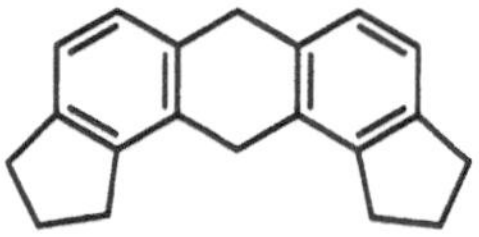 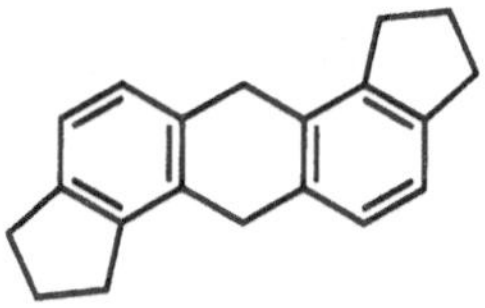 59:41	78
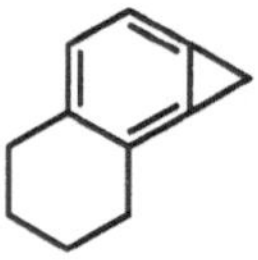	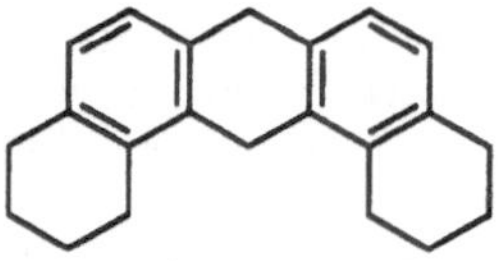 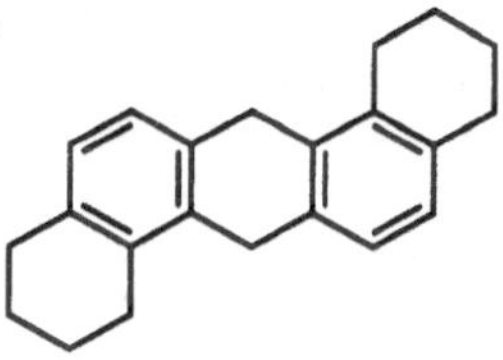 19:81	80
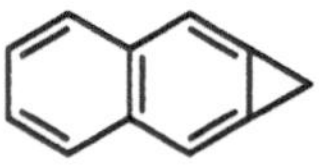	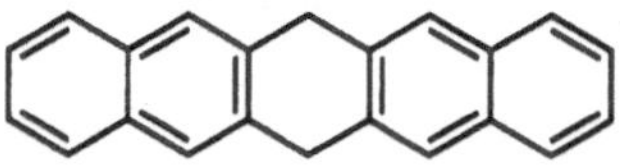	83
Cl Cl	Cl Cl Cl Cl	95

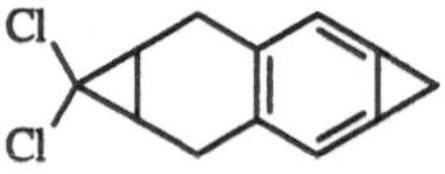

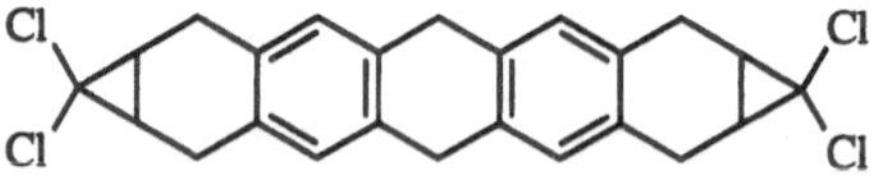

[a]Yields are isolated and are combined yields for those reactions which yield two products.

A possible mechanism is illustrated in Scheme 1.

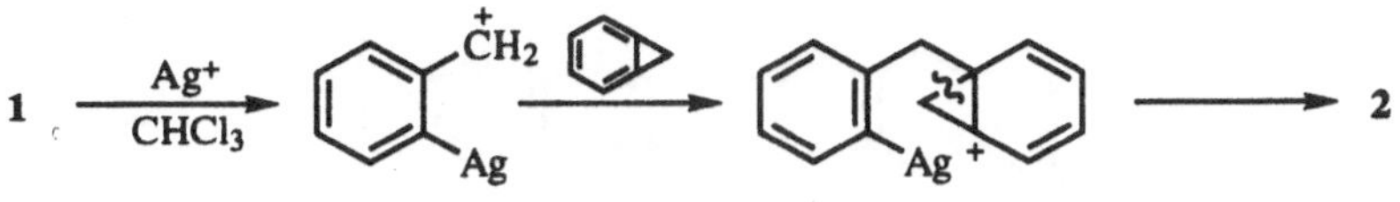

Scheme 1

The facile cleavage of the three-membered ring of benzocyclopropenes [11] by silver ion can be attributed to the strain energy associated with the cycloproparene ring system which has been estimated as ≈ 68 kcal/mole [12]. The regiochemistry of these cleavage reactions may be understood in terms of the arguments presented by Garratt and his co-workers [13] who have studied the reactions of several asymmetrically fused benzocyclopropenes with electrophiles. Their studies showed that fusion of a second strained ring leads to regioselective cleavage of the cyclopropene ring. The product ratios for the dimerization reactions presented in Table 1 may be explained, at least in part, in terms of these arguments. Thus fusion of either a four or a five-membered ring to the benzocyclopropene gives predominately *syn* products whereas fusion of the unstrained six-membered ring gives mainly the *anti* isomer. Steric effects may also be important in the predominant formation of the *anti* dimer when benzocyclopropene is fused to a six-membered ring.

The prospect that the polyacenes, or highly conjugated ladder polymers [14], might exhibit interesting electrical properties including unusual one-dimensional conductivity [15] has led to a flurry of activity in this area [14, 16-21]. Simple iterative chain-growing reactions using bicycloproparenes [22, 23] should lead to materials which upon aromatization would yield these polyacenes.

2. Synthesis and Oligomerization of di and tricycloproparenes

We have investigated the oligomerization of **5** [22] using silver ion and we find that polymerization does occur. The reaction can be monitored by ^1H NMR spectroscopy in $CDCl_3$ and the initially formed dimer **6** can be isolated readily by column chromatography.

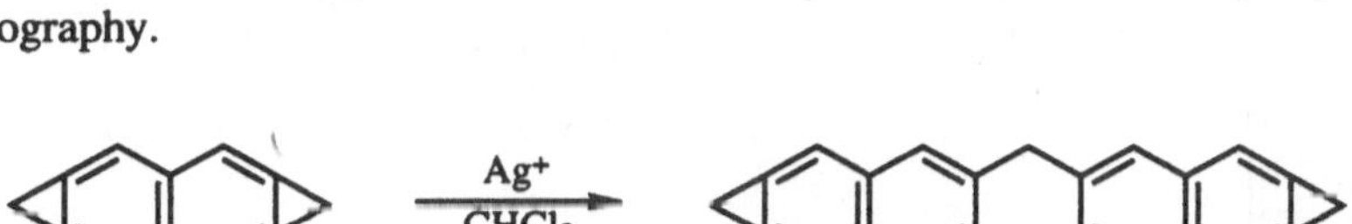

As the chain length increases in size, the oligomers begin to precipitate. Characterization of the resulting thermally stable oligomers has been achieved to some extent by solid state nuclear magnetic resonance spectroscopy. The cross-polarization experiment revealed aromatic signals at 132.3 and 126.1 ppm and one signal at 37.0 ppm which can be assigned to the saturated carbons.

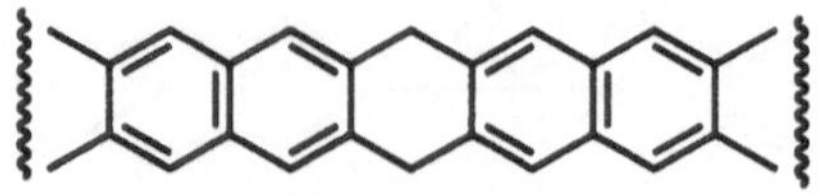

Unfortunately, the base-induced elimination-isomerization reaction of *gem*-dihalocyclopropanes used to synthesize **5** cannot be applied to the synthesis of cycloproparenes higher than the cyclopropanaphthalenes. However, we have found that the versatile cyclopropene synthon 1-bromo-2-chlorocyclopropene **7** [24] can be used to synthesize many of the higher cycloproparenes [25]. The cyclopropene itself can be synthesized readily from **8** either in solution or in the gas phase using fluoride salts to effect the elimination.

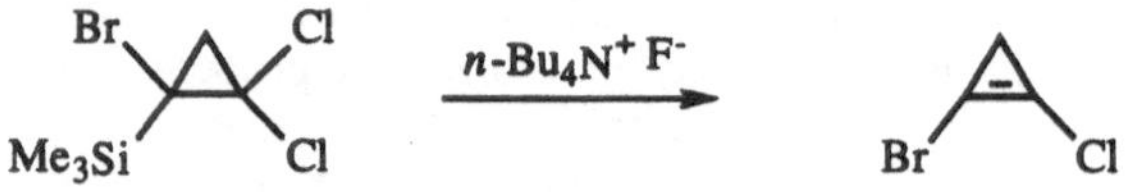

Br, Cl, Me_3Si, Cl — $n\text{-}Bu_4N^+ F^-$ → Br, Cl

8 **7**

A synthesis of the phenanthrene derivative **9** is illustrated in Scheme 2. The Diels-Alder adduct **10**, prepared by allowing **7** and the tetraene **11** [26] to stand at -20 °C for several days, could be converted to the starting material **12** by aromatization using DDQ. The elimination of **12** using potassium *t*-butoxide in tetrahydrofuran yielded the desired dicycloproparene.

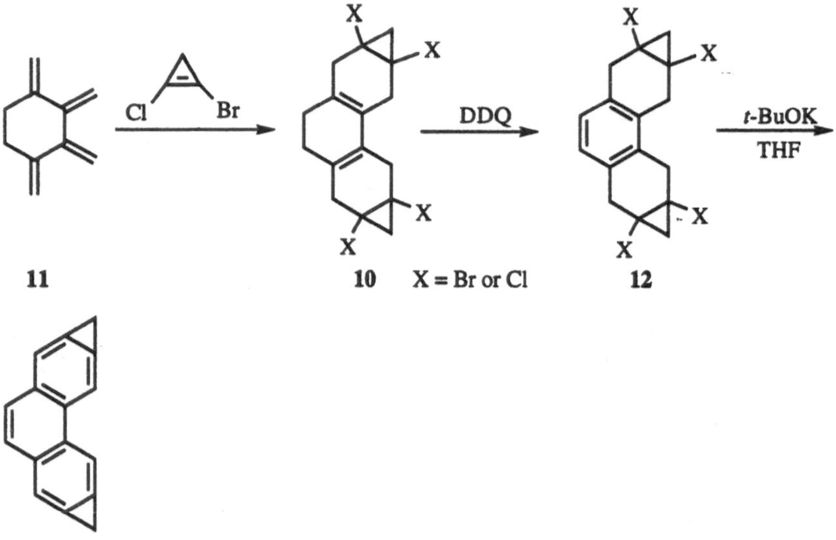

Cl, Br → X, X, X, X — DDQ → X, X, X, X — *t*-BuOK, THF →

11 **10** X = Br or Cl **12**

9

Scheme 2

Dimerization of **9** as described above for **5** was observed by nuclear magnetic resonance spectroscopy to give dimers **13** and **14** after only 5 minutes at 25 °C.

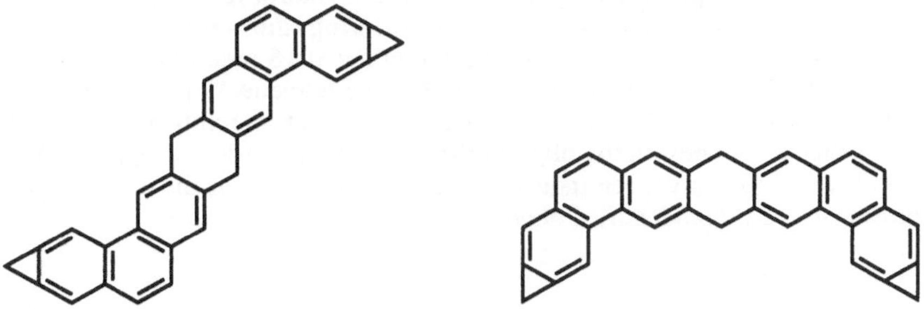

13 **14**

Prolonged reaction gave new materials provisionally identified as the ziz-zag structures (molecular trails) illustrated below.

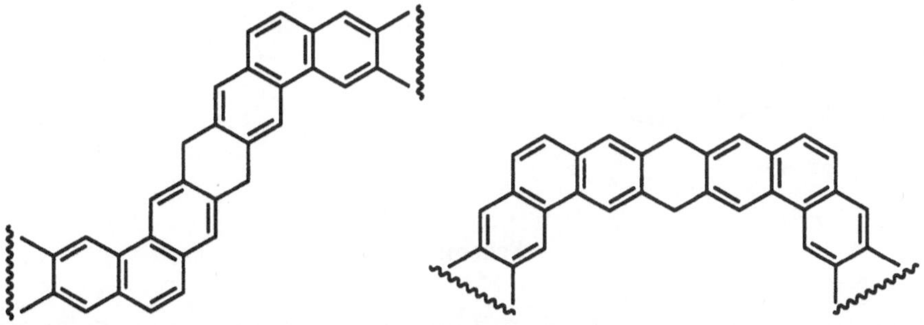

Dicyclopropanthracene **16** has been synthesized (Scheme 3) by dehydrohalogenation of the aromatized Diels-Alder adduct of **7** and 1,2,4,5-tetramethylenecyclohexane as described above for the phenanthrene derivative.

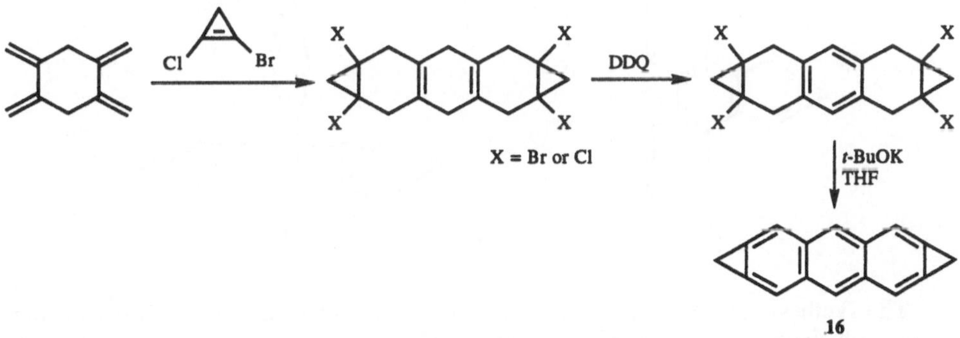

Cl Br

X X X X

DDQ

X X X X

X = Br or Cl

t-BuOK THF

16

Scheme 3

In contrast to **9**, dicyclopropanthracene gave only insoluble oligomers. No evidence for the expected dimer as observed for **5** and **9** could be found under the conditions used for the oligomerization reaction.

The versatile cycloproparene synthon 1-bromo-2-chlorocyclopropene [24] has also also been used as a reagent in a key step during the synthesis of the novel dicycloproparene **17** (Scheme 4). The starting material **18** required for this synthesis was prepared by the Diels-Alder addition of **7** to the tetraene **19** [27]. Elimination of **18** using potassium *t*-butoxide in tetrahydrofuran proceeded smoothly to yield the desired dicycloproparene. Studies on the oligomerization of **17** are still in progress. However, we have observed that silver ion induces rapid polymerization of the cycloproparene to yield dimers and as yet unidentified oligomers.

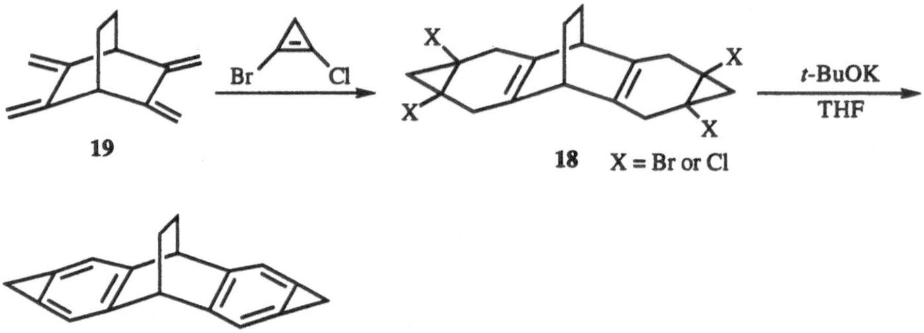

Scheme 4

Tricycloproparenes such as **20** and **21** have only recently been reported [28]. These compounds are of interest as precursors to graphitic-like networks with holes in the case of **20** or as honeycomb structures from **21**.

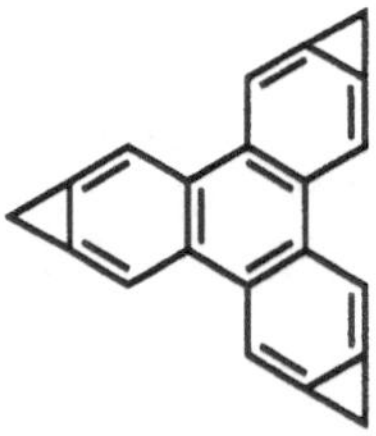

The synthesis of **20** is presented in Scheme 5. Conversion of the mixture of regio isomers obtained from the reaction of hexaradialene [29] with **7** was carried out at -50 °C. Purification of the white solid could be achieved readily by flash column chromatography on neutralized Florisil.

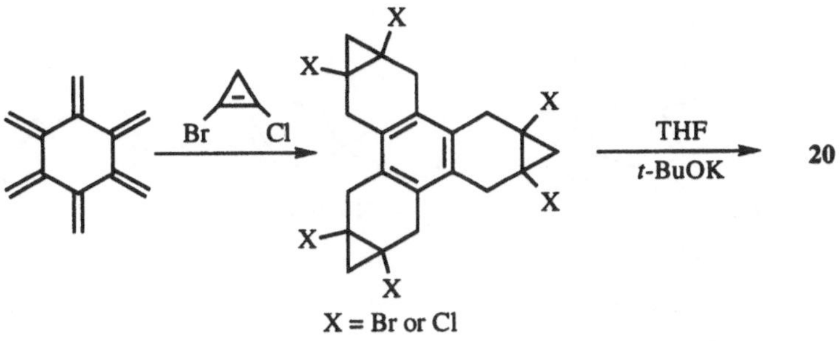

X
X
X
Br Cl
THF
t-BuOK
20
X
X
X
X = Br or Cl

Scheme 5

The synthesis of **21** is presented in Scheme 6. The starting material **22** could be prepared by repetitive addition of 1-bromo-2-chlorocyclopropene to hericene [30]. The elimination of **22** could not be effected using potassium *t*-butoxide in tetrahydrofuran as described for the synthesis of **20**; however, a mixture of N,N-dimethylformamide, hexamethylphosphoramide (5:1) and potassium *t*-butoxide at room temperature under an atmosphere of nitrogen provided a suitable medium to carry out this reaction. The cycloproparene could be isolated as a white solid in ≈ 50 % yield. Tri-cycloproparenes **20** and **21** undergo rapid oligomerization under the influence of silver ion in chloroform to yield insoluble polymers. Studies on the structures of these oligomers are under investigation.

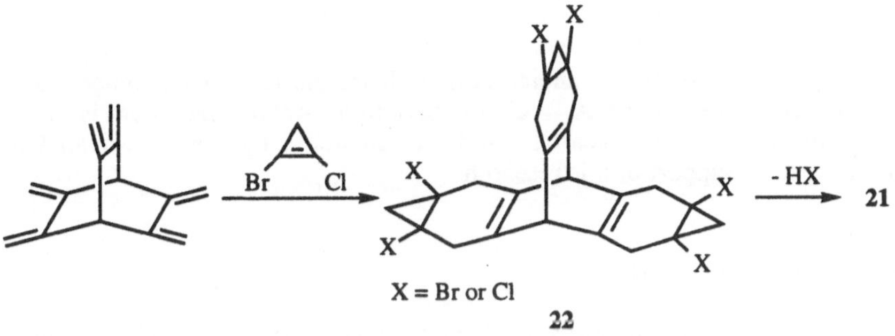

X X
Br Cl
X
X
- HX
21
X
X
X = Br or Cl
22

Scheme 6

The Diels-Alder adducts **23** and **24** were also isolated from the reaction of hericene with 1-bromo-2-chlorocyclopropene. These compounds could be converted to the cycloproparenes **25** and **26**, respectively, using potassium *t*-butoxide dissolved in the same N,N-dimethylformamide-hexamethylphosphoramide medium used to synthesize **21**.

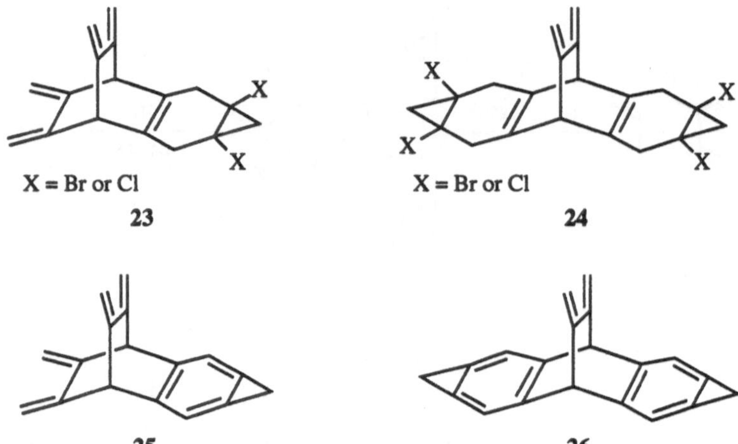

3. Summary

The above results demonstrate that bicycloproparenes and tricycloproparenes may be oligomerized readily using silver ion in chloroform. Characterization of these oligomers will be a long challenging task. The versatile cyclopropene synthon 1-bromo-2-chlorocyclopropene promises to have wide applicability in the synthesis of new cycloproparenes.

4. Acknowledgments

We gratefully acknowledge financial support from the National Science Foundation (CHE-906952) and the Robert A. Welch Foundation. Acknowledgment is made to the donors of the Petroleum Research Fund, administered by the American Chemical Society, for partial support of this research.

5. References and Notes

1. The publication of Feynman [*Saturday Review* **1960**, *43*, 45] "Wonders that Await" seems to have inspired much of the interest in this field of research.

2. Kenny, P. W., Miller, L. L. "Synthesis of Molecular Lines, Rigid Linear Molecules with Nanometer Scale Dimensions" *J. Chem. Soc. Chem. Commun.*, **1988**, *84*.

3. Kenny, P. W., Miller, L. L., Rak, S. F., Jozefiak, T. H., Christopfel, W. C. "Organized Monolayers of Polycyclic Aromatic Quinones" *J. Am. Chem. Soc.*, **1988**, *110*, 4445.

4. Schlüter, A. D. "Ladder Polymers: The New Generation" *Adv. Mater.*, **1991**, *3*, 282.

5. Löffler, M., Schlüter, A.-D., Gessler, K., Saenger, W., Toussaint, J.-M., Brédas, J.-L. "Synthesis of a Fully Unsaturated - Molecular Board" *Angew. Chem. Int. Ed. Engl.*, **1994**, *33*, 2209.

6. Kohnke, F. H., Stoddart, J. F. "The Evolution of Molecular Belts and Collars" *Pure and Appl. Chem.*, **1989**, *61*, 1581.

7. Kohnke, F. H., Mathias, J. P., Stoddart, J. F. "Structure-Directed Synthesis of New Organic Materials" *Angew. Chem. Int. Ed. Engl. Adv. Mater.*, **1989**, *28*, 1103.

8. Mathias, J. P., Stoddart, J. F. "Constructing a Molecular LEGO Set" *Chem. Soc. Revs.*, **1992**, *21*, 215.

9. Wegener, S., Müllen, K. "5, 6, 7, 8-Tetramethylenebicyclo[2.2.2]oct-2-ene as "Bis(diene)" in Repetitive Diels-Alder Reactions" *Chem. Ber.*, **1991**, *124*, 2101.

10. Billups, W. E., McCord, D. J., Maughon, B. R. "Dimerization of Cycloproparenes by Silver Ion" *Tetrahedron Lett.*, **1994**, *35*, 4493.

11. For other representative reactions of strained rings with silver ion see: Paquette, L. A., Wilson, S. E., Henzel, R. P. "Mechanistic Aspects of the Silver(I)-Promoted Rearrangements of Tricyclo[4.1.0.0^{2,7}]heptane Derivatives. Deuterium Isotope Effect Studies and Independent Generation of Argento Carbonium Ions" *J. Am. Chem. Soc.*, **1972**, *94*, 7771; Gassman, P. G., Atkins, T. J. "Transition Metal Complex Promoted Rearrangements. Tricyclo[4.1.0.0^{2,7}]heptane and 1-Methyltricyclo[4.1.0.0^{2,7}]heptane" *J. Am. Chem. Soc.*, **1972**, *94*, 7748; Gassman, P. G., Atkins, T. J. "Transition Metal Promoted Isomerizations of Highly Strained Polycyclic Systems. A Mechanistic Insight" *J. Am. Chem. Soc.*, **1971**, *93*, 4597; Sakai, M., Masamune, S. "Silver(I)-Catalyzed Rearrangement of Bicyclobutanes. Some Aspects of the Mechanism. I" *J. Am. Chem. Soc.*, **1971**, *93*, 4610; Sakai, M., Westberg, H. H., Yamaguchi, H., Masamune, S. "Silver(I)-Catalyzed Rearrangement of Bicyclobutanes. Some Aspects of the Mechanism. II" *J. Am. Chem. Soc.*, **1971**, *93*, 4611; Paquette, L. A., Allen, Jr., G. R., Henzel, R. P. "Silver(I) Ion Catalyzed Rearrangements of Strained σ Bonds. IV. The Fate of Tricyclo[4.1.0.0^{2,7}] heptane" *J. Am. Chem. Soc.*, **1970**, *92*, 7002; Cassar, L., Eaton, P. E., Halpern, J. "Silver(I)-and Palladium(II)-Catalyzed Isomerizations of Cubane. Synthesis and Characterization of Cuneane" *J. Am. Chem. Soc.*, **1970**, *92*, 6366.

12. Billups, W. E., Chow, W. Y., Leavell, K. H., Lewis, E. S., Margrave, J., Sass, R. L., Shieh, J. J., Werness, P. G., Wood, J. L. "Structure and Thermochemistry of Benzocyclopropenes. The Question of Bond Fixation and Strain Energy" *J. Am. Chem. Soc.*, **1973**, *95*, 7878.

13. Bee, L. K., Garratt, P. J., Mansuri, M. M. "Regioselective Ring Opening in Substituted Benzocyclopropenes. An Alternative or Complementary Mechanism for Electrophilic Substitution Involving Attack at a σ Bond" *J. Am. Chem. Soc.*, **1980**, *102*, 7076. See also: Billups W. E., Rodin, W. A. "Regioselective Ring Opening in Annelated Benzocyclopropenes" *J. Org. Chem.*, **1988**, *53*, 1312.

14. Bailey, W. J. *Encyclopedia of Polymer Science and Engineering 2nd ed.*, Mark, H. F., Bikales, N. M., Overberger, C. G., Menges, G., Kroschwitz, J. I. Eds., John Wiley and Sons, Inc: New York, pp. 158-235 (1990).

15. Kivelson, S., Chapman, O. L. "Polyacene and a New Class of Quasi-one-dimensional Conductors" *Phys. Rev. B.*, **1983**, *28*, 7236.

16. Tanaka, K., Ohzeki, K., Yamabe, T., Yata, S. "A Study on the Pristine and the Doped Polyacenic Semi-conductive Materials" *Synthetic Metals*, **1984**, *9*, 41.

17. Mishima, A., Kimura, K. "Superconductivity of the Quasi-one-dimensional Semiconductor Polyacene" *Synthetic Metals*, **1985**, *11*, 75; Aono, S., Nishikawa, K., Kimura, M., Kawabe, H. "Superconducting and Other Phases in Polyacenic Skeletons" *Synthetic Metals*, **1987**, *17*, 167.

18. Ozaki, M., Ikeda, Y., Nagoya, I. "An Approach to the Preparation of Polyacene by LB Method" *Synthetic Metals*, **1989**, 28, C801; Ozaki, M., Ikeda, Y., Nagoya, I. "An Approach to the Preparation of One-Dimensional Graphite." *Synthetic Metals*, **1987**, *18*, 485.

19. Marvel, C. S., Levesque, C. L. "The Structure of Vinyl Polymers: The Polymer from Methyl Vinyl Ketone" *J. Am Chem. Soc.*, **1938**, *60*, 280.

20. Bailey, W. J., Fetter, E. J., Economy, J. "Cyclic Dienes. XXVII. 1, 2, 4, 5-Tetramethylenecyclohexane" *J. Org. Chem.*, **1962**, *27*, 3479.

21. Kiji, J., Iwamoto, M. "Aromatization of Cyclized 1,2-Polybutadiene" *J. Polym. Sci. Polym. Lett. Ed.*, **1968**, *6*, 53.

22. Ippen, J., Vogel, E. "1,4-Dihydrodicyclopropa[b,g]naphthalene" *Angew. Chem. Int. Ed. Engl.*, **1974**, *13*, 736.

23. Billups, W. E., Haley, M. M., Claussen, R. C., Rodin, W. A. "Dicycloproparenes" *J. Am. Chem. Soc.*, **1991**, *113*, 4331.

24. Billups, W. E., Lin, L.-J., Arney, Jr., B. E., Rodin, W. A., Casserly, E. W. "1-Brono-2-Chlorocyclopropene - A New Cycloproparene Synthon. Synthesis of 1H-Cyclopropa[b]phenanthrene" *Tetrahedron Lett.*, **1984**, 3935.

25. Reviews: Billups, W. E., Rodin, W. A., Haley, M. M. "Cycloproparenes" *Tetrahedron*, **1988**, *44*, 1305; Halton, B. "The Cycloproparenes" *Ind. Eng. Chem.*

Prod. Res. Dev., **1980**, *19*, 349; "Developments in Cycloproparene Chemistry."*Chem. Rev.*, **1989**, *89*, 1161.

26. Hopf, H., Gottschild, D., Lenk, W. "Thermal Rearrangements, XIII. Thermal Isomerization of Exocyclic Allenes" *Isr. J. Chem.*, **1985**, *26*, 79.

27. Pilet, O., Birbaum, J.-L., Vogel, P. "Preparation and Diels-Alder Reactivity of 2,3,5,6,7,8-Hexamethylidenebicyclo[2.2.2]octane ('[2.2.2]Hericene'). Force-Field Calculations of Exocyclic Dienes as a Moiety of Bicyclic Skeletons" *Helv. Chim. Acta*, **1983**, 66, 19.

28. Billups, W. E., McCord, D. J., McMaughon, B. R. "Triscycloproparenes" *J. Am. Chem. Soc.*, **1994**, *116*, 8831.

29. Schiess, P., Heitzmann, M. "Hexakis(methylidene)-cyclohexane ("[6]Radialene")" *Helv. Chim. Acta*, **1979**, *61*, 844.

30. Hericene was prepared as described by Köhler and Steck. See: Köhler, F. H., Steck, A. "Ein- bis dreikernige Metall-π-Derivate von Hericen mit $(Me_3P)_3$ Fe-und CpCo-Fragmenten" *J. Organomet. Chem.*, **1993**, *444*, 165. For the first reported synthesis of hericene see: Pilet, O., Birnbaum, J.-L., Vogel, P. "Preparation and Diels-Alder Reactivity of 2,3,5,6,7,8-Hexamethylidenebicyclo[2.2.2]octane ('[2.2.2]Hericene'). Force-Field Calculations of Exocyclic Dienes as a Moiety of Bicyclic Skeletons" *Helv. Chim. Acta.*, **1983**, *66*, 19. For the origin of the name see: Nickon, A., Silversmith, E. (1987) The Name Game; Pergamon Press: New York 1987.

MULTIPLY ETHYNYLATED π-COMPLEXES OF IRON, COBALT, AND MANGANESE: MODULES FOR THE CONSTRUCTION OF RIGID ORGANOMETALLIC OBJECTS

U. H. F. BUNZ
MPI für Polymerforschung
Ackermannweg 10, 55021 Mainz, FRG

ABSTRACT. The synthesis and coupling reactions of multiply ethynylated π-complexes are discussed. The organometallic modules are prepared by a metalation/iodination/coupling (MIC)-sequence. The Stille and Hay type reactions of these ethynylated complexes gives rise to novel organometallic objects with unusual topologies, of which some show attractive lyotropic smectic or thermotropic nematic liquid crystalline behavior.

1. Introduction

In the context of preparative chemistry, modules are rigid, shape consistent molecular entities of defined lengths, fitted with two or more functional groups. These properties allow the connection of several modules into larger rigid and shape persistent molecular objects. Prominent examples of modules include Szeimies' [1] and Michls' oligobicyclo[1.1.1]pentanes (staffanes, tinkertoy-concept) [2a], Müllens' graphite sheets [3], Vollhardts cyclobuta-fused benzenes (phenylenes) [4], Moores' [5] and Tours' [6] phenylene ethynylene oligomers, Diederichs' tetraethynylethylenes [7], as well as Hawthornes' [8] and Michls' [2b] oligocarboranes. These approaches towards modular chemistry use mainly (but not exclusively) *covalent* bonds for the assembly of the modules into larger structures [9].

The progress in conceptualization, synthesis and assessment of properties in the field of modular chemistry has developed in a dramatic pace during the last few years. Yet all the above mentioned systems (with the exception of the carboranes) are purely *organic*. The organometallic branch of the enterprise still seems to be in its infancy [2c]. When starting out in 1992 we decided to use multiply ethynylated π-complexes for the construction of organometallic modules. Surprisingly only two examples of diethynylated π-complexes, i.e. 1,2- and 1,3-diethynylcyclobutadiene(cyclopentadienyl)cobalt (**1a**, **2a**) [10] were known; their synthesis was accomplished by dimerization of either 1,4-bis(trimethylsilyl)butadiyne (to give **2a**) or the codimerization of bis(trimethylsilyl)acetylene with 1,6-bis(trimethylsilyl)hexatriyne over $CpCo(CO)_2$ to obtain **1a**. The method used is elegant, but does not lead to a rational synthesis of other diethynylated π-complexes, which are the necessary building blocks for the construction of a variety of modular and/or supramolecular arrays (Figure 1).

J. Michl (ed.), Modular Chemistry, 473–484.

2. The Monomer Pool, The MIC-Process

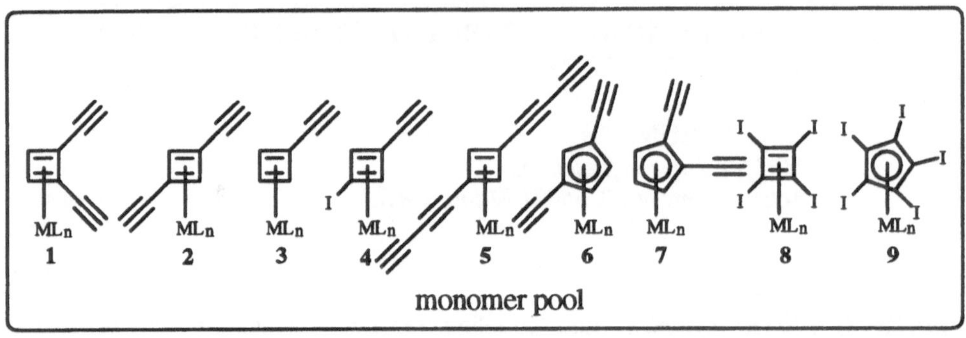

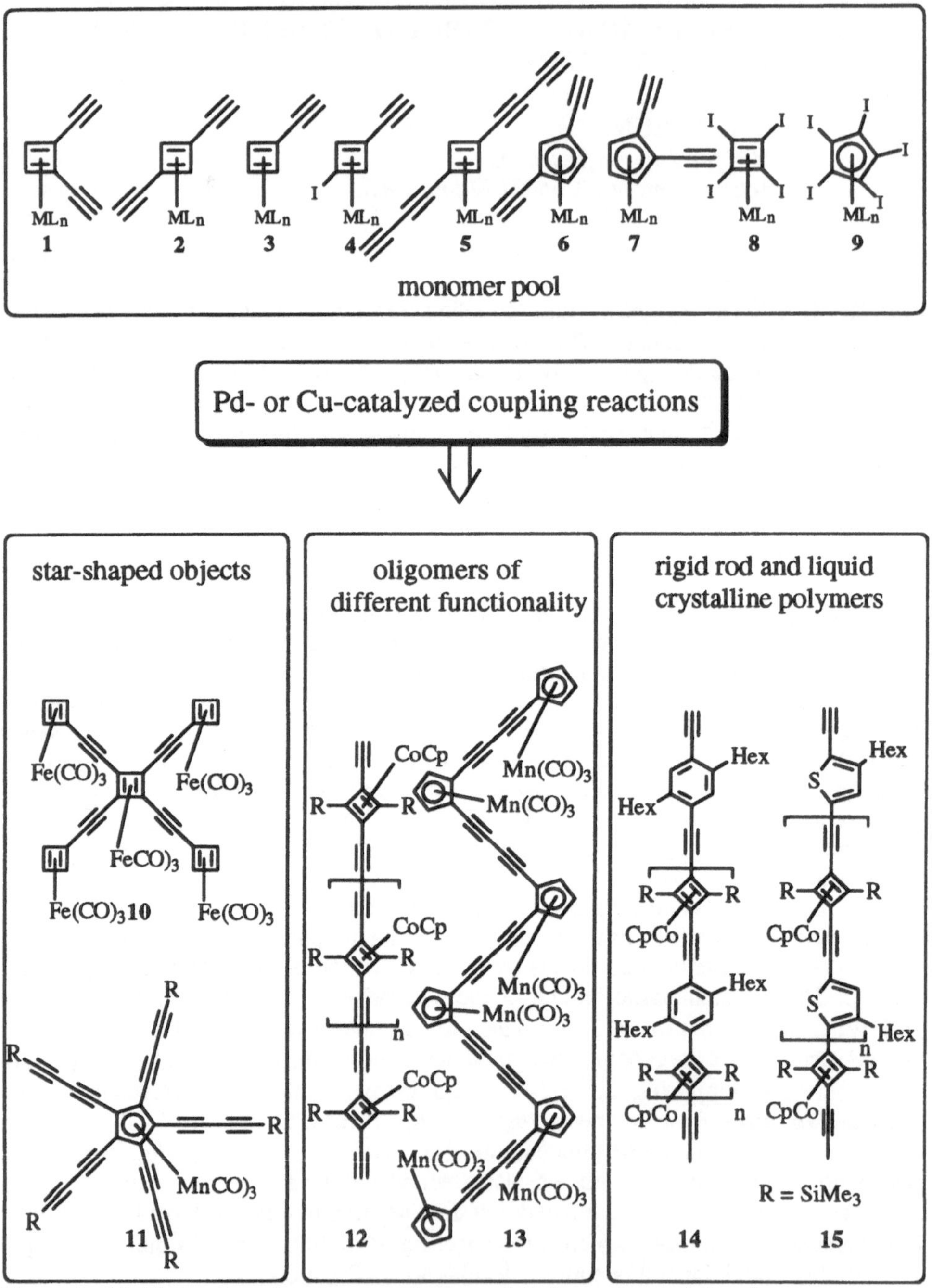

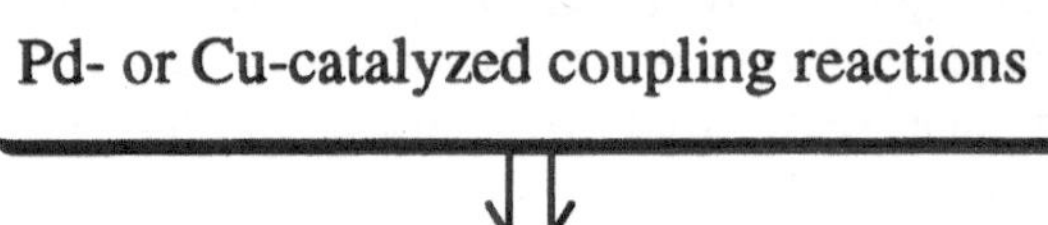

star-shaped objects

Fe(CO)3
FeCO)3
Fe(CO)3 10
R
MnCO)3
11

oligomers of different functionality

CoCp
R
n
Mn(CO)3
12
13

rigid rod and liquid crystalline polymers

Hex
S
R
CpCo
n
R = SiMe3
14
15

Figure 1. Multiply ethynylated π-complexes and their use in modular chemistry.

In order to use diethynylated π-complexes as modules, viable synthetic entries into this class of compounds (shown in Figure 1 as *monomer pool*) need to be developed. The literature showed that Sterzo and Stille [11] appended alkynes to the Cp ring of several half sandwich complexes (including the commercially available cymantrene, $CpMn(CO)_3$) by a metalation/iodination/coupling (MIC) sequence.

sec-BuLi, 1,2-diiodo-ethane; R_3Sn–≡, $PdCl_2(CH_3CN)_2$ — MIC strategy

$Mn(CO)_3$ **16**, $Mn(CO)_3$ **17**, $Mn(CO)_3$ **18a**

$Mn(CO)_3$ **18b** —MIC→ **7b** $Mn(CO)_3$, **6b** $Mn(CO)_3$

either only *ortho* or *ortho/meta*-mixture depending on reaction conditions

$Fe(CO)_3$ **19** —MIC→ $Fe(CO)_3$ **3b** —MIC→ $Fe(CO)_3$ **1**, $Fe(CO)_3$ **2**

either only *ortho* or *ortho/meta*-mixture depending on reaction conditions

1b —1,2-dichloro-ethylene, Pd-catalyst→ **19** (41 %) —*sec*-BuLi, Me_3SiCl→ **5b** (92 %)

This sequence works well because the electron withdrawing organometallic fragments facilitate the deprotonation of the ring as well as the palladium catalyzed Stille-type coupling of the iodide formed (such as **17**) with a tin alkyne. A good example is cymantrene, where the application of the MIC procedure leads to the formation of **18a** in a 82% overall yield [11]. Repetition of MIC using **18b** as a substrate gives rise to either pure **7a** or a mixture of **6a** and **7a** (ML_n = $Mn(CO)_3$, R =

TMS), depending upon the exact reaction conditions used [12]. The MIC process is not only useful for the synthesis of diethynylated Cp complexes, but was applied successfully to the tricarbonyl(cyclobutadiene)iron system as well. A series of monoalkynylated cyclobutadiene complexes with the general structure **3** was prepared using the MIC sequence [13]. To prepare diethynylated **1** (ML_n = $Fe(CO)_3$, R = TMS) starting from **3**, close scrutiny of the reaction temperature in the metalation step is necessary to obtain exclusively the *ortho* product **1** (ML_n = $Fe(CO)_3$ R^1 = TMS, R^2 = TMS, **1b**; Ph, **1c**; TIPS, **1d** *t*Bu-C≡C-C≡C-, **1e**,...).

Attempts to prepare modules with a different length increment have led to the synthesis of **5** [14]. Starting from **2a** (M_n = CpCo), **5** was prepared by application of a double chlorovinylation and subsequent double dehydrohalogenation. While the unsubstituted **5** is only stable in solution for several hours and decomposes upon concentration, silylation of the butadiyne termini leads to **5b**, which is stable under ambient conditions either in solution or in the solid state for at least some days.

3. Linear Oligomers and Polymers

3.1. CYCLOBUTADIENE OLIGOMERS

Iron: One of the goals was to synthesize the dumbbell **21** *via* extensive use of the MIC route. To this end MIC was repeatedly applied to **1c**, furnishing the iodide **20**, which was coupled in the last step to afford the ethynyl or butadiynyl-bridged dumbbells

R
R
$Fe(CO)_3$
MIC
3 steps,
20 %
yield
I
R
R
R
$Fe(CO)_3$
couple
n = 1, 76 %;
n = 2, 66 %
yield
R
R
R
R
R
R
n
$(CO)_3Fe$
$Fe(CO)_3$

19 **20** **21** R = $SiMe_3$

Me_3Si — I $Fe(CO)_3$ — Stille coupling, 40% → Me_3Si — $SiMe_3$ $Fe(CO)_3$ $Fe(CO)_3$

4 **22**

2, R = $SnMe_3$, ML_n = $Fe(CO)_3$ — Stille coupling, 31% → Me_3Si — $SiMe_3$ $Fe(CO)_3$ $Fe(CO)_3$ $Fe(CO)_3$

23

21 in 71 and 66% yield respectively [15].

The synthesis of linear oligomers [16] involves the use of either iodo and alkyne substituted (AB-difunctional, **4**) or diethynylated (AA-difunctional, **2**) monomers. Utilizing **4** and **2** afforded the dimer **22** and trimer **23**. Repetition of the sequence will furnish higher homologues (i.e., tetramer and pentamer). The application of the MIC procedure is possible with 1,2-diethynylated cyclobutadiene complexes as well, giving rise to the isolation of kinked oligomers **24** and **25**.

MIC is a powerful tool for the synthesis of modules based on cymantrene and tricarbonyl(cyclobutadiene)iron, due to the ease of the ring metalation. Using the 1,3-directing effect of TMS groups and the 1,2-directing effect of ethynyl groups in the metalation reaction allows the functionalization and control of virtually every position in the Cp or cyclobutadiene ring under consideration.

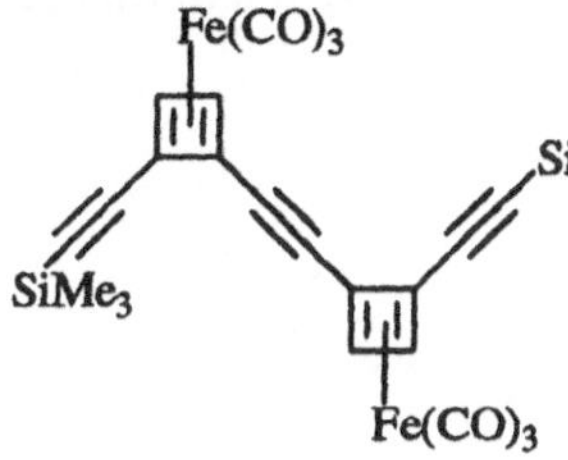

24 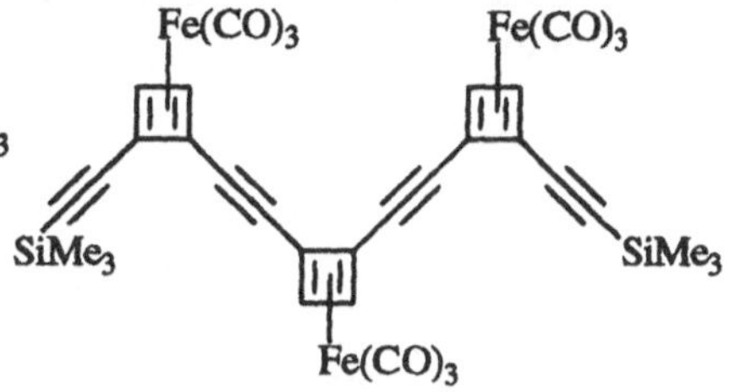 **25** 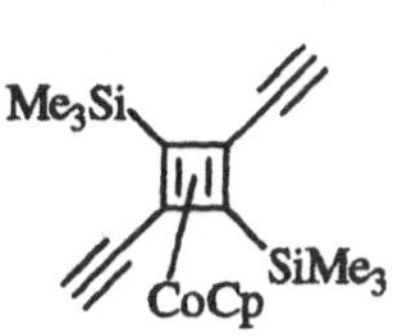 **2a**

Cobalt: 1,3-Diethynylcyclobutadiene(cyclopentadienyl)cobalt (**2a**) can be obtained on a 10 g scale. The availability renders **2a** an interesting module for rigid linear objects: a Hay type oligomerization of **2a** in boiling TMEDA led to the formation of a polymer of the structure **12** with $D_p \approx 15$ [17], as evidenced by ^{13}C NMR spectroscopy and analytical gel permeation chromatography (GPC). The GPC trace of **12** is informative, as it was possible to discern single oligomers up to the 17-mer, which was the most abundant species in the product. Despite the drastic conditions involved in the coupling, a monomodal distribution of *poly*-**12** with $M_w/M_n = 3.7$ was found. NMR spectroscopic investigation of *poly*-**12** evidenced as well that crosslinking of the butadiyne units did not occur. Comparison of the UV-vis spectra of *poly*-**12** and monomer **2a** (ML_n = CpCo) showed a considerable bathochromic shift of the observed bands and an increased ε-value in the polymer. To understand the optical properties, we decided to synthesize oligomers derived from **12**. To this end the coupling conditions were attenuated, the solvent TMEDA replaced with the lower boiling butanone (bp 80 °C). As consequence, a series of oligomers **12** (n = 0 - 7) was isolated.

In the UV-vis spectra of the oligomer series **12** the largest changes in λ_{max} and ε occur between monomer **2a** and dimer **21a** (n = 0). The spectra of the higher oligomers change noticeably up to the heptamer, but heptamer, octamer and nonamer display spectra very similar with respect to each other and to *poly*-**12**. The conclusions are that a) the coherence length of **12** is reached with the heptamer and b) λ_{max} does not change considerably from dimer to polymer, indicating that λ_{max} is a localized transition involving only the cyclobutadiene(cyclopentadienyl)cobalt unit.

The oligomers **12** (still bearing two terminal alkyne groups) are not only interesting as model compounds for spectroscopy but represent promising modules for the future connection of different redox active, mesogenic and/or photochemically active species utilizing Stille-type chemistry. It would be valuable though to find a method towards a more rational synthesis for discrete members of the *oligo*-**12** series. One of the problems here is that the synthesis of **2a** (*vide supra)* is not variable in differentiating the termini of the alkyne groups. But different alkyne termini, i.e. orthogonal protecting groups, would be necessary for the rational design of discrete oligomers.

Using the enlarged monomer **5a** (ML_n = CpCo) offered a possibility for the synthesis of tetrayne-linked polymers of type **26** by Hay coupling. Bisdiyne **5a** was completely consumed after 2 h at 30 °C. Reason for the enhanced coupling reactivity of **5a** compared to **2a** is the attenuated steric crowding in **5a**. But due to the diminished solubility of **26** in organic solvents, the polymer is relatively short; D_p does not exceed ≈ 7. In the solid state and in solution, **26** is deeply orange colored. It is thermally stable both in solution as well as in the solid state despite the octatetrayne bridge. The steric bulk of the organometallic fragment seems to be sufficient to supress any intermolecular crosslinking up to 200 °C in the solid state.

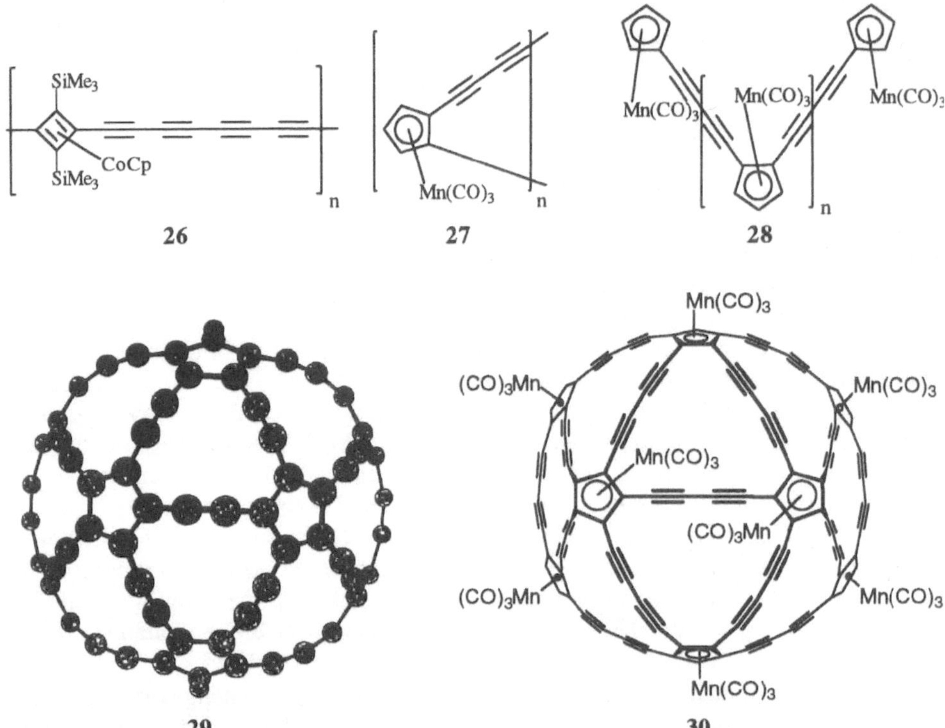

Figure 2. A fullerenyne and its organometallic $Mn(CO)_3$-stabilized derivative.

3.2. LINEAR FULLERENYNE SEGMENTS [12]

In a publication in 1993 Baughman *et al.* [28] speculated that the fullerenynes, a class of hollow and alkyne-expanded (exploded) fullerenes may be isolable. While it seems questionable that fullerenynes such as **29** will be stable enough to survive isolation (if successfully formed), an organometallic derivative of this particle (i.e. **30**) may be stable if synthesized. Even though if **30** is not attainable by synthesis at the moment, it is possible to prepare segments of it with relative ease: **6a** (ML_n = $Mn(CO)_3$) is active under the conditions of the Hay coupling, but it was not possible to isolate cyclic fullerenyne segments such as **36**. Instead the formation of **27** was observed ($D_P > 25$) in a 68% yield. It is not clear at the moment what governs the formation of cycles *vs.* polymer, but it may depend upon the steric crowding in the monomer. The rule would be, that an increased steric congestion of the monomer would lead to an increase in the formation of cycles, in accord with the Thorpe-Ingold effect exerted by *gem*-dimethyl groups upon the propensity of cyclisation [29].

In order to access linear oligomers **28** (one of which is **13**), we added **18a** as an end capping agent to **7** and performed Hay couplings. Preparative HPLC made the dimer through the heptamer accessible in yields ranging from 12.5 to 2.5%.

Diastereomer formation: The polymer **27** must be a mixture of different stereoisomers, for every $CpMn(CO)_3$ unit presents a stereocenter: surprisingly the stereoisomers can not be distinguished by conventional ^{13}C NMR spectroscopy and only six broadened signals are observed for **27**. The case of **28** shows a similar behavior. The single oligomer **28** typically does not show a split of the resonances in the ^{13}C NMR spectrum caused by diastereomers. But diastereomeric mixtures must occur starting with the tetramer. Attempts to separate discrete diastereomers of one oligomer by analytical HPLC were not met by success either. Apparently, the butadiyne bridges separate the stereocenters efficiently to prevent the spectroscopic and chromatographic discrimination of single diastereomers.

3.3. LIQUID CRYSTALLINE POLYMERS [18]

Polymers **14** and **15** [19] are prepared by the condensation of **2b** with 1,4-diiodo-1,4-dihexylbenzene or 2,5-diiodo-3-hexylthiophene in a Heck-Cassar-Sonogashira-Hagihara coupling [20] using the recent variant of Alami, Ferri, and Linstrumelle [21]. Typical D_p values fall between 12 and 20. Both polymers **14** and **15** are of rigid rod-like shape and were expected to show LC behavior as a consequence of their shape anisotropy. When a solution of **16** in dichloromethane was casted onto a glass slide and the solvent allowed to evaporate quickly, an isotropic yellow film was observed under crossed polarizers. Upon heating to 155 °C for a few minutes, a *schlieren* texture formed, typical for thermotropic nematic phases (Figure 2a). It is not possible to convert the LC-phase of **16** into the molten isotropic state, due to the onset of decomposition above 230 °C. The inaccessibility of the melt and the occurrence of a LC phase was corroborated by DSC results which showed that in the first scan (and only there!) a structural change occurred at ca. 165 °C, corresponding to the transition amorphous - liquid crystalline. When samples of carefully recrystallized **15** were subjected to powder diffractometry, several strong reflections were recorded, suggesting that **15** is partially crystalline. These reflexes disappeared upon heating the sample to 160 °C, indicating the loss of all *crystalline* order and the appearance of a nematic phase.

Polymer **15** is interesting in a different vein. It does not display a remarkable thermal behavior, but if solutions of **15** in dichloromethane are slowly evaporated, the occurrence of a texture similar to the ones displayed by low molecular smectic LC's is observed under crossed polarizers (Figure 2b). This is unusual, because main chain polymers are expected to show ordering only in one dimension, forming *nematic* phases. The smectic phase behavior was corroborated by electron microscopy (Figure 2c) showing brush like features with a lamellar ordering. The thickness of a single lamella thereby corresponds to the length of one macromolecule **15**. In accord with the above findings, powder diffractometry of concentrated solutions of **15** in trichlorobenzene showed a broad scattering feature in the region where in a crystallized sample the non-equatorial reflections appear. Quite interesting, the LC properties of **15** must be a consequence of the regular interchain ordering of the CpCo-units with respect to each other and not of a special ordering of the polymer chains.

Figure 3. a, b) LC-textures observed in **14** and **15** as evidenced by optical micrographs (crossed polarizers). *c)* Electron micrograph of polymer **15** showing the brush-like lamellar structures which are visible as „fans“ in the optical micrograph.

4. Stars

In our effort to produce shape persistent organometallic modules, we were interested in accessing star shaped derivatives as well. Starting with **8** (ML_n = $Fe(CO)_3$) [22], a series of tetraalkynylated cyclobutadiene complexes were produced in good to excellent yields by the Stille reaction [23, 24]. Not only simple alkynes but conjugated diynes and triynes were suited for the coupling reaction to furnish tetraalkynylated adducts. All of the stars **31** - **34** are stable brown yellow and crystalline materials. Introducing stannylated organometallic alkynes, a series of pentanuclear complexes were prepared by the same coupling reaction [25].

Pentaiodide **9** (ML_n = $Mn(CO)_3$) [26] coupled (catalyst: $[PdCl_2(CH_3CN)_2]$) [27] a series of alkynes and butadiynes to the cymantrene nucleus. The yields of **34** and **35** are less satisfying than those of the tricarbonyl(cyclobutadiene)irons. The reason for this behavior may be the enhanced steric crowding of the cymantrene nucleus, carrying five substituents, compared to the four substituents in the cyclobutadiene complexes. Still, the low yields of **34** and **35** do not pose a real problem to obtain up to 200 mg of the respective stars, due to the accessibility of **9** on a 10 g scale.

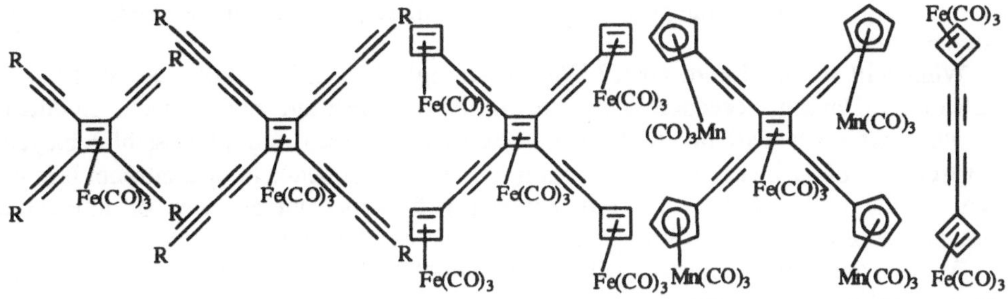

31 32 33 34 35

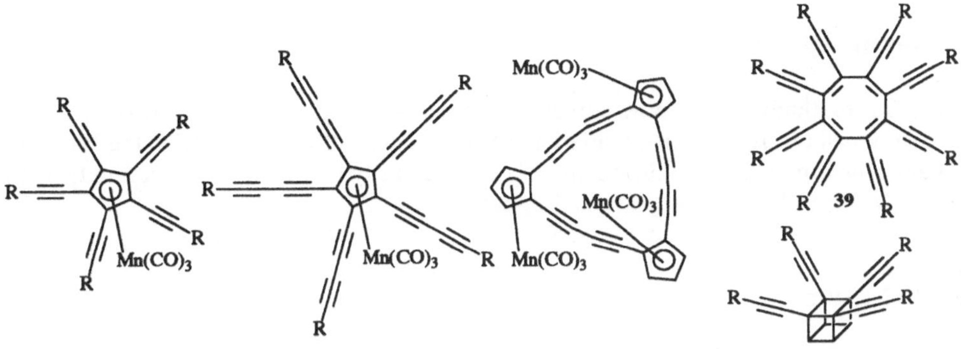

36 37 38 39 40

5. Conclusions and Outlook

Ethynylated and multiply ethynylated π-complexes are synthesized quickly by the MIC route. Electron withdrawing ML_n fragments thereby serve as convenient activating groups towards clean deprotonation (serving the ring substituted iodide) of the cyclobutadiene or Cp ring. Stille type couplingthen allows the efficient conversion of the iodo substituent(s) into alkyne and butadiyne groups for essentially all examined cases in respectable to excellent yields.

The synthesis of peralkynylated π-complexes by either the MIC-process (**21**) or by a one pot reaction (**29** - **32**) was achieved. The yield of the Pd-catalyzed coupling reactions is dependent upon a sensitive interplay of steric and electronic factors. The first rule is that the reaction goes the better, the less hindered the reaction partners are. Both the alkyne and the iodide are sensitive to steric hindrance. The second rule is that if stannylalkynes of structures similar to those of the used iodides are coupled, the yield of the cross coupling product decreases at the cost of the homocoupling product of the corresponding stannylalkyne (case of **31**). A third observation is that in going from monoynes to diynes and triynes, the yield of the coupling products diminishes. Instead an increasing amount of black and infusible material is obtained, indicating that the palladium catalyst may not be „innocent“ towards the product and starts to crosslink the extended alkyne arms, while protection by steric shielding is less and less feasible in the extended systems.

What will be our future goals? We are strongly inclined towards the synthesis of parts of **37**. Having succeeded in the preparation of linear and star-shaped segments of **37**, the next goals will lie in the synthesis of cyclic (**38**) and (if possible) bicyclic substructures of **37** by massive application of the MIC sequence to cymantrene. It is not clear at the moment, how far this concept will work, but a series of interesting cymantrene oligomers of differing topology should be obtained in this way.

Another important objective will be the demetalation of complexes **29** by cerium-ammonium nitrate, giving rise to the synthesis of either tetraethynylcubane **40** or octaethynylcyclooctatetraene **39**.

Acknowledgments

I wish to thank Jutta Wiegelmann, Markus Altmann and Mark Roos for their experimental efforts. Financial aid by the Volkswagen-Stiftung, the DFG, the Fonds der Chemischen Industrie and the BASF AG are gratefully acknowledged. I am indebted to Prof. Dr. Klaus Müllen for generous support.

6. References

1. Szeimies, G. (1992) Bridgehead olefins and small ring propellanes, in B. Halton, ed. *Adv. Strain in Organic Chemistry* **2**, 1.2
2. [a] Kaszynski, P.; Friedli, A. C.; Michl, J. (1992) Towards a molecular-size „tinkertoy" construction set. Preparation of terminally functionalized [n]staffanes, *J. Am. Chem. Soc.* **114**, 601. [b] Müller, J.; Base, K.; Magnera, T. F.; Michl, J. (1992) Rigid-rod oligo-*p*-carboranes for molecular tinkertoys. An inorganic Langmuir-Blodgett film with a functionalized outer surface, *J. Am. Chem. Soc.* **114**, 9721. [c] See also: Michl, J. these proceedings.
3. Müller, M; Petersen, J.; Strohmaier, R.; Günther, C.; Karl, N.; Müllen, K. (1996) Polybenzoid C_{54} hydrocarbons: synthesis and structural characterization in vapor-deposited ordered monolayers, *Angew. Chem., Int. Ed. Engl.* **35**, 886. Stabel, A.; Herwig, P.; Müllen, K.; Rabe, J. P. (1995) Diodelike current-voltage curves for single molecule-tunneling spectroscopy with submolecular resolution of an alkylated *peri*-condensed hexabenzocoronene, *ibid.* **34**, 1609.
4. Boese, R.; Matzger, A. J.; Mohler, D. L.; Vollhardt, K. P. C. (1995) C-3-Symmetric hexakis(trimethylsilyl)[7]phenylene [„tris(biphenylenocyclobutadieno)-cyclohexatriene"] - a polycyclic benzenoid hydrocarbon with slightly curved topology, *Angew. Chem., Int. Ed. Engl.* **34**, 1478 and refs.
5. Kawaguchi, T.; Walker, K. L.; Wilkins, C. L.; Moore, J. S. (1995) Double exponential dendrimer growth, *J. Am. Chem. Soc.* **117**, 2159 and refs. See also: Moore, J. S. these proceedings.
6. Tour, J. M. (1996) Conjugated macromolecules of precise length and constitution. Organic synthesis for the construction of nanoarchitectures, *Chem. Rev.* **96**, 537 and cited refs.
7. Diederich, F. (1994) Carbon scaffolding: building acetylenic all-carbon and carbon-rich compounds, *Nature* **369**, 199. Stang, P. J. Diederich, F. (eds.) (1995) *Modern Acetylene Chemistry*; Verlag Chemie, Weinheim. See also: Diederich, F. these proceedings.

8. Yang, X.; Knobler, C. B.; Zheng, Z.; Hawthorne, M. F. (1994) Host-guest chemistry of a new class of macrocyclic multidentate Lewis acids comprised of carborane-supported electrophilic mercury centers, *J. Am. Chem. Soc.* **116**, 7142 and refs.
9. For the use of weaker than covalent forces in modular chemistry see the articles by O. Yaghi, J. Wuest, M. Zaworotky and the therein cited literature in these proceedings.
10. Fritch, J. R.; Vollhardt, K. P. C. (1978) Bicyclobutadienylene(tricyclo[4.2.0.0^{2,5}]octa-1,3,5,7-tetraene)-η^5-cyclopentadienylcobalt. A possible intermediate in the remarkable gas phase rearrangement of 1,2-diethynyl-η^4-cyclobutadiene-η^5-cyclopentadienylcobalt, *J. Am. Chem Soc* **100**, 3643.
11. Lo Sterzo, C.; Stille, J. K. (1990) Use of palladium-catalyzed coupling reactions in synthesis of heterobimetallic complexes. Preparation of bis(cyclopentadienyl)acetylene heterodinuclear complexes, *Organometallics* **10**, 687 and cited refs.
12. Bunz, U. H. F.; Enkelmann, V.; Beer, F. (1995) *ortho-* and *meta-*(Diethynylcyclopentadienyl)tricarbonylmanganese: building blocks towards the construction of metal fragment substituted fullerenynes? *Organometallics* **14**, 2490.
13. Wiegelmann-Kreiter, J. E. C.; Bunz, U. H. F.; Schiel, P. (1994) Synthesis and characterizationn of diethynylated tricarbonyl(cyclobutadiene)iron complexes, the *ortho*-lithiation concept, *Organometallics* **13**, 4650. Wiegelmann, J. E. C.; Bunz, U. H. F. (1993) (Alkynylcyclobutadiene)tricarbonyliron: new organometallic alkynes, *ibid.* **12**, 3792. Bunz, U; (1993) Direct lithiation of (cyclobutadiene)tricarbonyliron and ((trimethylsilyl)-cyclobutadiene)tricarbonyliron with *sec*-butyllithium: selective *para*-metalation, *ibid.* 3594.
14. Altmann, M. Enkelmann, V. Bunz, U. H. F. (1996) Synthesis and characterization of a novel cyclobutadiene-octatetrayne polymer, *Chem. Ber.* **129**, 269.
15. Wiegelmann-Kreiter, J. E. C.; Bunz, U. H. F. (1995) Synthesis of dumbbell-shaped organometallics: synthesis of a peralkynylated dinuclear cyclobutadiene complex, *Organometallics* **14**, 4449.
16. Bunz, U. H. F.; Wiegelmann-Kreiter, J. E. C. (1996) Alkyne-substituted tricarbonyl(cyclobutadiene)iron complexes: Stille coupling of iodocyclobutadiene complexes with tin alkynes, *Chem. Ber.* **129**, *in print.*
17. Altmann, M. Bunz, U. H. F. (1994) Synthesis and characterization of a poly[(*para*-cyclobutadienylenecyclopentadienylcobalt)butadiynediyle, a butadiyne-cyclobutadiene copolymer, *Macromol. Rapid Commun.* **15**, 785. Altmann, M.; Enkelmann, V.; Bunz, U. H. F. (1996) Synthesis of linear oligomers of [1,3-diethynyl-2,4-bis(trimethylsilyl)-cyclobutadiene]cyclopentadienylcobalt: dimer to nonamer, *Organometallics* **15**, 394.
18. For a review see. Oriol, L.; Serrano, J.-L. (1995) Metallomesogenic polymers, *Adv. Mater.* **7**, 348.
19. Altmann, M.; Bunz, U. H. F. (1995) Polymers with complexed cyclobutadiene units in the main chain: the first example of a thermotropic, liquid crystalline organometallic polymer, *Angew. Chem., Int. Ed. Engl.* **34**, 569. Altmann, M. Enkelmann, V. Lieser, G. Bunz, U. H. F. (1995) Synthesis of novel polymers containing cyclobutadiene, thiophene, and alkyne units: polymeric organometallic mesogens, *Adv. Mater.* **7**, 726. Bunz, U. H. F. (1996) Synthesis of novel liquid crystalline organometallic polymers, *Pure Appl. Chem.* **68**, 309.
20. Dieck, H. A.; Heck, R. F. (1975) Palladium catalyzed synthesis of aryl, heterocyclic and vinylic acetylene derivatives, *J. Organomet. Chem.* **93**, 259. Cassar, I., Synthesis of aryl- and vinyl-substituted acetylene derivatives by the use of nickel and palladium complexes, *ibid.* 253. Sonogashira, K.; Tohda, Y.; Hagihara, N. (1975) A convenient synthesis of acetylenes: catalytic substitutions of acetylenic hydrogen with bromoalkenes, iodoarenes, and bromopyridines, *Tetrahedron Lett.* **16**, 4467.

21. Alami, M.; Ferri, F.; Linstrumelle, G. (1993) An efficient palladium-catalysed coupling reaction of vinyl and aryl halides or triflates with terminal alkynes, *Tetrahedron Lett.* **34**, 6403.
22. Amiet, G.; Nicholas, R.; Pettit, R. (1970) Mercuration of cyclobutadienyliron tricarbonyl, *J. Chem. Soc., Chem. Commun.* 161.
23. Farina, V.; Krishnan, B. (1991) Large rate accelerations in the Stille reaction with tri-2-furfurylphosphine and triphenylarsine as palladium ligands: mechanistic and synthetic implications, *J. Am. Chem. Soc.* **113**, 9585.
24. Stille, J. K. (1986) The palladium-catalyzed cross-coupling reactions of organotin reagents with organic electrophiles, *Angew. Chem., Int. Ed. Engl.* **25**, 508.
25. Bunz, U. H. F.; Enkelmann, V. (1993) The first complex with a tetraethynyl-cyclobutadiene ligand, *Angew. Chem., Int. Ed. Engl.* **32**, 1653. Bunz, U. H. F.; Enkelmann, V. (1994) Star-shaped tricarbonyl(cyclobutadiene)iron and cymantrene complexes: building blocks for carbon nets and organometallic construction nets? *Organometallics* **13**, 3823.
26. Bunz, U. H. F.; Enkelmann, V.; Räder, J. (1993) Tricarbonyl[η^5-(1-5)-pentapropyn-1-ylcyclopentadienyl]manganese, *Organometallics* **12**, 4745.
27. Beletskaya, I. P. (1983) The cross coupling reaction of organic halides with organic derivatives of tin, mercury, and copper catalyzed by palladium, *J. Organomet. Chem.* **250**, 551 and cited refs.
28. Baughman, R. H.; Galvao, D. S.; Cui, C.; Wang, Y.; Tomanek, D. (1993) Fullerenynes: a new family of porous fullerenes, *Chem. Phys. Lett.* **204**, 8.
29. Lightstone, F. C.; Bruice, T. C. (1994) Geminal-dialkyl substitution, intramolecular reactions, and enzyme efficiency, *J. Am. Chem. Soc.* **116**, 10789 and cited refs.

DIELS-ALDER OLIGOMERS OF BENZENE

WOLFRAM GRIMME, JOACHIM GOSSEL, JOHANN LEX
Institut für Organische Chemie der Universität zu Köln
Greinstr.4, 50939 Köln, Germany

ABSTRACT. Benzene, although comprising in its Kekulé structure the diene and ene functionality, does not form Diels-Alder oligomers. However, by an indirect route a Diels-Alder trimer of benzene, activated in its diene part by four chloro groups, can be obtained. This compound cycloadds to itself forming mixed Diels-Alder oligomers of benzene and tetrachlorobenzene.

The Kekulé structure of benzene comprises, in the most condensed form, both functionalities necessary for the Diels-Alder reaction, i.e. a diene and an ene. Benzene, therefore, might function as an AB module in a repetitive cycloaddition that forms three dimensional ladders of type **1** with an array of stacked double bonds.

1

a: n = 0

b: n = 1

Although the direct cycloaddition of benzene to itself is precluded by its high resonance energy, the Diels-Alder dimer **1a** [1,2] and trimer **1b** [3] have been synthesized by indirect routes. These structures represent AB modules for a repetitive Diels-Alder reaction, but the dimer is stable only at -60°C and the trimer does not cycloadd to itself up to 100°C and 8 kbar.

To achieve the Diels-Alder oligomerization of tribenzene **1b**, it was necessary to use the activated precursor **3**. **3** could be obtained in moderate yield by the cycloaddition of sesquibarrelene **2** [4] to tetrachlorothiophene dioxide. Since the stacked double bonds in

J. Michl (ed.), Modular Chemistry, 485–488.

2 possess a higher electron density, they react preferantially in this Diels-Alder reaction with inverse electron demand forming the isomeric tetrachloro-tribenzene **4.**

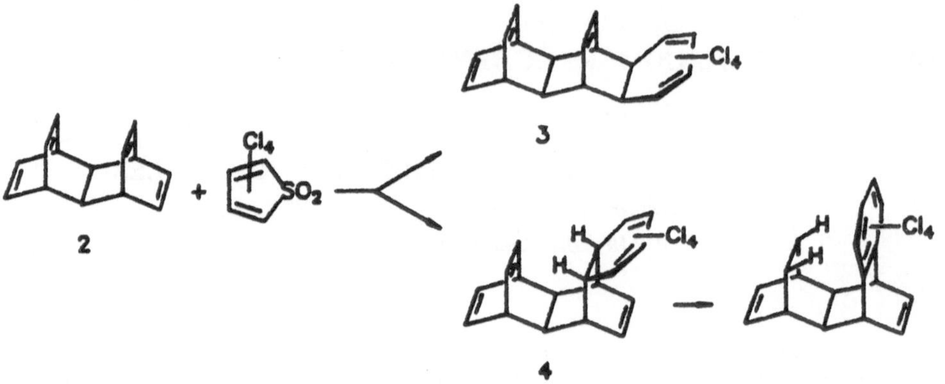

Tetrachloro-tribenzene **3** dimerizes at 80° and 8 kbar via a Diels-Alder reaction to give the hexabenzenes **5** and **6**. The more interesting isomer **5** with five stacked double bonds is the minor product, due to the higher dienophilicity of the stacked terminal double bond of **3**.

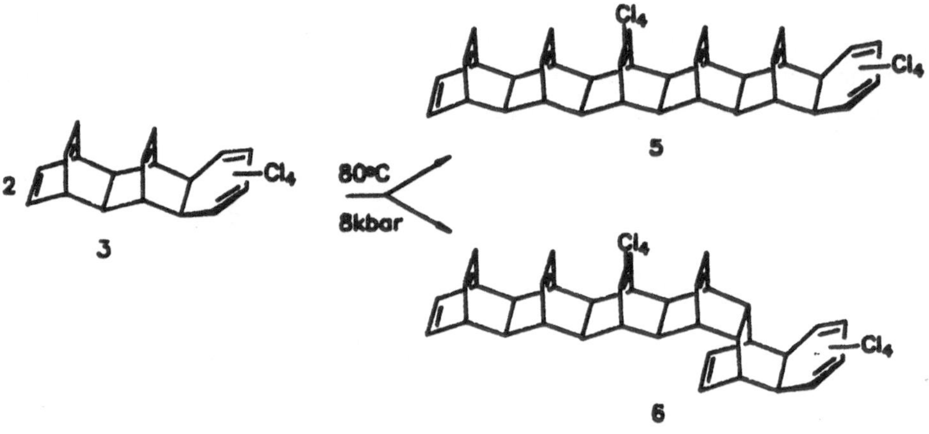

Both hexabenzenes **5** and **6** cyclorevert at 140 °C in a stepwise manner to benzene and tetrachlorobenzene.

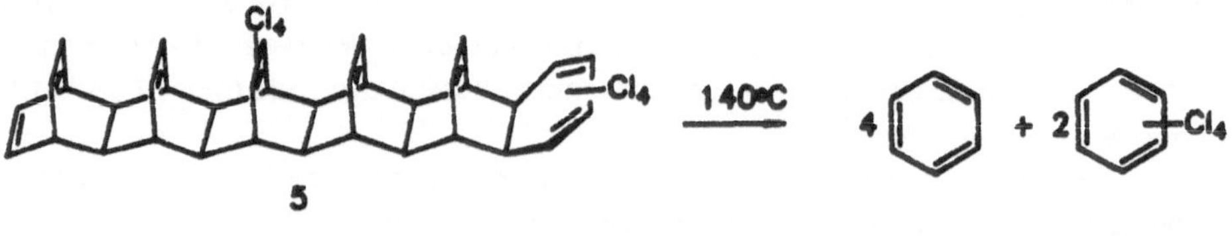

The geometry of **5** can be derived from the X-ray structure of the related all-syn-hemiundecabicyclo[2.2.2]octene **7**. This compound, which crystallizes with one molecule of methylene chloride, has the basal carbon atoms aligned along an arc of 90°.

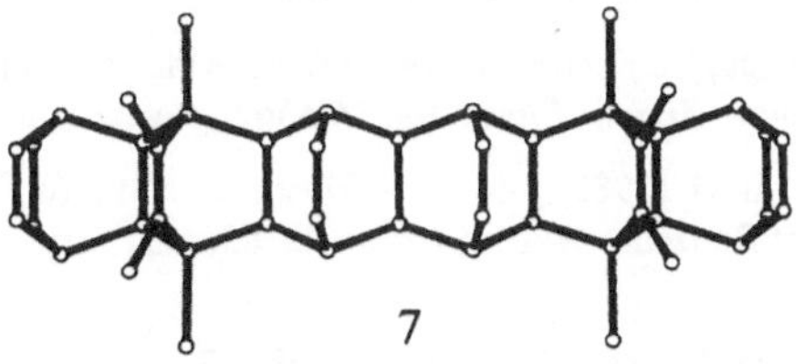

7

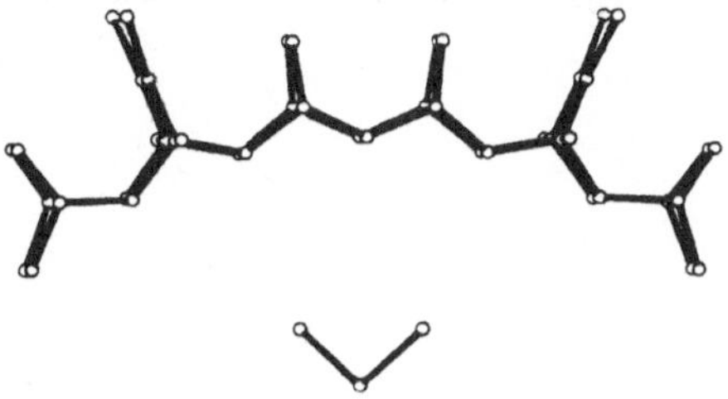

From this geometry it can be concluded that the all-syn octamer of **3** - a Diels-Alder tetracosamer of benzene - would close to **8**, a rigid macrocycle with a diameter of 20 Å and with 24 circularly stacked double bonds. In contrast to all other Diels-Alder oligomers of benzene, **8** should be stable at 140°C and - when doped - might show interesting electronic properties.

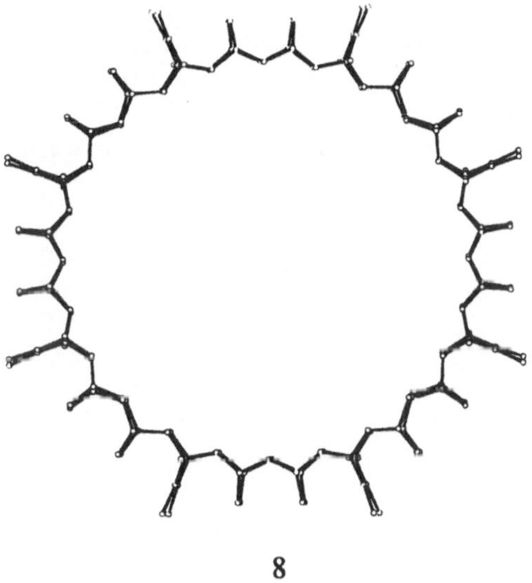

8 $-Cl_4$ $\xrightarrow[8kbar]{80°C}$? —

8

References

1. Braun, R., Kummer, M., Martin, H.-D. and Rubin, M.B. (1985) o,p-Dibenzene, *Angew. Chem.* **97**, 1054-1056; *Angew. Chem. Int. Ed. Engl.* **24**, 1059-1061.

2 Bertsch, A., Grimme, W. and Reinhardt, G. (1986) Diels-Alder Adducts of Benzene with Arenes and Their [4+2] Cycloreversion, *Angew. Chem.* **98**, 361-362; *Angew. Chem. Int. Ed. Engl.* **25**, 377-378.

3. Grimme, W. and Reinhardt, G. (1983) p-(o,p)-o-Tribenzene and p-(o,p)-(o,o)-p-Tetrabenzene, Diels-Alder Oligomers of Benzene, *Angew. Chem.* **95**, 636-637; *Angew. Chem. Int. Ed. Engl.* **22**, 617-618.

4. Melder, J.P., Wahl, F., Fritz, H. and Prinzbach, H. (1987) Synthesis and Photochemistry of Tetracyclo[6.2.2.$2^{3,6}$.$0^{2,7}$]tetradeca-4,9,11,13-tetraene Based Systems, *Chimia* **41**, 426-428.

Single Electron Tunneling in Molecular Nanostructures of Crystalline Gold Clusters Attached by Dithiols to Au [111]: Direct I(V) Measurements of Individual Surface Attached Gold Clusters by STM

CLIFFORD P. KUBIAK
Department of Chemistry
Purdue University
West Lafayette, IN 47907-1393
U.S.A.

RONALD P. ANDRES[†], THOMAS BEIN[‡], MATT DOROGI[§], SUE FENG[‡], JASON I. HENDERSON[‡], WILLIAM MAHONEY[†], RICHARD G. OSIFCHIN[†], RONALD G. REIFENBERGER[§]
[†] *School of Chemical Engineering*
[‡] *Department of Chemistry*
[§] *Department of Physics*
Purdue University
West Lafayette, IN 47907-1393
U.S.A.

ABSTRACT. Self assembled monolayers (SAMs) of the double-ended thiols 4, 4'-biphenyldithiol (BPD), and p-xylene-α,α'-dithiol (XYL) on gold substrates were used to covalently attach nanometer scale crystalline gold clusters to the exposed thiol surface. The positions of clusters remained constant after many repeated STM (Scanning Tunneling Microscopy) scans of the same region of the surface, allowing the current-voltage (I(V)) characteristics of individual gold clusters to be examined. The STM I(V) curve taken over a 1.8 nm diameter gold cluster deposited onto an (XYL) self-assembled monolayer on gold shows a gap at zero bias, and a "Coulomb staircase" progression at room temperature. The STM I(V) curve over the XYL SAM in a region without gold clusters shows an essentially linear response. The "Coulomb staircase" I(V) data are in good agreement with semiclassical theoretical predictions and allow the resistance of a single XYL molecule to be estimated at approximately 10MΩ.

J. Michl (ed.), Modular Chemistry, 489–501.

1. Introduction

We report the observation of room-temperature single-electron tunneling phenomena in the current-voltage responses of a scanning probe tip above molecular nanostructures consisting of crystalline (>200 atom) gold clusters attached by dithiols to gold surfaces. Self-assembled monolayers (SAMs) of alkane thiols on gold have been a subject of intense interest [1-13], and their orientation, packing, and stability have been investigated in some detail [14,15]. We demonstrate by results of reflection absorption infrared spectroscopy (RAIR), ellipsometry, and measurement of advancing contact angle (θ_a) that SAMs of double-ended thiols 4,4'-biphenyldithiol (BPD) and p-xylene α,α'-dithiol (XYL), can be formed in which only one end of the dithiol molecule attaches to the gold surface. Majda *et. al.* [16] and Whitesides *et. al.* [17] have employed $HS(CH_2)_nSi(Cl)_3$ with n=3 and 11 respectively, to show that thiol terminated films improve the adhesion of evaporated gold to a Si/SiO_2 substrate. Allison *et. al.* described the use of $HS(CH_2)_2NH_2$ or $HS(CH_2)_2N(CH_3)_2$ to functionalize gold substrates for the immobilization and subsequent imaging by STM of DNA [18-20]. We demonstrate that SAMs of dithiols can be used as chemically "sticky" surfaces for the attachment of crystalline, nanometer scale gold clusters. The cluster/dithiol/gold nanostructure so assembled is depicted in Figure 1. The key result is that individual gold clusters are chemically confined so that they can be reproducibly addressed by the scanning probe tip and their I(V) responses measured. The I(V) responses show a gap at zero bias and steps that are ascribed to theoretically predicted [21-23] single electron tunneling "Coulomb blockade" and "Coulomb staircase" phenomena at room temperature.

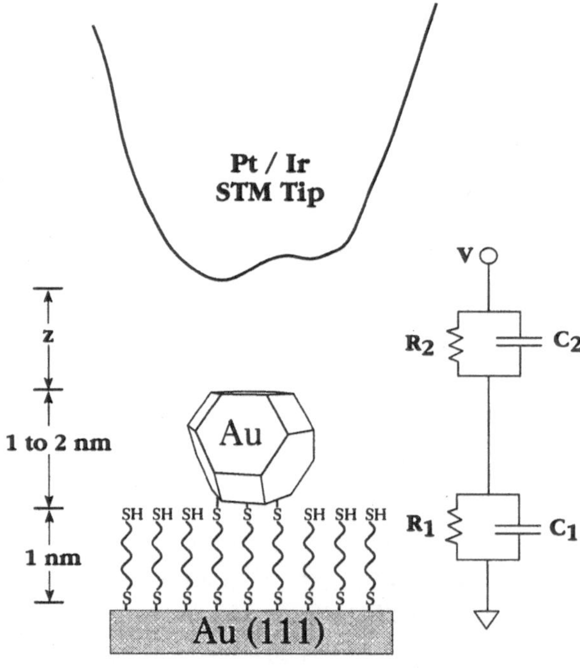

Figure 1. SAM of dithiols on gold with a crystalline gold cluster for the measurement of I(V) data of individual gold clusters by an STM tip. The equivalent circuit can be represented in terms of a pair of tunnel capacitors (R_1, C_1; R_2, C_2) in series.

2. Discussion

2.1 CHARACTERIZATION OF SELF-ASSEMBLED MONOLAYERS OF DITHIOLS

Detailed investigations into the orientation, packing, and stability of single component SAMs of alkane thiols have been reported [2,24-29]. The contact angle [2] and Reflection-Absorption Infared (RAIR) [15] data for these SAMs suggested that they were well ordered. SAMs of the aryl dithiols, BPD and XYL, have only been recently reported[30]. RAIR spectra of BPD/Au and XYL/Au SAMs show band positions similar to the IR spectra of the bulk solid in KBr [31]. All characterization was performed on Au substrates prepared as described elsewhere[32]. A comparison of the RAIR spectrum of BPD/Au with the bulk spectrum of BPD shows evidence that the BPD molecules are oriented with their long axes nearly normal to the surface (Figure 2). In particular, the band at 998 cm^{-1} in the bulk spectrum of BPD is assigned to the 18a normal mode of the phenyl rings [31]. The net dipole moment change associated with this mode is parallel to the long axis of the molecule. This mode shows enhanced intensity compared to the reduced intensity of the 18b normal mode at 1104 cm^{-1} with a dipole moment change perpendicular to the long axis of the molecule. RAIR spectra did not display a ν(S-H) mode, usually observed at approximately 2555 cm^{-1}. However, if the aryl thiols form SAMs with their long axes nearly normal to the gold surface, observation of the ν(S-H) mode would not be expected. For example, if the tilt angle of an aryl thiol is 7° [34], the S-H bond vector, based on typical C-S-H angles of 99° [35,36], would be only 2° off parallel to the underlying gold surface. The IR selection rules for a normal mode with a net dipole change parallel to the gold metal surface predict

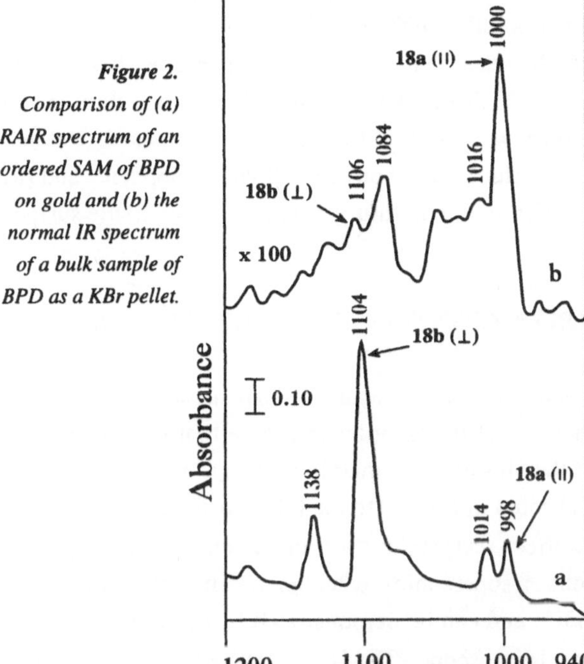

Figure 2. Comparison of (a) RAIR spectrum of an ordered SAM of BPD on gold and (b) the normal IR spectrum of a bulk sample of BPD as a KBr pellet.

that the ν(S-H) mode would not be observed. Another possible explanation for not observing the ν(S-H) mode is that the thiols may have been oxidized to disulfides upon transferring the sample from the inert atmosphere where the SAMs were formed to the RAIR chamber.

Ellipsometric data gave an experimental film thickness for BPD of 12.2 Å, in reasonable agreement with the expected film thickness (13 Å) for a perpendicular orientation. Recently, Chang et al [37] and others [34] suggested that aryl thiols pack with the long molecular axis essentially normal to the surface. Our experimental film thickness for XYL was 8.3 Å, consistent with a tilted molecular orientation (~24°). This tilt is apparently caused by the presence of the methylene carbon atoms in XYL. These orientations contrast with the 30° tilt proposed for alkane thiols such as octadecane thiol (ODT) [9,13,15]. Advancing water contact angle measurements also suggest that the aryl dithiols form SAMs with only one thiol attached to the surface. The advancing contact angle of water droplets on SAMs of BPD and XYL on gold lead to the same result, $\theta_a = 69\text{-}71°$, and thus indicate relatively hydrophilic surfaces. For comparison, contact angles on a $HS(CH_2)_{11}Si(Cl)_3$ modified Si/SiO_2 substrate, measured by Whitesides et. al., were 72-74° [17], in contrast to SAMs of a CH_3 terminated alkyl chain thiol like ODT ($\theta_a = 111°$).

The RAIR, ellipsometry film thickness, and contact angle data of SAMs of BPD and XYL on gold suggest that SAMs of BPD and XYL form with only one thiol group anchored to the gold surface and the other remote from the surface and exposed. Related studies concurrently establish that α, ω-dithiols based on aryl, alkynyl, thiophene, and [n] staffane spacers also form SAMs in which only one thiol group is attached to the surface [34,38].

2.2 SYNTHESIS AND DEPOSITION OF CRYSTALLINE GOLD CLUSTERS

The gold clusters are grown in the gas phase from the Multiple Expansion Cluster Source (MECS), described in detail elsewhere [39,40]. In order to assure that the clusters are single crystals, an aerosol mixture of the clusters entrained in argon gas is passed through a tubular oven in which the aerosol is heated to 1400°C to melt the clusters. The aerosol is then cooled to room temperature to effect recrystallization of the clusters, followed by expansion through a capillary to produce a supersonic cluster beam. The crystalline nature of the gold clusters is key to our studies and can be demonstrated by electron diffraction methods with a Tranmission Electron Microscope (TEM). Recrystallized gold clusters from the MECS were vacuum deposited onto the [0001] face of MoS_2. Figure 3 shows a plane view (perpendicular to substrate) bright field TEM micrograph of gold clusters on MoS_2. The crystallographic orientation of the clusters with respect to the substrate was determined using an indirect lattice imaging technique based on the parallel Moiré fringes apparent in Figure 3. The clusters are found to align heteroepitaxially, with the Au [111] facets of the clusters parallel to the [0001] face of MoS_2. Perpendicular to the MoS_2 basal plane are three equivalent MoS_2 [11$\bar{2}$0] planes. These, in turn, are parallel to one of a set of three equivalent Au [2$\bar{2}$0] planes [41].

Having established the crystalline nature of the MECS gold clusters by electron diffraction interference on crystalline MoS_2 substrates, the nanoscale clusters were deposited onto dithiol SAMs on Au [111] in a vacuum chamber held at 10^{-6} Torr as described previously [42,43]. The substrates were exposed to the cluster beam for fixed time intervals. After deposition, the samples are exposed to air and transferred into a UHV chamber for subsequent STM analysis.

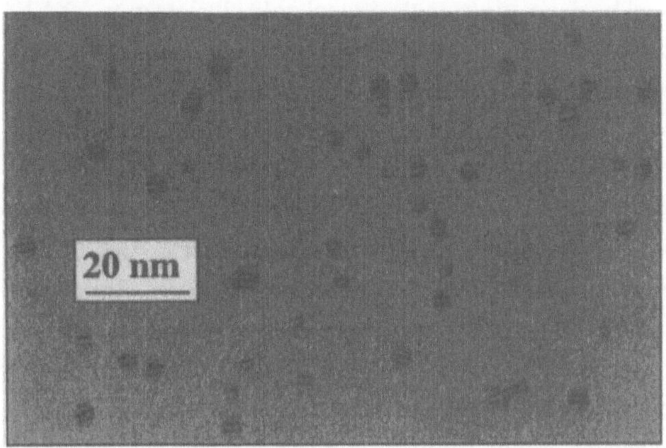

Figure 3. Bright field TEM of crystalline gold clusters deposited on MoS_2, exhibiting Moiré patterns. The Moiré fringes result when an electron beam passes though two superimposed crystal lattices having different lattice spacings and/or orientations. The Moiré fringes show both that the gold clusters are ordered crystals, and that they exhibit a preferential orientation with respect to the MoS_2 substrate. The spacing, D, of the Moiré fringes is

$$D = \frac{d_1 d_2}{d_1^2 + d_2^2 - 2d_1 d_2 \cos\phi)^{1/2}}$$

where d_1 (Au) and d_2 (MoS_2) are the lattice spacings of the two crystal planes that give rise to the pattern and ϕ is the angle of rotation of one plane with respect to the other [26,27]. The value of D obtained experimentally is 1.7 nm. This is consistent with interference of [2$\bar{2}$0] planes of Au and [11$\bar{2}$0] planes of MoS_2.

2.3 STM MEASUREMENTS

The STM measurements were made using a locally built system with a computer controlled digital feedback system for imaging [24] and spectroscopy [25]. The STM is housed in a UHV chamber with a base pressure of 3 x 10^{-10} Torr. STM tips were prepared by cutting 0.25 mm diameter Pt/Ir wire. After insertion into the STM chamber, tips were cleaned by resistive heating and field evaporation. Samples of Au [111] used in STM studies were prepared by evaporative deposition onto freshly cleaved mica.

2.4 SINGLE ELECTRON TUNNELING THROUGH CRYSTALLINE GOLD CLUSTER/DITHIOL SAM NANOSTRUCTURES

Crystalline gold clusters deposited onto dithiol SAMs on gold were found to be excellent systems for the study of room-temperature single-electron tunneling (SET) effects. In our

STM studies of gold clusters supported on both BPD and XYL dithiols, we found that both molecules tethered gold clusters at room temperature, but that the XYL dithiol provided less noise in both the imaging and spectroscopic modes; only these results will be discussed below. We observed Coulomb blockade behavior over the majority (~ 80%) of the clusters that were imaged. This, together with essentially linear I(V) responses over the "blank" XYL film, indicates that the SAM acts as a uniform, insulating layer between the gold clusters and the gold substrate. Molecular nanostructures for the study of room temperature SET effects were fabricated by depositing single crystal gold clusters onto a XYL/Au [111] SAM, Figure 1. Although other measurements of room-temperature Coulomb blockade and Coulomb staircase phenomena have been serendipitously observed and reported [44,45], this is believed to be the first time that the structure giving rise to the single electron tunneling behavior has been well characterized. If a cluster on XYL/Au [111] exhibits Coulomb blockade, suppression of the tunnel current centered about certain voltages is expected. As the tip moves closer to the cluster, the tip-cluster resistance R_2 should drop off rapidly, reflecting the exponential dependence of the tunnel current on the tip-cluster separation (Figure 1). The resistance and capacitance of the cluster-substrate junction, R_1 and C_1, should remain reasonably constant. The capacitances C_1 and C_2 can be estimated. When the tip is centered over a cluster, suppression of the tunnel current is found, suggesting room-temperature Coulomb blockade. Figure 4 compares STM I(V) data taken over the cluster and showing the suppression of tunnel current centered symmetrically about zero bias with the I(V) data taken over the XYL SAM in a region without gold clusters. In contrast to the I(V) data recorded over the cluster, the I(V) response over the XYL SAM is essentially linear. The two curves were taken with the same tunnel current and bias voltage set points (1 nA, -747 mV). Both the Coulomb blockade over the cluster and linear behavior over the XYL film were very reproducible.

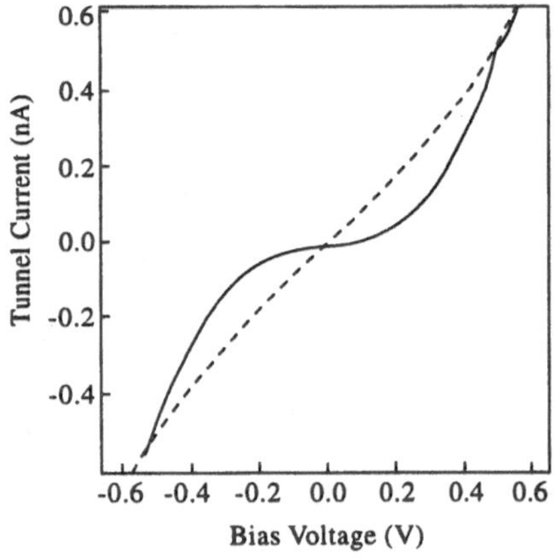

Figure 4. STM I(V) data taken over a 1.8 nm cluster (—) along with I(V) data taken over the XYL SAM in a region without gold clusters (——). The two curves were taken with the same tunnel current and bias voltage set points (1 nA, -747 mV). The I(V) curve taken over the cluster shows Coulomb blockade behavior, while the I(V) curve taken over the XYL film is essentially linear. Both I(V) curves are very symmetric.

The well characterized nature of the crystalline gold cluster/XYL SAM/Au [111] nanostructures provides an interesting opportunity to investigate whether the single-electron phenomena obey expectations of the semiclassical theory of "Coulomb blockade" [21-23].

In the semiclassical model, the two tunnel capacitors shown schematically in Figure 1 are biased by applying a voltage V_{bias} across the array. Each junction has a capacitance, an effective resistance, and a tunneling rate associated with it denoted by C_i, R_i, and Γ_i, with $i = 1, 2$. In principle, a fractional charge Q_o may exist on a cluster, permitting the entire I(V) data to be shifted positive or negative of the origin. Q_o may also add asymmetry to the detailed shape of the I(V) curve. Figure 5 shows the I(V,z) data taken as a function of the tip-cluster separation for a gold cluster on XYL/Au. These data were acquired for a cluster

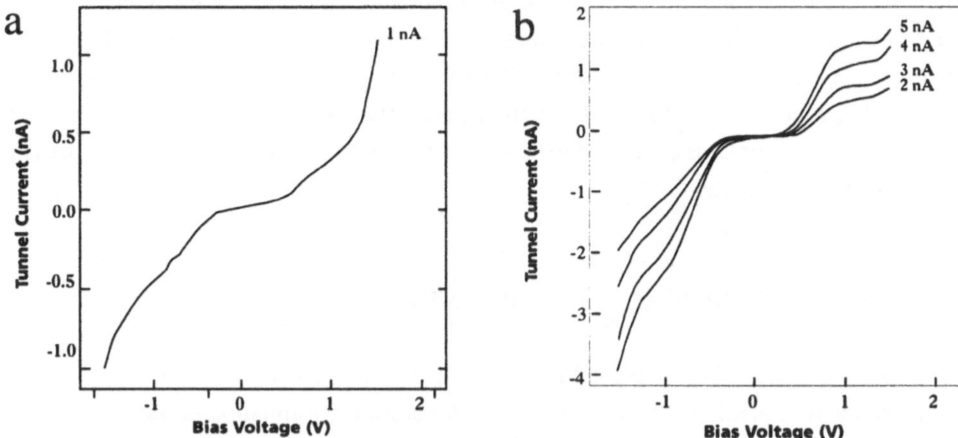

Figure 5. *I(V,z) data for a 1.8 nm gold cluster on XYL/Au, exhibiting Coulomb staircase behavior. The I(V) curves were taken with a bias voltage set point of -1.75 V and tunnel current set points of (a) 1nA; (b) 2 nA, 3 nA, 4 nA and 5 nA. Each of the I(V) curves is an average of 100 individual voltage sweeps.*

with a nominal height of ~1.8 nm as determined from the STM image. Distinct steps in the I(V) data are evident. The I(V) curves showing a Coulomb staircase were typically very asymmetric, indicating the presence of a fractional charge, Q_o, on the cluster.

Least squares fitting procedures were used to interpret the I(V,z) data shown in Figure 5 in the context of the semiclassical model [22]. Due to the asymmetric nature of the data, it was necessary to include the fractional charge Q_o in the fitting routine. It is interesting to see if the fitting parameters behave reasonably as the tip-cluster separation decreases. The rapid drop expected in the tip-cluster resistance R_2 is recovered from the fitting procedure, and shown in Figure 6. The cluster-substrate resistance R_1 varies, with an average value of 1.8 ± 1.2 MΩ. The

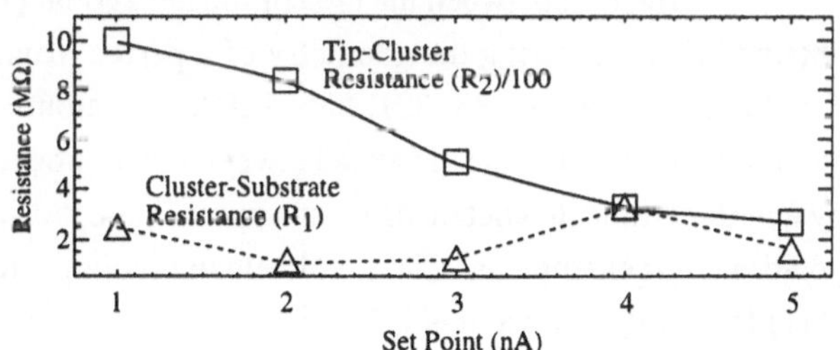

Figure 6. *Tip-cluster resistance (R_2) drop as a function of tip-cluster separation.*

tip-cluster capacitance C_2 behaves as expected, increasing slightly as the tip-cluster separation is decreased. The cluster-substrate capacitance C_1 decreases slightly as the tip-cluster separation is decreased. The average values of the parameters required to fit the data shown in Figure 5 are $C_1 = 0.8 \pm 0.5 \times 10^{-19}$ F, $C_2 = 1.3 \pm 0.3 \times 10^{-19}$ F, $R_1 = 1.8 \pm 1.2$ megohms, and $Q_o/e = -0.20 \pm 0.17$.

The capacitances calculated from the geometry indicated by the STM image agree well with the values calculated from the semiclassical model. For a cluster with its diameter found to be 1.8 nm and a tip diameter estimated to be 7.2 nm, the following capacitances can be estimated. If the cluster and tip are treated as spheres, estimates of C_1 and C_2 can be made using the method of image charges. These values give a tip-cluster capacitance C_2 of 1.4×10^{-19} F, which is very close to the average parameter value of 1.3×10^{-19} F. The cluster-substrate capacitance C_1 is estimated to be 1.7×10^{-19} F, which is a factor of 2 larger than the average fitting parameter for C_1 .

2.5 ESTIMATING THE ELECTRICAL RESISTANCE OF A MOLECULE

Molecular nanostructures of crystalline gold clusters adsorbed on SAMs of dithiols on gold provide reasonably well characterized systems for the estimation of the resistance of single molecules. Gold clusters prepared by the MECS source are crystalline. These clusters have been shown (*vide supra*) to orient with their [111] facets parallel to the substrate. The lowest energy structure of the crystalline clusters is believed to be that of a truncated octahedron. From the morphology of the cluster and the measurements of the electrical resistance of cluster/XYL/Au nanostructures, described in the previous section, it is possible to estimate the resistance of a single XYL dithiol molecule. Such an analysis may provide a better way to characterize the electrical properties of long-chain molecules compared to sandwiching a monolayer between two macroscopic conducting electrodes. Experiments using macroscopic contacts are often difficult to interpret because current flow through pin-hole imperfections can add unknown contributions to the measured conductivity [46].

A relation between the area of the hexagonal [111] facet and cluster height can be obtained by considering the geometry of a perfect truncated octahedron. For the case of Au clusters containing 38, 201, 586, 1289, etc. atoms, the area of the [111] facet can be analytically related to the distance between two opposite facets. For clusters not described by a perfect truncated octahedron, an approximate hexagonal area can be inferred by interpolation. For example, a 1.5 nm gold cluster having a truncated octahedral structure has a [111] face with an area of ~ 1.5 nm^2.

It has been shown by electron diffraction [47] that alkanethiol molecules have hexagonal symmetry on Au[111] with a S-S spacing of 0.497 nm and a calculated area per molecule of 0.214 nm^2. From the area of the [111] cluster face and the area per molecule, the number of molecules, N, between the cluster and substrate can be calculated. It follows that if the N molecules act as resistors in parallel, the resistance of a single XYL molecule will be $R_{molecule} \approx NR_1$, where R_1 is the cluster-substrate resistance from the semiclassical model fit.

From the Coulomb staircase data, the resistance of a single XYL molecule is estimated to fall between 7 MΩ and 23 MΩ, with the value around 10 MΩ deemed most reliable. These results are in agreement with a theoretical calculation of the resistance of a XYL molecule, in which $R_{molecule}$ was calculated to be 8.5 MΩ [48]. The resistance was calculated from the Landauer formula $R = h/2e^2T$. Here, T is the transmission coefficient through a XYL molecule between two gold surfaces, calculated using an extended Hückel molecular orbital description of the molecule.

3. Conclusions

This study demonstrates the ability to form well-characterized nanostructures for the study of room temperature single electron tunneling, and suggests an interesting method to study the conduction properties of individual molecules. The I(V) characteristics of the gold cluster/dithiol SAM assemblies investigated in this study show compelling evidence of Coulomb blockade and Coulomb staircase phenomena. These are the first observations of single electron tunneling phenomena at room temperature on well characterized systems.

4. Acknowledgments

Acknowledgment is made to the Army Research Office under grant DAAL03-G-0144. S. F. and J. I. H. gratefully acknowledge Purdue Research Foundation Graduate Fellowships. T. B. acknowledges funding from the National Science Foundation. We also wish to thank Tonja Henderson for preparing figures.

5. References

1. Bain, C. D.; Whitesides, G. M. (1989) Modeling organic surfaces with self-assembled monolayers, *Agnew. Chem., Int. Ed. Engl.* **28**, 506.
2. Bain, C. D.; Troughton, E. B.; Tao, Y.-T.; Evall, J.; Whitesides, G. M. (1989) Formation of monolayer films by the spontaneous assembly of organic thiols from solution onto gold, *J. Am. Chem. Soc.* **111,** 321.
3. Bain, C. D.; Evall, J.; Whitesides, G. M. (1989) Formation of monolayers by the coadsorption of thiols on gold: variation in the head group, tail group, and solvent, *J. Am. Chem. Soc.* **111**, 7155.
4. Biebuyck, H. A.; Whitesides, G. M. (1993) Interchange between monolayers on gold formed from unsymmetrical disulfide and solutions of thiols: evidence for sulfur-sulfur bond cleavage by gold metal, *Langmuir* **9**, 1766.
5. Collard, D. M.; Fox, M. A. (1991) Use of electroactive thiols to study the formation and exchange of alkanethiol monolayers on gold, *Langmuir* **7**, 1192.
6. Hickman, J. J.; Laibinis, P. E.; Auerbach, D. I.; Zou, C.; Gardner, T. J.; Whitesides, G. M.; Wrighton, M. S. (1992) Toward orthogonal self-assembly of redox active molecules on Pt and Au: selective reaction of disulfide with Au and isocyanide with Pt, *Langmuir* **8**, 357.
7. Kim, T.-Y.; Bard, A. J. (1992) Imaging and etching of self-assembled n-octadecanethiol layers on gold with the scanning tunneling microscope, *Langmuir* **8**, 1183.
8. Nuzzo, R. G.; Allara, D. L. (1983) Adsorption of bifunctional organic disulfides on gold surfaces, *J. Am. Chem. Soc.* **105**, 4481.
9. Sellers, H.; Ulman, A.; Shnidman, Y.; Eilers, J. E. (1993) Structure and binding of alkanethiolates on gold and silver surfaces: implications for self-assembled monolayers, *J. Am. Chem. Soc.* **115**, 9389.
10. Steinberg, S.; Rubenstein, I. (1992) Ion-selective monolayer membranes based upon self-assembling tetradentate ligand monolayers on gold electrodes. 3. application as selective ion sensors, *Langmuir* **8**, 1183.
11. Ulman, A. (1991)*An Introduction to Ultrathin Organic Films from Langmuir-Blodgett to Self-Assembly*; Academic Press: San Diego.
12. Weisshaar, D. E.; Lamp, B. D.; Porter, M. D. (1992) Thermodynamically controlled electrochemical formation of thiolate monolayers at gold: characterization and comparison to self-assembled analogs, *J. Am. Chem. Soc.* **114**, 5860.
13. Whitesides, G. M.; Laibinis, P. E. (1990) Wet chemical approaches to the characterization of organic surfaces: self-assembled monolayers, wetting, and the physical-organic chemistry of the solid-liquid interface, *Langmuir* **6**, 87.

14. Bain, C. D.; Whitesides, G. M. (1989) Formation of monolayers by the coadsorption of thiols on gold: variation in the length of the alkyl chain, *J. Am. Chem. Soc.* **111**, 7164.

15. Porter, M. D.; Bright, T. B.; Allara, D. L.; Chidsey, C. E. D. (1987) Spontaneously organized molecular assemblies. 4. structural characterization of n-electrochemistry, *J. Am. Chem. Soc.* **109**, 3559.

16. Goss, C. A.; Charych, D. H.; Majda, M. (1991) Application of (3-mercaptopropyl) trimethoxysilane as a molecular adhesive in the fabrication of vapor-deposited gold electrodes on glass substrates, *Anal. Chem.* **63**, 85.

17. Wasserman, S. R.; Biebuyck, H.; Whitesides, G. M. (1989) Monolayers of 11-trichlorosilylundecyl thioacetate: a system that promotes adhesion between silicone dioxide and evaporated gold, *J. Mater. Res.* **4**, 886.

18. Allison, D. P.; Warmack, R. J.; Bottomley, L. A.; Thundat, T.; Brown, G.M.; Woychick, R. P.; Schrick, J. J.; Jacobson, K. B.; Ferrel, T. L. (1992) Scanning tunneling microscopy of DNA: a novel technique using radiolabeled DNA to evaluate chemically mediated attachment of DNA to surfaces, *Ultramicroscopy* **42-44**, 1088.

19. Allison, D. P.; Bottomley, L. A.; Thundat, T.; Brown, G. M.; Woychik, R.P.; Schrick, J. J.; Jacobson, K. B.; Warmack, R. J. (1992) Immobilization of DNA for scanning probe microscopy, *Proc. Natl. Acad. Sci.* **89**, 10129.

20. Bottomley, L. A.; Haseltine, J. N.; Allison, D. P.; Warmack, R. J.; Thundat, T.; Sachleben, R. A.; Brown, G. M.; Woychik, R. P.; Jacobson, K. B.; Ferrell, T. L. (1992) Scanning tunneling microscopy of DNA: the chemical modification of gold surfaces for immobilization of DNA, *J. Vac. Sci. Technol. A* **10**, 591.

21. Kulik, I. O.; Shekhter, R. I. (1975) Kinectic phenomena and charge discreteness effects in granulated media, *Sov. Phys. JETP* **41**, 308.

22. Hanna, A. E.; Tinkham, M. (1991) Variation of the coulomb staircase in the two-junction system by functional electron charge, *Phys. Rev. B.* **44**, 5919.

23. Amman, M.; Wilkins, R.; Ben-Jacob, E.; Maker, P. D.; Jaklevic, R. C. (1991) Analytic solution for the current-voltage characteristic of two mesoscopic tunnel junctions coupled in series, *Phys. Rev. B.* **43**, 1146.

24. Piner, R.; Reifenberger, R. (1989) Computer control of the tunnel barrier width for the scanning tunneling microscope, *Rev. Sci. Instrum.* **60**, 3123.

25. Miller, T. G.; McElfresh, M. W.; Reifenberger, R. (1993) Density-of-states fine structure in the tunneling conductance of Y-Ba-Cu-O films: a comparison between experiment and theory, *Phys. Rev. B* **48**, 7499.

26. Hirch, P.; Howie, A.; Nicholson, R. B.; Pashley, D. W.; Whelan, M. J. (1977) *Electron Microscopy of Thin Crystals*; Robert E. Krieger Publishing Co.

27. Pashley, D. W.; Stowell, M. J.; Jacobs, M. H.; Law, T. J. (1964) The growth and structure of gold and silver deposits formed by evaporation inside an electron microscope, *Phil. Mag.* **10**, 127.

28. Rabolt, J. F.; Burns, F. C.; Schlotter, N. L.; Swalen, J. D. (1983) Anisotropic orientation in molecular monolayers by infrared spectroscopy *J. Chem. Phys.* **78**, 946.

29. Golden, W. G.; Snyder, C. D.;Smith, B (1982) Infrared reflection adsorption spectra of ordered and disordered arachidate monolayers on aluminum *J. Phys. Chem.* **86**, 4675.

30 Andres, R. P.; Bein, T.; Dorogi, M; feng, S.; Henderson, J. I.; Kubiak, C. P.; Mahoney, W.; Osifchin, R. G.; Reifenberger, R. (1996) "Coulomb Staircase" at room temperature in a self-assembled molecular nanostructure *Science* **272**, 1323.

31. The observed RAIR bands for BPD/Au were assigned using the system of substituted benzene normal modes, 1 - 20a,b, with symmetry labels appropriate for the local symmetry (C_{2v}) of the BPD ring system.[33] 3045 cm^{-1} (2 a_1), 3021 cm^{-1} (7b b_2), 1592 cm^{-1} (8a a_1), 1474 cm^{-1} (19a a_1), 1406 cm^{-1} (19b b_2), 1387 cm^{-1} (19b b_2), 1106 cm^{-1} (18b b_2), 1000 cm^{-1} (18a a_1). For XYL/Au, assignments again follow the system of substituted benzene normal modes, 1 - 20a,b, with symmetry labels appropriate for the local symmetry (D_{2h}) of the XYL ring system:[33] 3022 cm^{-1} (20b b_{2u}), 2921 cm^{-1} (ν_{as} - CH_2), 1511 cm^{-1} (19a $b_{1u)}$ 1423 cm^{-1} (19b b_{2u}), and 1263 cm^{-1} (13 b_{1u}).

32. Henderson, J. I.; Feng, S.; Ferrence, G.; Bein, T.; Kubiak, C.P. (1996) Self-assembled monolayers of dithiols, diisocyanides, and isocyanothiols on gold: 'chemically sticky' surfaces for covalent attachment of metal clusters and studies of interfacial electron transfer, *Inorg. Chim. Acta*, *242*, 115.

33. Varsanyi, G. (1974) *Assignments for Vibrational Spectra of Seven Hundred Benzene Derivatives*; Wiley: New York, NY.

34. Tour, J. M.; II, L. J.; Pearson, D. L.; Lamba, J. J. S.; Burgin, T.; Whitesides, G. M.; Allara, D. L.; Parikh, A. N.; Atre, S. (1995) Self-assembled monolayers and multilayers of conjugated thiols, α, ω-dithiols, and thioacetyl-containing adsorbates. Understanding attachments between potential molecular wires and gold surfaces, *J. Am. Chem. Soc.* **117**, *117*, 9529.

35. Iijima; Tsuchiya; Kimura (1977) The molecular structure of dimethyl sulfide, *Bull. Chem. Soc. Jap.* **50**, 2564.

36. March, J. (1985) Advanced organic chemistry reactions, mechanisms, and structure, Wiley: New York, NY.

37. Chang, S.-C.; Chao, I.; Tao, Y.-T. (1994) Structure of self-assembled monolayers of aromatic derivatized thiols on evaporated gold and silver surfaces: implications on packing mechanism, *J. Am. Chem. Soc.* **116**, 6792.

38. Obeng, Y.S.; Laing, M. E.; Friedl, A. C.; Yang, H. C.; Wang, D.; Thulstrup, E.W.; Bard, A. J.; Michl, J. (1992) Self-assembled monolayers of parent and derivatized [n] staffane-3,3$^{(n-1)}$ dithiols on polycrystalline gold electrodes, *J. Am. Chem. Soc.* **114**, 9943.

39. Park, S. B. Ph. D. Thesis, (1988) Optimal design of a reactor for gas-phase generation of metal microclusters, Purdue University.

40. Bowles, R. S.; Kolstad, J. J.; Calo, J. M.; Andres, R. P. (1981) Generation of molecular clusters of controlled size. *Surf. Sci.* **106**, 117.

41. Mahoney, W.; Lin, S. T.; Andres, R. P. (1995) Probing the nucleation of a thin metal film: atom deposition vs. cluster beam deposition, *MRS Symp.Proc.* **355**, 83.

42. Patil, A. N.; Paithankar, D. Y.; Otsuka, N.; Andres, R. P. (1993) The minimum-energy structure of nanometer-scale gold cluster, *Z. Phys. D.* **26**, 135.

43. Castro, T.; Li, Y. Z.; Reifenberger, R.; Choi, E.; Park, S. B.; Andres, R. P. (1989) Studies of individual nanometer sized metallic clusters using scanning tunneling microscopy, *J. Vac. Sci. Technol. A* **7**, 2845.

44. Anselmetti, D.; Richmond, T.; Baratoff, A.; Borer, G.; Dreier, M.; Bernasconi, M.; Guntherodt, H. J. (1994) Single-Electron tunneling observed at room temperature with adjustable double-barrier junction, *Europhys. Lett.* **25**, 297.

45. Schonenberger, C.; Houton, H. V.; Donkersloot, H. C. (1992) Single-electron tunneling observed at room temperature by scanning-tunneling microscopy, *Europhys. Lett.* **20**, 249.

46. Agarwal, V. K. (1975) Electrical behavior of langmuir films: a review, part 1, *Electrocomponent Sci. Tech.* **2**, 1.

47. Strong, L.; Whitesides, G. M. (1988) Structures of self-assembled monolayer films of organosulfur compounds adsorbed on gold single crystals: electron diffraction studies, *Langmuir* **4**, 546.

48. Samanta, M. P.; Tian, W.; Datta, S.; Henderson, J. I.; Kubiak, C. P. (1996) Electronic conduction through organic molecules *Phys. Rev. B.* **53**, R7626.

MOLECULAR OPTICAL RAILS BASED ON Aib

Modular Chemistry with Unusually Reliable Peptide Helices

ATSUO KUKI*, DEMETRIOS ANGLOS, JOSEPH D. AUGSPURGER, GAUTAM BASU, VANDANA A. BINDRA, MATTHEW KUBASIK, ADRIENNE PETTIJOHN

Department of Chemistry, Baker Laboratory, Cornell University
**Currently: Director of Computational Chemistry, ALANEX*
3550 General Atomics Court, San Diego, CA 92121-1194 USA

Abstract. Well established properties of peptide helices most often must be understood in the context of folding/unfolding temperatures and conformational equilibria. This limits their general usefulness as structurally well-defined modules. The conformational imperatives of the peptide backbone can be transformed, however, by the use of α-aminoisobutyric acid (Aib), whose geminal dimethyl functionality provides an unforgiving steric hindrance enforcing a well-defined helical backbone conformation. The corresponding unfolded state is unknown. The sequence composition and patterns required to attain this highly stable helix are described, as well as the 1D and 2D NMR evidence in solvents ranging from the benign $CDCl_3$ to the more aggressive DMSO. Temperature studies up to 150 ^{o}C are summarized. The utility of these helices has been considerably expanded in the direction of optical and electronic versatility by the design and synthesis of aromatic α-amino acids in which electron-rich and electron-deficient aromatic rings are fused to carbocylic structures incorporating the α-carbon. These aromatics possess the same α,α -dialkylation as the parent Aib and exhibit the same helical backbone propensity. These properties enable a modular strategy for the incorporation of electronic and optical functionality into variable sequence positions along an unchanging molecular optical rail.

J. Michl (ed.), Modular Chemistry, 503–516.

1. Introduction

In recent years the study of the conformational backbone propensities of α-aminoisobutyric acid (Aib) has drawn some attention; for example Aib has been reported in host-guest studies to possess a slightly higher intrinsic helix-forming propensity than alanine [1,2] . What is yet more distinctive, and more consequential, is the absence of the alternative β-sheet fold as a possibility for the Aib residue. In fact a considerable body of x-ray diffraction and 1–D NMR data had been accumulated during the 1980s on 100% pure Aib homo-oligomers, and the highly regular 3_{10}-helix formed reliably by such oligomers was thoroughly characterized [3] . The essential aspects of steric hindrance and crowding around the α-carbon, a nearly blank (energetically disallowed) Ramachandran map, the occurrence of conformational minima at the helical (ϕ,ψ) region, and the role of bond-angle bending (distortion away from tetrahedral) at the α-carbon [4] were all recognized and clearly expounded in earlier reviews and analyses [4-9] .

The natural role of Aib is limited to a class of membrane-inserting microbial peptide antibiotics, referred to as peptaibols, which possess distinctive hydrophobic sequences. A well-studied example is alamethicin [5,6,10] . In these peptides, Aib is a major constituent comprising typically 40-50% of the sequence composition. At this level of Aib content, the conformations are already invariably helical, though a wide variety of precise backbone hydrogen-bonding patterns is observed [7,11,12] . These include the α-helix, the 3_{10}-helix, and mixed α- / 3_{10}- helices. While these sequences are important as models of voltage-gated ion channels and of peptide/membrane association, we will not discuss them further here. Our attention was drawn to the comparatively unexplored region of >60% Aib content and the possibility that the excellent, regular, 3_{10}-helical structures of 100% Aib oligopeptides might well be retained as the Aib content was carefully diluted with guest residues down to lower, but still quite Aib-rich compositions.

2. Aib-Rich Helical Peptides

2.1 HELICES FROM COMMERCIALLY AVAILABLE AMINO ACIDS

The sequence pattern Ac-Aib-Aib-X-Aib-Aib-Y-Aib-Aib-NHMe was the first Aib-rich sequence studied in detail for solution conformational behavior

[13,14] . In these sequences the Aib residues serve as the host, and the precise N- and C-terminal protecting groups can be varied, but these were always selected such that the N-terminal group provided one additional carbonyl, and the C-terminal group provided one additional amide proton, for maximum intra-helical hydrogen bonding at both ends. The guest

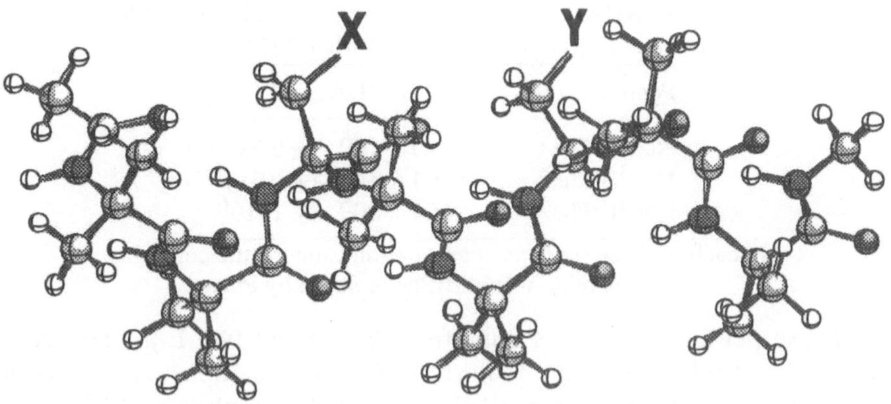

Figure 1. An Aib-rich octamer with two L-amino acid guest residues shown in the 3_{10}-helical conformation. In the 3_{10}-helix, the carbonyl of the i-th residue is hydrogen-bonded to the amide proton of residue i+3 . The pitch of the 3_{10}-helix is 6Å, with three residues per turn; it is thus longer, tighter, and narrower than the α-helix. Shown is a 3,6-substituted [6/8] peptide with Ac- and -NHMe terminating groups. Moving the insertion positions of these same guest L-amino acids leads to a family of *sequence permutation isomers* .

residues, "X" and "Y"were standard L-amino acids, that is mono-α-alkylated amino acids such as Phe, and Ala, or closely related synthetic amino acids such as β-(1-naphthyl)-alanine (Nap). This sequence pattern was verified by NMR methods in $CDCl_3$, CD_3CN, and DMSO to be a functionally ideal 3_{10}-helix [13-15] . In particular, the 2D NMR study (ROESY) was one of the earlier applications of this method to a peptide with so few α-protons [14] . and the structural assignment protocol used here has formed the basis of the later NMR ROESY work [16-18] .

The distance between the two β-carbons on the amino acids at positions 3 and 6 is close to 6Å (Fig. 1). This was exploited in a Remote Heavy Atom Effect (RHAE) study between a naphthyl fluorophore and a bromophenyl perturber unit mounted at these positions [19] . Fluorescence quenching by an exchange-mediated mechanism was implicated over this distance. The systematic comparison with the covalently closer 3,5-disubstituted and 4,5-

disubstituted sequence permutation isomers was then carried out [20-21] . Non-covalent electronic exchange interactions (successive *non-bonded* orbital contacts mediated by the electrons of interposed solvent or other intervening molecular groups) were observed to be significantly stronger than covalent (through-bond) interactions in these peptide helices [19-21] .

TABLE 1: RHAE: The enhancement in the rate constant for Nap (singlet S_1) -> Nap (triplet T_1) when BrPhe is present

Peptide	k(RHAE) s^{-1}
Nap^4BrPhe^5	0.5×10^6 $(\pm 0.3 \times 10^6)$
Nap^3BrPhe^5	1.1×10^6 $(\pm 0.3 \times 10^6)$
Nap^3BrPhe^6	14.1×10^6 $(\pm 1 \times 10^6)$

In each of these sequence permutation isomers, the control is the peptide w/ the BrPhe replaced by Phe.

These peptide sequences built from 6 Aib and two L-amino acids are designated as [6/8] to indicate their 75% Aib composition. As shown next, this designation is usefully extended to count the number of Aib *or* "Aib-class" amino acids in the sequence, out of the total number of amino acids in the peptide. The Aib or Aib-class amino acids are responsible for the distinctive and well-formed 3_{10}-helicity observed; such stable helicity would otherwise be quite unexpected in these relatively short peptide lengths.

2.2 CUSTOM SYNTHETIC Aib-CLASS AMINO ACIDS

All α,α–dialkylated amino acids are, structurally, side-chain extensions of Aib, and as expected, most (but not all) behave identically with regard to the backbone folding, We analyzed the set of those α,α–dialkylated-α-amino acids which do exhibit the same strong 3_{10}-helical backbone propensity as the parent molecule, Aib, and defined this set as the "*Aib-class*" amino acids [16,17] . Important classic examples are 1-aminocyclopentanecarboxylic acid and 1-aminocyclohexane-carboxylic acid (Ac_5c and Ac_6c, repectively) [9,40,41] . Such Aib-class amino acids may replace Aib in the sequence of an Aib-rich peptide [16] or replace Aib within an Aib homo-oligomer, with the reliable expectation that there will be no dilution of the "effective" Aib content. The distinctive 3_{10}-helical stability should then be fully preserved.

The synthesis of *new* Aib-class amino acids provides a clear benefit: we can install electronic and other chemically distinctive functionality into the sequence at positions selectable at will, yet insofar as the determinants of backbone helical conformation are concerned, these Aib-class *guest* residues will behave as *host* Aib residues.

The known Ac_5c and Ac_6c alicyclic amino acids can be considered as parents of higher aromatic homologues, and our research group has carried out the design and synthesis of a series of such electronically active cyclic amino acids, shown in Figure 2 [22-25] .

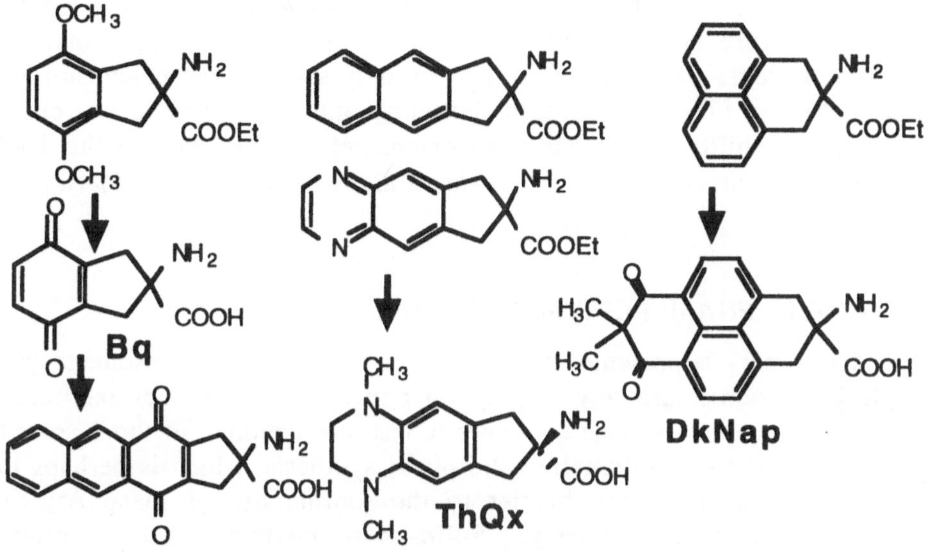

Figure 2: Eight synthetic Aib-Class cyclic amino acids. Due to the cyclic attachment to the backbone, these side chains are free from χ angle torsional flexibility. The arrows denote synthetic transformations (with suitable protection required): oxidation, Diels-Alder, reduction-reductive methylation, and Friedel-Crafts di-acylation. For example, the di-acylation is efficiently carried out [23] through phthalimide N-protection of the amino ester, $ClCOC(Me)_2COCl$ / $AlCl_3$ / ϕNO_2, and deprotection with MsOH / HCOOH. The reduction-reductive methylation from the quinoxaline to the N,N'-dimethyl-tetrahydroquinoxaline (ThQx) can furthermore be carried out upon a completed peptide pre-assembled from the quinoxaline precursor amino acid.

All Aib-class amino acids in Figure 2 contain the Aib frame within them, and are "tucked" or cyclically locked in their extensions beyond the β-carbons to retain 3_{10}-helical stability, unlike the non-cyclic and non-Aib-class α,α-diethylglycine. The two simple naphthyl-fused amino acids and the *p*-dimethoxyphenyl amino acid are fluorescent and can be employed as probes [26] , whereas the quinone amino acids and the diketo-naphthyl amino acid

(DkNap) are good electron acceptors, and the tetrahydroquinoxaline amino acid (ThQx) [25,27] is an excellent electron donor.

The conformational equivalence of DkNap, and of the heteroaromatic quinoxaline cyclic amino acid, to Aib was verified by solution phase NMR studies of the backbone helix at the [7/9] nonamer level [16,17] . The sequences contain two standard L-amino acids both to aid the NMR (two α-protons) and to bias the helix strongly into the right-handed form, e.g. ^{i}Boc-Aib-Aib-Aib-DkNap-Leu-ThQx-Ala-Aib-Aib-$NHCH_2CH_2OCH_3$ [16,25] . More such peptides can be synthesized for conformational analyses; however the evidence to date strongly supports the conclusion that these [6/8] and [7/9] sequences consistently and reliably yield the classic 3_{10}-helical stability which has been extensively documented in the 100% Aib homo-oligomers. This conformational evidence is summarized in section 3, below.

2.3 LIQUID PHASE PEPTIDE COUPLING

A brief summary is presented here on the current status of reasonable yield coupling protocols, drawing largely on experiences from our laboratory. This is warranted as the need continues to test new coupling methods so as to enable a transition to automatic solid-phase synthesis, which is perhaps the only significant remaining barrier to the routine use of these Aib-rich peptides in modular chemistry. Solid-phase methods (and the related ingenious C-terminal POE liquid-phase method) have been applied in the synthesis of Aib-containing peptides [28,29] but these efforts required the use of massive excess of amino acid which is unacceptable when the special synthetic Aib-class amino acids are considered. The problem is, of course, the high steric hindrance around the α-carbon, which is the price which must be paid in order to reap the low configurational entropy benefits of the desired Aib-rich product.

The most general methods are: for the coupling of a single Aib residue to the C-terminus of a peptide, the azirine method [30] ; and for the coupling of two or more Aib residues to the N-terminus of a peptide, the oxazolone method [3,31] . We have carried out both approaches. The synthesis of 1-dimethylamino-3,3-dimethylazirine on the one mole scale [30b], proceeds without incident, though an experienced chemist (Beth Secor) in our lab was required. We did not extend this method to the synthetic amino acids, which would have required the preparation of spiro-azirines. In general we found the application of the oxazolone approach to be most generally useful, and obtained yields typically on the 40%-80% range on peptide block couplings (CH_3CN reflux, ~8 hrs) [13,18,32,33] . These procedures are acceptable, and are based on pure acid-free oxazolone [16] .

It is expected that further improvement is likely to come from replacing, rather than refining, the oxazolone protocol. In 1994 an attractive alternative was published [34] , in which the efficient peptide coupling agent HBTU (or TBTU) [35] was altered to 7-aza-hydroxybenzotriazole derivatives, notably HATU [34] . We have employed both TBTU and HATU with good success in room temperature (liquid phase, 12-24 h) couplings of Aib and Aib-class amino acids to growing peptides. In solid-phase synthesis (single coupling cycle) with HATU, preliminary evidence indicates definitely improved yields, but still not up to the 95% range. This deserves further investigation, and the feasibility of routine solid-phase Aib-rich peptide synthesis may be on the verge of a significant upturn as methodologies offering consistently high yields may be anticipated in the near future.

3. Conformation and Stability

3.1 ROOM TEMPERATURE CONFORMATION: 1D and 2D NMR

Following the seminal work by the Balaram group [5] in India and the Toniolo group [3] in Italy, the solvent perturbation method [13,36,37,38] has been applied to these Aib-rich sequences to afford a clear-cut count of the number of intramolecularly hydrogen-bonded amide protons. The maximum number of amide protons will be protected by intrahelical hydrogen-bonding in the case of a stable 3_{10}-helix: for a fully protected n–mer, the number of amides is n+1 of which n-1 will be solvent-shielded, and so 2 amide protons will be exposed and solvent sensitive. In the alternative case of the α-helix - or, as is often overlooked, in the case of a 3_{10}-helix with a point defect [18] in the backbone hydrogen-bonding - there will be 3 amide protons exposed and hence keenly solvent sensitive [36] . The [6/8] peptides with Ala, Phe, Nap, or BrPhe at positions 3 and 6 show classic stable 3_{10}-helical hydrogen-bonding patterns (2 exposed amides, others highly protected) in all of these solvent perturbation experiments [13,17,32] .

The sequence permutation isomers in the [6/8] series displayed an unusual divergence from the expected pattern, when the two L-amino acids were positioned adjacent to each other. In this case, 3 rather than the expected 2 amide protons were observed to be solvent exposed [13,18] . Hence bringing the two non-Aib L–amino acid guests into contiguous positions can influence the hydrogen-bonding pattern even at 75% overall Aib content, Note that in the earlier review of Karle & Balaram (see their Figure 1) [7] , the uncertain behavior, in which it is difficult to predict a 3_{10} vs. α-helical outcome, typically occurs instead for Aib-containing peptides in the 25% to 50% range. Sequences with *contiguous* L-amino acids such as Ac-Aib-Aib-Aib-**Nap-Phe**-Aib-Aib-Aib-NHMe and ^{t}Boc-Aib-Aib-Aib-**Ala-Ala**-Aib-Aib-Aib-NHMe , however, display this *contiguity effect* in which the helix does not fold into the predicted ideal 3_{10}-helix. This is important in the present context as this discovery provided a clue to the clarification and refinement of the usual two-state 3_{10}-helix/α-helix viewpoint on such helices.

We examined the following two limiting perspectives: i) that the 3_{10}-helix/α-helix equilibrium of the entire peptide is controlled by the competing contributions from the Aib and L-amino acid components in a *sequence sensitive* way such that the *entire length* of the peptide shifted to an α-helix [13,15] , or ii) that the backbone instead responds at the local level to the Aib content of an *individual turn* [18] , and the one turn which necessarily becomes Aib-dilute in sequences with contiguous L-amino acids will suffer weakened hydrogen bonding, while the other turns are unaffected. This latter case of a point defect is diagrammed in Figures 8, 9 and 10 of ref. 18, and in the thesis of Dr. Pettijohn [39] . Both cases, the α-helix and the 3_{10}-helix with the point-defect in the Aib dilute turn, would match the 3 solvent sensitive amide protons observed in these contiguous L-amino acid sequences. A recent and extensive 2D NMR ROESY study settles this issue unequivocably by assigning all Aib amide resonances (otherwise undistinguished singlets which do *not* participate in COSY or TOCSY) and strictly ruling out the α-helix [18,39] . More than one type of defect is considered, and the *local point defect* at the hydrogen-bond which spans the Aib-dilute turn is implicated by the complete spectral evidence.

All possible defects, including point defects, dislocation defects (i.e. a mixed 3_{10}/α-helix), and others, can of course be included in a general Zimm-Bragg type theory for Aib-rich peptides. We emphasize, however, that calculating the energetics (and relative configurational entropies) of these helix/helix transitions and helical defects are very subtle, and very sensitive to one's precise model of crowded steric hindrance and bond-angle bending at the α-carbon. We have chosen instead the empirical route of consistent characterization of these short but highly stable Aib-rich helices by 2D NMR methods [14,17,18,32,33,37] .

By these methods, *all* Aib-rich peptides of [6/8] or [7/9] composition have been verified to produce stable regular 3_{10}-helices by all 1D (solvent perturbation) and 2D ROESY criteria, so long as contiguous L-amino acid sequences are avoided. This reliable helicity holds true both for peptides whose unique conformational character is provided by Aib, and for those peptides in which Aib-class amino acids are employed.

3.2 PROBING THE THERMAL STABILITY OF THE Aib-RICH HELICAL FOLD

Analysis of chemical shift behavior of the amide protons in DMSO up to 50 ^{o}C provided an intriguing result, that the 3_{10}-helix showed no sign of unfolding even for these remarkably short peptides (octamers and nonamers). Consequently the temperature-dependence NMR experiment for an octamer was pushed up to 150 ^{o}C in DMSO [17] . There was no break and absolutely no indication of an opening up of the helix to allow any significant solvent exposure of the 7 intrahelically hydrogen-bonded amide protons. Direct hydrogen-bond counting was performed for a nonamer at 120 ^{o}C by the solvent perturbation experiment, and again the result was the same: full 3_{10}-helical hydrogen bonding in the backbone was observed at 120 ^{o}C , just as seen at room temperature [17] . Both these peptides, the [6/8] and the [7/9], were fully sequence assigned and hence the two solvent exposed backbone amides were known unambiguously to be the two most N-terminal amides.

The thermal stability is in retrospect not surprising as the phenomenon of Aib-rich helical folding is based on severe and relatively unforgiving steric hindrance, which evidently serves to greatly reduce the configurational entropy of an alternative (and hitherto unobserved) unfolded state of an Aib-rich peptide.

4. The Molecular Optical Rail and An Outlook Upon Modular Chemistry Through Peptide Helices

Unique features of the helical rod, from which selected functionalized side chains radiate outwards, have been the focus of various *de novo* design peptide systems in the last ten years. But peptides in solution are not packed into higher order stable structures as are globular proteins, which typically require on the order of 70-100 amino acids as a minimum size to enable tertiary folding. High conformational stability is hence elusive among sequences in the 5-30 amino acid range. Even nominally helical structures inferred for such peptides designed from the more common high helical propensity amino acids suffer large hydrogen-bonding fluctuations in solution. The special sequences explored here, on the other hand, display unusual solution-phase helical stability. This reliable and stable fold implies that the Aib-based "Molecular Optical Rail" can be a practical and successful strategy for modular assembly.

A Molecular Optical Rail is a stable architecture in which electronically active components can be positioned at adjustable positions without concern for disrupting or refolding the structure. It is particularly relevant that the resultant thermal stability exceeds that of most proteins [17] . The essential feature for modularity, that chemical and/or electronic functionality [43] may be inserted at sequence variable positions without altering the architecture of the completed structure, is clearly met. Looking to the future, a key unexplored aspect is how to connect the resulting helices into yet larger scale constructs. Clearly, brute force assembly of very long Aib-rich helices, by application of methods such as segment condensation, is not likely to be the real vision. In the course of the work at Cornell, many ideas arose with respect to assembly of helices upward from planar solid supports of gold, or from optically transparent electrode materials such as alkyl-silylated indium tin oxide, and so on, mainly with the intent on high efficiency in vectorial charge separation. The solid spectroscopic foundation now established from room temperature to beyond 100 ^{o}C in sequence/structure data and understanding will, we hope, prove valuable to others interested in pursuing diverse and imaginative modular chemistry based on, or inspired by, Aib-rich 3_{10}-helices.

5. References

1. O'Neil, K. T. and DeGrado, W. F. (1990) A Thermodynamic Scale for the Helix-forming Tendencies of the Commonly Occuring Amino Acids, *Science*, **250**, 646–651.

2. Hermans, J., Anderson, A. G. and Yun, R. H. (1992) Differential Helix Propensity of Small Apolar Side Chains studied by Molecular Dynamics Simulations, *Biochemistry*, **31**, 5646–5653.

3. Toniolo, C., Bonora, G. M., Barone, V., Bavoso, A., Benedetti, E., Di Blasio, B., Grimaldi, P., Lelj, F., Pavone, V. and Pedone, C. (1985) Conformation of Pleionomers of α-Aminoisobutyric Acid, *Macromolecules*, **18**, 895–902; Pavone, V., Di Blasio, B., Santini, A., Benedetti, E.,, Pedone, C., Toniolo, C. and Crisma, M. (1990) The Longest, Regular Polypeptide 3_{10} Helix at Atomic Resolution, *J. Mol. Biol.*, **214**, 633-635; Toniolo, C., Crisma, M., Bonora, G. M., Benedetti, E., Di Blasio, B., Pavone, V., Pedone, C. and Santini, A. (1991) Preferred Conformation of the Terminally Blocked $(Aib)_{10}$ Homo-oligopeptide: A Long, Regular 3_{10} Helix, *Biopolymers*, **31**, 129–138.

4. Paterson, Y., Rumsey, S. M., Benedetti, E., Némethy, G. and Scheraga, H. A. (1981) Sensitivity of Polypeptide Conformation to Geometry. Theoretical Conformational Analysis of Oligomers of α-Aminoisobutyric acid, *J. Am. Chem. Soc.*, **103**, 2947–2955. See also the *dynamic* study of the Aib helix by modern computational approaches, in reference 42.

5. Prasad, B. V. V. and Balaram, P. (1984) The Stereochemistry of Peptides containing Aib, *CRC Crit. Rev. Biochem.*, **16**, 307–348.

6. Nagaraj, R. and Balaram, P. (1981) Solution Phase Synthesis of Alamethicin I, *Tetrahedron*, **37**, 1263-1270.

7. Karle, I. L. and Balaram, P. (1990) Structural Characteristics of α-Helical Peptide Molecules containing Aib Residues, *Biochemistry*, **29**, 6747–6756.

8. Marshall, G. R., Hodgkin, E. E., Langs, D. A., Smith, G. D., Zabrocki, J. and Leplawy, M. T. (1990) Factors Governing Helical Preferences of Peptides containing Multiple α,α-dialkyl Amino Acids, *Proc. Nat. Acad. Sci., USA*, **87**, 487–491.

9. Di Blasio, B., Pavone, V., Lombardi, A., Pedone, C. and Benedetti, E. (1993) Non-coded Residues as Building Blocks in the Design of Specific Secondary Structures: Symmetrically Disubstituted Glycines and β-Alanines, *Biopolymers*, **33**, 1037-1049.

10. Balasubramanian,T. M., Kendric, N. C. E. and Marshall, G. R. (1981) Synthesis and Characterization of the Major Component of Alamethicin, *J. Am. Chem. Soc.*, **103**, 6127-6132.

11. Karle, I. L., Flippen-Andersen, J. L., Uma, K., Balaram, H. and Balaram, P. (1990) α-Helix and Mixed 3_{10}/α-Helix in Cocrystallized Conformers of Boc-Aib-Val-Aib-Aib-Val-Val-Val-Aib-Val-Aib-OMe, *Proc. Nat. Acad. Sci., USA*, **86**, 765-769.

12. Slomczynska, U., Beusen, D. D., Zabrocki, J., Kociolek, K., Redlinski, A., Reusser, F., Hutton, W. C., Leplawy, M. T. and Marshall, G. R. (1992) Emericins III and IV and their Ethylalanine Epimers. Facilitated Chemical-enzymatic Synthesis and a Qualitative Evaluation of their Solution Structure, *J. Am. Chem. Soc.*, **114**, 4095–4106.

13. Basu, G., Bagchi, K. and Kuki, A. (1991) Conformational Preferences of Oligopeptides Rich in α-Aminoisobutyric Acid. I. Observation of a 3_{10}- / α-Helical Transition upon Sequence Permutation, *Biopolymers*, **31**, 1763-1774.

14. Basu, G. and Kuki, A. (1993) Evidence For a 3_{10}-Helical Conformation of an Eight Residue Peptide from ^{1}H-^{1}H Rotating Frame Overhauser Studies, *Biopolymers,* **33**, 995-1000.

15. Basu, G. and Kuki, A. (1992) Conformational Preferences of Oligopeptides Rich in α-Aminoisobutyric Acid. II. A Model for the 3_{10}- / α-Helical Transition with Composition and Sequence Sensitivity, *Biopolymers*, **32**, 61–71.

16. Bindra, V. A. and A. Kuki (1994) Conformational Preferences of Oligopeptides Rich in α-Aminoisobutyric Acid. III. Design, Synthesis and Hydrogen Bonding in 3_{10}-Helices, *Int. J. Pept. Prot. Res.*, **44**, 539-548.

17. Augspurger, J. D., Bindra, V. A., Scheraga, H. A. and Kuki, A. (1995) Helical Stability of *de novo* Designed Aib-rich Peptides at High Temperatures, *Biochemistry*, **34**, 2566-2576.

18. Pettijohn, A. and Kuki, A. (1996) A 3_{10}-Helical Peptide with a Point Defect in the Hydrogen-Bonding Pattern: Implications for the Role of Aib in Helix Stabilization, *Submitted for publication.*

19. Basu, G., Kubasik, M., Anglos, D., Secor, B. and Kuki, A. (1990) Long-Range Electronic Interactions in Peptides: The Remote Heavy Atom Effect, *J. Am. Chem. Soc.*, **112**, 9410-9411.

20. Basu, G., Anglos, D. and Kuki, A. (1993) Fluorescence Quenching in a Strongly Helical Peptide Series: The Role of Non-Covalent Pathways in Modulating Electronic Interaction, *Biochemistry*, **32**, 3067-3076 (1993).

21. Basu, G., Kubasik, M., Anglos, D. and Kuki, A. (1993) Spin-Forbidden Excitation Transfer and Heavy Atom Induced Intersystem Crossing in Linear and Cyclic Peptides, *J. Phys. Chem.*, **97**, 3956-3967.

22. Kotha, S. and Kuki, A. (1992) A Simple Method for the Synthesis of Cyclic α-Amino Acids, *Tetrahedron Lett.*, **33**, 1565-1568.

23. Kotha, S., Anglos, D. and Kuki, A. (1992) Friedel-Crafts Approach to Electron Deficient Cyclic α-Amino Acids, *Tetrahedron Lett.*, **33**, 1569-1572.

24. Kotha, S. and Kuki, A. (1993) Synthesis of a New Rigid Quinone-Amino Acid and Diels-Alder Extension to Higher Quinones, *Chem. Lett.*, 299-302.

25. Anglos, D., Bindra, V. and Kuki, A. (1994) Photoinduced Electron Transfer and Long-Lived Charge Separation in Rigid Peptide Architectures, *J. Chem. Soc. Chem. Commun.*, 213-216.

26. Kubasik, M. (1995) *Characterization of the Electronic and Dielectric Properties of Helical Peptides through Intramolecular Stark Spectroscopy*, Ph.D. Dissertation, Cornell University, Ithaca NY.

27. Anglos, D. (1994) *Photoinduced Intrapeptide Electron Transfer involving Novel Donor and Acceptor Amino Acids: a Triplet State Approach*, Ph.D. Dissertation, Cornell University, Ithaca NY.

28. Gisin, B. F., Kobayashi, S. and Hall, J. E. (1977) Synthesis of a 19-Residue Peptide with Alamethicin-like Activity, *Proc. Nat. Acad. Sci.,USA*, **74**, 115-119.

29. Jung, G., Bosch, R., Katz, E., Schmitt, H., Voges, K.-P. and Winter, W. (1983) Stabilizing Effects of 2-Methylalanine Residues on β-Turns and α-Helices, *Biopolymers*, **22**, 241-246.

30. a. Vitorelli, P., Heimgartner, H., Schmid, H., Hoet, P. and Ghosez, L. (1974) Addition of Carboxylic Acids and Cyclic 1,3-Diketones to 2-dimethylamino-3,3-dimethyl-1-azirine, *Tetrahedron*, **30**, 3737-3740; b. Haveaux, B., Dekoker, A., Rens, M., Sidani, A. R., Toye, J. and Ghosez, L. (1988) α-Chloro-Enamines, Reactive Intermediates for Synthesis: 1-chloro-N,N,2-trimethylpropenylamine, *Organic Syntheses Collective* **Vol VI**, W. E. Noland, Ed., John Wiley & Sons, New York, pp. 289-291.

31. Leplawy, M. T., Jones, D. S., Kenner, G. W. and Sheppard, R. C. (1960) Synthesis of Peptides derived from α-Methylalanine, *Tetrahedron*, **11**, 39-51.

32. Basu, G. (1993) *Remote Electronic Interactions and Conformational Transitions in a Helical Peptide Series: a Sequence Permutation Study*, Ph.D. Dissertation, Cornell University, Ithaca NY.

33. Bindra, V. A. (1994) *Design, Synthesis, and Conformational Analysis of Electronically Active 3_{10}-helical Peptides Rich in Aib-class Amino Acids* , Ph.D. Dissertation, Cornell University, Ithaca NY.

34. Carpino, L. A., El-Faham, A., Minor, C. A. and Albericio, F. (1994) Advantageous Applications of Azabenzotriazole (Triazolopyridine)-based Coupling Reagents to Solid Phase Peptide Synthesis, *J. Chem. Soc. Chem. Commun.*, 201-203.

35. Spencer, J. R., Antonenko, V. V., DeLaet, N. G. J. and Goodman, M. (1992) Comparative Study of Methods to Couple Hindered Peptides, *Int. J. Pept. Prot. Res.*, **40**, 282-293; Knorr, R., Trzeciak, A., Bannwarth, W. and Gillessen, D. (1989) New Coupling Reagents in Peptide Chemistry, *Tet. Lett.*, **30**, 1927-1931.

36. Bovey, F. A., Brewster, A. I., Patel, D. J., Tonelli, A. E. and Torchia, D. A. (1972) Determination of the Solution Conformation of Cyclic Peptides, *Accts. Chem. Res.*, **5**, 193–200.

37. Vijayakumar, E. K. S. and Balaram, P. (1983) Stereochemistry of α-Amino-isobutyric Peptides in Solution: Helical Conformations of Protected Decapeptides with Repeating Aib-L-Ala-Aib-L-Val Sequences, *Biopolymers*, **22**, 2133–2140.

38. Stevens, E. S., Sugawara, N., Bonora, G. M. and Toniolo, C. (1980) Conformational Analysis of Linear Peptides. 3. Temperature Dependence of NH Chemical Shifts in Chloroform, *J. Am. Chem. Soc.*, **102**, 7048–7050.

39. Pettijohn, A. (1995) *Investigation of 3_{10}-helical Conformations of Aib-rich Peptides using NMR, Resolution-enhanced FTIR, and CD Spectroscopies* , Ph.D. Dissertation, Cornell University, Ithaca NY.

40. Paul, P. K. C., Sukamar, M., Bardi, R., Piazzesi, A. M., Valle, G., Toniolo, C. and Balaram, P. (1986) Stereochemically Constrained Peptides. Theoretical and Experimental Studies on the Conformations of Peptides containing 1-Aminocyclohexanecarboxylic Acid, *J. Am. Chem. Soc.*, **108**, 6363–6370.

41. Toniolo, C. and Benedetti, E. (1988) Old and New Structures from Studies of Synthetic Peptides Rich in Cα,α,-Disubstituted Glycines, *ISI Atlas Sci. Biochem.*, **1**, 225-230.

42. Basu, G., Kitao, A., Hirata, F., Go, N. (1994) A Collective Motion Description of the 3_{10}- / α-Helix Transition: Implications for a Natural Reaction Coordinate, *J. Am. Chem. Soc.* **116**, 6307-6315.

43. Lang, K., Einarsdottir, O., Kotha, S., Kuki, A. (1997) Electron Transfer along Aib Helices: Long-lived Charge Separation in a Rigid Molecular Framework, *Submitted for publication.*

MODULAR DESIGN OF MULTI-PORPHYRIN ARRAYS FOR STUDIES IN PHOTOSYNTHESIS AND MOLECULAR PHOTONICS

JONATHAN S. LINDSEY
Department of Chemistry
North Carolina State University
Raleigh, NC 27695-8204 USA

ABSTRACT. Nature's modular chemistry paradigm involves small sets of building blocks that are joined covalently into larger structures, which subsequently self-assemble into supramolecular assemblies. We have been developing a modular chemistry employing synthetic porphyrin building blocks for creating bioorganic model systems and molecular photonic devices, an approach inspired by biology but differing both in implementation and in molecular structure. Design of a modular chemistry must consider issues including the linker properties, joining chemistry, solubility, diversity of components, and analytical methods for target molecule characterization. Combining porphyrins with different redox, photophysical, and catalytic functions into arrays enables the design of molecular devices for diverse functions such as light-harvesting, energy transduction, and molecular-scale information processing.

Introduction

The zenith of modular chemistries is found in Nature. All known living systems employ sets of small building blocks that are joined into larger covalent structures (Table 1). Many of these covalent structures subsequently undergo non-covalent assembly processes, forming supramolecular structures having long-range 3-dimensional organization. The assembly processes can involve self-assembly of like components or employ many different types of components. Thus modular covalent chemistry is often a prelude to modular non-covalent self-assembly [1].

There are three attractive aspects of modular chemistries. The size of the structures increases in a systematic and hierarchical manner, the chemistry for joining the modular components is of a single type and is employed repetitively irrespective of the number of building blocks constituting the set, and the incorporation of various numbers,

J. Michl (ed.), Modular Chemistry, 517–528.

combinations, and permutations of the building blocks can yield a diverse collection of target molecules.

Perhaps the best known example of Nature's modular chemistry paradigm is the amino acid → polypeptide → protein ascension, which occurs in all living organisms. In this chemistry, as with sugars and nucleotides, the sequence and nature of the building blocks determine the properties of the final products. A different building block approach occurs with the other examples in Table 1. Acetic acid is the sole building block used to form fatty acids of different lengths. Isoprene molecules together with other components are used in forming diverse natural products. Four porphobilinogen molecules combine to yield a single porphyrinogen structure, which undergoes various processing avenues to yield all of the known tetrapyrroles (heme, chlorophylls, vitamin B_{12}, etc.). The hemes and chlorophylls are then employed in ensembles to achieve diverse biological functions.

Photosynthetic systems employ collections of porphyrinic pigments, embedded in protein assemblies, to gather dilute sunlight and funnel photonic energy to reaction centers. The energy migration processes underlying light-harvesting phenomena are fast, efficient, and can involve the interplay of hundreds of pigments in a defined 3-dimensional architecture [2].

Table 1. Examples of molecular building blocks in biology.

Building block	#	Covalent Product	3-D aggregate
amino acids	20	polypeptide	proteins
sugars	few	polysaccharides	cellulose, starch
nucleotides	4	nucleic acids	nucleosomes, chromosomes
acetic acid	1	fatty acids	lipid membranes
isoprene	1	natural products	
porphobilinogen	1	porphyrins	light-harvesting antennae

We are interested in building synthetic model systems for studying photo-induced energy migration phenomena, both to mimic light-harvesting processes and to elicit new features in molecular photonic devices. This is very much a design and synthesis challenge, as a large number of pigments must be precisely arranged in close proximity. Among various pigments, synthetic porphyrins are attractive due to their strong absorption, their facile excited-state energy and electron-transfer reactions, and their close resemblance to the chlorophylls.

The construction of multi-porphyrin arrays should enable mimicry of photosynthesis and should provide access to a wide variety of molecular photonic devices. These include devices for light-harvesting, optical gating, photoinduced electron pumping, and electron or hole transport. The requirement that numerous porphyrins and ancillary molecules be arranged in a 3-D ensemble places a premium on synthetic approaches that have a modular building block nature. The approach we have been developing incorporates porphyrins as integral components of synthetic arrays, rather than as appendages to a molecular scaffolding. By incorporating porphyrins of various types, in principle a diversity of porphyrin-containing structures can be created for photochemical studies.

Original porphyrin building blocks

The earliest synthetic building block approach was developed by Collman and coworkers, who designed the picket fence porphyrins [3]. The amines provide ideal sites for further functionalization. This single compound launched a generation of enzyme and heme modeling studies (Figure 1).

An alternative approach employs metal ligation to construct a cofacial array of metalloporphyrins [4]. The metal-ligand self-assembled structures incorporate a large number of porphyrinic pigments and provide 1-dimensional conducting materials.

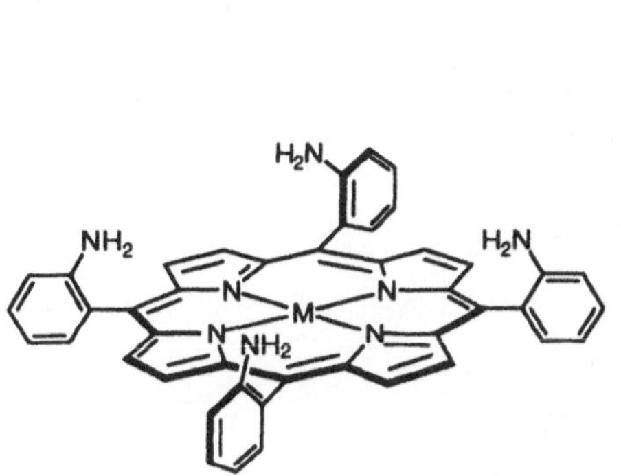

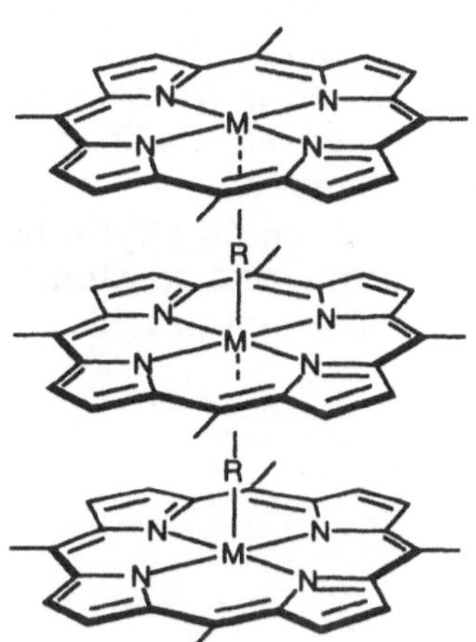

Figure 1. Classic building block approaches in porphyrin chemistry.

Each of these approaches is a powerful solution for the given design challenge. In heme modeling one is largely concerned with precise sculpturing of the environment of a single porphyrin, a feat ably accomplished with picket fence porphyrins and their congeners. In preparing anisotropic conductors one needs a large number of porphyrinic structures in a uniform assembly, for which indefinite association by ligation at the metal site is an ideal solution.

There are other domains where a large number of porphyrins need to be brought into close proximity, in well-defined architectures, with precise control over the metal and substituents of each porphyrin. This requires a building block approach that cannot be achieved exclusively with a picket fence porphyrin, or with a self-assembled stack of porphyrins.

Our porphyrin building blocks for preparing arrays

A viable building block approach requires a melding of molecular design, synthetic strategy, synthetic methodology, and analytical methods for chemical characterization. We have developed a modular strategy where a small set of porphyrin building blocks can be joined to make a diverse collection of synthetic arrays [5]. The porphyrin building blocks have three distinguishing features:

1) the porphyrins are facially encumbered to suppress cofacial aggregation, thereby providing enhanced solubility in organic solvents;
2) the porphyrins are in a defined state of metalation (i.e., have the metal that they will carry in the final product). Many of the metals in the periodic chart can be inserted into a porphyrin;
3) the porphyrins bear peripheral functional groups that are joined covalently in a controlled manner during the construction of the array.

Because the porphyrin building blocks bear multiple sites of attachment, the porphyrins constitute an integral part of the scaffolding of the array, much like amino acids constitute the integral components of proteins.

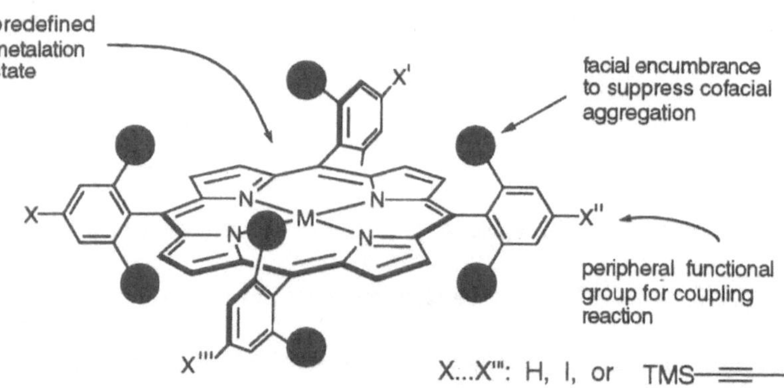

Figure 2. Schematic of a porphyrin building block.

The trimethylsilyl (TMS) group protects the ethyne during coupling reactions and other transformations, and can be removed with fluoride reagents.

In an integral building block approach (as opposed to one where appendages to a scaffolding are employed), the nature of the building block functional groups carries strong implications for the properties of the composite structure. For example, though the distinguishing feature of amino acids is their R group, the series of amide bonds in the polypeptide together with the R groups determines the properties and structure of the resulting protein. Furthermore, the chemistry for forming the linker joining the modular amino acids must be based on that of amines and acids. Similarly, the linker chosen for connecting the porphyrins can be expected to play an integral role in determining the properties of the multi-porphyrin arrays, and in turn imposes constraints on the type of functional groups in the building blocks and the available types of joining chemistry.

We have investigated various linkers in developing this building block approach [5]. For reasons involving solubility, coupling chemistries, distance and structure of the linker, we selected diphenylethynes to serve as linkers in the multi-porphyrin arrays. Iodo-phenyl and ethynyl-phenyl groups yield diphenylethyne linkers upon coupling under conditions that do not alter the remainder of the porphyrin building block (in particular the site of metalation). The diphenylethyne linkers are very attractive from a synthesis and photochemical standpoint, and other shorter or longer ethyne-based linkers can be employed.

This porphyrin building block approach is built around the following synthetic methods:

- a one-flask room temperature synthesis of *meso*-substituted porphyrins [6];
- identification of catalytic conditions for smoothly forming facially encumbered porphyrins bearing methyl, methoxy, or benzyloxy groups in the ortho-phenyl positions [7];
- a one-flask synthesis of dipyrromethanes, precursors for preparing *trans*-substituted porphyrins [8];
- a 9-step synthesis of porphyrins bearing 4 different *meso*-substituents [9];
- mild, efficient, clean, and rapid Pd-mediated coupling reactions for constructing diphenylethyne or diphenylbutadiyne linkers joining the porphyrins [10];
- gentle methods for metal insertion [11], including a room temperature method for inserting magnesium into porphyrins [12].

These synthetic methods provide access to a wide variety of porphyrin building blocks. For brevity we omit here any discussion of methods for preparing synthetic porphyrins and metalloporphyrins.

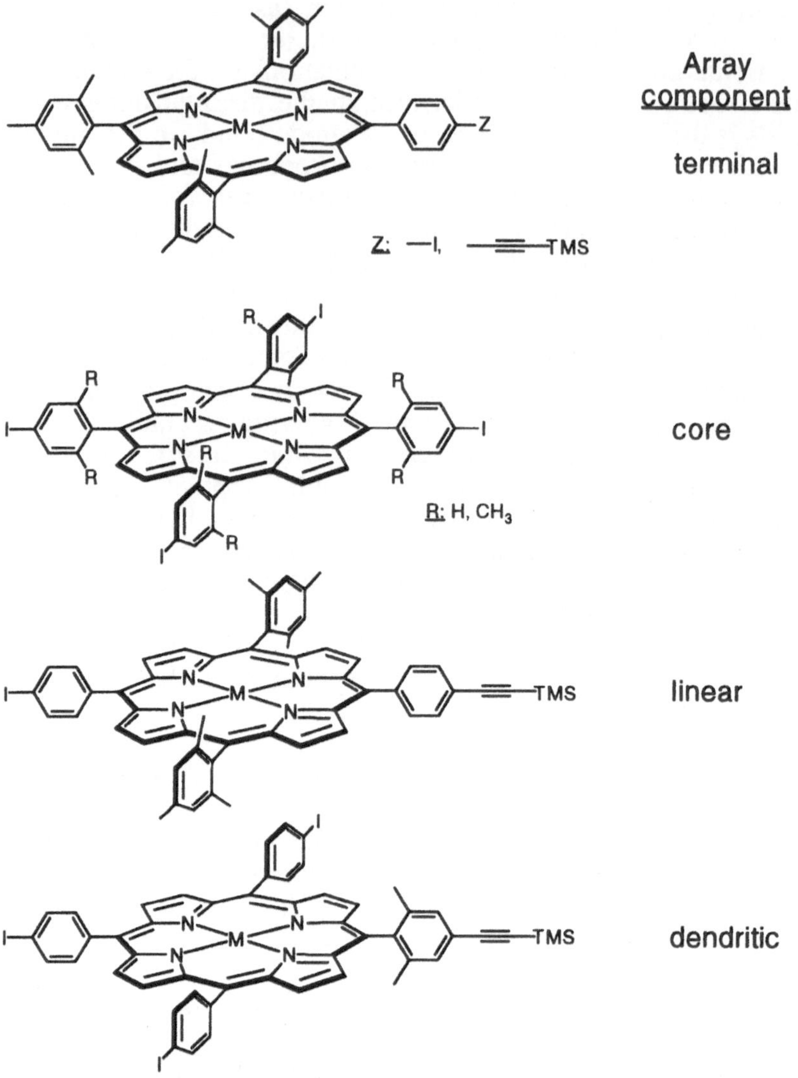

Figure 3. Small library of porphyrin building blocks, identified by the type of site they can occupy in a multi-porphyrin array.

Several examples of porphyrin building blocks that we have prepared are shown in Figure 3. In multi-porphyrin arrays, the porphyrins bearing one reactive group can occupy terminal sites, a porphyrin bearing 4 identical groups can serve as a symmetrical core site, a porphyrin bearing two reactive groups in a trans orientation can serve as a linear component, and a porphyrin bearing 3 groups of one type and one of another can serve as a component of a dendritic array. The recent development allowing placement of four different *meso*-substituents about the porphyrin periphery should broaden the scope of building block porphyrins that can be prepared.

Joining chemistry

The chemistry selected for joining any type of building blocks must meet two general criteria [5]:

1) The coupling reaction must have an inherent directionality (and be compatible with appropriate protecting groups) so that two different building blocks (A, B) can be joined selectively, forming the A-B product, not the A-A, B-B, or B-A side products.
2) The coupling reaction should be clean and high-yielding.

The coupling reaction for joining porphyrin building blocks must meet additional criteria:

3) The desire to incorporate diverse metalloporphyrins into arrays rules out acidic conditions that cause demetalation and forcing conditions that give transmetalation.
4) The desire to incorporate free base porphyrins into arrays rules out the presence of metals that can insert easily into porphyrins.
5) The intrinsic solubilities of the porphyrins require reactions to be performed in dilute solution, typically 0.001 - 0.01 M, which is as much as 100 times less concentrated than for typical coupling reactions. Maintaining solubility of the reactants, intermediates, and products is a key constraint since the coupling methods are to be extended to the synthesis of multi-porphyrin arrays.

The Pd-coupling reactions, involving the Cassar-Heck-Sonogashira reaction of an aryl iodide and an aryl ethyne, are especially attractive. The reaction is inherently directional. The Pd-coupling methods are performed in neutral to basic conditions where demetalation does not occur. Palladium does not insert into porphyrins under the mild conditions of the coupling reactions. Copper has frequently been employed as a cocatalyst in palladium-mediated couplings. Because copper readily inserts into free base porphyrins, we developed efficient Pd-mediated coupling conditions that are copper-free. Thus a diphenylethyne linkage can be constructed in 70% yield from mM reactants at 35 °C under simple anaerobic conditions for 1-2 h. The ability to synthesize multi-porphyrin arrays in dilute solution, under mild, low temperature, non-acidic, non-metalating conditions provides a rapid and efficient means for covalently joining free base and metalloporphyrin building blocks [11].

Overall synthetic approach

The modular synthetic approach is concise (Figure 4). An aldehyde is substituted with groups that will become the *meso*-substituents upon conversion to a porphyrin. Porphyrins then are joined in a covalent array using Pd-mediated coupling chemistry. One of the attractions of any modular chemistry, this one included, is the small number of different types of reactions that must be mastered. In the synthesis of multi-porphyrin arrays, there are generally 4 different types: aldehyde functionalization, porphyrin formation, porphyrin metalation, and Pd-mediated couplings of the porphyrin building blocks.

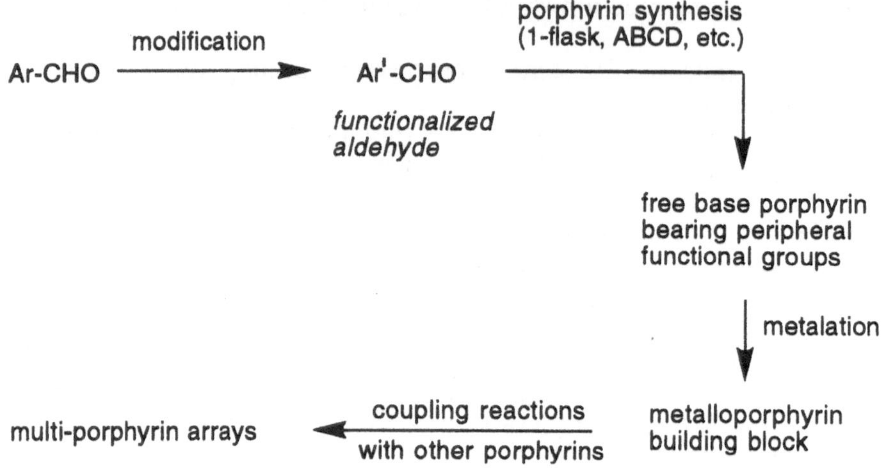

Figure 4. Outline of overall building block approach.

Initial fruits of modular chemistry

We have prepared a variety of multi-porphyrin arrays containing up to nine porphyrins that are joined via diphenylethyne linkers. Notable features of these arrays include relatively rigid intramolecular distances (though free rotation occurs about the ethynes), high solubility in solvents such as toluene, and incorporation of metals (Mg, Zn, Cu, or no metal) at defined sites. The structures formed include linear, star-like, or right-angle geometries, indicating the high level of architectural control accompanying the building block approach. The arrays are readily purified by size exclusion chromatography and can be characterized by absorption and fluorescence spectroscopies, matrix-assisted laser desorption mass spectrometry, and proton NMR spectroscopy.

As one example, we have constructed a linear array of pigments that functions as a molecular photonic wire [13]. The wire (90 Å from end-to-end) is built around a boron-dipyrromethene dye [14] as an input chromophore, three zinc porphyrins as a transmission element, and a free base porphyrin as an output chromophore. As suggested by the structure, the synthesis involved the use of only three building blocks (a boron-dipyrromethene dye, a zinc porphyrin, and a free base porphyrin). Absorption of a photon at the input results in emission of a different photon at the output with a signal transmission efficiency of 76% from input to output. This photonic wire provides one example of how light-driven signals can be manipulated in molecular-scale devices (Figure 5).

Other pigments can be incorporated into this building block design. A successful pigment must meet the photophysical design criteria and be compatible with the building block chemistry. The boron-dipyrromethene dye, which in essence is "half of a porphyrin", matches both sets of criteria rather well [14].

We are currently performing studies to characterize the mechanisms and dynamics of energy migration in the multi-pigment arrays. The high solubility and efficient building block syntheses suggests that longer arrays should be accessible.

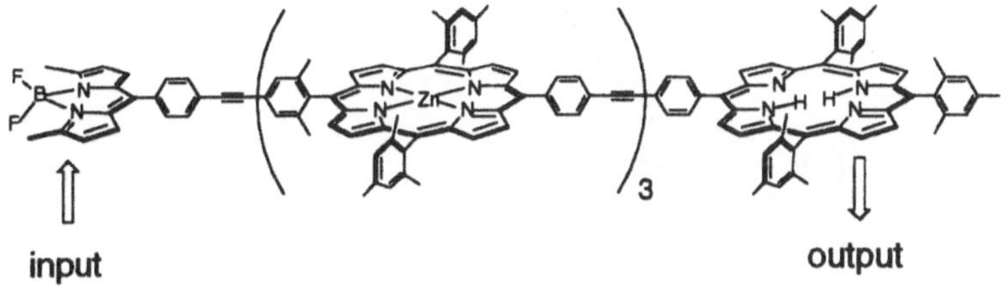

Figure 5. A molecular photonic wire.

A pentameric array of porphyrins has been constructed as a light-harvesting antenna [15]. Absorption of light by the peripheral zinc porphyrins is followed by energy transfer to the core free base porphyrin. The quantum efficiency of light-harvesting is high (>90%) and the energy transfer time from the zinc porphyrin to the free base porphyrin is several tens of psec (Figure 6).

Prof. David Bocian and his coworkers have performed studies to investigate the mechanisms of electronic communication in the arrays [16]. The visible absorption spectra (450-700 nm) show little electronic interactions (the spectra are the sum of the parts), yet the partially oxidized complexes exhibit rapid hole-hopping, indicating ground-state electron transfer via the diphenylethyne spacer. The resonance Raman spectra of neutral monomeric or dimeric porphyrins show an ethyne stretch upon excitation at 457 nm, which is interpreted as involving rotation of the diphenylethyne spacer toward coplanarity with the porphyrin, thereby mediating excited-state electronic energy transfer. Taken together these studies show that the linker plays an intimate role in the physical properties of the multi-porphyrin arrays.

These investigations have demonstrated that the arrays have interesting physical properties apart from their light-harvesting features. Electrochemical studies showed that up to 4 electrons can be removed from the pentameric array at nearly constant potential. In conjunction with their rapid hole-hopping properties, these results indicate that these or related arrays may also have application as hole-transport chains or as constant potential electron reservoirs [16].

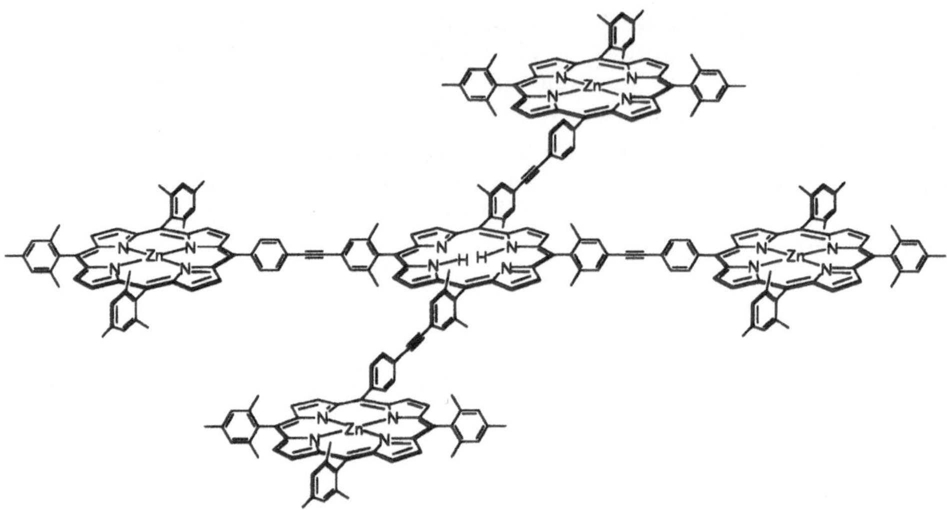

Figure 6. A pentameric light-harvesting array.

Recently we have prepared molecular optoelectronic gates [17]. These molecules are designed for electrochemical gating of the energy transfer process in a molecular photonic wire. The gates incorporate four pigments, including a boron-dipyrromethene dye, a zinc porphyrin, a magnesium porphyrin, and a free base porphyrin. Synthetic access to molecules of this complexity resides on a robust modular building block approach.

Outlook

Our work to date has focused on developing synthetic routes to porphyrin building blocks, and designing photochemical molecular devices using these building blocks. We are currently working to understand 1) how far the covalent modular building block approach can be pushed, 2) the diversity of architectures that can be produced, 3) how structural alterations in the linker affect the energy migration rates in the arrays, and 4) the photochemical features that can be elicited upon incorporation of various metalloporphyrins.

There are many facets of this building block chemistry that have not yet been explored. For example, in biological systems the number of building blocks varies from one modular chemistry to another (20 amino acids, 4 nucleotides, etc., see Table 1). It remains to be determined what the minimal basis set of porphyrin building blocks is (and indeed if there is one) with which any desired photochemical or catalytic objective can be achieved. We have not yet begun to explore the possibilities of supramolecular assembly of the covalent structures. The family of porphyrinic pigments extends beyond porphyrins into the fascinating menagerie of hydroporphyrins including chlorins, bacteriochlorins, isobacteriochlorins, corrins, corphins, etc. One can envisage a wide variety of molecular devices for catalysis and photochemistry built around these components, thus a family of "tetrapyrrole-building blocks" is sought that includes porphyrins and hydroporphyrins. Indeed, combining these different porphyrinic pigments, many of the metals in the periodic chart, a range of substituents to tune redox potentials and alter catalytic properties, and the length and structure of the linker joining adjacent pigments, should provide a lifetime of modular chemistry opportunities in this domain.

Acknowledgment

I thank my coworkers, named in the citations below, who have helped advance a modular building block chemistry of synthetic porphyrins. This work was funded primarily by the NIH (GM36238).

References

1. Kushner, D.J. (1969) Self-assembly of biological structures, *Bacteriological Rev.* **33**, 302-345. Lindsey, J.S. (1991) Self-assembly in synthetic routes to molecular devices. Biological principles and chemical perspectives, *New J. Chem.* **15**, 153-180.

2. Larkum, A.W.D. and Barrett, J. (1983) Light-harvesting processes in algae, *Adv. Bot. Res.* **10**, 1-219. Scheer, H., and Siegried, S., (eds.), (1988) *Photosynthetic Light-Harvesting Systems*, W. de Gruyter, Berlin.

3. Collman, J.P., Gagne, R.R., Reed, C.A., Halbert, T.R., Lang, G., and Robinson, W.T. (1975) "Picket fence porphyrins." Synthetic models for oxygen binding hemoproteins, *J. Am. Chem. Soc.* **97**, 1427-1439.

4. Segawa, H., Kunimoto, K., Susumu, K., Taniguchi, M., and Shimidzu, T. (1994) Synthesis of "wheel-and-axle"-type porphyrin arrays composed of phosphorous porphyrins, *J. Am. Chem. Soc.* **116**, 11193-11194.

5. Lindsey, J.S., Prathapan, S., Johnson, T.E., and Wagner, R.W. (1994) Porphyrin building blocks for modular construction of bioorganic model systems, *Tetrahedron* **50**, 8941-8968.

6. Lindsey, J.S. (1994) The synthesis of *meso*-substituted porphyrins, in F. Montanari and L. Casella (eds), *Metalloporphyrins-Catalyzed Oxidations*, Kluwer Academic Publishers, The Netherlands, pp. 9-86. Lindsey, J.S., Schreiman, I.C., Hsu, H.C., Kearney, P.C., and Marguerettaz, A.M. (1987) Rothemund and Adler-Longo reactions revisited: Synthesis of tetraphenylporphyrins under equilibrium conditions, *J. Org. Chem.* **52**, 827-836.

7. Lindsey, J.S. and Wagner, R.W. (1989) Investigation of the synthesis of ortho-substituted tetraphenylporphyrins, *J. Org. Chem.* **54**, 828-836.

8. Lee, C.-H., and Lindsey, J.S. (1994) One-flask synthesis of *meso*-substituted dipyrromethanes and their application in the synthesis of *trans*-substituted porphyrin building blocks, *Tetrahedron* **50**, 11427-11440.

9. Lee, C.-H., Li, F., Iwamoto, K., Dadok, J., Bothner-By, A.A., and Lindsey, J.S. (1995) Synthetic approaches to regioisomerically pure porphyrins bearing four different *meso*-substituents *Tetrahedron* **51**, 11645-11672.

10. Wagner, R.W., Johnson, T.E., Li, F., and Lindsey, J.S. (1995) Synthesis of ethyne-linked or butadiyne-linked porphyrin arrays using mild, copper-free, Pd-mediated coupling reactions, *J. Org. Chem.* **60**, 5266-5273.

11. Buchler, J.W. (1978) in Dolphin, D. (ed.), *The Porphyrins*, Academic Press, New York. Vol. I, pp. 389-483.

12. Lindsey, J.S. and Woodford, J.N. (1995) A simple method for preparing magnesium porphyrins, *Inorg. Chem.* **34**, 1063-1069.

13. Wagner, R.W. and Lindsey, J.S. (1994) A molecular photonic wire, *J. Am. Chem. Soc.* **116**, 9759-9760.

14. Wagner, R.W. and Lindsey, J.S. (1996) Boron-dipyrromethene dyes for incorporation in synthetic multi-pigment light-harvesting arrays, *Pure Appl. Chem.* **68**, 1373-1380.

15. Prathapan, S., Johnson, T. E., and Lindsey, J. S. (1993) Building-block synthesis of porphyrin light-harvesting arrays, *J. Am. Chem. Soc.* **115**, 7519-7520.

16. Seth, J., Palaniappan, V., Johnson, T.E., Prathapan, S., Lindsey, J.S., and Bocian, D.F. (1994) Investigation of electronic communication in multi-porphyrin light-harvesting arrays, *J. Am. Chem. Soc.* **116**, 10578-10592.

17. Wagner, R.W., Lindsey, J.S., Seth, J., Palaniappan, V., and Bocian, D.F. (1996) Molecular optoelectronic gates, *J. Am. Chem. Soc.* **118**, 3996-3997.

POLYIMIDE NANOFOAMS FROM PHASE SEPARATED TRIBLOCK COPOLYMERS

R. D. MILLER, B.-L. HSU, K. R. CARTER, H. J. CHA,
C. J. HAWKER, J. L. HEDRICK, R. A. DI PIETRO,
J. W. LABADIE, T. P. RUSSELL, M. I. SANCHEZ,
W. VOLKSEN AND D. YOON
IBM Research Division, Almaden Research Center,
650 Harry Road, San Jose, CA 95120-6099

J. E. McGRATH
Department of Chemistry, Virginia Polytechnic Institute and
State University Blacksburg, VA 24061 – 0212

ABSTRACT. Phase-separated block copolymers are capable of self-assembling into a variety of interesting morphological structures, the nature of which is determined primarily by the copolymer composition. A number of potential applications for such structures may be envisioned. In this paper, we describe the use of phase-separated structures in a polyimide matrix to produce low dielectric constant, closed-cell polyimide nanofoams by thermal decomposition of a labile polymeric block. The morphology in the cured triblock polyimide copolymer is maintained in the nanofoam produced.

1. Introduction

Block copolymers constitute an interesting class of materials which, under certain circumstances, are capable of supramolecular assembly [1,2]. Linear materials may be classified as diblock, triblock or multiblock depending on the block connectivity along the polymer backbone. More than two types of monomer blocks may also be incorporated in the polymer backbone. Other more complicated architectures, such as stars, grafts, dendrimers and hyperbranched systems are also possible [3].

$-(A)_n-(B)_m-$ $\quad$ $-(A)_n-(B)_m-(A)_n-$ $\quad$ $-(A)_n-(B)_m-(A)_n-(A)_n-(B)_m$

diblock $\quad$ triblock $\quad$ multiblock

In principle, block copolymers can be prepared in any living polymerization process (anionic, cationic, living free radical, etc.) where the polymer chain

J. Michl (ed.), Modular Chemistry, 529–542.

end(s) remain as active initiators for the polymerization of other active monomers. Polymers containing other types of reactive functionality may also be converted to block copolymers by reaction with suitably functionalized macromolecules.

Most polymers are incompatible when mixed as blends, resulting in macroscopic phase separation of the individual components. With block copolymers, phase separation of the component blocks can lead to the formation of nonrandom meso domains resulting in some very unusual polymer morphologies [1,2]. This is demonstrated in Figure 1 for the $\leftarrow AB \rightarrow_n$ diblock copolymer where the polymer morphology depends on the relative composition of the individual blocks. By altering the copolymer composition, it is possible to control the curvature of the interfaces as well as the domain shapes. In Figure 1, the composition of the diblock system progressively increases in the content of A in moving from left to right. The morphologies represented are idealized equilibrium structures for mutually incompatible blocks and would require substantial annealing in order to achieve the equilibrium morphologies. More recently, other more exotic morphologies have been discovered for diblock systems. The structures shown in Figure 1 represent equilibrium structures that would be obtained after annealing for long periods near the glass temperature of the high temperature block. Without annealing near the glass transition temperature, unusual nonequilibrium morphologies may be obtained. The structures shown represent self-assembling arrays with considerable potential for the preparation of nanostructures and devices depending on the nature of the component blocks, processing conditions, surface interactions, polymer functionality, etc. Obvious potential applications for materials with cylindrical morphologies would be anisotropic conductors, selective ion channels, catalyst supports, etc. Selective removal of the cylindrical phase in subsequent processing steps could lead to selective nanoporous membranes and filters [4]. Conversely, selective removal of the matrix material should result in regular nanostructures with potentially interesting electronic and spectral characteristics [5–7]. Lamellar structures could be used to modify surfaces, control surface topography, provide self-lubricating hard coatings, generate multilayer wave guides, confine two-dimensional conductors, etc. For the higher order morphologies such as cylinders and lamellae, it is important to not only optimize the extent of structural organization but also to be able to control the orientation of the macroscopic domains (e.g., parallel or perpendicular orientation to the surface). In the case of lamellar and cylindrical structures,

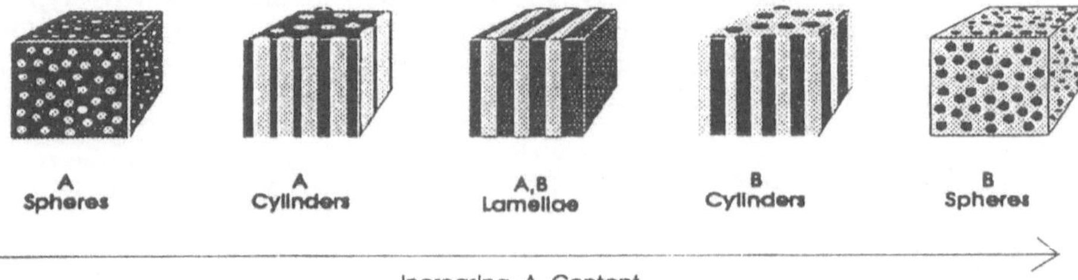

Figure 1. Idealized polymer morphologies of an $\leftarrow AB \rightarrow$ diblock copolymer as a function of increasing A content.

the copolymers form microdomains on the order of a few hundred angstroms which can assemble to grains as large as microns. Since these dimensions are usually small relative to the sample size, the overall orientation is usually random. Most attempts to orient higher order phase-separated structures to date have focused on surface interactions and modifications [9], special casting techniques including shear alignment [9 – 15] and the application of external fields [16,17].

Since spherical phase-separated morphologies avoid the need for domain orientation, applications for this type of structure are intrinsically easier to achieve. We will describe here the use of phase separation in block copolymers to prepare high temperature, low dielectric constant, organic polymeric nanofoams for use as dielectric insulators.

Although there are many techniques for the production of foamed polymers [18,19], most commercial processes result in open cell connecting pore structures and/or pores of large dimensions which would be unsuitable for dielectric applications. For such applications, the dielectric material must have excellent thermal and chemical stability, be processable in thin film form ($< 10\ \mu m$), display adequate mechanical properties, be impermeable to corrosive reagents and have pore dimensions which are small relative to the smallest electronic feature sizes (< 500 nm). As device dimensions continue to shrink causing wiring patterns for chip interconnects, transmission lines, module-board connects, etc., to become denser, access times are strongly effected by transmission delays and capacitance and inductive coupling, adversely effecting performance [20]. Significant reduction in the insulator dielectric constant can dramatically reduce transmission delays, electronic cross-talk and capacitance coupling between conductor lines, allowing ultimately the use of denser wiring patterns [20,21]. Depending on geometry and function, the delay times will scale according to the media dielectric constant raised to the $0.5 - 1.0$ power. The most common interlayer dielectric material (ILD) currently in use is SiO_2 with a dielectric constant which varies between $3.9 - 4.3$ depending on the mode of deposition. Although a plethora of organic polymers with dielectric constants ranging from $2.6 - 3.5$ have been auditioned and are under consideration [22], it seems unlikely that a suitable polymeric organic material will be found with a dielectric constant of less than 2.2 (poly(tetrafluoroethylene)) and acceptable thermal and mechanical properties.

Foamed polymers are potential low dielectric candidates, since the dielectric constant of air is low ($\varepsilon = 1.0$) and that of the foam will scale with the void volume fraction. It is noteworthy that highly porous aerogels ($90 - 98\%$ void volume fractions) have dielectric constants in the $1.03 - 2.0$ range, depending on the density [23]. The ultimate utility of foamed dielectric materials presupposes that the other requirements for a dielectric insulator can be satisfied.

2. Results and Discussion

Our scheme for the preparation of high temperature, low dielectric constant nanofoams, which is based on phase separation of block copolymers into spherical morphologies, is shown in Figure 2 [24]. In this example, the minor component block must be present in high enough concentration and have a sufficiently high molecular weight to produce the desired phase-separated morphology and at the same time should be thermally labile and decompose into monomer and/or small volatile fragments upon heating. The matrix material must be thermally and chemically stable with good mechanical properties and have a glass transition temperature (T_g) which is considerably higher than the labile block decomposition temperature in order to provide an adequate thermal processing window and prevent foam collapse. The labile block must decompose completely at a rate which allows the volatile fragments to be expelled by diffusion without pooling and significantly plasticizing the matrix host causing blowing effects which may lead to unacceptable increases in the pore sizes.

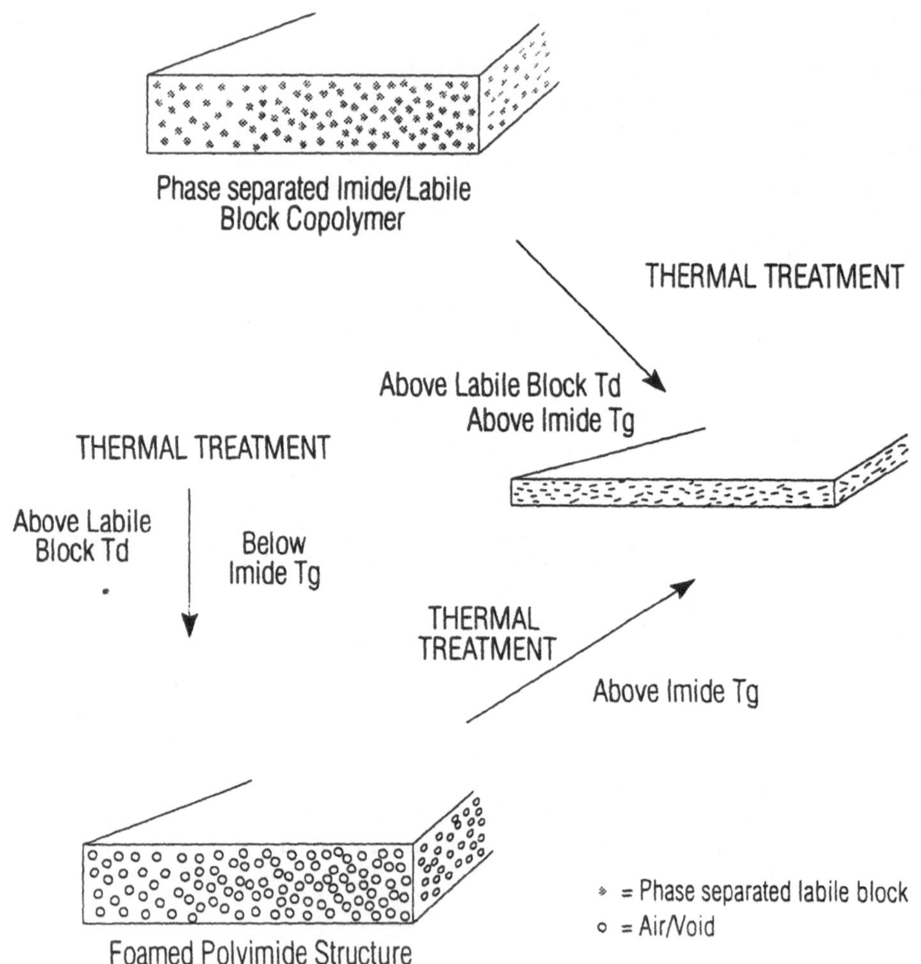

Figure 2. Schematic for the preparation of a polyimide nanofoam from a phase-separated block copolymer containing a thermally labile block.

As an appropriate matrix polymer we have selected polyimides, a class of thermally stable materials with numerous microelectronic applications [25]. These materials are usually prepared from the condensation of stoichiometric quantities of aromatic-bis-anhydrides with aromatic diamines. The soluble polyamic acids produced by ring opening polymerization can be spin-coated and cured thermally to the polyimides. Alternatively, the poly(amic acids) can be cured chemically by dehydration (e.g., acetic anhydride-pyridine) to yield polyimides directly. These materials are, however, often insoluble in common organic solvents and a processable precursor route is preferred.

A versatile precursor route is shown in Scheme 1. Pyromellitic dianhydride (PMDA) is opened by ethanol to yield a meta, para mixture of half acid esters which can be separated by fractional recrystallization and converted to the respective acid chlorides [25]. Reaction with a diamine, in this case, hexafluoroisopropylidene-bis-aniline (3FDA), yields the corresponding amic ester. The amic ester precursors offer a number of significant advantages over the more commonly employed poly(amic acids) in that they are soluble, stable and isolable. Thermal imidization also begins at higher temperatures (225 – 250 °C) than for the unstable amic acids, which allows some annealing of the polymer prior to imidization. Imidization of the amic ester at temperatures around 300 °C results in a polyimide homopolymer with a T_g of 435 °C in the case of PMDA-3FDA; the same value obtained either from thermal curing or chemical imidization of the corresponding amic acid. The amic ester route to polyimides is also applicable to a wide variety of diamines. Although many other materials have been surveyed, in this paper we report on nanofoams prepared from the PMDA-3FDA matrix material. This particular homopolymer shows excellent thermal stability (> 500 °C), a high T_g (435 °C), has a relatively low dielectric constant ($\varepsilon = 2.8$) and leads to isotropic films upon curing.

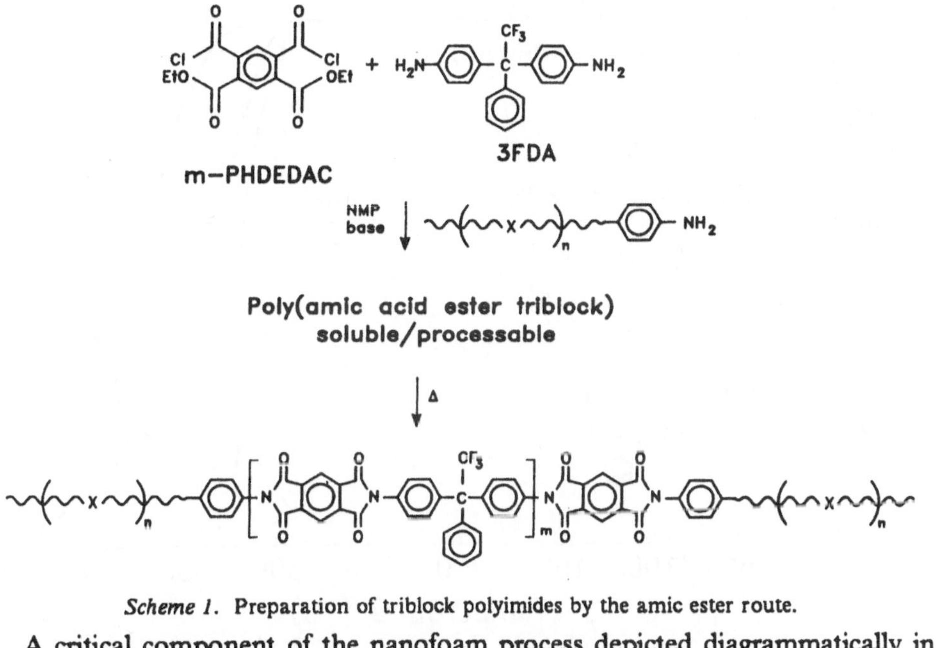

Scheme 1. Preparation of triblock polyimides by the amic ester route.

A critical component of the nanofoam process depicted diagrammatically in Figure 2 is the thermally labile block designed to ultimately produce the polymer voids. By utilizing a monofunctional aromatic amino substituted macromer with proper stoichiometric control in conjunction with PMDA and 3FDA, we can produce triblock copolymers as shown in Scheme 1. Vinyl addition polymers of styrene and α-methylstyrene [26−28] and copolymers were prepared anionically, and capped with ethylene oxide to produce hydroxyethyl end groups. Reaction of these reactively functionalized oligomers with p-nitrophenylchloroformate followed by catalytic reduction of the nitro group (Pd/C) yielded the terminally functionalized aryl amines. Hydroxyalkylated poly(methyl methacrylate), produced by group transfer polymerization, could be functionalized in the same manner [24,29]. Oligomers of poly(propylene oxide) (PO), generated by anionic ring opening polymerization, were functionalized directly as described [24,29,30]. In addition, oligomers derived from the ring opening polymerization of lactones, lactides and cyclic carbonates could also be prepared and functionalized similarly. Discussion of labile blocks derived from lactones and lactides will be deferred to another time. Aminophenyl and hydroxyethyl substituted polystyrene oligomers could also be generated by controlled free radical polymerization, with an appropriately functionalized initiator [31]. The possibility of preparing a wide variety of regular block and graft copolymers by controlled free radical polymerization techniques [31,32,33] should provide access to a large number of new macromers. Until recently, most regular block copolymers have been prepared by standard living polymerization techniques (anionic, cationic, etc.), procedures with limited functional group tolerance.

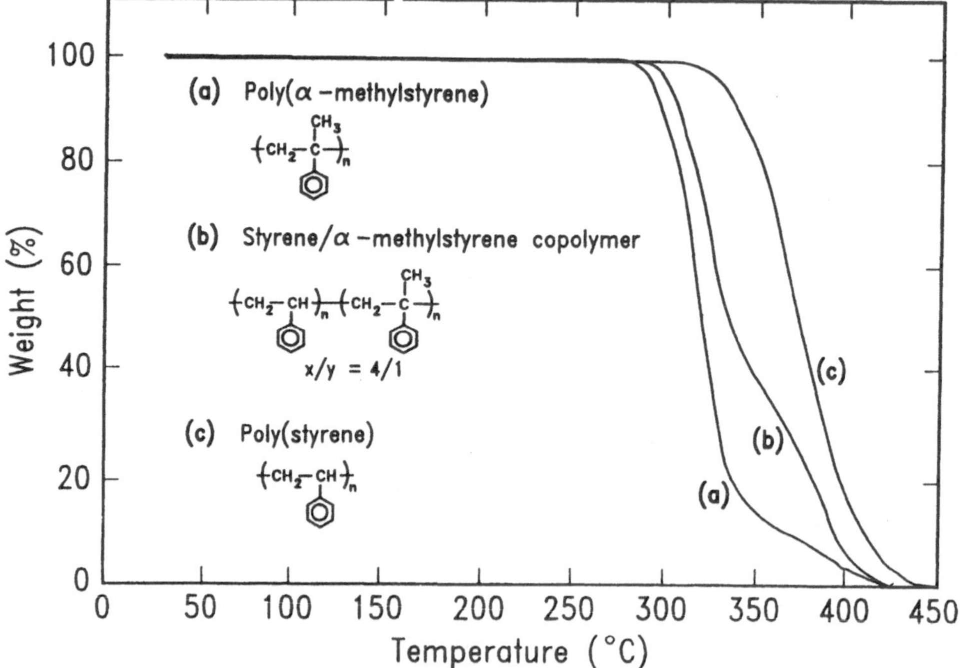

Figure 3. TGA analysis of a variety of styrene and α-methylstyrene homo- and copolymers (heating rate 10°/min).

Thermal gravimetric analysis (TGA) of the labile blocks showed that they cleanly decomposed upon heating to volatile fragments as shown in Figures 3 and 4. The styrenic oligomers undergo catastrophic decomposition above 300 °C to yield volatile products. In this regard, anionically polymerized polystyrene is the most stable while oligomers of α-methylstyrene are the least. The styrene oligomers prepared by the living free radical route were somewhat less stable than those prepared anionically. As expected, copolymers of styrene and α-methylstyrene show intermediate stabilities. Although functionalized oligomers of PO prepared by anionic polymerization are stable to ~ 300 °C in nitrogen, in the presence of air significant decomposition begins 40 – 60 degrees lower [24]. This fortuitous effect can be utilized to increase the effective processing window in the thermal conversion of the phase-separated morphology to a foam (vide infra). Pertinent characterization data for a number of functionalized thermally labile oligomers are shown in Table 1. The number average molecular weights of the amino functionalized oligomers were determined both by ^{1}H NMR analysis and by potentiometric titration (HBr-HOAc). The values derived from both procedures were consistent.

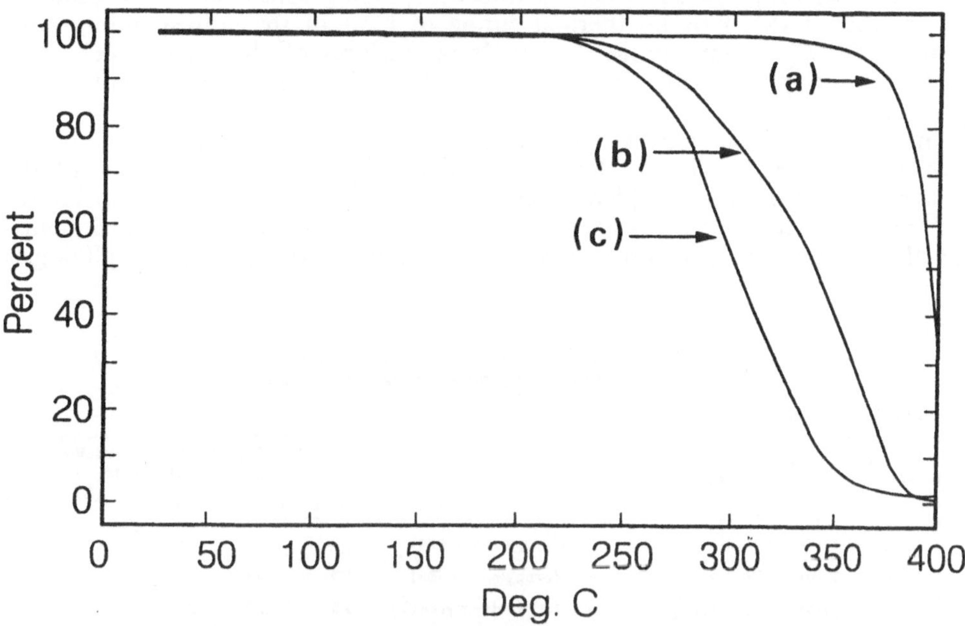

Figure 4. TGA study of poly(propylene oxide) (Mw = 8.5K) heating rate 10°/min; (a) air, (b) house nitrogen, (c) high purity nitrogen.

Triblock systems incorporating varying amounts of the thermally labile blocks were produced by condensation polymerization as shown in Scheme 1 by adjusting the reagent stoichiometries. The properties of a variety of the triblock copolymers derived from PMDA-3FDA are shown in Table 2. The wt.% of the labile block in the copolymers was determined both by ^{1}H NMR and by TGA analysis. Again, the two techniques yielded complementary results. The actual amount of the labile block incorporated into the copolymer was always

TABLE 1. Properties of the functionalized thermally labile blocks used to prepare the triblock polyimides.

Sample Entry	Thermally Labile Block Type	Polymerization Method	Molecular Weight, g/mol	T_g (°C)
1a	Poly(α-methylstyrene)	anionic	12,000	155
1b	Poly(styrene)	anionic	14,000	100
1c	Styrene/α-methylstyrene copolymer	(4/1) anionic	14,000	100
1d	Polystyrene	free radical	13,000	100
1e	Poly(propylene oxide)	anionic	5,600	-65

slightly lower than expected based on the reagent charge. Volume fractions were calculated based on the wt.% incorporation level and the respective experimental densities of PMDA-3FDA and the corresponding labile blocks. Films could be prepared either by directly spinning solutions of the imidized copolymers (NMP) or by thermal curing of films of the corresponding amic esters (300 °C) under nitrogen. No decomposition of the labile blocks was detected during the curing process (TGA, ^1H NMR). Phase separation in the triblock derivatives was indicated by differential scanning calorimetry and dynamic mechanical analysis, which yielded T_g values for the labile blocks which were very similar to those reported in Table 1. The T_g value of the matrix polyimide was considerably higher than the decomposition temperatures of the respective labile blocks as expected. Labile block content was maintained in the 15−30 wt.% regime to assure that noninterconnecting morphologies predominated.

TABLE 2. Properties of the triblock polyimide derivatives containing thermally labile polymer blocks.

Polymer Entry	Polyimide Type and Form	Thermally Labile Block Type	Thermally Labile Block Composition, wt. %			Volume Fraction Labile Block, (%)
			Charge	Incorporated ^1H NMR	Incorporated TGA	
2	PMDA/3FDA (imide)	Poly(propylene oxide)	15	9.9	9	11
3	PMDA/3FDA (imide)	Poly(propylene oxide)	25	23	22	27
4	PMDA/3FDA (imide)	Poly(α-methylstyrene)	25	20	24	27
5	PMDA/3FDA (alkyl ester)	Poly(α-methylstyrene)	25	--	24	27
6	PMDA/3FDA (alkyl ester)	Poly(styrene)	20	18	19	24
7	PMDA/3FDA (alkyl ester)	α-methylstyrene/styrene	20	15	14	18

Films of the triblock samples could be prepared by heating to 300 °C under nitrogen to either drive off residual solvent (NMP) from the fully imidized materials or to cure the poly(amic esters). The respective labile blocks were stable under the curing conditions. The samples containing PO were initially heated in argon or nitrogen during curing to prevent premature decomposition of the PO block. Subsequent decomposition of the labile blocks in the foaming process was initiated using a thermal protocol which was selected according to the labile block incorporated. For those polymers containing either styrene, α-methylstyrene or a copolymer mixture of these two monomers, a stepwise heating procedure terminating around 350 °C was employed. Following the heating period, decomposition of the thermally labile block was confirmed by ^{1}H NMR TGA analysis, and IR analysis. For cured copolymer samples containing PO labile blocks, the initial heating *in air* at 240 °C for 6 h was followed by 2 h at 300 °C to assure complete decomposition. TGA analysis of a triblock polymer containing ~ 15 wt.% of PO is shown in Figure 5. A weight loss equivalent to the weight of the labile block was observed. Densities of the foamed polymers were measured, whenever possible, using a density gradient column containing $Ca(NO_3)_2$ solutions. The column was calibrated with spheres of known densities. Foam porosities were also studied by small angle x-ray scattering [30] and by calibrated IR absorption [34]. The data obtained for a variety of PMDA-3FDA foams are shown in Table 3. In some cases, the void volume percentage was less than predicted by the volume percentage of the labile block in the initial triblock polymer. In these cases, it is possible that there is some collapse of the smaller voids due to capillary pressure during the foaming process. The significant decreases observed in the sample densities after processing are consistent with the generation of air-filled voids. Density studies on the foamed samples which originally contained α-methylstyrene labile blocks suggested that the pores were much larger and possibly interconnected. Samples of these polymers sunk continuously in the density gradient column suggesting liquid uptake by the sample was occurring.

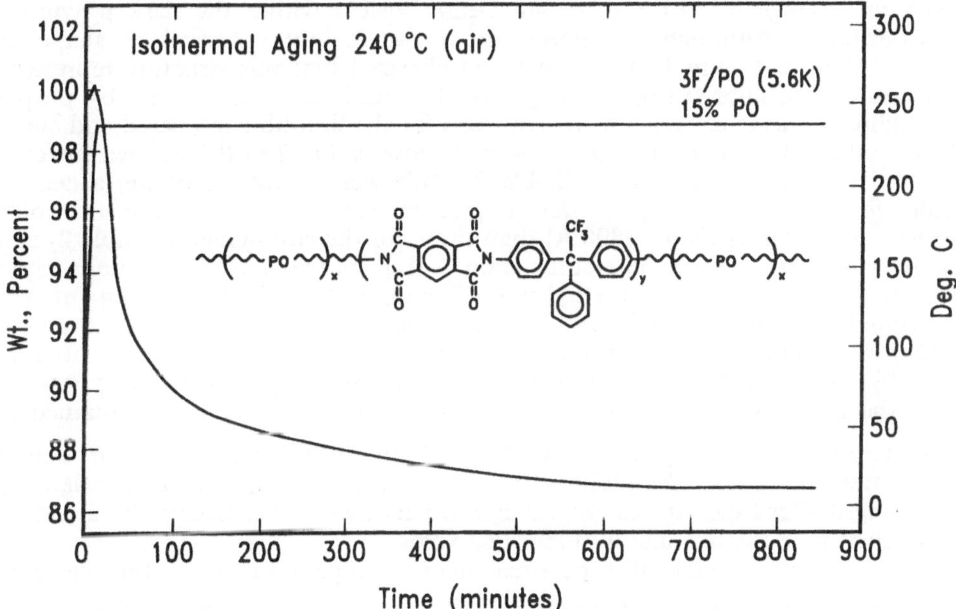

Figure 5. Weight loss in the triblock copolymer, PMDA-3FDA-PO (15 wt.%) upon heating in air at 240 °C.

TABLE 3. Foams prepared from a variety of PMDA-3FDA triblock polyimides. The entry numbers correspond to the copolymers listed in Table 2. The PO containing copolymers (entries 2 and 3) were foamed in air at 240 °C for 4 h and then heated to 300 °C for 2 h. The other samples were heated to 325 – 350 °C in nitrogen in the foaming process.

Sample Entry	Initial Labile Block Composition, Vol. %	Foam Density (g/cm^3)	Volume Fraction of Voids (Porosity), %
PMDA/3FDA Polyimide	--	1.35	--
2	11	1.17	13
3	27	1.11	18
4	27	1.13	16
5	27	Interconnecting Pores	30
6	24	1.17	14
7	18	Interconnecting Pores	19

Convincing evidence for foam formation from the triblock systems was obtained by transmission electron microscopy (TEM) studies. Figure 6 shows a TEM picture of the foam produced from PMDA-3FDA-PO (23 wt.%) (copolymer 3, Table 2). Noninterconnecting and randomly distributed pores with an average size of ~ 60 Å are clearly visible within the dark polyimide background. Although the pores are obviously not spherical in shape as suggested from Figure 1, it should be emphasized that this structure represents an idealized equilibrium morphology which would be obtained only by proper annealing. This is clearly not the situation for the thin films prepared and cured as described. On the other hand, Figure 7 shows a TEM of the foamed material derived from the copolymer 5 (Table 2). We were unable to obtain a density value for this sample using the density gradient column. In this case, the voids appear much larger (200 – 1800 Å) than those for the copolymer 3 (Table 2) and show evidence of some interconnection. It is possible that the more rapid decomposition of poly(α-methylstyrene) and/or the slow diffusion of the monomer fragments could lead to some plasticization of the matrix and a "blowing" effect leading to expanded pores. Small angle x-ray scattering (SAXS) studies [29] on PMDA-3FDA-PO before and after foaming suggest that the morphology present in the starting copolymer is largely maintained in the foam produced. Figure 8 shows the scattering profile ($Q = \frac{4\pi}{\lambda} \sin\theta$) before and after foaming. Although the scattering vector (Q) remains relatively unchanged after foaming, the scattering intensity increases dramatically, a feature consistent with the formation of air-filled voids.

The refractive index of a polymer gives a rough measure of the dielectric constant ($\varepsilon \sim n^2$ at optical frequencies). The measured refractive indices of the foamed materials are always substantially lower than for the matrix homopolymer as expected for a material of significantly lower density. For the foamed sample of copolymer 3 (Table 2), $n_{TE} = 1.46$ and $n_{TM} = 1.42$ were measured at 633 nm. The optical anisotropy (n_{TE}/n_{TM}) in the foam was only

Figure 6. A TEM picture of the triblock copolymer 3 (Table 2) after heating first at 240 °C, then at 300 °C in air. The light areas represent the thermally generated voids in the polyimide matrix (dark).

slightly larger (1.03 vs. 1.01) than in the fully cured homopolymer. These measurements suggest that the respective dielectric constants will be similarly isotropic. Preliminary dielectric studies indicate that the dielectric constant of the polyimide nanofoams may be empirically approximated by the expression [$\varepsilon =$ (refractive index)$^2 + 0.2$]. While the homopolymer PMDA-3FDA has a measured dry dielectric constant (out-of-plane) of ~2.8, that of the foam derived from PMDA-3FDA (23% PO) was only 2.30, a significant drop consistent with expectations.

In summary, phase separated block copolymers are capable of self-assembling into a variety of interesting polymer morphologies. The nature of these structures is a function of polymer composition. This characteristic could lead to a variety of interesting self-assembled structures with potential applications. We have demonstrated one such application by the conversion of phase-separated polyimide triblock copolymer nanostructures into low dielectric nanofoams. As expected, the morphologies present in the cured triblock copolymers were also maintained during the foaming process.

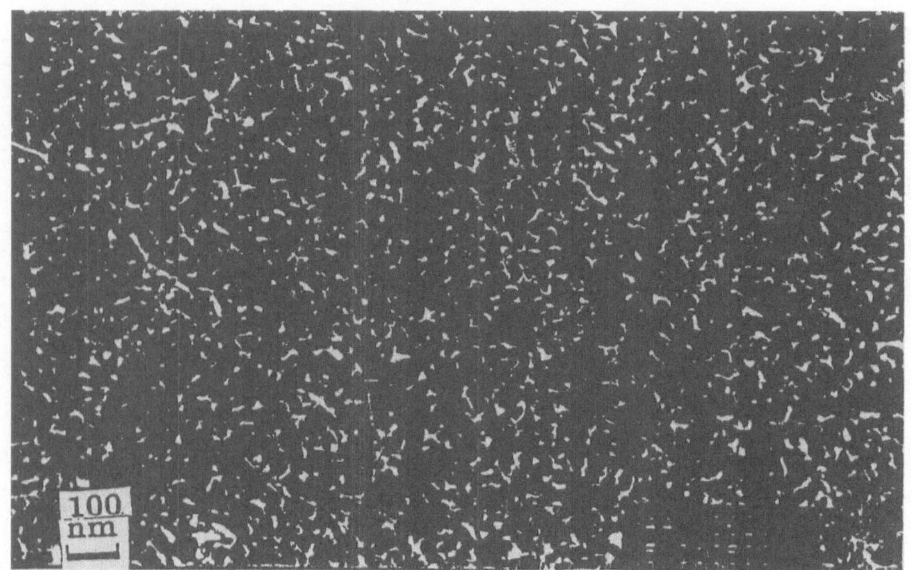

Figure 7. A TEM picture of the triblock copolymer 5 (Table 2) foamed by heating to 350 °C. The light areas are voids in the polymer matrix.

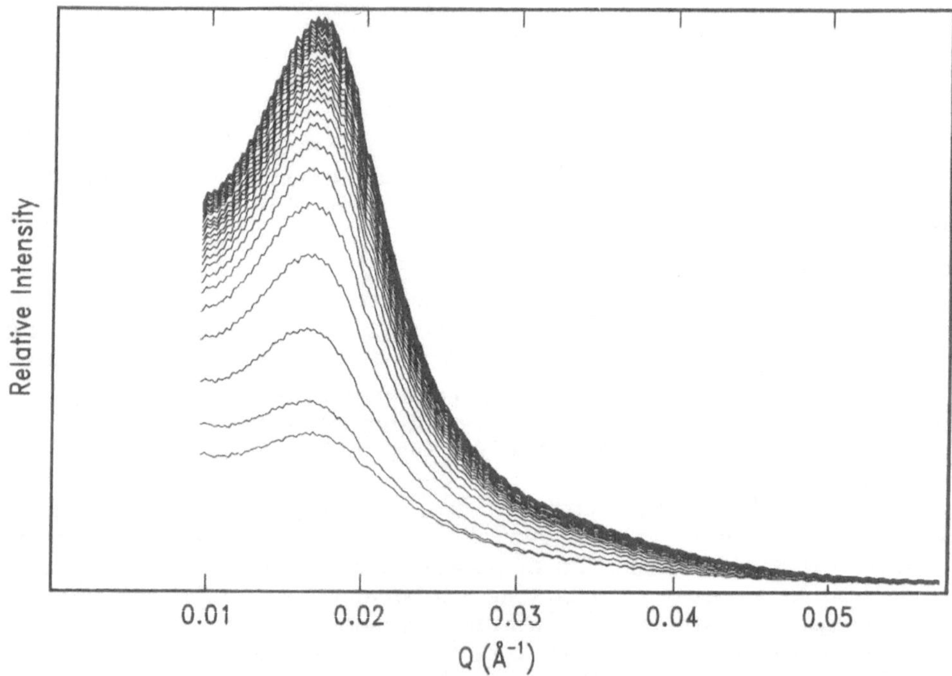

Figure 8. Small angle x-ray scattering of the copolymer PMDA-3FDA-PO as a function of the scattering vector Q($Å^{-1}$) upon heating from 60 – 250 °C in air after initially curing the sample to 300 °C under nitrogen.

3. References

1. Bates, F.S. and Fredrickson, G.H. (1990) Block copolymer thermodynamics: Theory and experiment, *Ann. Rev. Phys. Chem.* **41**, 525.
2. Leibler, L. and Fredrickson, G.H. (1995) The shape of polymers to come, *Chem. Brit.* **31(1)**, 42.
3. Stevens, M.D. (1990) *Reactions of Vinyl Polymers in Polymer Chemistry*, 2nd edition, Oxford University Press, Oxford, Chap. 9.
4. Lee, J.-S., Hirao, A. and Nakahama, S. (1989) Polymerization of monomers containing functional silyl groups: Porous membranes with controlled microstructures, *Macromolecules* **22**, 2602.
5. Canham, L.T. (1990) Silicon quantum wire array fabrication by electrochemical and chemical dissolution of wafers, *Appl. Phys. Lett.* **57**, 1046.
6. Steigerwald, M.L. and Brus, L.E. (1989) Synthesis, stabilization and electronic structure of quantum semiconductor nanoclusters, *Ann. Rev. Mater. Sci.* **19**, 471.
7. Efros, A.L. (1982) Interband absorption of light in a semiconductor sphere, *Sov. Phys. Semicond.* **16**, 772.
8. Sankaran, V., Yue, J., Cohen, R.E., Schrock, R.R. and Silbey, R.J. (1993) Synthesis of zinc sulfide clusters and zinc particles within microphase-separated domains of organometallic block copolymers, *Chem. Mater.* **5**, 1133.
9. Krausch, G. (1995) Surface induced self assemblies in thin polymer films, *Mater. Sci. Eng.* **B14(1)**, 1 and references cited therein.
10. Folkes, M.J., Keller, A. and Scalisi, F.P. (1973) Extrusion technique for the preparation of single crystals of block copolymers, *Colloid Polym. Sci.* **251**, 1.
11. Hadziioannou, G., Mathis, A. and Skoulios, A. (1979) Single crystals of styrene-isoprene-styrene three block copolymers with a cylindrical structure, I. Orientation study by small angle x-ray diffraction, *Colloid Polym. Sci.* **257**, 15.
12. Morrison, F.A. and Winter, H.H. (1989) Effect of unidirectional shear on the structure of triblock copolymers 1: Polystyrene-polybutadiene-polystyrene, *Macromolecules* **22**, 3533.
13. Almdal, K., Koppi, K.A., Bates, F.S. and Mortensen, K. (1992) Multiple ordered phases in block copolymer melts, *Macromolecules* **25**, 1743.
14. Albalah, R.J. and Thomas, E.L. (1993) Microphase separation of block copolymer solutions in a flow field, *J. Polym. Sci., Part B: Polymer Physics* **31**, 37.
15. Albalah, R.J. and Thomas, E.L. (1994) Roll-casting of block copolymers and of block copolymers homopolymer blends, *J. Polym. Sci., Part B: Polymer Physics* **32**, 341.
16. Morrison, F.A., Winter, H.H., Gronski, W. and Barnes, J.D. (1990) Effect of unidirectional shear on the structure of triblock copolymers 2: Polystyrene-polybutadiene-polystyrene, *Macromolecules* **23**, 4200.
17. Amundson, K., Helfand, E., Davis, D.D., Quan, X., Patel, S.S. and Smith, S.D. (1991) Effect of an electric field on block copolymer microstructure, *Macromolecules* **24**, 6546.
18. Aubert, J.H. and Sylvester, A.P. (April 1991) Microcellular foams: For what, *Chemtech*, p. 234.

19. Aubert, J.H. and Sylvester, A.P. (May 1991) Microcellular foams: Here's how, *Chemtech*, p. 290.
20. Tummala, R.R., Keyes, R.W., Grobman, W.D. and Kapur, S. (1989) Thin film dielectrics, in R. Tummala and E.J. Rymaszewski (eds.), *Microelectronics Packaging Handbook*, Van Nostrand Reinhold, New York, Chap. 9.
21. *Polymers in Microelectronics: Fundamentals and Applications*, D.S. Soane and Z. Mantynenko (eds.) (1989), Elsevier, Amsterdam.
22. Monk, D.J. and Soane, D.S. (1993) Interconnect Dielectrics, in C. P. Wong (ed.), *Polymers for Electronic and Photonic Applications*, Academic Press, Inc., New York, 119ff.
23. Hrubesh, L.W., Keene, L.E. and Latorre, V.R. (1993) Dielectric properties of aerogels, *J. Mater. Res.* **8(7)**, 1736.
24. Hedrick, J.L., Labadie, J.W., Russell, T.P., Hofer, D.C. and Wakharkar, V. (1993) High temperature polymer foams, *Polymer* **34**, 4717.
25. Volksen, W.M. (1994) Condensation polyimides: Synthesis, solution behavior and imidization characteristics, in *Advances in Polymer Science: High Performance Polymers*, Springer-Verlag, Berlin, Vol.117, 111ff.
26. Hedrick, J.L., Hawker, C.J., Di Pietro, R., Jerome, R. and Charlier, Y. (1995) The use of styrenic copolymers to generate polyimide nanofoams, *Polymer* **36(25)**, 4855.
27. Hedrick, J.L., Russell, T.P., Hawker, C., Sanchez, M., Carter, K., Di Pietro, R. and Jerome, R. (1995) Polyimide nanofoams prepared from styrenic block copolymers, *Microelectronics Technology: ACS Symposium Series* **614**, 425.
28. Hedrick, J.L., Di Pietro, R., Charlier, Y. and Jerome, R. (1995) Polyimide foams derived from poly(4,4′-oxydiphenyl pyromellitimide) and poly(α-methylstyrene), *High Perform. Polym.* **7**, 133.
29. Hedrick, J.L., Russell, T.P., Labadie, J., Lucas, M. and Swanson, S. (1995) High temperature nanofoams derived from rigid and semirigid polyimides, *Polymer* **36(14)**, 2685.
30. Charlier, Y., Hedrick, J.L., Russell, T.P., Jonas, A. and Volksen, W. (1995) High temperature polymer nanofoams based on amorphous, high Tg polyimides, *Polymer* **36(5)**, 987.
31. Hawker, C.J. and Hedrick, J.L. (1995) Accurate control of chain ends by a novel "living" free radical polymerization, *Macromolecules* **28**, 2993.
32. Georges, M.K., Veregin, R.P.N., Kazmaier, P.M. and Hamer, G.K. (1994) Taming the free radical polymerization process, *Trends Polym. Sci.* **2(2)**, 66.
33. Hawker, C.J. (1994) Molecular weight control by a "living" free radical polymerization process, *J. Am. Chem. Soc.* **116**, 11, 185.
34. Sanchez, M.I., Hedrick, J.L. and Russell, T.P. (1995) Nanofoam porosity by infrared spectroscopy, *J. Polym. Sci., Part B: Polym. Phys.* **33**, 253.

CONTROL OF THE MOLECULAR IN-PLANE ORIENTATION IN LANGMUIR-BLODGETT FILMS BY SHEARING

C. MINGOTAUD AND C. JEGO
Centre de Recherche Paul Pascal, Avenue A. Schweitzer,
F-33600 Pessac, France

ABSTRACT. Shearing of a Langmuir film by a rotating disk can induce orientation of molecules at the gas-water interface. Monolayers with such local anisotropy may be transferred onto a solid substrate by the Langmuir-Blodgett (LB) technique and give multilayer films which exhibit a preferred molecular orientation within the plane of the multilayers. This approach allows one, for the first time, to control in-plane anisotropy of Langmuir and LB films and could be a new way to build ordered architecture from molecular bricks using the LB technique.

1. Introduction

The Langmuir-Blodgett (LB) technique [1,2] has been widely used to generate systems exhibiting a high degree of 2D character. This LB method allows one to build new ordered molecular assemblies with specific architecture from modular components (ions, molecules, brickstone aggregates, etc). A large variety of molecules and polymers have been assembled into such well-organized multilayer films [3] which provide interesting physical or chemical properties in areas such as molecular electronics, optical devices and sensors. A classical anisotropy is found in the film between out-of-plane and in-plane properties. On the other hand, in-plane properties of the multilayers are often considered to be totally isotropic. For various physical properties and especially electrical conductivity, the lack of organization within the LB plane is a serious drawback which limits the usefulness of new materials obtained in this manner. In the research on LB multilayers, one of the new challenges is then to induce and control ordering of molecules within the plane of LB films. As shown in the following scheme, this new ordering process should lead to LB films where the molecules have a preferential direction instead of the original isotropic distribution.

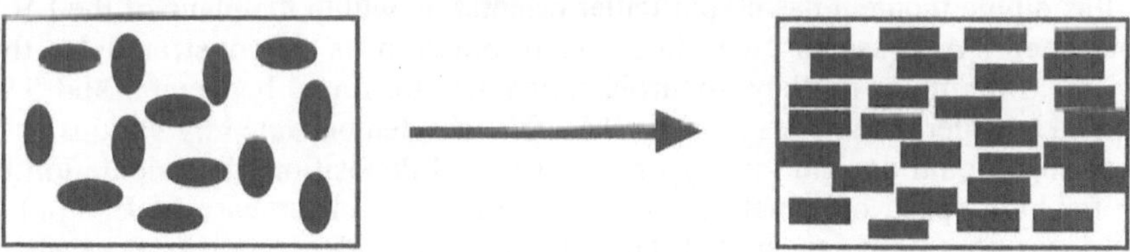

From this supramolecular organization, one expects to improve the properties of the material or to create new functionality thanks to this new ordering.

J. Michl (ed.), Modular Chemistry, 543–550.

2. Orientation due to the transfer

In-plane organization has been observed for various compounds in LB films[4-6]. Spontaneous orientation seems to occur during transfer of a Langmuir film onto a solid substrate for specific monolayers. Flow orientation models [7,8] have been proposed to explain this orientation during the transfer which then leads to anisotropic LB films. These models suggest that monolayer flow due to the transfer from the gas-water interface onto solid substrate induces orientation of the anisotropic molecules or aggregates of such compounds. Indeed, this phenomenon is mainly observed for polymers having a rod-like shape or for self-aggregating molecules (e.g., phthalocyanines). In particular, a substituted polyacrylate (**PA**) bearing "side-on fixed" mesogenic groups [9] (see Figure 1) presents this specific behavior in LB films and has been studied in detail.

PA

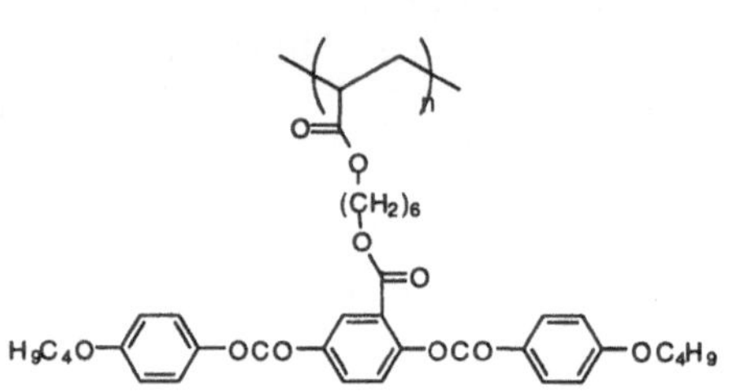

$(CH_2)_6$

H_9C_4O—⟨⟩—OCO—⟨⟩—OCO—⟨⟩—OC_4H_9

Figure 1. Chemical structure of the polyacrylate used in this study.

After transfer of twenty layers onto a solid substrate, the molecular orientation of this polymer has been analyzed by infrared dichroism. A dichroic ratio, ρ, can be defined for a particular IR peak by the relation

$$\rho = A_{\perp} / A_{//}$$

where $A_{\perp}$ and $A_{//}$ are respectively the absorption of the LB films when the light polarization is set perpendicular or parallel to the dipping direction. Using this dichroic ratio, one can define an order parameter, P_2, as:

$$P_2 = (1-\rho)/(1+\rho).$$

This parameter is equal to the average of $\cos(2\theta)$ over the distribution of molecular orientation, where θ is the angle of the dipole moment associated with the analyzed peak and the dipping direction [10,11]. P_2 is equal to 1 if all dipole moments are aligned with the dipping direction, equal to -1 if they are all perpendicular to this direction and null if this dipole moment has no particular orientation within the plane of the LB film. The polymer **PA** presents clear in-plane orientation as demonstrated by the IR dichroism. Maximum in-plane dichroic ratios $1/\rho$ are found between 5 and 5.9 and correspond to order parameters close to 0.6 - 0.7. As demonstrated by various authors, this ordering should depend mainly on the speed of deposition, the orientation being higher for high speed of transfer. To verify this point in the case of **PA** polymer, various depositions were done with different speeds. The order parameter associated with the 1605 cm^{-1} IR peak related to the main axis of the mesogenic core is plotted against the dipping speed in Figure 2. It increases quickly with the dipping speed and reaches a plateau corresponding to a maximum value of P_2. When the dipping speed is high enough, the orientation of the molecules or aggregates is then quasi-total and any increase of the speed does not change the order. The plateau in the plot of P_2 parameter corresponds to this saturation in the ordering. Generally, the range of dipping speeds one can use to transfer a given Langmuir film efficiently is very restricted.

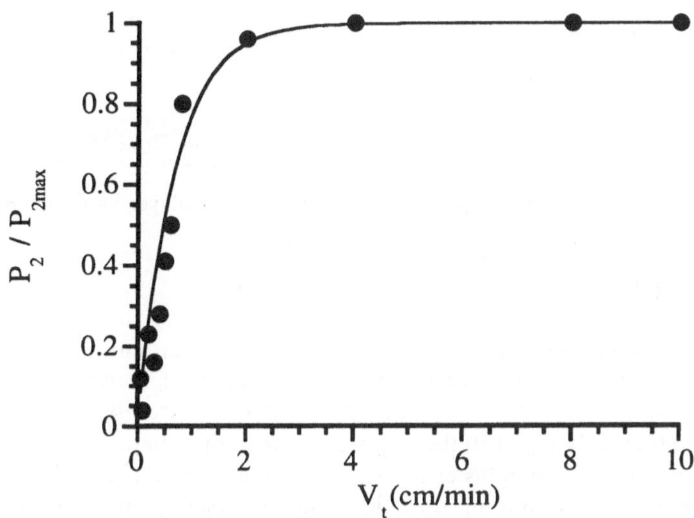

Figure 2. Dependence of the in-plane P_2 order parameter (associated with the 1605 cm^{-1} IR peak) in **PA** LB film with the deposition speed V_t.

Thus, the control of this in-plane orientation through a change of the dipping speed (or substrate size) is somewhat difficult. Fortunately, this control could be achieved by shearing as explained below.

3. Shearing of a monolayer by a rotating disk

3.1. EXPERIMENTAL SET-UP

A LB trough [12] was equipped with a rotating disk (made of Teflon, diameter 5 cm, height 10 mm) which can be put in contact with the monolayer (see Figure 3). Rotation speed and position of the disk compared to the substrate are mechanically set. The depth of the disk within the subphase is adjustable and generally set to 1 mm. Spreading, compression and deposition of the monolayer are carried out in a regular manner [13,14]. Thirty minutes before the transfer and during deposition, the disk is set in rotation. Transfer ratio and quality of the LB films seem to be unaffected by the shearing, even for high rotation speeds (>1.5 rps) where ripples are clearly produced at the interface by the disk rotation.

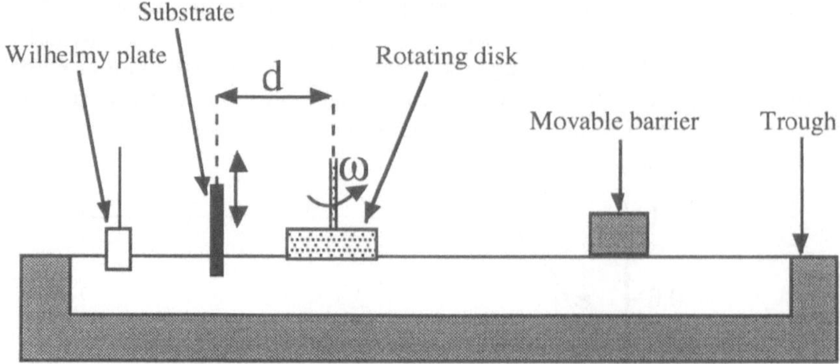

Figure 3. Experimental set-up for the shearing of a monolayer by a rotating disk.

3.2. CHANGE IN ORIENTATION WITH SHEARING

As shown in Figure 4, the shearing of the monolayer during deposition induces a decrease of the order parameter, P_2. For a small distance, d, between the substrate and the disk axis or for high rotation speed, ω, the parameter P_2 reaches a constant negative value of ca -0.6 exactly equal to the opposite of the initial value. The decrease in P_2 versus rotation speed is much faster for a small distance, d. The **PA** orientation, which is mainly parallel to the dipping direction without shearing, is then changed by the shearing stress; the preferential alignment under high shearing is perpendicular to the dipping direction. One should point out that shearing of the monolayer modifies the main orientation of the molecules within the LB film but not the anisotropy itself. The P_2 parameters are indeed equal in absolute value for high or null shearing.

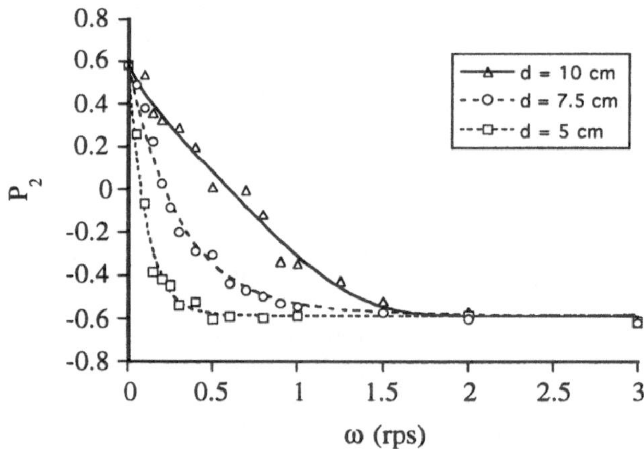

Figure 4. Order parameter P_2 in **PA** LB films obtained for various rotation speeds, ω, and distances, d, between the disk axis and the substrate.

This orientation process can be easily explained as shown in Figure 5. Without shearing, the flow due to the monolayer transfer from the interface to the substrate induces a preferential orientation of the molecules of elongated shape. These compounds should then lie close to the substrate with the long axis approximately parallel (see Figure 5a). On solid substrate, the molecules can then be found somewhat parallel to the dipping direction (see Figure 5b).

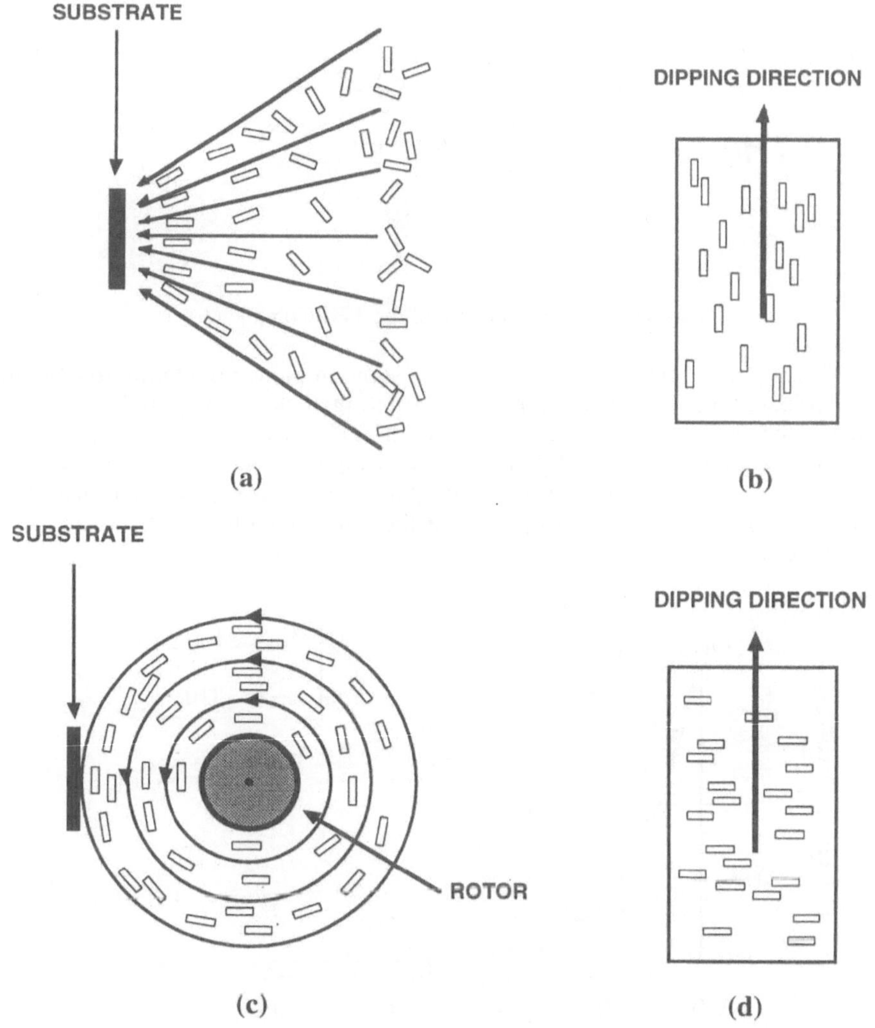

Figure 5. Scheme of the molecular orientation at the interface and in the LB films with or without shearing (see text for details).

When the rotating disk is in action, shearing forces the orientation of the molecules at the gas-water interface. The long axis of the molecules should be parallel to the flow stream, i.e., perpendicular to the orientation without shearing (see Figure

5c). If the flow due to the transfer is negligible compared to the shearing flow, the alignment at the gas-water interface will be maintained during deposition. The main axis of the molecules lies perpendicular to the dipping direction (see Figure 5d).

This orientation by shearing is clearly a general method to organize molecules at the gas-water interface. Indeed, comparable results were obtained for the **THP** compound (Figure 6) and similar change of preferential alignment in LB film was observed under shearing stress.

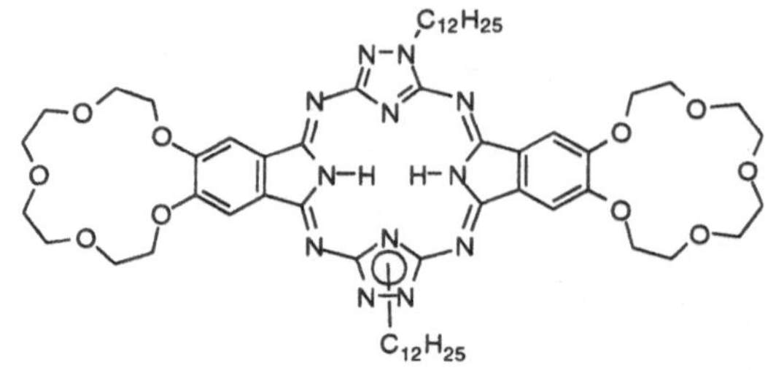

Figure 6. Chemical structure of the **THP** compound.

This suggests that all molecules having some in-plane orientation in LB films could be oriented in Langmuir films by a rotating disk. Compared to the **PA** experiments, the decrease of the order parameter in **THP** LB films versus rotation speed is very slow (see Figure 7). Smaller distance, d, or higher rotation speed, ω, had to be used in the case of **THP** to reach a given P_2 parameter. This qualitative difference should be related to the intrinsic properties of the Langmuir film and in particular its surface viscosity.

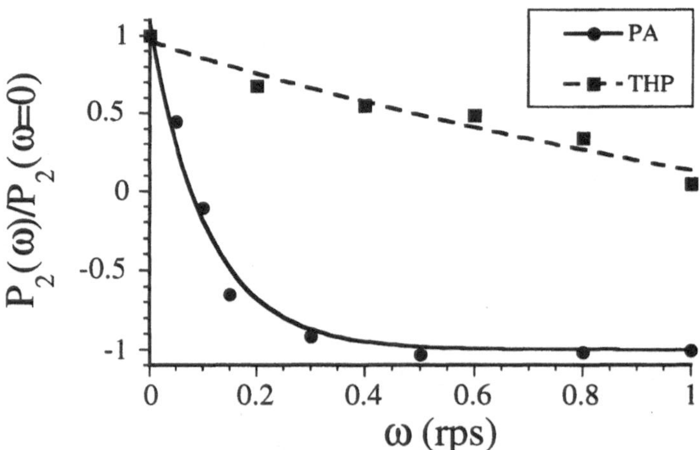

Figure 7. Change of the order parameter with rotation speed ω of the disk for **PA** or **THP** LB films for a distance d of 5 cm.

When neither of the two orientation processes (by shearing flow or by deposition flow) alone controls the alignment of the **PA** polymer, the P_2 parameter does not reach ± 0.6. Such values of P_2 could correspond either to a loss of anisotropy in the LB film due to the competition between the two processes or to a preferred orientation lying between the two main axes of the substrate (i.e., parallel and perpendicular to the dipping direction). In the case of **PA** molecules, more precise information about molecular orientation was obtained by recording the IR spectra of the LB films for different angles between the dipping direction and the polarized IR light. When the transfer is carried out without shearing, the maximum absorption is obtained for an electromagnetic field parallel to the dipping direction (see Figure 8).

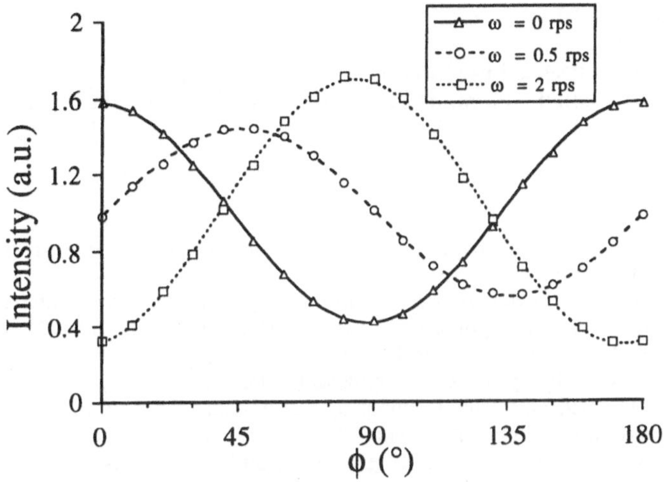

Figure 8. Intensity of the 1162 cm^{-1} IR peak of **PA** LB films obtained for a distance d of 7.5 cm and different rotation speeds ω. The intensity was measured for various angles, φ, between the polarized IR light and the dipping direction.

This demonstrates again that the mesogenic groups are preferentially oriented parallel to the dipping direction. For a LB film obtained with a high shearing (ω = 2 rps in Figure 8) and corresponding to a P_2 parameter equal to ca -0.6, the maximum absorption is observed when the light polarization is perpendicular to the dipping direction; the molecules are then mainly oriented around this axis. For an intermediate case (i.e., P_2 close to zero for ω = 0.5 rps in Figure 8), the intensity maximum is found around 45°, proving that molecules are still distributed around a preferential direction. This clearly indicates that all the anisotropy is not lost when the two orientation processes are in competition (even if this anisotropy is slightly decreased for such intermediate orientation as shown by the lower amplitude of the corresponding curve in Figure 8). The main molecular orientation within the LB films can then be set to various directions just by changing the rotation speed. For the **THP** compound, similar behavior has been observed. However, the lost of anisotropy for an intermediate orientation is much more important than in the case of **PA** monolayer.

4. Conclusion

Shearing of a Langmuir film by a rotating disk can induce orientation of molecules at the gas-water interface. Monolayers with such local anisotropy may be transferred onto a solid substrate by the Langmuir-Blodgett technique and give multilayer films which exhibit a preferred molecular orientation within the plane of the multilayers. This orientation depends mainly on the rotation speed and the distance between substrate and disk. This shearing enables control of in-plane ordering of molecules in LB films.

5. Acknowledgment

The authors are very grateful to Dr. N. Leroux for a kind gift of the **PA** polymer and to Pr. T. Torres (Universitad Autonoma de Madrid, Spain) for the **THP** compound.

6. References

(1) Gaines, G. L., Jr. (1966) *Insoluble Monolayers at Liquid-Gas Interfaces*, Interscience Publishers, New York.
(2) Ulman, A. (1991) *Introduction to Ultrathin Organic Films*, Academic Press, Boston.
(3) Mingotaud, A. F., Mingotaud, C. and Patterson, L. K. (1993) *Handbook of Monolayers*, Academic Press, San Diego.
(4) Vandevyver, M., Albouy, P.-A., Mingotaud, C., Perez, J., Barraud, A., Karthaus, O. and Ringsdorf, H. (1993) In-plane anisotropy and phase change in Langmuir-Blodgett films of disklike molecules, *Langmuir* **9**, 1561-1567.
(5) Penner, T. L., Schildkraut, J. S., Ringsdorf, H. and Schuster, A. (1991) Oriented films from polymeric amphiphiles with mesogenic groups: Langmuir-Blodgett liquid crystals, *Macromolecules* **24**, 1041-1049.
(6) Sauer, T., Arndt, T., Batchelder, D., Kalachev, A. A. and Wegner, G. (1990) The structure of Langmuir-Blodgett films from substituted phthalocyaninato-polysiloxanes, *Thin Solid Films* **187**, 357-374.
(7) Minari, N., Ikegami, K., Kuroda, S., Saito, K., Saito, M. and Sugi, M. (1988) Analytical model of flow dichroism in Langmuir-Blodgett Films, *Solid State Commun.* **65**, 1259-1262.
(8) Schwiegk, S., Vahlenkamp, T., Xu, Y. and Wegner, G. (1992) Origin of orientation phenomena observed in layered Langmuir-Blodgett structures of hairy-rod polymers, *Macromolecules* **1992**, 2513-2525.
(9) Leroux, N. (1993) Thesis, Université de Bordeaux I.
(10) Chollet, P.-A. and Messier, J. (1983) IR dichroism of anisotropic Langmuir-Blodgett multilayers, *Thin Solid Films* **99**, 197-204.
(11) Vandevyver, M., Barraud, A., Ruaudel-Teixier, A., Maillard, P. and Gianotti, C. (1982) Structure of porphyrin multilayers obtained by the Langmuir-Blodgett technique, *J. Colloid Interface Sci.* **85**, 571-585.
(12) ATEMETA, 35 Bd Anatole France, 93200 St Denis, FRANCE.
(13) Jego, C., Leroux, N., Agricole, B., Mauzac, M. and Mingotaud, C. (1994) Mesogenic pyrrole derivatives in Langmuir-Blodgett films, *J. Phys. Chem.* **98**, 13408-13413.
(14) Jego, C., Leroux, N., Agricole, B. and Mingotaud, C. (1996) Side-on liquid crystal polyacrylate in Langmuir-Blodgett films, *Liq. Cryst.* **20**, 691-696.

RODS, RINGS, BALLS AND STRINGS!

Structural Motifs in Carborane Chemistry

M. D. MORTIMER, W. JIANG, Z. ZHENG, R. R. KANE, I. T. CHIZHEVSKY, C. B. KNOBLER and M. F. HAWTHORNE

Department of Chemistry and Biochemistry, 405 Hilgard Avenue, University of California at Los Angeles, Los Angeles CA 90095-1569

Introduction

The three icosahedral carboranes, *ortho*, *meta* and *para* (Figure 1), have many desirable features as modular building blocks. All three isomers, although relatively expensive (the *para*-isomer especially), are commercially available and have an unlimited shelf life due to their remarkable thermal and chemical stability. Although carboranes are usually regarded as the province of inorganic chemists, their reactivity is firmly in the organic domain.

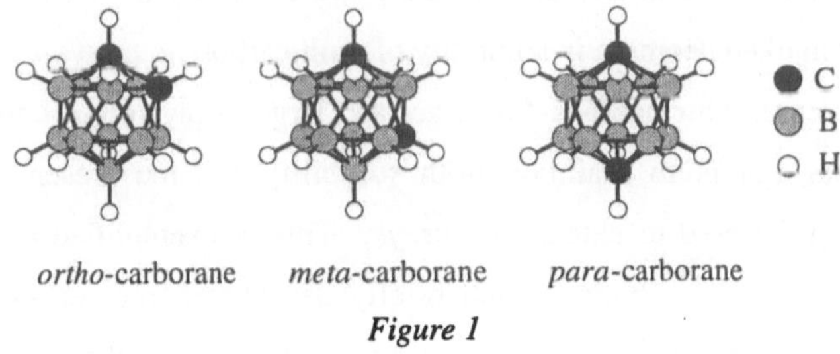

Figure 1

J. Michl (ed.), Modular Chemistry, 551–564.

The rigid three-dimensional icosahedron of the carborane isomers holds substituents in well-defined spatial relationships, and there are a number of relatively facile routes to substituted icosahedra. The two moderately acidic CH vertices are easily deprotonated and functionalized using organometallic reagents, and the ten BH vertices have a chemistry in many ways reminiscent of electrophilic substitution on aromatic rings. Most reactions at the boron vertices thus do not affect the carbon vertices, and *vice versa.* Due to the differing electron-withdrawing power of carbon and boron and the extremely delocalized molecular orbitals, there is a strong dipole moment in the *ortho* and *meta* isomers. *ortho*-Carborane has a dipole moment of 4.31 D, comparable with that of the HCl molecule. This has implications for the attachment of substituents to the boron vertices. Electrophiles are strongly directed to the boron vertices furthest away from the carbon atoms, while nucleophiles preferentially attack at the boron vertices closest to the carbon atoms. In *para*-carborane, of course, all the boron vertices are equivalent and the attachment of more than one substituent invariably leads to mixtures of isomers.

Most transformations maintain the integrity of the underlying geometry, with the notable exception of reaction with strong base, which allows the removal of one boron vertex adjacent to the carbon vertices in *ortho* or *meta* carborane. The icosahedron can then be "reconstructed" with either another boron atom, which can bear a substituent such as a functionalized alkyl or aryl, or with one of many transition metals or main group elements carrying other ligands. Thus, there is ample opportunity to "tailor-make" carboranes with substituents in one or more positions for use in subsequent reactions.

There are, however, some disadvantages common to all many vertex three-dimensional systems, particularly the limited regioselectivity of reactions at the boron vertices and the marked decrease in solubility of multicarborane arrays as the number of components increase. Our quest is for larger and larger arrays of carborane molecules, and our principal aim is to maintain both solubility and the presence of functional groups which can be used to extend the arrays. This is exemplified by several of the systems discussed in this article, which briefly describes some recent results in our ongoing efforts to build well-defined structural motifs based on carborane icosahedra, and illustrates the principal synthetic methods that we have found useful for controlled linkage and functionalization of carboranes.

Linear Arrays of *para*-Carborane Icosahedra: Carborods

Some time ago it was found that *para*-carborane serves as a useful monomer for the construction of rigid linear molecules by oxidative coupling of CLi groups using copper chloride (Scheme 1)[1]. Rods such as these can be used to study electron transport between donor and acceptor groups, or as rigid spacers between linking groups, which allows one to think in terms of framework assembly, or could be used to form highly organized monolayers on surfaces with thicknesses varying in a controlled fashion. The most important feature is an easily modified terminal group which can be used to attach groups at each end of the rod.

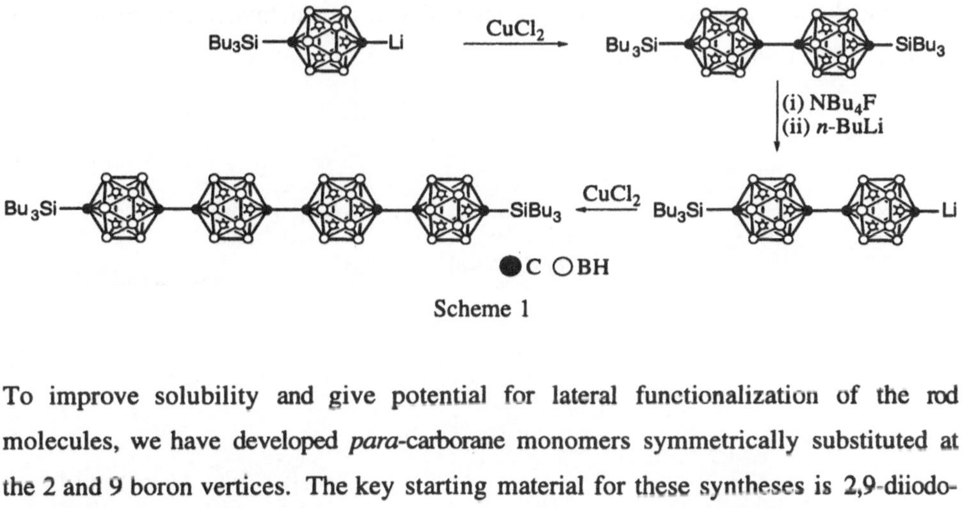

Scheme 1

To improve solubility and give potential for lateral functionalization of the rod molecules, we have developed *para*-carborane monomers symmetrically substituted at the 2 and 9 boron vertices. The key starting material for these syntheses is 2,9-diiodo-*para*-carborane, which is isolated in *ca.* 9 % yield from the reaction of *para*-carborane with two equivalents of iodine monochloride in the presence of a Lewis acid catalyst[2,3], along with the other four possible diiodo-*para*-carborane isomers (the total yield of this reaction approaches 90 %). The remaining isomers, which are difficult to separate, are converted on to deca-B-methyl-*para*-carborane (*vide infra*).

There are two principal reactions which we use to form B-C bonds at the 2 and 9 positions. The first is reaction of the BI vertices with a Grignard reagent in the presence of a palladium/copper catalyst system[3,4], which works well for substituents not carrying β-hydrogens (Scheme 2).

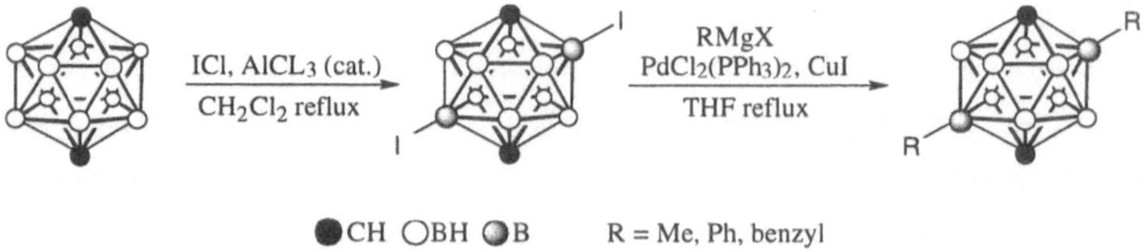

ICl, AlCL3 (cat.)
CH_2Cl_2 reflux
RMgX
$PdCl_2(PPh_3)_2$, CuI
THF reflux

●CH ○BH ◉B R = Me, Ph, benzyl

Scheme 2

The second route to B-substituted *para*-carboranes is reaction of a primary alkyne with a BI vertex using a similar catalyst system (Scheme 3)[3]. This is a route to boron vertices carrying long alkyl chains *via* hydrogenation of the triple bond, which are not accessible through Grignard reagents. Desilylation of a trimethylsilyl acetylene group gives a reactive acetylenic hydrogen.

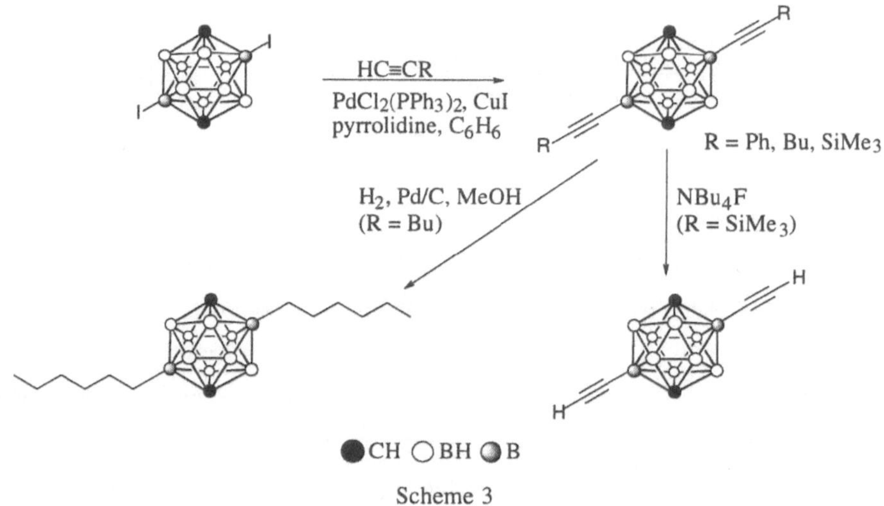

Scheme 3

We have prepared rods up to five monomers in length using 2,9-dimethyl-*para*-carborane and explored a variety of end groups. Use of the recently synthesized 2,9-dihexyl monomer is expected to significantly improve solubility and allow the preparation of much longer rods. We have also begun to explore the use of alkynyl groups as backbone elements in carborods. Using *para*-carborane derivatives with B-acetylene groups, two modes of coupling are available which maintain the linearity needed for rod synthesis. Glaser coupling of the acetylene groups gives a diacetylene linker, while reaction of a second BI vertex with the carboranyl acetylene gives a simple acetylene link (Scheme 4).

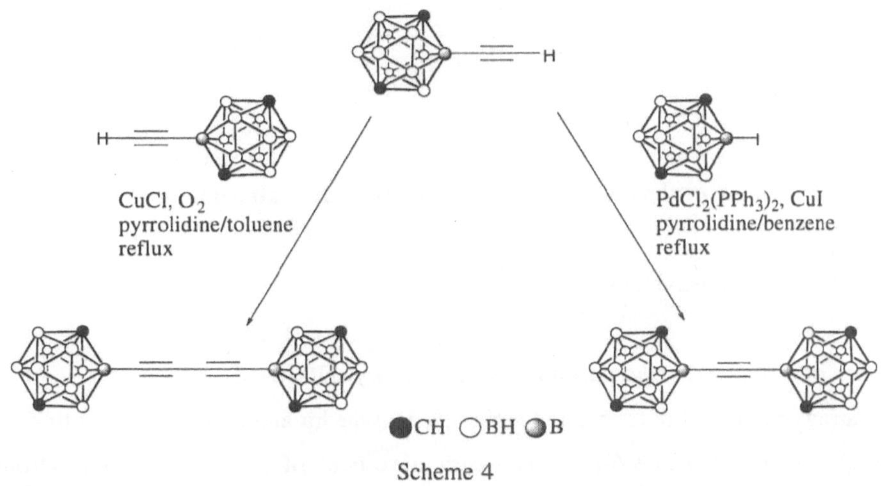

Scheme 4

Using 2,9-diiodo-*para*-carborane, rod-forming reactions can be propagated until the limits of solubility are reached. Our first hybrid rod molecule, shown in Scheme 5, is calculated to be approximately the same length as the tetramer carborod shown in Scheme 1 (discounting end groups), and is soluble in chloroform, whereas without capping silyl groups the tetramer rod is insoluble in any common solvent. Use of the boron vertices to propagate the rod backbone will allow a great deal of flexibility in lateral substitution at the carbon vertices. This will undoubtably be important for attaching solubilizing groups, and brings the possibility of infinitely extenable rigid rods a step closer. More speculatively, it may be possible to carry out rod forming reactions in two dimensions by using both opposite boron vertices and the carbon vertices. Thus the carboranes could act as crossing points in a framework.

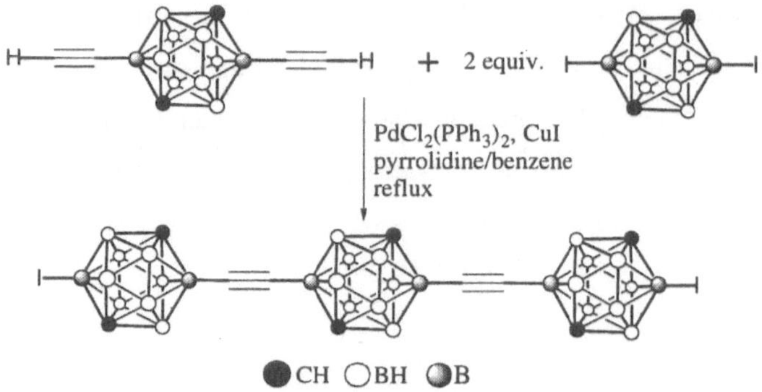

+ 2 equiv.

$PdCl_2(PPh_3)_2$, CuI
pyrrolidine/benzene
reflux

CH BH B

Scheme 5

Cyclic Arrays of *ortho*- and *meta*-Carborane Icosahedra

MERCURACARBORANDS

Reaction of dilithiated *ortho*-carborane with mercury salts results in the self-assembly of cyclic arrays composed of alternating carborane icosahedra and mercury atoms linked by C-Hg-C moieties (Scheme 6)[5]. The carbon vertices of the icosahedra are strongly electron-withdrawing and activate the mercury centers as Lewis acids, enabling the cycles to act as hosts to anionic guests. Anion template effects are found to play an important role in the nature of the products[6], allowing us to form rings of varying sizes. Thus, using a mercury salt with a non-coordinating anion such as acetate in the reaction with dilithio-*ortho*-carborane affords cyclic trimers with no guest. Halide ions promote tetramer formation with one or two centrally coordinated anion guests, and thiocyanate appears to give a pentameric product, as yet not structurally characterized. The halide ions in the tetrameric host can be removed easily with silver acetate leaving an empty host molecule. Substitution of the 9 and 12 boron vertices (those furthest from the carbon vertices) with alkyl groups[7] gives the cycles greater solubility in less polar solvents and allows a certain amount of design flexibility.

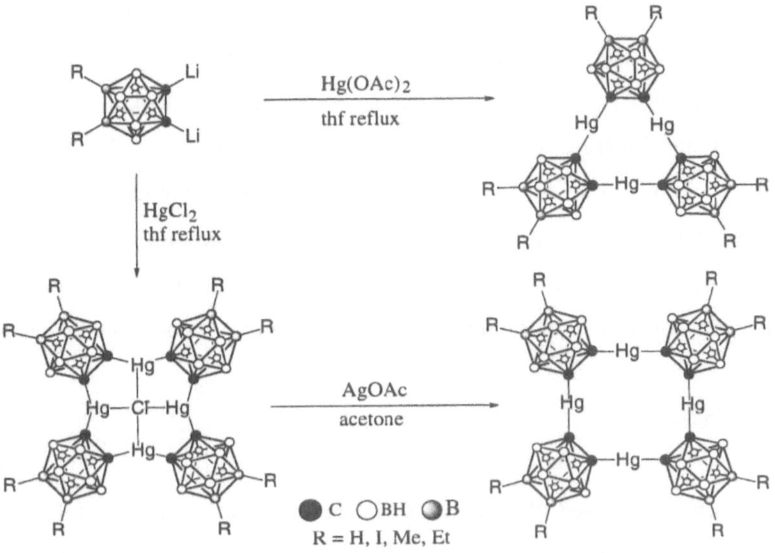

Scheme 6

By a procedure which is synthetically different, but results in a structurally analogous product, it is possible to synthesize an "inside-out" mercuracarborand in which the mercury atoms are bonded to boron vertices and the carbon vertices of the *meta*-carborane icosahedra are directed outwards (Scheme 7)[8].

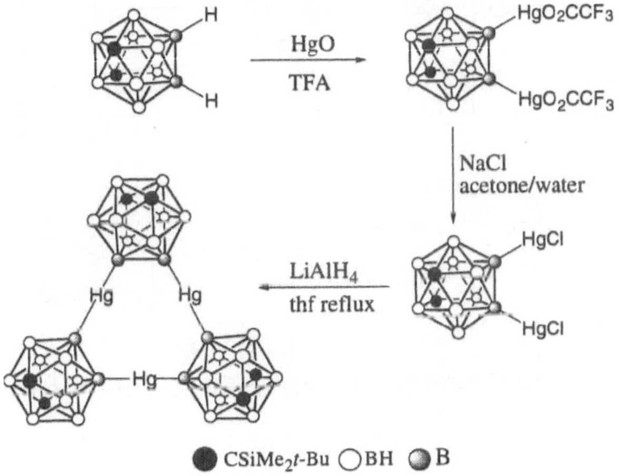

Scheme 7

As discussed in the synthesis of carborods, use of the boron vertices of the icosahedra for skeletal linkage within the array leaves the most easily functionalizable carbon vertices free for further reactions. The use of *meta*-carborane icosahedra also lends an interesting structural feature; substituents on the carbon vertices are above and below the plane of the cycle, whereas substituents on *ortho*-carborane lie in the plane of the cycle. Although structurally similar to the carbon-linked mercuracycles, the electronic properties of the "inside-out" trimer are very different. The boron vertices attached to the mercury atoms are electron donors, and the cycle does not coordinate guest anions.

CARBORACYCLES

Cyclic molecules composed of carboranes alternating with a variety of organic spacer groups may be synthesized by reaction of dilithiated carboranes with bifunctional alkyl or aryl halides[9,10]. Lacking the template-directed assembly and rigidity of the mercuracarborands, carboracycles are synthesized in a modular fashion from monoprotected carboranes. Depending on the linking group and mode of reaction, it is possible to isolate dimeric, trimeric or tetrameric cycles (Scheme 8).

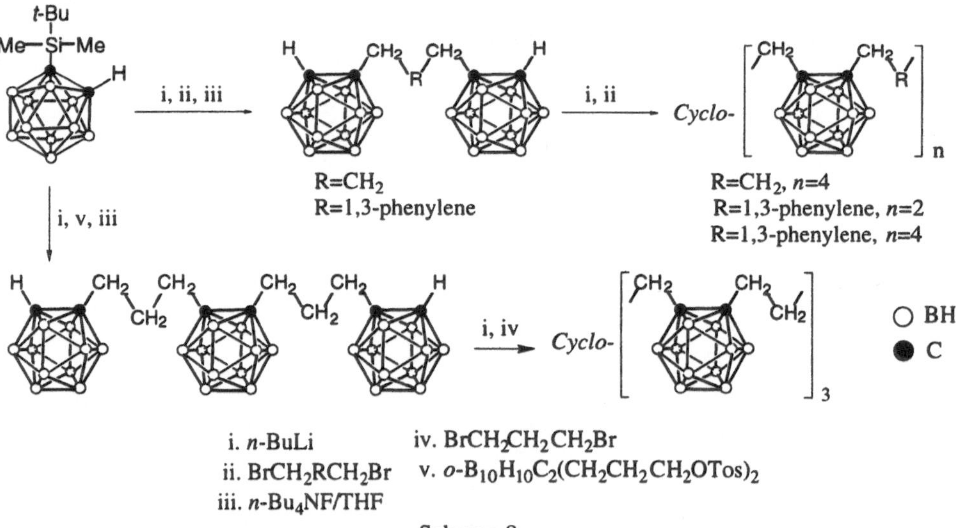

Scheme 8

We have made a wide variety of these "carboracycles" using both *ortho*- and *meta*-carboranes, and with propyl, 1,3-xylyl and 2,6-lutidyl spacer groups. As with other large carborane assemblies, solubility becomes a problem with increasing size, and methyl substitution of selected boron vertices of the icosahedra offers significant advantages in this respect.

Functionalization of carboracycles is possible by two principal routes; either by introduction of a functional group into the linking moiety, as with the introduction of lutidyl groups mentioned above, or by base degradation of the carborane icosahedra to *nido* eleven vertex species as shown in Equation 1. This opens the prospect of ligation to multiple metal centers and a fascinating extension of metallacarborane chemistry.

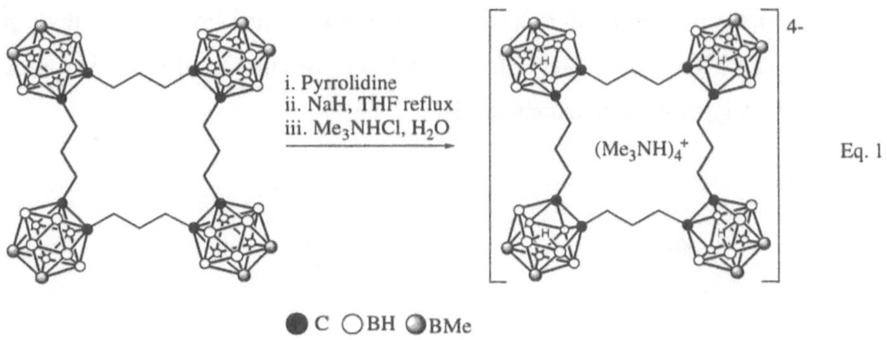

Multifunctionalized Derivatives of *para*-Carborane

In addition to our routes to functionalized carboranes *via* iodocarborane derivatives, we have explored the use of strong electrophilic reagents and catalysts to promote substitution at boron vertices. Electrophilic methylation of *para*-carborane using neat methyl triflate and triflic acid yields a species which is methylated at every boron vertex, deca-B-methyl-*para*-carborane (Scheme 9)[11]. Substituents on one or both carbon vertices of the icosahedra may be introduced prior to the methylation. Thus, it is possible to synthesize the novel "hydrocarbon ball", dodecamethyl-*para*-carborane (Figure 2). This compound has the rigidity and underlying structure of a carborane but presents only alkyl groups to its surroundings.

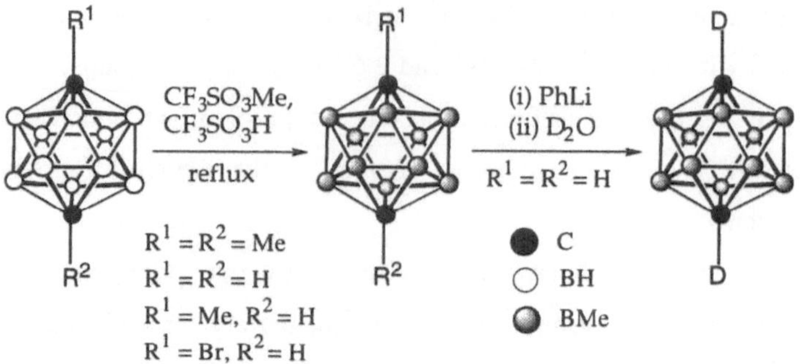

Scheme 9

Deca-B-methyl-*para*-carborane is susceptible to radical chlorination, giving a highly substituted molecule with a multitude of potential reactive sites. Reaction with chlorine gas under UV light introduces two chlorine atoms to each methyl group (Scheme 10 and Figure 3). Steric factors halt the reaction at this stage, allowing isolation of the Cl_{20} product in excellent yield.

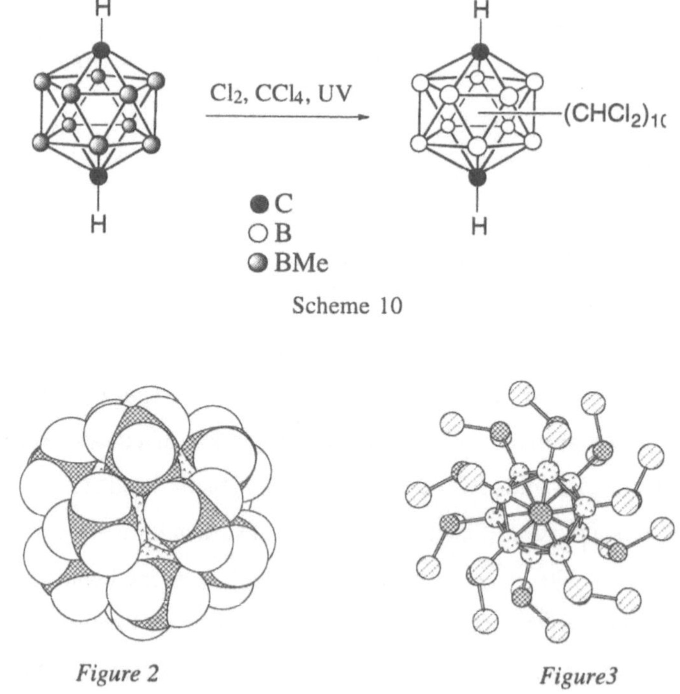

Scheme 10

Figure 2

Figure3

Oligomeric Carborane Phosphate Diesters

As part of our ongoing research into suitable boron-containing compounds for the treatment of cancer using boron neutron capture therapy (BNCT), we have developed methods for making "trailer" molecules containing a high proportion of boron which can be attached to cancer-specific monoclonal antibodies. Oligomeric carborane phosphate diesters can be synthesized quickly, easily and in high yield using established solid-phase techniques for the synthesis of DNA[12]. The starting point is one of many possible carborane-containing diols (Figure 4)[13]; one hydroxyl group of the monomer is protected with a dimethoxytrityl group while the other is activated with a phosphoramidite functionality.

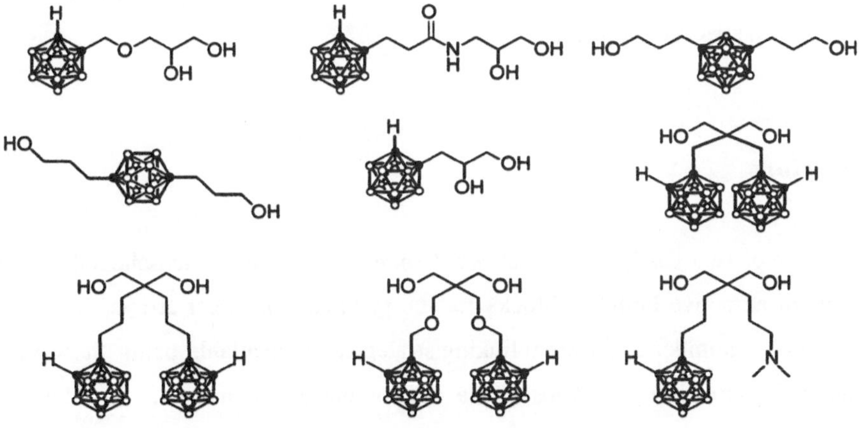

● C ○ BH

Figure 4

These monomers are then introduced directly to an automated DNA synthesizer, where coupling takes place at 98-99 % efficiency for each step (Scheme 11).

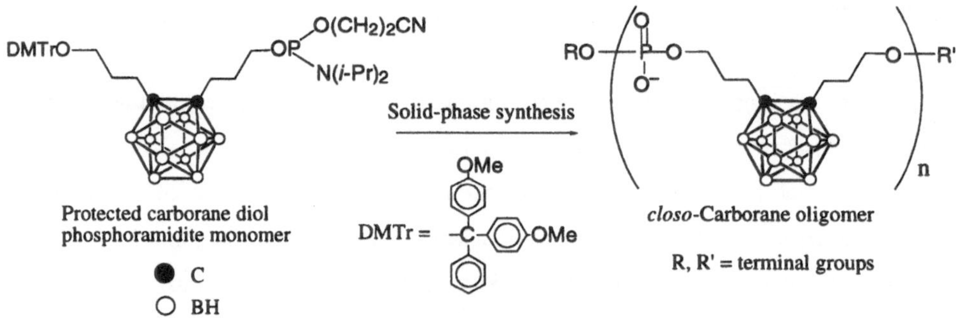

Solid-phase synthesis

Protected carborane diol phosphoramidite monomer

● C
○ BH

DMTr =

closo-Carborane oligomer

R, R' = terminal groups

Scheme 11

We have synthesized a variety of oligomers up to forty monomeric units in length which incorporate *ortho*-, *meta*- and *para*-carboranes either within the oligomer backbone or as pendant groups. In principle, nearly any diol may be included as a monomer for controlled oligophosphate synthesis, allowing the production of carefully tailored oligomers with specific properties.

Conclusions

The presence of two easily functionalized carbon vertices on the icosahedral carboranes makes them attractive building blocks for the synthesis of larger arrays. As we have demonstrated, a number of different linking strategies are available using these vertices, each lending interesting properties to the macromolecular construct. In addition, the boron vertices of the carboranes can be used to adjust the solubilities of the arrays, and we are beginning to extend our capabilities at these positions to more complex functions such as the introduction of reactive groups and the use of boron vertices for skeletal linkage. Further work will concentrate on three main areas; new linking motifs to join carborane subunits, methods to make larger and more complex arrays, and routes to specific properties of the arrays by selective functionalization of the constituent modules.

References

1. (a) Yang, X., Jiang, W., Knobler, C. B., Hawthorne, M.F. (1992) Rigid-rod molecules: carborods. Synthesis of tetrameric *p*-carboranes and the crystal structure of bis(tri-*n*-butylsilyl)tetra-*p*-carborane, *J. Am. Chem. Soc.* **114**, 9719-9721. (b) Muller, J., Base, K., Magnera, T. F., Michl, J. (1992) Rigid-Rod Oligo-*p*-Carboranes For Molecular Tinkertoys - an Inorganic Langmuir-Blodgett Film With a Functionalized Outer Surface, *J. Am. Chem. Soc.* **114**, 9721-9722.
2. Sieckhaus, J.F., Semenuk, N.S., Knowles, T.A., and Schroeder, H. (1969) Icosahedral Carboranes. XIII. Halogenation of *p*-Carborane, *Inorg. Chem.* **8**, 2452-2457.
3. Jiang, W., Knobler, C. B., Curtis, C. E., Mortimer, M. D., Hawthorne, M. F. (1995) Iodination Reactions of Icosahedral *para*-Carborane and the Synthesis of Carborane Derivatives with Boron-Carbon Bonds, *Inorg. Chem.* **34**, 3491-3498.
4. (a) Li, J., Logan, C. F., Jones, M. (1991) Simple Syntheses and Alkylation Reactions Of 3-Iodo-*ortho*-Carborane and 9,12-Diiodo-*ortho*-Carborane, *Inorg. Chem.* **30**, 4866-4868. (b) Zakharkin, L.I., Kovredov, A.I., Ol'shevskaya, V.A., and Shaugumbekova, Zh. H. (1982) Synthesis of B-organo-substituted 1,2-, 1,7-, and 1,12-Dicarba*closo*dodecaboranes(12), *J. Organomet. Chem.* **226**, 217-222.
5. (a)Yang, X., Knobler, C. B., Hawthorne, M. F. (1991) [12]Mercuracarborand-4, the 1st-Representative Of a New Class Of Rigid Macrocyclic Electrophiles - the Chloride Ion Complex Of a Charge-Reversed Analogue Of [12]Crown-4, *Angew. Chem., Int. Ed. Engl.* **30**, 1507-1508. (b) Yang, X., Knobler, C. B., Hawthorne, M. F. (1992) Macrocyclic Lewis acid host - halide ion guest species. Complexes of iodide ion, *J. Am. Chem. Soc.* **114**, 380-382. (c)Yang, X., Johnson, S. E., Khan, S. I., and ., Hawthorne, M. F. (1992) Multidentate Macrocyclic Lewis Acids. Release of "12-Mercuracarborand-4" from Its Iodide Complex and the Structure of Its Tetra(tetrahydrofuran) Dihydrate Complex, *Angew. Chem., Int. Ed. Engl.*, 1992, **31**, 893-896. (d) Yang, X., Zheng, Z., Knobler, C. B., Hawthorne, M. F. (1993) "Anti-crown" chemistry: synthesis of [9]mercuracarborand-3 and the crystal structure of its acetonitrile complexes, *J. Am. Chem. Soc.* **115**, 193-195. (e) Yang, X., Knobler, C. B., Hawthorne, M. F. (1993) Supramolecular chemistry: molecular aggregates of *closo*-$B_{10}H_{10}^{2-}$ with [12]mercuracarborand-4, *J. Am. Chem. Soc.* **115**, 4904-4905. (f) Yang, X., Knobler, C. B., Zheng, Z. Hawthorne, M. F. (1994) Host-Guest Chemistry of a New Class of Macrocyclic Multidentate Lewis Acids Comprised of Carborane-Supported Electrophilic Mercury Centers, *J. Am. Chem. Soc.* **116**, 7142-7159. (g) Zheng, Z., Knobler, C. B., Mortimer, M. D., Kong, G., Hawthorne, M. F. (1996) Hydrocarbon-Soluble Mercuracarborands: Syntheses, Halide Complexes, and Supramolecular Chemistry, *Inorg. Chem.* **35**, 1235-1243.
6. (a) Zheng, Z., Yang, X., Knobler, C. B., Hawthorne, M. F. (1993) An iodide ion complex of a hydrophobic tetraphenyl[12]mercuracarborand-4 having a sterically encumbered cavity, *J. Am. Chem. Soc.* **115**, 5320-5321. (b) Zheng, Z., Knobler, C. B., Hawthorne, M. F. (1995) Stereoselective Anion Template Effects: Syntheses and Molecular Structures of Tetraphenyl[12]Mercuracarborand-4 Complexes of Halide Ions, *J. Am. Chem. Soc.* **117**, 5105-5113.
7. Zheng, Z., Jiang, W., Zinn, A. A., Knobler, C. B., Hawthorne, M. F. (1995) Facile Electrophilic Iodination of Icosahedral Carboranes. Synthesis of Carborane Derivatives with Boron-Carbon Bonds via the Palladium-Catalyzed Reaction of Diiodocarboranes with Grignard Reagents, *Inorg. Chem.* **34**, 2095-2100.

8. Zheng, Z. P., Diaz, M., Knobler, C. B., Hawthorne, M. F. (1995) A Mercuracarborand Characterized By B-Hg-B Bonds - Synthesis and Structure Of Cyclo-$[(t\text{-}BuMe_2Si)_2C_2B_{10}H_8Hg]_3$, *J. Am. Chem. Soc.* **117**, 12338-12339.

9. Clegg, W., Gill, W. R., Macbride, J. A. H., Wade, K. (1993) $(1,7\text{-}C_2B_{10}H_{10}\text{-}1',3'\text{-}C_6H_4)_3$, a Cyclic Trimer From Meta-Carboranediyl and Meta-Phenylene Units - a New Category Of Macrocycle, *Angew. Chem., Int. Ed. Engl.* **32**, 1328-1329.

10. Chizhevsky, I. T., Johnson, S. E., Knobler, C. B., Gomez, F. A., Hawthorne, M. F. (1993) Carboracycles: a family of novel macrocyclic carborane derivatives, *J. Am. Chem. Soc.* **115**, 6981-6982.

11. Jiang, W., Knobler, C. B., Mortimer, M. D., Hawthorne, M. F. (1995) A camouflaged icosahedral carbone: dodecamethyl-1,12-dicarba-*closo*-dodecaborane(12) and related compounds, *Angew. Chem., Int. Ed. Engl.* **34**, 1332-1334.

12. Kane, R. R., Drechsel, K., Hawthorne, M. F. (1993) Automated syntheses of carborane-derived homogeneous oligophosphates: reagents for use in the immunoprotein-mediated boron neutron capture therapy (BNCT) of cancer, *J. Am. Chem. Soc.* **115**, 8853-8854.

13. (a) Drechsel, K., Lee, C. S., Leung, E. W., Kane, R. R., Hawthorne, M. F. (1994) Synthesis of new building blocks for boron-rich oligomers in boron neutron capture therapy (BNCT). I, *Tetrahedron Lett.* **35**, 6217-6220. (b) Kim, Y. S., Kane, R. R., Beno, C. L., Romano, S., Mendez, G., Hawthorne, M. F. (1995) Synthesis of new building blocks for boron-rich oligomers in boron neutron capture therapy (BNCT). II. Monomers derived from 2,2-disubstituted-1,3-diols, *Tetrahedron Lett.* **36**, 5147-5150.

ORGANIZATION OF A NON-AMPHIPHILIC SUPERMOLECULE IN MIXED MONOLAYERS

MARIA A. RAMPI*, CARLO A. BIGNOZZI*, PIERLORENZO CARUSO, DIETMAR MÖBIUS, and FRANCO SCANDOLA*
Max-Planck-Institut für biophysikalische Chemie, Postfach 2841, D-37018 Göttingen, Germany, and
**Dipartimento di Chimica, Università di Ferrara, via L. Borsari, 46, I-44100 Ferrara, Italy*

The metal complex dyad [fac-Re(phen)$(CO)_3$(BPA)Os(terpy)(bpy)]$(PF_6)_3$ has been organized in mixed monolayers at the air-water interface by co-spreading with the water insoluble negatively charged phospholipid dimyristoyl-phosphatidic acid (DMPA). The behavior of the mixed monolayers including the orientation of the dyad depends strongly on the molar mixing ratio of dyad:lipid. The dyad is incorporated in the mixed monolayer up to a limited molar fraction. The mixed monolayers have been transferred to solid substrates, and energy transfer experiments provide additional evidence for the structural model of the mixed dyad:DMPA monolayers.

1. Introduction

For the purpose of creating devices in molecular dimensions highly elaborate supermolecules have been synthesized that combine in a planned way subunits with different functions. Examples can be found in the fields of photoinduced charge separation with supermolecules mimicking the primary process in photosynthesis [1-3] or of light harvesting systems [4]. Future applications of such systems may be expected also in information storage and processing. The main obstacles to the use of these highly sophisticated molecules are the controlled organization of the molecules to large arrays and addressing the single molecule, although modern scanning techniques like scanning tunneling microscopy (STM), scanning force microscopy (SFM) and scanning near field optical microscopy (SNOM) may be able to solve the latter problem.

Various techniques may be used to obtain ordered phases incorporating supermolecules including the formation of organized monolayers

J. Michl (ed.), Modular Chemistry, 565–573.

[5]. We have investigated the possibility to organize a simple two-component supermolecule, the dyad [fac-Re(phen)$(CO)_3$(BPA)Os(terpy)(bpy)] $(PF_6)_3$ (structure see below) in phospholipid monolayers at the gas-water interface. The photophysical behaviour of this molecule has been characterized, and rapid energy transfer takes place from the Re(I) containing moiety to the Os(II) containing moiety [6]. The overall charge of the dyad is 3+, and therefore, the negatively charged lipid dimyristoyl-phosphatidic acid (DMPA) was chosen as lipid anchor. Mixed monolayers of the dyad and DMPA have been prepared at the gas-water interface by the method of co-spreading [7-9], i.e. delivering to the water surface a mixed solution of the two components in an organic volatile solvent. In the same way, organized monolayers of another metal (Rh and Rh) complex dyad have been formed recently [10]. The dyad as well as the two subunits abbreviated as Re and Os are non-amphiphilic and should be kept at the gas-water interface by electrostatic interactions with the lipid anchor. We report here on the behavior of the mixed dyad:lipid monolayers of different composition and propose structural models for the dense-packed monolayers. These mixed monolayers have been transferred to solid substrates and investigated by energy transfer from an excited cyanine dye donor. The organized dyad may be used as a component in designed functional structures of molecular dimensions.

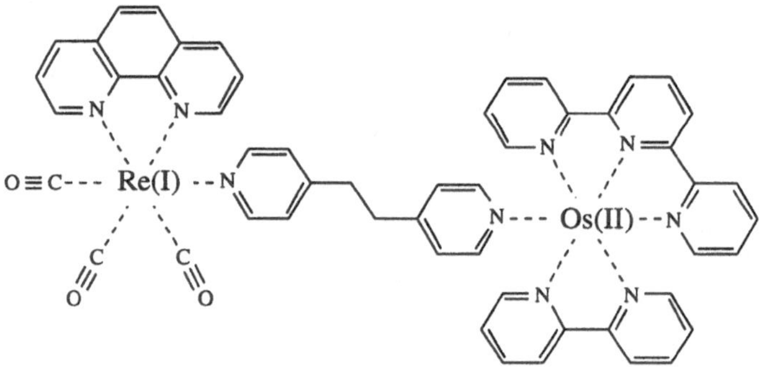

O≡C--- Re(I) --N N--- Os(II)--N

Structure of the dyad

2. Experimental

The synthesis of the dyad and of the subunits [fac-Re(phen)$(CO)_3$(BPA)](PF_6), "Re" and [Os(terpy)(bpy)(BPA)]$(PF_6)_2$, "Os", respectively, will be described elsewhere [6]. Here and in the structure of the dyad, the abbreviations are phen: o-phenanthroline, bpy: 2,2'-bipyridine, terpy: 2,2',2"-terpyridine, and BPA: 1,2-bispyridinyl-ethane. The dimyristoyl-phosphatidic acid (DMPA) was purchased from

Sigma and used without further purification. The stearic acid (for biochemical studies) was obtained from Merck and recrystallized from methanol. Eicosylamine was synthesized by W. Schulten of this laboratory. The cyanine dye dioctadecyl-oxacyanine perchlorate, "S9", was prepared by U. Lehmann of this laboratory. The solvents were chloroform (Baker, HPLC reagent), and methanol (Baker, HPLC reagent). Stearic acid (5 mM) and eicosylamine (1 mM), respectively, were spread from pure chloroform solutions, and the spreading solvent for DMPA (1 mM) and the mixed complex/DMPA solutions was a mixture of chloroform/methanol = 3:1 (v/v). The water used for the subphase was taken from a Milli-Q filtration system (Millipore Corp.). The glass plates (float glass) of dimensions 12x37x1 mm^3 were cleaned as described in a subsequent publication [11].

Surface pressure/area isotherms were measured at 21°C on a circular trough [12] with a Wilhelmy balance provided with a 15 mm wide filter paper. Reflection spectra were measured under normal incidence with a system equivalent to that described previously [13,14]. The reflection ΔR is the difference of the reflectivity of the monolayer covered water surface and that of the clean water surface.

Monolayers were transferred to glass plates at different constant surface pressures depending on the monolayer by vertical dipping of the plates using a circular trough. The first monolayer was eicosylamine. Details will be described elsewhere [11]. For energy transfer experiments we scanned the steady state fluorescence intensity at the maximum of the donor fluorescence band across the plate having a section with (I) and a section without (I_0) the acceptor layer to determine the intensity ratio I/I_0.

3. Results and Discussion

3.1 MONOLAYERS AT THE GAS-WATER INTERFACE

The lipid anchor DMPA forms insoluble monolayers at the gas-water interface as characterized by the surface pressure/area (π/A) isotherm shown in Figure 1, curve a. The isotherm has a plateau-like feature at a surface pressure of about 5 mN/m indicating a phase transition from the liquid to a liquid-crystalline phase. The π/A isotherm of the mixture dyad:DMPA, molar fraction of the dyad, $f_D = 0.44$, is plotted in Figure 1, curve b, as area per lipid molecule. This isotherm differs considerably from that of DMPA, indicating that at least part of the dyad is kept at the interface via interaction with the lipid anchor. The area per lipid molecule is larger in the whole surface pressure range than that of DMPA, and the phase transition is no longer seen.

The Re subunit shows a behavior very similar to that of the dyad, see Figure 1, curve c, measured for a molar fraction of Re, $f_{Re} = 0.5$. The

Os subunit behaves somewhat differently as shown by the π/A isotherm Figure 1, curve d, obtained with a molar fraction of Os, $f_{Os} = 0.5$. At high surface pressure above 40 mN/m the isotherm approaches that of DMPA and merges at 45 mN/m. Nevertheless, all isotherms shown in Figure 1 provide evidence for the presence of the complexes at the interface due to interaction with the lipid anchor.

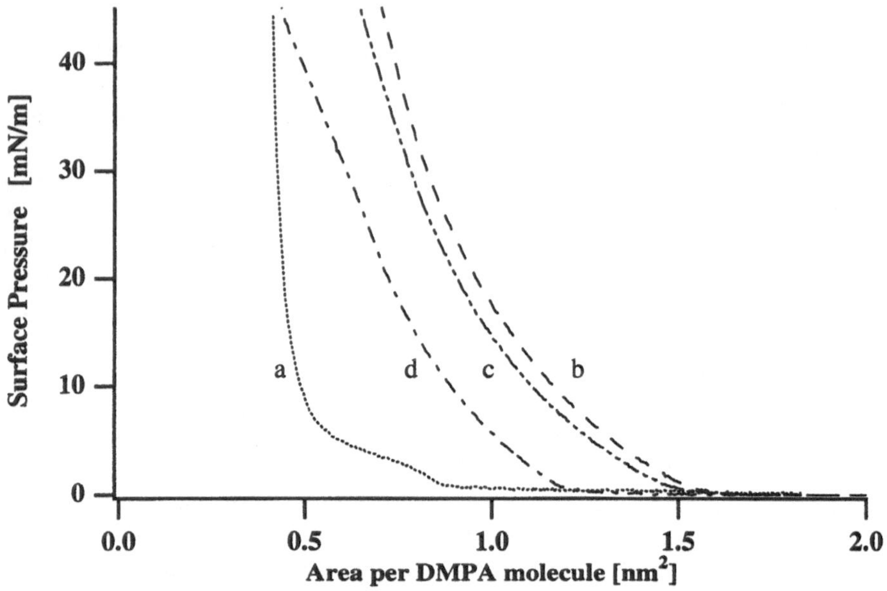

Figure 1. Surface pressure/area isotherms of the lipid DMPA (curve a) and of mixed monolayers of dyad:DMPA, $f_D = 0.44$ (b); Re:DMPA, $f_{Re} = 0.5$ (c); and Os:DMPA, $f_{Os} = 0.5$ (d).

This conclusion is reinforced by measurement of reflection spectra of the mixed monolayers at the gas-water interface. This technique is equivalent to the measurement of absorption spectra and gives a signal from the molecules at the interface only without contribution from molecules present in the bulk subphase. Figure 2 shows the reflection spectra taken at a surface pressure of 45 nmN/m of mixed monolayers of dyad:DMPA, $f_D = 0.5$ (a); Re:DMPA, $f_{Re} = 0.5$ (b); and Os:DMPA, $f_{Os} = 0.5$ (c). The spectra resemble closely the absorption spectra in solution (not shown here) and clearly prove the presence of the dyad and the subunits, respectively, at the gas-water interface. In particular, also the Os subunit is seen, although it does not require additional area with respect to the DMPA matrix at 45 mN/m, and therefore, must be located underneath the lipid head groups. The Re subunit, on the other hand, seems to be inserted between the lipid molecules.

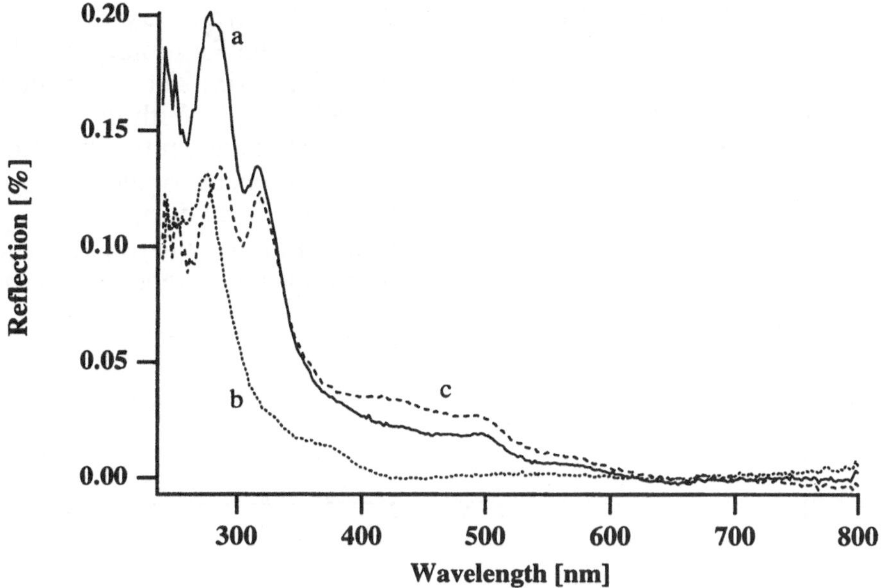

Figure 2. Reflection spectra of mixed monolayers at the gas-water interface π = 45 mN/m; dyad:DPMPA, $f_D = 0.5$ (a); Re:DMPA, $f_{Re} = 0.5$ (b); and Os:DMPA, $f_{Os} =$ 0.5 (c).

The dyad, due to its charge 3+, should require a minimum of 3 lipid molecules with 1- charge each, equivalent to a molar fraction of $f_D = 0.25$. At smaller values of f_D the surplus of lipid has its counterion in the double layer. In a series of π/A isotherms of mixed monolayers of dyad:DMPA with varying f_D (not shown here) an expansion with respect to pure DMPA is seen at small π, however, at a surface pressure of 45 mN/m a different behaviour is observed. The area per lipid molecule, A, is plotted vs. the molar fraction of the dyad in the spreading solution in Figure 3. At small values of f_D the area A increases only moderately up to $f_D \approx 0.1$, then increases stronger between $0.1 < f_D < 0.25$, and A is constant for $f_D \geq$ 0.25.

This behavior is rationalized on the basis of different molecular organization of the monolayer in the different parts of the plot. We assume that the dyad is lying flat underneath the dense-packed head groups of the matrix monolayer at small molar fractions of the dyad. Both parts of the dyad interact electrostatically with the negatively charged head groups of the DMPA matrix. In this arrangement the dyad occupies an area of 3.0 nm^2. If the electrostastic interactions do not markedly influence packing of the DMPA molecules at a surface pressure of 45 mN/m, we do not expect to see an expansion of the matrix monolayer with increasing fraction of the dyad until we reach dense packing of flat lying dyad at $f_D = 0.12$. When the fraction f_D is increased beyond this value, the dyad may resume a different orientation in the mixed monolayer since the Re moiety has the tendency to go into the hydrophobic part of the matrix. The Os moiety remains underneath the negatively charged head groups. With increasing f_D we shift the equilibrium between the two arrangements in the direction of the second until saturation is reached. This should occur for electrostatic reasons at the value $f_D = 0.25$. The line in Figure 3 has been calculated according a model which will be described in detail elsewhere [11] with reasonable parameters in the view of the dimensions and orientation of the dyad.

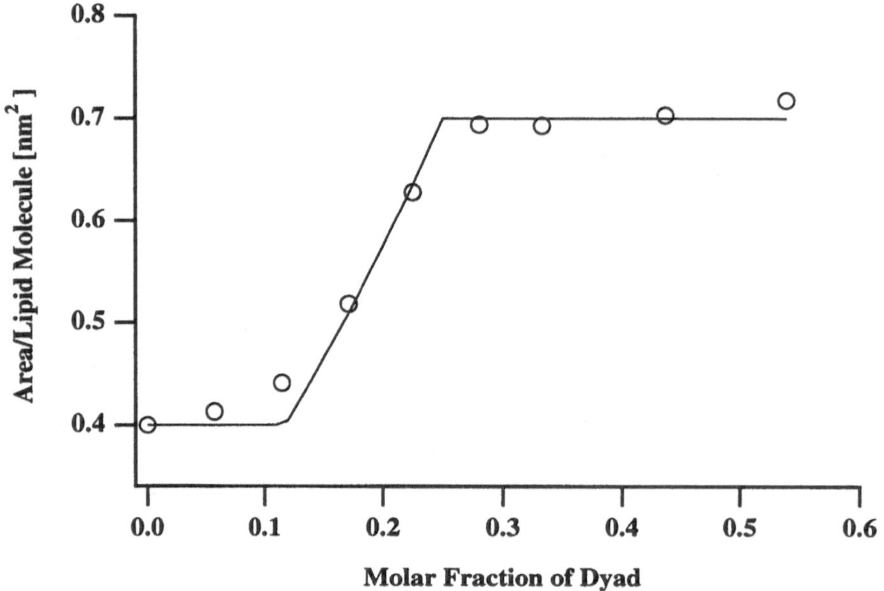

Figure 3. Mixed monolayers dyad:DMPA; area per lipid molecule at a surface pressure of 45 mN/m plotted vs. molar fraction of the dyad.

3.2 MONOLAYER SYSTEMS ON SOLID SUBSTRATES

Additional evidence for this structural model may be gained by energy transfer experiments in monolayer systems using an appropriate energy donor that may transfer its excitation energy to the Os moiety of the dyad only. The dye dioctadecyl-oxacyanine perchlorate, S9, which has been extensively used for energy transfer studies [5] has an emission with maximum around 420 nm. This band overlaps spectrally with the absorption of the Os moiety, a requirement for energy transfer, whereas the Re moiety has no absorption in this range. The Langmuir-Blodgett technique of monolayer transfer was used to construct systems on glass plates with the supermolecules organized in mixed monolayers.

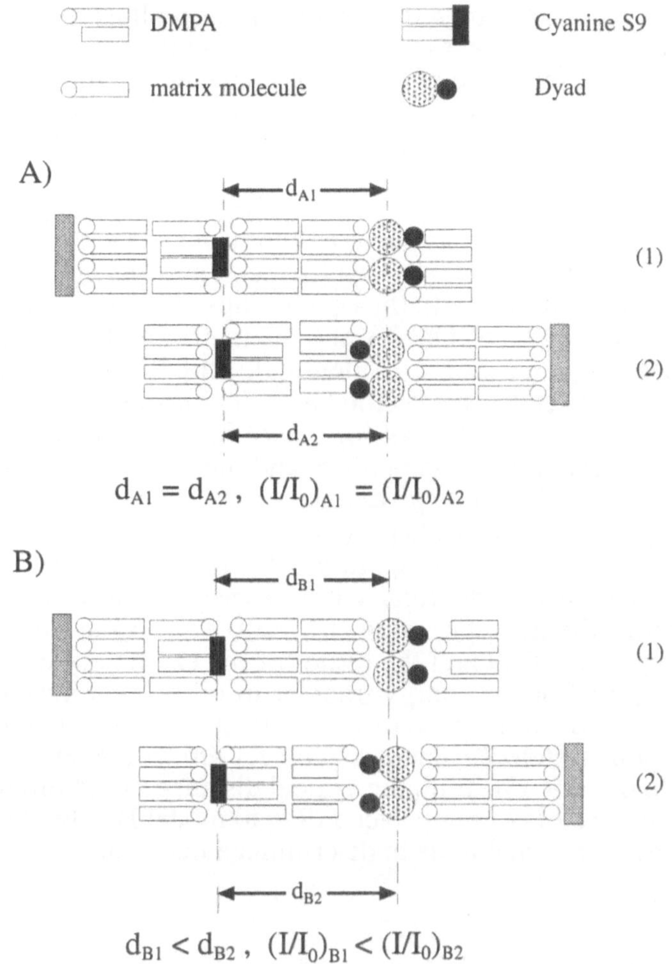

Figure 4. Structure of monolayer systems 1 and 2 (schematic) assembled by variation of the deposition sequence on glass plates for discrimination between the models A and B of organization of the mixed dyad:DMPA monolayer by energy transfer; donor: cyanine S9; acceptor: Os moiety (large disk); the ratio of fluorescence intensities of S9, I in the presence of the dyad, I_o in the absence of the dyad (determined experimentally), $(I/I_o)_1$ and $(I/I_o)_2$, should be identical for model A, but differ for model B.

The question we wanted to address is the location of the Os moiety of the dyad in the dense-packed mixed monolayer. Is the dyad organized such that the Re moiety is sticking between the head groups or in the hydrophobic part of the monolayer and the Os moiety directly underneath the lipid head groups (model A)? Or, alternatively, is the dyad entirely underneath the lipid head groups with the Re moiety in contact with the head groups and the Os moiety spaced away (model B)? From the π/A isotherm data we favor model A, since the Re subunit as well as the dyad show an addititonal area with respect to DMPA whereas the Os subunit does not. By variation of the sequence of monolayer transfer, the monolayer systems 1 and 2 have been assembled whose structures are schematically depicted with the symbols of the constituents in Figure 4 according to the two models A and B.

The chromophore of the cyanine dye (black rectangle) is located below the head groups of its matrix monolayer. The Re moiety (small disk) of the dyad is located between the head groups, and the Os moiety (large disk) directly underneath the head groups. Assuming the same thickness for the monolayers of DMPA and the matrix molecule (stearic acid) the distances d_{A1} and d_{A2} are identical, and the same fluorescence quenching of the cyanine dye by the Os moiety is expected, i.e. $(I/I_o)_{A1} = (I/I_o)_{A2}$. The situation is different for model B, since the distance between the plane of the cyanine chromophores and the Os moiety is increased by the thickness of the Re moiety (diameter approximately 1.13 nm). Consequently, $d_{B1} < d_{B2}$ and $(I/I_o)_{B1} < (I/I_o)_{B2}$.

Experimentally, we find the ratios $(I/I_o)_1 = 0.43 \pm 0.03$ and $(I/I_o)_2 = 0.39 \pm 0.02$, which are identical within experimental error, in favor of model A. From the average value $(I/I_o) = 0.41$ and the distance 5.8 nm between the planes of S9 and the center of the Os moiety in system 1 the critical radius for Förster energy transfer in monolayer systems [5] d_o = 6.4 nm is obtained. In the case of model B, the distance between S9 and the Os moiety in system 2 is increased by 1.1 nm with respect to system 1, and therefore d = 6.9 nm. With the critical distance of 6.4 nm the intensity ratio of $(I/I_o)_2$= 0.57 is expected according to model B. Clearly, the energy transfer experiments enable us to discriminate between the two models of monolayer organization.

4. Conclusion

The dyad of two metal (Re and Os) complex moieties has been successfully incorporated in monolayers using a lipid with negatively charged head group, DMPA, as anchor. The organization of the mixed monolayer depends on the molar fraction and the surface pressure. A structural model for the monolayer at high surface pressure (45 mN/m) with the Re moiety incorporated in the matrix monolayer and the Os moiety located directly underneath the head groups is supported by energy transfer experiments in monolayer systems assembled on glass plates using the Langmuir-Blodgett technique.

5. References

1. Closs, G.L. and Miller, J.R. (1988) Intramolecular Long-Distance Electron Transfer in Organic Molecules, *Science* **240**, 440-447.
2. Gust, D. and Moore, T.A. (1989) Mimicking Photosynthesis, *Science* **244**, 35-41.
3. Warman, J.M., Smit, K.J., Jonker, S.A., Verhoeven, J.W., Oevering, H., Kroon, J., Paddon-Row, M.N., and Oliver, A.M. (1993) Intramolecular Charge Separation and recombination in non-polar environment via long-distance electron transfer through saturated hydrocarbon barriers, *Chem. Phys.* **170**, 359-380.
4. Balzani, V. and Scandola, F. (1991) *Supramolecular Photochemistry*, Ellis Horwood, New York.
5. Kuhn, H. and Möbius, D. (1993) Monolayer Assemblies, in B. W. Rossiter and R. C. Baetzold (eds.), *Investigations of Surfaces and Interfaces*, John Wiley & Sons, Inc., New York, pp. 375-542.
6. Bignozzi, C.A. *manuscript in preparation.*
7. Hada, H., Hanawa, R., Haraguchi, A., and Yonezawa, Y. (1985) Preparation of the J aggregate of cyanine dyes by means of the Langmuir-Blodgett technique, *J. Phys. Chem.* **89**, 560-562.
8. Yonezawa, Y., Möbius, D., and Kuhn, H. (1986) Scheibe-Aggregate Monolayers of Cyanine Dyes without Long Alkyl Chains, *Ber. Bunsenges. Phys. Chem.* **90**, 1183-1188.
9. Ahuja, R.C., Caruso, P.-L., Möbius, D., Wildburg, G., Ringsdorf, H., Philp, D., Preece, J.A., and Stoddart, J.F. (1993) Molecular Organization via Ionic Interactions at Interfaces. 1. Monolayers and LB Films of Cyclic Bisbipyridinium Tetracations and Dimyristoylphosphatidic Acid, *Langmuir* **9**, 1534-1544.
10. Huesmann, H., Bignozzi, C.A., Indelli, M.T., Pavanin, L., Rampi, M.A., and Möbius, D. (1996) Organization of a metal complex dyad in monolayers, *Thin Solid Films* **284-285**, 62-65
11. Caruso, P.-L., Bignozzi, C.A., Rampi, M.A., Möbius, D., and Scandola, F. Organization of Non-Amphiphilic Metal Complexes in Monolayers by Co-Spreading with Lipid Anchor, *manuscript in preparation.*
12. Fromherz, P. (1975) Instrumentation for handling monomolecular films at an air-water interface, *Rev. Sci. Instrum.* **46**, 1380-1385.
13. Grüniger, H., Möbius, D., and Meyer, H. (1983) Enhanced light reflection by dye monolayers at the air-water interface, *J. Chem. Phys.* **79**, 3701-3710.
14. Orrit, M., Möbius, D., Lehmann, U., and Meyer, H. (1986) Reflection and transmission of light by dye monolayers, *J. Chem. Phys.* **85**, 4966-4979.

A MODULAR APPROACH TO LARGE FUNCTIONAL STRUCTURES

DAVID N. REINHOUDT,* DMITRY M. RUDKEVICH,
WILHELM T. S. HUCK, PAUL J. A. KENIS, REMKO H. VREEKAMP
*Laboratory of Organic Chemistry, University of Twente,
P.O. Box 217, 7500 AE Enschede, The Netherlands*

ABSTRACT. In this review a modular approach to design of functional (receptor) molecules, nanostructures, and molecular materials is described. Covalent combination of medium-sized functional building blocks (i.e., modules) results in highly effective bifunctional receptors, superstructured porphyrins and large hydrophobic surfaces. Noncovalently assembled bifunctional receptor systems and large well-defined aggregates (molecular objects) based on calixarene modules are discussed. Self-assembly of modules into NLO films and of large metallo-spheres into defined aggregates is described.

1. Introduction

Nature constructs receptors by the combination of a limited number of building blocks. Amino acids are combined to proteins, nucleotides to DNA and RNA, and monosaccharides to carbohydrates. Systematic variation of the monomer sequence gives access to an almost infinite number of functional bioreceptors. This is achieved at the expense of a high molecular weight which may be even larger when the biologically active species is composed of several sub-units that are connected by non-covalent bonds.

Most synthetic receptors have been prepared by *de novo* synthesis using modern synthetic methodologies, which allow an almost unlimited variation. This strategy focuses on the complementarity of functional groups of receptor and guest. The drawback is that for each individual guest molecule a new synthetic pathway needs to be developed.

Over the past years we have developed a new strategy for the synthesis of artificial receptors which is a compromise between the two extremes described above. We use medium-sized, relatively rigid molecules (i.e., *modules*) to which functional groups can be attached. Calix[4]- and -[6]arenes, resorcin[4]arenes, cyclodextrins, and porphyrins have the required structures for such building blocks. Previously we have covalently combined calix[4]arene with polyethylene glycols and with terphenyls to give calixcrown ethers and calixspherands, respectively [1a,b]. Calix[4]arene connected to cyclodextrins gives water-soluble receptors for organic substrates [1c]; metallosalens immobilized on a calixarene platform exhibit anion binding properties [2]. More recently, we have studied (covalent

J. Michl (ed.), Modular Chemistry, 575–585.

and noncovalent) combinations of different calixarenes with metallosalens, (metallo)porphyrins, and resorcin[4]arenes.

In this paper we describe three different approaches to combining modules into functional (receptor) molecules, nanostructures and molecular materials. The first approach is based on *covalent* combination of modules to (multi)functional molecules. The second approach deals with *noncovalent* (self-)assembly of modules into functional systems and/or well-defined molecular aggregates of nanosize dimensions. Finally, we demonstrate *self-organization* of modules into highly functional films and (nano)surfaces.

2. Covalently Linked Modules

For bifunctional recognition of inorganic salts we have covalently attached both cation and anion binding sites to the rigid lipophilic calix[4]arene platform. Bifunctional receptor **1** containing a UO_2-salen center for anion recognition and crown ether moiety for cation complexation has been prepared and used as a carrier to investigate the transport of hydrophilic salts such as cesium chloride across a supported liquid membrane [2].

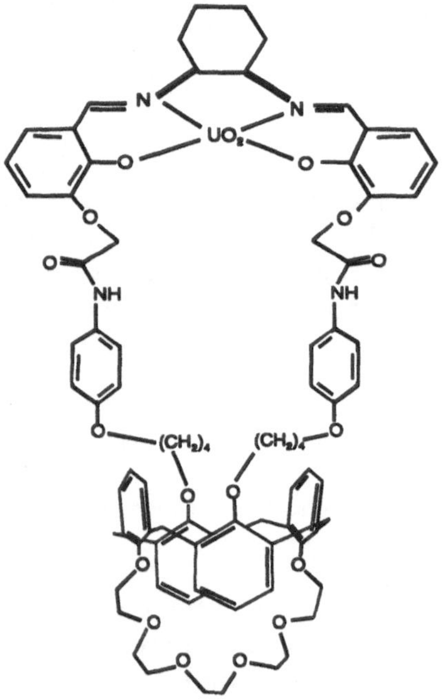

1 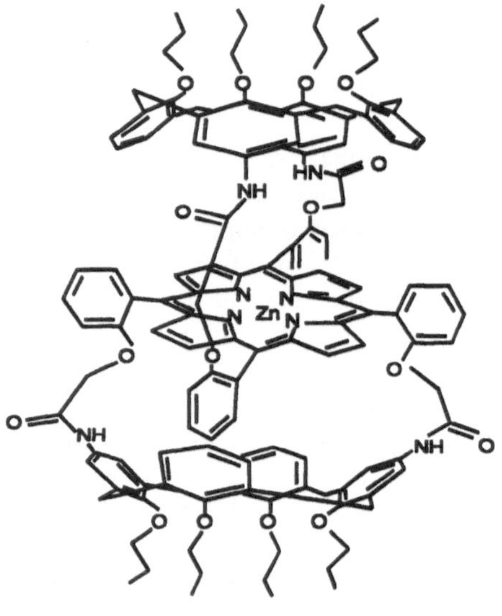 **2**

Multifunctional metalloreceptors have been prepared by covalent combination of two calix[4]arenes and one Zn tetraarylporphyrin core [3]. Porphyrins are being used in biomimetic chemistry for oxygen carrying and storage, and for recognition of biologically significant structures such as nucleic bases, amino acids, and anions.

Covalent combination of porphyrin and calix[4]arene building blocks in one preorganized molecular shape leads to receptors with a highly specific complexing ability. "Capped" metalloreceptor **2** in which the active Zn porphyrin fragment is for the first time covalently incorporated into a hydrophobic and rigid "egg-shaped" bis-calix[4]arene cavity has been prepared.

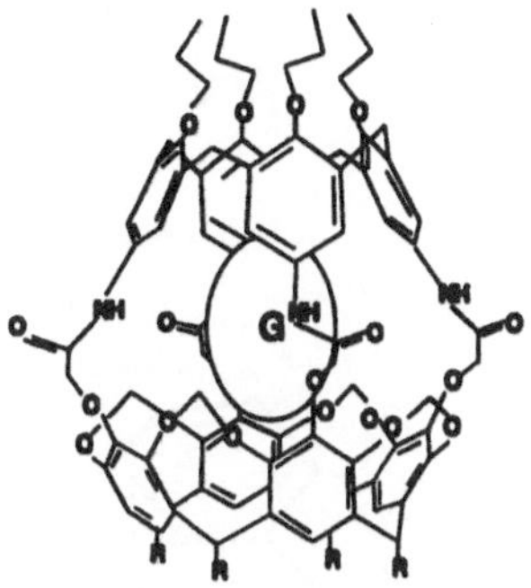

3

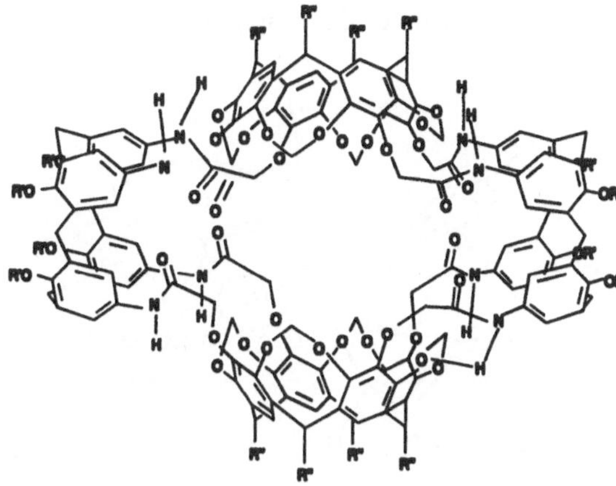

4

Searching for large hydrophobic surfaces of nanosize dimensions, we have been studying the combination of resorcin[4]arenes and calix[4]arenes in different ratios [4]. Combined in a 1:1 ratio we have obtained calix[4]arene-based carceplexes **3** which possess a non-symmetrical cavity that is completely closed. These molecules exist as two different diastereomers. NMR and computational studies have revealed a good relation between experimental and calculated activation energies of the isomerization process.

Reaction of two calix[4]arenes with one resorcin[4]arene gives three 2:1 coupling products having different configurations: endo-endo, endo-exo, and exo-exo. These molecules have a large hydrophobic surface on which steroids can be complexed; at least three functional groups are involved in complexation. The acetoxy group and the two hydroxyl groups are responsible for the complexation of steroids by these 2:1 addition products, most probably via CH-π and hydrogen bonding.

In a (2+2)-mode of coupling, the calix[4]arene and resorcin[4]arene give a rigid receptor **4** with a cavity of more than 1000 $Å^3$ [4].

3. Noncovalent Assembly of Modules into Functional Aggregates and Molecular Objects

Molecular self-assembly is an important part of life processes and biological systems. It results in a wide variety of complex structures, such as double-stranded DNA, viral protein coatings, lipid membranes, and globular proteins. Self-assembling processes have also found applications in the design of nanosize structures such as inorganic clusters, tubes and channels, monolayers, and hydrogen-bonded networks. The functions of such assemblies can be extrapolated from molecular properties.

In addition to coupling of the above described modules via covalent bonds, they can be also connected by non-covalent bonding to give *self-assembled functional receptor systems* and *objects* with molecular weights as large as 8000.

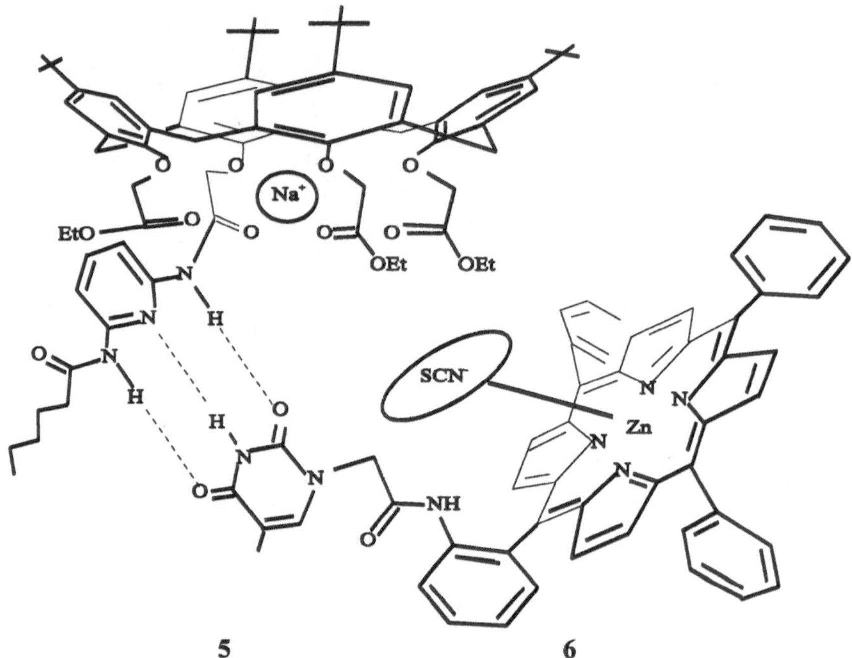

Figure 1. Noncovalently assembled bifunctional receptor **5•6** for NaSCN

In the *noncovalently* organized bifunctional receptor system (schematically represented in Figure 1) the Na^+ cation is complexed by calixarene **5**, the SCN^- anion coordinates to the Zn porphyrin **6**, and both cation and anion receptors **5** and **6** are connected via hydrogen-bonded aggregation (K_{ass} = 2.8 x 10^4 M^{-1} in toluene) [5]. The unique feature of this assembly is that all four components (calix[4]arene **5**, Zn porphyrin **6**, Na^+ cation, and SCN^- anion) are fixed in solution with their positions determined by specific ion-dipole interactions and hydrogen bonding. Moreover, the assembly can be "switched on" by complexation of the Na^+ cation with receptor **5**, and "switched off" by simply adding MeOH which will destroy the hydrogen bonds.

In the search for higher aggregates of functional molecules with well-defined geometry, Whitesides' cyanuric acid-melamine structural motif is currently being used. This "rosette" structure has the potential to function as a hydrogen-bonded platform for the aggregation of multiple calixarenes. The cyclic nature of the rosette provides substantial control over the degree of aggregation, and its highly ordered structure allows for the specific positioning of substituents and functional groups. Three features of the calixarenes are currently exploited in the construction of rosette-based assemblies. The first is the (almost) parallel orientation of two opposing aromatic rings, especially when the pinched cone conformation is adopted. Secondly, the bulky nature of the calix[4]arene can direct the association process towards the formation of the cyclic structure (*vs.* linear aggregates), and closes the space that is generated between the rosettes, thus creating a cavity. Finally, positions on both the upper and the lower rim can be employed to introduce functional groups.

The formation and stability of double rosettes from bis(melamine)calix[4]arene **7** calix-isocyanurates **8** and **9** (Figure 2) has been investigated by ^{1}H NMR titrations, NOE spectroscopy, and vapor pressure osmometry (VPO).

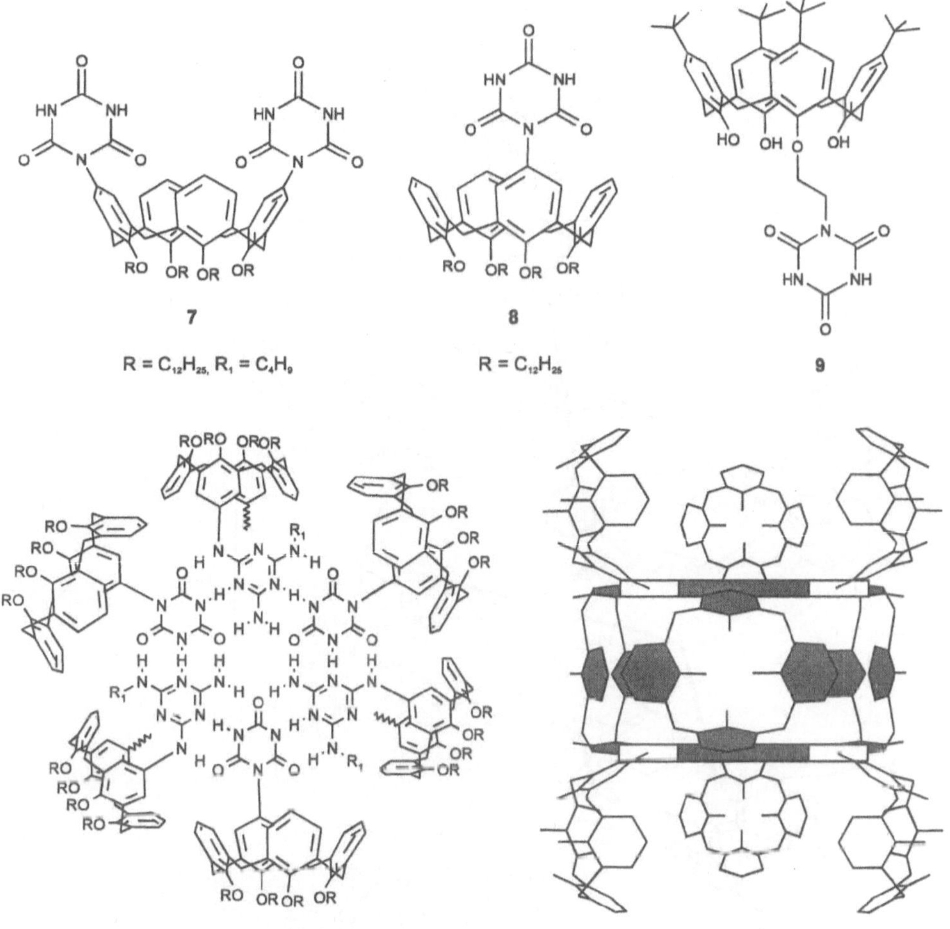

Figure 2. Schematic representation of the proposed structure of a double rosette consisting of **7** and **8** (for clarity, only the upper one of the two rosettes is drawn in the top view). In the side view the two rosettes are depicted as horizontal bars.

The results of the ^{1}H NMR titrations strongly support the formation of stable double rosettes as depicted in Figure 2. In these rosettes **7** and the cyanuric acid derivatives **8** and **9** are present in a ratio of 1:2. For the imido protons of the cyanuric acid derivatives two signals appear at very low field (14-16 ppm), indicating strong hydrogen bonding. Also the melamine NH's in the complexes are observed at lower field. The fact that only two signals are observed for the hydrogen bonded imido protons in situations where there is excess 7 indicates that the self-assembly is a highly cooperative process. NOE distances also are in accord with the proposed structures.

Mixing the bis-isocyanurate calix[4]arene **10** with flexible arms and tris-melamine calix[6]arene **11** with short and (rather) rigid arms in a ratio 3:2 in $CDCl_3$ also shows assembly; aggregates with expected masses of ca 8000 are formed (Figure 3).

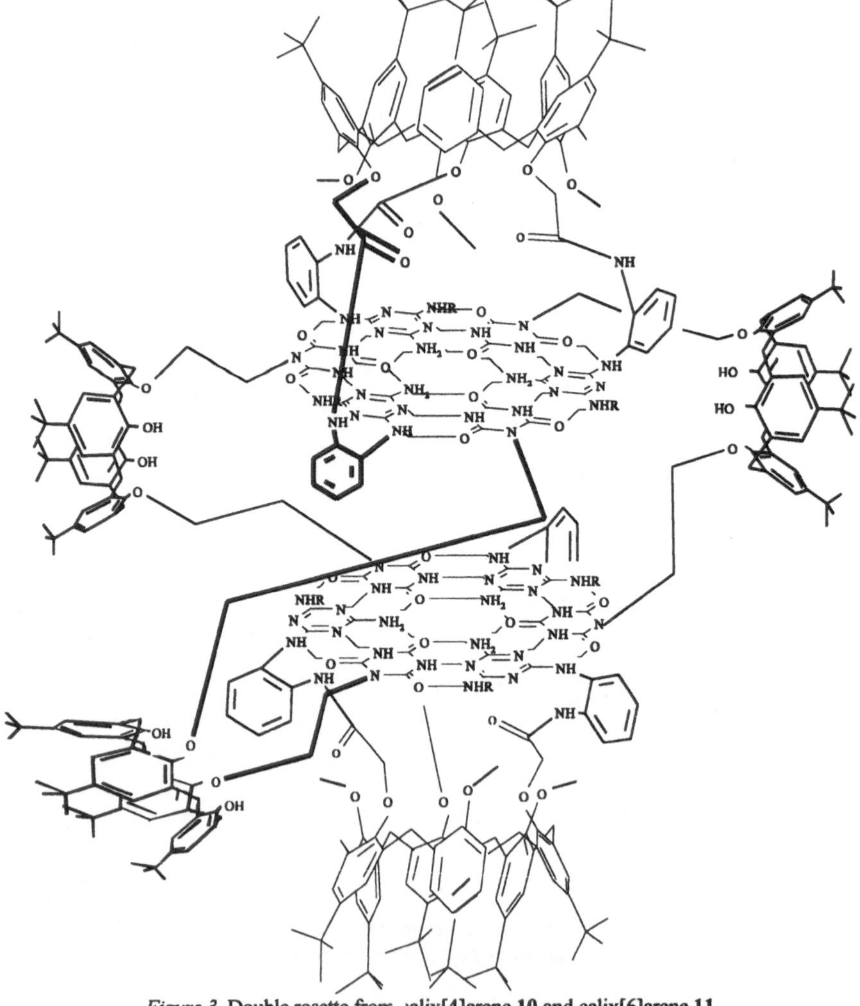

Figure 3. Double rosette from calix[4]arene **10** and calix[6]arene **11**

The ^{1}H NMR spectrum of the mixture **10** and **11** (ratio 3:2) is sharp already at room temperature. At the same time, only a very broad signal of isocyanurate NH protons can be found at ca 14.0 ppm. At 0 °C sharpening of the signal occurs, and at -30 – -60 °C there are two pronounced singlets at 16.1 and 14.0 ppm for both isocyanurate NH protons. This implies that the formation of a double rosette takes place.

4. Assembly of NLO-phoric Modules in a Strong Electric Field

Calix[4]arenes are also suitable as building blocks for nonlinear optical (NLO) active materials. Possible applications for those materials are frequency doubling (conversion of red laser light to blue) or electro-optic switching (optical computers). Macroscopic NLO activity requires an aligned arrangement of non-centrosymmetric building blocks. Organic molecules which are functionalized non-centrosymmetrically with electron donating (D) and accepting (A) groups which are connected via a π-conjungated system, so called D-π-A molecules, meet the requirements for NLO active building blocks.

NLO active calix[4]arenes **12a,b** have been obtained by selective functionalization of the upper rim with electron accepting groups like nitro or *p*-nitrostilbene, and the lower rim with an electron donating group (*n*-propoxy), as depicted in figure 4. In this way four D-π-A-units are combined within one molecule.

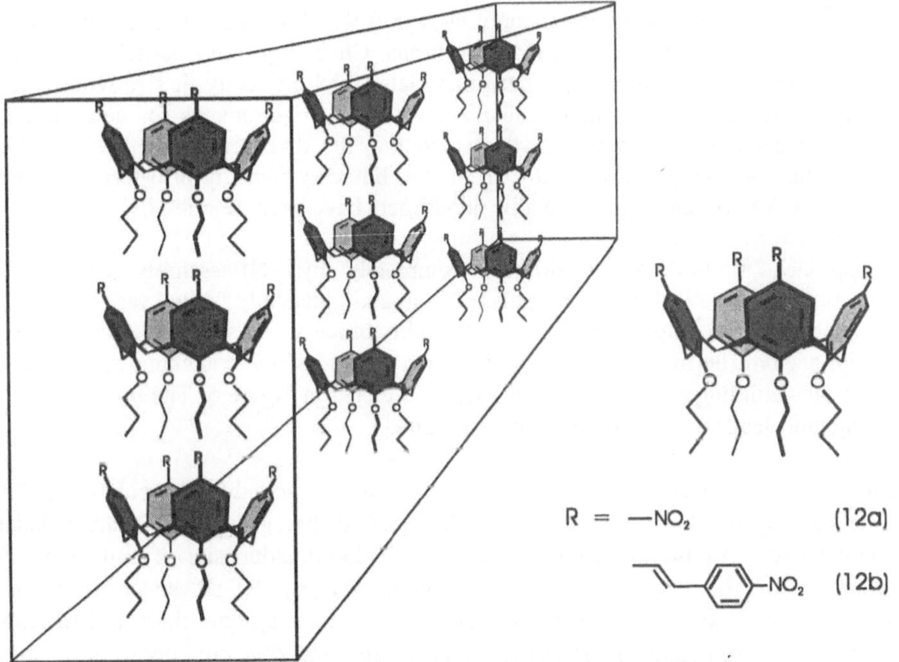

Figure 4. Assembly of calixarenes **12a** or **12b** after electrical field poling.

Spincoating of these building blocks (without a polymer matrix!) results in films (0.5-1.0 μm thick) which exhibit high NLO activity after poling [6]. The thermal stability of the d_{33}-values (frequency doubling parameter) of films consisting of neat **12a** and **12b** is high. At room temperature as well as at 80 °C (to which level the temperature can rise in photonic devices) no decay or less than 30 % decay of NLO activity is observed. After this initial decay a stable and still high d_{33}-value is reached. The origin of the high thermal stability of these NLO active films of calix[4]arenes alone is probably the stacking of these bowl-shaped molecules on top of each other (Figure 4). In this way very stable, large aggregates of columns are formed with very high dipole moments. Reorientation of these large aggregates which would mean decay of NLO activity, is difficult. Currently, dielectric measurements are in progress to prove the presence of these (columnar) aggregates of aligned molecules.

Another important property of NLO films used for frequency doubling, is transparancy at both the fundamental (1064 nm) and the frequency doubled wavelength (532 nm). The maxima of the absorbtion bands of **12a** and **12b** are well below 532 nm (291 and 370 nm, respectively). Furthermore, there is hardly any residual tailing absorbtion at that wavelength.

5. Large Self-assembled Metallospheres

Nanophysical fabrication methods (nanolithography) which operate from the bulk 'down' are close to having reached their diffraction limits. Chemistry builds structures from the atom 'up' and should in principle produce materials which perfectly fit size requirements. Dendrimers have the well-defined structure to meet the requirements for new materials. They are prepared by stepwise growth, which may be divergent or convergent. Until recently dendrimers were fully organic in nature, but very recently also transition metal containing derivatives and organometallic dendrimers have been described.

Our approach to building dendrimers completely by self-assembly employs the combination of a kinetically inert tridentate ligand and kinetically labile cyano groups in a square planar Pd(II) complex [7]. The self-assembly process is based on the substitution of the labile acetonitrile molecules for cyanomethyl groups of the tridentate ligand thereby adding a new building block as is schematically depicted in Figure 5. The branching in the repeating units leads to a metallodendritic structure.

When all solvent molecules, including acetonitrile, were removed from a solution of **13** in nitromethane by gentle heating *in vacuo*, the cyanomethyl group of the bis-palladated complex **13** replaces the acetonitrile at the fourth coordination site of palladium. This coordination of the cyano group can be monitored by FT-IR spectroscopy in a nitromethane solution and in the solid state (KBr disc), because of the characteristic shift of the C≡N stretch from 2250 cm^{-1} (free cyanomethyl group) to 2290 cm^{-1} upon coordination.

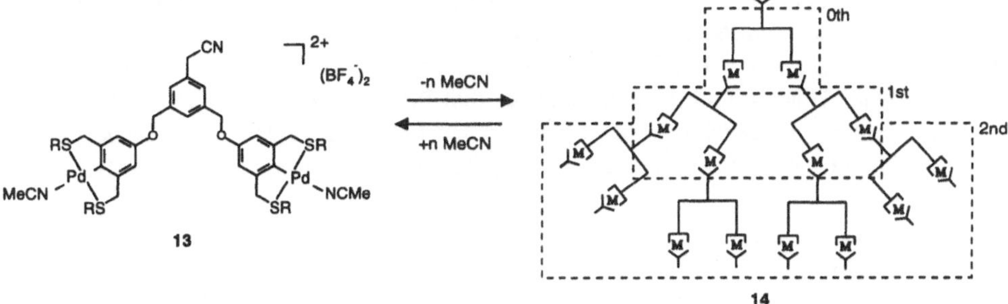

Figure 5. Schematic self-assembly of a second generation metallodendrimer.

The ^{1}H NMR spectrum in CD_3NO_2 of the acetonitrile-free compound **14** shows very broad signals, indicating association in solution. When small amounts of acetonitrile were added to this solution, all signals became sharp. This proves that the self-assembly of **13** and disassembly of **14** is a completely reversible process. QELS (Quasi Elastic Light-Scattering) measurements of nitromethane solutions showed particles with a modal hydrodynamic diameter of 200 nm.

The size of the aggregates was measured by Atomic Force Microscopy (AFM), after evaporation of the nitromethane solution on a gold surface. A representative part of the surface is shown in Figure 6. The globular shaped assemblies are clearly visible on a relatively rough gold substrate. The average diameter, as found by the grain size analysis routine of the instruments software of these aggregates is 205 nm, with a relatively narrow distribution: 80% (95%) of the diameters are found within s_d (2 s_d) of the mean value. Grazing angle FT-IR on the gold surface covered with the spheres showed the characteristic C≡N signal of coordinated cyanogroups at 2290 cm^{-1}. When a glass substrate was used instead of gold, the same spheres were observed, of roughly the same size. This indicates that the interaction of the spheres with the surface does not significantly influence their size.

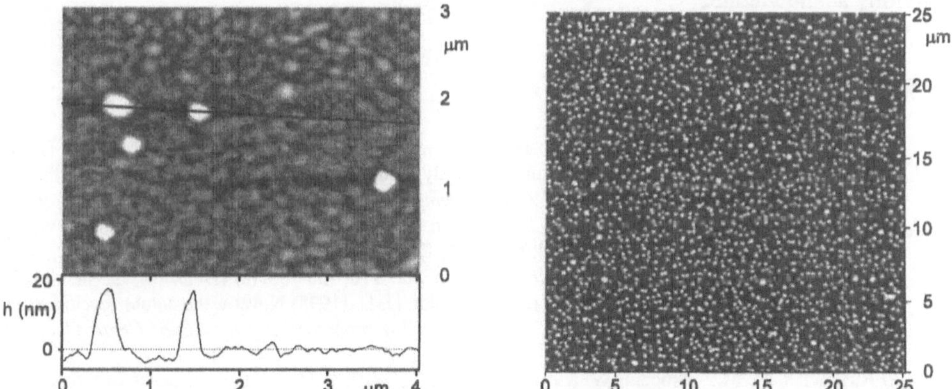

Figure 6. 25x25 μm AFM scan of self-assembled spheres on a gold substrate (right) An enlargement of a representative 3x4 μm area (left) is some isolated spheres (white dots). (Left): A line scan over two of the aggregates. The grey scale is 40 nm.

Samples for Transmission Electron Microscopy (TEM) were prepared by slow evaporation of a nitromethane solution on a carbon-coated copper grid. The pictures clearly show globular aggregates in the 150-200 nm range. This is in good agreement with the diameter measured with AFM (Figure 6) and indicates that Pd is present throughout the spherical assembly. Energy Dispersive X-ray Spectrometry (EDX) revealed the presence of the elements Pd and S in these aggregates, proving that these structures are no artifacts.

6. Conclusions

A modular approach to large functional structures has been described which applies either covalent or noncovalent combination of medium-size functional building blocks. This approach has produced bifunctional receptors, capped porphyrins, and large hydrophobic surfaces by covalent combination of the appropriate modules (e.g. calix[4]- and -[6]arenes, resorcin[4]arenes, cyclodextrins, porphyrins, and metallosalens). Noncovalent assembly of bifunctional receptor systems and large well-defined aggregates (molecular objects) based on calixarene modules have been 'synthesized'. Transparent and highly NLO active thin films, consisting of calix[4]arenes only, have been assembled. The perfect combination of high NLO activity and transparency makes these calix-films suitable materials for frequency doubling. By proper design of a branched repeating unit, large (metallo)spheres with well-defined dimensions can be formed by self-assembly. Self-assembly occurs not only in the solid state, but also in solution.

7. Acknowledgements

The work described in this paper is the result of the efforts of more than 10 coworkers whose names appear in the references. Financial support from the Technology Foundation (STW), the Technical Science Branch of the Netherlands Organization for Scientific research (SON/NWO), and of AKZO-NOBEL Research, Arnhem, The Netherlands is gratefully acknowledged.

8. References

1. (a) Ghidini, E.; Ugozzoli, F.; Ungaro, R.; Harkema, S.; Abu El-Fadl, A. and Reinhoudt, D. N. (1990) Complexation of alkali metal cations by conformationally rigid, stereoisomeric calix[4]arene crown ethers: a quantitative evaluation of preorganization, *J. Am. Chem. Soc.* **112**, 6979-6992. (b) Iwema Bakker, W. I.; Haas, M.; Khoo-Beattie, C.; Ostaszewski, R.; Franken, S. M.; den Hertog, Jr., H. J.; Verboom, W.; de Zeeuw, D.; Harkema, S. and Reinhoudt, D. N. (1994) Kinetically stable complexes of alkali cations with calixspherands: an evaluation of shielding, *J. Am. Chem. Soc.* **116**, 123-133. (c) van Dienst, E.; Snellink, B. H. M.; von Piekartz, I.; Engbersen, J. F. J. and Reinhoudt, D. N. (1995) Novel water-soluble cyclodextrin-calix[4]arene host molecules with strongly enhanced binding properties, *J. Chem. Soc., Chem. Commun.* 1151-1152.
2. (a) Rudkevich, D. M.; Verboom, W. and Reinhoudt, D. N. (1994) Calix[4]arene salenes: a bifunctional receptor for NaH_2PO_4, *J. Org. Chem.* **59**, 3683-3686. (b) Rudkevich, D. M.; Mercer-Chalmers, J. D.;

Verboom, W.; Ungaro, R.; de Jong, F. and Reinhoudt, D. N. (1995) Bifunctional recognition: simultaneous transport of cations and anions through a supported liquid membrane, *J. Am. Chem. Soc.* **117**, 6124-6125.

3. (a) Rudkevich, D. M.; Verboom, W. and Reinhoudt, D. N. (1995) Capped biscalix[4]arene-Zn-porphyrin metalloreceptor with a rigid cavity, *J. Org. Chem.* **60**, 6585-6587. (b) Rudkevich, D. M.; Verboom, W. and Reinhoudt, D. N. (1994) Biscalix[4]arene-Zn-tetraarylporphyrins, *Tetrahedron Lett.* **35**, 7131-7134.
4. (a) Timmerman, P.; Verboom, W.; van Veggel, F. C. J. M.; van Duynhoven, J. P. M. and Reinhoudt, D. N. (1994) A novel type of stereoisomerism in calix[4]arene-based carceplexes, *Angew. Chem., Int. Ed. Engl.* **33**, 2345-2348. (b) Timmerman, P.; Brinks, E. A.; Verboom, W. and Reinhoudt, D. N. (1995) Synthetic receptors with preorganized cavities that complex prednisolone-21-acetate, *J. Chem. Soc., Chem. Commun.* 417-418. (c) Timmerman, P.; Verboom, W.; van Veggel, F. C. J. M.; van Hoorn, W. P. and Reinhoudt, D. N. (1994) An organic molecule with a rigid cavity of nanosize dimensions, *Angew. Chem., Int. Ed. Engl.* **33**, 1292-1295.
5. Rudkevich, D. M.; Shivanyuk, A. N.; Brzozka, Z.; Verboom, W. and Reinhoudt, D. N. (1995) A self-assembled bifunctional receptor, *Angew. Chem., Int. Ed. Engl.* **34**, 2124-2126.
6. (a) Kelderman, E.; Derhaeg, L.; Verboom, W.; Engbersen, J. F. J.; Harkema, S.; Persoons, A. and Reinhoudt, D. N. (1993) Calix[4]arenes as molecules for second order nonlinear optics, *Supramolecular Chem.* **2**, 183-190. (b) Kelderman, E.; Heesink, G. J. T.; Derhaeg, L.; Verbiest, T.; Klaase, P. T. A.; Verboom, W.; Engbersen, J. F. J.; van Hulst, N. F.; Clays, K.; Persoons, A. and Reinhoudt, D. N. (1993) Highly ordered films of neat calix[4]arenes for second order nonlinear optics, *Adv.Mater.* **5**, 925-930.
7. Huck, W. T. S.; van Veggel, F. C. J. M.; Kropman, B. L.; Blank, D. H. A.; Keim, E. G.; Smithers, M. M. A. and Reinhoudt, D. N. (1995) Large self-assembled organopalladium spheres, *J. Am. Chem. Soc.* **117**, 8293-8294.

ITERATIVE BUILDING BLOCK APPROACHES TO DISCRETE POLYSTANNANE OLIGOMERS

LAWRENCE R. SITA* AND KAZUSATO SHIBATA
Searle Chemistry Laboratory
Department of Chemistry
The University of Chicago
5735 South Ellis Avenue
Chicago, Illinois 60637
U.S.A.

ABSTRACT. New synthetic methodology has been developed that can be utilized for the iterative construction of several new families of discrete linear polystannane oligomers of the general formula, X-$(R_2Sn)_n$-Y (X = Y and X = R). A full analysis of these compounds by a variety of techniques, including variable temperature UV-Vis and ^{119}Sn NMR spectroscopies, has been performed, and the results of these investigations provide additional information regarding the structure/property relationships of this unique class of compound.

1. Introduction

Linear catenation of group 14 heavy atoms generates novel optical/electronic properties that are associated with the corresponding polymetallane structures, R-$(R_2M)_n$-R (M = Si, Ge, Sn and Pb) [1 - 4]. In this regard, the synthesis and characterization of discrete group 14 linear polymetallane oligomers have been instrumental in providing data that can be used to either refine existing theoretical models or develop new ones. Towards this goal, we have recently developed new synthetic methodology that allows one to carry out the controlled, rational synthesis of various families of polystannane oligomers. Key to the success of this work has been the design and implementation of a new class of "building block" that can be used to construct a desired linear polystannane sequence in an iterative, pre-programmed fashion. This new modular chemistry now provides an important mechanism through which the novel structure/property relationships of the heavier group 14 polymetallanes, the polystannanes, can be probed.

J. Michl (ed.), Modular Chemistry, 587–599.

2. Synthesis

From the outset, certain requirements that had to be met for the controlled synthesis of linear polystannane oligomers were recognized. These consisted of the needs for: (1) a facile and directional method for the construction of Sn-Sn bonds, (2) a set of "protecting groups" that can be used to prevent undesirable side reactions and which can be selectively and quantitatively removed to "unmask" a new site of reactivity, and (3) new methods of purification and analysis to ascertain that the air-sensitive polystannane products are homogeneously and homologously pure. With regard to the first of these, the hydrostannolysis reaction, which can be viewed as the metathetical exchange between an organotin hydride and an organotin amide according to Eq. 1, has previously been used to prepare tri- and tetrastannanes in high yield (Eq. 2) [4b]. However, in order to apply this synthetic methodology for the intended purpose, a protecting group was

$$R_3Sn\text{-}NMe_2 + R'_3SnH \longrightarrow \left[R_3Sn \cdots H \cdots Sn R'_3 \cdots NMe_2 \right]^{\ddagger} \longrightarrow R_3Sn\text{-}SnR'_3 + HNMe_2 \qquad (1)$$

$$H\text{-}(R_2Sn)_n\text{-}H + 2\ R'_3SnNMe_2 \xrightarrow[n\,=\,1\text{ or }2]{} R'_3Sn\text{-}(R_2Sn)_n\text{-}SnR'_3 + 2\ HNMe_2 \qquad (2)$$

required that could be removed, after hydrostannolysis, to reveal a new functionality that could then serve as the site for extension of the polystannane backbone. This was ultimately achieved by developing the β-alkoxy substituent as a masked synthon for the Sn-H functionality. As shown in Eq. 3, the key building block for polystannane synthesis, compound **1**, can be easily prepared through the hydrostannation of ethyl vinyl ether with *in situ* generated Bu_2SnHCl, followed by amide/halide

$$Bu_2SnH_2 + Bu_2SnCl_2 \rightleftharpoons [\,2\ Bu_2SnHCl\,] \xrightarrow[45^\circ\text{ C, 18h}]{\text{CH}_2\text{=CHOEt}} 2\ Bu_2Sn(Cl)CH_2CH_2OEt \xrightarrow[\text{THF, -78° C}]{LiNMe_2} 2\ Bu_2Sn(NMe_2)CH_2CH_2OEt \qquad (3)$$

1

(55% yield)

exchange of the intermediate product with lithium dimethylamide (LDMA). In this latter process, it is critical that the addition of the LDMA reagent be performed slowly at low temperature (-78° C) in order to minimize the formation of $Bu_2Sn(NMe_2)_2$ which must subsequently be removed by careful fractional distillation of crude **1**.

To unveil a new Sn-H functional group, it was determined that the β-alkoxy substituent can be quantitatively removed with either (a) diisobutylaluminum hydride (DIBAL-H) or (b) *in situ*-generated alane (AlH_3), prepared from zinc dibromide and lithium aluminum hydride, according to Eq. 4. This latter method has been found to be preferable for the synthesis of longer chain polystannane oligomers (e.g. n ≥ 5).

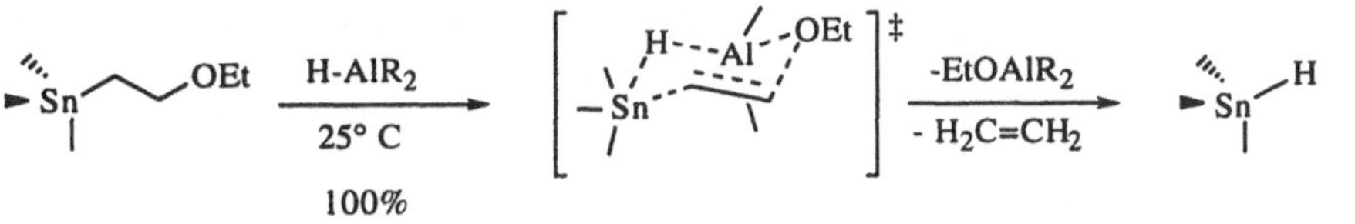

$$\ce{>Sn-CH2CH2-OEt} \xrightarrow[\text{25° C, 100\%}]{\text{H-AlR}_2} [\text{transition state: H--Al--OEt / Sn}]^{\ddagger} \xrightarrow[\text{- H}_2\text{C=CH}_2]{\text{-EtOAlR}_2} \ce{>Sn-H} \qquad (4)$$

R = isobutyl, H

Compound **1** has proven to be a versatile building block for the iterative stepwise and divergent syntheses of linear polystannane oligomers. Thus, as Scheme 1 reveals, if one starts with a triorganotin hydride, then repetition of the reaction sequence, (1) hydrostannolysis, followed by, (2) deblocking, provides for a stepwise growth of the polystannanes [3b]. Here, it is important to mention that the polar nature of the β-alkoxy substituent allows for the rigorous preparative purification of the compounds by conventional column chromatography on silica gel and for their analysis by reverse-phase (C_{18}) high pressure liquid chromatography (HPLC).

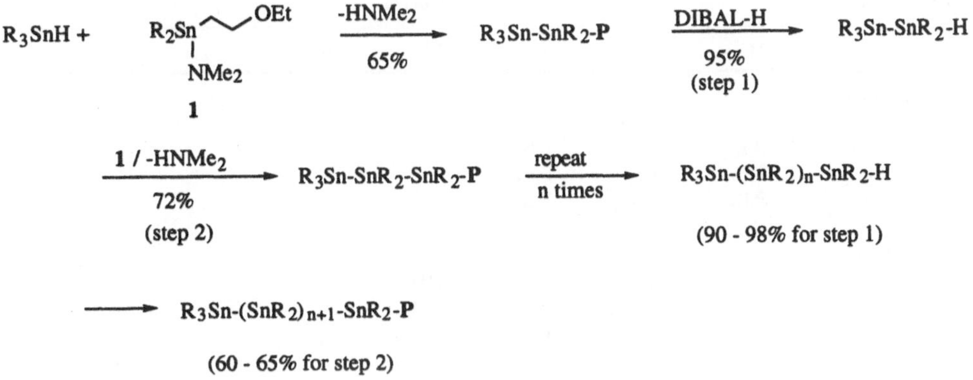

$$R_3SnH + R_2Sn(NMe_2)CH_2CH_2OEt\ (\mathbf{1}) \xrightarrow[65\%]{-HNMe_2} R_3Sn\text{-}SnR_2\text{-}P \xrightarrow[95\%\ (\text{step } 1)]{\text{DIBAL-H}} R_3Sn\text{-}SnR_2\text{-}H$$

$$\xrightarrow[72\%\ (\text{step } 2)]{\mathbf{1}\ /\ -HNMe_2} R_3Sn\text{-}SnR_2\text{-}SnR_2\text{-}P \xrightarrow[\text{n times}]{\text{repeat}} R_3Sn\text{-}(SnR_2)_n\text{-}SnR_2\text{-}H \quad (90 - 98\%\ \text{for step } 1)$$

$$\longrightarrow R_3Sn\text{-}(SnR_2)_{n+1}\text{-}SnR_2\text{-}P \quad (60 - 65\%\ \text{for step } 2)$$

P = 2-ethoxyethyl; R = Bu

Scheme 1

Complementary to the stepwise approach of Scheme 1, if one begins with a diorganotin dihydride, then repetition of the same reaction sequence generates an iterative divergent route to the odd series of linear polystannane oligomers as shown in Scheme 2. Unfortunately, the intermediate polystannane dihydrides that are produced in this process appear to be much less stable and a decomposition pathway that now appears is butyl group migration to a terminal tin atom. Subsequent hydrostannolysis with **1** then produces a mixture of peralkyl, mono-functional and α,ω-difunctional polystannane oligomers. Although this side reaction lowers the overall yield,

$R_2SnH_2 + 2\ Bu_2Sn(NMe_2)CH_2CH_2OEt$ (**1**) $\xrightarrow{-2\ HNMe_2}$ **P**-Bu_2Sn-SnR_2-$SnBu_2$-**P** 75 - 80%

$\xrightarrow{2\ AlH_3}$ H-Bu_2Sn-SnR_2-$SnBu_2$-H $\xrightarrow[-2\ HNMe_2]{2\ \mathbf{1}}$ **P**-$(Bu_2Sn)_2$-SnR_2-$(SnBu_2)_2$-**P** 40 - 50%

$\xrightarrow{repeat}$ **P**-$(Bu_2Sn)_3$-SnR_2-$(SnBu_2)_3$-**P** 40 - 50%

P = 2-ethoxyethyl; R = n-Bu or t-Bu

Scheme 2

the mixture of products can usually be separated to provide oligomers that one could not otherwise obtain. Hence, it has now been possible to obtain the number of new polystannane derivatives shown in Table 1. In this

TABLE 1. Linear Polystannane Oligomers Obtained in Pure Form.

Compound	Substituents						
	Sn_1	Sn_2	Sn_3	Sn_4	Sn_5	Sn_6	Sn_7
tristannane							
2a	Bu_3	t-Bu_2	Bu_3				
2b	Bu_3	Bu_2	P,Bu_2				
2c	P,Bu_2	Bu_2	P,Bu_2				
2d	P,Bu_2	t-Bu_2	P,Bu_2				
tetrastannane							
3a	Bu_3	Bu_2	Bu_2	P,Bu_2			
3b	P,Bu_2	Bu_2	Bu_2	P,Bu_2			
3c	Bu_3	t-Bu_2	Bu_2	P,Bu_2			
pentastannane							
4a	Bu_3	Bu_2	Bu_2	Bu_2	P,Bu_2		
4b	P,Bu_2	Bu_2	Bu_2	Bu_2	P,Bu_2		
4c	P,Bu_2	Bu_2	t-Bu_2	Bu_2	P,Bu_2		
hexastannane							
5a	Bu_3	Bu_2	Bu_2	Bu_2	Bu_2	P,Bu_2	
5b	Bu_3	Bu_2	t-Bu_2	Bu_2	Bu_2	P,Bu_2	
heptastannane							
6	P,Bu_2	Bu_2	Bu_2	t-Bu_2	Bu_2	Bu_2	P,Bu_2

P = 2-ethoxyethyl

regard, it is interesting to note that the heptastannane 6 is the longest linear polystannane oligomer that has been prepared to date in pure form.

3. Properties

3.1. ELECTRONIC SPECTROSCOPY

The electronic spectra collected for the linear polystannanes listed in Table 1 all show a low energy transition which red shifts and increases in intensity with increasing chain length as exemplified by the comparison between the spectra for the pentastannane **4c** and the heptastannane **6** shown in Figure 1. Of interest is the observation that virtually no perturbation is made in the position of this transition in going from n-butyl to t-butyl substituents on the central tin atom in the two pentastannanes represented by compounds **4b** and **4c** [*cf.* λ_{max} 296 nm (ε_{max} 49 700) for **4b** vs 297 nm (54 700) for **4c**].

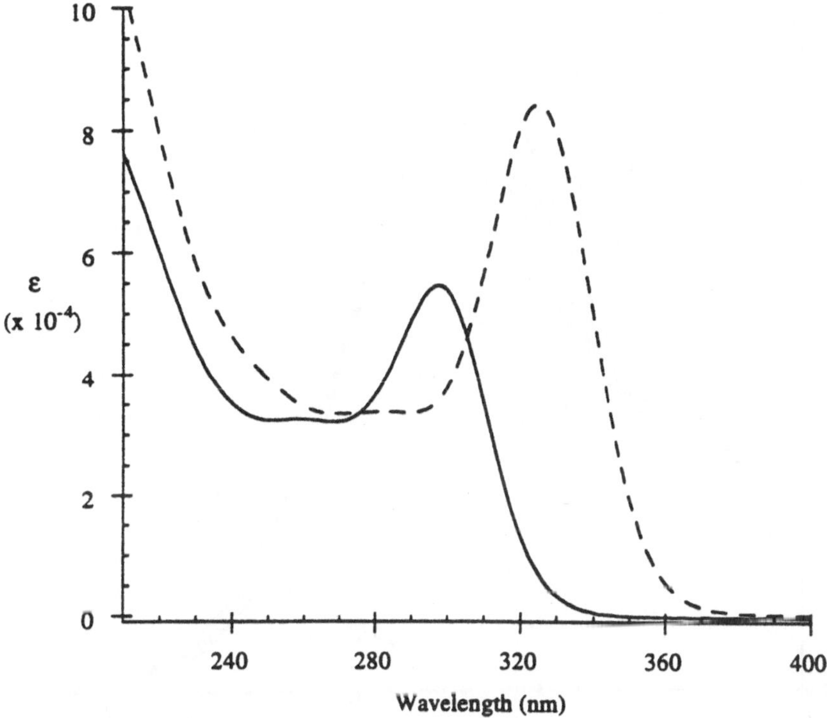

Figure 1. A comparison between the electronic spectra of the pentastannane **4c** (solid line) and the heptastannane **6** (dashed line).

Previously, it has been demonstrated that high molecular weight linear polysilane and polystannane materials exhibit reversible thermochromic behavior, characterized by a bathochromic shift of the low energy transition with decreasing temperature, that is attributed to an increase, at lower temperatures, of the population of the all-*trans* chain segments that define the chromophoric length in the polymetallane backbone [3d, 5]. In contrast, the polystannane oligomers listed in Table 1, that are longer than three tin atoms in length, all display a uniquely different kind of reversible thermochromic behavior. As shown in Figure 2 for the heptastannane **6**, this behavior is characterized by a red shift and decrease in intensity with increasing temperature. At the present time, we believe the origin of this behavior to be electronic rather than structural (or conformational) in nature. More specifically, given the small force constants for Sn-Sn bond stretching, it is possible that a thermal population of close lying vibrational states of the polystannanes shifts the equilibrium Sn-Sn bond lengths in the ground state structure which, in turn, provides for a better Franck-Condon overlap with the excited state.

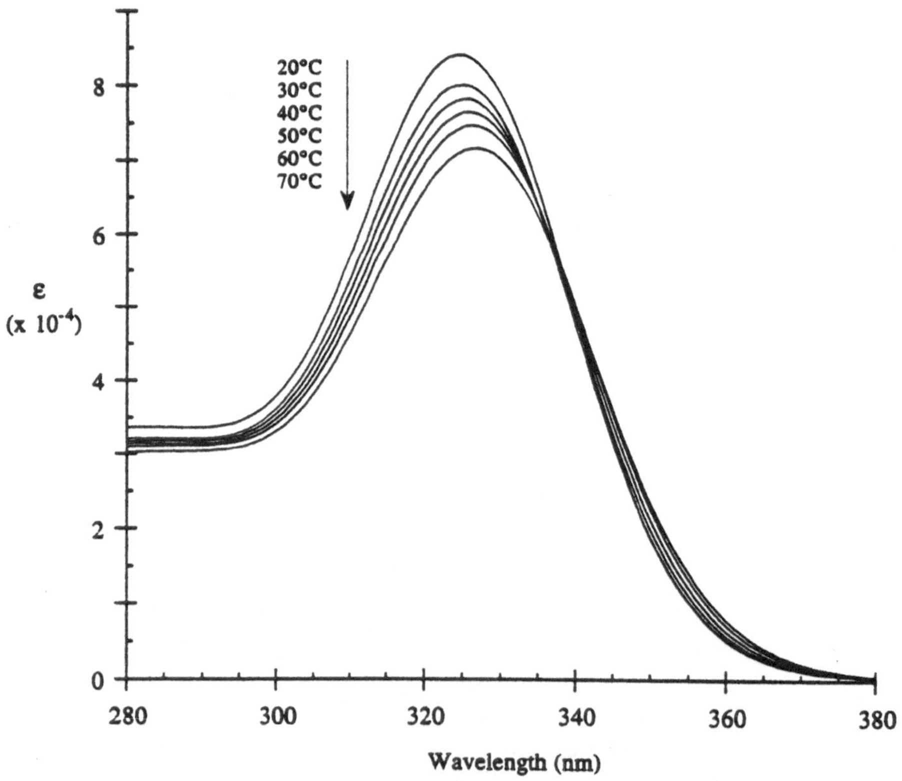

Figure 2. Temperature dependence of the electronic spectra of the heptastannane **6**.

One question that arises is: what is the maximum red shift in the low energy transition that one can expect for a given series of linear polystannane oligomers? Unfortunately, the methods outlined in Schemes 1 and 2 are not practical for the synthesis of long-chain oligomers with lengths greater than eight tin atoms. Accordingly, we developed the alternative

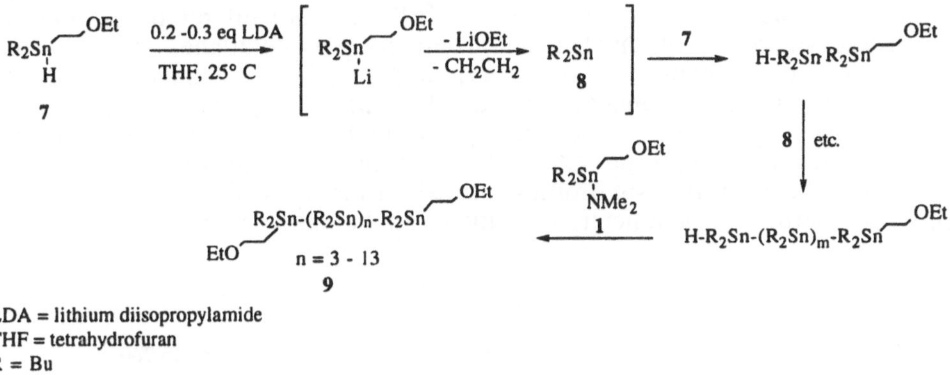

$R_2Sn(H)CH_2CH_2OEt$ (**7**) —(0.2 -0.3 eq LDA, THF, 25° C)→ [$R_2Sn(Li)CH_2CH_2OEt$ —(- LiOEt, - CH_2CH_2)→ R_2Sn (**8**)] —(**7**)→ $H\text{-}R_2Sn\text{-}R_2SnCH_2CH_2OEt$ —(**8**, etc.)→ $H\text{-}R_2Sn\text{-}(R_2Sn)_m\text{-}R_2SnCH_2CH_2OEt$ —($R_2Sn(NMe_2)CH_2CH_2OEt$, **1**)→ $EtOCH_2CH_2\text{-}R_2Sn\text{-}(R_2Sn)_n\text{-}R_2SnCH_2CH_2OEt$, n = 3 - 13 (**9**)

LDA = lithium diisopropylamide
THF = tetrahydrofuran
R = Bu

Scheme 3

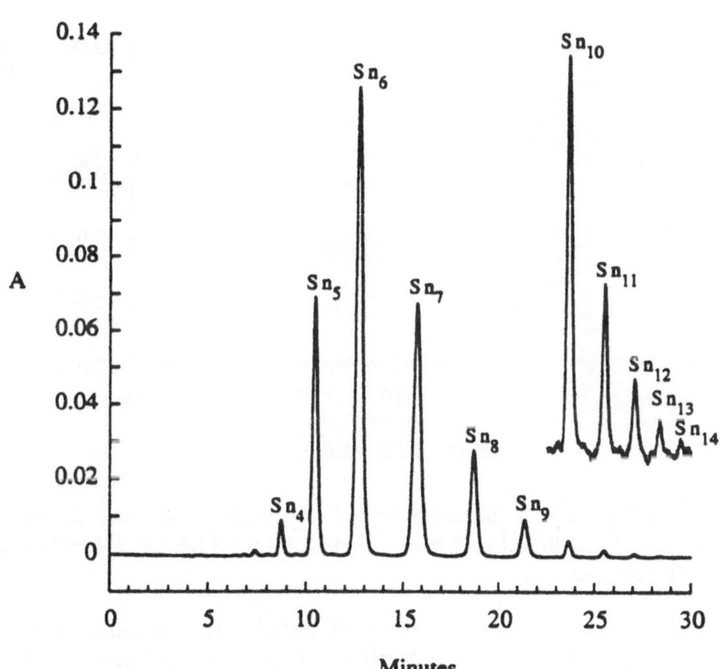

Figure 3. HPLC chromatogram for **9** (peak detection at 322 nm).

oligomerization process shown in Scheme 3 which allows one to generate, in one step, a family of oligomers that can then be separated by reverse-phase HPLC as demonstrated in Figure 3 [6]. By using a diode-array UV–Vis detector, it is then possible to accurately obtain λ_{max} for the series and in this fashion, values for polystannanes with lengths of up to 15 tin atoms have been recorded for the series **9** shown in Scheme 3. This data was subsequently used to show that the semiempirical Sandorfy Hückel molecular orbital approximation (model C) [7] works well for describing the electronic structures of the heavier group 14 polymetallanes [6]. In this model, the polystannane backbone is treated as a linear chain of interacting sp^3 orbitals with two resonance integrals, β_{vic} and β_{gem}, now being introduced to describe, respectively, the strong (vicinal) interaction between two overlapping sp^3 orbitals on adjacent tin atoms and the weaker (geminal) interaction between two sp^3 orbitals located on the same tin atom. The ratio, $\beta_{gem} / \beta_{vic}$ defines a parameter, *m*, which can be taken as a measure of the

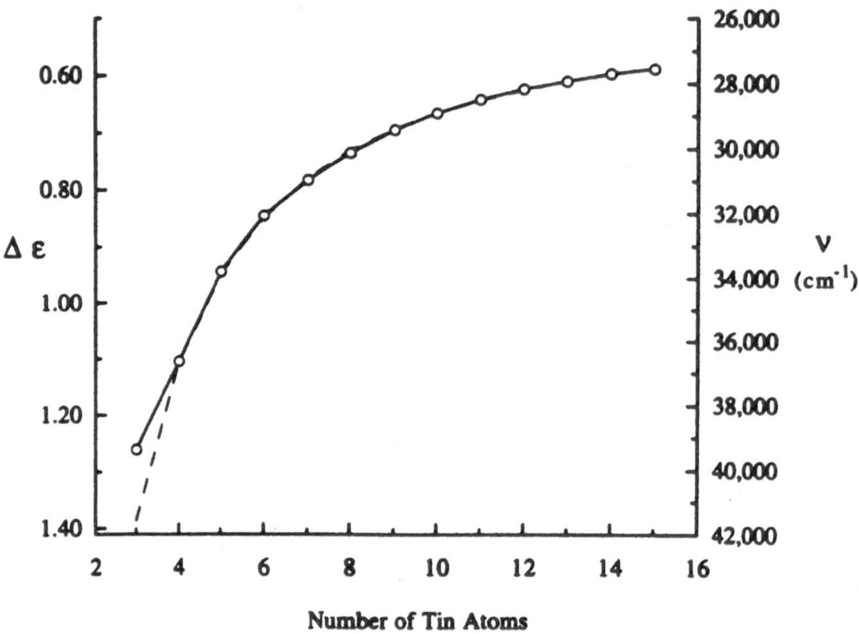

Figure 4. Curves for the calculated (Sandorfy model C) energy differences (LUMO-HOMO) (dashed line) and observed transition frequencies (circles and solid line) as a function of chain length in linear polystannanes.

degree of delocalization along the σ-bonded backbone. As shown in Figure 4, setting $m = 0.75$ produces a very high degree of correlation between the curve generated by the observed chain length behavior of the transition frequencies (cm^{-1}) of **9** and that determined by a plot of the calculated energy difference between the highest occupied molecular orbital (HOMO) and lowest unoccupied MO (LUMO), $\Delta\varepsilon$, as a function of polystannane chain length. A similarly high value of *m* has previously been obtained for all-trans polysilane oligomers [8], thereby indicating, that electrons in polystannanes are also anticipated to be highly delocalized along the σ-bonded backbones. Returning now back to the original question of what is the maximum red shift in the low energy transition that one can expect for a given series of linear polystannane oligomers, extrapolation of the plot in Figure 4 to n = ∞ in **9** provides a predicted λ_{max} of 382 nm that is in surprisingly good agreement with the λ_{max} value of 384 nm obtained by Tilley and co-workers [3d] for the high molecular weight polystannane material, H-$(Bu_2Sn)_n$-H ($M_n \geq 10{,}000$). This close agreement between experiment and theory for polystannanes supports the hypothesis presented by Michl and co-workers [9] that the simple Sandorfy model C approximation should be adequate for describing the electronic structure of polystannane and polyplumbane oligomers due to the absence in these systems of significant perivalent interactions that are responsible for the conformational effects observed for polysilanes.

3.2 ^{119}Sn NMR SPECTROSCOPY

Due to the occurrence of two isotopes of tin with I = 1/2, ^{119}Sn and ^{117}Sn, of relatively high natural abundance (8.58% and 7.61%, respectively) and high receptivity (natural abundance times sensitivity) (25.6 and 19.9, respectively) relative to ^{13}C (=1), ^{119}Sn NMR spectroscopy is a powerful tool for investigating polystannane structures [4b]. In this regard, we were interested in determining whether the relative magnitudes of the various coupling constants, $^{n}J(^{119}Sn-^{119}Sn)$ and $^{n}J(^{119}Sn-^{117}Sn)$, in linear polystannanes might provide useful information pertaining to the nature of their electronic structures. A typical high resolution ^{119}Sn NMR spectrum of a polystannane recorded at 186 MHz is shown in Figure 5. From this data, the magnitudes of almost all the various coupling constants that are observed can be successfully extracted through simulations of the spectra, and as Table 2 reveals, this can be achieved even for the heptastannane **6**. An important feature of linear polystannanes that we have discovered through ^{119}Sn NMR spectroscopy is that, for chain lengths greater than three tin atoms, the absolute magnitudes of the various coupling constants, and most notably,

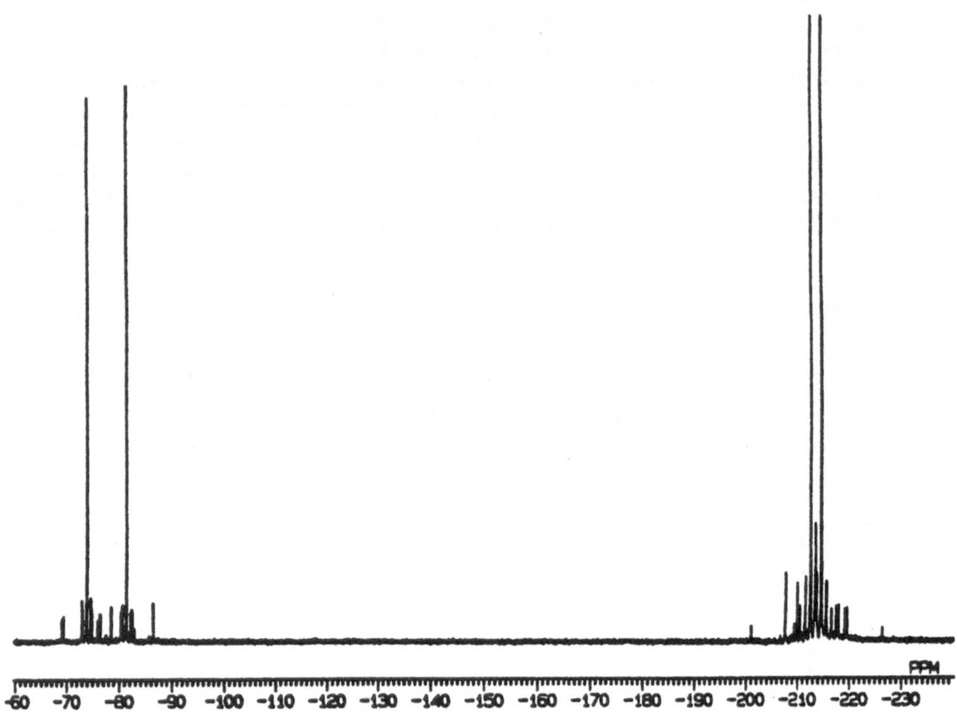

Figure 5. A $^{119}Sn\{^{1}H\}$ NMR (C_6D_6, 40° C) spectrum of the tetrastannane **4a**.

TABLE 2. Absolute Magnitudes of the $^{n}J(^{119}Sn\text{-}^{117}Sn)$ Coupling Constants (Hz) of the Heptastannane **6**.[a,b]

n = 1			n = 2			n = 3		n = 4		n = 5
AB	BC	CD	AC	BD	CC	AD	BC	AC	BB	AB
1315	257	401	233	< 5	233	163	172	52	57	18

[a] Sn_1,Sn_7 = A; Sn_2, Sn_6 = B; Sn_3, Sn_5 = C; Sn_4 = D
[b] Recorded at 40° C.

those of the direct one-bond constants, are temperature dependent. This unprecedented behavior is exemplified by the curves shown in Figure 6 for the tetrastannane **4a**. In previous work, we have shown that the direct one-bond coupling constants in a stereochemically rigid polycyclic polystannane are also dramatically temperature dependent and we attributed this behavior to a shift in the equilibrium structural parameters of the polystannane framework at higher temperatures due to an increase in nonbonded steric

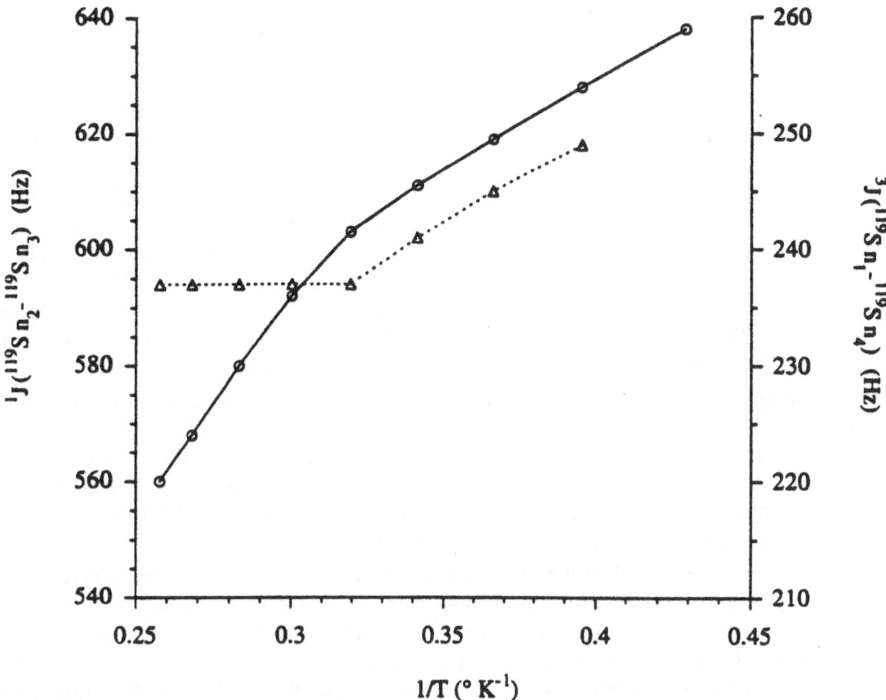

Figure 6. Temperature dependence of $^1J(^{119}Sn_2$-$^{119}Sn_3)$ (circles) and $^3J(^{119}Sn_1$-$^{119}Sn_4)$ (triangles) of the tetrastannane **4a**.

interactions amongst the organic substituents [10]. Given that linear polystannanes also exhibit temperature dependent one-bond coupling constants we now believe that this phenomenon may be a general one for heavy-atom frameworks and its origin might be associated with a thermal population of either low-lying vibrationally excited states that make negative contributions to the magnitude of the coupling constant [11]. Clearly more investigations are required to elucidate both the nature and generality of this temperature dependent phenomenon and studies along these lines are now in progress.

4. Conclusion

Linear group 14 polymetallanes remain an interesting class of structure for study and our understanding of their electronic structures is far from being complete. In this regard, the new methodology described herein should prove useful for providing access to a variety of different families of discrete polystannane oligomers. An elucidation of the origins of the unique temperature dependent physical properties of these compounds should then lead to a further refinement of our present theories regarding the nature of structure and bonding in group 14 catenates.

5. References

1. For M = Si, see: Miller, R. D. and Michl, J. (1989) Polysilane High Polymers, *Chem. Rev.* **89**, 1359-1409 and references cited therein.
2. For M = Ge, see: (a) Trefonas, P., III, and West, R. (1985) Organogermane Homopolymers and Copolymers with Organosilanes, *J. Polym. Sci. Ed.* **23**, 2099-2107. (b) Miller, R. D. and Sooriyakumaran, R. (1987) Soluble Alkyl Substituted Polygermanes: Thermochromic Behavior, *J. Polym. Sci. Ed.* **25**, 111-125.
3. For M = Sn, see: (a) Adams, S. and Dräger, M. (1987) Polystannanes $Ph_3Sn\text{-}(t\text{-}Bu_2Sn)_n\text{-}SnPh_3$ (n = 1-4): A Route to Molecular Metals?, *Angew. Chem. Int. Ed. Engl.* **26**, 1255-1256. (b) Sita, L. R. (1992) A New Strategy for the Synthesis of Homologously Pure Linear Polystannane Oligomers, *Organometallics* **11** 1442-1444. (c) Imori, T. and Tilley, T. D. (1993) High Molecular Mass Polystannanes via Dehydropolymerization of Di(n-butyl)stannane, *J. Chem. Soc., Chem. Commun.* 1607-1609. (d) Imori, T., Lu, V.; Cai, H., Tilley, T. D. (1995) Metal-Catalyzed Dehydropolymerization of Secondary Stannanes to High Molecular Weight Polystannanes, *J. Am. Chem. Soc.* **117**, 9931-9940.
4. For recent reviews of polystannanes, see: (a) Sita, L. R. (1994) Heavy-Metal Organic Chemistry: Building with Tin, *Acc. Chem. Res.* **27**, 191-197. (b) Sita, L. R. (1995) Structure / Property Relationships of Polystannanes, in R. West (ed), *Advances in Organometallic Chemistry*, Academic Press, Boston, pp. 189-243.
5. (a) Trefonas, P., III, Damewood, J. R., Jr., West, R., Miller, R. D. (1985) Organosilane High Polymers: Thermochromic Behavior in Solution, *Organometallics* **4**, 1318-1319. (b) Balaji, V. and Michl, J. (1991) Singlet Excitation in Polysilanes: *Ab Initio* Calculations on Oligosilane Models, *Polyhedron* **10**, 1265-1284.
6. Sita, L. R, Terry, K. W., Shibata, K. (1995) Sandorfy Hückel Molecular Orbital Approximation for Modeling the Electronic Structures of Long-Chain Polystannanes, *J. Am. Chem. Soc.* **117**, 8049-8050.
7. Sandorfy, C. (1955) LCAO MO Calculations on Saturated Hydrocarbons and Their Substituted Derivatives, *Can. J. Chem.* **33**, 1337-1351.
8. Herman, A., Dreczewski, B., Wojnowski, W. (1985) The Degree of σ-Bond Delocalization in Polysilanes and Its Influence on Reactivity by Through-Bond Interactions. The PES Scaled Sandorfy C Approach, *Chem. Phys.* **98**, 475-481.
9. (a) Plitt, H. S., Downing, J. W., Raymond, M. K., Balaji, V., Michl, J. (1994) Photophysics and Potential-energy Hypersurface of Permethylated Oligosilanes, *J. Chem. Soc., Faraday Trans.* **90**, 1653-1662. (b) Albinsson, B., Teramae, H., Downing, J. W., Michl, J. (1996) Conformers of Saturated Chains: Matrix Isolation, Structure, IR and UV Spectra of $n\text{-}Si_4Me_{10}$, *Chem. Eur. J.* **2**, 529-538. (c) Albinsson, B., Teramae, H., Plitt, H., Goss, L. M., Schmidbaur, H., Michl, J. (1996) Matrix-Isolation IR and UV Spectra of Si_3H_8 and Si_4H_{10}: Isomers and Conformers of Oligosilanes, *J. Phys. Chem.* **100**, 8681-8691.

10. Sita, L. R., Kinoshita, I. (1991) Contribution of Nonbonded Interactions to the Destabilization of a Group 14 Bicyclo[1.1.1]pentane, *J. Am. Chem. Soc.* **113**, 5070-5072.

11. (a) Pople, J. A. and Santry, D. P. (1964) Molecular Orbital Theory of Nuclear Spin Coupling Constants, *Mol. Phys.* **8**, 1-18. (b) Raynes, W. T. and Panteli, N. (1983) The Nuclear Spin-Spin Coupling in the Hydrogen Molecule: Its Equilibrium Value and Bond-Length Dependence, *Chem. Phys. Lett.* **94**, 558-560.

FULLERENE TINKER TOYS

DANIEL T. COLBERT and RICHARD E. SMALLEY
Center for Nanoscale Science and Technology, Rice Quantum Institute, and Departments of Chemistry and Physics
Rice University
P.O. Box 1892, M.S. 100
Houston, TX 77251, USA

ABSTRACT. Fullerenes comprise an infinite class of pure carbon cages. The variety of fullerene structures, their strength and stability, and their ability to undergo chemical derivatization make them ideal building blocks for a nanoscale architecture. Prospects for assembling such fullerene "tinker toys" are discussed with emphasis on their form and growth. Formation of single-walled carbon nanotubes, which are likely to be the ultimate nanoscale building beams, are discussed in detail.

1. Introduction: Fullerene Zoology and Synthesis

Carbon is unique in the periodic table in its ability to form pure covalently bonded extended structures of great strength. Metals also form pure crystals, but the extreme delocalization of their valence electrons favors weak, less directional bonding with many neighbors. Other non-metals, e.g., N and O, form bonds with one another that are so strong that their free energy is minimized by pairing up (witness the explosive nature of many nitrogen compounds). Carbon has characteristics of both these extremes: it forms very strong directional bonds with itself and has a high enough valence that it enjoys bonding to two, three or four partners.

The variety of pure carbon structures would be limited to diamond, graphite, and exotic and highly reactive sp-hybridized molecules if appeal could be made only to the various ways carbon atoms can hybridize. Organizing carbon compounds in a Linnaean hierarchy, the diversity of the "kingdom" of carbon is infinitely enriched in the "phylum" *graphite* (i.e., 3-coordinated carbon) by the inclusion of pentagons and heptagons into the predominantly hexagonal graphene lattice. The curvature conferred by non-hexagonal subunits enables a wealth of structures built on a common theme: Fullerenes. The unifying fullerene theme is the elimination of energetic dangling bonds

J. Michl (ed.), Modular Chemistry, 601–607.

by curving, which is accomplished by the incorporation of pentagons, according to the "Pentagon Road" mechanism [1], discussed shortly.

There is an infinitely rich diversity of structures made possible by the use of pentagons and heptagons together with hexagons. Under the "class" *fullerene* of the "phylum" *graphite* are several "orders" including *spheroidal single-shell fullerenes, fullerene onions*, and *single- and multi-wall nanotubes.* This somewhat arbitrary classification scheme concludes with a miscellaneous class of fullerenes that employ heptagons, yielding a wealth of strange and fanciful objects (cf. Fig. 1). In fact, the use of hexagonal, pentagonal, and heptagonal substructures are sufficient to produced caged carbon structures of any topology.

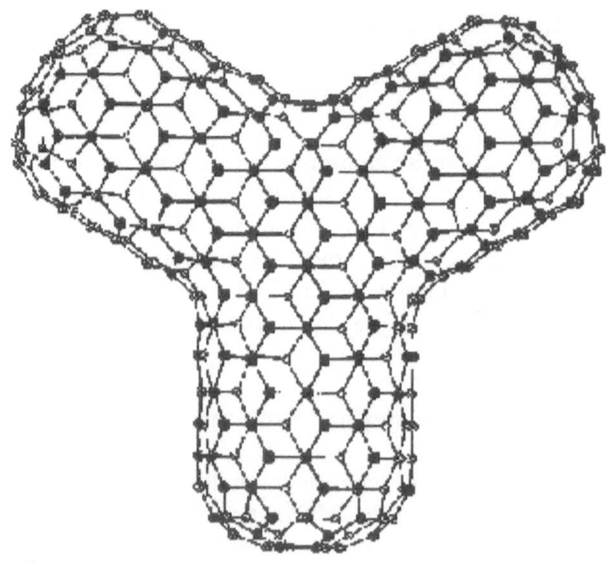

Figure 1. A fanciful fullerene that incorporates hexagons, pentagons (positive curvature), and heptagons (negative curvature). [Used with permission from G. Scuseria.]

2. The Pentagon Road

The pentagon road is a gas-phase mechanism, appropriate for present discussion in light of the fact that no condensed-phase synthesis of a complete fullerene has yet emerged. It is built on partial annealing of carbon clusters in the gas phase, requiring the breaking of C-C bonds at the high temperatures at which fullerenes are known to form. The mechanism is based on the assumptions that the energetically favored form of any *open* graphitic sheet is one that (a) is made up of only hexagons and pentagons, (b) has as many pentagons as possible, while (c) avoiding adjacent pentagons [1]. The resulting fullerene structure does not correspond to a free energy minimum under growth temperatures and pressures, but it *is* kinetically constrained. That is, if the condensing vapor is allowed to anneal to closed fullerenes, then the transformation of this metastable state to graphite, the free-energy minimum, is extremely slow, just as for the transition from diamond to graphite at ordinary temperatures and pressures. The only difference is that fullerenes probably do not correspond to the equilibrium structure *anywhere* in the phase diagram of carbon, but this is an academic point not relevant to this discussion.

The favored terminus of the Pentagon Road under ordinary conditions is C_{60}. In fact, yields of C_{60} up to 40% of all the condensing carbon vapor have been observed [2,3]. There is a long stretch along the Pentagon Road on the way toward C_{60} during which ten dangling bonds persist on the cluster periphery. Under annealing conditions of pure carbon vapor, elimination of these dangling bonds provides the energetic incentive to curve and close as a fullerene. However, one might ask whether it is possible to take a detour off the Pentagon Road by stabilizing the open structure against closing, and lengthen the fullerene-shell precursor into a fullerene tube. As discussed below, we believe this is precisely what happens when carbon vapor co-condenses with a small percentage of metal vapor. At a later stage of growth, once the tubular form is irreversible, and the metal particle is larger, the metal catalyzes the insertion of additional carbon between the metal cluster and the growing tube.

3. Catalytic Growth of Single-Wall Nanotubes by Laser Vaporization

Single-wall nanotubes (SWNTs) are never found with defects along their sides, in contrast to their multi-wall brethren (MWNT), which always contain numerous defects, whether made catalytically from hydrocarbons, or from a carbon arc discharge. These defects are stress points that can lead to breakage, diminishing their tensile strength. They will also form sites for electron scattering, reducing the nanotubes' expected special electron transport properties.

A MWNT that has suffered a bite out of its outer layer(s) can survive such a defect due to the integrity of the inner layers, and due to stabilizing "spot-welds" between adjacent layers. (These spot welds are now also thought to form at the edges of growing nanotubes, and are crucial for their growth by helping to keep the tip open [4].) In contrast, SWNTs should find living with defects much more difficult. Because of the greater strain energy of these thinner tubes, even a few missing atoms in the tube wall are likely to lead to breakage quickly at any temperature high enough to form them in the first place. In this natural selection of the fittest, only SWNTs with perfect walls would remain. Thus, while defects are inevitable for multi-walled nanotubes, they are highly discriminated against in SWNTs. Indeed, no transmission electron micrographs of defective SWNTs have been published. Therefore, SWNTs should represent the idealization of nanotube material.

It was recently discovery in our laboratory that SWNTs were produced catalytically using a simple extension of the laser vaporization/heated oven technique developed earlier in this group to produce C_{60} and other small fullerenes in 30-40% yield [2,3]. SWNTs were formed in much higher yield and with less amorphous overcoating than those made in the arc, the only other known method for producing SWNTs. The attempt to make SWNTs in this way grew from a prior discovery in our laboratory that multi-walled nanotubes were formed by laser-vaporization with a slight modification in operating conditions [4]. Loading the graphite target with about 1 atom percent transition metal (as in the arc production of SWNTs) was an obvious extension. Only the yield was surprising.

Figures 2 and 3 show transmission electron micrographs (TEM) of high-yield Co/Ni-catalyzed (Co:Ni:C in atomic ratios 0.6:0.6:98.8) nanotube material deposited on the cold collector. Single-walled nanotubes typically were found grouped in structures in which many tubes run together aligned in van der Waals contact over most of their length. This morphology requires a very high density of SWNTs in the aerosol for so many tubes to have formed aligned prior to depositing on a cold collector, with very little other carbon available to coat the SWNTs prior to their alignment. The high nanotube yield (over 80% of all the carbon vaporized becomes nanotubes) is especially remarkable considering that the soluble fullerene yield from the same run was found to be about 5%, with much of the remaining carbon consisting of giant fullerenes and onions.

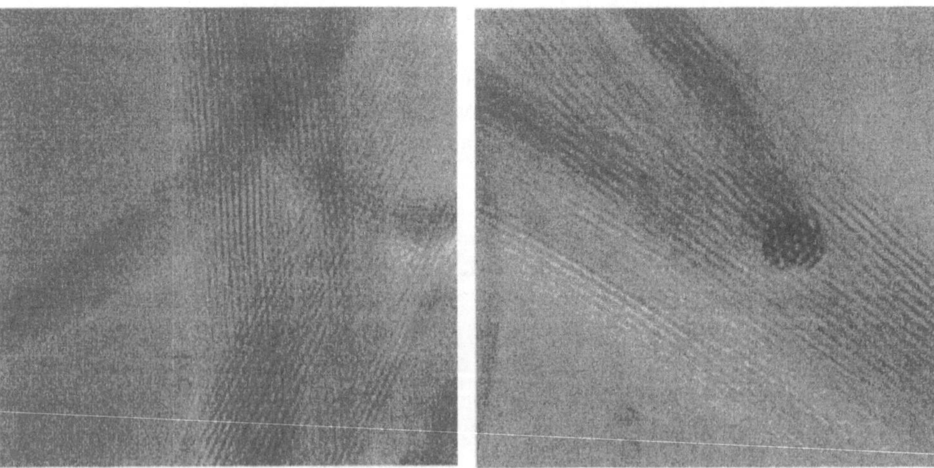

Figure 2. Single-walled nanotube ropes. *Figure 3.* A cross-sectional view of a rope. Fringes are spaced by about 1.3 nm in both figures.

Single-wall nanotubes produced in this way are particularly clean, as might be expected given such high yields. Typical arc-produced SWNTs are covered with a thick layer of amorphous carbon, limiting their usefulness. Much less coating is seen with the present tubes, the bundling leaving very little surface area exposed for overcoating.

The remainder of this paper describes a possible mechanism for formation of SWNTs in both the laser-vaporization apparatus and in the arc. In both cases, essentially *all* catalytically produced nanotubes have a single wall with a very narrow distribution of diameters (although it is somewhat broader in the arc [5]; this is understood to result from a competing mechanism that occurs primarily in the arc method in which pre-formed metal particles nucleate SWNTs). Formation of *multiwalled* nanotubes by metal-catalyzed chemical vapor deposition is widely thought to proceed via solvation of carbon vapor in the metal particle, followed by nanotube precipitation. The particle size is assumed in these models to control the outer diameter of the associated nanotube. Under the present ~1 at% metal loading conditions, however, too few metal atoms will have clustered to form a particle of the appropriate size (~100 atoms for a 1 nm diameter particle) by the time significant carbon clustering occurs. Even if early metal clustering could occur, though, it would somehow have to produce particles of highly uniform diameters to account for the uniform diameter tubes produced from them.

An alternative to nucleation of a nanotube from a pre-existing (large) metal particle is that the tube diameter is *intrinsic* to the condensation of carbon. In this view, the role of metal in nucleation is only to prevent closure of a partially curved carbon cluster to a fullerene, i.e., to cause a detour from the Pentagon Road to the Tube Road (although later the metal assumes the important role of catalyzing the addition of carbon feedstock to grow the tube). Here, a few (perhaps only one!) metal atoms chemisorbed onto the open edge of a fullerene precursor prevent closure into a fullerene. Like the legendary Sisyphus, each time the would-be fullerene climbs the energetic hill toward closure by incorporating the seventh, eighth, ninth, etc., pentagon, the rapidly diffusing metal atom(s) locally anneal the pentagons to hexagons, thus returning the edge to the fully open starting point where it must begin its ascent anew.

Under these circumstances, the relevant question becomes: what is the most likely configuration of a carbon cluster that is constrained to be *open*? A spheroidal fullerene of increasing diameter has the enormous energetic disadvantage of an increasingly large edge circumference. The only reasonable alternative that avoids large open edges is a tubular shape, the diameter of which will be determined by an energetic competition between the strain of bending the graphene sheet, favoring larger diameters, and the number of dangling bonds at the open edge, favoring smaller diameters. Detailed energetic calculations predict that the lowest energy open tublet of 250-600 carbon atoms (the range for which we assume the tube diameter is irreversibly fixed) is the (10,10) tube, denoting the zero-helicity armchair tube of D_5 symmetry [6]. The calculated (10,10) tube diameter is 1.36 nm, in excellent agreement with diameters observed by TEM and by X-ray diffraction studies of SWNT material [7]. Once the open shell has lengthened by a few nm, it is kinetically locked into the tube conformation.

We note that continued growth of the single wall nanotube will still be favored even when the metal particle at its tip grows beyond the initial 1-2 nm diameter. While the lower strain energy of a second layer precipitated from a larger metal particle would be energetically favorable compared with adding to the inner layer, the unavoidable open edges introduced as this second layer nucleates would present a high energetic barrier. This view thus places emphasis on the constrained metal particle size only during initial lengthening of the nanotube. This explains why there has been no observed correlation between nanotube diameter and catalytic particle size.

A final clue to the growth of SWNTs is found in the ends of the "ropes" of aligned individual tubes. All rope ends observed show that the individuals end within a few tube diameters of one another, even though these individuals are thousands of diameters long. Either the ropes break cleanly during sample preparation (sonication), or else they grow together with their growth fronts running together. We find it difficult to believe that individual tubes adhere and align *after* they have grown very many diameters in length, so we envisage a growth mechanism whereby a tube nucleus collides with and adheres to a longer tube, quickly aligning to maximize their van der Waals contact. Both tubes grow by addition of feedstock condensing onto their sides and diffusing to their open tips. However, whereas the longer tube receives reactive flux only from behind its tip, the shorter tube receives additional flux from its front, that is, feedstock diffusing on the longer tube directly toward the open tip of the shorter tube. This additional flux, which is likely to be more reactive than the backside flux, the rate of growth of shorter tubes will always be greater than that of longer tubes. This adherence and catch-up continues as long as individual tube nuclei are available in the condensing vapor. In this way, growing tips will quickly catch up to one another, accounting for the observed uniform rope ends.

4. Future Prospects

The most intriguing aspect of these results is that the yield is so high. This must mean that the "live" end of a growing nanotube is capable of efficiently using large carbon clusters as a feedstock. In particular, it is quite possible that C_{60} itself is consumed by the catalytic Ni/Co cluster at the end of the growing nanotube, adding 60 atoms to extend the tube length. This then suggests the possibility of vastly protracting the growth stage of the SWNT. For example, suppose we insert a network of tungsten wires into the downstream gas flow after the carbon/Ni/Co target but well before the end of the heated zone of the oven. Suppose further that this wire network is arranged so that many of the still-growing nanotubes are caught with their live ends trailing out in the blowing wind of the inert carrier gas flow. It may then be possible to feed the growing "live" ends with an appropriate pure carbon feedstock, e.g., hydrocarbons, carbon vaporized from a pure graphite target, or even C_{60} vapor. Since further laser vaporization of the graphite/Ni/Co target is unnecessary it should be possible to continue growth indefinitely without overcoating the sides of the nanotube with amorphous carbon, or growing the catalytic particle too big by further addition of Ni and Co vapor. Variations on this theme may ultimately permit very clean single wall nanotube "ropes" to be grown of many cm length, all aligned in the flowing wind.

Acknowledgments: This work was supported by the Office of Naval Research, the National Science Foundation, the Texas Advanced Technology Program, the Robert A. Welch Foundation, and used equipment designed for study of fullerene-encapsulated catalysts supported by the Department of Energy, Division of Chemical Sciences.

References

1. Smalley, R.E. (1992) Self-Assembly of the Fullerenes, *Accts. Chem. Res.* **25**, 98-105.
2. Smalley, R. E. and Colbert, D. T., Self-Assembly of Fullerene Tubes and Balls, Proceedings of the Robert A. Welch Foundation 39th Conference on Chemical Research: Nanophase Chemistry (1996).
3. Haufler, R.E., Chai, Y., Chibante, L. P. F., Conceicao, J., Jin, C., Wang, L.-S.,Maruyama, S., and R. E. Smalley (1991) Carbon Arc Generation of C_{60}, *Mat. Res. Soc. Symp Proc.* **206**, 627-638.
4. Guo, T., Nikolaev, Rinzler, A. G., Tománek, D., Colbert, D. T., Smalley, and R. E. (1995) Self-Assembly of Tubular Fullerenes, *Journal of Physical Chemistry* **99**, 10694-10697.
5. Bethune, D. S., Kiang, C. H., De Vries, M. S., Gorman, G. Savoy, R., Beyers, R., Cobalt-Catalyzed Growth of Carbon Nanotubes With Single Atomic Layer Walls, *Nature* **363**, 605 (1993).
6. Xu, C., Colbert D. T., Scuseria G., and Smalley, R. E., in preparation.
7. Thess, A., Lee, R., Nikolaev, P., Dai, H., Petit, P., Robert, J., Xu, C., Lee, Y.-H., Kim, S.-G., Rinzler, A. G., Colbert, D. T., Scuseria, G., Tománek, D., Fischer, J. E., Smalley, R. E., Crystalline Ropes of Metallic Carbon Nanotubes, *Science* **273** (July 26, 1996 issue).

NEW MODULES — NEW FAMILIES OF INTERLOCKED MOLECULES

PETER T. GLINK, J. FRASER STODDART
School of Chemistry, The University of Birmingham
Edgbaston, Birmingham B15 2TT, United Kingdom.

ABSTRACT. Secondary dialkylammonium ions form inclusion complexes — with pseudorotaxane geometries and in varying stoichiometries — with large ring crown ethers such as dibenzo[24]crown-8 and bis-*p*-phenylene[34]crown-10. The complexes are stabilised primarily by [N-H···O] and [C-H···O] hydrogen bonding interactions. The self-assembly of three new rotaxanes — and the solid state structures of eight pseudorotaxanes and a rotaxane — based on this supramolecular system is descibed.

1. Introduction

Supramolecular chemistry [1] — chemistry beyond the molecule — has, at its very heart, the concept of a modular approach to the construction of multicomponent molecular assemblies and supramolecular arrays. The piecing together of small complementary molecular building blocks into ordered superstructures allows for the preparation of novel materials which would be virtually impossible to produce by conventional synthetic chemistry. One of the goals of contemporary science is the production of functioning materials [2-4] which operate at a molecular level, but express

J. Michl (ed.), Modular Chemistry, 609–622.

themselves in a macroscopic manner. It is becoming apparent that such materials probably will not be realised using the techniques of conventional synthetic chemistry on its own, but rather through a marriage of its power with that of supramolecular science. Whereas synthetic chemistry has reached a stage where relatively small molecules of extreme structural complexity can be synthesised almost at will, supramolecular chemistry is still in its infancy. For supramolecular chemistry to flourish, a greater understanding of noncovalent bonding interactions and how they rely upon complementary functional groups and then preside over self-assembly, is necessary so that the chemist can add new building blocks or modules to the molecular 'tool kit'. Our own research group has been successful, during the last ten years or so, in developing a versatile template-directed approach to the designed self-assembly of complex molecular assemblies, such as catenanes [5], rotaxanes [6] and pseudorotaxanes [7] The approach is based on the mutual cooperative noncovalent bonding interactions between π-electron rich aromatic units (*e.g.* hydroquinone rings) and π-electron deficient aromatic units (*e.g.* bipyridinium dications). These components comprise a so-called 'molecular meccano set' [5,6]. To progress to more complex multicomponent systems, the number of pieces in the meccano set — that is, the number of different modules with designated properties — must be increased. To this end, we have uncovered [8-10] very recently a new self-assembling system which is extremely simple in its composition, yet is capable of providing intricate and fascinating supramolecular arrays and molecular assemblies.

The binding of ammonium (NH_4^+) and primary alkylammonium (RNH_3^+) ions with macrocyclic polyethers has received much attention from supramolecular chemists since the discovery by Pedersen [11] of the crown ethers and the complexing ability they exhibit towards a wide range of cations. Complexes of ammonium and primary alkylammonium ions with crown ethers usually occur [12] in a face-to-face manner and are stabilised predominantly by hydrogen-bonding and other ion-dipole interactions. It has been noted [13], from solid state investigations, that the NH_3^+ centres of these ions prefer to lie as deeply as possible within the plane of the macrocycle. By contrast, the binding of secondary dialkylammonium ions ($R_2NH_2^+$) has received relatively little attention [14-20] — presumably, because their association constants for face-to-face complexation with those crown ethers — that bind NH_4^+ and RNH_3^+ ions well — are relatively small. It occurred to us that secondary dialkylammonium ions ($R_2NH_2^+$) should form stable complexes with crown ethers if the ion were given the option of *threading* through the macrocyclic ring, rather than being constrained to form a face-to-face complex. The supramolecular architectures formed by this type of complexation are those of pseudorotaxanes [7]. For simple alkyl groups (R) in a $R_2NH_2^+$ ion, threading through the crown ether requires that the macro-ring be constituted of at least 24 atoms. This much could be gleaned from some preliminary molecular modelling studies on the computer. Thus, we began our study of secondary dialkylammonium binding with the well-known crown ethers, dibenzo[24]crown-8 (**DB24C8**) [11], asymmetric-dibenzo[24]crown-8 (***asym*DB24C8**) [11] and bis-*p*-phenylene[34]crown-10 (**BPP34C10**) [5,6]. The salts we have been studying are simple in their constitutions — *e.g.* the monocationic, dibenzylammonium (**1**$^+$), di-*n*-butylammonium (**2**$^+$) and benzyl-*n*-butylammonium (**3**$^+$) hexafluorophosphates, and the dicationic, α,α'-bis(benzylammonium)-*p*-xylene (**4**$^{2+}$) bis(hexafluorophosphate). To date, we have been successful in obtaining solid state structures of eight complexes derived by mixing these modules [8-10]. We have utilised the understanding gained in studying these complexes in order to self-assemble three new rotaxanes [21]. In this short review, we will give a brief overview of these fascinating supramolecular complexes and molecular compounds.

2. Pseudorotaxanes

The first hint that complexation occurs between the three monocations, **1**$^+$, **2**$^+$ and **3**$^+$, and **DB24C8** and ***asym*DB24C8** is that their insoluble hexafluorophosphates are readily taken up into $CHCl_3$ and CH_2Cl_2 solutions in the presence of either of these macrocyclic polyethers. The three salts, **1**·PF_6, **2**·PF_6 and **3**·PF_6, cocrystallise with **DB24C8** to produce [8,10] inclusion complexes of 1:1 stoichiometry with

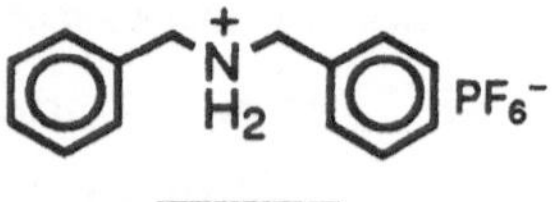

1.PF6   2.PF6   3.PF6

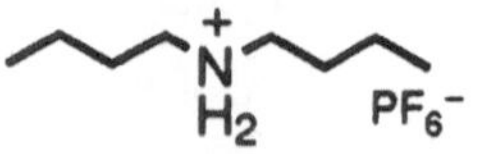

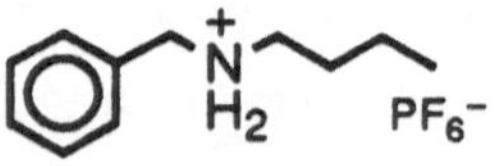

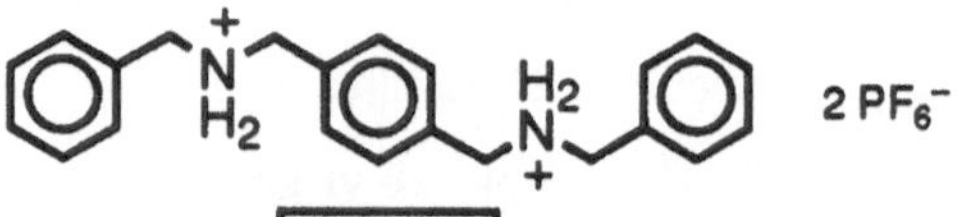

4.2PF6

DB24C8 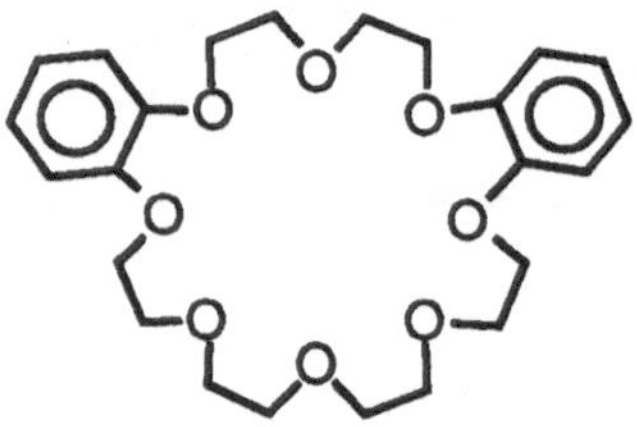   *asym*DB24C8 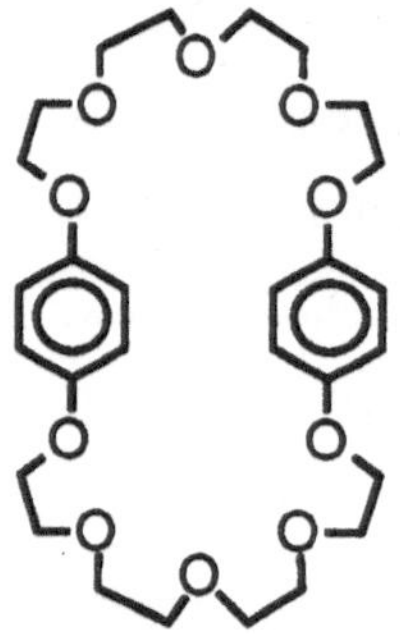   BPP34C10

pseudorotaxane geometries (Figure 1a-c). These complexes are all stabilised by a series of hydrogen bonds — two [N-H···O] contacts involving the hydrogen atoms of the NH_2^+ group and oxygen atoms of the crown ether, and either one or two [C-H···O] contacts involving hydrogen atoms on the methylene group adjacent to the formally cationic nitrogen atom. In the case of the $[\mathbf{DB24C8{\cdot}1}]^+$ and $[\mathbf{DB24C8{\cdot}3}]^+$ complexes (Fig. 1a and 1c), two independent 1:1 complexes — in which the oxygen atoms involved in hydrogen-bonding are different, as are the conformations of the dialkylammonium ions — are observed in each unit cell, indicating that more than one hydrogen bonding motif is possible. These 1:1 complexes also exist in solution as indicated by ^1H NMR spectroscopic techniques. In a variety of solvents, the $[\mathbf{DB24C8{\cdot}1}][PF_6]$ complex exists as a discrete species — that is, sharp signals for the resonances of the protons of the 1:1 complex are observed in $CDCl_3$, CD_3CN and CD_3COCD_3 solutions along with

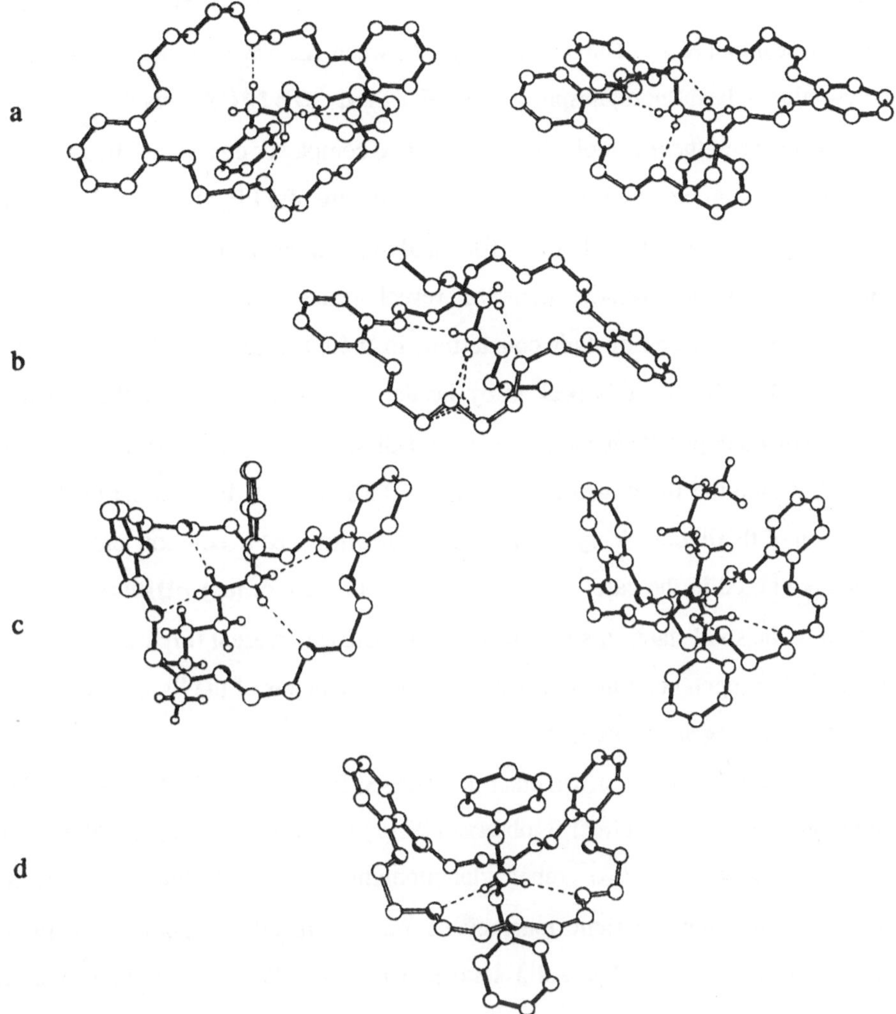

Figure 1. Solid state structures of the 1:1 complexes (a) [**DB24C8·1**]$^+$ (two independent 1:1 complexes), (b) [**DB24C8·2**]$^+$, (c) [**DB24C8·3**]$^+$ (two independent 1:1 complexes), and (d) [***asym*DB24C8·1**]$^+$.

signals associated with the uncomplexed host and guest. This situation arises because of the physical difficulty of threading a phenyl ring of **1**$^+$ through the cavity of **DB24C8** and, consequently, association and dissociation of the two species are slow in solution on the ^{1}H NMR timescale at 300 or 400 MHz. The di-*n*-butylammonium ion (**2**$^+$) exhibits faster kinetics for these processes, as indicated by the sets of sharp, time-averaged signals observed in $CDCl_3$, CD_3CN and CD_3COCD_3 for protons in both the host and guest molecules. The 1:1 complex between the asymmetric salt, benzyl-*n*-butylammonium hexafluorophosphate (**3**·PF_6) and **DB24C8** displays kinetics intermediate between these exhibited by the first two complexes discussed. Broad signals are observed at room temperature in $CDCl_3$ for a mixture of **3**·PF_6 and **DB24C8**. Mass spectrometry is another useful tool for studying these complexes. The three supramolecular systems already discussed reveal peaks which correspond to 1:1 complexes having lost one of their counterions in their respective FAB mass spectra. Indeed, for [**DB24C8**·**1**]$^+$, this peak is by far the most intense one in the spectrum, indicating strong complexation and a slow rate of dissociation of the components in the 'gas phase'. Changing the crown ether to ***asym*DB24C8** has little effect [10] on the complexation of these ions. Very similar results are witnessed in solution and in the gas phase as were noted in the case of **DB24C8**. The structure of [***asym*DB24C8**·**1**]$^+$ was solved [10] in the solid state. A single [2]pseudorotaxane is present (Figure 1d) in which the [N-H···O] hydrogen bonding occurs with the oxygen atoms of the 'longer' of the two polyether loops acting as the acceptors.

When the size of the crown ether is increased from **DB24C8** to **BPP34C10**, a fascinating supramolecular effect is observed [9,10] in the solid state, in solution and in the 'gas phase' — two dibenzylammonium ions (**1**$^+$) thread themselves simultaneously through the same macrocyclic ring to produce a complex with 1:2 (host:guest) stoichiometry. The solid state structure of [**BPP34C10**·(**1**)$_2$]$^{2+}$ reveals (Figure 2) a double-stranded [3]pseudorotaxane. This superstructure is stabilised by four [N-H···O] hydrogen-bonds and two aromatic-aromatic edge-to-face [C-H···π] interactions. A complex of 1:2 stoichiometry exists not only in the solid state — two equivalents of **1**·PF_6 are taken up into a solution of **BPP34C10** in CD_2Cl_2 (this stoichiometry is readily determined by integration of the relevant probe protons in the ^1H NMR spectrum). Mass spectrometry indicates that such a complex exists in the 'gas phase' as well: there is a peak in the FAB mass spectrum corresponding to the [**BPP34C10**·(**1**)$_2$][PF_6] ion.

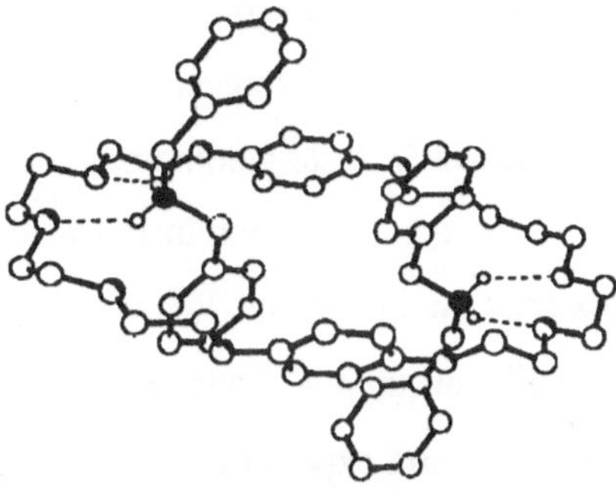

Figure 2. A double-stranded [3]pseudorotaxane — the solid state structure of the 1:2 complex $[\mathbf{BPP34C10}\cdot(\mathbf{1})_2]^{2+}$.

We have noted the effect that increasing the size of the crown ether has on the stoichiometry of the complex — namely, two secondary dialkylammonium ions can be accommodated side-by-side in the cavity of **BPP34C10**. The next question was — what would happen if we added more secondary dialkylammonium NH_2^+ centres to the thread-like component? The bis(dialkylammonium) salt $\mathbf{4}\cdot[PF_6]_2$ was cocrystallised with **BPP34C10**. Remarkably, a superstructure [9,10] of 2:2 stoichiometry is formed (Figure 3) in which two dications $\mathbf{4}^{2+}$ are threaded through two **BPP34C10** that are positioned immediately adjacent to one another. This double-threaded doubly encircled [4]pseudorotaxane is stabilised by a total of nine [N-H···O] and three [C-H···O]

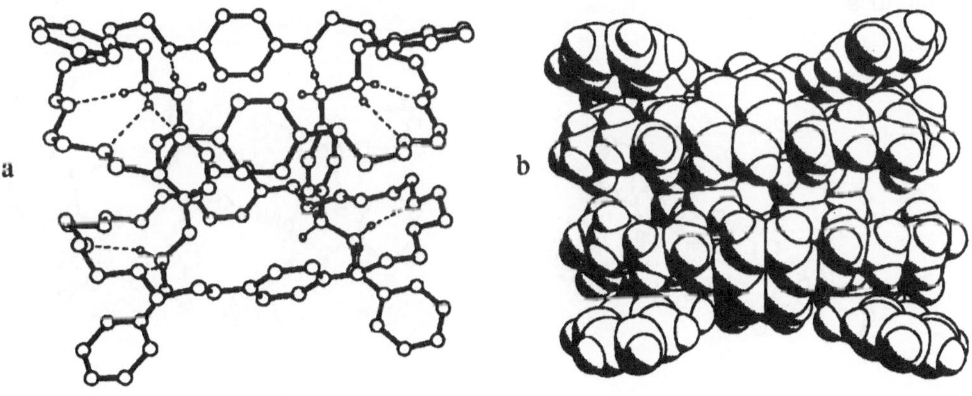

Figure 3. A doubly encircled double-stranded [4]pseudorotaxane — (a) ball-and-stick and (b) space-filling representations of the solid state structure of the 2:2 complex $[(\mathbf{BPP34C10})_2\cdot(\mathbf{4})_2]^{4+}$.

hydrogen-bonds between the hydrogen atoms of the NH_2^+ centres and the adjacent benzylic methylene groups and some of the oxygen atoms of the crown ether rings. However, evidence for complexes with 2:2 stoichiometries has not been found in either the solution or gas phases. In the gas phase, for instance, the peak of highest mass corresponds to a 1:1 complex after the loss of one PF_6^- counterion.

If we retain the thread-like dication $\mathbf{4}^{2+}$ and return to the smaller crown ethers, **DB24C8** and ***asym*DB24C8**, we would expect these species to form stable complexes with 2:1 (host:guest) stoichiometries — that is, each NH_2^+ centre of $\mathbf{4}^{2+}$ should be able to complex one of the crown ethers. This is indeed the case. In the solid state, both $[(\mathbf{DB24C8})_2\cdot\mathbf{4}]^{2+}$ and $[(\boldsymbol{asym}\mathbf{DB24C8})_2\cdot\mathbf{4}]^{2+}$ exist (Figure 4) as double-threaded [3]pseudorotaxanes [9,10] which are centrosymmetric about the central *p*-xylyl unit of $\mathbf{4}^{2+}$. Both complexes are stabilised by a series of [N–H···O] and [C–H···O] hydrogen-bonding interactions, and, additionally, $[(\mathbf{DB24C8})_2\cdot\mathbf{4}]^{2+}$ is stabilised by an almost parallel π-π stacking interaction between the central *p*-xylyl spacer and one of the catechol rings of each of the **DB24C8** rings. In solution, the stoichiometries of these 2:1 complexes are determined readily — since the slow kinetics of complexation and decomplexation are once again operative in these complexes — by integration of the signals of the complexed hosts and guest. Both 2:1 complexes are observed in their respective FAB mass spectra as the peaks of highest mass, having lost one of their counterions.

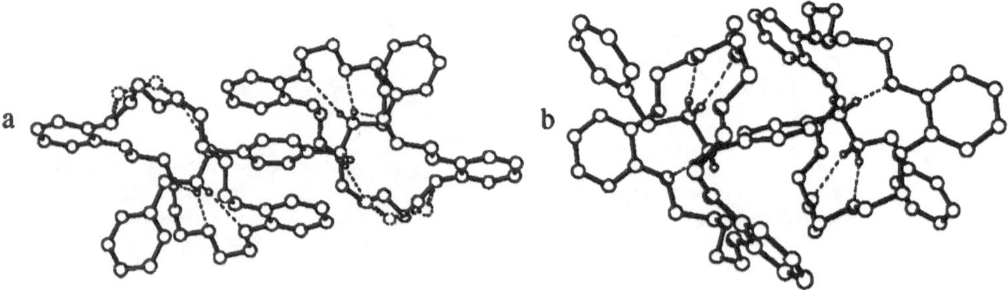

Figure 4. Doubly encircled [3]pseudorotaxanes — the solid state structures of the 2:1 complexes (a) $[(\mathbf{DB24C8})_2\cdot\mathbf{4}]^{2+}$ and (b) $[(\boldsymbol{asym}\mathbf{DB24C8})_2\cdot\mathbf{4}]^{2+}$.

The types of complexes we have been studying are quite neatly summarised with the aid of the cartoons shown in Figure 5. By expanding or reducing the size of the cavity of the crown ether or by adding or removing NH_2^+ centres, we have identified at least one example of the four types of complexes of varying stoichiometry. No doubt the general idea represented in this Figure can, and will, be expanded on as we explore further the limits we may place on this very simple self-assembling system. The ease with which the modular approach to self-assembly produces such wonderfully diverse structures could also herald some fascinating developments in macromolecular science.

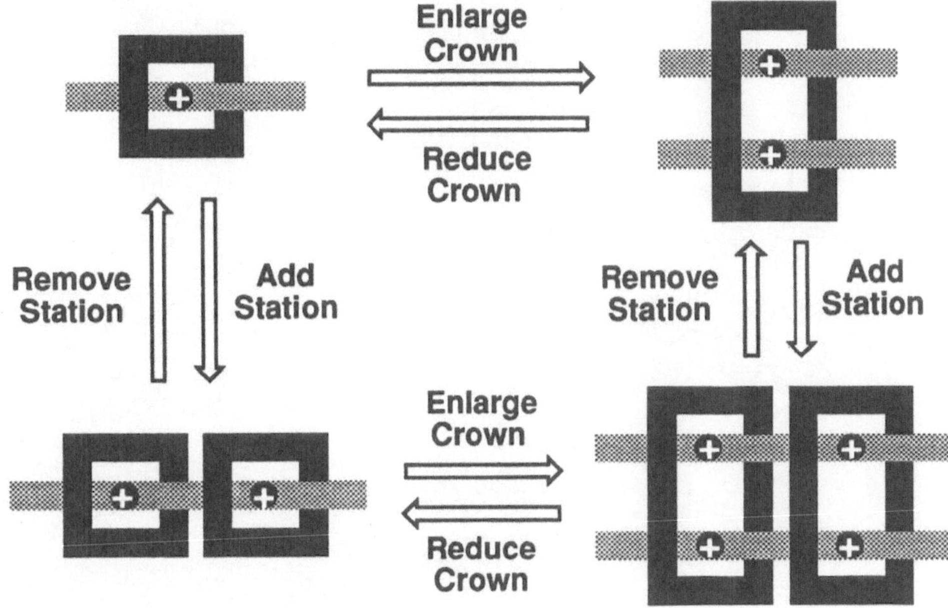

Figure 5. Cartoon representation of the types of pseudorotaxanes we have prepared.

3. Rotaxanes

As a first step towards extending these supermolecules of a pseudorotaxane architecture into the domain of mechanically–interlocked molecules, we have prepared [21] three new rotaxanes based on binding between secondary dialkylammonium ions and crown ethers. The key to their self-assembly lies in the synthesis of the terminally-functionalised

Bu^tO_2C–C≡C–$CO_2{}^tBu$ + N_3–CH_2–Ar–CH_2–$\overset{+}{N}H_2$–CH_2–Ar–CH_2–N_3 PF_6^- + Bu^tO_2C–C≡C–$CO_2{}^tBu$

5.PF_6

CH_2Cl_2 Δ

DB24C8

asym **DB24C8**

PF_6^-

7·PF_6 $m = 2$, $n = 1$ (**DB24C8**), 31%

8·PF_6 $m = 1$, $n = 2$ (***asym*DB24C8**), 24%

in CH_2Cl_2 or CH_2Cl_2/MeCN solutions and result in two [2]rotaxanes, **7**·PF_6 and **8**·PF_6, and a [3]rotaxane **9**·$2PF_6$, respectively, being formed in moderate yields. The stoichiometries of the rotaxanes are determined readily by ^1H and ^{13}C NMR spectroscopies. In addition, FAB mass spectrometry produces spectra in which the peaks of highest mass and greatest intensity are those characteristic of the rotaxanes.

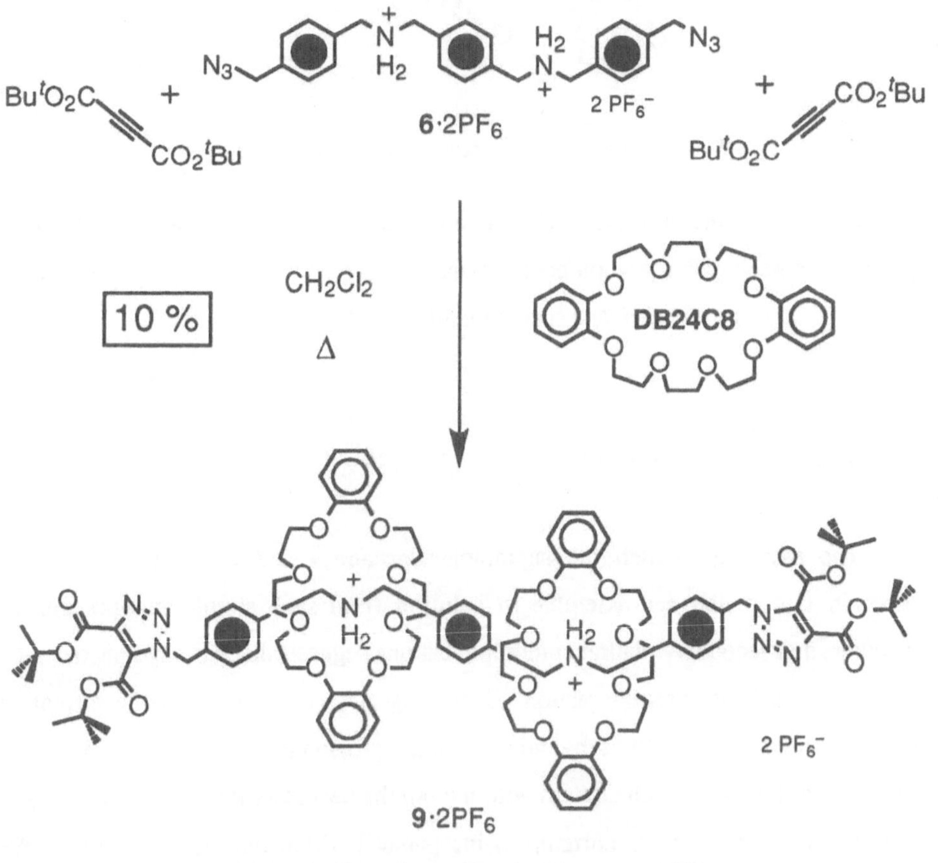

Scheme 2. The self-assembly of a [3]rotaxane, **9**·$2PF_6$.

The solid state structure of one of these rotaxanes — namely **7**·PF_6 — was solved (Figure 6) by X-ray crystallography [21]. It shows quite clearly the interlocked nature of the two components of this molecule. The [N–H···O] hydrogen-bonding interactions, which are the basis for the self-assembly of this rotaxane, are evident in the

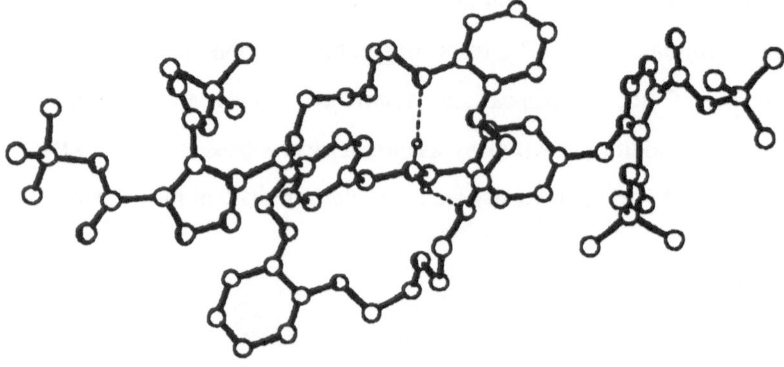

Figure 6. The solid state structure of the [2]rotaxane 7^+.

crystal structure, *i.e.*, the 'information' contained in the rotaxane precursors 'lives on' in the [2]rotaxane itself. This phenomenon has been observed repeatedly in the solid state for a range of catenane and rotaxane superstructures [5,6].

4. Conclusions and Outlook

The ease with which the supramolecular arrays and molecular assemblies discussed in this review self-assemble in solution from such simple components as crown ethers and secondary dialkylammonium cations suggests that we have uncovered a quite fundamental host-guest interaction. Exploiting these interactions in the formation of more complicated assemblies by mixing small programmed modules — from an expanded 'tool kit' — which contain within them the necessary information to support the self-assembly processes, is currently being pursued within our research group. We believe that, although this self-assembling system has taken almost thirty years to be uncovered — since the observation that primary alkylammonium ions bind to crown ethers in a face-to-face manner — in the laboratory, the availability of the components and the intricacy and beauty of some of these complexes should result in considerable and renewed interest in the interactions between these complementary molecules and ions.

5. References

[1] Lehn, J.-M. (1995) *Supramolecular Chemistry: Concepts and Perspectives*, VCH, Weinheim.

[2] Whitesides, G.M., Mathias, J.P., and Seto, C.T. (1991) Molecular self-assembly and nanochemistry: a chemical strategy for the synthesis of nanostructures, *Science* **254**, 1312-1319.

[3] Whitesides, G.M., Simanek, E.E., Mathias, J.P., Seto, C.T., Chin, D.N., Mamusen, M., and Gordon, D.M. (1995) Noncovalent synthesis — using physical-organic chemistry to make aggregates, *Acc. Chem. Res.* **28**, 37-44.

[4] De Silva, A.P. and McCoy, C.D. (1994) Switchable photonic molecules in infomation technology, *Chem. Ind. (London)*, 992-996.

[5] Amabilino, D.B., Ashton, P.R., Brown, C.L., Córdova, E., Godínez, L.A., Goodnow, T.T., Kaifer, A.E., Newton, S.P., Pietraszkiewicz, M., Philp, D., Raymo, F.M., Reder, A.S., Rutland, M.T., Slawin, A.M.Z., Spencer, N., Stoddart, J.F., and Williams, D.J. (1995) Molecular meccano. 2. Self-assembly of [*n*]catenanes, *J. Am. Chem. Soc.* **117**, 1271-1293.

[6] Anelli, P.L., Ashton, P.R., Ballardini, R., Balzani, V., Delgado, M., Gandolfi, M.T., Goodnow, T.T., Kaifer, A.E., Philp, D., Pietraskiewicz, M., Prodi, L., Reddington, M.V., Slawin, A.M.Z., Spencer, N., Stoddart, J.F., Vicent, C., and Williams, D.J. (1992) Molecular meccano. 1. [2]Rotaxanes and a [2]catenane made to order, *J. Am. Chem. Soc.* **114**, 193-218.

[7] Ashton, P.R., Philp, D., Spencer, N., and Stoddart, J.F. (1991) The self-assembly of [*n*]pseudorotaxanes, *J. Chem. Soc., Chem. Commun.* 1677-1679.

[8] Ashton, P.R., Campbell, P.J., Chrystal, E.J.T., Glink, P.T., Menzer, S., Philp, D., Spencer, N., Stoddart, J.F., Tasker, P.A., and Williams, D.J. (1995) Dialkylammonium ion/crown ether complexes: the forerunners of a new family of interlocked molecules, *Angew. Chem. Int. Ed. Engl.* **34**, 1865-1869.

[9] Ashton, P.R., Chrystal, E.J.T., Glink, P.T., Menzer, S., Schiavo, C., Stoddart, J.F., Tasker, P.A., and Williams, D.J. (1995) Doubly encircled and double-stranded pseudorotaxanes, *Angew. Chem. Int. Ed. Engl.* **34**, 1869-1871.

[10] Ashton, P.R., Chrystal, E.J.T., Glink, P.T., Menzer, S., Schiavo, C., Spencer, N., Stoddart, J.F., Tasker, P.A., White, A.J.P., and Williams, D.J. (1996) Pseudorotaxanes formed between secondary dialkylammonium salts and crown ethers, *Chem. Eur. J.* **2**,709-728.

[11] Pedersen, C.J. (1967) Cyclic polyethers and their complexes with metal salts, *J. Am. Chem. Soc.* **89**, 7017-7036.

[12] Goldberg, I. (1984) Complexes of crown ethers with molecular guests, in Atwood, J.L., Davies, J.E.D., and MacNicol, D.D. (eds.), *Inclusion Compounds, Vol. 2* Academic Press, London, pp. 261-335.

[13] Trueblood, K.N., Knobler, C.B., Lawrence, D.S., and Stevens, R.V. (1982) Structures of the 1:1 complexes of 18-crown-6 with hydrazinium perchlorate, hydroxylammonium perchlorate and methylammonium perchlorate, *J. Am. Chem. Soc.* **104**, 1355-1362.

[14] Metcalfe, J.C., Stoddart, J.F., and Jones, G. (1977) Complexation of primary alkylammonium salts and secondary dialkylammonium salts by *N,N*-dimethyl-1,7-diaza-4,10-dioxacyclododecane, *J. Am. Chem. Soc.* **99**, 8317-8319.

[15] Izatt, R.M., Lamb, J.D., Izatt, N.E., Rossiter, Jr., B.E., Christensen, J.J., and Haymore, B.L. (1979) A calorimetric titration study of the reaction of several organic ammonium cations with 18-crown-6 in methanol, *J. Am. Chem. Soc.* **101**, 6273-6276.

[16] Krane, J., and Aune, O. (1980) Complexation of dibenzylammonium salt by *N,N*-dimethyl-1,7-diaza-4,10-dioxacyclododecane. A reinvestigation, *Acta Chem. Scand.* **B34**, 397-401.

[17] Tsukube, H. (1984) Characteristic transport properties of diaza-crown ethers for primary and secondary ammonium cations, *Bull. Chem. Soc. Jpn.* **57**, 2685-2686.

[18] Abed-Ali, S.S., Brisdon, B.J., and England, R. (1987) Functionalised polyorganosiloxanes as selective liquid extractants, *J. Chem. Soc., Chem. Commun.* 1565-1566.

[19] Vögtle, F., and Hoss, R. (1992) Synthetic molecular receptor for piperazine and related diamines, *J. Chem. Soc.*, *Chem. Commun.* 1584-1585.

[20] Misumi, S. (1993) Recognitory coloration of cations with chromoacerands, *Top. Curr. Chem.* **165**, 163-192.

[21] Ashton, P.R., Glink, P.T., Stoddart, J.F., Tasker, P.A., White, A.J.P., and Williams, D.J. (1996) Self-assembling [2]- and [3]-rotaxanes from secondary dialkylammonium salts and crown ethers, *Chem. Eur. J.* **2**, 729-736.

ZIRCONOCENE "MOLECULAR ZIPPERS":

Implements for Construction of Well-Defined Polymers and Macrocycles

T. DON TILLEY AND SHANE S. H. MAO
Department of Chemistry
University of California at Berkeley
Berkeley, CA 94720-1460

ABSTRACT. The zirconocene-coupling of diynes may be employed as a carbon-carbon bond-forming reaction for construction of macromolecular structures. Diynes of the type RC≡C-C_6H_4-C_6H_4-C≡CR (R = alkyl) lead to polymers $[C_6H_4C_6H_4C_4R_2\text{-}ZrCp_2]_n$, with zirconacyclopentadiene units in the polymer backbone. Via reactions of the Zr-C bonds of this polymer, introduction of various unsaturated groups into the polymer chain is achieved. Polymers are also obtained by the coupling of diynes containing silicon substituents in the spacer group, such as $MeC{\equiv}CSiMe_2C_6H_4SiMe_2C{\equiv}CMe$. Interestingly, the resulting polymers quantitatively degrade upon mild heating in solution to macrocycles that contain zirconacyclopentadiene units. The coupling of diynes with terminal silyl substituents also provides high-yield routes to macrocycles.

1. Introduction

Major efforts in materials chemistry focus on the molecular engineering of new supramolecular structures with tailored properties [1-3]. Such structures form the basis for "molecular devices" which may have a number of possible functions (e.g., electronic [1f,g], chemical [1a,b], optical [1b,d,e], and/or mechanical [1a,b]). Many of these functions rely on charge-transport capabilities in the structure, which are often incorporated via electron-delocalized segments [1a]. In principle, the electronic properties for such components are intrinsically tunable, via synthetic manipulation of band gaps, ionization potentials and electron affinities [1a,d]. Thus, it is important to develop synthetic methods that allow the systematic variation of electronic properties for polymer chains, oligomers, macrocycles, etc. A potentially useful strategy for fine-tuning these properties involves use of versatile "synthetic intermediate" structures which may be derivatized in a variety of ways [4]. With this modular approach, specified components in a delocalized structure may be interchanged at will. Here we describe macromolecules and macrocycles which may be modified by the introduction of modules via synthetically versatile zirconacyclopentadiene groups.

Over the past two decades, zirconacyclopentadienes have been studied intensively as synthetic intermediates in organic synthesis [5]. These versatile synthons are readily converted to other derivatives including dienes [5], aromatics [6], and heterocycles [7], and are formed by the coupling reaction of alkynes with "zirconocene", Cp_2Zr, which is generated in solution as a reactive intermediate [8]. For example, photolysis of dimethylzirconocene in the presence of alkynes affords the corresponding zirconacyclopentadiene complex in 35-50 % yield [8a]. Thermolysis of $Cp_2Zr(CO)_2$ in the presence of alkynes also generates zirconacyclopentadienes, in 41-51 % yield [8b], but higher yields are typically obtained by the reductive coupling of Cp_2ZrCl_2/alkyne

J. Michl (ed.), Modular Chemistry, 623–635.

mixtures with magnesium [8c]. Finally, Negishi has described a simple procedure for generation of zirconocene [8d], stabilized as a 1-butene adduct [8e], by addition of 2 equivalents of n-BuLi to a THF solution of Cp_2ZrCl_2 at -78 °C (eq 1). By this method, zirconacyclopentadienes are obtained in nearly quantitative yield.

$$Cp_2ZrCl_2 \xrightarrow[\text{-78 °C, -BuH}]{\text{2 n-BuLi}} \left[Cp_2Zr(\text{1-butene}) \right] \xrightarrow[-\text{1-butene}]{2\,R-C\equiv C-R} Cp_2Zr(C_4R_4) \quad (1)$$

Zirconocene-mediated alkyne coupling appeared to us to represent a potentially powerful method for the construction of macromolecular structures. For example, the intermolecular coupling of diynes should in principle lead to polymers incorporating zirconacyclopentadiene groups into the main chain [9]. In related work, two groups have recently reported the synthesis of cobaltacyclopentadiene polymers by a similar strategy [10]. Thus, we initiated an investigation into the zirconocene coupling reactions of diynes RC≡C-X-C≡CR possessing various rigid spacer groups X [4]. Previous work on the zirconocene-coupling of diynes with flexible spacer groups has afforded bicyclic compounds [9] and oligomers [9c,d]. In this contribution, we provide an overview of the results we have obtained on diyne couplings by zirconocene.

2. Diyne Syntheses

Our investigations have revealed three types of diyne coupling reactions, which are distinguished by the nature of the diyne substituents [4]. In particular, silicon substituents in the alkyne play a very important role in determining the regiochemistry of metallacycle formation, by directing the carbon atoms to which they are bound toward zirconium [11]. Also, as discussed below, silyl substituents lead to reversible coupling reactions which produce macrocycles [4a,b,g]. Thus, the pattern of diyne substitution plays an important role in determining the final structure for the coupled product.

2.1. DIYNES WITHOUT SILICON SUBSTITUENTS

Diynes **1** and **2** were synthesized in high yield via palladium catalyzed cross-coupling [12] (eq 2). The reactions go smoothly at room temperature for 4,4'-diiodobiphenyl, but elevated temperatures were required for reactions with dibromobiphenyl.

$$X-C_6H_4-C_6H_4-X + 2\,HC\equiv CR \xrightarrow[Et_3N,\ THF]{PdCl_2(PPh_3)_2/CuI} RC\equiv C-C_6H_4-C_6H_4-C\equiv CR \quad (2)$$

X = Br, I

1, R = Me, 94 %

2, R = $(CH_2)_4Me$, 95 %

2.2. DIYNES WITH INTERNAL SILICON SUBSTITUENTS

Diynes with internal silicon substituents were synthesized in three steps (eq 3). The Grignard reaction afforded silyl ethers $(EtO)Me_2Si$-X-$SiMe_2(OEt)$ in moderate yield (ca. 65%), and the following conversion to the corresponding silyl chlorides is quantitative.

The final step to produce the diynes **3-5** proceeds in high yield, but it is worth noting that the direct reactions of $(EtO)Me_2Si\text{-}X\text{-}SiMe_2(OEt)$ with LiCCMe were not successful.

$$Br{-}X{-}Br + (EtO)_2SiMe_2 + 2\,Mg \xrightarrow{THF} (EtO)Me_2Si{-}X{-}SiMe_2(OEt) \quad (3)$$

$$(EtO)Me_2Si{-}X{-}SiMe_2(OEt) \xrightarrow{CH_3COCl} ClMe_2Si{-}X{-}SiMe_2Cl$$

$$ClMe_2Si{-}X{-}SiMe_2Cl \xrightarrow[THF]{MeC{\equiv}CLi} MeC{\equiv}CMe_2Si{-}X{-}SiMe_2C{\equiv}CMe$$

3, X = 1,4-phenylene — 60% total yield

4, X = 1,3-phenylene — 52% total yield

5, X = 4,4'-biphenylene — 48% total yield

2.3. DIYNES WITH TERMINAL SILICON SUBSTITUENTS

Diyne **6** was also synthesized via palladium catalyzed Heck-coupling (eq 4). Protiodesilylation of **6** afforded the terminal diyne, which was then lithiated and reacted with a silyl chloride to give the corresponding diynes **7** and **8**.

$$X{-}C_6H_4C_6H_4{-}X + 2\,HC{\equiv}CSiMe_3 \xrightarrow[Et_3N,\ THF]{PdCl_2(PPh_3)_2/CuI} Me_3SiC{\equiv}C{-}C_6H_4C_6H_4{-}C{\equiv}CSiMe_3\ (\mathbf{6})$$

X = Br, I

$$\mathbf{6} \xrightarrow{KOH/MeOH} HC{\equiv}C{-}C_6H_4C_6H_4{-}C{\equiv}CH \quad (4)$$

$$HC{\equiv}C{-}C_6H_4C_6H_4{-}C{\equiv}CH \xrightarrow[(2)\ RSiMe_2Cl]{(1)\ LiNMe_2} RMe_2SiC{\equiv}C{-}C_6H_4C_6H_4{-}C{\equiv}CSiMe_2R$$

7, R = $(CH_2)_9CH_3$

8, R = $(CH_2)_4CH{=}CH_2$

3. Zirconocene Coupling of Diynes Without Silicon Substituents

The coupling reaction of **1** with zirconocene (generated by addition of 2 equiv of *n*-BuLi to Cp_2ZrCl_2 at -78 °C) afforded an insoluble red polymer, which was characterized by infrared spectroscopy and elemental analysis as having the formula $[C_6H_4C_6H_4C_4Me_2ZrCp_2]_n$. Soluble derivatives of this polymer may be obtained by incorporation of alkyl groups into the Cp rings, or as diyne substituents. Thus a soluble polymer (**9**) with a monomodal molecular weight distribution (by gel permeation chromatography, gpc) was obtained from the reaction of zirconocene with **2** (eq 5). The method used to generate zirconocene has a strong influence on the molecular weight and polydispersity of the polymer. Negishi's method afforded a higher number-average molecular weight polymer, with a polydispersity close to 2 (M_w/M_n = 37 000/18 000). The magnesium reduction method afforded polymer with a higher weight-average molecular weight and a larger polydispersity (M_w/M_n = 80 000/14 000). It was observed that the polymerization proceeded in a homogeneous fashion in the former case, but heterogeneously with the magnesium reagent. Therefore the two molecular weight distributions appear to result from different polymerization mechanisms. Polymer **9** is soluble in THF and toluene, and processible into thin films. It is sensitive to moisture, but relatively stable in dry air.

$C\equiv C(CH_2)_4CH_3$ … **2** … $C\equiv C(CH_2)_4CH_3$ —"Cp_2Zr", THF→ **9** (5)

To help elucidate the polymer structure, we prepared and completely characterized a molecular species for comparison of spectroscopic data, by reaction of zirconocene with $PhC\equiv C(CH_2)_4CH_3$. This reaction gave about 80% 2,4-diphenyl-3,5-dipentyl-zirconacyclopentadiene and 20% 2,5-diphenyl-3,4-dipentylzirconacyclopentadiene. By comparison of the NMR spectra for these compounds with polymer **9**, the distribution of substitution patterns given in eq 5 was established.

As a versatile polymer synthon, **9** undergoes a variety of reactions to form new types of conjugated chains. For example, it reacts with HCl to form a conjugated biphenylene-butadienyl polymer (**10**, M_w/M_n = 22 000/11 000) in quantitative yield (eq 6).

9 —HCl, $- n\ Cp_2ZrCl_2$→ **10** (6)

Air-stable polymer **10** is soluble in common hydrocarbon solvents, and can be separated from Cp_2ZrCl_2 and purified by precipitation from methanol. The spectroscopic data for this polymer are consistent with the assigned structure in equation 6. Polymer **10** is very stable, as shown by the thermogravimetric analysis (TGA) trace, which reveals an onset decomposition temperature of 403 °C (under nitrogen), and a weight decrease to 20% after thermolysis.

A regular, biphenylene-phosphole copolymer was produced by reaction of **9** with $PhPCl_2$ (Scheme 1). The yellow polymer **11** (M_w/M_n = 13 000/7 000) is soluble in common organic solvents. Its $^{31}P\{^1H\}$ NMR spectrum (Figure 1) consists of two peaks which appear in an approximate integrated ratio which is consistent with the distribution of 2,5- vs. 2,4-substitution patterns in the precursor polymer.

Other new polymers which have been derived from **9** are shown in Scheme 1. Polymer **9** reacts cleanly with iodine to give the iodo-derivatized polymer **12**, and with S_2Cl_2 to give the biphenylene-thiophene polymer **13**. In addition, **9** reacts with $MeO_2CC\equiv CCO_2Me$, in the presence of two equivalents of CuCl and LiCl, to form a functionalized polyphenylene (**14**) [6]. The electronic properties of all the new polymers described above are currently under investigation.

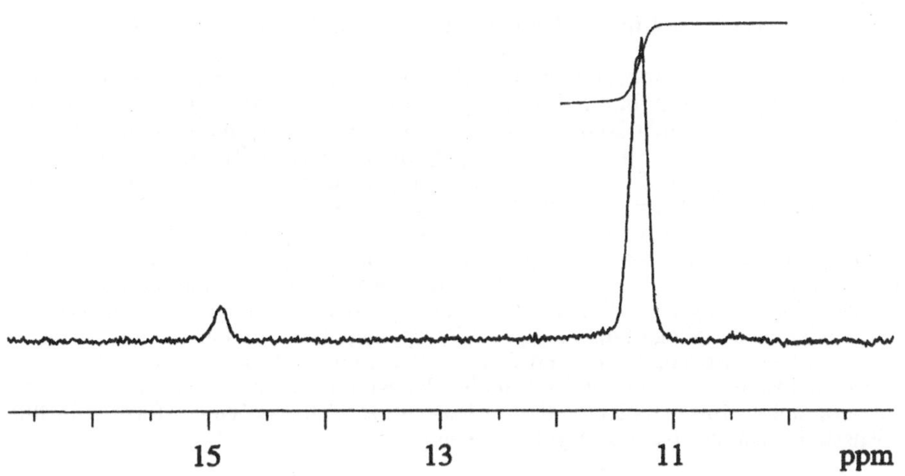

Figure 1. $^{31}P\{^{1}H\}$ NMR spectrum for **11**

Scheme 1

4. Zirconocene Coupling of Diynes with Internal Silicon Substituents

In contrast to the zirconocene-coupling of diynes without silicon substituents, the coupling of diynes with internal silicon substituents affords highly regiospecific polymers. Yellow polymer **15** (M_w/M_n = 13 000/4 600 vs. polystyrene standards) was generated from diyne **3** in 90 % yield with Negishi's reagent (eq 7). Refractionation of the polymer by addition of a benzene solution to pentane separated polymer of higher molecular weight (M_w/M_n = 17 000/8 000; 62% yield). Spectroscopic data are consistent with the structure shown, and reflect regiospecific coupling to afford 2,5-silyl-zirconacyclopentadiene rings in the polymer backbone. In particular, the ^1H and ^{13}C NMR shifts for **15** closely correspond to those for the model complex 2,5-bis(dimethylphenylsilyl)-3,4-dimethylzirconacyclopentadiene, which was synthesized in 95% yield by the coupling of dimethylphenyl(propynyl)silane. End group analysis of **15** by ^1H NMR spectroscopy provided a molecular weight of ca. 28 000. The great disparity between this value and the molecular weight obtained by gel permeation chromatography (4600) indicates that at least some of the polymers formed in this condensation reaction have macrocyclic structures.

(7)

Polymer **15** may be converted in high yield to other σ-π conjugated structures. For example, hydrolysis of **15** (with M_w/M_n = 13 000/4 600) generated the colorless polymer $[Me_2SiC_6H_4SiMe_2CH{=}CMeCMe{=}CH]_n$ (M_w/M_n = 12 000/4 200) in 94% isolated yield, with little or no degradation of the polymer backbone. The reaction of **15** with iodine produced the vinyl iodide polymer $[Me_2SiC_6H_4SiMe_2C(I){=}CMeCMe{=}C(I)]_n$ (M_w/M_n = 8 800/5 000) in 74 % isolated yield. Interestingly, the ^1H NMR spectrum of $[Me_2SiC_6H_4SiMe_2C(I){=}CMeCMe{=}C(I)]_n$ indicates the presence of inequivalent SiMe groups, which do not coalesce to 90 °C in toluene-d_8. This inequivalence, also observed for the "monomeric" compound $PhMe_2Si(I){=}CMe\text{-}CMe{=}C(I)SiMe_2Ph$, is presumably due to restricted rotation about the Si-C(I) bonds.

The gpc trace for **15** reveals a primarily monomodal molecular weight distribution in addition to a small peak (at ca. M = 1000, Scheme 2), which appeared to correspond to a cyclic oligomer. The yield of this oligomer increases under more dilute reaction conditions (e.g., 50% for a monomer concentration of ca. 0.1 M whereas at 0.6 M, only 3% of this compound is formed). In investigating the influence of reaction temperature on the molecular weight properties of **15**, we discovered that elevated temperatures favor production of the oligomeric species. Furthermore, it can be shown that this species (**16**) is formed by the thermolytic degradation of polymer **15**. Thus in refluxing tetrahydrofuran, **15** is quantitatively converted to **16** (by gpc; Scheme 2). Compound **17** was characterized as a trimeric macrocycle by mass spectroscopy and X-ray crystallography. Its formation is undoubtedly due to reversibility of the alkyne-coupling reactions in this system, which provides a low-energy pathway to the preferred thermodynamic product, in this case a trimer.

The X-ray crystal structure of **16** revealed a planar Si_6 macrocyclic core, with an average transannular Si···Si distance of ca. 13.2 Å. Like polymer **15**, **16** may be converted in high yield to derivatized macrocycles, including the hydrolyzed cyclophane $[Me_2SiC_6H_4SiMe_2CH{=}CMeCMe{=}CH]_3$ and the iodo-derivatized cyclophane $[Me_2SiC_6H_4SiMe_2C(I){=}CMeCMe{=}C(I)]_3$. The cyclophane $[Me_2SiC_6H_4SiMe_2$-

CH=CMeCMe=CH]3 adopts a chair conformation, with the rings stacked in columns along the *a* direction. For these somewhat irregular rings, the inequivalent transannular Si···Si distances are 14.2, 11.8, and 12.8 Å.

Scheme 2

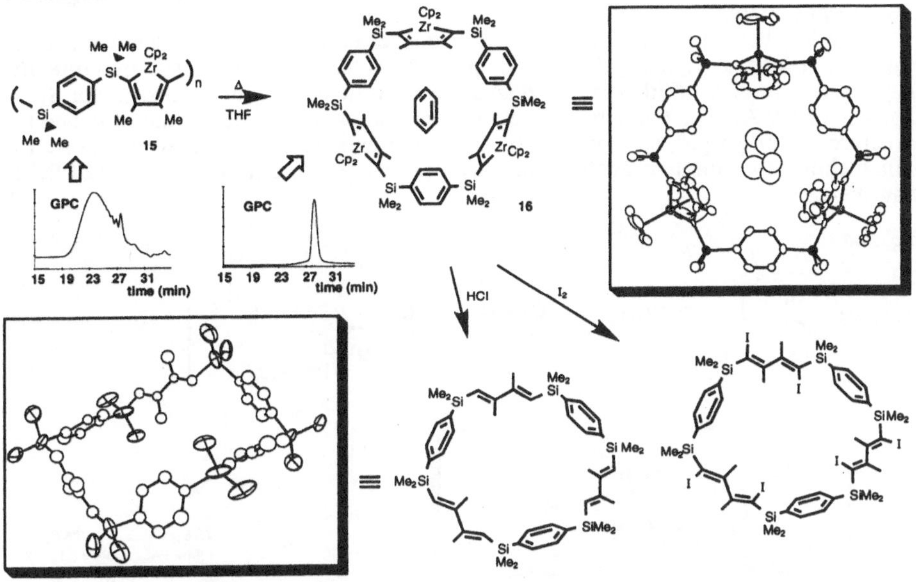

A larger ring was obtained by the zirconocene-coupling of diyne **5**, which possesses the longer 1,4-biphenylene spacer group. Initially, this coupling reaction gives a yellow polymer **17** in 90% isolated yield with M_w/M_n = 18 000/8 500) (Scheme 3). The polymer is converted in refluxing tetrahydrofuran to an equilibrium mixture of the polymer and the trimeric macrocycle **18**. This equilibrium depends on temperature and concentration, and in dilute solution (0.5 g / 70 mL THF) the equilibrium lies heavily to the side of the macrocycle. With larger macrocycles, then, there is less of a thermodynamic bias in favor of the macrocyclic product. We are currently exploring the limits of this method for producing macrocycles with very large dimensions.

Scheme 3

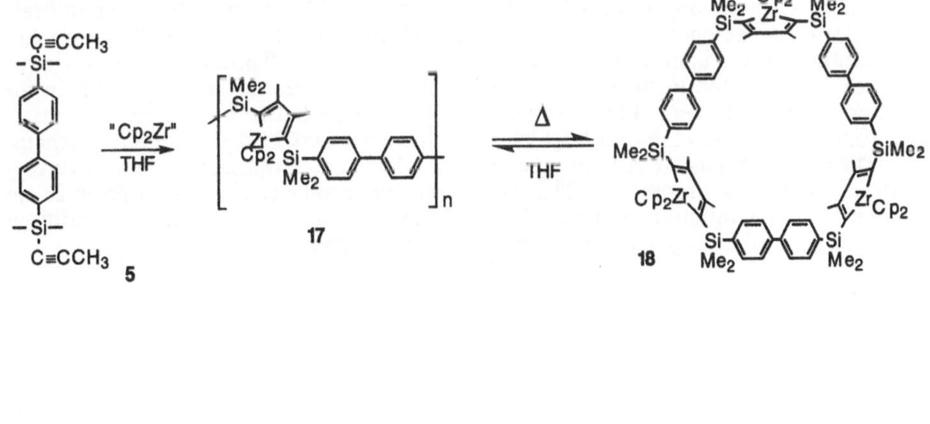

Initial experiments indicate that the composition of macrocycles formed by "equilibrium zirconocene-coupling" may be controlled to at least some degree by the geometry of the spacer group. Diyne **4**, with a 1,3-phenylene spacer group, is coupled to a regiospecific polymer (**19**; M_w/M_n = 5 800/2 600) in 50% yield (Scheme 4). The lower molecular weight and lower yield of **19** (vs. **15**) may be attributed to the relatively short spacer group, which should lead to less extensive coupling due to crowding at the reacting Zr centers.

When heated in THF, polymer **19** is converted to a *dimeric* macrocycle **20** in high yield (Scheme 4), and yellowish single crystals formed from the solution directly. The crystal structure of **20** reveals a planar macrocyclic core, with a transannular Si···Si distance of 8.9 Å. The two zirconocene units are on opposite sides of the ring, and the structure is fairly rigid, as indicated by variable temperature ^1H NMR spectroscopy, which shows that the two cyclopentadienyl resonances do not coalesce to 120 °C in toluene-d_8.

Scheme 4

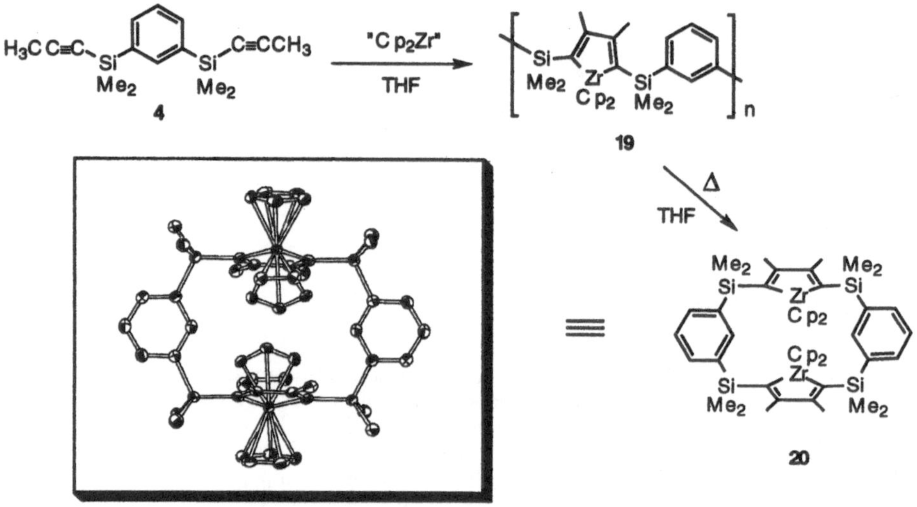

5. Zirconocene Coupling of Diynes With Terminal Silicon Substituents

With diynes possessing terminal silyl substituents, it is not possible to incorporate "linear" zirconocyclopentadiene units into a chain, since the silyl groups will enforce 3,4-substitution for the connection points to the spacer groups. Thus, couplings with this type of diyne by the Negishi reagent ($Cp_2ZrCl_2/^nBuLi$) do not usually result in the production of polymer. For example, diyne **6** is coupled to a mixture of trimeric macrocycle **21** and tetrameric **22**, which are formed in a combined yield of 85% (Scheme 5). These compounds are readily separated by solvent extractions, since they have very different solubility properties. The thermodynamic product of the reaction is **21**, as shown by its quantitative conversion from **22** in refluxing THF. Reaction conditions have been modified to give a high yield of either compound.

Scheme 5

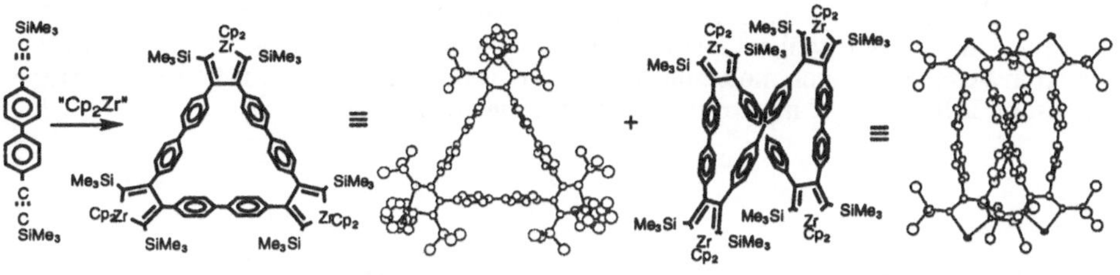

6 21 22

Molecules of **21** have a disk-like shape with coplanar zirconacyclopentadiene rings, and they are packed in a herringbone fashion in the crystal. This molecular shape is similar to that observed for related hexadehydro[18]-annulenes [13a]. Compound **22** has a folded, COT-like geometry similar to that proposed for Sondheimer's octadehydro-[24]annulene [13], however the folding is rather extreme such that the angle defined by the Zr_3 planes is 31.3°. Further distortion of this structure away from D_{2d} symmetry (to crystallographic D_2 symmetry), and buckling of the biphenylene groups by 5.6 and 9.2 °, appears to result from steric interactions between biphenylene substituents on the ZrC_4 rings. Tetramer **22** exhibits interesting fluxional behavior in solution, such that the inequivalent Cp rings coalesce at 30 °C ($\Delta G^{\ddagger}$ = 15.8 kcal mol^{-1}). The rate of this process is unaffected by the presence of THF, which is expected to promote reversibility for the alkyne-coupling steps. Therefore it appears that exchange of the Cp rings ("inner" and "outer") occurs by inversion of the macrocycle through a planar D_{4h} structure [13b].

Reactions of **21** and **22** with HCl in THF afford the corresponding hydrolyzed macrocycles **23** and **24**, respectively, in high yield. Single crystals of **23** were obtained from acetonitrile/toluene solution, and the molecules contain one equivalent of acetonitrile per macrocycle, located near the center of the macrocycle cavity (Scheme 6).

Scheme 6

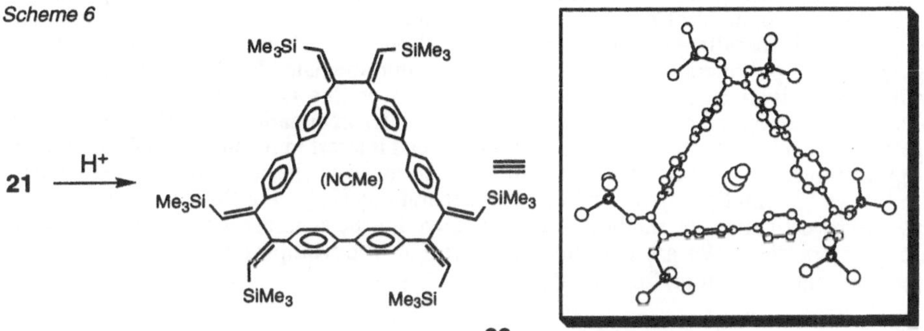

23

Generation of zirconocene in a different manner, by reduction of Cp_2ZrCl_2 with Mg, produces more of the kinetic product (7:3 ratio of **22** to **21**), but also higher oligomers (ca 30% by weight) which have estimated degrees of polymerization from 5 to ca. 100 (by gel permeation chromatography; M_w/M_n = 4 400/1 800). These higher oligomers appear to be cyclic, since the infrared spectrum of the mixture contains no stretching vibrations assignable to -C≡CSiMe$_3$ end groups.

Initial results indicate that the macrocycle synthesis outlined in Schemes 5 and 6 is versatile. In a fairly trivial extension of this method, we have obtained the macrocycles shown in Scheme 7. Such molecules may serve as building blocks for liquid crystalline phases, or 3-dimensional networks with well-defined porosity.

Scheme 7

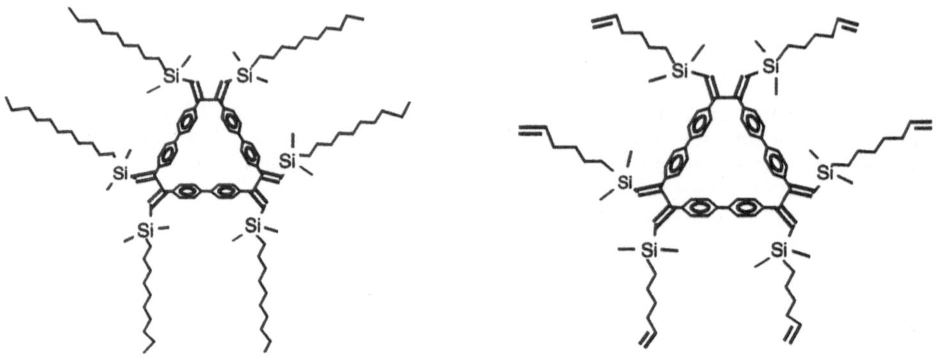

6. Conclusion

The zirconocene coupling of alkynes is a versatile synthetic technique which can be applied to the construction of macromolecular objects. Our initial studies on reactions of this type, summarized above, indicate that this method may prove to be a very powerful tool in the development of new carbon-based materials. In one application, conjugated polymers with metallacyclic rings in the backbone have been synthesized. These organometallic polymers act as versatile synthetic intermediates for the synthesis of new families of derivatized polymers. Perhaps more surprising is the facile conversion of polymers containing silicon substituents to macrocycles in quantitative yield. This result is particularly relevant, given the current wide-spread interest in macrocycles for a variety of applications, and the fact that synthetic routes to macrocycles are generally associated with low yields, multiple steps, and difficult separations [2,3]. For these reasons, we are currently investigating the methods outlined above for the synthesis of macrocycles with tailored properties. We are particularly interested in developing routes to very large rings, and rings containing functional groups.

7. Acknowledgments

We thank the Division of Chemistry and Materials Research at the National Science Foundation for support of this work. We thank Dr. Fred Hollander of the department X-ray facility (CHEXRAY) for determination of the crystal structures.

8. References

1. Selected references on conjugated polymers: (a) Skotheim, T.A. (1986) *Handbook of Conductive Polymers*; Marcel Dekker: New York, Vol. 1 and 2. (b) Bredas, J.L. and Silbey, R. (1991) *Conjugated Polymers: The Novel Science and Technology of Highly Conducting and Nonlinear Optically Active Materials*, Kluwer Academic Publishers: Dordrecht, The Netherlands. (c) Miller, R.D. and Michl, J. (1989) Polysilane High Polymers, *Chem. Rev.* **89**, 1359-1410. (d) Burroughes, J.H., Bradley, D.D.C., Brown, A.R., Marks, R.N., Mackay, K., Friend, R.H., Burns, P.L., Holmes, A.B. (1990) Light-Emitting Diodes Based on Conjugated Polymers, *Nature* **347**, 539-541. (e) Bradley, D.D.C. (1993) Conjugated Polymer Electroluminescence, *Synth. Met.* **54**, 401-415. (f) Sailor, M.J., Ginsburg, E.J., Gorman, C.B., Kumar, A., Grubbs, R.H., Lewis, N.S. (1990) Thin Films of N-Si/Poly-$(CH_3)_3$Si-Cyclooctatetraene-Conducting-Polymer Solar Cells and Layered Structures, *Science* **1990**, *249*, 1146-1149. (g) Sailor, M.J., Klavetter, F.L., Grubbs, R.H., Lewis, N.S. (1990) Electronic Properties of Junctions between Silicon and Organic Conducting Polymers, *Nature* **346**, 155-157.
2. Selected references on modular construction of macrocycles: (a) Vogtle, F. (1993) *Cyclophane Chemistry*, John Wiley & Sons Ltd.: Chichester, West Sussex. (b) Diederich, F. and Rubin, Y. (1992) Synthetic Approaches toward Molecular and Polymeric Carbon Allotropes, *Angew. Chem. Int. Ed. Engl.* **31**, 1101-1123. (c) Diederich, F. (1994) Carbon Scaffolding - Building Acetylenic All-Carbon and Carbon-Rich Compounds, *Nature* **69**, 199-207. (d) Moore, J.S. and Zhang, J.S. (1992) Efficient Synthesis of Nanoscale Macrocyclic Hydrocarbons, *Angew. Chem. Int. Ed. Engl.* **31**, 922-924. (e) Zhang, J.S. and Moore, J.S. (1992) Nanoarchitectures 3. Aggregation of Hexa(phenylacetylene) Macrocycles in Solution - a Model System for Studying π-π Interactions, *J. Am. Chem. Soc.* **114**, 9701-9702. (f) Zhang, J., Pesak, D.J., Ludwick, J.L. and Moore, J.S. (1994) Nanoarchitectures 5. Geometrically Controlled and Site-Specifically-Functionalized Phenylacetylene Macrocycles, *J. Am. Chem. Soc.* **116**, 4227-4239. (g) Venkataraman, D., Lee, S., Zhang, J. and Moore, J.S. (1994) An Organic Solid with Wide Channels Based on Hydrogen Bonding between Macrocycles, *Nature* **371**, 591-593. (h) Bissell, R.A., Cordova, E., Kaifer, A.E. and Stoddart, J.F. (1994) A Chemically and Electrochemically Switchable Molecular Shuttle, *Nature* **369**, 133-137. (i) Romero, M.A. and Falis, A.G. (1994) Revolveneynes - Novel Eneyneparacyclophanes by Sequential Palladium Coupling, *Tetrahedron Lett.* **35**, 4711-4714. (j) Walter, C.J. and Sanders, J.K.M. (1995) Free-Energy Profile for a Host-Accelerated Diels-Alder Reaction - The Sources of Exo Selectivity, *Angew. Chem. Int. Ed. Engl.* **34**, 217-219.
3. Selected references on modular construction of networks: (a) Lehn, J.M. (1990) Perspectives in Supramolecular Chemistry - From Molecular Recognition towards Molecular Information Processing and Self-Organization, *Angew. Chem. Int. Ed. Engl.* **29**, 1304-1319. (b) Kaszynski, P., Friedli, A.C. and Michl, J. (1992) toward a Molecular-Size Tinkertoy Construction Set - Preparation of Terminally Functionalized [n]Staffanes from [1.1.1]Propellane, *J. Am. Chem. Soc.* **114**, 601-620. (c) Whitesides, G.M., Simanek, E.E., Mathias, J.P., Seto, C.T., Chin, D.N., Mammen, M. and Gordon, D. M. (1995) Noncovalent Synthesis - Using Physical-Organic Chemistry to Make Aggregates, *Acc. Chem. Res.* **28**, 37-44. (d) Simard, M., Su, D. and Wuest, J.D. (1991) Use of Hydrogen Bonds to Control Molecular Aggregation - Self-Assembly of 3-Dimensional Networks with Large Chambers, *J. Am. Chem. Soc.* **113**, 4696-4698. (e) Garcia-Tellado, F., Geib, S.J., Goswami, S. and Hamilton, A.D. (1991) Molecular Recognition in the Solid State - Controlled Assembly of Hydrogen-Bonded Molecular Sheets, *J. Am. Chem. Soc.* **113**, 9265-9269. (f) Abrahams, B.F., Hoskins, B.F., Michail, D.M. and Robson, R. (1994) Assembly of Porphyrin Building Blocks into Network Structures with Large Channels, *Nature*, **369**, 727-729. (g) Swager, T.M. and Marsella, M.J. (1994) Molecular Recognition and Chemoresistive Materials, *Adv. Mater.* **6**, 595-597. (h) Beer, P.D. (1994) Charged Guest Recognition by Redox Responsive Ligand Systems, *Adv. Mater.* **6**, 607-609. (i) Bunz, U.H.F. (1994) Polyynes-Fascinating Monomers for the Construction of Carbon Networks, *Angew. Chem. Int. Ed. Engl.* **33**, 1073-1076. (j) Amabilino, D.B., Stoddart, J.F. and Williams, D.J. (1994) From Solid-State Structures and Superstructures to Self-Assembly Processes, *Chem. Mater.* **6**, 1159-1167.
4. (a) Mao, S.S.H. and Tilley, T.D. (1995) A New Route to Unsaturated Organosilicon Polymers and Macrocycles Based on Zirconocene-Coupling of 1,4-MeC≡C$(Me_2Si)C_6H_4(SiMe_2)$C≡CMe, *J. Am. Chem. Soc.* **117**, 5365-5366. (b) Mao, S.S.H. and Tilley, T.D. (1995) Efficient, Zirconocene-Mediated Cyclotrimerization and Cyclotetramerization of $Me_3SiC{\equiv}CC_6H_4C_6H_4C{\equiv}CSiMe_3$ to Unsaturated Macrocycles, *J. Am. Chem. Soc.* **117**, 7031-7032. (c) Mao, S.S.H. and Tilley, T.D.

(1996) Polymers with Linked Macrocyclic Rings in the Main Chain: Zirconocene Coupling of 1,8-Cyclotetradecadiyne. *Macromolecules.*, in press. (d) Mao, S.S.H.; Tilley, T.D. (1996) Zirconocene Coupling of 1,8-Cyclotetradecadiyne to Polymers Containing Macrocyclic Rings in the Main Chain. *Polymer Preprints. 37(1)*, 443. (e) Mao, S.S.H. and Tilley, T.D. Cross-Conjugated Polymers via Condensation of a Zirconocene (Alkynylbenzyne) Derivative Generated by Thermolysis of $Cp_2ZrMe(C_6H_4C{\equiv}CSiMe_3)$ *J. Organomet. Chem.*, in press. (f) Mao, S.S.H. and Tilley, T.D. (1996) Conjugated Polymers Containing Zirconacyclopentadiene in the Backbone - A Versatile Synthon for Polyphenylene, Polybiphenylenebutadiene, and Conjugated Polymers with Thiophene, Phosphole, and Borole Groups in Their Backbones, manuscript in preparation. (g) Mao, S.S.H. and Tilley, T.D. (1996) Zirconocene-Coupling of Diynes with Internal Silicon Substituents as an Efficient Route to Unsaturated Organosilicon Polymers and Macrocycles, manuscript in preparation.

5. (a) Broene, R.D. and Buchwald, S.L. (1993) Zirconocene Complexes of Unsaturated Organic Molecules - New Vehicles for Organic Synthesis, *Science* **261**, 1696-1701, and references therein. (b) Negishi E. and Takahashi T. (1994) Patterns of Stoichiometric and Catalytic Reactions of Organozirconium and Related Complexes of Synthetic Interest, *Acc. Chem. Res.* **27**, 124-130. (c) Cardin, D.J., Lappert, M.F., Raston, C.L. (1986) *Chemistry of Organozirconium and Hafnium Compounds*, John Wiley & Sons: New York.
6. Takahashi, T., Kotora, M. and Xi, Z.F. (1995) Cycloaddition of Zirconacyclopentadienes to Alkynes Using Copper Salts - Formation of Benzene Derivatives, *J. Chem. Soc., Chem. Commun.* 361-362.
7. Fagan, P.J., Nugent, W.A. and Calabrese, J.C. (1994) Metallacycle Transfer from Zirconium to Main Group Elements - A Versatile Synthesis of Heterocycles, *J. Am. Chem. Soc.* **116**, 1880-1889.
8. (a) Alt, H. and Rausch, M.D. (1974) Photochemical Reactions of Dimethyl Derivatives of Titanocene, Zirconocene, and Hafnocene, *J. Am. Chem. Soc.* **96**, 5936-5937. (b) Demerseman, B., Bouquet, G. and Bigorgne, M. (1977) Etude Comparee Des Dicyclopentadienyl-Titane et-Zirconium Dicarbonyle : Reactivite Chimique et Spectre Infra-Rouge Basses Frequences, *J. Organomet. Chem.* **132**, 223-229. (c) Thanedar, S. and Farona, M.F. (1982) A One-Step Synthesis of Bis(η^5-Cyclopentadienyl)Zirconacyclopentadiene Compounds, *J. Organomet. Chem.* **235**, 65-68. (d) Negishi, E., Cederbaum, F.E. and Takahashi, T. (1986) Reaction of Zirconocene Dichloride with Alkyllithiums or Alkyl Grignard Reagents as a Convienient Method for Generating a "Zirconocene" Equivalent and Its Use in Zirconium-Promoted Cyclization of Alkenes, Alkynes, Dienes, Enynes, and Diynes, *Tetrahedron Lett.* **27**, 2829-2837. (e) Buchwald, S.L., Watson, B.T. and Huffman J.C. (1987) The Synthesis, Reaction, and Molecular Structure of Zirconocene-Alkyne Complexes, *J. Am. Chem. Soc.* **109**, 2544-2546.
9. The coupling of diynes with a flexible spacer group generally results in bicyclic compounds or low molecular weight oligomers: (a) 8d. (b) Negishi, E., Holmes, S.J., Tour, J.M., Miller, J.A., Cederbaum, F.E., Swanson, D.R., and Takahashi, T. (1989) Novel Bicyclization of Enynes and Diynes Promoted by Zirconocene Derivatives and Conversion of Zirconabicycles into Bicyclic Enones via Carbonylation, *J. Am. Chem. Soc.* **111**, 3336-3346. (c) Nugent, W.A, Thorn, D.L. and Harlow, R.L. (1987) Cyclization of Diacetylenes to e,e-Exocyclic Dienes. Complementary Procedures Based on Titanium and Zirconium Reagents, *J. Am. Chem. Soc.* **109**, 2788-2796. (d) RajanBabu, T.V., Nugent, W.A., Taber, D.F. and Fagan, P.J. (1988) Stereoselective Cyclization of Enynes Mediated by Metallocene Reagents, *J. Am. Chem. Soc.* **110**, 7128-7135. (e) Takahashi, T., Kasai, K., Xi, Z.F. and Denisov, V. (1995) Reaction of Zirconocene Ethylene Complex With Diynes - Formation of Bridged Zirconacyclopentenes, *Chem. Lett.* **5**, 347-348. (f) Du, B., Farona, M.F., McConnville, D.B. and Youngs, W.J. (1995) Synthesis and Structure of a Tricyclic Bis(Zirconacyclopentadiene) Compound, *Tetrahedron* **51**, 4359-4370.
10. (a) Ohkubo, A., Fujita, T., Ohba, S., Aramaki, K., Nishihara, H. (1992) A Novel Intramolecular Metallacycle-Phosphine Reaction of an Electrochemically Oxidized Cobaltacyclopentadiene Complex, *J. Chem. Soc,. Chem. Commun.*, 1553-1555. (b) Ohkubo, A., Aramaki, K., Nishihara, H. (1993) Synthesis of a Novel π-Conjugated Organometallic Polymer, Poly(4,4'-Biphenylene-2,5-Cobaltacylopenta-2,4-Dienylene), by Metallacycling Polymerization (MCP), *Chem. Lett.* 271-274. (c) Nishihara, H., Shimura, T., Ohkubo, A., Matsuda, N., Aramaki, K. (1993) Poly(Arylene Cobaltacyclopentadienylene) - A New Class of Organometallic π-Conjugated Conducting Polymers, *Adv. Mater.* **5**, 752-754. (d) Nishihara, H., Ohkubo, A., Aramaki, K. (1993) Synthesis of π-Conjugated Organometallic Polymers Based on Metallacyclization Reactions, *Synth. Met.* **55**, 821-826. (e) Fang, M.C., Watanabe, A. and Matsuda, M. (1994) Synthesis of σ-π Conjugated Alternating Silylene-Diacetylene Copolymers and Their Optical and Electrical Properties, *Chem.*

Lett. 13-16. (f). Shimura, T.; Ohkubo, A.; Matsuda, N.; Matsuoka, I.; Aramaki, K.; Nishihara, H. (1996) Synthesis of Poly(arylene cobaltacyclopentadienylene)s, a New Class of Organometallic π-Conjugated Polymers, by Metallacycling Polymerization and Their Physical Properties. *Chem. Mater. 8*, 1307.

11. In the formation of zirconacyclopentadienes, aryl and silyl substituents in the alkyne prefer the α position. See : Negishi E. and Takahashi T. (1994) Patterns of Stoichiometric and Catalytic Reactions of Organozirconium and Related Complexes of Synthetic Interest, *Acc. Chem. Res.* **27**, 124-130 and references therein.
12. (a) Tao, W.J., Nesbitt, S. and Heck, R.F. (1990) Palladium-Catalyzed Alkenylation and Alkynylation of Polyhaloarenes, *J. Org. Chem.* **55**, 63-69. (b) Tao, W.J., Silverberg L.J., Rheingold A.L., Heck R.F. (1989) Alkyne Reactions with Arylpalladium Compounds, *Organometallics* **8**, 2550-2559. (c) Takahashi, S., Kuroyama, Y., Sonogashira, K., Hagihara, N. (1980) A Convenient Synthesis of Ethynylarenes and Diethynylarenes, *Synthesis* 627-629.
13. (a) Diederich, F., Rubin, Y., Knobler, C.B., Whetten, R.L, Schriver, K.E., Houk, K.N. and Li, Y. (1989) All-Carbon Molecules - Evidence for the Generation of Cyclo[18]Carbon from a Stable Organic Precursor, *Science* **245**, 1088-1090. (b) Li, Y., Rubin, Y., Diederich, F. and Houk, K.N. (1990) Electronic and Structural Properties of the Cyclobutenodehydroannulenes, *J. Am. Chem. Soc.* **112**, 1618-1623. (c) Rubin, Y., Knobler, C.B. and Diederich, F. (1990) Synthesis and Crystal Structure of a Stable Hexacobalt Complex of Cyclo[18]Carbon, *J. Am. Chem. Soc.* **112**, 4966-4968.

DIRECTING NUCLEATION AND GROWTH OF MOLECULAR CRYSTALS ON ORDERED SUBSTRATES: THE ROLE OF EPITAXIAL INTERACTIONS

M. D. WARD, S. J. BONAFEDE and A. C. HILLIER
Department of Chemical Engineering and Materials Science
University of Minnesota
Amundson Hall, 421 Washington Ave. SE
Minneapolis, MN 55455

ABSTRACT. Much of the interest in molecular materials, whether in thin film or bulk form, stems from the ability to employ molecular-level "crystal engineering" strategies, which aim to rationally manipulate crystal packing and control physical and electronic properties. However, understanding the assembly of molecules, particularly on surfaces, during nucleation of these phases is imperative if the crystallization process is to be directed rationally. We describe herein strategies for directing the organization of molecular assemblies during nucleation and growth of thin films and bulk crystals of organic materials. The formation of these nuclei and their supramolecular structure depend upon a subtle balance of intermolecular forces within the nuclei and nuclei-substrate epitaxial interactions. In some cases, the formation of these nuclei have been observed directly, and their molecular level structure characterized, using scanning tunneling microscopy (STM) and in-situ real-time atomic force microscopy (AFM). Organic crystals can be used as substrates for directing nucleation of other organic crystalline materials by "ledge-directed epitaxy." The examples presented illustrate the influence of the substrate on nucleation rate, orientation of nuclei, monolayers, multilayers, and crystals with respect to the nucleating substrate, polymorphism, and quality.

1. Crystallization and Modular Self-Assembly

Over the last several decades substantial effort has been expended on the design of molecular crystals using "crystal engineering" approaches [1], in which the solid state framework is built from carefully selected molecules (which can be considered as "modules" in the spirit of this symposium). This approach typically involves the use of molecules with topologies and functional groups which are capable of directing the self-assembly of the molecules into a desired solid state motif. Such directing forces may include hydrogen bonding, charge-transfer interactions, metal-ligand coordination, electrostatic interactions, or simply shape-directing van der Waals forces.

While crystal engineering has been successful when small systematic changes are made, these efforts commonly are frustrated by the numerous alternative ways in which molecules can assemble in the solid state to form polymorphs. This is largely a consequence of van der Waals interactions, which exist between all atoms and account for a significant portion of the lattice energy, but are individually weak and relatively

J. Michl (ed.), Modular Chemistry, 637–649.

isotropic. Although polymorphs have the same chemical composition, they can differ dramatically with respect to their physical, chemical, and electronic properties. This includes the optical properties of organic dyes, the conductivity of one-dimensional conductors, the magnetism of organic ferromagnets, and the frequency doubling characteristics of non-linear optical materials (where noncentrosymmetric packing is required). Polymorphism is also critical in the pharmaceutical industry, particularly with respect to kinetic solubility, mechanical properties, purity, and mass flow characteristics. The issue of polymorphism also extends to thin molecular films and modular assemblies where structure is, and will continue to be, important with respect to function.

Polymorphism generally is difficult to control and predict. Indeed, there are numerous documented and anecdotal accounts of "disappearing polymorphs" or at least problems with reproducibility in processing of a polymorphic system [2]. Relatively minor changes in temperature, impurities or solvent composition can influence the crystal form isolated in a given process. The common occurrence of polymorphism can be attributed to different trajectories describing the multistep self-assembly of molecules crystals. Nucleation is the most energetically demanding step, and the rate determining step, in crystallization. Consequently, a thermodynamically less preferred polymorph can form if its activation energy for nucleation is less than that of the others. It is our belief that polymorphism can be controlled by guiding the self-assembly of molecules into a supramolecular motif that resembles the desired polymorph during the formation of prenucleation aggregates that form crystal nuclei.

Homogeneous nucleation is energetically unfavorable owing to the creation of phase boundaries with their corresponding surface energies. However, if the number of molecules in an incipient nucleus reaches a critical value when enthalpic contributions from bulk intermolecular interactions exceed the energetic penalty of creating new surface area, crystal growth on the nucleus will be energetically favored. Hence, crystallization is an activated process in which the activation energy depends upon the surface energy, γ, and the volume free energy ΔG_V, according to eq. (1).

$$\Delta G^*_{hom} = \frac{16\pi}{3}\left(\frac{\gamma^3}{\Delta G_V^2}\right) \tag{1}$$

However, the activation energy for <u>heterogeneous</u> nucleation on substrates is smaller than homogeneous nucleation due to lowering of the surface energy of the nucleus and the substrate upon interfacial contact (eq. 2). The activation energy is modified by a function $f(\Theta)$, which describes the interaction of the nucleus with the substrate surface in terms of a classical wetting contact angle Θ (eq. 3). Inspection of eq. (6) reveals that $f(\Theta) < 1$ for values of $\Theta < 90^o$ (that is, favorable wetting); $\Delta G^*_{het} < \Delta G^*_{hom}$ under these conditions. The wetting analogy allows nucleation to be described in terms of the accompanying change in total surface free energy, $\Delta\gamma_{het}$ (eq. 4), where γ_{32}, γ_{21} and γ_{31} are the individual interfacial energies between the nutrient phase and aggregate, the aggregate and substrate and the nutrient phase and substrate, respectively. The terms S_{32} and S_{21} are the interfacial contact areas of the nutrient phase and aggregate, and of the aggregate and substrate, respectively. Larger values of Θ correspond to smaller values of S_{21} for a given nucleus volume.

$$\Delta G^*_{hom} = \frac{16\pi}{3}\left(\frac{\gamma^3}{\Delta G_v^2}\right) \tag{2}$$

$$f(\Theta) = \frac{1}{4}(2+\cos\Theta)(1-\cos\Theta)^2 \tag{3}$$

$$\Delta\gamma_{het} = \gamma_{32}S_{32} + (\gamma_{21} - \gamma_{31})\, S_{21} \tag{4}$$

The dominance of heterogeneous nucleation suggests that one strategy for controlling polymorphism is the rational design of substrate surfaces that can guide the assembly of molecules into supramolecular motifs resembling those of specific, desired polymorphs.

We describe herein two approaches to directing polymorph selectivity. One approach involves molecular crystal substrates having crystallographically well-defined ledge sites that provide a "three-dimensional" surface capable of "shape matching" to a desired incipient nucleus, thereby increasing the interfacial area of contact between the substrate and the prenucleation aggregate. The other relies on heteroepitaxial nucleation on a two-dimensional substrate driven by lattice matching between the substrate and ordered overlayers that serve as incipient nuclei, but in which commensurism is achieved only for large overlayer areas. Both approaches illustrate the important role of interfacial structure in controlling the supramolecular motif of molecular assemblies at surfaces.

2. Polymorph Selectivity by Ledge-Directed Epitaxy

It is evident from eq. (4) that under conditions where aggregate-substrate interactions are favorable, that is, negative values of $(\gamma_{21} - \gamma_{31})$, increasing S_{21} lowers the activation energy for nucleation. The value of S_{21} for nucleation on a single substrate plane is defined by a single interface. However, if a nucleus of identical volume is able to contact two substrate planes, such as intersecting terrace and step planes, S_{21} will be larger and the activation energy for nucleation will be lowered (for simplicity, we assume here that γ_{21} and γ_{31} are identical for both planes). This condition, which is related to conventional models of crystal growth where attachment of atoms or molecules occurs preferentially at step sites [3] is the basis for nucleation driven by ledge-directed epitaxy (LDE) on surfaces of single crystals recently demonstrated in our laboratory [4].

The LDE mechanism relies on the presence of molecular crystal substrates which have crystallographically well-defined ledges, which are defined by the intersection of step and terrace planes. A substrate ledge is described by a crystal direction $[uvw]_{sub}$, and will have a crystallographically defined dihedral angle, θ_{sub} (Figure 1). The free energy of a prenucleation aggregate, whose supramolecular stucture resembles the solid state structure of the crystalline phase, will be lowered by interaction with the substrate ledge site if (1) its contacting planes are dispersive in nature so that interfacial interactions are attractive and repulsive interactions are minimized, (2) its contacting planes are close-packed surfaces in order that interfacial interactions are maximized, and (3) the angle between the contacting aggregate planes, θ_{agg}, has approximately the same value as θ_{sub}. If these conditions are met the substrate can induce nucleation by "shape matching." Aggregate planes can be deduced from inspection of the crystal structure of the compound under investigation. If two planes in the crystal structure have a dihedral angle similar to that of the substrate ledge, the condition $\theta_{agg} \approx \theta_{sub}$ will lead to a larger S_{21} term, and consequently, to enhanced interfacial stabilization of the aggregate. Conversely, if θ_{agg} is grossly different than θ_{sub}, S_{21} will be limited to a single interface and the activation barrier for nucleation will be larger than that for interaction at two interfaces at the ledge.

The LDE concept suggested a strategy for selective nucleation of polymorphs. Polymorphs of a given composition are described by different supramolecular motifs in the solid state. Therefore, crystal planes of different polymorphs differ with respect to their two-dimensional molecular structures and dihedral angles, which correspond to θ_{agg} of the prenucleation aggregate. If a particular polymorph of a compound has two well-defined, close-packed crystal planes whose dihedral angle matches that of the substrate ledge site, nucleation of that phase will be favored. Conversely, nucleation of a polymorph which lacks a pair of intersecting close-packed planes with $\theta_{agg} = \theta_{sub}$, will be less preferred.

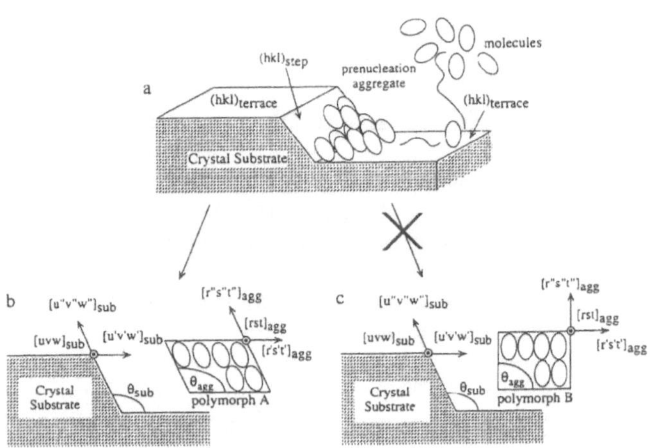

Figure 1. Conceptual representation of polymorph selectivity by ledge-directed epitaxy (LDE). The critical feature is stabilization of an incipient nucleus by interfacial interaction between the ledge site and two planes of a prenucleation aggregate (i.e., the incipient nucleus) which correspond to planes in a particular polymorph. LDE is realized when the dihedral angles of the ledge and the two contacting planes of the nucleus are equivalent, and when the two contacting planes of the aggregate are close packed. These conditions lead to a greater lowering of the free energy of the aggregate relative to those of other polymorphs, thereby directing the selectivity.

We have demonstrated this principal by selectively growing a thermodynamically less preferred polymorph of the organic dye $(DMTC^+)(TMO^-)\cdot CHCl_3$ (DMTC = 3,3'-dimethylthiacarbocyanine; TMO = 3,3',5,5'-tetramethyltrimethine oxonol) on freshly cleaved single crystals of succinic acid (SA). Single crystals of SA can be cleaved parallel to the $(0\bar{1}0)_{SA}$ plane, which contains hydrogen-bonded succinic acid chains along $[10\bar{1}]_{SA}$ organized into $(0\bar{1}0)_{SA}$ layers held together by dispersive interactions. Cleaving produces distinct linear features attributable to $[10\bar{1}]_{SA}$ ledges when the crystals are viewed normal to the $(0\bar{1}0)_{SA}$ surfaces. These ledges are formed by the intersection of $(0\bar{1}0)_{SA}$ terrace and $(1\bar{1}1)_{SA}$ step planes, and have a crystallographically defined dihedral angle of $\theta_{sub} = 112.6°$ Both planes can be considered as low-energy, close-packed, dispersive planes. The strong hydrogen bonding along $[10\bar{1}]_{SA}$ stabilizes the $(0\bar{1}0)_{SA}$ and $(1\bar{1}1)_{SA}$ planes, thereby favoring their formation when SA crystals are cleaved while also inhibiting their reconstruction.

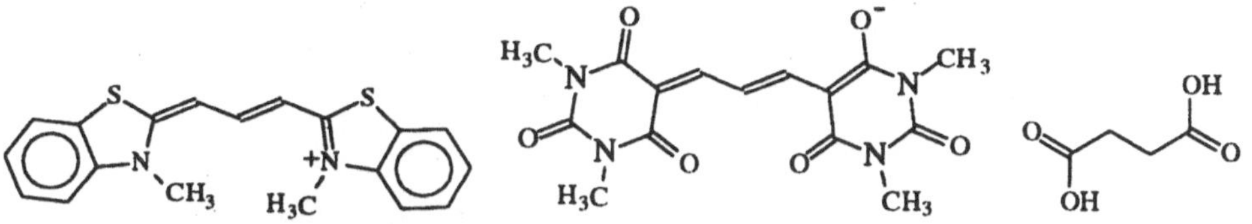

DMTC TMO SA

The $DMTC^+$-TMO^- salts crystallize into seven polymorphs and seven pseudopolymorphs (i.e., solvates) which can be readily distinguished by their colors, morphologies and, for some phases, by their x-ray powder diffraction patterns. Two polymorphs were obtained in sufficient quantity and quality for single crystal x-ray analysis, and were designated **Ia** and **Ib** (**Ia** = space group $P2_1/c$, $a = 7.660$ Å, $b = 23.950$ Å, $c = 20.201$ Å, $\beta = 82.35°$; **Ib** = space group $P2_1/a$, $a = 14.528$ Å, $b = 13.676$ Å, $c = 18.548$ Å, $\beta = 90.54°$) [5]. Polymorph **Ia** exhibits a red color by transmission but a gold color by reflection, and **Ib** exhibits a red color by both transmission and reflection. These phases formed simultaneously during evaporation of $CHCl_3$ solutions containing the ions, but **Ib** was clearly the minor phase in all attempts using conventional crystallization procedures.

Crystallization of $DMTC^+$-TMO^- salts from $CHCl_3$ solutions in the presence of freshly cleaved SA crystals afforded a dramatically different polymorph distribution than that observed in the absence of SA [6]. After approximately 24 hours, red rectangular crystals were evident on the $[10\bar{1}]_{SA}$ ledges, with their long axis parallel to the ledge. The color, spectroscopic characterization, atomic force microscopy, and the thermal phase behavior of these crystals enabled their identification as polymorph **Ib**, the thermodynamically less preferred form.

Comparison of the crystal packing of **Ia** and **Ib**, and its relationship to the geometry of the $[10\bar{1}]_{SA}$ ledge site reveals the reason for the selectivity toward **Ib**. In both structures, the $DMTC^+$ and TMO^- ions stack in an alternating manner along the *b*-axis, but in **Ia** the ions are mutually perpendicular while in **Ib** they are mutually parallel (Figure 2). This difference in solid state packing leads to different packing densities in the various crystal planes. The solid state structure of **Ia** reveals only one plane with a close-packed surface, namely $(10\bar{1})_{\mathbf{Ia}}$, and no reasonable crystal plane combinations with a dihedral angle approaching that of the $[10\bar{1}]_{SA}$ ledge. Therefore, strong interfacial interaction between a prenucleation aggregate of **Ia** with both planes of the $[10\bar{1}]_{SA}$ ledge is not possible, consistent with the absence of growth of this polymorph in the presence of SA.

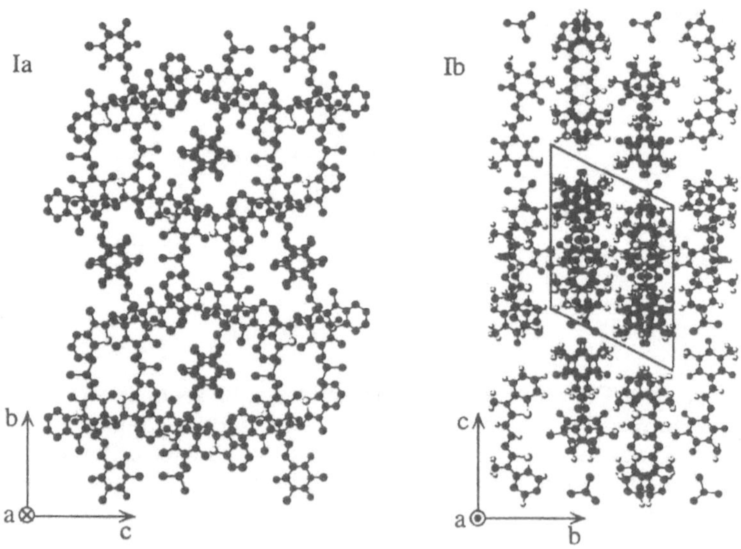

Figure 2. Crystal structure of polymorphs **Ia** and **Ib** of (DMTC)(TMO)·$CHCl_3$ the outlined region in **Ib** corresponds to an aggregate of DMTC and TMO molecules viewed down their stacking direction. This aggregate has strong intermolecular interactions along the stacking direction and has close-packed, intersecting (014) and (010) planes. In contrast, analysis of the structure of **Ia**, reveals no such pair of intersecting planes.

In contrast, analysis of the crystal structure of **Ib** reveals the existence of a one-dimensional aggregate of $DMTC^+$ and TMO^- ions stacked along the *a*-axis. This aggregate has locally close-packed $(010)_{\mathbf{Ib}}$ and $(014)_{\mathbf{Ib}}$ planes (the $(014)_{\mathbf{Ib}}$ plane can be generated by removal of a $CHCl_3$ solvate molecule from the $(001)_{\mathbf{Ib}}$ plane; see outline, Figure 3). The dihedral angle of $(010)_{\mathbf{Ib}} \cap (014)_{\mathbf{Ib}}$ corresponds to $\theta_{agg} = 113.0^o$, nearly identical to the value of $\theta_{sub} = 112.6^o$ for the $[10\bar{1}]_{SA}$ ledge. The $(010)_{\mathbf{Ib}}$ and

$(014)_{Ib}$ planes share the $[100]_{Ib}$ direction; therefore $[100]_{Ib}$ would coincide with $[10\bar{1}]_{SA}$ if these aggregate planes contact the SA ledge planes (Figure 3). Indeed, this is the orientation deduced from the morphology and AFM of **Ib** single crystals grown at the SA ledges. We surmise that interfacial contact of the $(014)_{Ib}$ and $(010)_{Ib}$ planes with the ledge and π–π stacking interactions between $DMTC^+$ and TMO^- ions along $[001]_{Ib}$ favor growths of a one-dimensional aggregate *along the ledge* until the aggregate reaches critical size, this one-dimensional aggregate having a cross section corresponding to the area outlined in Figure 2. Aggregate assembly in this manner preserves the dihedral angle match and close-packed planes at the ledge interface, thereby maximizing dispersive contact during nucleation.

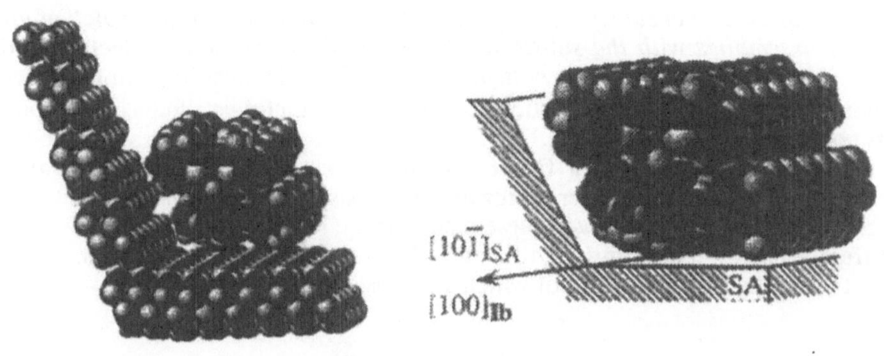

Figure 3. (left) Schematic representation of an incipient nucleus of **Ib** interacting with the $[10\bar{1}]_{SA}$ ledge site, in which the supramolecular structure of the nucleus is identical to that of the bulk crystal. The nucleus is depicted in an orientation identical, with respect to the $[10\bar{1}]_{SA}$ ledge, to that of mature crystals grown on the SA substrate. The stacking axis of the DMTC and TMO molecules is parallel to the ledge. The selectivity toward this less thermodynamically preferred phase and the obsesrved orientation is the result of good agreement between the dihedral angles of the $[10\bar{1}]_{SA}$ ledge (θ_{sub} = 112.6°) and the pair of intersecting (014) and (010) planes (θ_{sub} = 113°) in the nucleus. (right) Schematic representation of the growth of **Ib** along the $[10\bar{1}]_{SA}$ ledge. Strong intermolecular interactions in the nucleus along the stacking direction also favor growth along the ledge.

3. Two-Dimensional Heteroepitaxial Growth of Organic Overlayers and Crystals

It is likely that a significant amount of modular chemistry will be directed toward the construction of molecular assemblies in the form of thin monolayer or multilayer films on surfaces, as it is conceivable that such systems may lead to novel device structures. These modular systems may be assembled by covalent bonds, or by non-covalent interactions such as hydrogen bonds, charge-transfer interactions, or simply van der Waals bonding. However, fabrication of these systems will require not only control of supramolecular structure based on intermolecular interactions between the "modules," but also an understanding of the interactions between the molecular assemblies and the substrate surfaces upon which they form. The role of substrates in the assembly of molecules has become evident in recent studies of ordered organic thin films prepared by molecular beam epitaxy [7,8]. These studies have revealed that the structure and orientation of molecular overlayers can be driven by epitaxy two-dimensional substrates.

In the course of studies aimed toward understanding the solution nucleation and growth of organic crystals, we have found that on appropriate two-dimensional ordered substrates crystal nuclei actually consist of ordered, molecularly thick overlayers. We have examined the formation of these overlayers in real time with *in situ* atomic force microscopy (AFM), with particular attention to the influence of the substrate on the structure and orientation of these overlayers. *A key conclusion of these studies is that overlayer structure and orientation is determined not by the interaction between individual molecules and substrate sites, but rather by coincident lattice match of large molecular assemblies with the substrate.* Based on the discussion in Section 1, this leads to a better "wetting" of the overlayer so that ΔG^*_{het} is lower than in the absence of epitaxy. We describe here a particular example in which two-dimensional epitaxy directs polymorph selectivity.

Charge-transfer salts of ET (ET = bis(ethylenedithiolo)tetrathiafulvalene), which can be prepared by electrocrystallization methods, form numerous conducting and superconducting phases. However, these materials are plagued by polymorphism, with the different phases exhibiting significantly different electronic properties. This is exemplified by the ET-I_3 system, for which 14 phases are known.

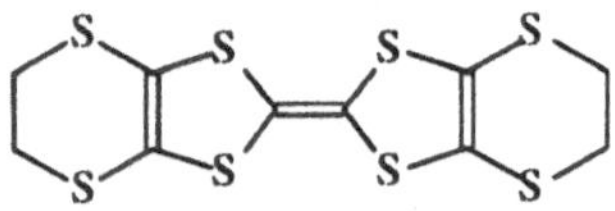

ET

The synthesis of these materials by electrochemical oxidation of ET in the presence of I_3^- typically occurs with simultaneous growth of several different polymorphs at the electrode surface. The α-$(ET)_2I_3$ and β-$(ET)_2I_3$ forms, which both contain layers of ET and I_3^- ions but with slightly different packing of ET molecules, generally predominate (Figure 4). Notably, when electrocrystallization is performed on freshly cleaved highly oriented pyrolytic graphite (HOPG) surfaces only the β form is observed, whereas on electrochemically roughened HOPG surfaces only α is formed. This suggested that epitaxy with the ordered substrate was important for directing the selectivity toward the β form.

The origin of this polymorph selectivity became evident when the electrochemical growth of these crystals was examined with *in situ* AFM. Upon application of an anodic potential to a freshly cleaved HOPG electrode in an acetonitrile solution of ET and n-$Bu_4N^+I_3^-$, large islands formed on the terraces of the HOPG substrate, eventually coalescing into a single monolayer (Figure 5). The height of these islands and the monolayer was 15.5 Å, nearly identical to the layer thickness of β-

$(ET)_2I_3$. High resolution imaging of the monolayer revealed a structure and periodicity of contrast resembling that of the (001) plane of β-$(ET)_2I_3$, with lattice constants of $\mathbf{b_1}$ = 12.0 Å, $\mathbf{b_2}$ = 8.5 Å and γ = 108°, very similar to the crystallographic parameters for β-$(ET)_2I_3$ (a = 6.6 Å (= 1/2$\mathbf{b_1}$), b = 9.1 Å, γ = 110°). These β-like monolayers serve as nuclei for large crystals of β-$(ET)_2I_3$, which form as macroscopic plates on the HOPG electrode with their (001) faces parallel to the substrate. These observations, as well as real time studies of the crystal growth from the monolayers, clearly indicated that macroscopic growth is dictated by the initial events.

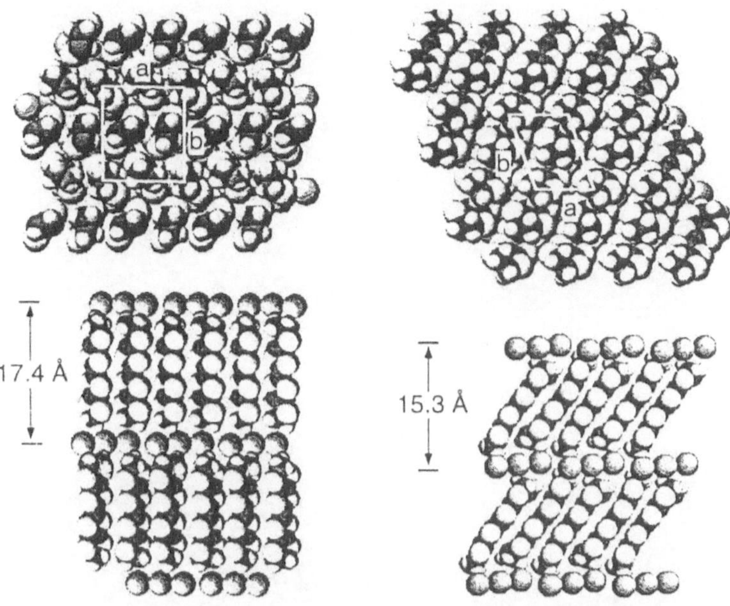

Figure 4. Layer motifs of α-$(ET)_2I_3$ (left) and β-$(ET)_2I_3$ (right), as observed perpendicular (top) and parallel (bottom) to the (001) plane. The two polymorphs are similar in that they contain stacks of ET molecules arranged into (001) layers separated by layers of I_3^- anions. However, the two forms differ with respect to the declination of the ET molecules. Consequently, the layer thicknesses of α- and β-$(ET)_2I_3$ are 17.4 Å and 15.3 Å, respectively.

If the AFM tip was rastered over a small region of the monolayer at a force which was slightly higher than that used for imaging, the monolayer could be removed and the substrate could be imaged. This allowed direct determination of the azimuthal orientation of the monolayer with respect to the substrate. The average orientation of the overlayer was such that the overlayer vector $\mathbf{b_1}$ was rotated approximately 10° from the $\mathbf{a_1}$ lattice vector of HOPG, although angle as large as 20° were measured. Using the AFM lattice constants and the value of 10°, the azimuthal orientation can be described by overlayer latttice vectors having magnitude and direction in terms of the HOPG lattice vectors of $\mathbf{b_1} = 4.9\mathbf{a_1} + 1.1\mathbf{a_2}$ and $\mathbf{b_2} = 3.86\mathbf{a_2} - 3.86\mathbf{a_1}$. Macroscopic crystals grown from these monolayers were also oriented, but the a-axes of the crystals (corresponding to $\mathbf{b_1}$ of the monolayer) always were oriented 18° from $\mathbf{a_1}$ of HOPG.

These observations were consistent with lattice matching calculations designed to find the optimum epitaxial orientations of the unreconstructed (001) plane of β-$(ET)_2I_3$. These calculations predicted a coincident lattice with $\mathbf{b}_1$ rotated 19° from $\mathbf{a}_1$, such that $\mathbf{b}_1 = 4\mathbf{a}_1 + 2\mathbf{a}_2$ and $\mathbf{b}_2 = 3.3\mathbf{a}_2 - 4\mathbf{a}_1$. This coincident lattice can also be described as a $\mathbf{b}_3 \times 3\mathbf{b}_2$ supercell whose vertices are commensurate with the HOPG substrate (Figure 6). The discrepancy between the monolayer orientation, and the observed bulk crystal and calculated orientation is attributed to measurement error.

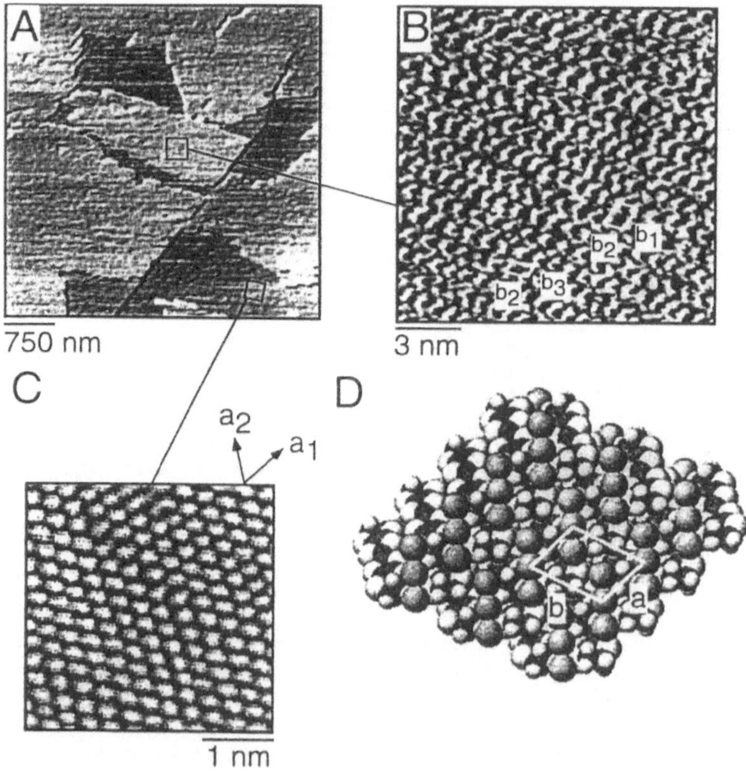

Figure 5. (A) An atomic force microscopy image acquired *in situ* during anodic oxidation of ET at a freshly cleaved highly oriented pyrolytic graphite (HOPG) electrode in 0.1 M n-$Bu_4N^+I_3^-$. This particular image shows a partially formed monolayer of $(ET)_2I_3$ which has a height of 15.5 Å, identical within experimental error to the thickness of the (001) planes in β-$(ET)_2I_3$. (B) High resolution imaging of a region of the monolayer reveals contrast periodicity identical to that of the (001) plane of β-$(ET)_2I_3$. The cell $\mathbf{b}_1 \times \mathbf{b}_2$ is evident from the real space data, while the cell $\mathbf{b}_2 \times \mathbf{b}_3$ depicted here, which corresponds to the unit cell of the bulk crystal, is evident in the Fourier space data (not shown). (C) Image of the underlying HOPG substrate exposed after mechanical removal of the overlayer with the AFM tip. (D) Molecular packing of the (001) surface of β-$(ET)_2I_3$. The lattice constants *a* and *b* correspond to $\mathbf{b}_3$ and $\mathbf{b}_2$ of the monolayer, respectively.

It is important to note that an epitaxial analysis of possible α-$(ET)_2I_3$/HOPG interfaces did not reveal commensurate or coincident lattice. This suggests that the polymorph selectivity for β-$(ET)_2I_3$ is driven by the epitaxy described above. Our calculations also indicate the orientation of the molecules in the overlayer is not identical to the minimum energy orientation of a single ET molecule. Rather, it is the in-phase relationship of the overlayer with the substrate at *supramolecular length scales* that stabilizes the β-like monolayer and determines its orientation.

We have also calculated the potential energy of an isolated β-$(ET)_2I_3$ (001) layer as a function of layer dimensions in order to evaluate its stiffness, which is a measure of its tendency to resist reconstruction that may be necessary to achieve epitaxy with the substrate. In this case, strong charge-transfer and S-S interactions within the (001) layer results in a rather deep minimum in the potential function at lattice dimensions nearly identical to that observed in the crystal structure. Conversely, the potential energy dependence on the overlayer-substrate separation or on the azimuthal orientation is rather shallow. As a consequence, the elastic constant of this layer is large relative to the elastic constant describing the overlayer-substrate interactions (the elastic constant is the second derivative of the energy). The presence of aperiodic interactions within the length scale defined by the $\mathbf{b}_3$ x $3\mathbf{b}_2$ supercell in Figure 6 can be described as a strain in terms of the displacement of the molecules from preferred sites. However, the low elastic constant of the overlayer-substrate interface leads to small interfacial stresses (stress = elastic constant x strain). This favors the formation of relatively defect-free monolayers with large domains, as observed for the β-like monolayer.

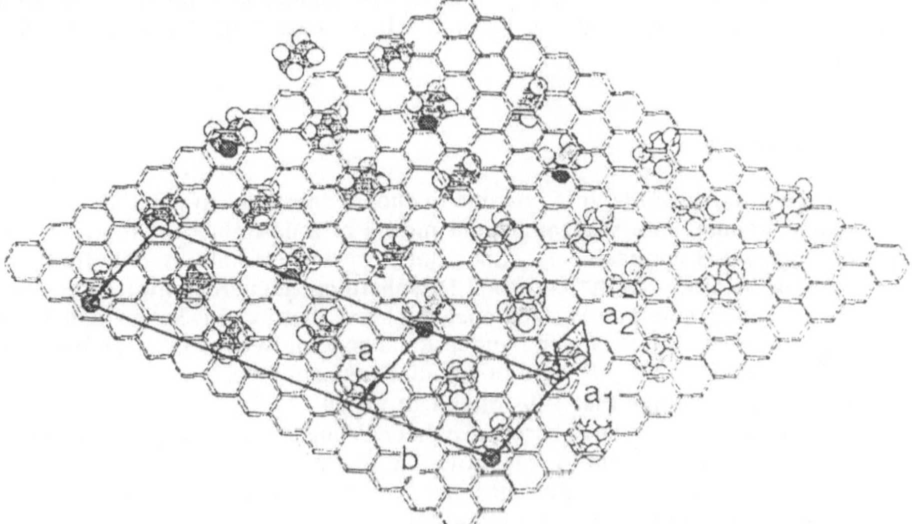

Figure 6. Schematic representation of the coincident lattice relationship between the (001) plane of β-$(ET)_2I_3$ and the HOPG substrate. The coincident overlayer can be described as having a $\mathbf{b}_1$ x $3\mathbf{b}_2$ supercell which is commensurate with the HOPG substrate. For purposes of clarity, only the ethylene groups of the external rings of the ET molecules are shown.

In the absence of epitaxy with the (001) layer, it is likely that a layer having a smaller elastic constant would have formed on the HOPG surface, as the energetic penalty for reconstructing a layer in order to achieve epitaxy under these conditions. Indeed, we have observed this behavior with $(perylene)_2ClO_4$, for which no epitaxial match between crystal planes and HOPG was found by our calculations. In this case, AFM has revealed the formation of a molecularly thick overlayer with perylene molecules lying flat (or nearly flat) on the substrate. The overlayer has lattice constants and an azimuthal orientation that lead to a coincident lattice with the HOPG substrate. The structure of this overlayer is best described as a reconstructed (100) plane, in which the perylene molecules interact only by van der Waals forces, resulting in a small elastic constant for the layer. Under these conditions, the energetic benefit of epitaxial substrate-overlayer interactions outweighs the penalty of reconstruction.

These and other observations have established that crystalline overlayers with structures resembling bulk crystal packing will form if layers having strong intralayer interactions can be epitaxially matched to the substrate. If such an epitaxy is not present, it is likely that an alternative orientation will be observed in which the energetic penalty of reconstructing the layer to achieve epitaxy is small. Epitaxial growth is likely to lead to low-defect thin films, particularly when the substrate-overlayer interactions are relatively weak and the corresponding interfacial stresses are small. Conversely, reconstruction of overlayers may lead to defects in subsequent layers due to the mismatch between the reconstructed layer and the ideal bulk structure. Alternatively, appropriately chosen substrates may induce the formation of reconstructed overlayers which can serve as nuclei for otherwise unattainable polymorphic forms. Clearly, the substrate must not be ignored when fabricating thin films based on molecular modules.

4. Conclusions

These examples illustrate that it is possible to choose substrates to guide the assembly of molecular modules into supramolecular motifs of a desirable polymorphic form. These principals are applicable to either bulk organic crystals or thin films with molecular dimensions. In the case of LDE, there are numerous molecular crystal substrates with cleavable low surface energy planes that expose ledges suitable for nucleation; substrates with strong one-dimensional bonding vectors that stabilize ledges are the best candidates. In the case of two-dimensional heteroepitaxial growth, we conclude from our studies that substrates can be chosen by simply analyzing the epitaxial interface for an overlayer based on either existing crystal structures or calculated layer structures. It is important to note, however, that it is not yet possible to design a substrate for a polymorph which has not yet been discovered or structurally characterized. However, we anticipate that libraries of molecular crystal substrates or two-dimensional substrates can be used with combinatorial methods to screen for new polymorphs.

Acknowledgments. The authors gratefully acknowledge the financial support of the Office of Naval Research and the National Science Foundation.

References

1. Desiraju, G. R. (1989) *Crystal Engineering: The Design of Organic Solids*, Elsevier, New York.
2. Dunitz, J. D., and Bernstein, J. (1995) Disappearing Polymorphs *Acc. Chem. Res.* **28**, 193 - 200.
3. Tiller, W. A. (1991)*The Science of Crystallization: Microscopic Interfacial Phenomena*; Cambridge University Press, New York.
4. Carter, P. W. and Ward, M. D. (1993) Topographically Directed Nucleation of Organic Crystals on Molecular Single-Crystal Substrates *J. Am. Chem. Soc.* **115**, 11521 - 11535.
5. Etter, M. C., Kress, R. B., Bernstein, J., and Cash, D. J. (1984) Solid State Chemistry and Structures of a New Class of Mixed Dyes. Cyanine-Oxonol *J. Am. Chem. Soc.* **106**, 6921 - 6927.
6. Bonafede, S. J. and Ward, M. D. (1995) Selective Nucleation and Growth of an Organic Polymorph by Ledge-Directed Epitaxy on a Molecular Crystal Substrate *J.Am. Chem. Soc.* **117**, 7853 - 7861.
7. Nebesny, K. W., Collins, G. E., Lee, P. A., Chau, L.-K., Danziger, J., Osburn, E., and Armstrong, N. R. (1991) Organic/Inorganic Molecular Beam Epitaxy: Formation of an Ordered Phthalocyanine/SnS2 Heterojunction *Chem. Mater.* **3**,829 - 838.
8. Forrest, S. R., Burrows, P. E., Haskal, E. I., and So, F. F. (1994) Ultrahigh-vacuum quasiepitaxial growth of model van der Waals thin films. II. Experiment *Phys. Rev. B* **49**, 11309-11321.

NANOCOMPOSITES, MOLECULAR COMPOSITES AND THEIR FIELDS OF APPLICATION

GERHARD WEGNER
Max-Planck-Institut für Polymerforschung. Postfach 31 48,
D-55021 Mainz, Germany

ABSTRACT. An important thrust in modern macromolecular chemistry is to construct complex architectures either directly from monomer units, or to design macromolecules capable of forming a desired superstructure by self-organization. Chain rigidity is an example of a structural principle leading to self organization depending on the aspect ratio of the molecular objects and their mutual interactions. These molecularly defined objects serve as the building blocks of systems for which ultrathin layers composed of hairy rod macromolecules (HRM) are an example. Depending on architectural details layered architectures of HRM are models for molecular nanocomposites, exhibit interesting optical, electrochemical, dielectric or barrier properties. They are also designed to be used in devices the performance of which depends on molecular interactions. Thus, the design of novel macromolecules comprises the design of supramolecular architectures and functions based on architectural principles from the very beginning. A systems approach is emphasized as part of the synthetic strategy.

1. Design principles of Nanocomposites

1.1. RATIONAL DESIGN OF SUPRAMOLECULAR MATERIALS

The target of our research is the rational step-by-step construction of molecular solids useful for a variety of advanced technologies in which control of the molecular and supramolecular architecture is a prerequisite to function. The philosophy and strategy is outlined in scheme I. Molecularly defined objects of the same or different kind are assembled in a controlled manner. Supramolecular architectures of materials arise which have specific functions by design. The targets are materials for micro (and nano-) mechanics, membranes, catalytic systems, photonic devices, LC displays etc.

J. Michl (ed.), Modular Chemistry, 651–661.

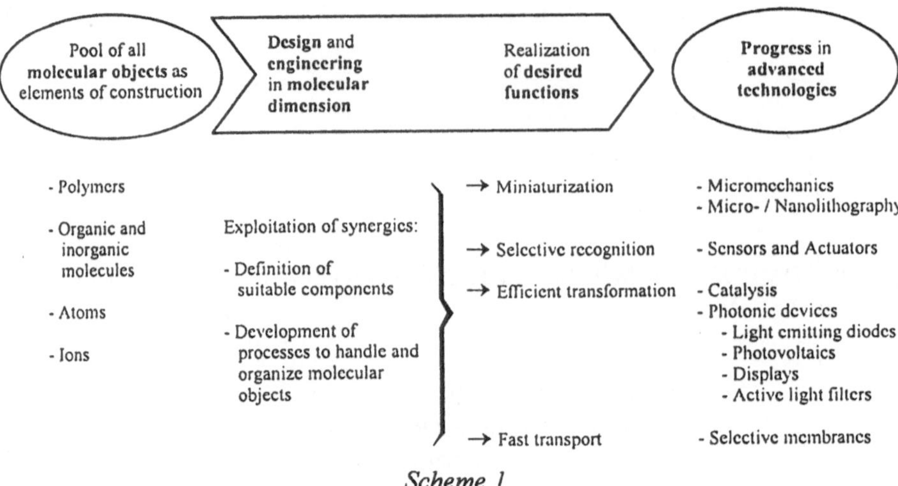

Scheme 1

1.2 HAIRY-ROD MACROMOLECULES (HRM) AND THE ASSEMBLY PROCESS

Our approach to the problem starts by definition and design of shape persistent macromolecules, Fig. 1 [1-8]. They serve as objects to be assembled in a first step as monolayers at the air-water-interface of a Langmuir trough as shown in Fig. 2 [9-11]. These are transferred to solid substrates by a dipping process. Flow alignment of the rod-like element occurs in the course of the transfer which has been shown to work as a two-dimensional analogue of an extrusion [9].

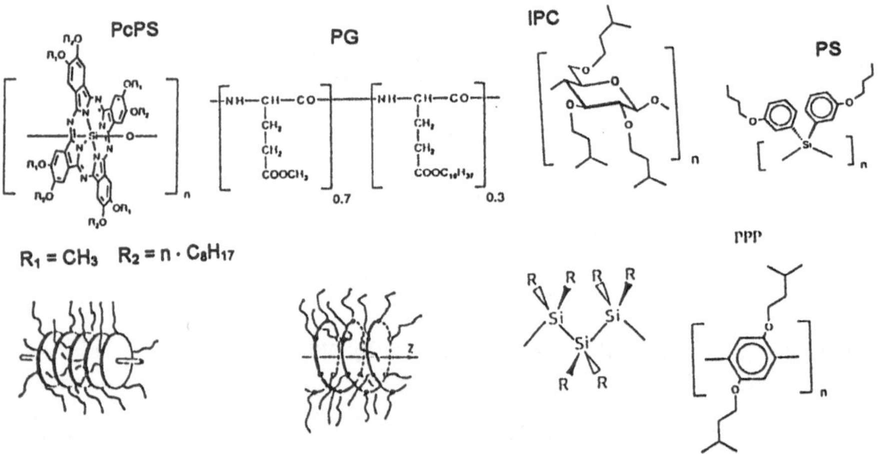

Figure 1 · Typical hairy rod molecules; PcPS = phthalocyaniato-(poly)siloxanes; PG = polyglutamate; IPC = Isopentyl cellulose; PPP = poly-(*p*-phenylene)

Rod-like macromolecules with their backbones decorated by numerous short and flexible hydrophobic side chains [see Fig. 1] have been thus designed for construction of layered films by this technique [1-9]. The shape persistent backbones are surrounded by a liquid-like skin of side chains. The layered films are best described as molecularly reinforced liquids where the backbone elements are embedded in a continuous matrix of interdigitating side chains. The build-up process results in a homogeneous orientation of all backbone elements within a given layered assembly [9, 10]. In general, the layered assemblies of these hairy-rod macromolecules (HRM) have a liquid-crystalline texture and are without any pinholes at all [11, 12]. When rods of different chemical composition are used, a series of different architectures is achieved, e.g. nanocomposites composed of layers alternating in chemical structure and direction of the rods. Such systems show interesting optical, electrochemical and electrical properties when properly composed. Furthermore, guest molecules can be placed ('doped') into the matrix of the side chains in which the rod backbones are embedded. Thus, when ionophores are chosen as guests, ion sensitive membranes and ion sensitive FET-devices can be achieved [1-3, 15].

2. Properties and Applications

The excellent mechanical and optical properties of such composites are demonstrated by Fig. 3. Here the architecture (simplified) of a multilayer film of copolyglutamates is shown together with the components of the mechanical tensor determined from a layered assembly of total thickness of ca. 1 μm (500 layers) by a light scattering technique (Brillouin spectroscopy). The anisotropic character of the composite is thus proven [13].

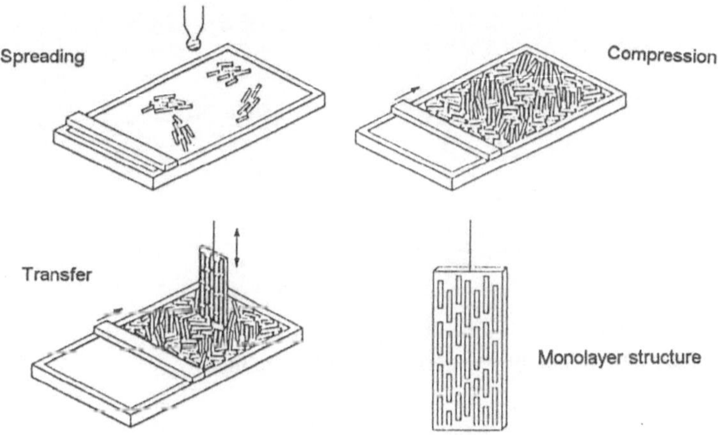

Figure 2 Construction of monolayers [1-4]

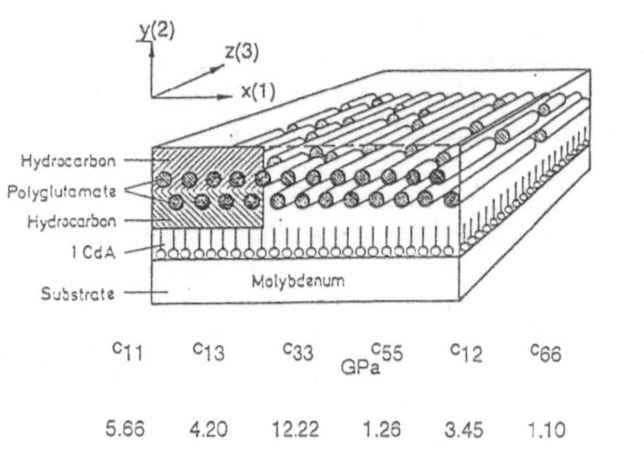

Figure 3 Mechanical modulus of a multilayer assembly of copolyglutamate

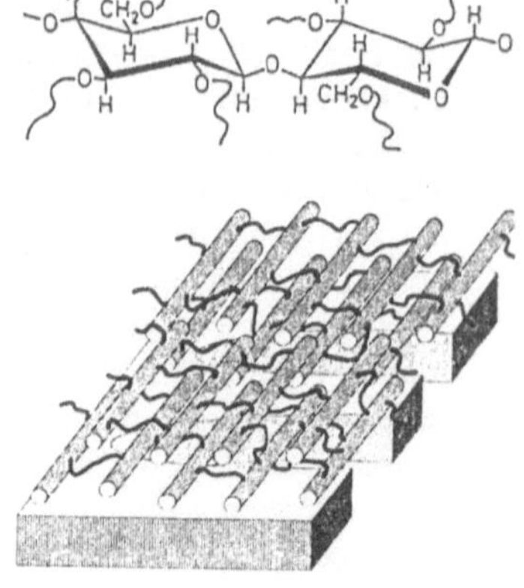

Figure 4a Crosslinked multilayer of cellulose derivatives on top of porous substrate

2.1 ULTRATHIN MEMBRANES

The fact that multilayers of photocrosslinkable HRM are readily formed and that the introduction of crosslinks gives rise to solvent swellable stable ultrathin films supports the idea to use these systems as liquid separation and/or transport membranes. HRM based on cellulose alkylethers or copoly(glutamate)s can be used to construct membranes if transferred by the LB technique to a porous Celgard 2400® film as the substrate [5].

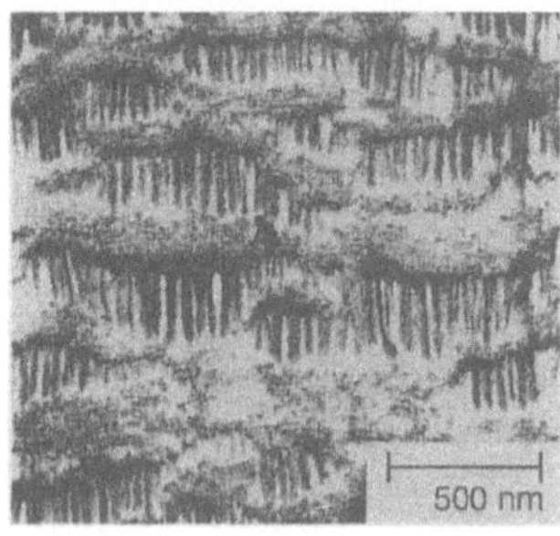

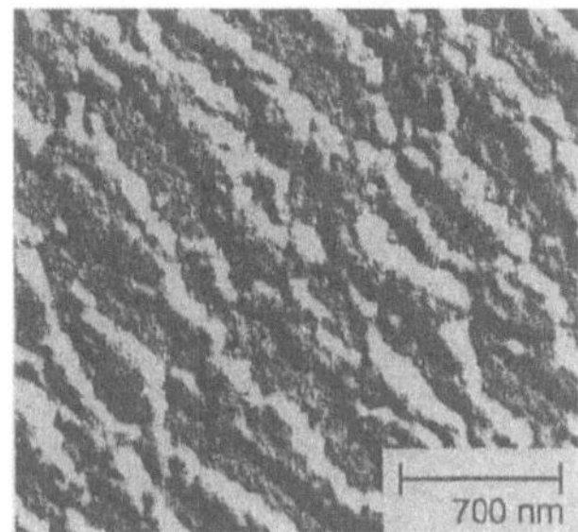

Figure 4b Celgard 2400 Film before and after coverage by 40 ML of crosslinked cellulose derivative (TEM replica)

Because of the small layer thickness of less than 10Å for each individual layer [6] and the possibility to introduce functional groups, cellulose ethers are most promising for the design of membranes. The ability to undergo crosslinking is achieved by attaching a few cinnamoyl residues to the cellulose backbone; all residual OH groups are alkylated by isopentyl residues. Fig. 4a shows schematically the architecture of such a crosslinked multilayer assembly on top of a porous substrate. Fig. 4b shows the actual porous Celgard 2400®-film before and after coverage. These membranes serve as size exclusive separation devices or osmotic membranes, since they are constructed like a sieve: only molecules of smaller diameter as the distance between the bars can penetrate [5].

2.2 DESIGN OF POLARIZED LIGHT EMITTERS

Another application is demonstrated by Figs. 5a, b. Here HRM's based on poly-p-phenylene substituted with alkoxy side chains are used to construct an electroluminiscent device which emits polarized light in the visible region of the spectrum. LED's based on polymers have been pioneered by the Cambridge group. in the recent years and have become a rapidly growing field of research. The device shown here is the first rational approach to create polarized light as is necessary for background illumination of LCD-devices and computer screens.

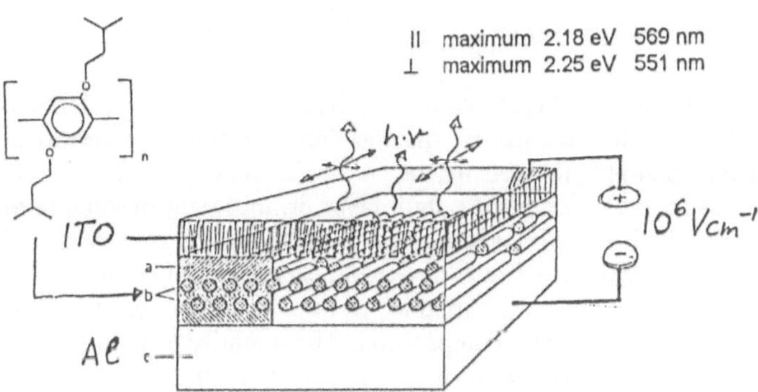

Figure 5a Light emitting device (LED) [see ref. 14]

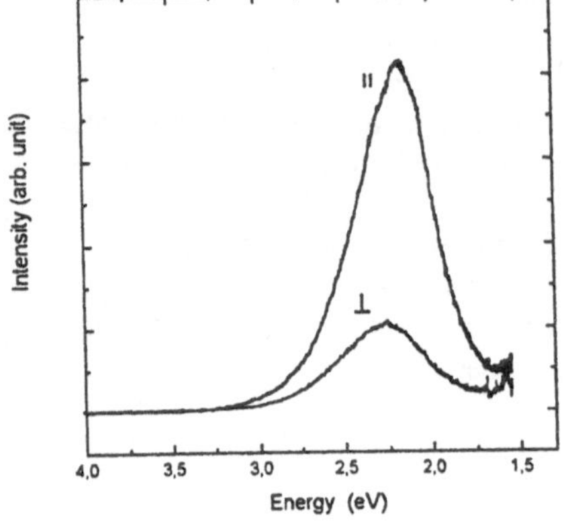

Figure 5b Polarized light emission from the device shown in Fig. 5a (100ML, ext. quantum eff. $4 \cdot 10^{-3}$)

Other phenomena related to charge carrier transport have been described, among others the construction of spatially and electronically defined tunneling barriers consisting of a single monolayer of an insulating HRM (cellulose derivative) placed in the middle of a multilayer assembly of semiconducting phthalocyanine polymers (PCPs, see Fig. 1) [16]. Furthermore, the same polymers (PCPs) have been evaluated for their electrochemical properties. Thin films of 1-100 ML show facile electron and ion transport in electrochemical oxidation and reduction in aqueous and non-aqueous media [11].

2.3 CONSTRUCTION OF A CHEMFET

A final example is given in Fig. 6 describing the set-up of a field effect based ion sensor (CHEMFET) for monitoring the ion concentration in electrolytes. This set-up was developed [15] to measure the Na or Ca-ion concentration in aqueous media and relies on the presence of a multilayer arrangement of ionophore doped HRM's on top of the MOS-FET which is in contact with the electrolyte. The time and environmental stabilities of such devices is excellent and compares well with conventional glass electrodes. As explained in Fig. 6 the acronym ELBOS device was used to describe the special composition. The quantity to be measured is the change in capacitance which in turn is controlled by the interfacial potential between electrolyte and LB membrane and LB membrane and MOS surface resp. The LB membrane itself consists of 10-30 ML of a copolyglutamate doped with an ionophore with selectivity for the ion in the aqueous electrolyte to be monitored. In order to prevent loss of the ionophore to the electrolyte the layered structure is covered by a few (ca. 2-6) ML of PCPs which works as a diffusion barrier for the ionophore but not for the ions to be detected.

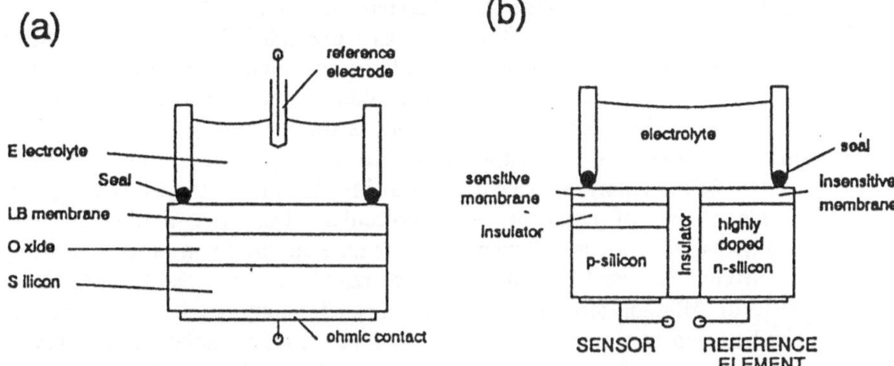

Figure 6 ELBOS-CHEMFET to monitor ion concentration in aqueous electrolytes (15) a) principle of the set-up; b) complete device layout

Figure 7 shows the exceptionally fast response of the ELBOS device to instantaneous changes of the ion concentration. This is of course due to the fact that the sensitive layer which controls the surface potential is only a few nm thick.

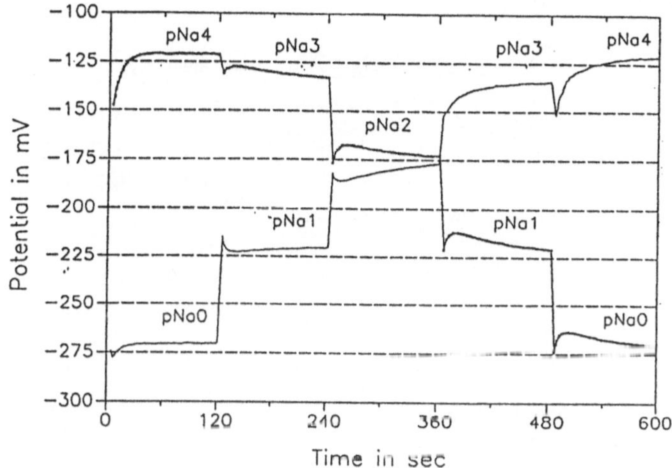

Figure 7 Time dependent response of the ELBOS device in Fig. 6b (the LB layer was doped by a Na^+ ionophore) to instantaneous change in Na^+ ion concentration.

2.4 TOPOCHEMICAL ASPECTS AND SCAFFOLDING [4]

All reactions occurring inside the layered assemblies of HRM can be characterized by a "dimensionality" as exemplified by Figs. 8a-c. "Dimensionality of the reaction" should be differentiated from "product dimensionality". The latter refers to the spatial structure of the product, i.e. whether a two- or three-dimensional network is the reaction product. Chemical reactions between HRMs within a given layer or in adjacent layers are controlled by the extent of the reaction zones defined in Fig. 8c, that is the overlap regions of the side chains of adjacent macromolecules. The cross-linking reaction discussed above which leads to membranes is an example for a reaction leading to a 3D product. Since LB multilayers are readily constructed layer-by-layer from HRM of different. structure or chemical functionality reactive layers can be placed in between non-reactive layers to achieve "reaction isolation" and layers having corresponding functions can be placed in adjacent layers. An example for a Diels-Alder reaction used to "stitch up" two adjacent layers in a multilayer assembly is shown in Fig. 8c. Details have been published recently [4]. In terms of a topochemical description the stitching reaction occurs in z-direction [Fig. 8a] but the product is a strictly 2D network composed of 2 adjacent layers.

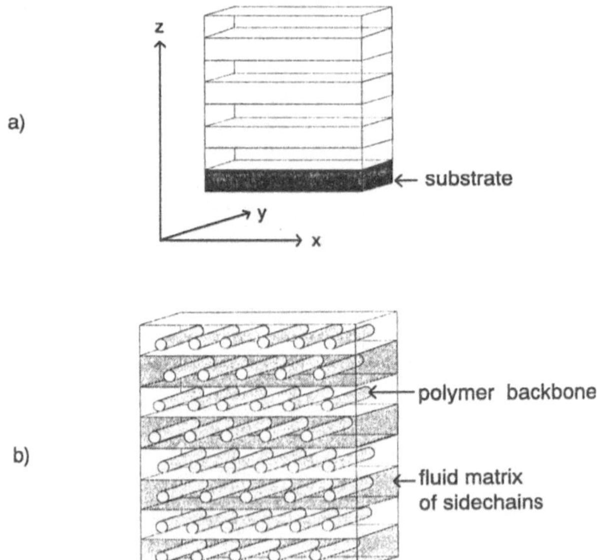

Figure 8 a) General architecture of a LB assembly and assignment of directions.
b) Layered assembly of HRM, where the side chain matrix has fluid characteristics.

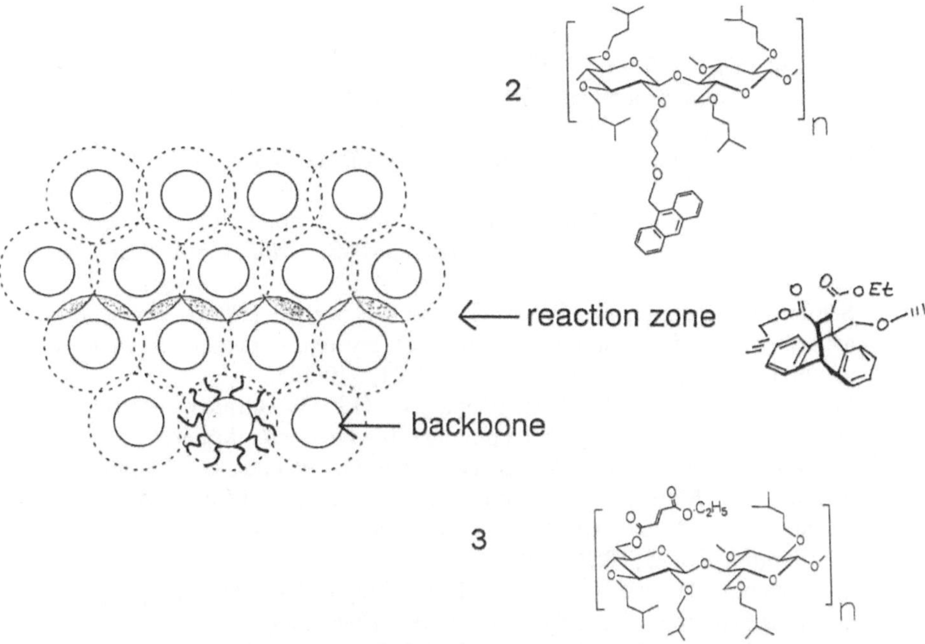

Figure 8 c) Definition of reaction zones and example for a chemical reaction of Diels-Alder type occurring between adjacent functionalized layers embedded in non-reactive layers (isopentyl cellulose). The reaction occurs thermally initiated and selectively between the layers [4] as indicated.

Acknowledgement

The background of the work reported here was laid in the context of the program "Ultrathin layers of polymers as the basis of new optical, electronic or biologically active systems" supported by the German Ministry of Research and Technology (BMFT) 1986-1992. The concept of hairy rod macromolecules was developed mainly in the Theses of E. Orthmann and Th. Sauer (PcPS), G. Duda (PG), F. Embs (polysilanes), Th. Vahlenkamp (PPP), M. Schaub (cellulose derivatives) and Th. Arndt (spectroscopy). Outstanding contributions came from a number of staff scientists of the MPI-P, notably by Dr. Ch. Bubeck (optics), Dr. D. Neher (device physics), Dr. M. Schulze (synthetic chemistry), Dr. G. Lieser (microscopy) and by postdoctoral students Dr. A. Ritcey, Dr. M. Suzuki, Dr. S. Iida and visiting scientists Dr. K. Yase and Prof. Y. Xu. More recently, among others the following students contributed significantly: Th. Schwiegk, K. Mathauer, M. Seufert, Ch. Fakirov, A. Ferencz and M. Remmers. Intensive interaction with Prof. W. Knoll and his students at the MPI-P and with Prof. B. Hoffmann (University of Karlsruhe) is also greatly acknowledged.

References

1. Wegner, G. (1992) Ultrathin films of polymers: Architecture, characterization, properties, *Thin Solid Films* **216,** 105-116.
2. Wegner, G. (1992) Photochemistry and photophysics of nanocomposites prepared from rod-like macromolecules by LB-technique, *Mol. Cryst. Liq. Cryst.* **216**, 7-12.
3. Wegner, G. (1993) Control of molecular and supramolecular architecture of polymers, polymer systems and nanocomposites, *Mol. Cryst. Liqu. Cryst.* **235**, 1-34.
4. Seufert, M., Schaub, M., Wenz, G., and Wegner, G. (1995) Topochemical aspects of the formation of networks in layered Langmuir-Blodgett (LB) assemblies, *Angew. Chem. Int. Ed. Engl.* **34,** 340-343.
5. Seufert, M., Fakirov, Ch., and Wegner, G. (1995) Ultrathin membranes of molecularly reinforced liquids on porous substrates, *Adv. Mater.* **7,** 52-55.
6. Schaub, M., Fakirov, Ch., Lieser, G., Wenz, G., Wegner, G., Albouy, P.-A., Schmidt, A., Majrkzak, C., Satija, S., Wu, H., and Foster, M.D. (1995) Ultrathin layers and supramolecular architecture of isopentyl cellulose, *Macromolecules* **28,** 1221-1228.
7. Vahlenkamp, Th., and Wegner, G. (1994) Poly (2.5-dialkoxy-p-phenylene)s - synthesis and properties, *Makromol. Chem. Phys.* **195,** 1933-1952.
8. Ferencz, A., Ries, R., and Wegner, G. (1993) Polymer hemiporphyrazines as a new class of rod-like macromolecules for the LB-technique, *Angew. Chem. Int. Ed. Engl.* 32, 1184-1187.
9. Schwiegk, S., Vahlenkamp, T., Xu, Y., and Wegner, G. (1992) On the origin of orientation phenomena observed in layered LB structures of hairy rod polymers, *Macromolecules* **25,** 2513-2525.
10. Suzuki, M., Ferencz, A., Iida, S., Enkelmann, V., and G. Wegner, (1993) Effects of surface topology and highly anisotropic polymer LB-films on liquid crystal alignment, *Adv. Mater.* **5,** 359-364.
11. Ferencz, A., Armstrong, N.R., and Wegner, G. (1994) Electrochemical characterization of the oxidation of LB-films of polyphthalocyaninatosiloxane, *Macromolecules* **27,** 1517-1528.
12. Yase, K., Schwiegk, S., Lieser, G., and Wegner, G. (1992) Molecular orientation of substituted phthalocyaninatopolysiloxane in Langmuir-Blodgett-films, *Thin Solid Films* **210/211,** 22-25.
13. Johannsmann, D., Mathauer, K., Wegner, G., and Knoll, W. (1992) Viscoelastic properties of thin films probed with a quartz-crystal resonator, *Phys. Rev.* B **46,** 7808-7815.

14. Cimrova, V., Neher, D., and Wegner, G. (1996) Polarized light. Emission from LEDs prepared by the Langmuir-Blodgett-technique, *Adv. Mater.* **8**, 146-148.
15. Erbach, R., Hoffmann, B., Schaub, M., and Wegner, G. (1992) Application of rod-like polymers with ionophores as Langmuir-Blodgett membranes for Si-based sensors, *Sensors and Actuators* **B6**, 211-216.
16) Wegner, G. (1991) Ultrathin films of polymers, *Ber. Bunsenges. Phys. Chem.* **95**, 1326-1333.

DESIGNING POROSITY IN COORDINATION SOLIDS

O. M. YAGHI* AND HAILIAN LI
Department of Chemistry and Biochemistry
Goldwater Center for Science and Engineering
Arizona State University
P.O. Box 871604
Tempe, Arizona 85287-1604

ABSTRACT. Two strategies for producing porous coordination solids are described. The first deals with the construction of rigid metal-organic frameworks using ligands having multidentate functionalities, and shapes that prevent interpenetration. These ligands are chosen to enhance the selective inclusion of certain guest molecules. The second strategy addresses the character of the guest species that occupy the channels of the metal-organic framework. Here, hydrogen-bonded aggregates have been designed as guest species not only to allow the formation of large voids but also to facilitate the removal of inclusions, thus permitting access into the channels by other molecular species.

1. Introduction

The development of strategies for the construction of coordination solids with open-framework structures is at a stage where it is becoming increasing possible to realize target topologies and compositions [1]. Recently, a number of extended open frameworks have been prepared by linking organic ligands to metal ions to give covalent structures that have the same topologies as basic inorganic structures such as SiO_2 [2], rutile [3], PtS [4], CaCuP [5], and $ThSi_2$ [6]. The larger size of the building units in these coordination solids compared to those of their inorganic counterparts creates open space where guest solvent molecules or counterions reside in the crystal. At the outset of this study, attempts to liberate the neutral guests or to exchange the counterions has resulted in the destruction of the metal-organic framework, thus precluding its use as a porous material. We believe that in these cases the energy required to remove the inclusions from the channels by breaking the guest-framework interactions is sufficient to destroy the metal-organic architecture.

To provide solutions to this problem, we have focused our efforts on using building units that are capable of generating strong bonding interactions within the backbone structure of the framework, and on employing guest molecules or ions that are expected to self-aggregate within the channels without forming extensive interactions to the

J. Michl (ed.), Modular Chemistry, 663–670.

framework. We have demonstrated success in these two areas by using a multi-bidentate ligand, 1,3,5-benzenetricarboxylate (BTC), with cobalt(II) to form a stable open framework, $CoC_6H_3(COO)_2(COOH)(NC_5H_5)_2 \cdot 2/3NC_5H_5$, and by employing large hydrogen-bonded aggregates of nitrate ions and water as guests in the structure of copper(I) bipyridine, $Cu(4,4'\text{-}bpy)_{1.5} \cdot NO_3(H_2O)_{1.25}$. In both of these compounds, situations were created where operations such as the removal of the neutral guests or ion exchange of the counterions leave the framework intact. This contribution represents a summary of the synthetic routes developed for the production of these frameworks. The key structural features responsible for the inclusion properties of these solids are also presented.

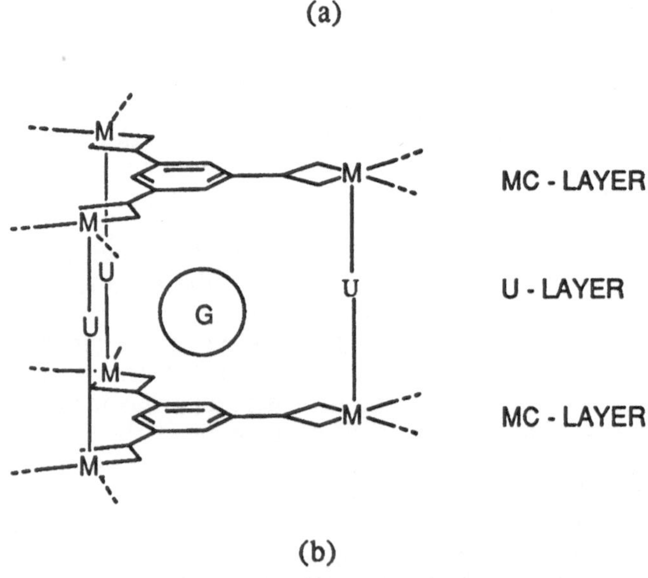

Figure 1. (a) The structural formula of 1,3,5-benzenetricarboxylate, and (b) a schematic representation of its coordination to a metal ion to form alternating metal-carboxylate and spacer unit (U) layers leading to voids where guest inclusions (G) reside.

2. Selective Binding in $CoC_6H_3(COO)_2(COOH)(NC_5H_5)_2 \cdot 2/3NC_5H_5$

Given that carboxylate groups are known to bind to transition metal ions within the same plane as either bidentate or monodentate ligands, we postulated that the reaction of 1,3,5-benzenetricarboxylate (BTC) (Figure 1a) with a first-row transition metal may yield a material composed of metal-carboxylate (MC) layers as shown in Figure 1b. Here, U groups hold adjacent layers and create channels where inclusions reside. Depending on the character of U, suitable environments for accommodating various inclusions (G) could be generated.

2.1. SYNTHESIS

Solid $Co(NO_3)_2 \cdot 6H_2O$ (0.582 g, 2 mmol) and the acid form of BTC (BTCH3) (0.420 g, 2 mmol) were dissolved in 15 mL ethanol, then added to a solution of polyethylene oxide (PEO) (0.250 g, MW = 100,000) in 1,2-dichloroethane (5 mL). The mixture was stirred until a clear solution was obtained. Pyridine was diffused into this solution resulting in the formation of large pink crystals. These were collected and washed successively with 1,2-dichloroethane, ethanol, and acetone to yield 0.85 g of product. (Found; C, 54.99; H, 3.82; N, 7.44; Co, 12.43%: Calculated for $CoC_6H_3(COO)_2(COOH)(NC_5H_5)_2 \cdot 2/3NC_5H_5$; C, 56.11; H, 3.66; N, 7.82; Co, 12.33%.) This material is insoluble in water and common organic solvents. The homogeneity of the bulk product was confirmed by comparison of the observed and calculated powder diffraction pattern, which was produced using the single crystal data.

2.2. SINGLE CRYSTAL X-RAY STRUCTURE

An X-ray crystal analysis study was performed [at 20±1 °C, hexagonal, space group $P6_3/mcm$ - D^3_{6h} (no. 193) with a = 16.711(4) Å, c = 14.189(2) Å, V = 3423(1) Å^3, and Z = 6 for x = 1, d_{calcd} = 1.464 g cm^{-3}, d_{found}= 1.48 ± 0.01 g cm^{-3}, R_1 (unweighted, based on F) = 0.068 for 552 independent reflections, μ_a (MoK_α^-) = 0.80 mm^{-1}] to reveal the presence of a neutral porous network represented by the formula, $CoC_6H_3(COO)_2(COOH)(NC_5H_5)_2 \cdot 2/3NC_5H_5$. A single layer of the structure is shown in Figure 2a, where two $BTCH_3$ and one BTC units are coordinated to each of the cobalt(II) centers. The remaining positions on the metal centers are occupied by pyridine molecules which were treated as being statistically disordered with two preferred orientations around the Co-N bond. This arrangement generates extended sheets along the xy plane. In the crystal, these sheets stack along the z axis to give alternating cobalt-carboxylate and pyridine layers as shown in Figure 2b. Close examination of the remaining space between the sheets reveals the presence of uncoordinated pyridine molecules which occupy the rectangular channels (7 × 10 Å) resulting from the tight fit between the layers. This gives the structure a three-dimensional character in that these closely interacting pyridines running along the third dimension create a rigid channel system; as the layer-layer separation

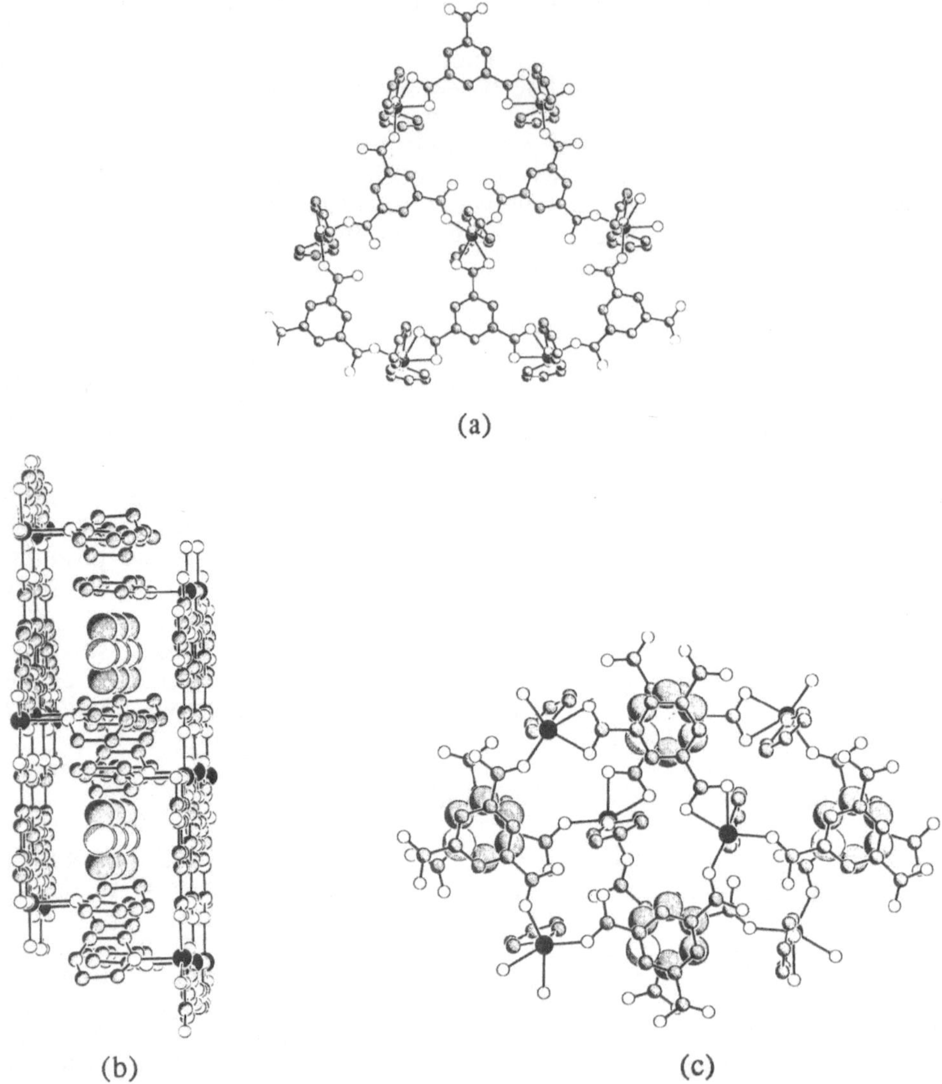

Figure 2. (a) A single layer of the extended porous network of $CoC_6H_3(COO)_2(COOH)(NC_5H_5)_2 \cdot 2/3\ NC_5H_5$. The metal-carboxylate layer is shown approximately along the *xy* plane with the spheres: dark, Co; open, O; partially shaded, N; and shaded, C. Only one of the two preferred orientations about the Co-N bond is shown for the statistically disordered coordinated pyridine molecules. The hydrogen atoms on the pyridines and btc units are omitted for clarity. (b) A perspective drawing perpendicular to the *z* axis. Here, the pyridine guest molecules are represented in the space-filling van der Waals radii to show their occupation of the channels within the structure. (c) A view of the structure along the *z* axis showing the π-stacking of the pyridine guests with the benzene rings of the btc units.

distance and the channel structure remain unaltered upon removal or inclusion of guests.

2.3. SELECTIVE BINDING STUDIES

Thermal gravimetric analysis performed on a crystalline sample showed cleanly at 190 °C a weight loss of 11.7%, corresponding to the loss of the pyridine guests occupying the channels (2/3 NC_5H_5 per formula unit), followed at 350 °C by another weight loss for a total 45.5%, corresponding to the remaining pyridine molecules bound to the metal centers within the framework (2 NC_5H_5 per formula unit). This material does not lose its crystallinity upon removal of the pyridine inclusions. Heating to 200 °C for six hours showed that the positions and intensities of the most intense lines (002), (003), (004), (005) and (006) remain unchanged relative to those observed for the unheated sample of the solid. Elemental microanalysis for the heated material (Found; C, 52.38; H, 3.46; N, 6.84%: Calculated for $CoC_6H_3(COO)_2(COOH)(NC_5H_5)_2$; C, 53.66; H, 3.32; N, 6.59) point to the absence of pyridine guests.

Examination of the single crystals with an optical microscope at 60 °C under 0.001 mm Hg pressure showed that they retain their morphology and crystallinity upon loss of the pyridine guests. Based on infrared data, we found that solid samples of this material, where the pyridine guests have been removed, selectively absorb aromatic molecules such as benzene, nitrobenzene, cyanobenzene, and chlorobenzene but not acetonitrile, nitromethane, or 1,2-dichloroethane. We believe that the remarkable selectivity of this open framework towards aromatic molecules is a direct consequence of their π-stacking with the carboxylate units present within the sheets as shown in Figure 2c. The use of multidentate ligands, coupled with our ability to functionalize the framework contribute greatly to the unusual thermal stability exhibited by this material.

3. Hydrothermal Synthesis of $Cu(4,4'\text{-bpy})_{1.5}\cdot NO_3(H_2O)_{1.25}$

Hydrogen-bonded aggregates of nitrate-water can be used as inclusion to support the formation of open structures with copper(I) and 4,4'-bipyridine(4,4'-bpy). However, hydrothermal conditions have to be used due to the poor solubility of 4,4'-bpy in water at room temperature.

3.1. SYNTHESIS

A mixture of $Cu(NO_3)_2\cdot 2.5H_2O$ (0.17 g, 0.74 mmol), 4,4'-bpy (0.17 g, 1.11 mmol), and 1,3,5-triazine (0.040 g, 0.49 mmol) in deionized water (15 mL) was transferred to a stainless steel bomb, which was sealed and placed in a programmable furnace. The temperature was raised to 140 °C at 5°/min and held at that temperature for 24 h, then cooled at 0.1°/min to 90 °C and held for 12 h, followed by further cooling at the same rate to 70 °C and held for another 12 h, and finally cooled down to room temperature at 0.1 °C/min. The resulting rectangular parallelepiped orange crystals of Cu(4,4'-

bpy$)_{1.5}$·$NO_3(H_2O)_{1.25}$ were collected and washed with deionized water (3 × 5 mL) and ethanol (2 × 5 mL), then air-dried to give 0.25 g of product (87% yield based on bpy). This compound is stable in air and is insoluble in water and common organic solvents.

3.2. SINGLE CRYSTAL X-RAY STRUCTURE

X-ray structure determination performed on a single crystal [at 20±1 °C, orthorhombic, space group *Fddd* (No. 70) with a = 18.272(5) Å, b = 23.498(5) Å, c = 29.935(7) Å, V = 12853(6) Å^3, and Z = 32, d_{calcd} = 1.581 g cm^{-3}, R_1 (unweighted, based on F)=0.036 for 1404 independent reflections, μ_a (MoK_α^- = 1.39 mm^{-1}] isolated from the reaction mixture reveal an extended cationic framework constructed from the building unit shown in Figure 3a. Here, symmetry-equivalent and slightly-distorted trigonal-planar copper(I) centers (N7-Cu-N1 = 125.4 (2)°, N13a-Cu-N1 = 125.7 (2)°, and N13a-Cu-N7 = 108.5 (2)°) are linked by rod-like 4,4′-bpy ligands to form six porous and identical interpenetrating 3-D networks with each single network (shown in Figure 3b) having three different size channels running along the [100], [010] and [001] crystallographic axes with dimensions 26 × 20 Å, 10 × 12 Å and 43 × 18 Å, respectively. The overall structure contains six-interpenetrating frameworks. The extent of interpenetration in this structure does not fill all the available voids but leaves a significant portion of it in the form of two rectangular extended channels (8 × 6 Å and 4 × 5 Å) which are occupied by nitrate anions that are hydrogen-bonded to solvent water molecules.

3.3. ION-EXCHANGE STUDIES

The lability of the hydrogen bonded inclusions was examined by thermal gravimetric analysis of a 21.631 mg microcrystalline sample, which revealed that the water guests are liberated at 120 °C as indicated by a weight loss of 6% corresponding to 1.25 H_2O per formula unit. Infrared and elemental microanalysis data reveal that most of the nitrate ions can be exchanged in aqueous media with other simple ions such as SO_4^{2-} and BF_4^-.

4. Summary

This work demonstrates the importance of using multi-bidentate ligands in the formation of stable metal-organic frameworks capable of maintaining their porous structure in the absence of inclusions. The selective binding of aromatic solvents to the cobalt-carboxylate framework indicates that specific features can be built into the structure to generate environments capable of accommodating with discrimination a variety of guests. The large hydrogen-bonded aggregates observed as inclusions in the open structure of copper(I) bipyridine point to a possible relationship between the size of the guest and the size of the channels ultimately produced in the crystal. It appears that liberation of water from these aggregates allows easier access to

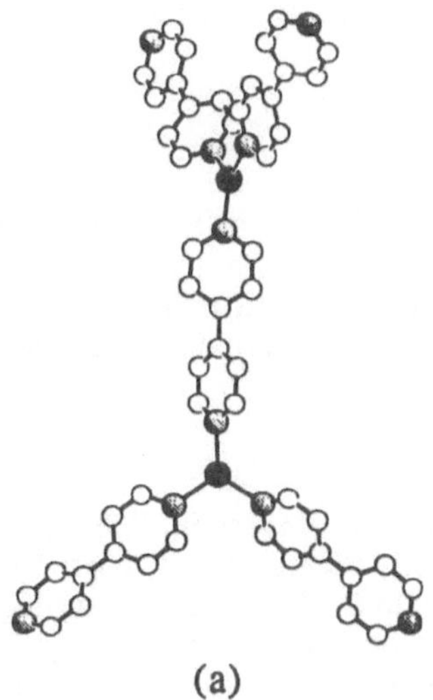

(a)

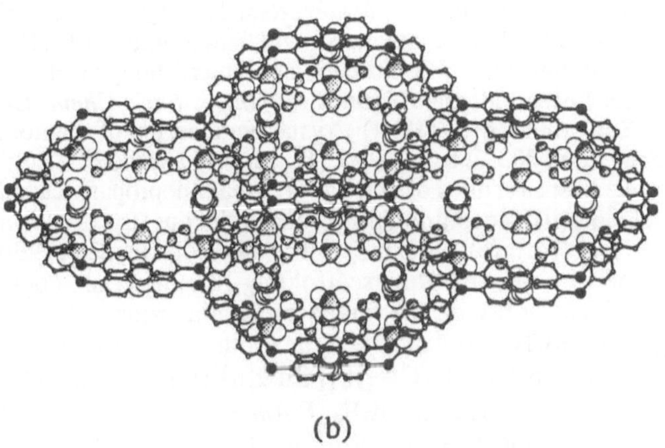

(b)

Figure 3. (a) The building block unit present in crystalline Cu(4,4'-bpy)$_{1.5}$·$NO_3(H_2O)_{1.5}$ (open spheres, C; shaded spheres, N; dark spheres, Cu). (b) The structure of a single framework shown approximately down the crystallographic direction [100]. The framework is represented as a line drawing and the nitrate anions and the water guest inclusions are depicted with large open spheres, O; large shaded spheres, N; small shaded spheres, H. The 4,4'-bpy hydrogens are omitted and all atoms are assigned arbitrary sizes for clarity.

the channels and thus more facile ion-exchange of the nitrate ions residing within those channels.

5. Acknowledgments

The financial support of the National Science Foundation, the donors of the Petroleum Research Fund administered by the American Chemical Society, and the Exxon Education Foundation is acknowledged. I am grateful for the efforts of my coworkers T. L. Groy, G. Li and H. Li.

6. References

1. (a) Yaghi, O.M. (1995) in *Access in Nanoporous Materials*, Pinnavaia, T.J. and Thorp, M.F. (eds.), Plenum press, New York, 111-121. (b) Stein, A., Keller, S.W., and Mallouk, T.E. (1993) Turning down the heat: design and mechanism in solid-state synthesis, *Science* **259**, 1558-1564. (c) Fagan, P.J. and Ward, M.D. (1992) Building molecular crystals, *Sci. Am.* **267**, 48-54. (d) Bein, T. (ed.) (1992) *Supramolecular Architecture: Synthetic Control in Thin Films and Solids*, American Chemical Society, Washington DC. (e) Dagani, R. (1991) Building complex multimolecule assemblies poses big challenges, *Chem. Eng. News*, May 27, 24-30.
2. (a) Yaghi, O.M., Richardson, D.A., Li, G., Davis, C.E., and Groy, T.L. (1994) Open-framework solids with diamond-like structures prepared from clusters and metal-organic building blocks, *Mater. Res. Soc. Symp. Proc.* **371**, 15-19. (b) MacGillivray, L.R., Subramanian, S., and Zaworotko, M.J. (1994) Interwoven two- and three-dimensional coordination polymers though self-assembly of Cu(I) cations with linear bidentate ligands, *J. Chem. Soc., Chem. Commun.*, 1325-1326. (c) Zaworotko, M.J. (1994) Crystal engineering of diamondoid networks, *Chem. Soc. Rev.*, 283-288. (d) Park, K.-M. and Iwamoto, T. (1992) Urea- and thiourea-like host structures of *catena*-[(1,2-diaminopropane)cadmium(ii) tetra-μ-cyanonickelate(ii)] accommodating aliphatic guests, *J. Chem. Soc., Chem. Commun.*, 72-74. (e) Hoskins, B.F. and Robson, R. (1990) Design and construction of a new class of scaffolding-like materials comprising infinite polymeric frameworks of 3D-linked molecular rods. A reappraisal of the $Zn(CN)_2$ and $Cd(CN)_2$ structures and the synthesis and structure of the diamond-related frameworks $[N(CH_3)_4][Cu^IZn^{II}(CN)_4]$ and $Cu^I[4,4',4'',4'''$-tetracyanotetraphenyl-methane$]BF_4$, *J. Am. Chem. Soc.* **112**, 1546-1554.
3. Batten, S.R., Hoskins, B.F., and Robson, R. (1991) 3D knitting patterns. Two independent, interpenetrating rutile-related infinite frameworks in the structure of $Zn[C(CN_3]_2$, *J. Chem. Soc., Chem. Commun.*, 445-447.
4. Abrahams, B.F., Hoskins, B.F., Michail, D.M., and Robson, R. (1994) Assembly of porphyrin building blocks into network structures with large channels, *Nature* **369**, 727-729.
5. Gardner, G.B., Venkataraman, D., Moore, J.S., and Lee, S. (1995) Spontaneous assembly of a hinged coordination network, *Nature* **374**, 792-795.
6. Yaghi, O.M. and Li, H. (1995) Hydrothermal synthesis of a metal-organic framework containing large rectangular channels, *J. Am. Chem. Soc.* **117**, 10401-10402.

INDEX